D1191319

THE OCEAN BASINS AND MARGINS

Volume 5

The Arctic Ocean

THE OCEAN BASINS AND MARGINS

THE OCEAN BASINS AND MARGINS

Edited by

Alan E.M. Nairn
Department of Geology
University of South Carolina
Columbia, South Carolina

Michael Churkin, Jr.
U.S. Geological Survey
Menlo Park, California

and

Francis G. Stehli
University of Florida
Gainesville, Florida

Volume 5

The Arctic Ocean

PLENUM PRESS · NEW YORK AND LONDON

Library of Congress Cataloging in Publication Data

Nairn, A E M
    The ocean basins and margins.

    Includes bibliographies.
    CONTENTS: v. 1. The South Atlantic.–v. 2. The North Atlantic.–v. 3. The Gulf of
Mexico and the Caribbean.–v. 4A. The eastern Mediterranean.–v. 4B. The western
Mediterranean.–v. 5. The Arctic Ocean.
    1. Submarine geology. 2. Continental margins. I. Stehli, Francis Greenough, joint
author. II. Churkin, Michael, 1932-   III. Title.
QE39.N27   551.4′608   72-83046
ISBN 0-306-37775-6 (V. 5)

© 1981 Plenum Publishing Corporation
A Division of Plenum Publishing Corporation
227 West 17th Street, New York, N.Y. 10011

Printed in the United States of America

CONTRIBUTORS TO THIS VOLUME

K. Birkenmajer
Institute of Geology
Polish Academy of Sciences
Cracow, Poland
and Institute of Historical Geology
    and Paleontology
Copenhagen, Denmark

Claire Carter
U.S. Geological Survey
Menlo Park, California

N. Z. Cherkis
Naval Research Laboratory
Washington, D.C.

Michael Churkin, Jr.
U.S. Geological Survey
Menlo Park, California

David L. Clark
Department of Geology and
    Geophysics
University of Wisconsin
Madison, Wisconsin

Peter R. Dawes
Geological Survey of Greenland
Copenhagen, Denmark

J. Thomas Dutro, Jr.
U.S. Geological Survey
Washington, D.C.

Stephen Eittreim
U.S. Geological Survey
Menlo Park, California

R. H. Feden
Naval Research Laboratory
Washington, D.C.

H. S. Fleming
Naval Research Laboratory
Washington, D.C.

Arthur Grantz
U.S. Geological Survey
Menlo Park, California

J. Wm. Kerr
J. Wm. Kerr and Associates
Calgary, Canada
(Formerly Geological Survey of
    Canada, Calgary, Canada)

Y. A. Kosygin
Institute of Tectonics and Geophysics
Khabarovsk, USSR

D. K. Norris
Geological Survey of Canada
Calgary, Alberta, Canada

L. M. Parfenov
Institute of Tectonics and Geophysics
Khabarovsk, USSR

John S. Peel
Geological Survey of Greenland
Copenhagen, Denmark

R. K. Perry
Naval Research Laboratory
Washington, D.C.

Rhoda Robinson
U.S. Geological Survey
Menlo Park, California

N. A. Shilo
Northeast USSR Scientific Research
    Institute
Siberian Division, Academy of
    Sciences of the USSR
Magadan, USSR

George Soleimani
U.S. Geological Survey
Menlo Park, California

S. M. Til'man
Northeast USSR Scientific Research
    Institute
Siberian Division, Academy of
    Sciences of the USSR
Magadan, USSR

James H. Trexler, Jr.
U.S. Geological Survey
Menlo Park, California

P. R. Vogt
Naval Research Laboratory
Washington, D.C.

Olive T. Whitney
U.S. Geological Survey
Menlo Park, California

C. J. Yorath
Geological Survey of Canada
Vancouver, British Columbia,
    Canada

CONTENTS

Chapter 4. Evolution of the Canadian Arctic Islands: A Transition between the Atlantic and Arctic Oceans

J. Wm. Kerr

Chapter 5. The Northern Margin of Greenland from Baffin Bay to the Greenland Sea

Peter R. Dawes and John S. Peel

Chapter 6. The Geology of Svalbard, the Western Part of the Barents Sea, and the Continental Margin of Scandinavia

K. Birkenmajer

Chapter 7.   Geology of the Soviet Arctic: Kola Peninsula to Lena River

Michael Churkin, Jr., George Soleimani, Claire Carter,
and Rhoda Robinson

Chapter 8.   Tectonics of the Soviet Far East

Y. A. Kosygin and L. M. Parfenov

Chapter 9.   The Tectonic Zones of Northeastern USSR and the Formation of Its Continental Crust

N. A. Shilo and S. M. Til'man

Chapter 10.   Geology and Physiography of the Continental Margin North of Alaska and Implications for the Origin of the Canada Basin

Arthur Grantz, Stephen Eittreim, and Olive T. Whitney

Chapter 11. **The Greenland–Norwegian Sea and Iceland Environment:
             Geology and Geophysics**

P. R. Vogt, R. K. Perry, R. H. Feden, H. S. Fleming,
and N. Z. Cherkis

Chapter 12. **Geology and Geophysics of the Amerasian Basin**

David L. Clark

Chapter 1

CONTINENTAL PLATES AND ACCRETED OCEANIC TERRANES IN THE ARCTIC

Michael Churkin, Jr., and James H. Trexler, Jr.

U.S. Geological Survey
Menlo Park, California

I. INTRODUCTION

The geology of the Arctic Ocean Basin is in many ways less well understood than that of the other ocean basins discussed in preceding volumes of this series. Very broad shelf seas conceal large parts of the continents fronting on the Arctic. This geography plus ice cover and difficult working conditions has retarded the accumulation of data and progress in resolving the controversial origin and history of the Arctic Ocean (Churkin, 1973; Churkin *et al.*, this volume).

In the other major ocean basins, bathymetric features are much better known and have been studied by closely spaced seismic and magnetic lines and, in many areas, by deep-sea drilling. In the Arctic, however, there has been very little direct sampling of the main submarine features. As a result, its first-order submarine features can be assigned very different roles in the various proposed schemes of Arctic evolution. A prime example is the Alpha–Mendeleyev Ridge System, which has been identified both as an extinct spreading center (Vogt and Ostenso, 1970; Hall, 1973; Kerr, this volume) and as a collision zone (Herron *et al.*, 1974), two diametrically opposed viewpoints. Without new data, the controversy of the ridge's origin will probably not be resolved.

Understanding of the Arctic is a key to the understanding of the relations between the Eurasian and North American continental landmasses. For global tectonics, the Arctic evolution serves as an important link uniting the geologic histories of the Pacific with the North Atlantic.

This chapter briefly analyzes the geology covered in the succeeding chapters by outlining the major tectonostratigraphic terranes. The geologic history of each terrane is summarized by the use of columnar sections. Correlation diagrams are used to show the regional relations of terranes. On the basis of this terrane analysis, a map showing the major continental plates and accreted terranes has been constructed. Finally, a pre-late Mesozoic reconstruction of the Arctic, including paleomagnetic data, suggests that the Pacific extended into the Arctic. Northward drift of major continental plates and accreted terranes formed the modern connection between North America and Siberia.

For this chapter we have followed the tectonostratigraphic terminology used by Jones *et al.* (1977), Churkin and Eberlein (1977), and Berg *et al.* (1978). A terrane is a sequence of rocks forming a discrete block of geology that can be distinguished from adjoining blocks by a characteristic stratigraphy and structure, resulting in a geologic history substantially different from that of adjacent terranes. Each terrane is bounded by known or inferred faults that

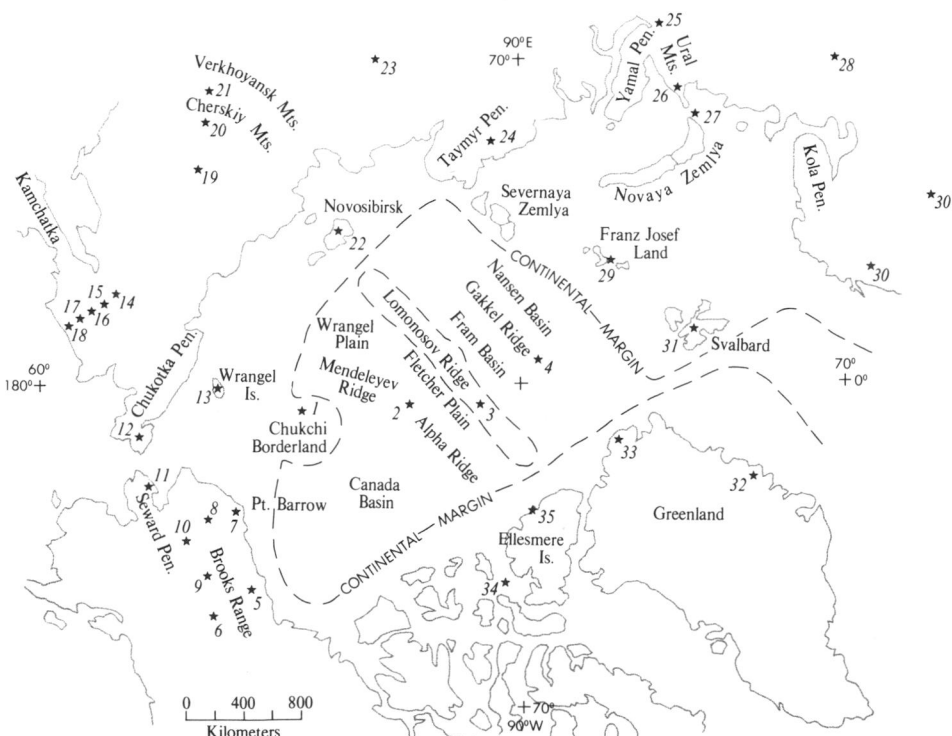

Fig. 1.   Index map of the Arctic showing some major physiographic features and the locations (★) of the columnar sections shown in Figs. 2–4.

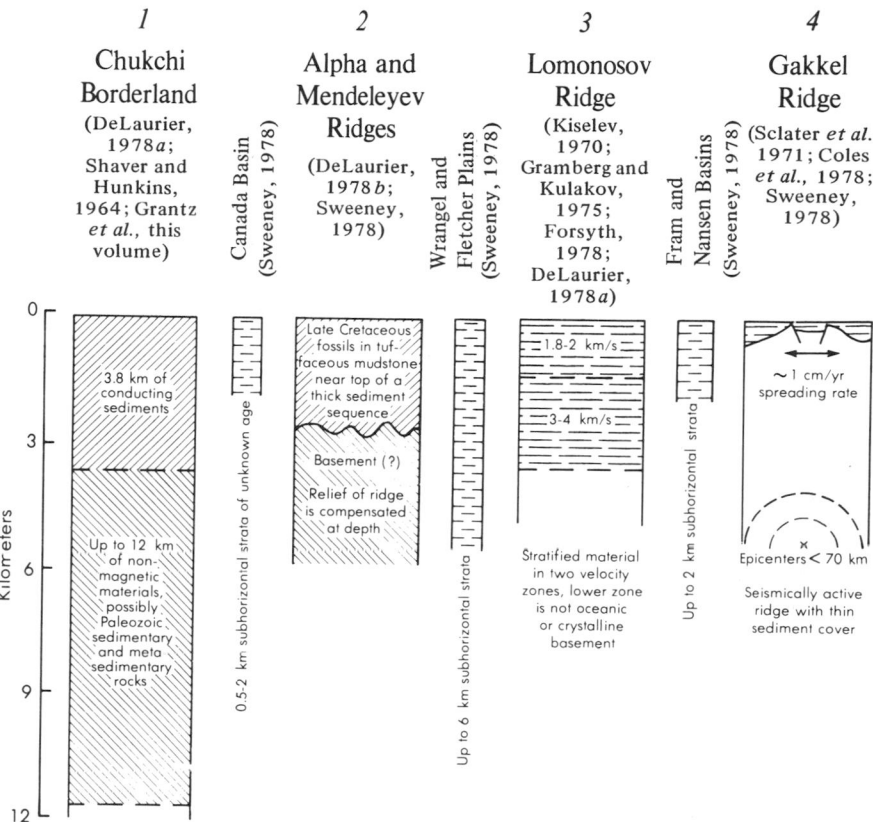

Fig. 2.   Columnar sections of the major bathymetric features of the Arctic Ocean Basin.

mark sutures and zones of collision. When a small terrane can be shown to have its own basement and to be capable of acting as a discrete lithospheric plate, we call it a microplate.

II. TECTONOSTRATIGRAPHIC TERRANES

Within the foldbelts rimming the Pacific Ocean Basin, numerous tectonostratigraphic terranes have been recognized as being discrete oceanic and microcontinental terranes that were successively accreted to the larger continental plates (Monger *et al.*, 1972; Jones *et al.*, 1972; Churkin and Eberlein, 1977). Some of these allochthonous terranes are true microplates with their own basement in the form of ophiolites or crystalline complexes.

A comparable analysis of the Arctic focusing on the basement, stratigraphic succession, and structure of key features (Fig. 1) has identified a preliminary set of tectonostratigraphic terranes. The basic geologic data from

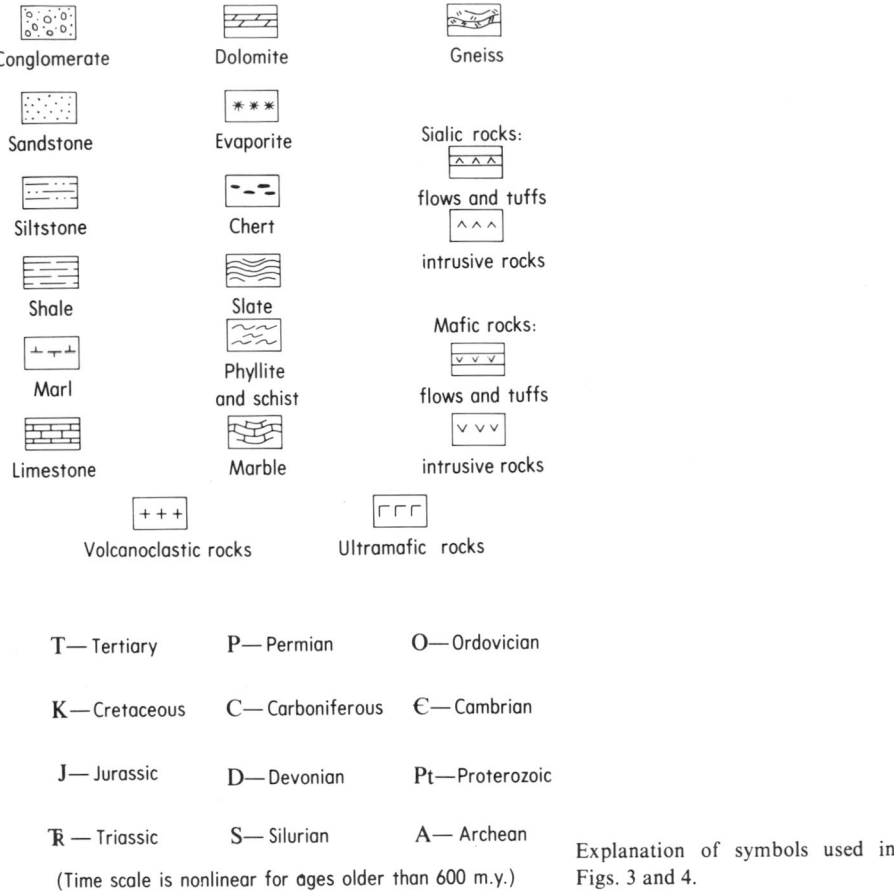

Conglomerate	Dolomite	Gneiss
Sandstone	Evaporite	Sialic rocks:
Siltstone	Chert	flows and tuffs
Shale	Slate	intrusive rocks
Marl	Phyllite and schist	Mafic rocks: flows and tuffs
Limestone	Marble	intrusive rocks
Volcanoclastic rocks		Ultramafic rocks

T—Tertiary   P—Permian   O—Ordovician

K—Cretaceous   C—Carboniferous €—Cambrian

J—Jurassic   D—Devonian   Pt—Proterozoic

Ŗ—Triassic   S—Silurian   A—Archean

(Time scale is nonlinear for ages older than 600 m.y.)

Explanation of symbols used in Figs. 3 and 4.

each terrane that are critical to any interpretation of regional relations are shown in the form of generalized stratigraphic columns (Figs. 2–4).

The resulting terrane map (Fig. 5), when compared to a generalized tectonic map (Churkin, 1973, p. 492), shows that the large platform areas are major continental plates composed mainly of shelf or continental sedimentary sequences laid down on some type of crystalline basement (for example, Fig. 4, columns 30–34). The foldbelts separating the major plates from one another, either continent to continent or continent to ocean, are composed of smaller terranes down to the scale of individual fault slices (Fig. 3, columns 7–9, 15–18). The narrow foldbelt terranes, unlike the broader continental terranes, are generally rich in volcanic, plutonic, and ocean-floor continental-margin sequences that lie on various basement rocks, including ophiolites, metaplutonic, and volcanic (remnant arc) rocks (Fig. 3, columns 5, 8–10, 14–18).

III. RELATIONS BETWEEN TERRANES

Placement of the boundaries of the Arctic terranes is preliminary, and many contacts, especially between the older terranes, are obscured by younger overlapping sedimentary sequences (for example, Verkhoyansk Foldbelt and Sverdrup Basin) or intrusive rocks (Cherskiy Foldbelt), or volcanic rock cover (Okhotsk–Chukotka magmatic arc across southern Chukotka) (Fig. 5). In other places, the original terrane boundaries are sharply defined along younger faults that reactivated the old sutures (Tintina and Farewell–Denali Fault Systems). Blueschists (Fig. 3: Penzhina Foldbelt, column 14, and Koryak Mountains, columns 15–18) along some of these sutures indicate low-temperature/high-pressure tectonics at plate boundaries. Relations between the major continental plates are shown in Fig. 6.

In Alaska, continental sequences make up the Yukon–Porcupine terrane of east-central Alaska, much of the Brooks Range and the North Slope, and the Seward Peninsula. The Yukon–Porcupine terrane (Fig. 3, column 6), with its well-developed late Precambrian section (Tindir Group) and overlying,

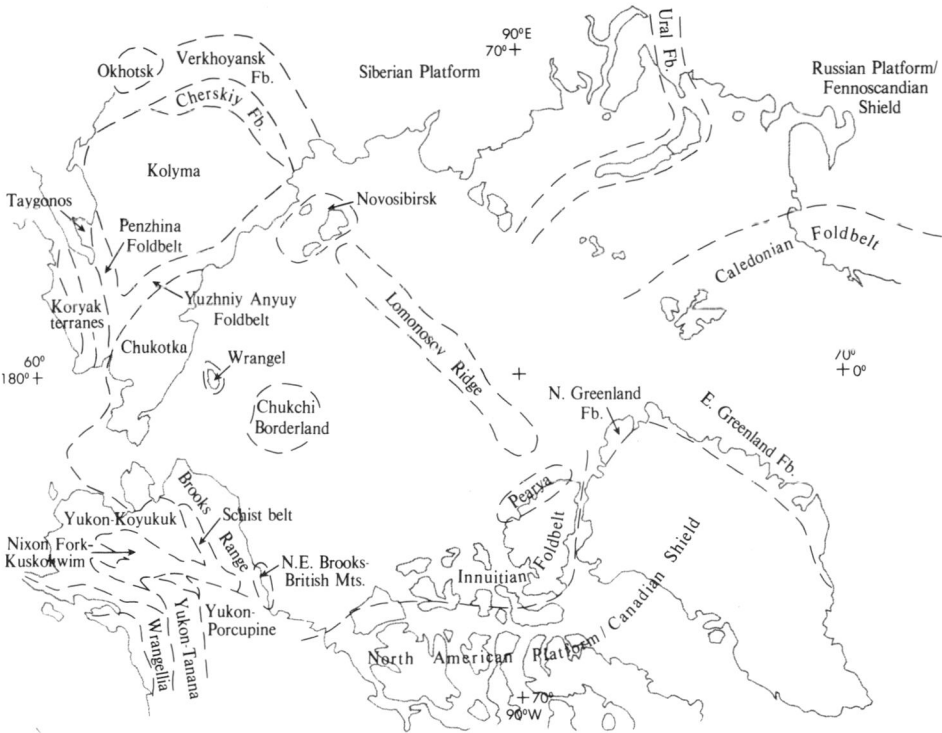

Fig. 5.   Tectonostratigraphic terranes of the Arctic.

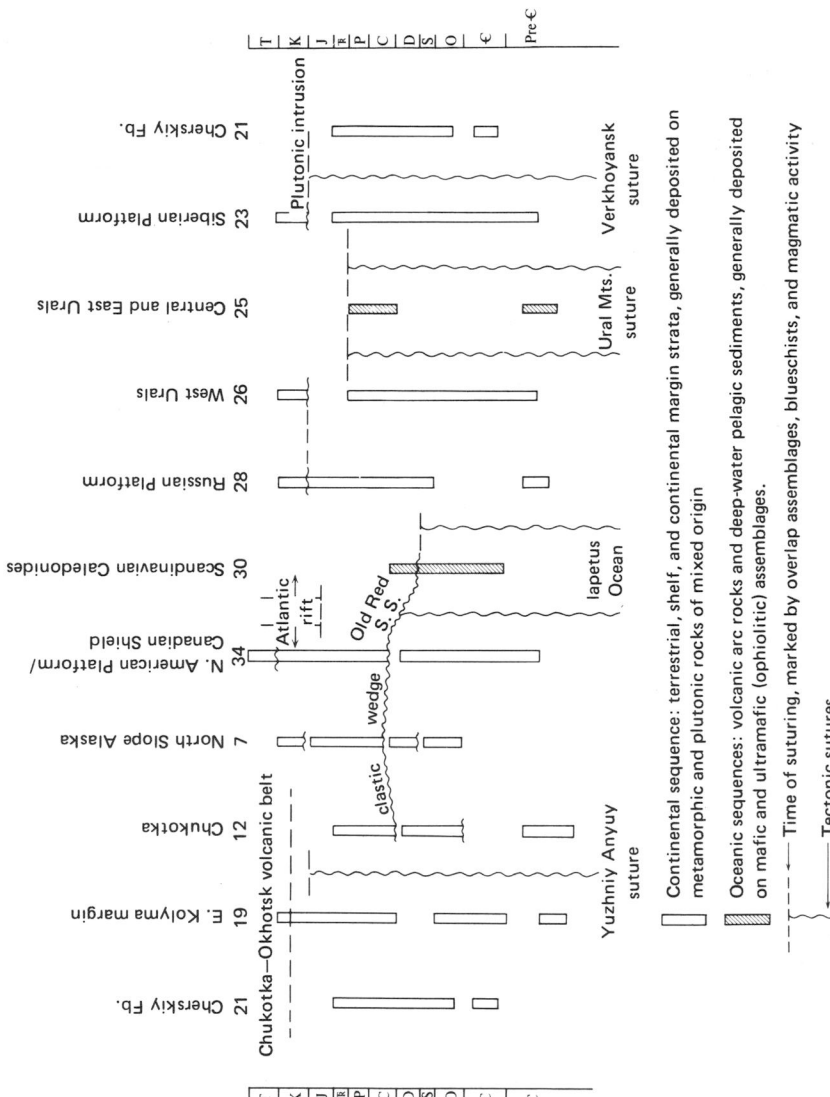

Fig. 6.   Regional tectonic relations of major terranes rimming the Arctic.

nearly complete Paleozoic sequence of platform rocks, correlates well with the late Precambrian and Paleozoic sequences in Yukon Territory that overlap eastward onto the Canadian Shield. Recognition of bona fide North American geology in other parts of Alaska is less certain, and terrane analyses, coupled with paleomagnetic results (Richards, 1974; Jones *et al.*, 1977; Churkin and Eberlein, 1977; Berg *et al.*, 1978; Churkin and Carter, 1979; Rob Van der Voo and Meridee Jones, 1979, written communication), indicate that much of the rest of Alaska is made of allochthonous terranes of uncertain origin.

Paleozoic rocks like those in the Yukon–Porcupine terrane occur in southwestern Alaska in the Kuskokwim–Nixon Fork region, but there they overlie a Precambrian crystalline basement (Fig. 3). The Paleozoic carbonate rocks in Nixon Fork and in the Yukon–Porcupine area (Fig. 3, column 6) become more shaly and attenuated to the south. A series of mainly younger and mainly oceanic terranes has been accreted to their southern margin (Churkin and Carter, 1979). Reconstruction of a similar facies shift in thrust slices along the Brooks Range (Fig. 3, columns 7 and 8) suggests that another collision-deformed Paleozoic continental margin extends along the southern Brooks Range and eastern Seward Peninsula (Churkin *et al.*, 1979). Reconstruction of these continental margins indicates that the ancient continental backbone of Alaska extends in a narrow arcuate belt across the center of Alaska to the Nixon Fork–upper Kuskokwim region, where it terminates near the Aniak–Thompson Creek Fault (Fig. 9). The presence of a similar continental margin along the Brooks Range suggests that the continental and accreted microplate framework of southwestern Alaska may have been offset right-laterally northeastward to the present-day Brooks Range (Fig. 9) (Churkin and Carter, 1979). Distribution of carbonate platform rocks in Seward Peninsula (Fig. 3, column 11), St. Lawrence Island, and the Chukotka Peninsula (Fig. 3, column 12) shows that this Arctic continental platform continued with oroclinal bends from Alaska across the Bering Sea to Chukotka, establishing a western extension of the North American plate into Eurasia (Churkin, 1970; Patton and Tailleur, 1977).

If these correlations are correct, then the Yuzhniy Anyuy Foldbelt marks the western margin of the North American plate (Fujita, 1978), where it is in contact with the Kolyma plate (Figs. 5 and 11). The Yuzhniy Anyuy suture appears to be a continent–continent collision zone that closed in the Late Cretaceous (Fig. 12) (Fujita, 1978). However, there is considerable controversy concerning the Kolyma plate (Fig. 3, columns 19 and 20) (Puscharovskiy, 1977). Originally it was considered to be an entirely continental plate with Precambrian basement and Paleozoic platform cover, but new data suggest that its interior may include a Mesozoic eugeosynclinal sequence (Shilo *et al.*, 1973).

The Cherskiy Foldbelt (Fig. 3, column 21; Fig. 5) along the west side of the Kolyma plate contains a thrust-faulted Paleozoic sedimentary sequence that can be interpreted as representing a collision-deformed continental margin separated from the Siberian platform by both a broad foldbelt (Verkhoyansk Mts.–Ilya-Debin Synclinorium) and a suture zone containing a Late Jurassic or Early Cretaceous magmatic arc (Fig. 6) (Churkin, 1972). Farther west, collision during the Permian and Triassic of the Siberian Platform and the Fennoscandian Shield formed the Ural Foldbelt (Hamilton, 1970). Originally the Siberian Platform may have been part of North America that broke off in the late Precambrian (Sears and Price, 1978).

The history of interaction between the Fennoscandian and North American Shields is long and complex but has been worked out in some detail (for example, Gee, 1975; Siedlecka, 1975). Briefly, the modern Atlantic rift has been a zone of activity since at least the late Precambrian when it opened to form the Iapetus Ocean (Fig. 6). This ocean closed in the Early Silurian, and continued compression during the Silurian and Devonian formed the Caledonian Foldbelt (Dewey, 1969). Reopening of the northern Atlantic commenced approximately 81 m.y. ago, and spreading on the Gakkel Ridge during the past 63 m.y. formed the Eurasian Basin (Pitman and Talwani, 1972).

This assembly of continental plates that front on the Arctic indicates a more or less continuous relative motion of the continental plates, resulting mainly in collision but also rifting (Fig. 6). Besides these major continental plates, small tectonostratigraphic terranes rimming the Arctic have geologic histories so different from their neighbors that they suggest allochthonous borderland terranes similar to those rimming the northern Pacific Basin (Figs. 5 and 7).

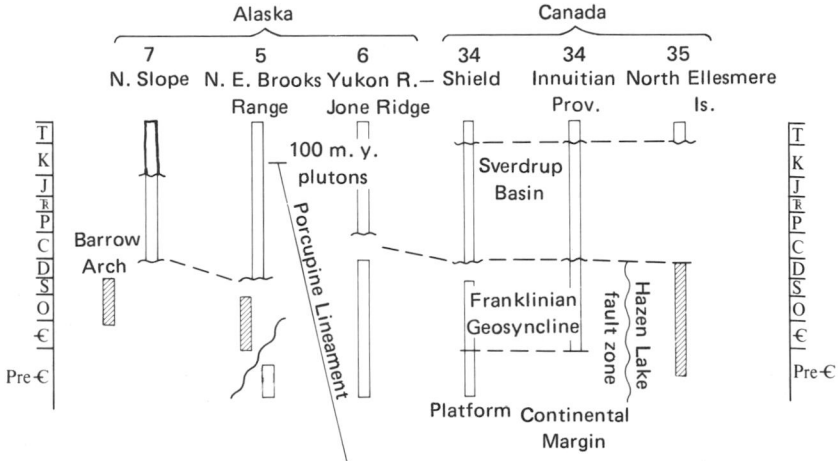

Fig. 7. Regional tectonic relations of the major plates and accreted terranes in the North American Arctic. See Fig. 6 for explanation.

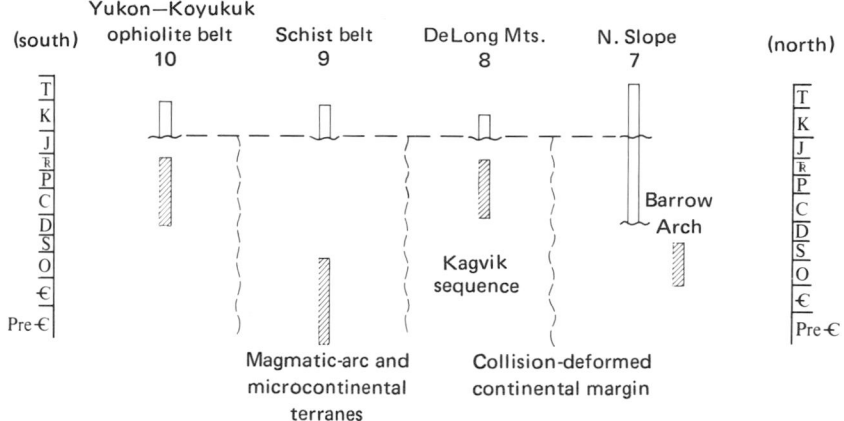

Fig. 8. Regional tectonic relations of the major terranes of the western Brooks Range, Alaska—Kobuk River to Point Barrow. See Fig. 6 for explanation.

In the Arctic Basin, the submarine Lomonosov Ridge (Fig. 2, column 3; Fig. 5) appears to be a microcontinental plate that rifted from the Barents and Kara Sea margin about 63 m.y. ago as the Mesozoic Atlantic opening continued. More controversial is the hypothesis of Mesozoic rifting and rotation of Alaska from the Canadian Arctic (Carey, 1955; Tailleur, 1973). Whatever the ultimate origin of the Canada Basin (Fig. 2), the geologic history of early Paleozoic and Precambrian rocks in the northern Ellesmere Island terrane (Fig. 4, column 35) and in the pre-Mississippian sequence of the northeastern Brooks Range and British Mountains (Fig. 3, column 5; Fig. 7), including perhaps the pre-Mississippian rocks of the Barrow Arch and Wrangel Island (Fig. 3, column 13; Fig. 7), suggests that they too are borderland terranes or magmatic arcs, and may be allochthonous.

A much larger number of enigmatic borderland terranes rims the northern Pacific Basin. Most of these contain sequences of oceanic origin rich in volcanic and plutonic rocks. In northern Alaska these are accreted to a collision-deformed Paleozoic continental margin along the south side of the Brooks Range (Fig. 8) and farther south along a similar continental margin that is subparallel to the Tintina and Denali–Farewell Fault Systems (Fig. 9). Outward to the Pacific, the terranes successively range in age from Precambrian and early Paleozoic to Cenozoic, a time span implying a long history of accretion (Churkin and Carter, 1979).

West of the Nixon Fork–Kuskokwim region of southwestern Alaska, the continental plate of Precambrian schist and overlying Paleozoic carbonate platform strata with its south-facing continental margin terminates and does not reappear in the Koryak Mountains, a correlative terrane across the Bering Sea (Fig. 5). Instead, west of the Nixon Fork area lies a mainly oceanic ter-

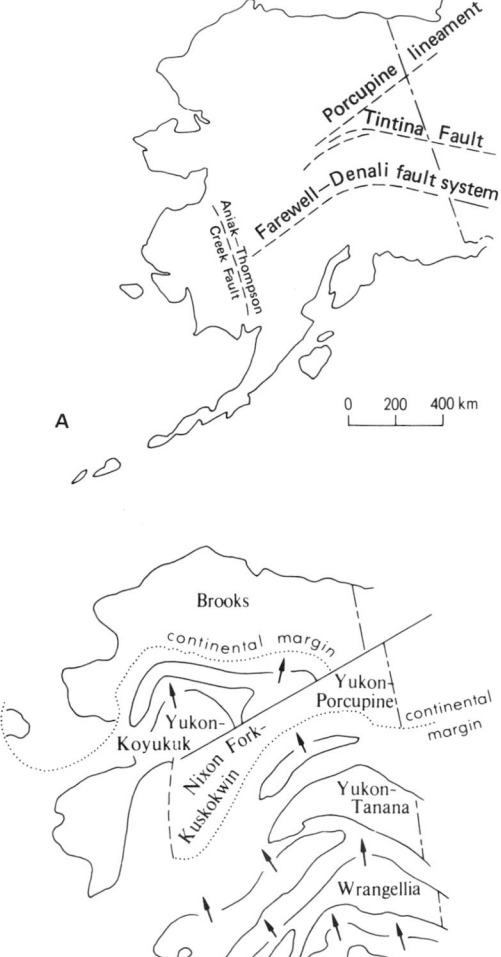

Fig. 9. Alaska; showing (A) the major fault systems, and (B) some of the larger terranes pulled apart to show the sequence of accretion (modified after Churkin and Carter, 1979).

rane of late Paleozoic and early Mesozoic age that extends northward to form the large Yukon–Koyukuk Basin containing a late Paleozoic ophiolite complex (Patton and Tailleur, 1977) (Fig. 3, column 10; Fig. 9).

This Yukon–Koyukuk oceanic terrane in turn fronts on the Brooks Range and Seward Peninsula on its north and west sides (Figs. 1 and 8). The Brooks Range–Seward Peninsula terrane, like the Nixon Fork–Kuskokwim terrane, has a Paleozoic carbonate platform cover that seems to overlie a Precambrian basement. Also, a collision-deformed continental margin with associated microcontinental and volcanic arc terranes comparable to that found in the Nixon Fork–Kuskokwim terrane has been recognized in the western Brooks

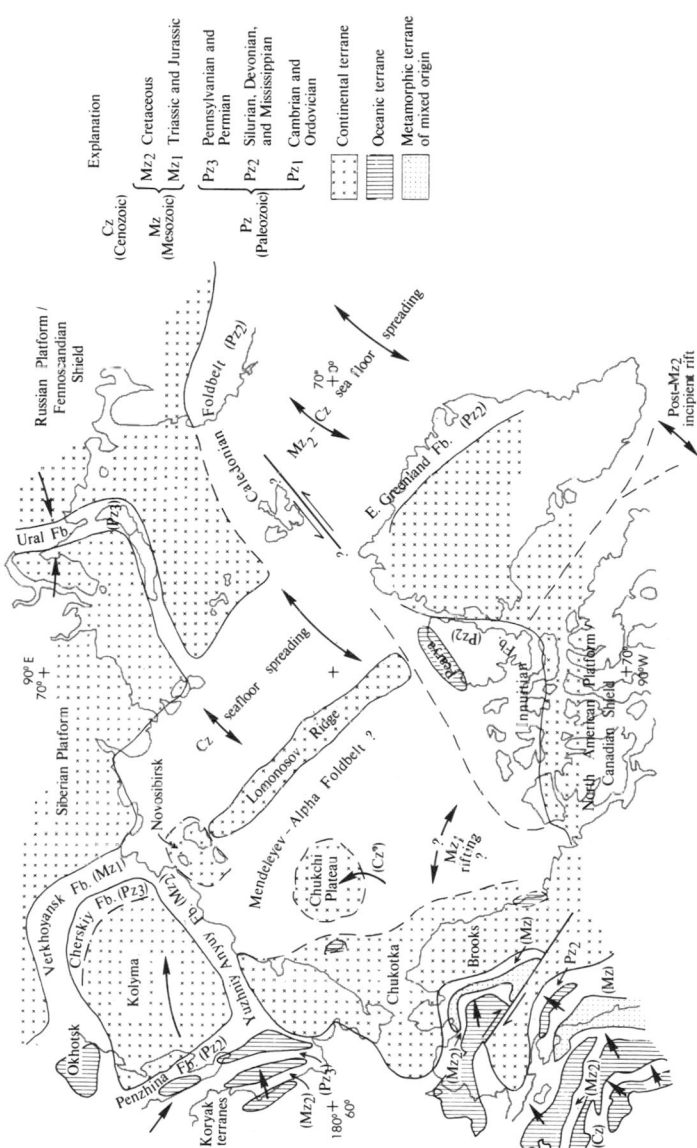

Fig. 10. Continental plates and accreted terranes of the Arctic with minor paleogeographic reconstructions to show trajectories of drift, accretion, and rifting.

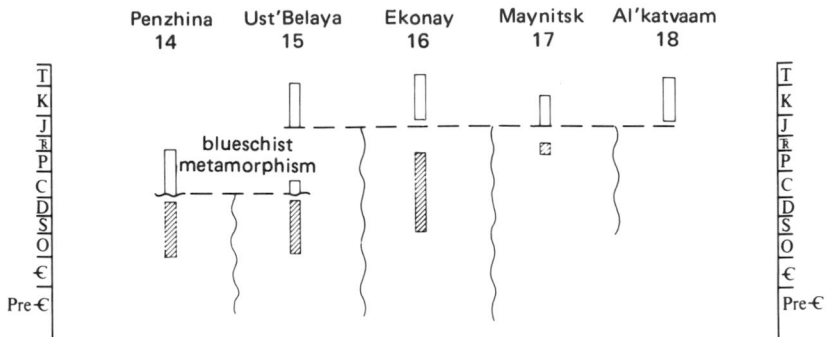

Fig. 11.   Regional tectonic relations of accreted terranes in the Koryak Mountains. See Fig. 6 for explanation.

Range (Churkin et al. 1979). These similarities between the Brooks Range and Nixon Fork–Kuskokwim terranes suggest that the Brooks Range, together with the Seward Peninsula and Chukotka, formed a plate that may have been cut from the west edge of the Nixon Fork terrane and translated north-eastward to its present position.

The Porcupine lineament appears to mark the suture between the Brooks Range and the Yukon–Porcupine terranes (Fig. 9). The enigmatic Yukon–Koyukuk oceanic terrane nearly surrounded by continental terranes is thought to be a product of mid-Paleozoic rift or aulacogene development within a continental framework (Patton *et al.*, 1977). We believe instead that the Yukon–Koyukuk terrane is a captured fragment of a proto-Pacific plate (perhaps part of the Kula plate) as described below. Early and mid-Paleozoic sequences rich in volcanic and plutonic rocks that occur in the southern Brooks Range and particularly in the Donnerak and Ambler River areas can be interpreted as magmatic arc rocks. These oceanic rocks are difficult to account for if the Yukon–Koyukuk terrane is interpreted as an intracratonic aulacogen.

The continental backbone of Alaska to which terranes from the Pacific have been accreted continues from Seward Peninsula to Chukotka Peninsula across the Bering Strait via St. Lawrence Island (Patton and Tailleur, 1977). The west margin of the Chukotka plate is sutured to the Kolyma plate along the Yuzhniy Anyuy Foldbelt (Figs. 6 and 10).

On the Pacific side of the Chukotka and Kolyma plates that make up the continental backbone of northeastern USSR, there are, as in Alaska, accreted terranes mainly of oceanic origin (Figs. 1 and 11). The Penzhina Fold-belt characterized by Devonian or Mississippian ophiolites and blueschist (Dobretsov, 1974) marks the suture between the Chukotka continental plate and successively accreted oceanic microplates in the Koryak Mountains and in the Taygonos Peninsula (Nekrasov, 1976; Fujita, 1978). The record of early

Paleozoic, late Paleozoic, and late Mesozoic ophiolities and associated blue-schists (Firsov and Dobretsov, 1969; Dobretsov, 1974) indicates a long history of oceanic terrane accretion in the Soviet northeast resembling that in the North American Cordillera.

IV. REVIEW OF PALEOMAGNETIC DATA

Paleomagnetic data for terranes of the Arctic are sparse, but polar wandering paths have been established for a few terranes, primarily utilizing results from the USSR (McElhinny, 1973; Irving, 1977). Hamilton (1970) used paleomagnetic data to show that the Siberian and Russian Platforms sutured in the late Paleozoic and afterward traveled together in a roughly northern direction to their present position (Fig. 12). The Kolyma plate, which makes up most of the northeast USSR, moved north from the central Pacific since the Paleozoic and collided with the Siberian plate in the mid-Mesozoic (McElhinny, 1973). This travel path, based on paleomagnetic data for Kolyma, does not support the proposal by Herron et al. (1974) that Kolyma came across the Arctic Basin to collide with North America in the Paleozoic and then moved back to its present position. Irving (1977) reconstructed the drift of the major continental plates from the Devonian to the present and showed that in the beginning of the Mesozoic continental plates amassed in the mid-Atlantic (Laurasia). In the Jurassic, Laurasia broke up and major plates drifted north into the Arctic. The Siberian–Russian Platform moved around the Arctic to the east, while the North American plate moved around and into the Arctic to the west—the overall movement being pincerlike with the jaws of the pincers closing the ocean between Alaska–Chukotka and Siberia at the Yuzhniy Anyuy suture (Fig. 12).

In addition to this general northward movement of major continental plates, northward-moving smaller plates, including some of oceanic origin, have been recognized in the Cordillera (Jones et al., 1972). Northward displacement of some of these terranes has been supported by paleomagnetic data (Packer and Stone, 1971; Beck, 1976). The Wrangellia terrane, a unique lava and sedimentary sequence of Triassic age, has been shown to be displaced about 30° northward (Jones et al., 1977). The Alexander terrane, a volcanic arc terrane with rocks of Ordovician through Permian age in southeastern Alaska, has been shown to be displaced about 15° northward (Jones et al., 1972; Rob Van der Voo and Meridee Jones, 1979, written communication).

Paleomagnetic results from the high Arctic are less conclusive. Preliminary data from the Brooks Range have been interpreted to support the hypothesis of counterclockwise rotation of the Brooks Range from the Canadian Arctic Islands (Newman et al., 1977). However, the Brooks Range paleomagnetic sampling has been on allochthonous thrust slices whose orientation is suspect. Furthermore, a widespread Cretaceous or younger thermal remagnetization event that has not been taken into account may have obliterated any older paleomagnetic features.

V. DISCUSSION

Figure 10 outlines the major plates and terranes discussed above and shows by arrows and with minor paleogeographic reconstruction (not time specific) the former positions and relations of the Arctic terranes. The long history of successive collisions and terrane accretions around the rim of the northern Pacific suggests that a continental framework existed as a backstop or docking facility for migrating terranes, preventing them from bypassing Alaska and sweeping into the Arctic Basin. The collision and suturing of terranes along the Pacific margin of the Brooks plate had certainly started by Cretaceous time and possibly much earlier (Churkin et al., 1979) (Fig. 8). The Penzhina Foldbelt (Fig. 11) separating Pacific terranes from the Chukotka and Kolyma plates has a record of collision and suturing in the late Paleozoic (Dobretsov, 1974). These data indicate that the continental framework of plates separating the Arctic from the Pacific had developed over a long period of time, culminating in a continental assembly resembling the present configuration by the Late Cretaceous (Fig. 6).

In the Paleozoic, the northern continental blocks must have formed a very open framework with wide oceans between some of them (Fig. 12). Paleomagnetic data suggest that the continental blocks started to converge with the suturing of the Urals and continued converging as the Proto-Pacific plates, including the Kolyma block, moved north and collided with Siberia. Closure of the strait separating Kolyma from Chukotka (associated with Atlantic rifting) sealed off the Arctic Basin in the early Cretaceous, trapping the leading edge of the proto-Pacific "Kula" plate to form part of the modern Amerasian Basin. Our reconstructions show:

1. Continental plates reach the polar region and begin circumpolar drift that narrows the ocean separating Chukotka–Alaska from Siberia (Fig. 12A). This Early Jurassic reconstruction does not show the counterclockwise rotation of the Chukotka–Alaska Peninsula from the Canadian Arctic Islands as suggested by Carey (1955) and Tailleur (1973). If Chukotka–Alaska was not a peninsula at this time, an even wider ocean would have separated North America and Siberia, providing more direct access to the Arctic for the Kula plate but excluding the Kula plate from the Canada Basin proper.

2. Continued circumpolar drift of continents in combination with

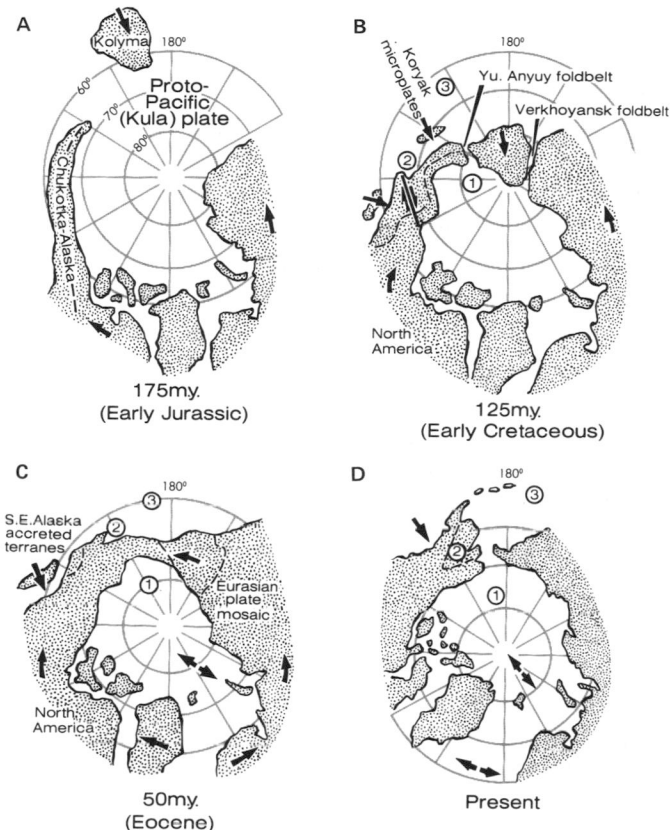

Fig. 12.   Paleogeographic reconstructions of the Arctic showing northward drift, collision, and accretion. (Based on paleomagnetic reconstructions by Irving, 1977, and interpretations presented in this chapter.)

northward drift of Kula and Kolyma plates results in collsion and suturing of Kolyma and Siberia, and suturing of Alaska–Chukotka with the Eurasian plate mosaic (Fig. 12B). Multiple collisions and accretion of microplates and terranes against the Pacific margin of Alaska–Chukotka result in its oroclinal deformation and shear. Part of the Pacific (Kula) plate, isolated behind the newly formed continental mosaic, forms the nucleus for the Arctic Basin.

3.   Circumpolar drift continues with the opening of the Atlantic and Eurasian basins, deforming the continental connection between North America and Eurasia (Fig. 12C).

4.   Modern polar configuration shows the present position of three possible remnants of the Kula plate (Fig. 12D): (1) part of the Amerasian Basin, (2) the Yukon–Koyukuk Basin, and (3) the Bering abyssal plain (Cooper et al., 1976).

VI. CONCLUSIONS

An analysis of the stratigraphy and tectonics of the key features of Arctic geology provides a preliminary outline of the major northern continental plates. Foldbelts that developed along the edges of these plates, particularly those fronting on the Pacific and, to a lesser degree, the Arctic Basin are in turn composed of a number of smaller terranes and microplates. Most of these terranes appear to be oceanic, but some are microcontinental blocks or metamorphic terranes of uncertain origin. Paleomagnetic data indicate a general northward movement of these plates during the Paleozoic and early Mesozoic, resulting in an open framework of continental plates separated around the pole by wide oceans. During this time the proto-Pacific plate extended into the Arctic. Atlantic and then Eurasian Basin opening and general northward drift resulted in collision and suturing in Eurasia. Northward-moving microplates and terranes also were accreted to the growing continental mosaic, bridging North America with Eurasia and thereby sealing off a part of the proto-Pacific plate (Kula plate) to form the nucleus of the modern Arctic Basin.

ACKNOWLEDGMENTS

The U.S. and Soviet Academies of Sciences sponsored exchange visits for Churkin to the USSR in 1967 and 1975. Special thanks are given to Canadian geologists H. Gabrielse, J. W. H. Monger, J. W. Kerr, and H. P. Trettin; to many Soviet geologists, including N. L. Dobretsov, G. I. Kameneva, A. M. Karasik, V. M. Merzlyakov, L. M. Parfenov, and S. M. Tilman; and to northern European geologists P. R. Dawes, M. B. Edwards, D. G. Gee, N. Henricksen, David Roberts, A. Siedlecka, and A. Simonen for discussion and much information.

REFERENCES

Aleksandrov, A. A., 1973, Ophiolites of the Ust'Belaya Mountains, *Dokl. Akad. Nauk SSSR*, v. 219(1), p. 171–174.

Aleksandrov, A. A., Bogdanov, N. A., Byalobzheskiy, S. G., Markov, M. S., Til'man, S. M., Khain, V. E., and Chekhov, A. D., 1975, New data on the tectonics of the Koryak Highlands, *Geotectonics (USSR)*, v. 9(5), p. 292–299.

Baragar, W. R. A., and McGlynn, J. C., 1976, Early Archean basement in the Canadian Shield: A review of the evidence, *Geol. Surv. Can. Pap.* 76-14, 21 p.

Beck, M. E., Jr., 1976, Discordant paleomagnetic pole positions as evidence of regional shear in the western Cordillera of North America, *Am. J. Sci.*, v. 276, p. 694–712.

Berg, H. C., Jones, D. L., and Coney, P. J., 1978, Map showing pre-Cenozoic tectonostratigraphic terranes of southeastern Alaska and adjacent areas, *U.S. Geol. Surv. Open-File Rep.* 78-1085, 2 sheets, scale 1:1,000,000.

Bird, K. J., Connor, C. L., Tailleur, I. L., Silberman, M. L., and Christie, J. L., 1978, Granite on the Barrow Arch, northeast NPR-A, in: *The U.S. Geological Survey in Alaska: Accomplishments during 1977*, Johnson, K. M., ed., *U.S. Geol. Surv. Circ.* 772-B, p. B24–B25.

Birkelund, T., Perch-Nielsen, K., Bridgwater, D., and Higgins, A. K., 1974, An outline of the geology of the Atlantic coast of Greenland, in: *The Ocean Basins and Margins*, v. 2: *The North Atlantic*, Nairn, A. E. M., and Stehli, F. G., eds., New York: Plenum Press, p. 125–155.

Bogdanov, N. A., and Til'man, S. M., 1964, Obshchie cherty razvitye Paleozoiskikh struktur ostrova Wrangelia i zapadnoi chasti Khrebta Bruksa (Aliaska), in: *Soveshchanie po Problem Tektoniki, Moskva 1963, Skladchatye Oblasti Evrazii, Materialy*, Moscow: Nauka, p. 219–230.

Brabb, E. E., and Churkin, M., Jr., 1969, Geological map of the Charley River quadrangle, east-central Alaska, *U.S. Geol. Surv. Miscell. Geol. Invest. Map* I-573, 1 sheet, scale 1:250,000.

Brosgé, W. P., and Reiser, H. N., 1964, Geologic map and section of the Chandalar quadrangle, Alaska, *U.S. Geol. Surv. Miscell. Geol. Invest. Map* I-375, scale 1:250,000.

Bütler, H., 1961, Continental Carboniferous and Lower Permian in central East Greenland, in: *Geology of the Arctic*, V. I, Raasch, G. O., ed., Toronto: Toronto Universtiy Press, p. 205–213.

Callomon, J. H., 1961, The Jurassic system in East Greenland, in: *Geology of the Arctic*, v. I, Raasch, G. O., ed., Toronto: Toronto University Press, p. 258–268.

Carey, S. W., 1955, Orocline concept in geotectonics, part I. *Pap. Proc. R. Soc. Tasmania*, v. 89, p. 255–288, Tasmania University Department of Geology Publication 28.

Carter, C., and Laufeld, S., 1975, Ordovician and Silurian fossils in well cores from the North Slope of Alaska, *Am. Assoc. Petrol. Geol. Bull.* v. 59(3), p. 457–464.

Churkin, M., Jr., 1970, Fold belts of Alaska and Siberia and drift between North America and Asia, in: *Proceedings of the Geological Seminar on the North Slope of Alaska, Palo Alto, California, 1970*, Adkison, W. L., and Brosgé, M. M., eds., Los Angeles, California: American Association of Petroleum Geologists, Pacific Section, p. G1–G17.

Churkin, M., Jr., 1972, Western boundary of the North American continental plate in Asia, *Geol. Soc. Am. Bull.*, v. 83, p. 1027–1036.

Churkin, M., Jr., 1973, Geologic concepts of Arctic Ocean Basin, in: *Arctic Geology*, Pitcher, M. G., ed., *Am. Assoc. Petrol. Geol. Mem.* 19, p. 485–499.

Churkin, M., Jr., and Carter, Claire, 1979, Collision-deformed Paleozoic continental margin in Alaska—A foundation for microplate accretion [abs.], *Geol. Soc. Am. Abstr. Progr. Cordill. Sect.*, v. 11(3), p. 72.

Churkin, M., Jr., and Eberlein, G. D., 1977, Ancient borderland terranes of the North American Cordillera: Correlation and microplate tectonics, *Geol. Soc. Am. Bull.*, v. 88, p. 769–786.

Churkin, M. J., Jr., Nokleberg, W. J., and Huie, C., 1979, Collision-deformed Paleozoic continental margin, Western Brooks Range, Alaska, *Geology*, v. 7, p. 379–383.

Coles, R. L., Hannaford, W., and Haines, G. V., 1978, Magnetic anomalies and the evolution of the Arctic, in: *Arctic Geophysical Review*, Sweeney, J. F., ed., *Can. Dep. Energy Mines Resources*, v. 45(4), p. 51–66.

Cooper, A. K., Scholl, D. W., and Marlow, M. S., 1976, Plate tectonic model for the evolution of the eastern Bering Sea Basin, *Geol. Soc. Am. Bull.*, v. 87, p. 1119–1126.

Dawes, P. R., 1971, The north Greenland foldbelt and environs, *Geol. Soc. Denmark Bull.*, v. 20, p. 197–239.

Dawes, P. R., 1973, The north Greenland foldbelt: A clue to the history of the Arctic Ocean Basin and the Nares Strait lineament, in: *Implications of Continental Drift to the Earth Sciences*, Tarling, D. H., and Runcorn, S. K., eds., New York: Academic Press, v. 2, p. 925–948.

DeLaurier, J. M., 1978a, Arctic Ocean sediment thicknesses and upper mantle temperatures from magnetotelluric soundings, in: *Arctic Geophysical Review*, Sweeney, J. F., ed., *Can. Dep. Energy Mines Resources*, v. 45(4), p. 35–50.

DeLaurier, J. M., 1978*b*, The Alpha Ridge is not a spreading center, in: *Arctic Geophysical Review*, Sweeney, J. F., ed., *Can. Dep. Energy Mines Resources*, v. 45(4), p. 87–90.

Dewey, J. F., 1969, Evolution of the Appalachian/Caledonian orogen, *Nature (London)*, v. 222, p. 124–129.

Dillon, J. T., and Pessel, G. H., 1979, Tectonic significance of Late Devonian and Late Proterozoic U/Pb zircon ages from metaigneous rocks, Brooks Range, Alaska [abs.], *Geol. Soc. Am. Abstr. Progr. Cordill. Sect.*, v. 11(3), p. 75.

Dobretsov, N. L., 1974, Lavsonit-glaukofanovye slantsy Penzhinskogo poyasa i drugie Paleozoiskie glaukofanslantsevye poyasa, in: *Glaukofanslantsevye i Eklogit-Glaukofanslantsevye Komleksy SSSR*, Sobolev, V. S., ed., Novosibirsk: Nauka, Siberskoe otdelenie, p. 26.

Firsov, L. V., and Dobretsov, N. L., 1969, Age of glaucophane metamorphism at the northwestern fringe of the Pacific Ocean, *Dokl. Akad. Nauk SSSR*, v. 185(4), p. 883–886.

Forsyth, D. A., 1978, Review of Arctic crustal studies, in: *Arctic Geophysical Review*, Sweeney, J. F., ed., *Can. Dep. Energy Mines Resources*, v. 45(4), p. 75–86.

Fujita, K., 1978, Pre-Cenozoic tectonic evolution of northeast Siberia: *J. Geol.*, v. 86(2), p. 159–172.

Gale, G. H., and Roberts, D., 1974, Trace element geochemistry of Norwegian Lower Paleozoic basic volcanics and its tectonic implications, *Earth Planet. Sci. Lett.*, v. 22, p. 380–390.

Gee, D. G., 1975, A tectonic model for the central part of the Scandinavian Caledonides, *Am. J. Sci.*, v. 275-A, p. 468–515.

Gnibidenko, G. S., 1969, *Metamorficheskie Kompleksy v Struktyrakh Severo-Zapadnogo Sektora Tikho Okeanskogo Poyasa*, Novosibirsk: Nauka, Siberskoe otdelenie, 134 p.

Gramberg, I. S., and Kulakov, Yu. N., 1975, General geological features and possible oil and gas provinces of the Arctic Basin, in: *Canada's Continental Margins*, Yorath, C. J., Parker, E. R., and Glass, D. J., eds., *Can. Soc. Petrol. Geol. Mem.* 4, p. 525–529.

Grinberg, G. A., Gusev, G. S., Milanovskiy, E. E., Mokshantsev, K. B., Slavin, V. I., Khain, V. E., 1977, Constitution and development of the Kolyma massif in the light of new data, *Geotectonics (USSR)*, v. 11(4), p. 260–268.

Hall, J. K., 1973, Geophysical evidence for ancient sea-floor spreading from Alpha Cordillera and Mendeleyev Ridge, in: *Arctic Geology*, Pitcher, M. G., ed., *Am. Assoc. Petrol. Geol. Mem.* 19, p. 542–561.

Haller, J., 1961, Account of Caledonian orogeny in Greenland, in: *Geology of the Arctic*, v. 1, Raasch, G. O., ed., Toronto: Toronto University Press, p. 170–187.

Hamilton, W., 1970, The Uralides and the motion of the Russian and Siberian platforms, *Geol. Soc. Am. Bull.*, v. 81, p. 2553–2576.

Herron, E. M., Dewey, J. F., and Pitman, W. C., 1974, Plate tectonics model for the evolution of the Arctic, *Geology*, v. 2(8), p. 377–380.

Irving, E., 1977, Drift of the major continental blocks since the Devonian, *Nature (London)*, v. 270(5635), p. 304–309.

Jones, D. L., Irwin, W. P., and Ovenshine, A. T., 1972, Southeastern Alaska—A displaced continental fragment?, *U.S. Geol. Surv. Prof. Pap.* 800-B, p. 211–217.

Jones, D. L., Silberling, N. J., and Hillhouse, J., 1977, Wrangellia—A displaced terrane in northwestern North America, *Can. J. Earth Sci.*, v. 14(11), p. 2565–2577.

Kameneva, G. I., 1977, The tectonic setting of Wrangel Island and its Paleozoic structural bonds with Alaska [in Russian]: *Tektonika Arktiki Skladchatiy Fundament Shel-fovikh Sedimentatsionnikh Basseinov*, Leningrad: NIIGA, p. 122–129.

Katz, H. R., 1961, Late Precambrian stratigraphy in East Greenland, in: *Geology of the Arctic*, v. 1 Raasch, G. O., ed., Toronto: Toronto University Press, p. 299–328.

Kiselev, Yu. G., 1970, Some of the features of the present morphotectonic structure of the Lomonosov Ridge based on seismic data [in Russian]: *Morskaya Geologiya i Geofizika*, v. 1, p. 123–128.

Maync, W., 1961, The Permian of Greenland, in: *Geology of the Arctic*, v. 1, Raasch, G. O., ed., Toronto: Toronto University Press, p. 214–223.

McElhinny, M. W., 1973, Paleomagnetism and plate tectonics of eastern Asia, in: *The Western Pacific: Island Arcs, Marginal Seas, Geochemistry*, Coleman, P. J., ed., New York: Crane, Russak, p. 407–414.

Merzlyakov, V. M., 1972, *Stratigrafia i Tektonika Omulevskogo Podnyatiya*, Akademiya Nauk SSSR, v. 19, 151 p.

Miller, T. P., and Sainsbury, C. L., 1975, Alaska: Zone 8, in: *Mesozoic–Cenozoic Orogenic Belts, Data for Orogenic Studies*, Spencer, A. M., ed., Edinburgh: Scottish Academic Press, p. 587–588.

Monger, J. W. H., Souther, J. G., and Gabrielse, H., 1972, Evolution of the Canadian Cordillera; a plate-tectonic model, *Am. J. Sci.*, v. 272, p. 577–602.

Nalivkin, D. V., 1973, Geology of the U.S.S.R. [translated from Russian by N. Rast], Edinburgh: Oliver and Boyd, 855 p.

Nekrasov, G. E., 1976, *Tektonika i Magmatism Taigonosa Severo-Zapadnoi Kamchatki*, Akademiya Nauk SSSR, 150 p.

Newman, G. W., Mull, C. G., and Watkins, N. D., 1977, Northern Alaska paleomagnetism, plate rotation and tectonics [abs.], in: Program and abstracts: Alaska Geological Society Symposium, Anchorage, p. 16–19.

Packer, D. R., and Stone, D. B., 1971, An Alaskan Jurassic paleomagnetic pole and the Alaskan orocline, *Nature (London)*, v. 237, p. 25–26.

Patton, W. W., Jr., and Tailleur, I. L., 1977, Evidence in the Bering Strait region for differential movement between North America and Eurasia, *Geol. Soc. Am. Bull.*, v. 88, p. 1298–1304.

Patton, W. W., Jr., Tailleur, I. L., Brosgé, W. P., and Lanphere, M. A., 1977, Preliminary report on the ophiolites of northern and western Alaska, in: *North American Ophiolites*, Coleman, R. G., and Irwin, W. P., eds., *Oreg. Dep. Geol. Miner. Ind. Bull.* 95, 183p.

Pitman, W. C., and Talwani, Manik, 1972, Sea-floor spreading in the North Atlantic, *Geol. Soc. Am. Bull.*, v. 83, p. 619–646.

Pushcharovskiy, Yu. M., 1977, The problem of the Kolyma Massif, *Geotectonics (USSR)*, v. 11(4), p. 243–244.

Reiser, H. N., Brosegé, W. P., Detterman, R. L., and Dutro, J. T., Jr., 1978, Geologic map of the Demarcation Point quadrangle, Alaska, *U.S. Geol. Surv. Open-File Rep.* 78-526, 1 sheet, scale 1:250,000.

Richards, G. H., 1974, Tectonic evolution of Alaska, *Am. Assoc. Petrol. Geol. Bull.*, v. 58(1), p. 79–105.

Rutten, M. G., 1969, *The Geology of Western Europe*, Amsterdam: Elsevier Publishing Co., 520 p.

Sclater, J. G., Anderson, R. N., and Bell, M. L., 1971, Elevation of ridges and the evolution of the central eastern Pacific: *J. Geophys. Res.*, v. 76, p. 7888–7915.

Sears, J. W., and Price R. A., 1978, The Siberian connection: A case for Precambrian separation of the North American and Siberian cratons, *Geology*, v. 6, p. 267–270.

Shaver, R., and Hunkins, K., 1964, Arctic Ocean geophysical studies; Chukchi Cap and Chukchi Abyssal Plain, *Deep-Sea Res. Oceanogr. Abstr.* v. 11, p. 905–916.

Shilo, N. A., Merzlyakov, V. M., and Terekhov, M. I., 1973, The Alazey–Oloy eugeosynclinal system, a new Mesozoic unit in the northeast USSR: *Dokl. Acad. Sci. USSR Earth Sci. Sect.*, v. 210(5), p. 99–100.

Siedlecka, A., 1975, Late Precambrian Stratigraphy and Structure of the Northeastern Margin of the Fennoscandian Shield (East Finnmark–Timan Region), *Nor. Geol. Unders. Publ.* no. 316, p. 313–348.

Sweeney, J. F., ed., 1978, *Arctic Geophysical Review*, Can. Dep. Energy Mines Resources, v. 45(4), 108 p.

Tailleur, I. L., 1973, Probable rift origin of Canada Basin, Arctic Ocean, in: *Arctic Geology*, Pitcher, M. G., ed., *Am. Assoc. Petrol. Geol. Mem.* 19, p. 526–535.

Thorsteinsson, R., 1961, Lower Paleozoic stratigraphy of the Canadian Arctic Archipelago [abs.], in: *Geology of the Arctic*, v. 1, Raasch, G. O., ed., Toronto: Toronto University Press, p. 380.

Trettin, H. P., 1973, Early Paleozoic evolution of northern parts of Canadian Arctic Archipelago, in: *Arctic Geology*, Pitcher, M. G., ed., *Am. Assoc. Petrol. Geol. Mem*. 19, p. 57–75.

Trettin, H. P., Frisch, T. O., Sobczak, L. W., Weber, J. R., Niblett, E. R., Law, L. K., DeLaurier, J. M., and Whitham, K., 1972, The Innuitian province, in: *Variations in Tectonic Styles in Canada*, Price, R. A., and Douglas, R. J. W., eds., *Geol. Assoc. Can. Spec. Pap*. 11, p. 83–179.

Trümpy, R., 1961, Triassic of East Greenland, in: *Geology of the Arctic*, v. 1, Raasch, G. O., ed., Toronto: Toronto University Press, p. 248–254.

Vol'nov, D. A., Voitzekhovskiy, V. N., Ivanov, O. A., Sorokov, D. S., Yashin, D. S., 1970, Glava X-Novosibirskie ostrova, in: *Geologiya SSSR: Tom XXVI, Ostrova Sovetskoi Arktiki*, Sidorenko, A. A., ed., Moscow: NEDRA, p. 324, 374.

Vogt, P. R., and Ostenso, N. A., 1970, Magnetic and gravity profiles across the Alpha Cordillera and their relation to Arctic sea-floor spreading, *J. Geophys. Res*., v. 75, p. 4925–4937.

Wenk, E., 1961, Tertiary of Greenland, in: *Geology of the Arctic*, v. 1, Raasch, G. O., ed., Toronto: Toronto University Press, p. 278–284.

Chapter 2

GEOLOGY OF ALASKA BORDERING THE ARCTIC OCEAN

J. Thomas Dutro, Jr.
U.S. Geological Survey
Washington, D.C.

I. INTRODUCTION

The geology of Alaska bordering the Arctic Ocean can be visualized as a succession of depositional episodes separated by distinct tectonic or orogenic events that leave identifiable patterns in the geologic framework of the northern third of the state (Fig. 1). These six episodes are: (1) Recent to early Cretaceous, including the Gubik and Sagavanirktok Formations, the Colville and Nanushuk Groups, the Fortress Mountain Formation, Torok Formation, Okpikruak Formation, and Bathtub Graywacke; (2) Jurassic through early Permian, including the Kingak Shale, Shublik Formation, Ivishak Formation, Echooka Formation, and Siksikpuk Formation; (3) Middle Pennsylvanian through early Mississippian, including various carbonate units of the Lisburne Group, the Kayak Shale, and the Kekiktuk Conglomerate; (4) Upper and Middle Devonian, including the Kanayut Conglomerate, Hunt Fork Shale, and associated Frasnian units, and unnamed Middle Devonian sandstone and limestone units; (5) Middle and early Devonian, Silurian, Ordovician, and Cambrian, including the Skajit Limestone, Nanook Limestone, Katakturuk Dolomite, and unnamed early Paleozoic formations; and (6) Precambrian, including the Neruokpuk Quartzite and several unnamed Upper Precambrian sedimentary units in northeasternmost Alaska.

J. Thomas Dutro, Jr.

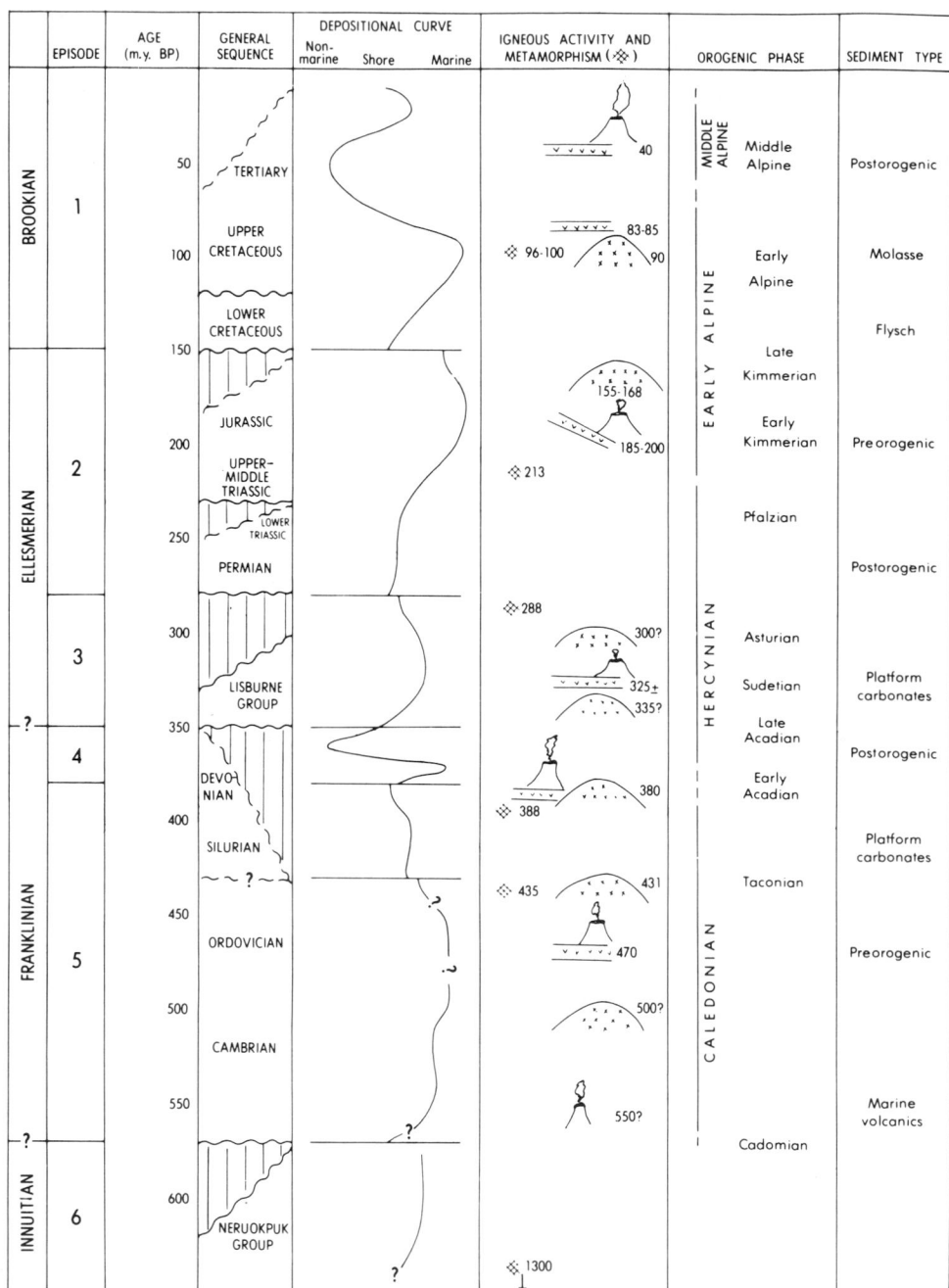

Fig. 1.   Graphic summary of the geology of northern Alaska, showing the depositional episodes, unconformities, transgressive–regressive curve, igneous activity, and approximate correlations.

As one approaches the shoreline from the icy vastness of the Arctic Ocean, the low coastline and its outlying sandy islands, spits, and bars barely rise above sea level. On a fogless day, if one is as far east as the mouth of the Canning River, glacier-covered mountains can be seen on the distant horizon. For hundreds of miles to the west, however, only a few low hills and ridges break the monotonous flatness of the Arctic Coastal Plain. The inland mountains cannot be viewed again until the western capes have been rounded, although bedrock cliffs of the Lisburne Peninsula are reminders of the geologic events preceding the depositional episode that built the Arctic plain.

Enormous quantities of sand, silt, and clay are dumped into the Arctic Ocean by the Colville, Sagavanirktok, Canning, Kongakut, and other rivers that flow northward from the Brooks Range during the short Arctic summers. Great wedges of sediment have built out into the Beaufort and Chukchi Seas, coalescing into a wide shallow marine shelf. Similar processes built the entire coastal plain, beginning in early Cretaceous time when the coastline was as much as 300 km to the south. The geologic history from the earliest Cretaceous can be seen as a backward projection in time of processes that are continuing in the Arctic regions today.

II. ANALYSIS OF DEPOSITIONAL EPISODES

Cretaceous and younger strata reflect the history of the shifting strandline as it retreated northward from the cliffy headlands of the ancient Brooks Range (Fig. 2). Earliest Cretaceous rocks (the Okpikruak and related formations) in places contain great boulders and blocks of strata as old as the Carboniferous that appear to have fallen into the edge of the sea, much as they do today from the cliffs of the Lisburne Peninsula.

Away from these ancient seacliffs, both along the shore and offshore, marine sands and gravels were deposited and, in areas of low terrigenous supply far from the river mouths, marine limestones with coquinas of *Buchia* shells accumulated.

Younger conglomerates of the Fortress Mountain Formation indicate a continuing influx of coarse material from the mountainous terrain, but a general decrease in grain size upward and northward across the coastal plain suggests that these rocks were deposited farther from the highlands as the shoreline advanced northward and eastward. Two large deltaic systems developed in the late early Cretaceous (Nanushuk Group), one in the western foothills near the Arctic coast and the other in the central foothills south of the Colville River.

Complexly intertonguing marine and nonmarine facies, with the development of coal in the Upper Cretaceous, suggest a low fluvial plain with a much

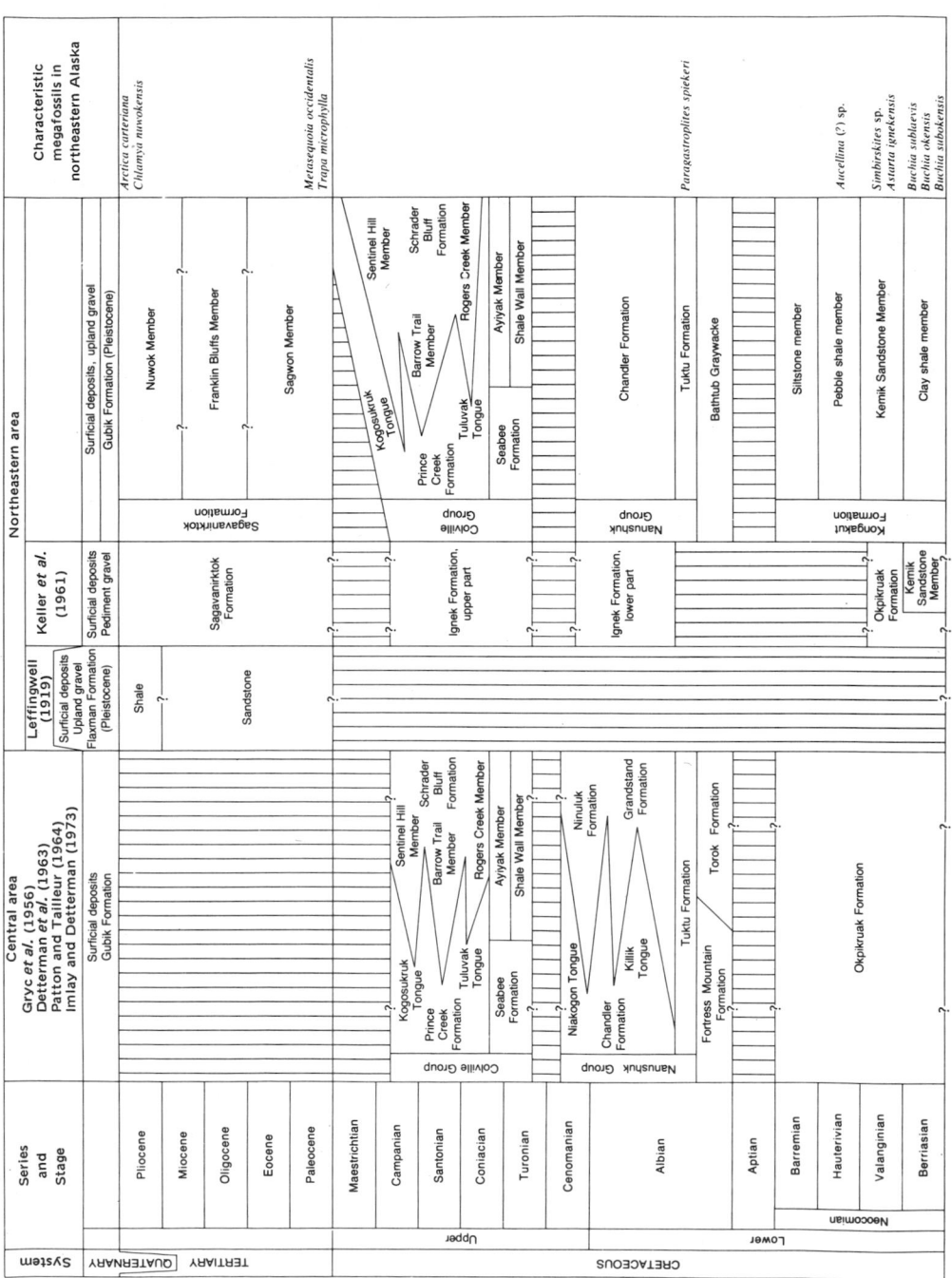

Fig. 2.   Stratigraphic sequence in episode 1 (Recent–Neocomian), from Detterman *et al.* (1975).

more temperate climate than that of the present day. Plant fossils recovered from these Cretaceous rocks suggest the temperate nature of the climate during the late Cretaceous and early Tertiary (Detterman *et al.*, 1975).

Thick gravels of the Lower Tertiary Sagavanirktok Formation remind the observer that a rejuvenation of the ancestral Brooks Range occurred near the end of Cretaceous time. Folded Cretaceous strata in the foothills and faulted terranes in the Brooks Range provide clues to the timing of this tectonic event.

The glaciers glimpsed on approaching the mountains are remnants of more extensive Pleistocene glaciation that covered high areas of the Brooks Range and carved the great U-shaped valleys that contain most of the major rivers (Porter, 1966). Morainal lobes extend far out into the foothills and outwash deposits cover large areas of the southern part of the Arctic lowlands. From the Canning River eastward, these glacial gravels extend all along the mountain front and piedmont glaciers may actually have reached the sea during glacial maxima, with icebergs calving into the Arctic Ocean as they do today off northern Ellesmere Island (Detterman *et al.*, 1958; Hamilton, 1978).

Perhaps the most significant event in the geologic history of northern Alaska was the building of the ancient Brooks Range that provided all the sediment during Cretaceous and later times. Before the Cretaceous, no southern mountain source was available to build seaward a great complex of marine and nonmarine rocks. Jurassic through Permian strata were deposited in entirely different settings (Detterman *et al.*, 1975).

Jurassic and Triassic rocks reflect deposition in relatively deep water far from shore (Fig. 3). The Jurassic Kingak Shale is made up mostly of silt and clay-sized sediment and the main fossils found in it are nektonic ammonoids. The Triassic Shublik Formation is composed of thin-bedded phosphatic limestone, shale and chert with flat monotid clams and ammonites as the major fossils. These Triassic strata comprise a thin, reduced sequence in which very little sediment represents comparatively long periods of time.

In the position of the present-day Brooks Range, the Lower Mesozoic, especially the Jurassic, consists of chert, oil shale, and tuffaceous sedimentary and mafic igneous rocks. South of the Brooks Range, an ophiolitic terrane apparently was emplaced during Jurassic time (Patton *et al.*, 1977).

The oldest rocks of this second sequence are included in the Sadlerochit Group, which crops out in the northeastern Brooks Range. Middle Permian conglomeratic calcareous sandstones and limestones of the Echooka Formation are basal strata that indicate a northern source for the clastic material. Similarly, the Lower Triassic Ivishak Formation had a northern source for delta-front clastics that grade southward into finer laminated beds containing many ironstone and limestone concretions.

Thus, the strata in the second depositional package are predominantly deep-water beds from a northern source, flanked on the south by an ophiolitic terrane.

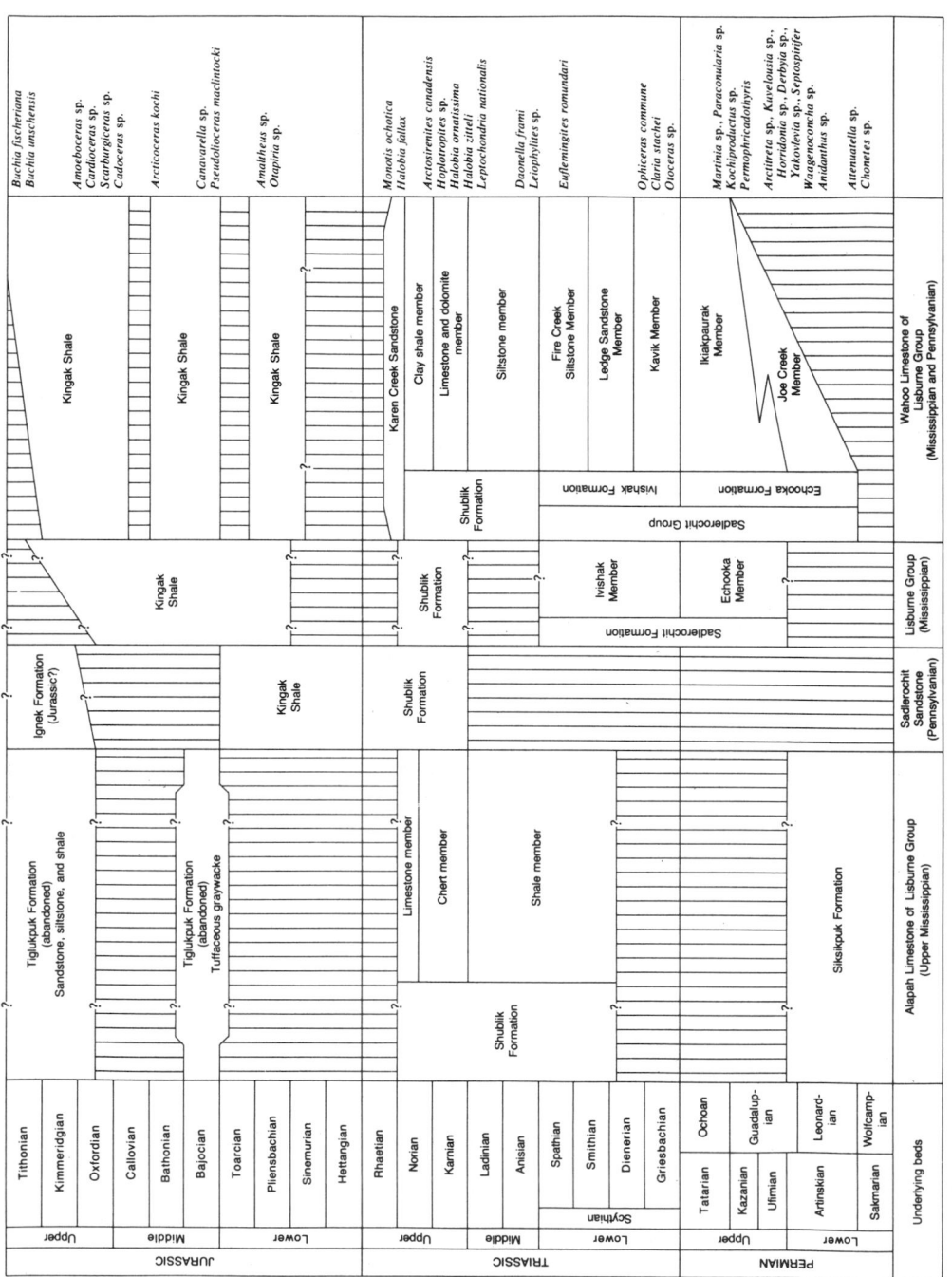

Fig. 3.   Stratigraphic sequence in episode 2 (Jurassic–middle Permian), from Detterman *et al.* (1975).

The orogenic event that built the Brooks Range in the late Jurassic profoundly altered the course of geologic history in northern Alaska. The precise relationship of the ophiolite emplacement to the early part of the Brooks Range Orogeny is not understood. Most of the thrusting in the range itself is considered to reflect later events that were related to the folding of Upper Cretaceous rocks north of the mountains. The youngest strata involved in the range thrusting are probably late early Cretaceous in age.

Intrusive granitoid rocks were emplaced near the southern edge of the Brooks Range during the Jurassic and, as mentioned above, associated mafic intrusive and extrusive strata are of about the same age.

There is no record of the late Pennsylvanian in northern Alaska. For that matter, strata of this age are rarely found in Alaska. Because rocks of early and late Permian age were deposited on various parts of the Lisburne Group, it is evident that earlier Carboniferous beds were regionally eroded during the latter part of the Pennsylvanian.

The third depositional episode involves the Carboniferous, a time of widespread marine transgression onto an early Paleozoic northern landmass (Fig. 4). Basal clastic rocks, in places conglomeratic and including nonmarine redbeds and some coal, lie with angular unconformity on all older rocks in the northeastern part of the Brooks Range (Brosgé et al., 1962). Further west in the central Brooks Range, littoral marine sandstone succeeds, with no structural unconformity, a thick sequence of Upper Devonian nonmarine clastic rocks. In the western Brooks Range, fully marine calcareous sandstone and sandy limestone lie directly on uppermost Devonian marine shales and graywackes. Thus, the transgression apparently proceeded from southwest to northeast, the basal beds climbing in time from Mamet foram zone 6 in the DeLong Mountains to zone 15 in the Sadlerochit Mountains of the extreme northeastern Brooks Range (Brosgé and Dutro, 1973). In all areas there was an upward progression into the shallow-water carbonates of the Lisburne Group, with the youngest part of that sequence also in the northeast where Middle Pennsylvanian (Atokan) strata occur (Armstrong and Mamet, 1977).

This thick complex of intertonguing carbonate facies has been the object of extensive stratigraphic and paleontologic study, and the Lisburne is one of the prime targets for oil and gas exploration in the subsurface of northern Alaska (Bird and Jordan, 1977).

The northern landmass that shed sediments southward into the Paleozoic and earliest Mesozoic seas has been explored by drilling on the Barrow Arch. The pre-Carboniferous cores of the large anticlinoria of the eastern Brooks Range are extensions of this complex Precambrian to Middle Paleozoic terrane (Reed, 1968; Dutro, 1970). Much of this older positive area is composed of intricately folded and faulted Cambrian and Ordovician strata that unconformably overlie mid-Precambrian quartz-mica schists and quartzites

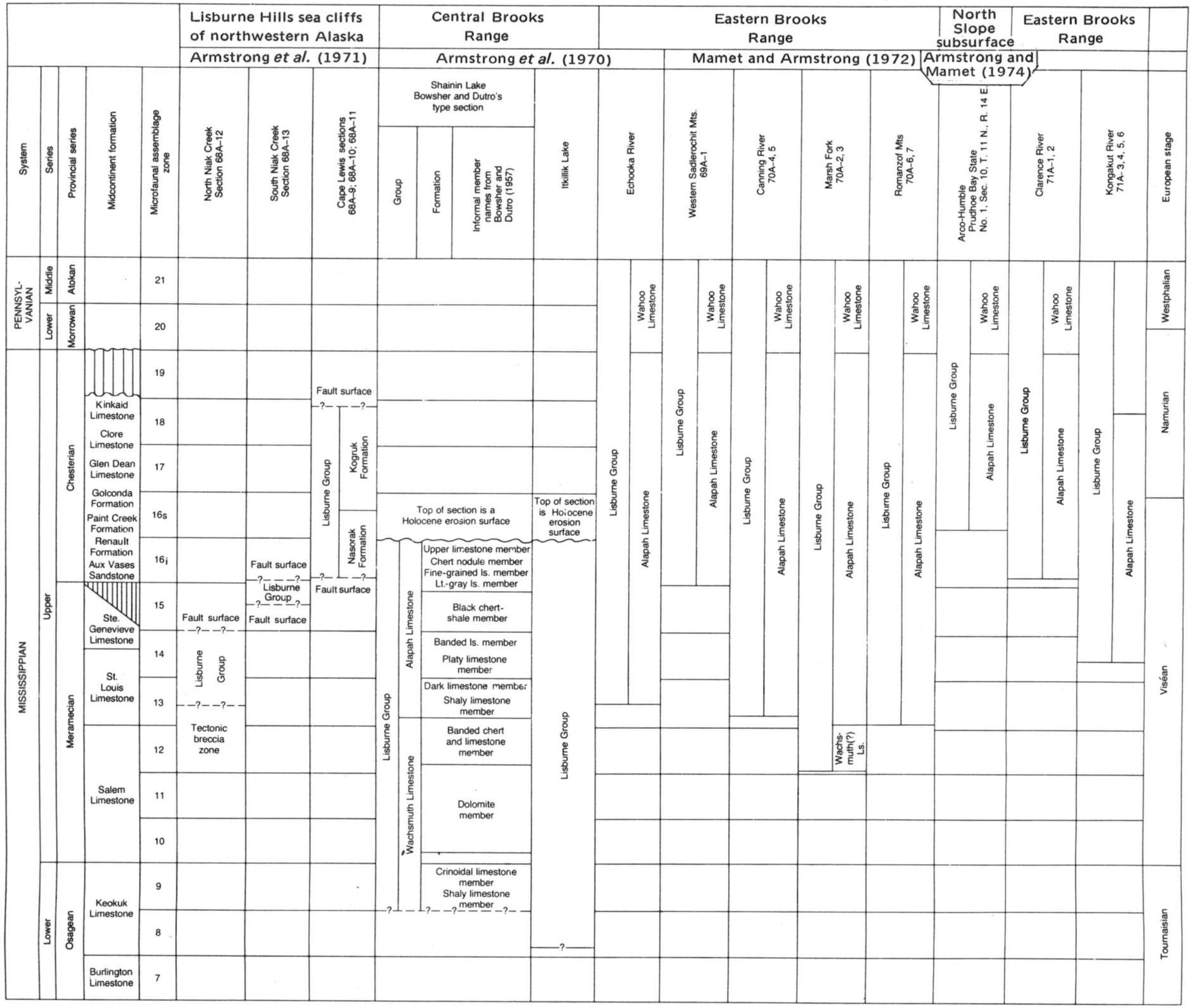

Fig. 4. Carboniferous sequence of episode 3 (middle Pennsylvanian–early Mississippian), from Armstrong and Mamet (1975).

(Reiser *et al.*, 1971, 1974). In a few places, Middle Devonian shallow-water marine sandstones lie with angular unconformity on Ordovician laminated deep-water cherts and argillites. Intrusive granitoid rocks in the eastern Brooks Range, dated at about 380 m.y., are additional evidence of this Devonian phase of orogeny.

An older orogenic episode is indicated by 431-m.y.-old granitic intrusives in the Upper Jago River country. A thick pile of mafic marine volcanic rocks, lying above Upper Cambrian outer shelf or upper slope strata, are possibly related to this orogeny (Reiser, 1970).

The Doonerak anticlinorium represents a westward extension of these older rocks into the central Brooks Range where probable Ordovician through Devonian strata are also unconformably overlain by Carboniferous rocks (Dutro *et al.*, 1976).

In the northern part of the central Brooks Range, rocks of the fourth depositional cycle reach their maximum development. Sands and gravels transported southward and westward from the northern landmass were deposited along river valleys and, ultimately, in deltaic environments that graded seaward to the southwest (Fig. 5). An extremely complex series of terrestrial, deltaic, and littoral deposits comprises the latest Devonian Kanayut Conglomerate (Gryc *et al.*, 1967). These rocks grade downward into another complex of marine facies in the Hunt Fork Shale and the "Beaucoup Formation." Although most of these strata are deeper-water marine shales and graywackes, organic buildups developed during the early late Devonian (Frasnian) in areas adjacent to pre-late Devonian structural highs (Dutro *et al.*, 1977; Dutro, 1978).

Rocks as old as Middle Devonian, probable equivalents of the thin sandstone wedge to the northeast, lie unconformably on Silurian carbonate and older metamorphic and clastic rock sequences. Precambrian radiometric ages in blue amphiboles from some of these metamorphic rocks suggest that they could be approximately equivalent to the older schists of the northeastern Brooks Range (Grybeck *et al.*, 1977).

The oldest sequence did not consist entirely of metamorphic rocks, however. In the northeast, particularly in the Demarcation Point quadrangle, a thick series of relatively undisturbed sedimentary rocks is apparently of late Precambrian age (Fig. 6). The exact relationships to the older schists is not clear, but the juxtaposition with Cambro-Ordovician strata suggests that they are indeed late Precambrian, like the strata of the Tindir Group in east-central Alaska (Dutro *et al.*, 1972).

III. OROGENIC HISTORY

Igneous events that are dated radiometrically fall roughly into five periods, each of which can be correlated with similar events elsewhere.

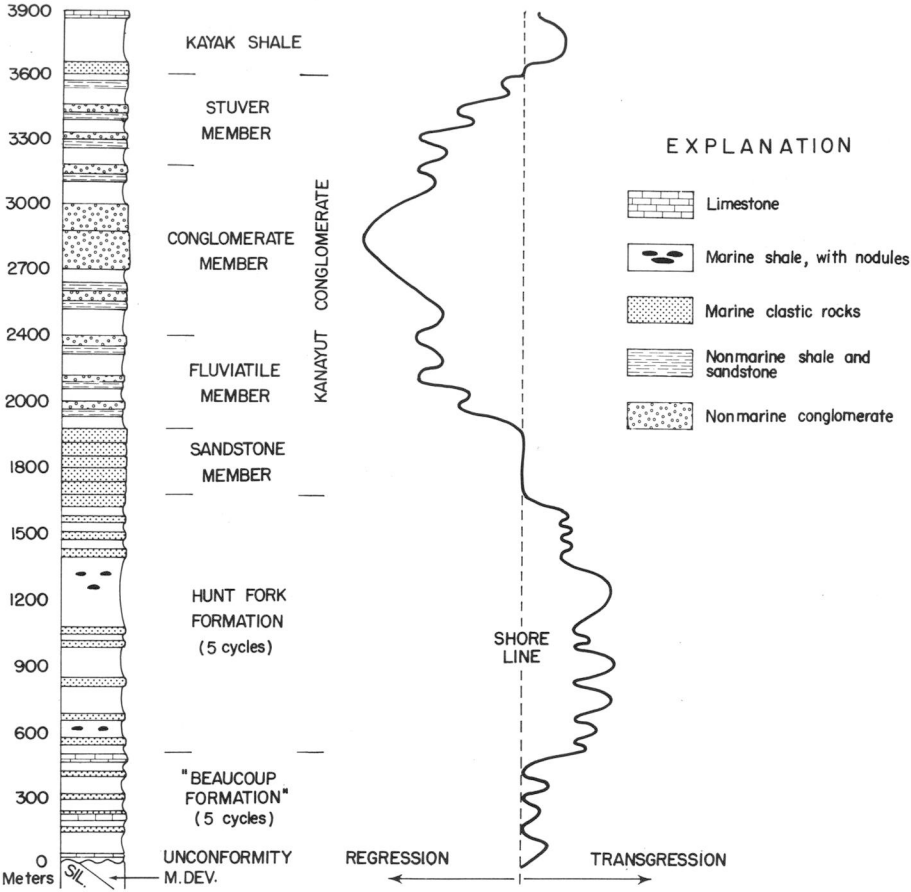

Fig. 5. Cyclic deposition in episode 4 (late Devonian).

The oldest of these is about 450 m.y. old, within the Taconian period of orogeny (or early Caledonide of some authors). In the Romanzof Mountains, a quartz monzonite pluton that invades the Precambrian Neruokpuk Quartzite has been dated as 431 ± 13 m.y. (Reiser 1970) and in the central Brooks Range at Mt. Doonerak mafic dikes intruding an early Paleozoic terrane are dated at about 470 m.y. (Dutro *et al.*, 1976). A few tens of kilometers to the southeast of the Romanzofs, Middle Devonian near-shore sandstone overlies the Ordovician chert sequence with angular unconformity. Elsewhere in northeastern Alaska, this early Paleozoic event is obscured by the pervasive pre-Carboniferous unconformity that characterizes the region.

In several places, granitic rocks have yielded radiometric ages that relate them to the Acadian. The Okpilak Granite in the Romanzof Mountains has been variously dated as 330 ± 35 to 405 ± 45 m.y., spanning much of

Devonian time. As the granite is unconformably overlain by early Carboniferous rocks, it most likely represents an Acadian intrusive event (Sable, 1977). The Chandalar Pluton in the central Brooks Range has been dated at 380 ± 40 m.y. (Brosgé and Reiser, 1964) and a recent report of granite at the bottom of East Teshekpuk well (Bird *et al.*, 1978) suggests a similar interpretation. A K–Ar age of 332 ± 10 m.y. is given for this granite, which is unconformably overlain by late Mississippian sedimentary rocks.

Two plutons in the northern Yukon are also of probable Acadian age. The Mt. Sedgwick Pluton was dated by Wanless *et al.* (1965) at about 355 m.y.

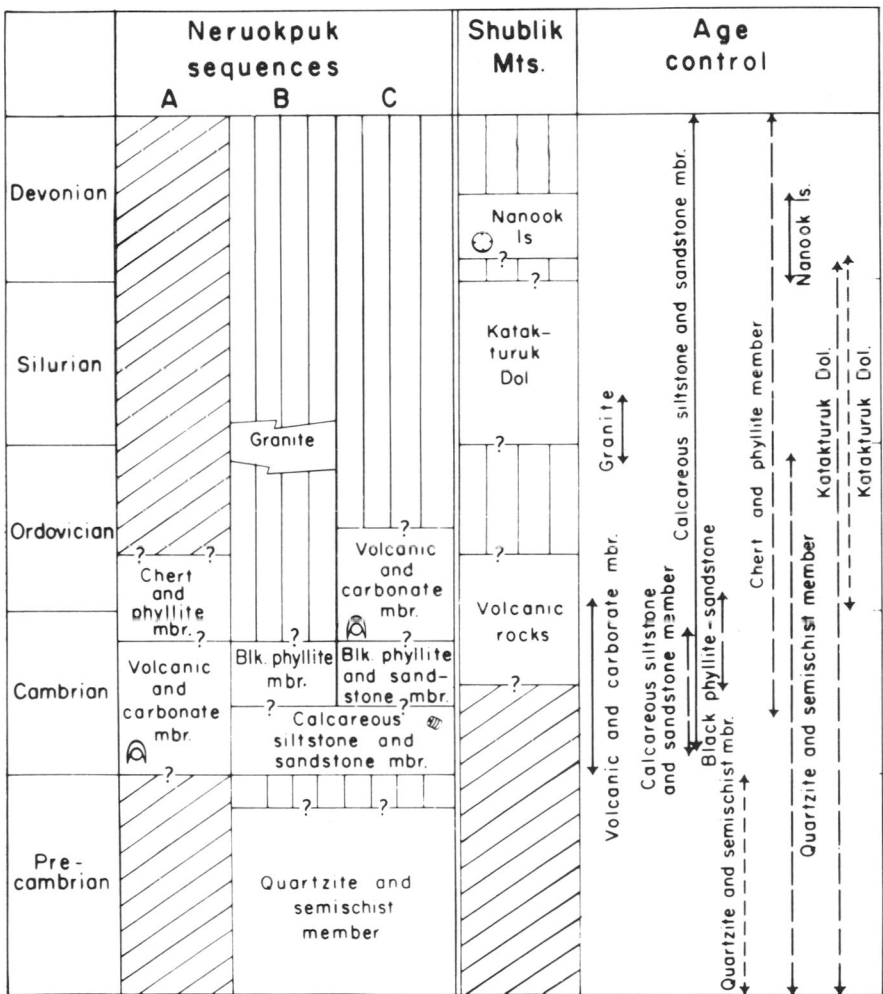

Fig. 6.   Lower Paleozoic and Precambrian sequences in episodes 5 and 6, from Dutro *et al.* (1972).

and the Mt. Fitton Pluton was reported by the same authors to be 370 ± 16 m.y. old.

Several samples from mafic dikes that intrude Lower Paleozoic rocks of the Doonerak antiform have been dated at about 380 m.y. (Dutro *et al.*, 1976). Like the older Taconic volcanic rocks, they seem to be slightly older than the ages of the plutons of the same orogeny.

Both intrusive and extrusive volcanic activity occurred during the late early Carboniferous in the Ivishak River area. Because the stratiform volcaniclastic rocks are bracketed by dated marine beds, the age can be estimated at about 325 m.y. Two possibly related plutonic events are less well dated. A pluton in the Coleen quadrangle is dated as 335 m.y. (Brosgé and Tailleur, 1970) and a hornblende granodiorite in the Chandalar quadrangle may be about 300 m.y. old, although there are difficulties with interpretation of conflicting dates on that body. This mid-Carboniferous activity might be assigned to the Sudetian orogenic event.

IV. HYPOTHESIS OF ARCTIC PLATE MOVEMENT

The tectonic evolution of Alaska involves an interplay between the development of the Arctic Ocean Basin and the evolution of the northern Pacific rim. There are several published views of the Arctic's tectonic history, and most of them involve southward transport of what is now northern Alaska, either by rotation away from the Canadian Arctic Islands or by direct movement south from the region of the Alpha Rise (Tailleur and Brosgé, 1970; Churkin, 1969, 1975; Pushcharovskiy, 1976).

In my view, the ancient Paleozoic landmass north of Alaska has a history similar to that of northern Ellesmere Island and the North Atlantic Caledonides. The several fragments of the orogen, breaking up and moving away from the old Fennoscandian core, took different routes across the Arctic region during the Mesozoic. Both Spitzbergen and Ellesmere may have rotated through about 90°, while Alaska, the Arctic Islands, and Greenland were moving at different rates on straighter courses.

As the leading edge of the Alaskan portion reacted with a northward-moving portion of the ancient north Pacific, the rocks of that oceanic segment may have been obducted onto what is now the southern flank of the Brooks Range. The resulting collision produced, by late Jurassic time, an ancestral Brooks Range that could then begin shedding sediment into the Cretaceous basin north of the range.

Continued southward transport produced thrust shuffling in the range itself so that both northward and southward directed movements are observed

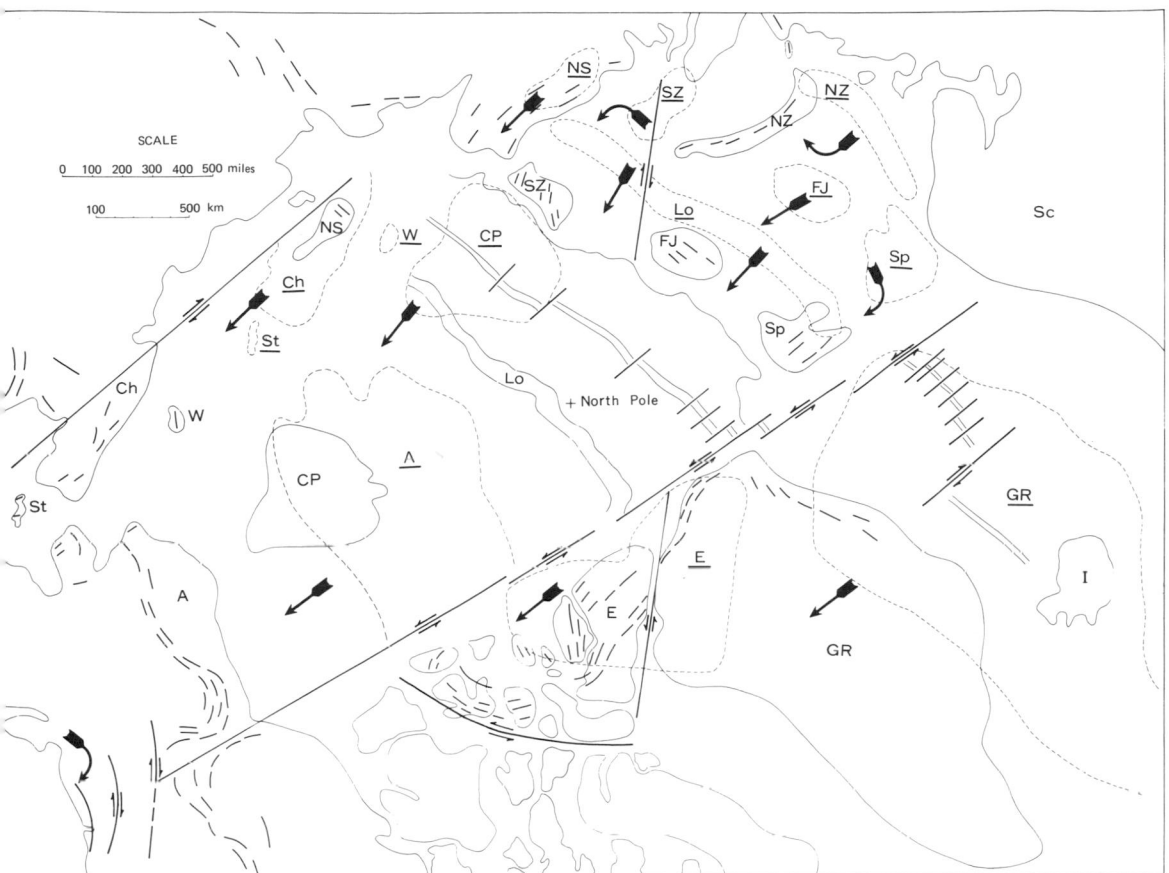

Fig. 7. Hypothetical development of the Arctic Ocean Basin and adjacent regions in post-Paleozoic time. A, northern Alaska; Ch, Chukchi Peninsula; CP, Chukchi Plateau; E, Ellesmere Island; FJ, Franz Josef Land; GR, Greenland; I, Iceland; Lo, Lomonosov Ridge; NS, New Siberian Islands; NZ, Novaya Zemlya; Sc, Scandinavia; Sp, Spitzbergen; St. St. Lawrence Island; SZ, Severna Zemlya; W, Wrangel Island. Dashed outlines with underlined letters indicate approximate positions during early Mesozoic time. Large arrows show directions of relative transport. Small arrows show relative transcurrent movements.

in the various fault slices. Folding of the molasse basin rocks through the Tertiary may be the expression of continuous, though slowed, southward underthrusting of the entire northern part of Alaska. It must be recognized that by Tertiary time the major part of Alaska was consolidated into a landmass of its own, and further southward movement in the north would have been seriously hampered.

If this scheme has validity, the suture between the Alaskan and Canadian Arctic parts should occur along the northern edge of the Arctic Islands, intersecting the coastline near the mouth of the Mackenzie River. This possible tectonic scenario is sketched in Fig. 7.

V. SUMMARY

This trip backward in time capsulizes in a general way the geologic development of the area that is now northern Alaska. The Cretaceous through Recent history is predominantly one of erosion of the ancient Brooks Range and filling of the Colville Geosyncline under processes like those in operation today. All the rocks older than Jurassic, however, were deposited far from their present sites. One of the intriguing tectonic puzzles involves the geographic positions and relative timing of events that occurred during the orogeny that built the Brooks Range.

I suggest that pre-Cretaceous deposition took place far to the north along the western margin of the Caledonides. Subsequently, the region that is now northern Alaska moved southward and impinged upon a northward moving plate, or plates, from the Pacific region. Interactions between these two major parts of the earth's crust produced the Brooks Range during the Jurassic. Later developments reflect continuing readjustments of the northern third of Alaska as southward movement was dissipated in diverse systems of thrust faulting and lateral displacement.

ACKNOWLEDGMENTS

I am indebted to many colleagues who, during the past 30 years, shared with me the excitement and pleasures of exploring the geology of northern Alaska. Among these are W. P. Brosgé, H. N. Reiser, R. L. Detterman, A. K. Armstrong, A. L. Bowsher, E. G. Sable, I. L. Tailleur, C. G. Mull, M. D. Mangus, A. H. Lachenbruch, M. C. Lachenbruch, R. L. Morris, C. J. Gudim, B. L. Mamet, T. H. Nilsen, T. D. Hamilton, and W. E. Yeend. I am especially grateful for the enlightened leadership of George Gryc, P. E. Cloud, Jr., George Gates, A. T. Ovenshine, and R. L. Miller, who provided the research environment in which geological serendipity could effect workable geologic syntheses.

REFERENCES

Armstrong, A. K., and Mamet, B. L., 1974, Carboniferous biostratigraphy, Prudhoe Bay State 1, Arctic Alaska, *Am. Assoc. Petrol. Geol. Bull.*, v. 58(4), p. 646–660.
Armstrong, A. K., and Mamet, B. L., 1975, Carboniferous biostratigraphy, northeastern Brooks Range, Alaska, *U.S. Geol. Surv. Prof. Pap.* 884, 29 p.
Armstrong, A. K., and Mamet, B. L., 1977, Carboniferous microfacies, microfossils, and corals, Lisburne Group, Arctic Alaska, *U.S. Geol. Surv. Prof. Pap.* 849, 144 p., 47 pls., 19 figs.

Armstrong, A. K., Mamet, B. L., and Dutro, J. T., Jr., 1970, Foraminiferal zonation and carbo-
    nate facies of Carboniferous (Mississippian and Pennsylvanian) Lisburne Group, central and
    eastern Brooks Range, Arctic Alaska, *Am. Assoc. Petrol. Geol. Bull.*, v. 54(5), p. 687–698.
Armstrong, A. K., Mamet, B. L., and Dutro, J. T., Jr., 1971, Lisburne Group, Cape Lewis–Niak
    Creek, northwestern Alaska, in: *Geological Survey Research, 1971, U.S. Geol. Survey Prof.
    Pap.* 750-B, p. B23–B34.
Bird, K. J., and Jordan, C. F., 1977, Lisburne Group (Mississippian and Pennsylvanian), potential
    major hydrocarbon objective of Arctic Slope, Alaska, *Am. Assoc. Petrol. Geol. Bull.*, v.
    61(9), p. 1493–1512.
Bird, K. J., Connor, C. L., Tailleur, I. L., Silberman, M. L., and Christie, J. L., 1978, Granite on
    the Barrow Arch, northeast NPRA, in: *The United States Geological Survey in Alaska:
    Accomplishments during 1977*, Johnson, K. M., ed., *U.S. Geol. Surv. Circ.* 772-B, p.
    B24–B25.
Bowsher, A. L., and Dutro, J. T., Jr., 1957, The Paleozoic section in the Shainin Lake area,
    central Brooks Range, Alaska, *U.S. Geol. Survey Prof. Pap.* 303-A, 39 p.
Brosgé, W. P. and Dutro, J. T., Jr., 1973, Paleozoic rocks of northern and central Alaska, in:
    Arctic Geology, Pitcher, M. G., ed., *Am. Assoc. Petrol. Geol. Mem.* 19, p. 361–375.
Brosgé, W. P., Dutro, J. T., Jr., Mangus, M. D., and Reiser, H. N., 1962, Paleozoic sequence in
    eastern Brooks Range, Alaska, *Am. Assoc. Petrol. Geol. Bull.*, v. 46(12), p. 2174–2198.
Brosgé, W. P. and Reiser, H. N., 1964, Geologic map and section of the Chandalar quadrangle,
    Alaska, *U.S. Geol. Surv. Misc. Geol. Inv. Map* I-375, scale 1:250,000.
Brosgé, W. P. and Tailleur, I. L., 1970, Depositional history of northern Alaska, in: *Proceedings
    of the Geological Seminar on the North Slope of Alaska, Palo Alto, California, 1970*, Adkison,
    W. L., and Brosgé, M .M., eds., Los Angeles: American Association of Petroleum Geologists,
    Pacific Section, p. D1–D17.
Churkin, M. Jr., 1969, Paleozoic tectonic history of the Arctic basin north of Alaska, *Science*, v.
    165, p. 549–555.
Churkin, M., Jr., 1975, Basement rocks of Barrow Arch, Alaska, and circum-Arctic Paleozoic
    mobile belt, *Am. Assoc. Petrol. Geol. Bull.*, v. 59(3), p. 451–456.
Detterman, R. L., Bowsher, A. L., and Dutro, J. T., Jr., 1958, Glaciation on the Arctic Slope of
    the Brooks Range, northern Alaska, *Arctic*, v. 11(1), p. 43–61.
Detterman, R. L. Bickel, R. S., and Gryc, G., 1963, Geology of the Chandler River region,
    Alaska, *U.S. Geol. Survey Prof. Pap.* 303-E, p. 223–324.
Detterman, R. L., Reiser, H. N., Brosgé, W. P., and Dutro, J. T., Jr., 1975, Post-Carboniferous
    stratigraphy, northern Alaska, *U.S. Geol. Surv. Prof. Pap.* 886, 46 p.
Dutro, J. T., Jr., 1970, Pre-Carboniferous carbonate rocks, northeastern Alaska, in: *Proceedings
    of the Geological Seminar on the North Slope of Alaska, Palo Alto, California, 1970*,
    Adkison, W. L., and Brosgé, M. M., eds., Los Angeles: American Association of Petroleum
    Geologists Pacific Section, p. M1–M8.
Dutro, J. T., Jr., 1978, Potential strata-bound lead-zinc mineralization, Philip Smith Mountains
    quadrangle, Alaska, in: *The United States Geological Survey in Alaska: Accomplishments
    during 1977*, Johnson, K. M., ed., *U.S. Geol. Surv. Circ.* 772-B, p. B9–B11.
Dutro, J. T., Jr., Brosgé, W. P., Lanphere, M. A., and Reiser, H. N., 1976, Geologic significance
    of Doonerak structural high, central Brooks Range, Alaska, *Am. Assoc. Petrol. Geol. Bull.*,
    v. 60, p. 952–961.
Dutro, J. T., Jr., Brosgé, W. P., and Reiser, H. N., 1972, Significance of recently discovered Cam-
    brian fossils and reinterpretation of the Neruokpuk Formation, northeastern Alaska, *Am.
    Assoc. Petrol. Geol. Bull.*, v. 56(4), p. 808–815.
Dutro, J. T., Jr., Brosgé W. P., and Reiser, H. N., 1977, Upper Devonian depositional history,
    central Brooks Range, Alaska, in: *The United States Geological Survey in Alaska: Accom-
    plishments during 1976*, Blean, K. M., ed., *U.S. Geol. Surv. Circ.* 751-B, p. B16–B18.
Grybeck, D., Beikman, H. M., Brosgé, W. P., Tailleur, I. L., and Mull, C. G., 1977, Geologic
    map of the Brooks Range, Alaska, *U.S. Geol. Surv. Open-File Map* 77-166B, 2 sheets.
Gryc, G. Bergquist, H. R., Detterman, R. L., Patton, W. W., Jr., Robinson, F. M., Rucker, F. P.,

and Whittington, C. L., 1956, Mesozoic sequence in Colville River region, northern Alaska, *Am. Assoc. Petrol. Geol. Bull.*, v. 40(2), p. 209–254.

Gryc, G., Dutro, J. T., Jr., Brosgé, W. P., Tailleur, I. L., and Churkin, M., Jr., 1967, Devonian of Alaska, in: *International Symposium on the Devonian System, Calgary, 1967*, Oswald, D. H., ed., Calgary: Alberta Society of Petroleum Geologists, v. 1, p. 703–716.

Hamilton, T. D., 1978, Late Cenozoic stratigraphy of the south-central Brooks Range, in: *The United States Geological Survey in Alaska: Accomplishments during 1977*, Johnson, K. M., ed., *U.S. Geol. Surv. Circ.* 772-B, p. B36–B38.

Imlay, R. W., and Detterman, R. L., 1973, Jurassic paleobiogeography of Alaska, *U.S. Geol. Surv. Prof. Pap.* 801, 34 p.

Keller, A. S., Morris, R. H., and Detterman, R. L., 1961, Geology of the Shaviovik and Sagavanirktok Rivers region, Alaska, in: *Exploration of Naval Petroleum Reserve No. 4 and Adjacent Areas, Northern Alaska, 1944–53*, Part 3: *Areal Geology*, *U.S. Geol. Surv. Prof. Pap.* 303-D, p. 169–222.

Leffingwell, E. de K., 1919, The Canning River region, northern Alaska, *U.S. Geol. Survey Prof. Pap.* 109, 251 p.

Mamet, B. L., and Armstrong, A. K., 1972, Lisburne Group, Franklin and Romanzof Mountains, northeastern Alaska, in: *Geological Survey Research, 1972*, *U.S. Geol. Survey Prof. Pap.* 800-C, p. C127–C144.

Patton, W. W., Jr., and Tailleur, I. L., 1964, Geology of the Killik–Itkillik region, Alaska, *U.S. Geol Survey Prof. Pap.* 303-G, p. 409–500.

Patton, W. W., Jr., Tailleur, I. L., Brosgé, W. P., and Lanphere, M. A., 1977, Preliminary report on the ophiolites of northern and western Alaska, in: *North American Ophiolites*, Coleman, R. G., and Irwin, W. P., eds., *Oreg. Dep. Geol. Miner. Ind. Bull.* 95, 183 p.

Porter, S. C., 1966, Pleistocene geology of Anaktuvuk Pass, Central Brooks Range, Alaska, *Arctic Inst. North Am. Tech. Pap.* 18, 100 p.

Pushcharovskiy, Yu. M., 1976, Tectonics of the Arctic Ocean Basin, *Geotectonics (USSR)*, v. 10(2), 85–91 [in English].

Reed, B. L., 1968, Geology of the Lake Peters area, northeastern Brooks Range, *U.S. Geol. Surv. Bull.* 1236, 132 p.

Reiser, H. N., 1970, Northeastern Brooks Range—A surface expression of the Prudhoe Bay section, in: *Proceedings of the Geological Seminar on the North Slope of Alaska, Palo Alto, California, 1970*, Adkison, W. L., and Brosgé, M. M., eds., Los Angeles: American Association of Petroleum Geologists, Pacific Section, p. K1–K13.

Reiser, H. N., Brosgé W. P., Dutro, J. T., Jr., and Detterman, R. L., 1971, Preliminary geologic map, Mt. Michelson quadrangle, Alaska, *U.S. Geol. Surv. Open-File Map* 490, 2 sheets.

Reiser, H. N., Brosgé, W. P., Dutro, J. T., Jr., and Detterman, R. L., 1974, Preliminary geologic map of the Demarcation Point quadrangle, Alaska, *U.S. Geol. Surv. Mineral Inv. Map* MF-610.

Sable, E. G., 1977, Geology of the western Romanzof Mountains, Brooks Range, northeastern Alaska, *U.S. Geol. Surv. Prof. Pap.* 897, 84 p.

Tailleur, I. L., and Brosgé, W. P., 1970, Tectonic history of northern Alaska, in: *Proceedings of the Geological Seminar on the North Slope of Alaska, Palo Alto, California, 1970*, Adkison, W. L., and Brosgé, M. M., eds., Los Angeles: American Association of Petroleum Geologists, Pacific Section, p. E1–E19.

Wanless, R. K., Stevens, R. D., Lachance, G. R., and Rimsaite, R. Y. H., 1965, Age determinations and geological studies; Part 1, Isotopic ages, Report 5, *Can. Geol. Surv. Pap.* 64-17, Part 1, 126 p.

Chapter 3

THE NORTH AMERICAN PLATE FROM THE ARCTIC ARCHIPELAGO TO THE ROMANZOF MOUNTAINS*

D. K. Norris
Geological Survey of Canada
Calgary, Alberta, Canada

and

C. J. Yorath
Geological Survey of Canada
Vancouver, British Columbia, Canada

I. INTRODUCTION

The part of the North American plate under consideration lies north of latitude 68°N and extends from the Coppermine Arch at longitude 120°W to the Romanzof Uplift at longitude 142°W. It embraces portions of Canada's continental shelf, the Cordilleran Foldbelt, and the Interior Platform of mainland Canada and Banks Island of the Arctic Archipelago. The project area includes segments of several physiographic (after Bostock, 1970) and tectonic (after Norris, D. K., 1973) elements shown in Figs. 1 and 2, respectively.

The treatise is intended to complement Chapter 2 on the geology of Arctic Alaska by J. T. Dutro and Chapter 4 on the Arctic Archipelago by J. Wm. Kerr. Although consistent with one another on the bedrock geology of this flank of Canada Basin, the chapters are not necessarily in agreement on matters of speculation on the origin and tectonic evolution of the Canada Basin. Collectively, however, the several contributions provide the reader with up-to-date information as a foundation for a more meaningful interpretation of the history of the Arctic Ocean floor and of the continental shelves that flank it. Insofar as the evolution of the Arctic Ocean Basin is intimately and inextricably linked to the relative motions of the North American and Eurasian plates, it is apparent that the Arctic Ocean Basin is the Rosetta Stone of plate tectonics. A fuller understanding of its history is essential to a definitive statement of the evolution and interaction of the two plates. Many highly speculative papers have been written on the tectonic evolution of the Arctic Ocean Basin, but too few have respected the constraints imposed by the geological facts of its margins. This chapter will present, therefore, new data on the bedrock geology of a part of the southern rim of Canada Basin and will use these data to assess current theories of the geological history of the region.

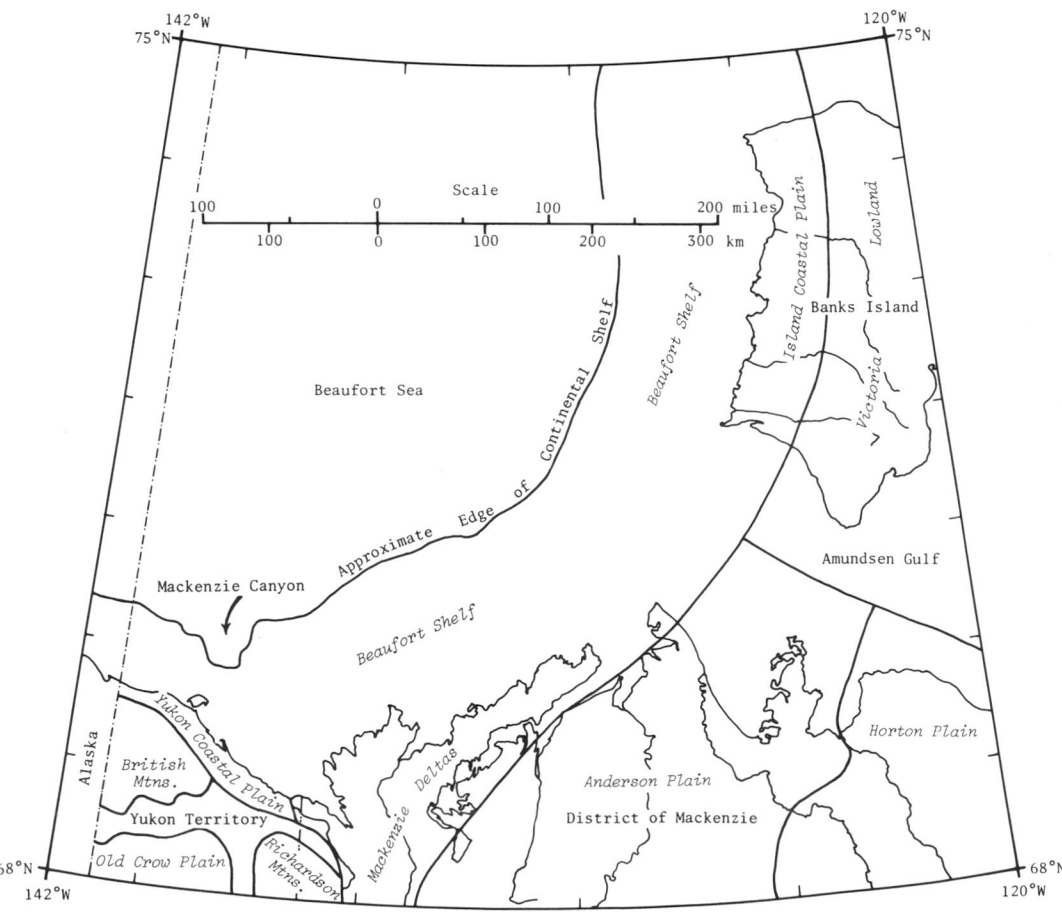

Fig. 1.   Physiographic divisions within the project area of northwestern Canada (after Bostock, 1970).

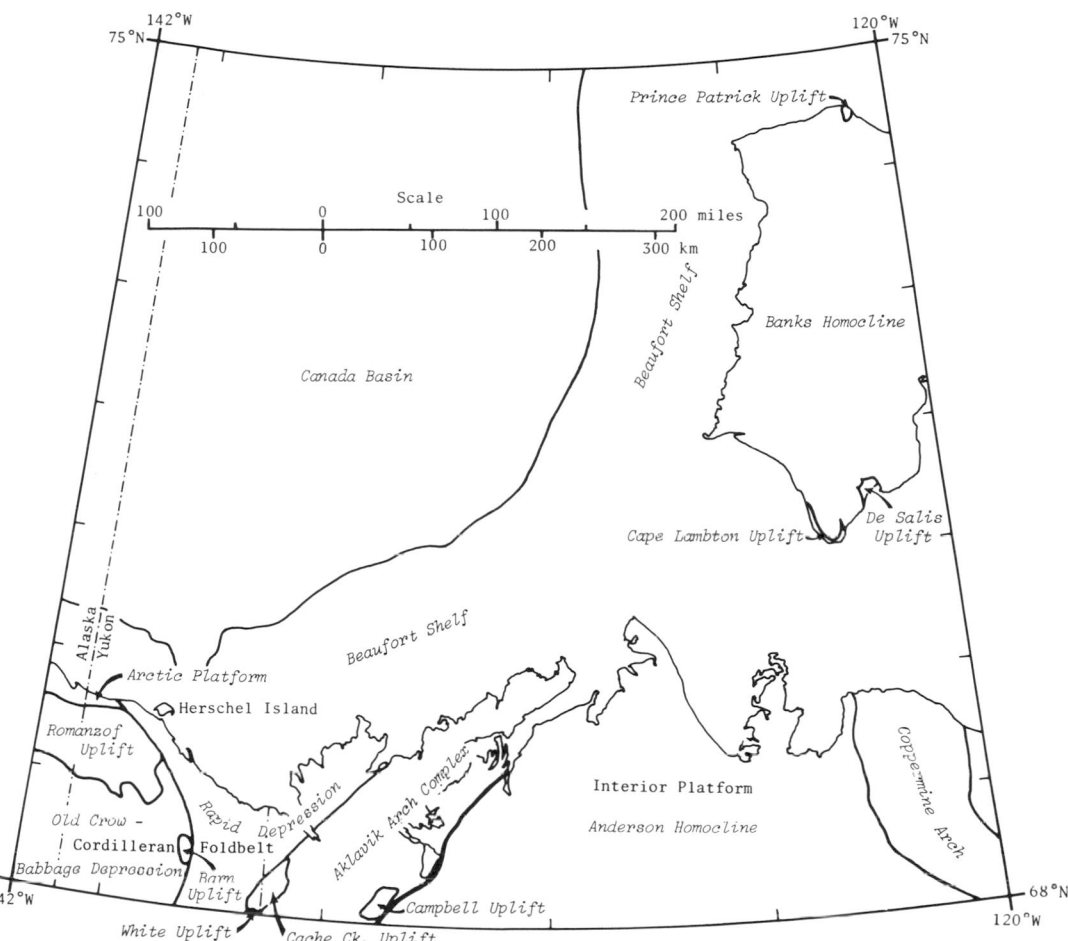

g. 2.   Tectonic elements within the project area of northwestern Canada. The boundary between the Interior Platform
d the Cordilleran Orogen is defined by the heavy solid line on the southeast flank of the Aklavik Arch Complex.

D. K. Norris contributed the basic data on the geology of northern Yukon Territory and adjacent District of Mackenzie, prepared the geological map (Fig. 3a) and synthesized the structural and stratigraphic framework of the region. C. J. Yorath and H. R. Balkwill contributed the data for Anderson Homocline and Coppermine Arch, and A. D. Miall for Banks Homocline. The structure sections were constructed jointly by Norris and Yorath, with the latter generally being responsible for interpretation of the Beaufort Shelf using onshore geology and available offshore geophysics. We view this chapter as a progress report but we are nonetheless responsible for the geological interpretations and conclusions drawn therefrom.

II. STRATIGRAPHIC SUCCESSION

A. Introduction

The stratigraphic succession of the northern Cordillera and adjacent Interior Platform of the mainland and Arctic Archipelago has the form of an eastward- and northward-tapering wedge. It ranges in age from Proterozoic to Paleogene and in thickness from zero at the edge of the Canadian Shield to approximately 7000 m at the junction between the Cordilleran Foldbelt and the Interior Platform. The wedge comprises two superposed, genetically and compositionally distinct, stratigraphic assemblages. The lower, Proterozoic to Lower Cretaceous platformal, miogeoclinal, and eugeoclinal assemblage, spanning roughly 1300 m.y., is the lateral continuation of the relatively undeformed foreland sequence of the Interior Platform and Beaufort Shelf. The upper, Lower Cretaceous to Paleogene exogeoclinal assemblage, spanning approximately 85 m.y., is the syn- and postorogenic suite of rocks derived from the deformed and uplifted regions farther into the Cordilleran Orogenic System.

Of the many unconformities punctuating the stratigraphic succession, a few are regional as well as interregional, occurring in several tectonic elements. They divide the rocks into five major, discrete, lithostratigraphic sequences that are continuous from the northern mainland to the southwestern Arctic Archipelago. The sequences are founded largely on the concept of tectonic environments as used by Sloss et al. (1949) rather than primarily on source area location as defined by Lerand (1973, p. 373). Only one new term (Inuvikian) is introduced; others, now established in the literature, are amended.

The stratigraphic column (Table I and Fig. 4) comprises four sequences termed, from oldest to youngest, the Inuvikian, Franklinian, Ellesmerian, and Brookian. They are assemblages of formations and groups and, in the definition of Sloss et al. (1949, p. 110-111), "are simply the strata which are included between objective, recognizable horizons, and are without specific time significance." The horizons separating them are either unconformities representing major hiati or are stratigraphic contacts separating depositional suites that represent fundamentally distinct origins. Certain genetic relations to source areas are implied because the sequences formed primarily in response to

──➤

Fig. 3. (a) Geological map, (b) structure sections, and (c) sources of information for the geological map of the project area of northwestern Canada. In Fig. 3a (1) standard structural geological symbols are used, (2) contours in offshore areas represent generalized free air gravity anomalies at intervals of 10 mgals (after Sobczak et al., 1973; Wold et al., 1970), (3) boreholes for hydrocarbons are solid circles, and (4) formational groupings only in subsurface are bracketed. Structure section C''–C' is in part interpreted from seismic profile A of Hofer and Varga (1972). Note vertical exaggeration in structure sections A–A' and B–B'.

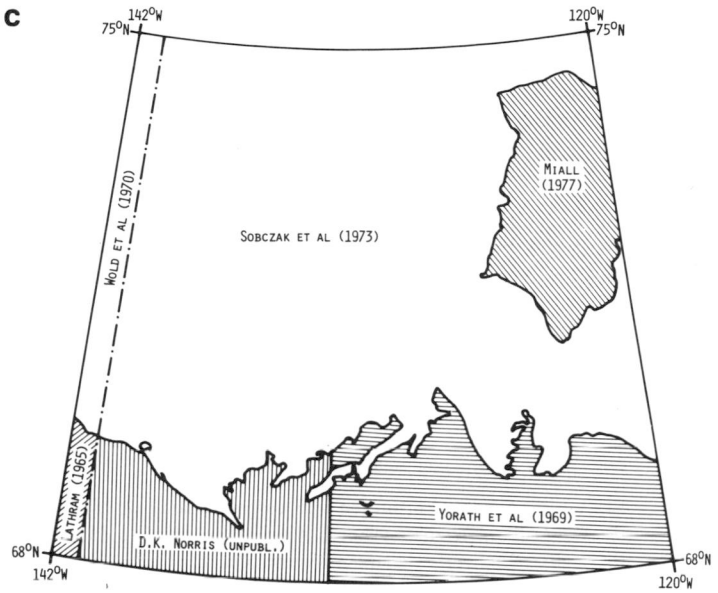

Fig. 3 (cont.)

epeirogeny in the relatively stable masses of the craton and Beaufort Shelf or to episodic orogeny in the flanking mobile belt of the Cordilleran Orogenic System. The basement (Hu), deformed in the Hudsonian Orogeny, embraces the crystalline rocks of the Canadian Shield and their westward extension beneath the Interior Platform, miogeocline, and Beaufort Shelf. It is shown only in the structure sections (Fig. 3b) at the base of the Proterozoic or Phanerozoic succession and will not be dealt with further.

B. The Inuvikian Sequence (I)

The name Inuvikian is proposed for the lowest, relatively unaltered sedimentary sequence of Proterozoic (Helikian?) age, resting nonconformably upon the Hudsonian basement (see Table I). It comprises undifferentiated clastic and carbonate rocks (Imh) in the Melville Hills, tentatively assigned to the Shaler Group (Balkwill and Yorath, 1970, p. 11), rocks correlated with the Glenelg Formation (Ig) at the southern tip of Banks Island (Thorsteinsson and Tozer, 1962, p. 26), the unnamed clastic and carbonate succession (Ic) exposed in the core of Campbell Uplift (Dyke, 1975, p. 525), and the undifferentiated Neruokpuk Formation (In) in Romanzof Uplift (Norris, D. K., 1974, Table I). All are intruded locally by basic sills and dikes.

Few positive correlations among Inuvikian successions have been made.

TABLE I
Table of Formations

Sequence	Symbol	Formational groupings[a,b]
Brookian	Bf	Fluviatile deposits of the modern Mackenzie Delta.
	Bq	Undifferentiated Quaternary deposits of Old Crow Plain, Eskimo Lakes area, and Beaufort Shelf.
	Bb	Beaufort Formation and seaward equivalents on Beaufort Shelf.
	Br	Reindeer Formation; Eureka Sound Formation.
	Bmc	Moose Channel Formation.
	Bk	Boundary Creek and Tent Island Formations; Amundsen Gulf Group; Kanguk Formation.
	Bdb	"Upper shale–siltstone division"; "upper sandstone division" and flyschoid equivalents; Darnley Bay Group; Isachsen, Christopher, and Hassel Formations.
		Interregional unconformity
Ellesmerian	Ek	Kingak Formation and eastward equivalents; "white and coaly quartzite division"; Wilkie Point and Mould Bay Formations.
	El	Endicott Group; Lisburne Group; Sadlerochit Formation and eastward equivalents; Shublik Formation.
		Interregional unconformity
Franklinian	Fi	Canol and Imperial Formations; Melville Island Group.
	Fr	Road River Formation; Hare Indian Formation.
	Fb	Clastics and carbonates including Old Fort Island, Mount Cap, and Saline River Formations; Franklin Mountain and Mt. Kindle Formations; Gossage, Bear Rock, Vunta, and Hume Formations; Katakturuk and Nanook Formations.
	Fg	Sedgwick, Fitton, Hoidahl, and Ammerman granites.
	Fv	Unnamed volcanics and carbonates.
		Interregional unconformity
Inuvikian	I	Unnamed Precambrian formations in Melville Hills (Imh); Glenelg Formation (Ig); unnamed formations in Campbell Uplift (Ic); undifferentiated Neruokpuk Formation (In); Tindir Group (It)
		Interregional unconformity
Unnamed	Hu	Crystalline rocks deformed by the Hudsonian Orogeny (structure sections only).

[a] Listing of formational groupings does not necessarily imply stratigraphic order.
[b] Terms in quotation marks are informal.

The Glenelg Formation of Victoria Island has been identified in southern Banks Island and possible equivalents of the Shaler Group occur in the Melville Hills on the Coppermine Arch. The Proterozoic sediments exposed in the core of the Campbell Uplift immediately east of the Mackenzie Delta, moreover, resemble the red and gray siltstone and gray dolomite of the type

Glenelg Formation of Victoria Island (Thorsteinsson and Tozer, 1962, p. 28) but individual rock units have not been correlated. The Inuvikian Sequence in Campbell Uplift, however, yielded a well-defined virtual magnetic pole indicating a Neohelikian age of about 1100 m.y. (Norris, D. K., and Black, 1964, p. 50). Paleomagnetic dating of the unfossiliferous Neruokpuk Formation in Romanzof Uplift has been unsuccessful because of strong, unstable remanent magnetic components in the rocks and the age of this succession can be said only to be older than the Lower Cambrian volcanics and carbonates that overlie it. The Neruokpuk could be either Helikian or Hadrynian in age. Moreover, the argillites, poorly sorted feldspathic sandstones, limestones, and basic igneous rocks comprising the Neruokpuk (In) differ markedly from the orthoquartzites and algal dolomites of the Inuvikian sequence in the extreme southwest corner of the project area (It) as well as east of the Mackenzie Delta (Ic), suggesting not only a different provenance, but also different environments of deposition.

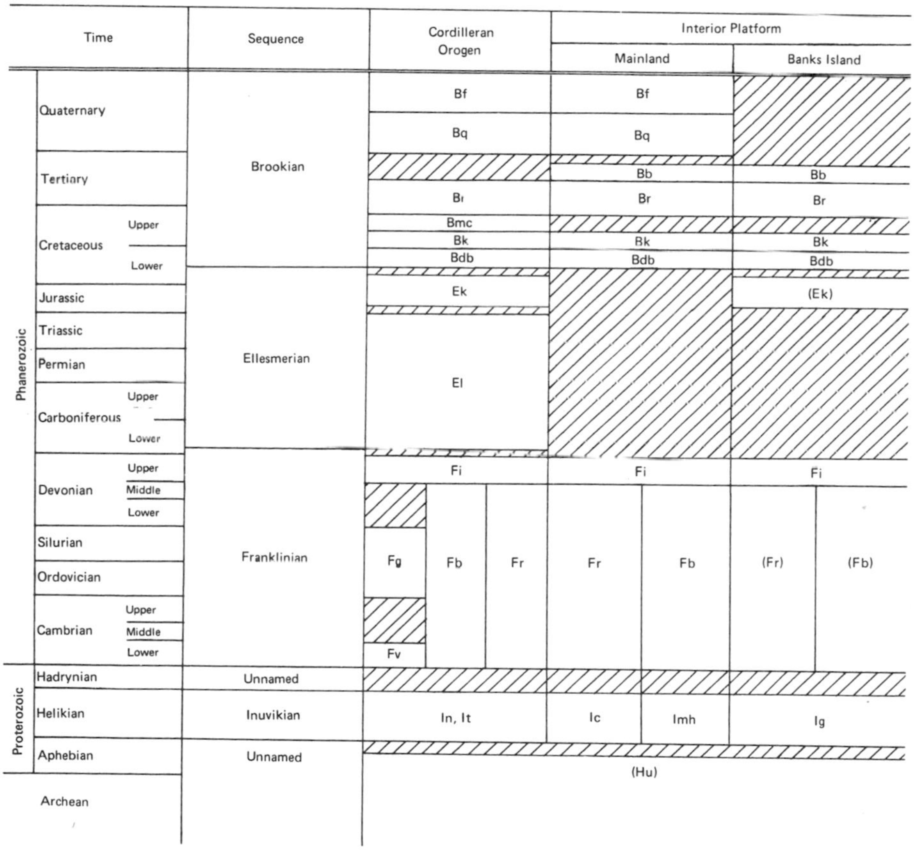

Time			Sequence	Cordilleran Orogen			Interior Platform		
							Mainland		Banks Island
Phanerozoic	Quaternary		Brookian	Bf			Bf		
				Bq			Bq		
	Tertiary						Bb		Bb
				Br			Br		Br
	Cretaceous	Upper		Bmc					
				Bk			Bk		Bk
		Lower		Bdb			Bdb		Bdb
	Jurassic		Ellesmerian	Ek					(Ek)
	Triassic								
	Permian								
	Carboniferous	Upper		El					
		Lower							
	Devonian	Upper	Franklinian	Fi			Fi		Fi
		Middle							
		Lower							
	Silurian			Fg	Fb	Fr	Fr	Fb	(Fr) (Fb)
	Ordovician								
	Cambrian	Upper							
		Middle							
		Lower		Fv					
Proterozoic	Hadrynian		Unnamed						
	Helikian		Inuvikian	In, It			Ic	Imh	Ig
	Aphebian		Unnamed						
							(Hu)		
	Archean								

Fig. 4.   Sequences on the southern rim of Canada Basin.

C. The Franklinian Sequence (F)

The Franklinian Sequence embraces the lithostratigraphic succession from the base of the Cambrian to the top of the Devonian System. The term Franklinian is adopted from Lerand (1973, p. 373) but its lower limit has been adjusted to include the Lower and Middle Cambrian Series. Thus the succession includes the sedimentary, intrusive, and volcanic rocks between the late Proterozoic Racklan and the late Devonian to early Mississippian Ellesmerian tectonism. It is bounded below and above by regional and interregional unconformities. The sequence contains three spatially and genetically distinct assemblages: a suite of basic volcanics and carbonates (Fv) confined to Romanzof Uplift, a widespread assemblage of carbonates and clastics (Fb, Fr, Fi) and a suite of acid igneous intrusions (Fg), which within the project area is confined largely to Barn and Romanzof Uplifts.

1. *Lower Paleozoic Volcanics and Carbonates (Fv)*

In the core of Romanzof Uplift is a 1000-m-thick assemblage of volcanic conglomerate, discontinuous mafic flows, gray argillite, and beds and blocks of gray limestone. It occurs as a spectacularly castellated erosional remnant resting upon the Proterozoic Neruokpuk Formation. No younger sequences are observed to overlie it in Canada. It is continuous with the "volcanic and carbonate member" in northeastern Alaska where it was reported (Dutro *et al.*, 1972, p. 813) to contain indigenous Lower as well as Upper Cambrian trilobites. There the unit is lithologically gradational into an overlying, thick, chert and phyllite sequence (Dutro *et al.*, 1972, p. 809), which is probably the Lower and Middle Paleozoic Road River Formation.

2. *Lower and Middle Paleozoic Carbonates and Clastics (Fb, Fr)*

In the Interior Platform of the northern mainland and Banks Island, a basal, shallow-water clastic unit is overlain by a widespread blanket of stable, shallow-water platform carbonate deposits. Together, the clastics and carbonates comprise unit Fb and they rest unconformably upon the Inuvikian Sequence. Their limited presence in the relatively undeformed rocks of southern Beaufort Shelf is conjectural (see Fig. 3b, structure sections). West of the Mackenzie Delta, however, they occur as an inlier in White Uplift (Dyke, 1971). There the Lower and Middle Paleozoic carbonate suite is a monotonous sequence of pelletoid limestones which contrasts with the coeval succession of distinctive, mappable carbonate units of the platform. The suite reflects a marked difference in the epeirogenic regime of that time (Macqueen, 1975, p. 297). Shales intervening laterally between the limestones of White Uplift and the continuous carbonate blanket of the Interior Platform suggest, moreover, that the White Uplift succession represents a shallow-water carbonate buildup of restricted lateral extent (Norford, 1964, p. 10). West of the project area in Brooks Range, a thick succession of Middle Devonian or older carbonate rocks comprising the Katakturuk and Nanook Formations (Dutro, 1970, p. M1) is exposed locally. It indicates that stable shelf carbonate sedimentation also occurred there in the early and mid-Paleozoic in regions now within the Cordilleran Foldbelt. Because the two formations can be divided into several informal members (Dutro, 1970, p. M2), the depositional regime may have been more like that of the Interior Platform rather than that of the eastern flank of the Cordilleran Foldbelt.

The Richardson Trough (Lenz, 1972, p. 329) was a narrow, north-trending, structurally controlled site of graptolitic shale and limestone deposition adjacent to the Interior Platform at least from the Middle Cambrian to the Lower Devonian. These sediments comprise the Road River and Hare Indian Formations (Fr). The northern end of the trough is within the project area. It flares approximately where it is intersected by the Aklavik Arch Complex (Fig. 2), which the northern limit of the carbonate shelf (Fb) west of the trough forms Clastic rocks equivalent to the carbonate banks, therefore, are widespread on the northern mainland west of the Mackenzie Delta and may occur in the Beaufort Shelf. The lateral transition from the Lower and Middle Paleozoic shale and limestone (Fr) to the platform carbonates (Fb) on the east flank of the trough occurs both on the mainland and on Banks Island in the subsurface. It is postulated to take place along a northeast-trending, curvilinear line in the approximate location of the northwest flank of the Aklavik Arch Complex. As suggested by Miall (1976), the Richardson Trough may link with the Hazen Trough of the Arctic Archipelago. The Katakturuk and Nanook Formations beyond the project area in the northeast Brooks Range therefore lie within the arms of the flare and they may extend seaward onto Beaufort Shelf (see Lerand, 1973, Fig. 28). Clastic equivalents of these formations may occur in the region of the seaward end of all structure sections (Fig. 3b). It is possible, however, that the entire Lower and Middle Paleozoic section was removed by erosion prior to deposition of the Ellesmerian Sequence in the outer reaches of the Beaufort Shelf.

The carbonate–shale interfaces on the two sides of the Richardson Trough oscillated laterally with time and resulted in complex intertonguing of the two facies. Moreover, there was progressive encroachment of the clastics upon the carbonate banks flanking the trough (Norford, 1964, p. 6, 7) so that in some areas the banks were ultimately blanketed with shale. Thus, in the eastern half of the project area, unit Fr interfingers with as well as overlies unit Fb (see Fig. 3b, structure sections). In the western half of the area, on the other hand, the remaining Lower and Middle Paleozoic sequence is in the shale facies. If carbonate banks formed at all in the west, they were limited to the eastward extension of the Katakturuk and Nanook Formations in the region of Romanzof Uplift and were removed at the sub-Ellesmerian unconformity.

3. *Upper Devonian Clastic Deposits (Fi)*

On the northern mainland east of Mackenzie Delta, the base of the Upper Devonian succession is marked by the transgressive Canol Formation, a distinctive, thin, black, siliceous shale locally with interbeds of calcareous siltstone. The formation is not recognized on Banks Island. Conformably overlying the Canol is the Imperial Formation, an assemblage of alternating dark gray and brown shales, and micaceous, green siltstone and sandstone with many of the attributes of turbidity current sedimentation. It is interpreted to represent a deep-water, flyschoid deposit. Small areas of unnamed, Upper Devonian, plant-bearing, fine-grained clastics occur on the margins of Old Crow Plain in the southwest corner of the project area (see Fig. 3a). Coeval strata on northeastern Banks Island are undifferentiated Melville Island Group representing "a spectrum of depositional environments ranging from marine shelf, with occasional development of carbonate reefs, through nearshore and coastal plain" (Klovan and Embry, 1971, p. 722–724). None of the succession there is reported to be flyschoid in character like its counterpart on the northern mainland.

4. *Lower Paleozoic Plutons (Fg)*

Acid igneous intrusions within and adjacent to the project area (see Table II) range from granites through quartz monzonites to syenodiorites. They occur as isolated stocks or cupolas exposed over areas of only a few square kilometers (e.g., Sedgwick, Fitton, and Hoidahl) and as batholiths outcropping intermittently over hundreds of square kilometers (e.g., Old Crow). All are variably weathered and hydrothermally altered so that some apparent radiometric ages may have little tectonic significance. They range from 95 to 431 m.y. and it is most likely that the maximum value of 431 m.y. is the approximate minimum intrinsic age of plutonism within the project area. This coincides approximately with the Ordovician–Silurian boundary and the time gap between this event and the next youngest orogenic phase (Ellesmerian Orogeny, 350 m.y. ago) was at least 80 m.y.

D. The Ellesmerian Sequence (E)

The Ellesmerian Sequence (Lerand, 1973, p. 373) is herein redefined to include the rock succession from the Visean Stage of the Mississippian to the upper part of the Hauterivian Stage of the Lower Cretaceous. Its upper

TABLE II
Acid Igneous Intrusions within and Adjacent to the Project Area

Intrusion	Radiometric age (m.y.)	Kind	Rock type	Mineral	Exposed area (km²)	Longitude	Latitude	Reference(s)
Okpilak	310 ± 35	Pb–α	Qtz. monzonite	Zi	370	144°00'W	69°15'N	Sable (1965), p. 168
	128	K–Ar	Qtz. monzonite	Bi		144°00'W	69°15'N	Sable (1965), p. 169
	405 ± 45	Pb–α	Qtz. monzonite	Zi	370	144°00'W	69°15'N	Sable (1965), p. 168
	125	K–Ar	Qtz. monzonite	Bi		144°00'W	69°15'N	Sable (1965), p. 169
Brooks Rg.	380 ± 40	Pb–α	Granite	Zi				Brosgé and Reiser (1964)
	125	K–Ar	Granite	Bi				Brosgé and Reiser (1964)
Romanzof	431 ± 13	K–Ar	Granite	Hb	15	143°43'W	69°09'N	Reiser (1970), p. K4
Mt. Sedgwick	95	K–Ar	Granite	Bi	65	139°07'W	68°51'N	Baardsgaard et al. (1961), p. 459
	355	K–Ar	Granite	Hb	65	139°08'W	68°52'N	Wanless et al. (1965), p. 23
	341 ± 14	K–Ar	Granite	Hb	65	139°04'W	68°51'N	Wanless et al. (1974), p. 19–20
	312 ± 11	K–Ar	Granite	Bi		139°04'W	68°51'N	Wanless et al. (1974), p. 19–20
Mt. Fitton	353	K–Ar	Qtz. monzonite	Bi	8	138°01'W	68°30'N	Baardsgaard et al. (1961), p. 459
	370 ± 16	K–Ar	Granite	Bi	8	138°03'W	68°28'N	Wanless et al. (1965), p. 22
Hoidahl	265 ± 10	K–Ar	Granite	Or	1	138°09'W	68°33'N	Bell (1974), p. 26; Shell (unpublished)
	300 ± 20	K–Ar	Granite	Or	1	138°09'W	68°33'N	Bell (1974), p. 26; Shell (unpublished)
	406 ± 30	Rb–Sr	Granite	Bi	1	138°09'W	68°33'N	Bell (1974), p. 26; Shell (unpublished)
Old Crow	265 ± 12	K–Ar	Granite	Bi	1700	140°44'W	67°34'N	Wanless et al. (1965), p. 22
	354 ± 10	Rb–Sr	Granite	Bi	1700	140°48'W	67°40'N	Ziegler (1969), p. 15
	295 ± 10	K–Ar	Granite	Bi	1700	140°48'W	67°40'N	Ziegler (1969), p. 17
	299 = 9	?	Granite	?	1700	140°48'W	67°40'N	Bell (1974), p. 26
Schaeffer	220	K–Ar	Granite	Bi	60	139°50'W	67°44'N	Baardsgaard et al. (1961), p. 459
Dave Lord	364 ± 15	K–Ar	Syenodiorite	Hb	6	139°15'W	67°36'N	Norris, D. K. (unpublished)

boundary is adjusted to include much of the texturally and mineralogically mature, epicontinental Jurassic and Lower Cretaceous clastic rock exposed on the mainland and present in the subsurface of Banks Island (see Table I and Fig. 4).

The sequence is bounded below by the regional unconformity at the base of the Mississippian System. It is bounded above by the Hauterivian and Barremian basal transgressive "upper shale–siltstone division" of the northern mainland and by the coal-bearing clastics of the Isachsen Formation on Banks Island.

The Ellesmerian Sequence is divided into two temporally distinct assemblages (El and Ek). In the region west of Mackenzie Delta, a Mississippian through Triassic carbonate and coal-bearing clastic assemblage (El) occurs at the base of the sequence (see Fig. 4). It is overlain disconformably by Jurassic and Lower Cretaceous clastics (Ek), which occur in the subsurface of Banks Island. In the northern mainland east of the delta, the entire Ellesmerian Sequence is missing.

1. *Upper Paleozoic and Triassic Clastics and Carbonates* (*El*)

West of Mackenzie Delta, the Carboniferous, Permian, and Triassic Systems comprise a diachronous, generally paraconformable succession of clastic and carbonate rocks. In the vicinity of Romanzof and Barn Uplifts (Fig. 2), the Mississippian Kekiktuk Formation, a quartzite pebble and cobble conglomerate, occurs locally at the base. It is overlain by plant- and coal-bearing shales which are gradational upward and laterally into marine, calcareous shales and skeletal micritic limestones. These shales and limestones belong to the Mississippian Kayak Formation and together with the Kekiktuk they comprise the Endicott Group (Table I), a northeasterly transgressive, predominantly clastic interval that rests with profound angular unconformity on the deformed argillites and quartzites of the Proterozoic Neruokpuk Formation. The Endicott is overlain gradationally by resistant, gray, skeletal limestones and dolomites of the Lisburne Group, ranging in age from late Mississippian (Visean) to mid-Pennsylvanian (Moscovian) (Bamber and Waterhouse, 1971, p. 81). Clastics and thin carbonates of the Lower and Middle Permian Sadlerochit Formation of the project area rest disconformably upon the Lisburne Group on the crest and northeast flank of Romanzof Uplift. Correlative rocks are extensively exposed in the core of Cache Creek Uplift (Fig. 2), where they rest disconformably upon Pennsylvanian (Moscovian) dolomites (Bamber and Waterhouse, 1971, p. 86) assigned to the Lisburne Group. There the unnamed Permian succession is many times thicker than the Sadlerochit Formation in the Sadlerochit Mountains west of the project area. Moreover, the occurrence of an early Middle Permian (Ufimian) brachiopod fauna (Norris, D. K., 1976a, Fig. 97.3) close to its upper, disconformable contact with Jurassic rocks in Cache Creek Uplift strongly suggests that only strata coeval with the Permian Echooka Member of the type area are present in northwestern Canada. The overlying Ivishak Member of the Sadlerochit Formation, if deposited, was removed at one or other of the unconformities at the base of the Upper Triassic and Lower Jurassic Series.

The thin sequence of dark gray carbonates and clastics comprising the Shublik Formation is both middle and late Triassic in age west of the project area (Detterman, 1970, p. 5). In the Romanzof and Barn Uplifts, however, it may be late Triassic because only Norian fossils have been found (Mountjoy, 1967a, p. 11). The formation is bounded both above and below by regional unconformities. The Shublik is observed to overstep the Lisburne and Endicott Groups northeastward and rest upon the Neruokpuk Formation on the north flank of Romanzof Uplift (Norris, D. K., 1972b, p. 96). It rests upon shale and coal of the Kayak Formation on the north flank of Barn Uplift. The Shublik in turn is overstepped northeastward by the Jurassic and Lower Cretaceous Kingak Formation southwest of Herschel Island (Fig. 2) so that at least locally the Kingak rests unconformably upon deformed Neruokpuk (Norris, D. K., 1972b, p. 98).

2. Jurassic and Lower Cretaceous Epicontinental Deposits (Ek)

The epicontinental sequence in the Mackenzie Delta area is an alternating succession of texturally mature, quartz arenites and concretionary shale, ranging in age from early Jurassic to mid-early Cretaceous (Hauterivian). A variety of names, both formal and informal, identifies the many rock units there (see Jeletzky, 1960, 1967) and in most places fossils are essential to a meaningful interpretation of the stratigraphic and structural framework of the region. The sandstones intertongue with the shale and are presumed to represent seaward migrating sandbars or barrier islands (Young, 1974b, p. 188) in a generally north- and northwest-prograding clastic sequence derived from a cratonic landmass to the east and southeast. The coarser clastics overstep the shales and thin out in the direction of progradation so that in the vicinity of Herschel Island the sequence is dominated by dark gray, marine shales of the Kingak Formation. There, a persistent, dark gray quartzite correlated with the mid-lower Cretaceous (Valanginian to Hauterivian) "white and coaly quartzite division" (see Table I) occurs at the top of the Jurassic and Lower Cretaceous epicontinental deposits. Like the "white and coaly quartzite division," it marks the top of the Ellesmerian Sequence. On southern Beaufort Shelf, these prograding clastics are interpreted to pass laterally into shale in the same way that Lerand (1973, Figs. 21, 22) suggested for the region north of the Mackenzie Delta. The predominance of shale in the Jurassic and Lower

Cretaceous section assigned to the Wilkie Point and Mould Bay Formations in the Orksut I-44 well (Fig. 3b, Section A–A'; Table III; Miall, 1975a, p. 258) in south-central Banks Island, moreover, would suggest that seaward equivalents in Beaufort Shelf (Fig. 3b, structure section A–A') are also in shale facies.

E. The Brookian Sequence (B)

The Brookian Sequence embraces the stratigraphic succession from the upper part of the Neocomian (Upper Hauterivian) to the Holocene. It includes, therefore, the entire sedimentary record of the fundamental redistribution of source areas and depositional basins or troughs concurrent with onset of early Cretaceous tectonism in the northern Cordillera.

The sequence is divided into a number of widespread rock units (see Table I). Its base is the Lower Cretaceous, transgressive "upper shale–siltstone division" of the northern mainland, which is overstepped east of the Mackenzie

TABLE III
Boreholes Included in Figure 3b, Structure Sections

Structure section	Borehole name	T.D. (m)	Status
A–A'	Columbia et al. Amoco Ikkariktuk M-64	1722	Dry and abandoned
	Deminex et al. Orksut I-44	3060	Dry and abandoned
	Elf et al. Storkerson Bay A-15	2048	Dry and abandoned
B–B'	Elf Horton River G-02	2478	Dry and abandoned
	Texcan C & E Nicholson N-45	863	Dry and abandoned
C–C'	Amoco Ulster Scurry Inuvik D-54	1562	Dry and abandoned
	Gulf East Reindeer P-60	1920	Dry and abandoned
	Gulf Mobil East Reindeer A-01	2954	Dry and abandoned
	Gulf Mobil Ikhil I-37	4704	Dry and abandoned
	IOE BA Shell Tununuk K-10	3757	Dry and abandoned
	BA Shell IOE Reindeer D-27	3861	Dry and abandoned
	Gulf Mobil Ya Ya P-53	3033	Suspended gas
	Gulf Mobil Toapolok O-54	2786	Dry and abandoned
	Shell Kumak J-06	3481	Suspended oil
	IOE Taglu C-42	4895	Suspended gas
	IOE Tablu G-33	2994	Gas
	IOE Taglu F-43	4555	Suspended gas
	IOE Taglu West P-03	3310	Suspended gas
	IOE Tablu D-55	3706	Dry and abandoned
	Imperial Immerk B-48	2708	Dry and abandoned
D–D'	Shell Aklavik A-37	2584	Dry and abandoned
	Shell Beaverhouse Creek H-13	3748	Dry and abandoned
	IOE Blow River YT E-47	4267	Dry and abandoned
	IOE Spring River YT N-58	2136	Dry and abandoned
	Pacific Imp et al. Roland Bay YT L-41	2752	Dry and abandoned

Delta by the Darnley Bay Group (Brideaux and McIntyre, 1975, p. 3), on Banks Island by the Isachsen Formation, and possibly on Beaufort Shelf. Intermediate rocks include the Lower and Upper Cretaceous flyschoid and molassoid deposits of the northern mainland (Young, 1974b, Fig. 2). Its top comprises the Tertiary and Quaternary clastic rocks identified on Banks Island, on Beaufort Shelf, in Mackenzie Delta, and in Old Crow Flats.

1. *Upper Neocomian Transgressive Deposits and the Aptian–Albian Flyschoid Rocks (Bdb)*

Throughout the project area, the northward-prograding clastic Jurassic and Lower Cretaceous epicontinental deposits were transgressed in the late early Cretaceous (late Neocomian). West of the Mackenzie Delta, a blanket of marine muds comprising the "upper shale–siltstone division" was spread across the region. East of the delta (Yorath *et al.*, 1975, p. 17) as well as on Banks Island, however, coeval rocks of the late Neocomian appear to be absent.

This basal unit of the Upper Neocomian is overlain by regressive, marine, coarse clastics of the Aptian "upper sandstone division" and in turn by Albian, flyschoid sandstones, mudstones, and conglomerates. The appearance of carbonate and slaty particles and feldspar in the upper part of the "upper sandstone division" heralded the change in source area and dispersal patterns during the transition from the epicontinental to the flyschoid phase (Young, 1974b, p. 200) and reflected the beginning of major uplift and shift of provenance to the west. The fundamental reversal in direction of dominant source area from eastern cratonic to western uplifted segments of the northern Cordilleran Mobile Belt would appear, therefore, to have occurred approximately in the Aptian. In the southern part of the Canadian Cordillera, on the other hand, it occurred in the late Jurassic (Norris, D. K., and Bally, 1972, p. 2) or about 30 million years earlier. How this time difference may relate to possible progressive northward migration of the deformation front remains to be studied.

On Anderson Homocline east of Mackenzie Delta, the Aptian–Albian succession comprises sandstone, shale, and coal of the Darnley Bay Group (Yorath *et al.*, 1975, p. 9). They rest unconformably upon Middle Paleozoic carbonates and shales of the Franklinian Sequence. On Banks Island, moreover, a somewhat similar assemblage of coeval, clastic rocks is differentiated into the Isachsen, Christopher, and Hassel Formations (Miall, 1977). They rest upon Jurassic shales (Wilkie Point and Mould Bay Formations) of the Ellesmerian Sequence; equivalents of the "upper shale–siltstone division" are absent at the unconformity separating these Jurassic and Lower Cretaceous rocks.

2. *Upper Cretaceous and Lower Tertiary Molassoid Deposits (Bk, Bmc, Br)*

The Lower Cretaceous flyschoid depositional phase was followed in the early late Cretaceous (Cenomanian) by a regional, marine transgression and the deposition of two successive homogeneous blankets of mudstone (Bk) separated by an unconformtiy. On the northern mainland the older is the Boundary Creek Formation and the younger the Tent Island Formation (Young, 1975, p. 5, 7). They correspond both lithologically and in age to the Smoking Hills and Mason River Formations, respectively, of the Amundsen Gulf Group (Table I) on Anderson Plain (Yorath *et al.*, 1975, p. 18) and to the undifferentiated Kanguk Formation on Banks Island (Miall, 1977).

West of Mackenzie Delta, these Upper Cretaceous shales are overlain by a molassoid, terrigenous, seaward-thickening, generally progradational clastic wedge (Bmc) of the Moose Channel Formation (Young, 1975, p. 20). It is late Cretaceous (Maestrichtian) in age and marks the first major regression in the pattern of large-scale marine transgressions and nonmarine progradations of the molassoid phase (Young, 1974*b*, p. 192) of the northern Cordillera. The wedge is overlain by the delta-plain facies of the Lower Tertiary (Paleocene) Reindeer Formation (Br), segments of which occur here and there west of the Mackenzie Delta as well as in Caribou Hills immediately east of it (Mountjoy, 1967*b*, p. 10) on the northwest flank of the Aklavik Arch Complex (see Fig. 2). Correlative, nonmarine, clastic rocks and coal on Banks Island were assigned to the uppermost Cretaceous and Lower Tertiary Eureka Sound Formation (Thorsteinsson and Tozer, 1962, p. 65).

3. *Upper Tertiary and Quaternary Unconsolidated Deposits (Bb, Bq, Bf)*

The bedrock on western Banks Island and beneath the Old Crow Plain and the Mackenzie Delta is almost completely covered by unconsolidated fluvial, lacustrine, and glacial clastic sediments. They have been divided into three map units (Fig. 3a,b): the Upper Tertiary and Lower (?) Quaternary fluvial sands and gravels of the Beaufort Formation (Bb) on Banks Island, on Anderson Homocline, and their seaward equivalents on Beaufort Shelf; the undifferentiated Quaternary lacustrine, fluviatile, marine, and glacial sediments (Bq) of the Old Crow Basin, the Eskimo Lakes area, and the Beaufort Shelf; and the Holocene fluviatile deposits of the modern Mackenzie Delta (Bf). Because of their areal extent, these deposits lie on various older formations and effectively mask important geological relationships within the Cordillera as well as between the Cordilleran and Innuitian Orogenic Systems. Interpretation of structural and stratigraphic continuity in these critical areas, therefore, depends largely upon available geophysical data.

III. TECTONIC DEVELOPMENT

A. Geotectonic Elements

1. *Introduction*

The project area contains several of the tectonic elements comprising the Interior and Arctic Platforms, and the Cordilleran Orogenic System (Norris, D. K., 1974). The boundaries of these elements are defined or redefined on the basis of known structural and stratigraphic discontinuities. Where extensively covered by Quaternary sedimentary deposits or by the Beaufort Sea, however, the boundaries are arbitrarily generalized along the limit of outcrop.

There are basically four classes of elements: platforms, fault arrays, depressions, and uplifts or arches, each being characterized by specific structural and stratigraphic attributes. The limits of each element are defined in terms of attributes acquired during and since the Laramide Orogeny, although some elements may have originated in the Proterozoic. The Interior Platform of the northern mainland of Canada and its structural continuation into the Arctic Archipelago, termed the Arctic Platform, are integral parts of the craton. They are composed of gently inclined, relatively undeformed, paraconformable, layered successions of sedimentary rocks, with minor basic intrusions confined to the Proterozoic sequences. They are noted for their regional stability and in the extreme appear to have sustained only broad arching, such as occurred, for example, in the Coppermine Arch and in minor uplifts in both southern and northern Banks Island.

Fault arrays are bundles of generally parallel or subparallel faults of a common type. They usually have strike lengths measured in hundreds of kilometers. The faults or strands comprising them may overlap in any dimension, bifurcating or converging upward, downward, or along strike. Collectively they may constitute a closely knit association of structures of more or less constant width that maintains the sense and magnitude of tectonic transport of the array, or they may diverge, splay and die out through dissipation of displacement on a greater number of strands. Moreover, the arrays may change direction with respect to the principal deformative stresses with consequent change in the nature of the displacement on individual faults, as seen for example in the Richardson Fault Array where it passes southward from the Richardson Anticlinorium behind or structurally west of the Mackenzie Foldbelt (Norris, D. K., 1973, p. 27).

Depressions and their counterparts as uplifts or arches are structurally negative and positive regions, respectively, commonly juxtaposed and complementing one another. They may be formed by faults or folds or by combinations of faulting and folding. Arches are structurally simple, broad, positive, anticlinal features largely confined to the relatively stable, cratonic regions. Depressions and uplifts in the Cordilleran Orogenic System, on the other hand, are intimate associations of faults and folds that commonly have undergone cycles of uplift and depression, cumulative deformation, and differential erosion or sedimentation. Thus a particular tectonic element may have been an uplift at one time and a depression at another. It may possess spectacular angular unconformities locally in addition to the regional and interregional unconformities identified on the platform. It is important to appreciate, therefore, that these elements are defined by their present features, which developed through their entire history. Although their Holocene trend may be similar to that of their ancestors, their areal and linear extent may be quite different from those in the geological past.

2. *Romanzof Uplift*

The Romanzof Uplift, defined by Payne (1955) in northeast Brooks Range, originally was part of the Colville Geanticline, uplifted in Tertiary time along east-striking reverse faults. Mostly Paleozoic and Proterozoic rocks are exposed with the Mesozoic Systems removed by erosion. The Uplift's northern limit both in Alaska and immediately adjacent Canada is the Arctic Platform (Fig. 2), a pre-Albian tectonic feature, according to Payne (1955), which occupied the northern part of the present Arctic Coastal Plain. Its southern limit is the Brooks Range Geanticline, a Mesozoic and early Tertiary tectonic element in which mostly Paleozoic rocks are exposed. The boundary between the Brooks Range Geanticline and the Romanzof Uplift, as interpreted by Payne, coincides approximately with the northern limit of the Old Crow–Babbage Depression. D. K. Norris (1974, Fig. 1, p. 19) included the Barn Uplift as part of the Romanzof Uplift, and thereby obscured an important structurally low area, which is basically a northward continuation of the Old Crow–Babbage Depression (see Fig. 2).

By identifying the Barn Uplift and Old Crow–Babbage Depression as separate tectonic elements, the writers wish to amend the boundary of Romanzof Uplift, as defined by Payne, to include only the area of uplifted and exposed Proterozoic Neruokpuk Formation (In) along with included outliers of Phanerozoic rocks (see Fig. 3a, geological map). Thus the Old Crow–Babbage Depression is expanded to include the area of folded and faulted Upper Paleozoic and Triassic clastics and carbonates (El) on the south flank of the British Mountains and originally embodied in the Brooks Range Geanticline. The northern limit of the Romanzof Uplift, on the other hand, remains the boundary with the Arctic Platform and its seaward continuation as Beaufort Shelf. South of Herschel Island, however, the Romanzof Uplift and Old Crow–Babbage Depression are bounded by the Rapid Depression (Fig. 2).

Two angular unconformities and two disconformities partition the sedimentary succession within the amended Romanzof Uplift. Moreover, the Mount Sedgwick Stock, with a maximum measured radiometric age of 355 m.y. (Table II), is the only known acid igneous intrusion within the uplift. There are, however, many minor, basic sills and dikes. A prominent Cambrian volcaniclastic assemblage (Fv) in the uplift straddles the Alaska–Canada border.

The angular unconformity at the base of the Mississippian Kekiktuk Formation forms the southern and southeastern boundary of the Romanzof Uplift as well as the base of the outliers of the Paleozoic Lisburne Group. It is recognized also to the south of the project area in the Aklavik Arch Complex (Norris, D. K., 1974, p. 32). Beneath the unconformity are deformed argillites, quartzites, and limestones of the Neruokpuk Formation (In), repeated on a megascopic scale by a family of southeast-trending, high-angle reverse faults (see Fig. 5) that dip both toward the interior of the uplift and toward the Beaufort Shelf. Because many of these faults appear to have large stratigraphic separations within the Neruokpuk Formation and only minor or negligible separations within the Mississippian and younger strata, they were kinematically active for the most part prior to the deposition of the Kekiktuk

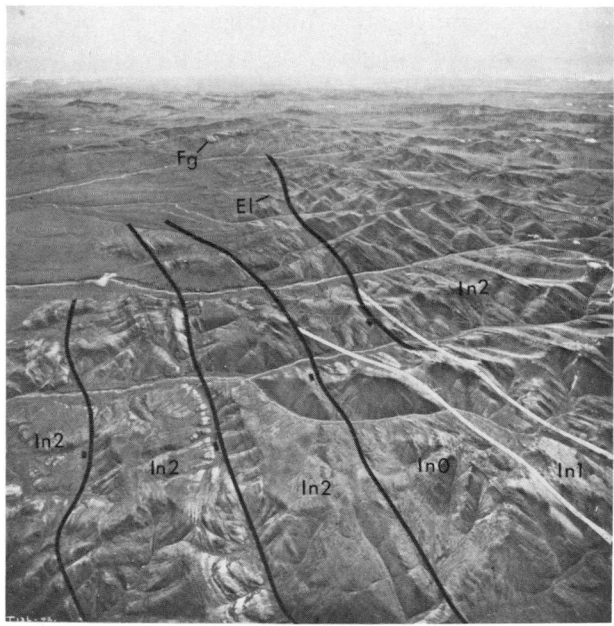

Fig. 5. Northeast flank of Romanzof Uplift, showing repetition of the Inuvikian Neruokpuk Formation (In0, In1, In2), the Mount Sedgwick stock (Fg), and the Shublik Formation (El). NAPL Oblique Photo T13L-42. View is to the southeast. Depositional contacts are identified by white lines and faults by black lines. Solid black dashes indicate downthrown walls.

Formation in Mississippian (Visean) time. There is, in fact, good evidence to suggest that this deformation occurred prior to the Cambrian, as outlined below.

In the British Mountains, a major unconformity appears to separate the relatively undeformed Cambrian volcanic, carbonate, and shale succession (Fv) from the underlying Neruokpuk Formation (Norris, D. K., 1976a, p. 457). Differences in dip and in intensity of deformation as well as the occurrence of a variety of lithologic units of the Neruokpuk in contact with the Cambrian support the thesis that the Neruokpuk was deformed principally prior to the Cambrian. Moreover, the Cambrian succession rests unconformably on the hanging wall of a major, reverse fault in the uplift. Its structural position on one of the older rock units of the Neruokpuk suggests that the latter formation was already faulted at the time the Cambrian succession was deposited and that the bulk of the deformation in Romanzof Uplift took place prior to the early Cambrian.

The stratigraphic succession in the Romanzof Uplift would appear to have undergone differential elevation and depression repeatedly in the Phanerozoic. The absence of the Cambrian unit (Fv) beneath the Mississippian Endicott and Lisburne Groups (El) suggests that much of the uplift may have been elevated prior to the Mississippian in order that the Cambrian strata were removed by erosion over all but a limited area. Immediately west of the project area, Mississippian rocks rest with angular unconformity upon an undated chert and phyllite assemblage in gradational contact with the trilobite-bearing Lower Cambrian volcanics, carbonates, and clastics (Dutro et al., 1972, p. 809). The authors would like to suggest that the cherts and phyllites are coeval with some part of the Lower and Middle Paleozoic Road River Formation. The latter may have been widespread in the northern Cordillera and served as a principal northern source of Upper Devonian sediments during epeirogenic uplift in the late Devonian and earliest Mississippian (Tournasian). Thus the angular unconformity at the base of the Endicott Group may not represent the Ellesmerian Orogeny but rather a much earlier event.

Two tectonically important overstepping relationships are present on the north flank of Romanzof Uplift, the one at the base of the Triassic Shublik Formation and the other at the base of the Jurassic and Lower Cretaceous Kingak Formation (Norris, D. K., 1974, p. 35). Differential elevation in early or mid-Triassic time was documented there by the occurrence of Upper Triassic Shublik Formation in depositional contact with the Proterozoic Neruokpuk. The full stratigraphic succession of Carboniferous and Permian rocks on the southwest flank of Romanzof Uplift has been overstepped northward in the direction of the Beaufort Shelf. The second important overstepping relationship on the north flank of the uplift is documented (Norris, D. K., 1974, p. 35) by the occurrence of the Jurassic and Lower Cretaceous

Kingak Formation unconformably upon the Neruokpuk. The northern part of the uplift was elevated in earliest Jurassic time, then depressed and stabilized from the Jurassic (Toarcian) through the early Cretaceous to accumulate a seaward-thickening clastic wedge (see Fig. 3b, structure sections) that recorded the onset and progress of the Laramide Orogeny (s.1.). Regionally significant, fundamental and widespread restructuring of the epicontinental seas and land masses then took place. With orogeny and uplift in the interior of the Cordillera, the principal sources of clastics shifted from the east and northeast to the west and southwest (Norris, D. K., 1974, p. 34).

Romanzof Uplift comprises two contrasting structural styles stacked one upon the other at the unconformity separating the Neruokpuk Formation from younger rocks. The lower (and older) style is confined to the Neruokpuk and comprises open to collapsed, cylindrical, flexural-slip folds with axial surfaces either vertical or dipping steeply to the southwest or northeast. Southeast-trending, high-angle reverse faults divide the region into a number of elongate blocks. Although exposure is poor, the style is clearly and abundantly displayed in the canyons of the Malcolm and Firth Rivers, which head in or cut across the uplift. The contorted, cleaved, intruded, and erosionally truncated Neruokpuk Formation is the basement for the upper (and younger) style displayed in the Phanerozoic rocks, which were deformed principally by the late Cretaceous and Tertiary (Laramide) orogeny.

On the crest and north flank of the Romanzof Uplift, the Cambrian (Fv), the Carboniferous, Permian, and Triassic (El), and the Jura-Cretaceous (Ek) rocks commonly occur as isolated, synclinal remnants or as slabs bounded on one side by high-angle reverse faults extending upward from the Neruokpuk basement. The faults were initially Proterozoic structures, rejuvenated by compression in the late Cretaceous or early Tertiary, with the concurrent development of open folds. On a mesoscopic scale, the carbonate and volcanic rocks are characteristically closely fractured, and the Jura-Cretaceous shales are acutely folded along northwest-trending axial surfaces. Although the structural trends of the two styles are parallel and northwest, the deformation of the Phanerozoic rocks is mild compared to that in the Proterozoic basement.

3. *Old Crow–Babbage Depression*

The Old Crow–Babbage Depression lies on the north flank of the Brooks Range Geanticline as defined by Payne (1955). Within the project area it is flanked on the north by the amended Romanzof Uplift, and on the east by the Rapid Depression and Barn Uplift. South of the area it is bounded by the Kaltag–Porcupine Fault and the Aklavik Arch Complex. Sedimentary rocks, ranging from the Proterozoic Tindir Group (It) to the upper Lower Cretaceous "upper shale–siltstone division" (Bdb), occupy the depression. Three small granitic intrusions (Fg) have been identified, one on Mount Fitton with cupolas to the south and northwest, another on Ammerman Mountain, straddling the International Boundary, between Firth and Old Crow Rivers, and a third 40 km west of the boundary (see Fig. 3a, geological map). Over a large area in the center of the depression the bedrock is masked by Quaternary lacustrine and fluviatile deposits (Bq).

The profound unconformity at the base of the Mississippian Endicott Group can be mapped on the north flank of the depression as well as locally within it. Unnamed Devonian, plant-bearing, slaty argillites and quartzites (Fi) occur locally between the Endicott (El) and Tindir Groups (It) (Norris, D. K., 1972b, p. 93) on the north side of the Old Crow Plain at least as far west as Ammerman Mountain. The Endicott and Lisburne Groups cover a large area of the depression between the Old Crow Plain and Romanzof Uplift. The Triassic Shublik Formation, on the other hand, is confined for the most part to the headwaters of Babbage River and around the southeast plunge of the Romanzof Uplift. Immediately east of the Mount Sedgwick Stock, however, the Shublik was removed by erosion at the unconformity at the base of the Jura-Cretaceous Kingak Formation (Ek).

The incompetent shales of the Kingak are preserved between Barn and Romanzof Uplifts as well as in the prominent structural depression between Old Crow Plain and British Mountains. They are overlain by the Lower Cretaceous "upper shale–siltstone division" (Bdb), which is preserved locally in synclinal depressions west of Barn Uplift and south of Romanzof Uplift.

The Old Crow–Babbage Depression appears to be underlain by a complex association of structural highs and lows, with the area of Old Crow Plain (Fig. 1) interpreted to contain sedimentary rocks locally as young as Cretaceous beneath a cover of Tertiary nonmarine and Holocene lacustrine deposits (Lawrence, 1973, p. 311). The gently dipping Endicott Group (El) exposed in the middle of the Old Crow Plain and the apparently flat-lying Mesozoic and older strata indicated nearby on the two short seismic sections published by Lawrence (1973, Figs. 5 and 6) suggest, moreover, that the rocks in the plain may not be folded as acutely as they are peripheral to it. Rather they are faulted and differentially uplifted without significant rotation. Thus the isolated outcrop of Endicott Group may be part of an uplifted block, fault-bounded on the west, gently inclined to the east, and overlain by Upper Paleozoic carbonate rocks of the Lisburne Group that connect with extensive exposures of the group at the headwaters of Black Fox Creek (see Fig. 3a, geological map). The age, areal extent, and continuity of the youngest bedrock beneath the Tertiary clastics in the vicinity of the Old Crow River are uncertain. It is suggested that, if resistant strata as young as the "upper shale–siltstone division" (Bdb) were present there, they would be exposed.

Recessive shales coeval with the Jura-Cretaceous Kingak Formation may be present locally as a thin veneer upon Lisburne and Endicott rocks. Both of the latter groups would appear, however, to have been removed southwestward because strata correlated with the Neocomian "white and coaly quartzite division" (Ek) rest upon the Proterozoic Tindir Group on the flanks of the Old Crow Batholith (D. K. Norris, unpublished data).

A few southeast-trending, reverse faults paralleling the long dimension of the Romanzof Uplift occur north of the Old Crow Plain. They are the continuation of the belt of southwest-dipping thrust faults of the Brooks Range west of the project area that repeats the Upper Paleozoic succession (Lathram, 1965) and appears to glide in a major detachment in shale and coal of the Endicott Group. In eastern Brooks Range, however, the faults are observed to decrease progressively in number and in stratigraphic separation so that, toward the 141st meridian, only a few discontinuous strands remain (Fig. 3a, geological map). They commonly can be mapped for just a few tens of kilometers. Rarely do they possess the great lateral continuity that characterizes the thrust belts in Alaska and the eastern Cordillera of Canada in the vicinity of the 49th parallel of latitude. The few reverse faults that extend beyond the Romanzof Uplift turn abruptly south and are transformed into right-lateral strike-slip faults. One such fault merges with the Rapid Fault Array and defines the eastern flank of Barn Uplift. Another bifurcates and repeats the carbonate rocks of the Lisburne Group (El) on the west flank of the uplift (see Norris, D. K., 1973, p. 31). The sense of relative motion on the reverse faults was mechanically congruent with that on the strike-slip faults and caused relative displacement of the Old Crow–Babbage Depression northward toward Beaufort Shelf (Norris, D. K., 1974, p. 36).

4. Barn Uplift

Barn Uplift lies along the eastern margin of the Old Crow–Babbage Depression (Fig. 2). It is defined herein as the elongate area of exposure of Lower Paleozoic rocks (Fr) in Barn Mountains (Fig. 6). The term Barn Uplift was introduced informally by D. K. Norris (1973, p. 24) in a summary account of the geology of northernmost Yukon Territory and adjacent District of Mackenzie. Martin (1959, p. 2414) first reported the presence of Silurian graptolites in the uplift and paved the way for the correlation of these unnamed rocks with coeval graptolitic shales and limestones in the Richardson Trough (Lenz, 1972, p. 329) south of the project area.

The stratigraphic succession comprising Barn Uplift is acutely folded and faulted. It is an assemblage of dark gray to black, red, and green shales, black and green chert, ridge-forming gray quartzite and argillaceous siltstone, commonly with load casts (Dyke, 1974, p. 303) and light gray weathering,

Fig. 6. Barn Uplift, showing block faulting of unnamed Franklinian clastics and carbonates (Fr), overlain unconformably by Ellesmerian Endicott and Lisburne Groups (El) on the west (left), and the Kingak Formation (Ek) on the north. The contact with the "upper shale–siltstone division" and younger flyschoid rocks (Bdb) on the north and northeast is a fault. Fg, Franklinian granite; MB, MacKenzie Bay. Solid circles locate two fossil localities of Ordovician/Arenigian (OAr) and Silurian/Ludlovian (SLu) ages. NAPL Oblique Photo T13L-107. View is to the north. Depositional contacts are identified by white lines and faults by black lines. Solid black dashes indicate downthrown walls.

arenaceous limestone. According to Dyke, the quartzite interbeds are commonly massive, are usually fine to medium grained, and are rarely conglomeratic. The lithologies of this unnamed sucession contrast markedly with the black, silty shales and limestones of the type Road River Formation in the Richardson Trough. They are, however, similar to the slaty, gray, black, and red argillites, gray quartzites, and gray limestones of the Neruokpuk Formation of the Romanzof Uplift, although they lack the interbedded basic volcanics and dikes. A minor granitic cupola (Hoidahl) of the Mount Fitton Stock occurs in the northwest quarter of the uplift (see Fig. 3 and Table II).

Because some lithologies are common to both the Romanzof and Barn Uplifts, their successions were initially considered equivalent (Norris, D. K., *et al.*, 1963) and the term Neruokpuk was unfortunately extended to include the pre-Mississippian rocks in Barn Mountains. The rocks there were interpreted to be largely Proterozoic in age and the graptolitic shales within them were considered to have been preserved in structurally depressed blocks (Norford, 1964, p. 137).

Detailed stratigraphic and structural studies by geologists of Atlantic Richfield Canada Ltd. (Lenz, 1972) and by the Geological Survey of Canada (Dyke, 1974) indicate, however, that the pre-Mississippian Barn Uplift assemblage is most probably entirely of early Paleozoic age and that none of it may be Precambrian. Graptolites collected in the uplift as well as in inliers to the south (Martin, 1959, p. 2414; Norford, 1964, p. 137; Norris, D. K., 1970, p. 231; Norris, A. W., 1971, p. 106; Lenz and Perry, 1972, p. 1131) represent many stages from Lower Ordovician to Lower Devonian. Paucity of fossils, structural omissions due to faulting, and differential erosion prior to Mississippian deposition can account for the apparent absence of some stages in this part of the Cordilleran Orogenic System. It seems reasonable, however, to suggest a complete or nearly complete succession of Lower Ordovician to at least Upper Silurian strata (Lenz and Perry, 1972, p. 1131) in the Barn Uplift.

The Barn Uplift is divided into many elongate blocks bounded by nearly vertical, curvilinear, north-trending faults (Figs. 3a and 6). Repetition of stages identified by graptolites in adjacent blocks indicates, moreover, that the rock succession from east to west across the uplift is a series of highly deformed, generally steep-dipping panels. Strata as young as middle to late Silurian occur in the outermost blocks and as old as early Ordovician in the middle of the uplift. Of the five blocks shown in Fig. 6, the long, narrow central one contains Lower Ordovician (Arenigian, OAr) graptolites near its eastern boundary and Upper Silurian (Ludlovian, SLu) graptolites near its western boundary (paleontological control from Lenz and Perry, 1972, Fig. 2). These observations together with evidence from vertical air photographs indicate that a relatively undeformed, west-facing panel, representing the maximum observed number of faunizones in the uplift, is approximately 1000 m thick. This appears to be approximately the minimum stratigraphic thickness of Phanerozoic rocks exposed in the uplift, because even lower beds may occur at depth and higher strata may have been removed at the sub-Mississippian unconformity.

Because the unnamed strata (Fr) comprising the Barn Uplift and the

Endicott and Lisburne Groups (El) both strike north, the contact between the two successions is a paraconformity on the west flank of the uplift. On the south flank, however, the unconformity is essentially right angular. Clearly the Ordovician and Silurian beds were deformed, uplifted, and bevelled prior to the transgression of the late Mississippian (Visean) clastics of the Endicott Group. Stratigraphic separations of up to 1000 m appear to have been accomplished on the various components of the north-trending set of nearly vertical faults prior to the Visean. Minor reactivation of movement along them, possibly in conjunction with oblique-slip faulting in the Rapid Depression in the late Cretaceous and early Tertiary, accounts for the markedly reduced offsets on the extensions of these faults into the Mississippian and younger rocks (see Figs. 3a and 6).

5. Rapid Depression

Rapid Depression is a northeast-trending, structurally controlled low between the Old Crow–Babbage Depression and Romanzof and Barn Uplifts on the west, and the Aklavik Arch Complex and Cache Creek Uplift on the east (see Fig. 2). On Beaufort Shelf the low is identified by the −40- and −50-mgal free air gravity anomalies (Fig. 3a) in Mackenzie Bay. It terminates up plunge just south of latitude 68°N.

The rocks filling much of the depression are siltstones, turbiditic sandstones, and conglomerates (Bdb) of early Cretaceous age. This is predominantly the flyschoid sequence (Norris, D. K., 1974, p. 34; Young, 1974b, p. 188) that signaled a fundamental and widespread restructuring of landmasses and depositional sites. Beginning in the mid-early Cretaceous, more than 3000 m of immature, flyschoid clastics were deposited in the axial region of the depression in the Blow River area (see Fig. 3a), with coarse conglomerates predominating. On the flanks, flyschoid rocks are observed to thin markedly (Fig. 3b, structure section D–D' and Fig. 7) and to rest unconformably on a variety of older units because of the interregional unconformity at the base of the Brookian Sequence.

More than 1200 m of Upper Cretaceous and Tertiary molassoid rocks (Bk and Bmc) were deposited axially in the Rapid Depression on top of the flysch. They were penetrated in the IOE Blow River YT E-47 well (Fig. 3b, Section D–D'; Table III). Still younger continental deposits (Br) occur in a prominent syncline in the depression 25 km northwest of the well (Norris, D. K., 1972b, p. 97).

The general eastward-fining trend, paralleling downcurrent paleodispersal directions in the flyschoid sequence (Young, 1974a, p. 295), and the marked contrast in mineralogical maturity between the epicontinental (Ek) and flyschoid (Bdb) phases support the concept of a redistribution of landmasses

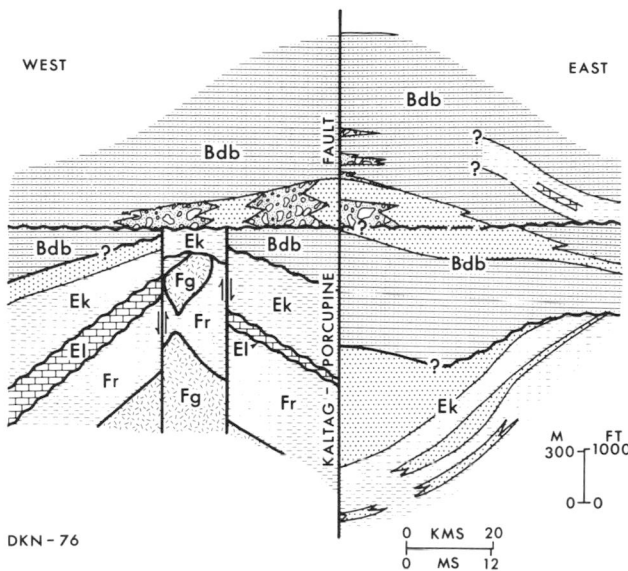

Fig. 7. Schematic cross section of Rapid Depression, showing stratigraphic and structural relations of Franklinian, Ellesmerian, and Brookian Sequences across the Kaltag–Porcupine Fault. Datum is the unconformity at the base of the prominent flyschoid conglomerates and its postulated eastward continuation. Diagram is based upon Fig. 3 of Young (1974a, p. 294), amended to allow for the lateral equivalence of the conglomerates and the "upper sandstone division."

and depositional sites. During the Jurassic and early early Cretaceous, the widespread epicontinental, marine, mineralogically mature sandstone and shale succession (Ek) was derived from the craton to the east and southeast. In the late early Cretaceous, texturally and mineralogically immature, chert, lithic sandstones, and conglomerates (Bdb) began to be shed from the deforming and uplifting regions deeper into the orogenic system to the west and southwest. The change from the Ellesmerian to the Brookian Sequence marks the initiation of the earliest phase of the Laramide Orogeny (s.l.) in the northern Canadian Cordillera.

The presence of the transgressive and flyschoid rocks resting upon various stratigraphic units in the axial region of the trough and upon progressively older formations away from the axis (Fig. 7) indicates not only that the depression was structurally controlled, but also that it had a long antecedent history. On the east flank of Barn Uplift the flyschoid rocks rest upon a thin succession of Lower Cretaceous (Berriasian) shales and sandstones (Triad Oil Co., personal communication, 1959), whereas it rests upon the slightly younger (Valanginian) "white and coaly quartzite division" and upon shales and siltstones of the Kingak Formation in immediately adjacent areas. Eastward in the vicinity of Cache Creek Uplift, moreover, the flyschoid rocks are observed again to lie unconformably upon sandstone of the "white and coaly quartzite division" and in turn upon shales of the Kingak (Norris, D. K., 1976a, p. 458, 459) on the crest of the uplift. Westward from the axis of Rapid Depression, contact relations between the flyschoid sequence and older strata are generally poorly exposed and obscured by faulting. On the west flank of Barn Uplift, however, flyschoid shales also lie unconformably upon the "white and coaly quartzite division" and farther northwest, somewhat parallel to the flank of the depression, upon the Kingak Formation. Both flanks of the depression would appear to have been relatively uplifted and to have contributed detritus prior to their being denuded and overrun with sediment as the trough filled. Indeed, some paleocurrent measurements (Young, 1974b, Fig. 6) suggest transport both to the northwest and east away from the axis of the Cache Creek Uplift for the flyschoid and younger clastics.

Evidence from stratigraphic omissions support still earlier tectonic activity in the position of the depression. In the early Cretaceous, resistant sandstones and quartzites ("white and coaly quartzite division") comprising the unnamed, prominent, north-trending range of mountains a few kilometers east of Bonnet Lake (see Fig. 8) were eroded from the block immediately west of the Kaltag–Porcupine Fault (Fig. 7). They recur west of the Barn Uplift as a 200-m, ridge-forming unit outlining much of the structure in the depression between there and the southeast plunge of Romanzof Uplift. Immediately south of latitude 68°N, and up the plunge of Rapid Depression, moreover, Lower Jurassic sandstones rest unconformably upon Lower Paleozoic shales (Fr) of the Road River Formation (D. K. Norris, unpublished). And finally, toward the axis of the depression, a spectacular unconformity separates the Franklinian and Ellesmerian Sequences (Fig. 8). Folded and faulted, chert pebble and cobble conglomerates at the base of the Mississippian Endicott Group rest upon vertically dipping, strongly cleaved shales coeval with the Lower Paleozoic Road River Formation.

The nature and timing of the structural events controlling the Rapid Depression are revealed by the family of nearly vertical, curviplanar faults within and bounding the depression. The Kaltag–Porcupine Fault (K-P), a major, right-lateral transcurrent fault, trends northeast across Old Crow–Babbage Depression and merges with the Yukon Fault (Y), a thrust approximately at the southern limit of Rapid Depression (see Fig. 9). Together they turn northward, anastomose, and flare out as a family of nearly vertical faults (Rapid Fault Array) ranging in trend from northwest to northeast. They cut the Cretaceous flyschoid and molassoid successions into a number of blocks and depress them systematically so that the blocks in the axial region of the depression are structurally lowest. The latest vertical displacement of the Rapid Fault Array was in the early Tertiary or later.

Fig. 8.   Unnamed uplift in the axial region of Rapid Depression showing the uncomformity between the Franklinian (Fr) and Ellesmerian (Ek) Sequences, the Kaltag-Porcupine (K–P), Skull (SF), and other faults of the Rapid Fault Array, Blow Syncline (BS), and Bonnet Lake (BL). NAPL Oblique Photo T13R-114. View is to the north. Depositional contacts are identified by white lines and faults by black lines.

The fact that these younger movements are along faults in the Rapid Depression would strongly suggest reactivation of older faults. Indeed, the Kaltag–Porcupine Fault defines the western limit of the "white and coaly quartzite division" comprising the mountains east of Bonnet Lake, and a splay from it can account for the major stratigraphic omission at the base of the Lower Jurassic sandstones along the west flank of the depression just south of latitude 68°N. Moreover, the omission of the lower part of the flyschoid suc-

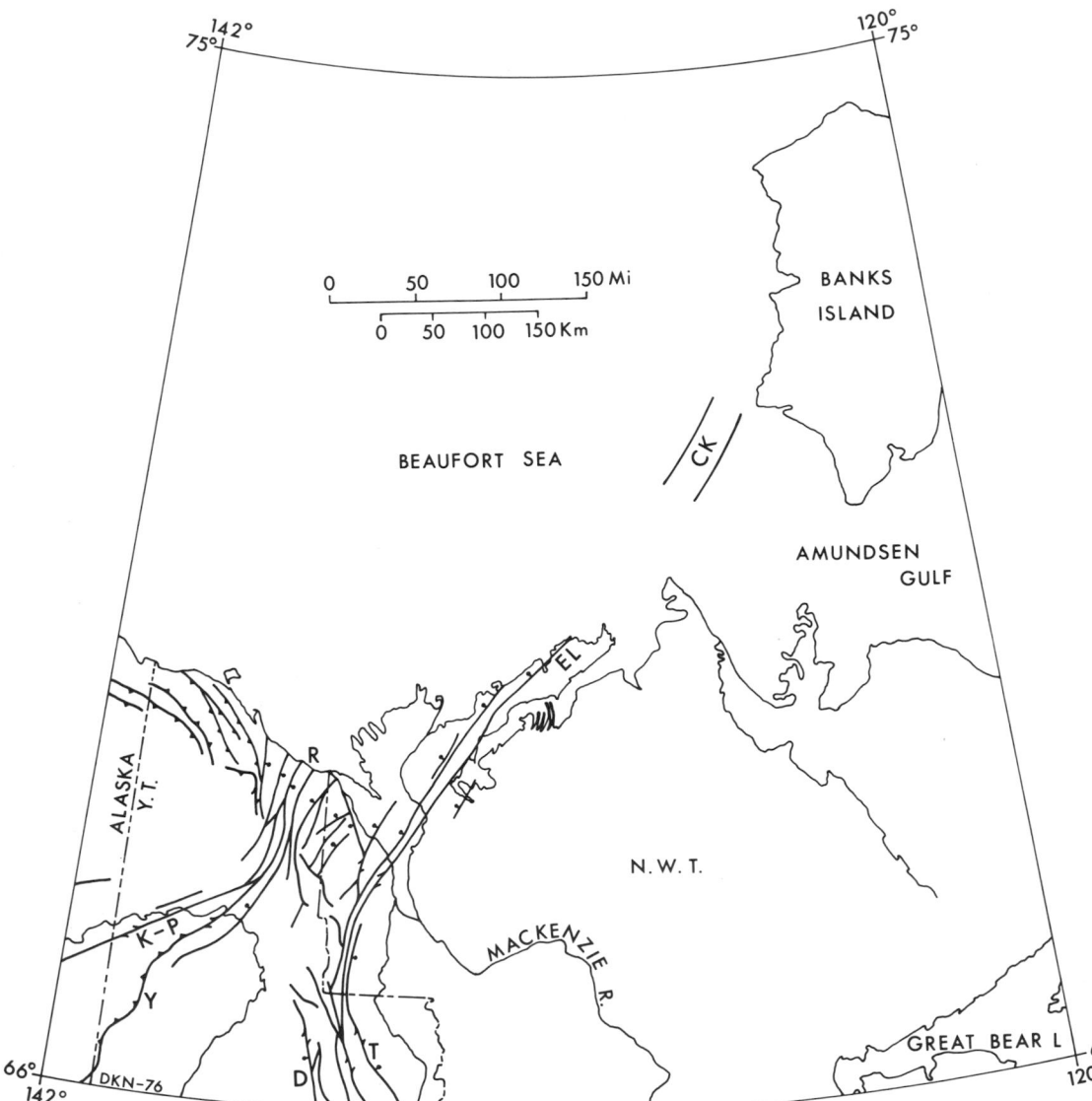

Fig. 9.   Spatial relations between the Richardson (D–T) and Rapid (R) Fault Arrays, and Cape Kellett (CK) Fault Zone
The Kaltag–Porcupine (K-P), Yukon (Y), Deception (D), Trevor (T), and Eskimo Lakes (EL) Faults are identified. Soli
circles indicate downthrown walls, teeth upthrown, and barbs the sense of displacement on most strands of the arrays.

cession at Mount Fitton suggests syndepositional faulting in the trough (see
Fig. 7), perhaps in conjunction with regional upwarping of the flanks to form
Barn and Cache Creek Uplifts. Mechanically, right-lateral displacement on the
curviplanar Kaltag–Porcupine Fault requires uplift of the general Barn–
Romanzof region where the fault zone turns north and the motion is

transformed from one principally of strike-slip to one of high-angle reverse displacement.

It appears, therefore, that the position, orientation, and timing of the Rapid Depression were controlled by a combination of major fault systems in the Cretaceous and Tertiary. The Rapid Depression is the Mesozoic analogue of the Paleozoic Richardson Trough (Lenz, 1972, p. 329) and is a fundamental tectonic feature of the outer part of the Cordilleran Orogenic System.

6. Aklavik Arch Complex

The term Aklavik Arch was first used by Jeletzky (1961, p. 538) for a northeast-trending "mid-Valanginian anticlinal structure" extending from Bell Basin (northeastern Eagle Foldbelt of this paper) south of the project area, across Mackenzie Delta to the vicinity of Sitidgi Lake. Moreover, he applied the name "Dave Lord Creek Arch" (Dave Lord Uplift of this paper) to one of the late Middle Devonian "broad anticlinal arches" (Martin, 1959, p. 2442 and Fig. 12b), recognized by Martin and identified by Knipping (1960, p. 1 and Fig. 5) as "Dave Lord Ridge . . . the general location of the late Palaeozoic orogenic belt." The term "ancestral Aklavik Arch" subsequently was introduced by Jeletzky (1962, p. 66) for a regional arch which he interpreted to be the predecessor of the "Dave Lord Creek" and "Aklavik" arches. According to Jeletzky "It could be that these two arches . . . only arose in Cretaceous time, perhaps because of the lateral offset of parts of ancestral Aklavik Arch by north-trending strike-slip faults." He postulated, therefore, that the Dave Lord Uplift along the north flank of Eagle Foldbelt and Campbell Uplift immediately east of Mackenzie Delta were one physically continuous structural high prior to Cretaceous time. Because the Aklavik Arch is now recognized as a composite structural feature embracing among its several parts Dave Lord and Campbell Uplifts, it was renamed the Aklavik Arch Complex (Yorath and Norris, 1975, p. 590) and recognized geologically as one of the most complex tectonic elements in the northern Canadian Cordillera.

The Aklavik Arch Complex is a composite, northeast-trending element extending in Canada from Keele Range adjacent to the Alaska border to the east side of the Mackenzie Delta (Fig. 10). From southwest to northeast it embraces Keele Platform, Dave Lord Uplift, White and Cache Creek Uplifts, Canoe Depression, and Rat and Campbell Uplifts. Some of the structural depressions separating these uplifts are unnamed. Individually the long axes of the uplifts and depressions trend approximately 25° counterclockwise from the long axis of the complex. Collectively they are arranged in a systematic, right-hand en echelon pattern that trends obliquely to the regional structural grain of this northern part of the Cordilleran Orogenic System.

The rocks comprising the Aklavik Arch Complex range from Proterozoic

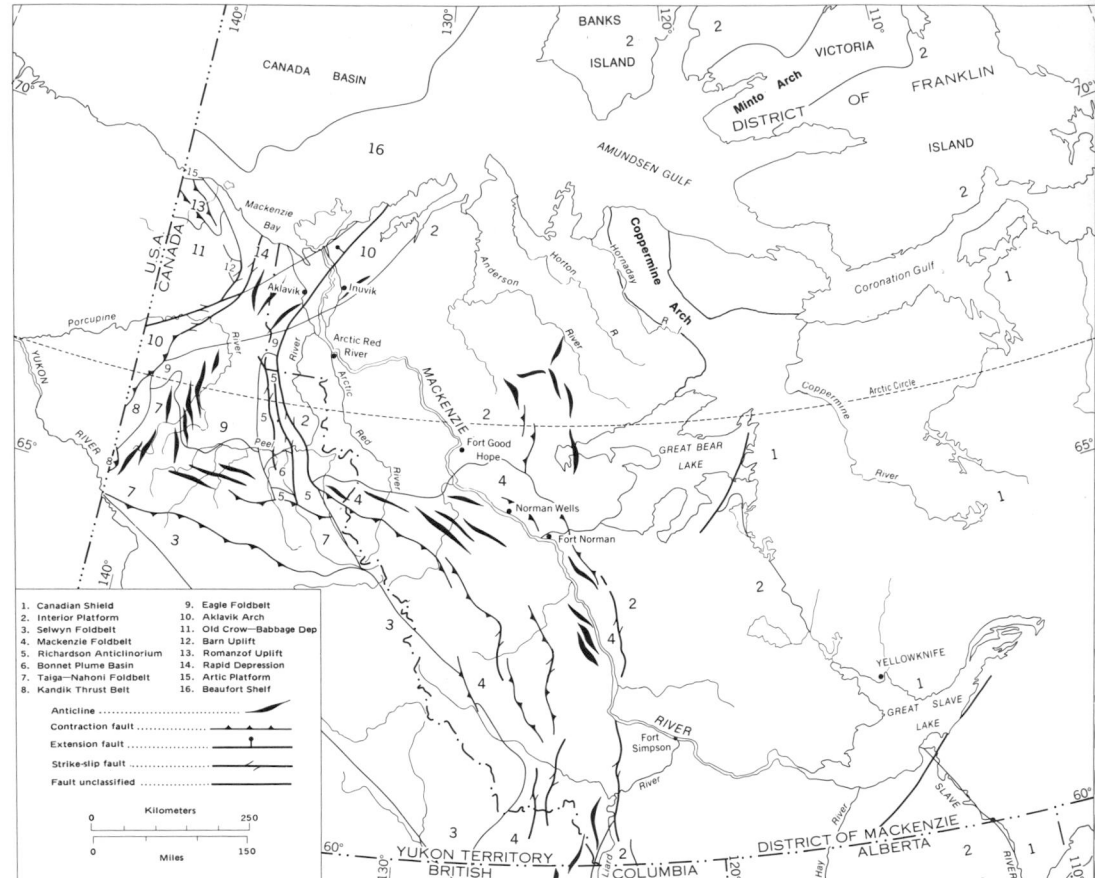

Fig. 10. Tectonic elements of the northern Cordillera and Interior Platform of Canada (after Norris and Hopkir 1977).

to Tertiary although the successions are by no means complete. Many unconformities, some spectacularly angular, are present in the complex and attest to
intermittent and prolonged activity. South of the project area in Keele Platform, for example, the unconformity between the Proterozoic succession and the
shoreline sandstones and shales assigned to the Middle Cambrian indicates
uplift and erosion in the late Proterozoic or early(?) Cambrian. Moreover, the
Proterozoic dolomite and clastic succession (Ic) in the core of Campbell Uplift
(see Fig. 3a, geological map) at the northeast limit of surface expression of the
complex is overlain unconformably by gently dipping to flat-lying, Middle and
Lower Paleozoic carbonates and clastics (Fb). Although the contact has not
been seen, it is presumed to be angular as well, because of the generally steeper
dips in the Proterozoic rocks.

The next episode of deformation in the Aklavik Arch Complex appears to have taken place between the Lower Devonian and the Middle Permian. It may have coincided in part with the Ellesmerian Orogeny of the Arctic Archipelago (Thorsteinsson and Tozer, 1960). In Rat Uplift, an unnamed clastic unit of Lower or Middle Permian age rests with angular unconformity on deformed Upper Devonian shales and sandstones (Norris, D. K., 1974, p. 30) as well as on Lower Paleozoic Road River Formation (Knipping, 1960, p. 1). Moreover, on the east flank of White Uplift (Dyke, 1975), Lower Permian carbonates and clastics rest disconformably upon Carboniferous (Bashkirian–Moscovian) limestones and dolomites (Bamber and Waterhouse, 1971, Fig. 10). The Carboniferous succession in turn rests disconformably on Middle Devonian dolomites (Norris, A. W., 1968, p. 233). Farther southwest on the crest of Dave Lord Uplift where it is crossed by the Porcupine River (Fig. 10), deformed Lower Devonian and older graptolitic shales and chert are overlain with angular unconformity by undated chert-pebble conglomerates that may be as old as Mississippian and as young as early Cretaceous (Norris, D. K., 1974, p. 30). The presence in Middle and Upper Paleozoic rock of angular unconformities in some places and of disconformities in others suggests, moreover, that the intensity of this deformation varied from place to place within the arch complex.

Early Mesozoic deformation of part of the Aklavik Arch Complex is indicated by an angular unconformity at the base of the Lower Jurassic Series. Northwest and southeast of the arch complex, as well as in Dave Lord Uplift, the base of the Lower Jurassic succession is a disconformity. In Rat Uplift, however, Lower Jurassic sandstones rest upon Lower Permian clastics and in turn upon the Upper Devonian. Additional uplift and erosion in the region of Rat Uplift is indicated, moreover, by the gradual wedging out and overstepping of Lower Cretaceous sandstones on the crest of the arch by still younger Lower Cretaceous clastics (Jeletzky, 1961, p. 538). The fact that, adjacent to the Aklavik Arch Complex, the beds below and above these unconformities are paraconformable, that they are locally angularly unconformable in the complex, and that Upper Triassic carbonates and clastics are preserved here and there away from it, again attests to deformation and uplift confined to specific components of the complex.

Another angular unconformity occurs between upper Lower Cretaceous (Albian) conglomerates and folded and faulted Middle and Upper Paleozoic formations on the southeast flank of the Dave Lord Arch (Norris, D. K., 1974, p. 31). The deformation is confined to the interval from the Middle Permian through earliest Cretaceous. In Campbell Uplift (Fig. 2), however, Albian shales (Dyke, 1975, p. 527) lie unconformably upon Lower and Middle Paleozoic formations and in turn upon Proterozoic clastics and carbonates in the core of the uplift. The Albian transgression of the northern Interior Platform included the region of Campbell Uplift and the gentle truncation of underlying formations is attributed to deformation between the late Devonian and the late early Cretaceous.

Evidence of late Cretaceous or early Tertiary Laramide deformation of the Aklavik Arch Complex is widespread. Rocks as young as early Cretaceous are intimately involved in the structure of Rat Uplift. Locally, moreover, mid-Upper Cretaceous formations dip gently off the flanks of the complex as, for example, on the northwest side of Dave Lord Uplift (Norris, D. K., 1972b, p. 97). Furthermore, on the southeast flank of Campbell Uplift, gently folded, rusty shales and siltstones, dated as Aptian or Albian (W. S. Hopkins, personal communication, 1971), are overlain by flat-lying, unconsolidated to semiconsolidated sand and silt presumed to be the Upper Tertiary Beaufort Formation. The Laramide deformation of the Aklavik Arch Complex appears, therefore, to have taken place in the latest Cretaceous or early Tertiary in the region of Mackenzie Delta.

It is apparent that the Aklavik Arch Complex is a tectonic association of uplifted and depressed components that were active independently and intermittently, beginning possibly in the late Proterozoic (or earlier?) and continuing through to the latest Cretaceous or early Tertiary. Trending obliquely to the regional structural grain, it is demonstrably a mobile element that persisted through at least two major orogenies affecting the northern Canadian Cordillera, one in the late Paleozoic and the other in the late Cretaceous or early Tertiary. Both structurally and stratigraphically it is one of the most complex elements in the Canadian Cordillera and doubtless it has played a fundamental role in the evolution of the Cordilleran Orogenic System.

The Aklavik Arch Complex is cut transversely by the Richardson Fault Array (Fig. 9), a family of curvilinear, nearly vertical faults trending north and northeast from the Richardson Anticlinorium (Fig. 10) to the Beaufort Shelf. It is flanked by a family of faults extending northeast from the Alaska border to form an integral part of the Richardson Fault Array (see Fig. 9). The juxtaposition of the Keele Platform and Dave Lord–Cache Creek Uplift, and of Rat and Campbell Uplifts may, therefore, be due to right-lateral displacement on some of the faults of these two strike-slip systems (Norris, D. K., 1974, p. 32).

7. *Anderson Homocline*

Anderson Homocline (Fig. 2) was defined by D. K. Norris (1975b) to embrace the gentle west- and northwest-dipping homocline of Cretaceous and Paleozoic rocks (Yorath, 1973, p. 42) between the Coppermine Arch and Campbell Uplift. It is part of the Interior Platform of North America.

Physiographically it includes parts of Anderson and Horton Plains (see Fig. 1) and is relatively flat and featureless.

The Lower and Middle Paleozoic succession (Fb) is composed for the most part of dolomites and limestones with a basal, varicolored siltstone and sandstone sequence resting with angular unconformity upon Proterozoic dolomites and clastics (Imh). These platform carbonates are overlain by black, calcareous shales of the Hare Indian Formation (Fr), and in turn by shales and sandstones of the Upper Devonian Canol and Imperial Formations (Fi).

Lower Cretaceous (Aptian and Lower Albian) shales (Bdb) rest unconformably upon successively older Paleozoic formations from west to east, indicating progressive overstepping of the west flank of the ancestral Coppermine Arch (see Fig. 3a). They lie upon the Imperial Formation (Fi) adjacent to Campbell Uplift, in turn upon the Canol (Fi), Hare Indian (Fr), the Hume, Bear Rock, Mount Kindle, and Franklin Mountain Formations (Fb), and finally upon the unnamed Proterozoic units (Imh) exposed on the crest of the arch. The Carboniferous to Lower Cretaceous Ellesmerian Sequence is absent from the Anderson Homocline. Upper Cretaceous shales (Bk) are confined to the structurally lower part of the homocline adjacent to Campbell Uplift. They are overlain by small patches of Upper Tertiary Beaufort Formation (Bb).

The moderately to gently angular unconformities at the base and top of the Paleozoic System in Anderson Homocline indicate differential uplift prior to early Paleozoic and Cretaceous sedimentation (Yorath *et al.*, 1969, p. 22). There are, moreover, disconformities with varying amounts of local relief within the Proterozoic succession, at the base of the Bear Rock and Canol Formations and within the Cretaceous System. Collectively, these unconformities document only epeirogenic instability in this part of the Interior Platform since the late Proterozoic. They contrast markedly with those observed in the mobile belt of the adjacent Cordilleran Orogenic System.

8. *Coppermine Arch*

The term Coppermine Arch (Fig. 2) was introduced by Douglas *et al.* (1963, Fig. 1) for an anticlinal feature involving Proterozoic and Lower Paleozoic rocks in the extreme northwest corner of the Canadian Shield. Although defined only in outline, the arch clearly embraced not only a part of the shield but also the inlier of Proterozoic carbonates and clastics adjacent to Amundsen Gulf (Fig. 3a). The structure includes the saddle containing Lower Paleozoic Platform carbonates (Fb) that separates the two areas of Proterozoic rocks. The Coppermine Arch trends northwest and plunges beneath Amundsen Gulf. Water depths of 600 m suggest an even more pronounced saddle or other structural depression there. If the arch continues northwestward to Nelson Head on the southern tip of Banks Island (Fig. 3a), the Glenelg Formation (Ig) there represents a stratigraphically younger part of the Proterozoic succession in the Coppermine Arch.

The rocks comprising the arch are dolomite, limestone, quartzite, and shale of late Proterozoic age. They are intruded by northwest-trending, Proterozoic, diabasic dikes and sills. Lower Cretaceous rocks are preserved locally in structural depressions on the flanks of the arch (see Fig. 3a).

The epeirogenic history of Coppermine Arch appears to have begun in the late Proterozoic (Cook and Aitken, 1969; Balkwill and Yorath, 1970). Stratigraphic thinning and omission of formations in fault-bounded blocks suggest movements during and after deposition of the Proterozoic rocks. Tilting toward the northeast and deep bevelling occurred prior to the initiation of Paleozoic sedimentation. Cambrian through Lower Silurian rocks display no change of facies suggestive of a topographic high along the arch (Cook and Aitken, 1969). Pre-Middle Devonian uplift is indicated, however, by the absence of Upper Ordovician and Lower Silurian rocks beneath the Hume Formation. Corals and stromatoporoids in the Hume, moreover, demonstrate the return to stable shelf conditions. Regional northwest tilting and bevelling before the Cretaceous resulted in Lower Cretaceous shales overstepping onto successively older formations to the southeast. Like Campbell Uplift to the west, positive movement and extension of the arch in the late Cretaceous dropped these Lower Cretaceous rocks in grabens on the flanks of the arch where they are preserved today.

9. Banks Homocline

Banks Homocline (Figs. 2 and 11) is defined herein as the gently west-dipping, Brookian Cretaceous and younger clastic sequence underlying Victoria Lowland and Island Coastal Plain in western Banks Island (Fig. 1). It includes, therefore, the west flank of Prince Albert Homocline (Thorsteinsson and Tozer, 1960, p. 5 and Fig. 2), which is defined to embrace the sequence of Ordovician to Upper Devonian rocks on the northwest flank of Minto Arch (Fig. 10). It merges with the Beaufort Shelf without break in slope or structure.

Beneath the homocline is a succession of Jurassic (Ek) and older rocks deformed into several doubly plunging, curvilinear uplifts and depressions (Miall, 1974, Fig. 1). Surface mapping in conjunction with gravity data (Miall, 1977) suggests that none of these structures has great length and it is not surprising that uplifts at the northern and southern tips of the island are demonstrably discontinuous. Noteworthy is the Lower and Middle Paleozoic carbonate bank (Fb) grading westerly into deep-water shale (Fr) on Beaufort Shelf (Fig. 11 and Miall, 1976, Fig. 8). Implicitly these shales may connect with those interpreted to be present farther southwest on the shelf (see Fig. 3b,

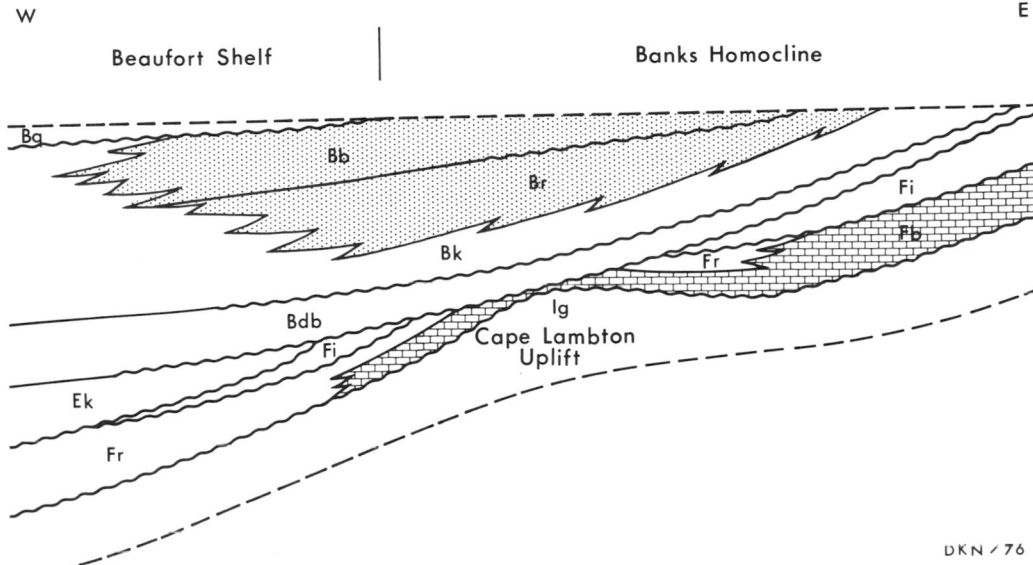

Fig. 11. Schematic cross-section of Banks Homocline and Beaufort Shelf along the line of structure section A–A', showing the stratigraphic setting of the western margin of the stable Franklinian shelf (Fb) and of the regressive Brookian clastics (Br and Bb). All stratigraphic units are identified by abbreviations included in Table I.

structure sections), with those reported on the north flank of Romanzof Uplift (Norris, D. K., 1976a, p. 457), and with the northward continuation of Richardson Trough.

The rocks comprising Banks Homocline range in age from mid-early Cretaceous to early Tertiary. The nonmarine Isachsen Formation at their base rests unconformably upon a depositional surface of considerable relief, with remnants of the Isachsen occurring in possible exhumed valleys in a Devonian plateau (Miall, 1975b, p. 567). The Isachsen Formation is overlain by the Albian marine, transgressive and regressive Christopher and Hassel Formations. Collectively these three formations constitute unit Bdb of Table I and Fig. 3. They are overlain unconformably by the transgressive Kanguk shale (Bk), which by the end of the Cretaceous had covered much of central and western Banks Island (Miall, 1975b, Fig, 10). Renewed uplift in the craton to the east as well as locally within the Banks Island area in latest Cretaceous time, however, caused an influx of terrigenous detritus and initiated a gradual change to deltaic sedimentation. Eureka Sound Formation (Br), a westward progradational sequence, was deposited across western Banks Island and adjacent Beaufort Shelf in the early Tertiary. It was interrupted in the mid-Tertiary because of regional uplift accompanied by faulting (Miall, 1975b, p. 583). Sedimentation presumably continued uninterrupted in the offshore areas of what is now Beaufort Shelf. Finally, fluvial and deltaic clastics and peat were deposited on the homocline, beginning in the Middle Miocene (Hills and Fyles, 1973, p. 11), to comprise the Beaufort Formation (Bb), and probably continuing into the Holocene to add to the late Tertiary and Quaternary sedimentary record on Beaufort Shelf.

Banks Homocline, although generally west dipping, is broadly warped into arcuate anticlines and synclines, some of which appear to be linked right-hand en echelon (see Fig. 3a, geological map). The folding involved the Upper Tertiary Beaufort Formation and therefore the homocline may embrace some of the youngest deformation in the report area.

10. *Beaufort Shelf*

The structural feature defined as the Beaufort Shelf (Fig. 2) embraces that segment of the North American plate that underlies the continental shelf of the Beaufort Sea. It coincides approximately with the physiographic feature referred to as the Beaufort Shelf (Fig. 1), a segment of the Arctic Continental Shelf (Bostock, 1970). Its areal extent is delimited by bathymetry and is defined arbitrarily because its outer margin differs from that of many other continents. It lacks the pronounced physiographic break at about the 200 m isobath. According to Sobczak (1975b, p. 745), the depth of the break increases gradually from 60 m north of Alaska to 650 m northwest of Meighen Island in the Arctic Archipelago. Only in the region north of Mackenzie Delta is the physiography typical of that of other shelves. The shelf widens from about 70 km north of Alaska to about 150 km adjacent to Banks Island (Wold et al., 1970, p. 851-852). Within the report area (see Fig. 3a), its northern limit beneath Beaufort Sea is drawn at the 200 m isobath opposite the mainland, swinging to the 500 m isobath opposite Amundsen Gulf and Banks Island. The Mackenzie Canyon (Fig. 1), therefore, can be identified from this limit, whereas the canyon beneath Amundsen Gulf cannot. Beaufort Sea has one of the narrowest continental shelves found anywhere in the Arctic (Wold et al., 1970, p. 851).

The southern boundary of Beaufort Shelf is a Holocene geographic accident determined by modern sea level. Thus the stratigraphy and structure of Rapid Depression, Aklavik Arch Complex, and Anderson and Banks Homoclines continue into the shelf (see Fig. 3b, structure sections). However, the stratigraphic sequences, the presence and nature of unconformities, and the continuity and style of the structures shown there are speculative. Only the uppermost part of the shelf has been reached with the drill. Interpretation lower down is based upon available geological and geophysical data.

The sedimentary formations comprising the Beaufort Shelf are extrapolated from onshore. They appear to occur throughout the shelf within the report area and to constitute a relatively simple stratigraphic package comprising the whole or parts of the Inuvikian, Franklinian, Ellesmerian, and Brookian Sequences. Resting upon the crystalline rocks of the basement (Hu) is a variable thickness of undifferentiated Inuvikian clastics and carbonates (I), locally cut by both basic and possibly acidic intrusions. They are overlain in turn by Franklinian, Lower and Middle Paleozoic, stable platform carbonates and clastics (Fb) and their seaward equivalents in shale (Fr). As indicated earlier, this relationship implies a seaway in the position of Beaufort Shelf that connected with the Richardson Trough in the early and middle Paleozoic. Upper Devonian and younger Paleozoic sedimentary rocks are presumed to be limited to the southernmost fringe of Beaufort Shelf immediately northwest of Mackenzie Delta. They are exposed onshore (see Fig. 3b, section D–D') but are presumed to be truncated northward at the unconformity beneath the Ellesmerian Sequence. Marine Kingak shales (Ek) and their lateral equivalents form a widespread, thin blanket throughout Beaufort Shelf and probably occur locally in the subsurface of the Banks Homocline (Fig. 3b, section A–A'). They may comprise the bulk of the cores of diapirs and folds in the offshore areas of southern Beaufort Sea west of Mackenzie Delta (Fig. 3b, sections C''–C and C–C'') and hence may have played an important role in localizing closures in the younger, gas-bearing sedimentary rocks. They are overlain unconformably by the Lower Cretaceous transgressive and flyschoid sequence (Bdb). The latter in turn is covered by a complex intertonguing succession of transgressive and regressive marine shales (Bk) and prograding deltaic deposits (Bmc, Br, Bb) and their offshore equivalents of late Cretaceous to Tertiary age. Unconsolidated Quaternary sediments (Bq, Bf) cover all older formations as they prograde from the mouths of rivers and are distributed laterally by longshore currents. The similarity of stratigraphic successions in the Beaufort Shelf, the northern mainland, and contiguous parts of the Arctic Archipelago implies likeness and continuity in the geological history throughout the Phanerozoic and possibly much of the late Proterozoic. Unconformity-bounded marine successions common to the mainland and archipelago imply intermittent, ancestral seas in the position of Beaufort Shelf at least since the late Proterozoic.

The free-air gravity field (Wold et al., 1970; Sobczak et al., 1973) over southern Beaufort Sea reveals three prominent, negative anomalies on and adjacent to the southern limit of the continental shelf, in addition to the string of elliptical, positive anomalies approximately over the northern shelf edge (Fig. 3a, geological map). The −50-mgal anomaly centered over Kugmallit Bay and Richards Island was interpreted (Hornal et al., 1970, p. 9) to identify a Cretaceous and Tertiary clastic succession more than 6000 m thick. Drilling and geophysical data confirm this conclusion. In Mackenzie Bay, moreover, a bipartite anomaly, with lows of −40 and −50 mgals opposite the mouths of

Babbage River and Shallow Bay, respectively, as well as the −40-mgal anomaly opposite the mouth of Kongakut River in Alaska, suggests economically important depocenters in the Beaufort Shelf in the late Cretaceous and Tertiary.

As pointed out by Wold *et al.* (1970, p. 849), the predominant feature in the free-air gravity field is the series of elliptical, positive anomalies (100 mgals) over the northern edge of the continental shelf parallel to the coastlines of the mainland and Banks Island. They concluded the the "more or less continous 100 mgal high" reflects a seaward thinning of the crust from 35 km under the continent to less than 20 km beneath the continental slope, combined with thinner sedimentary cover over a linear ridge in the basement. Saddles between the highs occur approximately opposite the mouths of the major river and marine channels draining into southern Beaufort Sea. They interpreted the saddles to define accumulations of low density sediments transported by the major river and channel systems, in the same way that the nearshore negative anomalies define depocenters for the late Cretaceous and Tertiary on and adjacent to the southern limit of the shelf. Alternatively many of the large positive free-air anomalies can be explained by a prograded wedge of Tertiary and Quaternary sediments acting as an uncompensated load on the crust (Sobczak, 1975*b*, p. 751).

The geological structure of Beaufort Shelf is dominated by diapirs in the Mesozoic and Tertiary rocks. Beneath the shelf in the vicinity of Herschel Island, the diapirs are probably composed of shale, with cores of Kingak Formation and Lower Cretaceous flyschoid clastics that were intruded into the younger succession consequent upon movement along the seaward- and landward-dipping reverse faults comprising the northwesterly trending splays of the Rapid Fault Array. These relationships are illustrated in the most seaward parts of profiles C–C' and C–C''. Beneath Mackenzie Bay, Richards Island, and northward to the edge of the continental shelf similar diapirs may occur in the more deeply depressed part of Rapid Depression. However, in this latter region, sedimentary growth fault-induced diapirs composed of Upper Cretaceous and Tertiary prodeltaic shales are interpreted to intrude their coarser equivalent clastics. The lutokinetic diapirs (Yorath and Norris, 1975) occur at relatively higher structural levels and have been the targets of considerable petroleum exploration.

Beneath Tuktoyaktuk Peninsula and seaward of it, a family of northeast-trending extension faults, termed the Eskimo Lakes Fault Zone (Cote *et al.*, 1974), has been documented by drilling and geophysics. These down-to-basin faults, comprising the northeasterly extension of the Richardson Fault Array, depress progressively the stratigraphic succession beneath Beaufort Shelf and ultimately merge with the northeasterly splays of the Rapid Fault Array. They have had a long history of periodic movements, the last and most dramatic being during Laramide tectonism. As such they were syntectonic with late Cretaceous and Tertiary deposition, which resulted in substantial increases in stratigraphic thicknesses of several units in their footwalls. Some of the strain in individual fault blocks was manifest in roll-over anticlines such as at Atkinson Point where oil was recovered from epicontinental fluvial Aptian sandstones.

11. *Mackenzie Delta*

The most striking Holocene depositional feature along the southern rim of Canada Basin is Mackenzie Delta, a maze of modern lakes and channels masking important structural and stratigraphic relationships on the northeast flank of the Cordilleran Orogenic System. Mackay (1963, p. 98) likened the delta to the shape of an upturned left palm and fingers of a hand with the fingers pointing northward. The indentation between the thumb and first finger corresponds to Shallow Bay, with the fingers representing the distributary channels. From south to north the subaerial delta is about 210 km in length, and from west to east about 65 km in width. The area, including land and water, is approximately 12,000 km². From its rather blunt southern extremity, its two flanks trend curvilinearly northward and at about latitude 68°20′N they turn abruptly northwest in dogleg fashion to terminate at Mackenzie Bay. The bulk of the delta has been built by sediments from Mackenzie River but the Peel River drainage network has contributed significantly.

Outcrops of Lower Devonian and older limestones and dolomites in Campbell Uplift in the modern delta near its east margin (Norris, D. K., 1975b) suggests that the sedimentary fill in the upper delta was thin. In the lower delta and offshore, however, there is a buried, flat-bottomed ice scour channel or canyon that probably followed a pre-Wisconsin fluvial valley (Shearer, 1972, Fig. 1). This is Mackenzie Canyon (Fig. 1). It slopes gently seaward, is aligned with the northwest-trending, lower subaerial delta, and is presumed to intersect the continental slope north of Herschel Island at a depth of about 350 m. The ancestral Mackenzie River, on the other hand, may have flowed northeast in a graben now occupied by Campbell and Sitidgi Lakes to enter the Arctic Ocean in the general region of Eskimo Lakes.

The size, shape, and orientation of Mackenzie Delta may be structurally controlled but there are few data to support this thesis. Westward migration has been prevented by the Richardson Mountains. Moreover, the straight, eastern scarp of the Aklavik Range (Fig. 12) cuts obliquely across the local structural grain (Norris, D. K., 1975b) and is probably of erosional origin (Mackay, 1963, p. 97). Around the dogleg bend, in the region of the Yukon–Northwest Territories boundary, the west side of the delta cuts orthogonally across the strike of the bedrock geology, and in the subaqueous delta

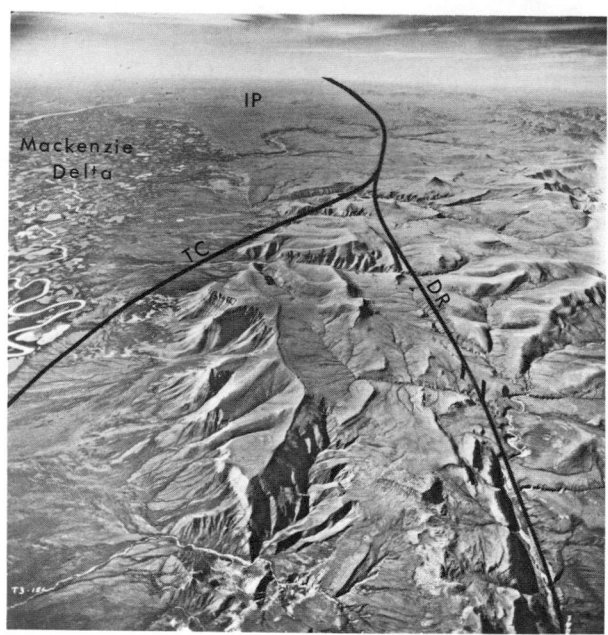

Fig. 12.   Aklavik Range and the western margin of Mackenzie Delta, showing the Treeless Creek (TC) and Donna River (DR) strands of the Richardson Fault Array, and the Interior Platform (IP). NAPL Oblique Photo T3-15L. View is to the south. Barbs indicate sense of relative displacement.

both walls of Mackenzie Canyon parallel the structural grain of Romanzof Uplift and the west flank of Rapid Depression. The canyon may be following structurally weak or incompetent bedrock units offshore. The buried scour channel of the subaerial delta on the northeast side, however, cuts obliquely across the Upper Cretaceous and Tertiary succession of Caribou Hills on the northwest flank of the Aklavik Arch Complex and there appears to be no underlying structural control.

B. Geotectonic Patterns

1. *Introduction*

A distinctive feature of the northern part of the Cordilleran Orogen is the limited lateral continuity of stratigraphic units and of faults and folds. It can be explained by the truncation of formations and structures by the plethora of unconformities within the succession. The sedimentary units do not have a consistent direction of thickening, but instead thicken and thin laterally with important local facies changes. To complicate matters, this truncation can parallel or transect any of the numerous strike directions. Thus the sedimentary pile is composed of a mosaic of interlocking and overlapping bodies of rocks of irregular size and shape, and of contrasting mechanical properties. It was constructed progressively, and was deformed and eroded intermittently so that some structures are extensive and others are localized within the pile. Still others have served to position and orient later and more extensive deformations. Only the youngest (Laramide) structures can be recognized throughout all the tectonic elements comprising the area. Because earlier deformations are more readily decipherable, they provide important information about the evolution of this part of the Cordillera.

The stratigraphic and structural style of the region contrasts with that in the outer part of the orogen from well south of the project area. There a relatively simple eastward-tapering wedge of sedimentary rocks embraces an orderly stack of formations whose lithological and mechanical characteristics are uniform over vast areas, and which includes a family of eastward-converging disconformities. The wedge is cut by bundles of Laramide thrust and high-angle reverse faults, some components of which extend for hundreds of kilometers along strike and presumably root deeply in the Cordilleran Orogenic System (Bally *et al.*, 1966). The earlier history of those equivalent parts of the orogen farther afield was not nearly so complicated, in spite of its being masked, as in the southern Rocky Mountains and Foothills of Canada, through extreme telescoping by thrust faulting and folding during the Laramide Orogeny.

Two basic geotectonic patterns emerge for the northern part of the Canadian Cordillera. The one relates to the Laramide folding and the other to the systems of strike-slip faults, ranging in age from at least the late Proterozoic to the early Tertiary. These patterns identify a regionally systematic response to orogenic forces throughout considerable geological time and shed some light on the role of the northern Cordillera in the tectonic development of the North American plate.

2. *En Echelon Folding*

The eastern margin of the Cordilleran orogenic belt in northern Canada is characterized by bundles of folds that are arcuate in plan and form prominent salients and re-entrants in the mountain front (Norris, D. K., 1972*a*, p. 634). Arrays of cylindrical to conical folds within these bundles are arranged systematically right- and left-hand en echelon, and identify a regional, basic, Laramide kinematic pattern. From south to north the folds are commonly right-hand en echelon in the Mackenzie Foldbelt from below latitude 60°N to the apex of the arcuate mountain front opposite Great Bear Lake. There they revert to a left-hand pattern that persists in generally west-trending fold bundles to the great re-entrant in the Ogilvie Mountains southwest of Eagle

Plain. Around the arc the north-trending folds reciprocate to right-hand en echelon much as in eastern Mackenzie Foldbelt. They terminate abruptly against the Aklavik Arch Complex, which is constructed of northeast-trending fault blocks and structural highs and depressions arranged systematically right-hand en echelon. On the north flank of Romanzof Uplift the fold pattern is again left-hand en echelon (see Norris, D. K., 1972a, Fig. 1).

The arcuate form of the fold bundles may be due primarily to initial curvature of the eastern margin of the miogeocline, and the systematic alternation from right- to left-hand en echelon arrays, to differential transport of the sedimentary veneer in conjunction with minor rotation of the fold bundles about vertical axes (Norris, D. K., 1972a, p. 642). The association of these arrays with major faults having apparent dextral movement in Mackenzie Foldbelt (Gabrielse et al., 1973, p. 111) is predictable and indeed appears to be an integral part of the Laramide evolution of the northern Cordillera.

3. *Richardson Fault Array*

The Richardson Fault Array (Fig. 9) is a curvilinear system of nearly vertical faults traceable for 1000 km from Beaufort Shelf south of Banks Island to the core of the Mackenzie Foldbelt. Offshore, at the mouth of Amundsen Gulf, it is a zone of large normal faults, which Lerand (1973, p. 334) called the Cape Kellett Fault Zone. According to Lerand, it marks a hinge to the west and north of which the dip and thickness of the Phanerozoic section increase rapidly. It is postulated that the zone continues southwest across the Beaufort Shelf to connect with the Eskimo Lakes Fault Zone (Cote et al., 1974, p. 73) beneath Tuktoyaktuk Peninsula. There, seismic and well data identified Eskimo Lakes Fault (Fig. 9), one strand of the zone, to have a vertical separation of up to about 2500 m at the pre-Mesozoic level (Cote et al., 1974, p. 73). The Richardson Fault Array then crosses the Aklavik Arch Complex as it turns gently southward to become an anastomosing and braided system of faults with marked vertical separations in northern Richardson Mountains. The Donna River Fault (Fig. 12; Jeletzky, 1960) with associated gypsum intrusions (Kent and Russell, 1961) in Aklavik Range is a component part of the array (Norris, D. K., 1975b).

The complicated pattern of faults in the Richardson Anticlinorium continues south through Bonnet Plume Basin, (Norris D. K., and Hopkins, 1977) to the front of Wernecke Mountains. There the array turns southeastward and changes in style from one of nearly vertical faults with horizontal separations in Richardson Anticlinorium, to one of steeply dipping, high-angle reverse faults in the core zone of Mackenzie Foldbelt (Fig. 10).

Extrapolations of Richardson Fault Array to the north and south are purely speculative. There are no data to support the northward continuation of the Cape Kellett Fault Zone west of Banks Island and hence to provide a structural link between the Innuitian and Cordilleran Orogens. Southward along the boundary between Mackenzie and Selwyn Foldbelts (Norris, D. K., 1972a, Fig. 1), it would appear that the Laramide, high-angle reverse faults may have inherited their position and orientation from earlier lines of weakness and mask the true (strike-slip?) nature of older faults.

The formations comprising individual blocks within Richardson Fault Array range in age from Proterozoic (Helikian or older) to early Tertiary. Horsts of slaty argillite and quartzite, the oldest stratigraphic assemblage in the northern Cordillera (Norris, D. K., 1975c), are extensively exposed in Wernecke Mountains and they occur locally in the core of Richardson Anticlinorium. In both regions they are overlain unconformably by Lower Cambrian rocks. Semiconsolidated clastics and coal of early Tertiary age, on the other hand, are found in structural depressions such as Bonnet Plume Basin (Norris, D. K., and Hopkins, 1977) and in down-faulted blocks beneath Tuktoyaktuk Peninsula and offshore areas of southern Beaufort Sea. The contact relations of these formations do not establish a maximum time range for displacement in the array but they do confirm activity from before the early Cambrian to well into the Tertiary.

An important characteristic of the Richardson Fault Array is the stratigraphic contrast between Proterozoic and Lower and Middle Paleozoic formations from east to west across it in the general region of the Richardson Anticlinorium and Bonnet Plume Basin. Both the juxtaposition of contrasting depositional regimes and the presence of major stratigraphic omissions emphasize the fundamental importance of the array in the geological evolution of the outer part of the Cordilleran Orogen.

In the Mackenzie Foldbelt and in nearby parts of the platform east and north of there, the Helikian succession is made up of a large number of formations (see Aitken et al., 1973, Table I) that extend over wide areas. It includes an unnamed, gray dolomite and chert unit at the base, overlain by a sequence of variously colored clastic and carbonate units, both named and unnamed. Almost all of these units have been mapped from Mackenzie Foldbelt westward into the array, but none can be recognized west of it.

In addition to the occurrences in the array cited above, the oldest slaty argillites and quartzites in the Proterozoic succession are exposed here and there west of the array in the cores of structural highs in eastern Taiga–Nahoni Foldbelt (Norris, D. K., 1974, Fig. 1). There they are overlain by a banded gray and orange dolomite formation (Green, 1972, p. 15), which is assumed to be of Helikian age. An angular unconformity between these dolomites and the Phanerozoic rocks introduces considerable uncertainty about the correlation of the dolomites with possible counterparts on the opposite side of the array. Barring facies changes, the Helikian formation on the two sides does not appear to be equivalent.

The conglomerates, iron formation, carbonates, and clastics comprising the Hadrynian sequence east of the array (Aitken *et al.*, 1973, Table I) is bounded above and below by regionally angular unconformities. Only the lower part of the conglomerate (Rapitan) formation can be recognized on the west flank of the array (Norris, D. K., 1975*a*). The remainder of the Hadrynian sequence has been removed at the unconformity at the base of the Phanerozoic section.

East of Richardson Fault Array, both in the Interior Platform and in the Mackenzie Foldbelt, the Lower and Middle Paleozoic interval is made up of a number of distinctive carbonate formations (Macqueen, 1975, p. 291). They comprise the east flank of Richardson Trough (Lenz, 1972, p. 329), the curvilinear, north-trending, graptolite-bearing shale trough extending at least from the core of Mackenzie Foldbelt to Beaufort Shelf in the position of the Richardson Fault Array.

West of the array is an unnamed, coeval, monotonous succession of gray, platform carbonates with shale tongues, which is not readily divisible into regionally mappable units (Norris, D. K., 1975*a*). Macqueen (1975, p. 293 and 297) stated that it accumulated on shallow-water carbonate banks under remarkably stable conditions, reflecting a real difference in epeirogenic regime from its counterparts east of the array. The overlying, Upper Devonian clastic sequence thickens and coarsens westward into the Richardson Trough but is abruptly truncated near the western margin of the array at the unconformity beneath the Carboniferous System. Formations comprising the latter appear to have overlapped the trough without being influenced by it.

The Richardson Fault Array controlled the position, timing, and life span of the Richardson Trough and the Richardson Anticlinorium. As a fault-bounded intracratonic depression, Richardson Trough is a taphrogeosyncline (Kay, 1945, p. 1172). It is also an aulacogen (Shatsky and Bogdanoff, 1960), as first suggested by Churkin (1975, p. 454), and it persisted for a period of at least 200 m.y. from the beginning of the Middle Cambrian to the end of the Devonian. The thick argillite and quartzite assemblage at the base of the Proterozoic section in the general region of the trough is strongly suggestive, moreover, that it may well have been in existence intermittently through the Proterozoic.

Three observations of right-lateral separation across strands of the Richardson Fault Array indicate intermittent and cumulative displacement from the Proterozoic to about the end of the Devonian Period (Norris, D. K., and Hopkins, 1977). On the north plunge of Richardson Anticlinorium (Norris, D. K., 1975*c*), Upper Devonian clastics are separated right laterally approximately 10 km. Secondly, on the west flank of Bonnet Plume Basin,

Middle Devonian and older carbonates are separated dextrally by approximately 20 km. And thirdly, an isolated mass of conglomerate (diamictite), unique to the Hadrynian Rapitan Formation on the south flank of Bonnet Plume Basin, is offset a minimum of 40 km right laterally from the main mass of conglomerate at the headwaters of Snake River (Norris, D. K., and Hopkins, 1977). Carboniferous and younger formations, on the other hand, appear to have been laid down in the absence of activity on the array, although they have been deformed and uplifted in later geological times by the array. The relatively small strike separation of Upper Devonian formations and the lack of conclusive evidence of sizeable, dextral displacement in younger rocks would suggest that strike slip on the Richardson Fault Array may have terminated by the early Carboniferous (see Fig. 13a). It is clear, however, from the faulting of Upper Cretaceous and Lower Tertiary formations that the inversion of the early and mid-Paleozoic Richardson Trough into the Richardson Anticlinorium must have taken place in the middle or late Tertiary.

In late Devonian and Mississippian time, the stable depositional regime in the eugeocline and miogeocline to the west changed drastically (Tempelman-Kluit, 1976, p. 1354). The emplacement of intermediate volcanics, flyschoid graywackes and shales, and granodiorite batholiths there suggests a westward transfer of tectonic instability (and strike-slip faulting?) in the direction of an ancestral margin of the North American plate.

4. *Tintina Fault*

Tintina Fault (Fig. 13b; Roddick, 1967) is one of the great right-lateral, transcurrent faults in western North America. It is marked by Tintina Trench, a valley feature that is traceable for 1000 km from Liard Plain on the northern border of British Columbia through Yukon Territory to eastern Alaska. Throughout its length, the Tintina Fault separates sharply contrasting geological regimes. Southwest of it are the metamorphic and intrusive rocks of the Yukon Crystalline Terrane, and northeast of it are the deformed, but relatively unmetamorphosed, sedimentary and intrusive rocks of the Selwyn Foldbelt (Fig. 10).

The linearity and length of the Tintina Fault suggest not only that it is vertical and deep (Roddick, 1967, p. 28), but also that there has been large-scale shearing and strike-slip displacement along it. Extensive greenstone bodies trending obliquely to the fault appear to be structurally controlled. Moreover, placer mining and road cuts have exposed gouge and crushed rock for several kilometers along the strike of the fault (Aho, 1959, p. 337–338).

The main evidence for the great right-lateral displacement on the Tintina Fault is from the matching of major geological units on the two sides. One of

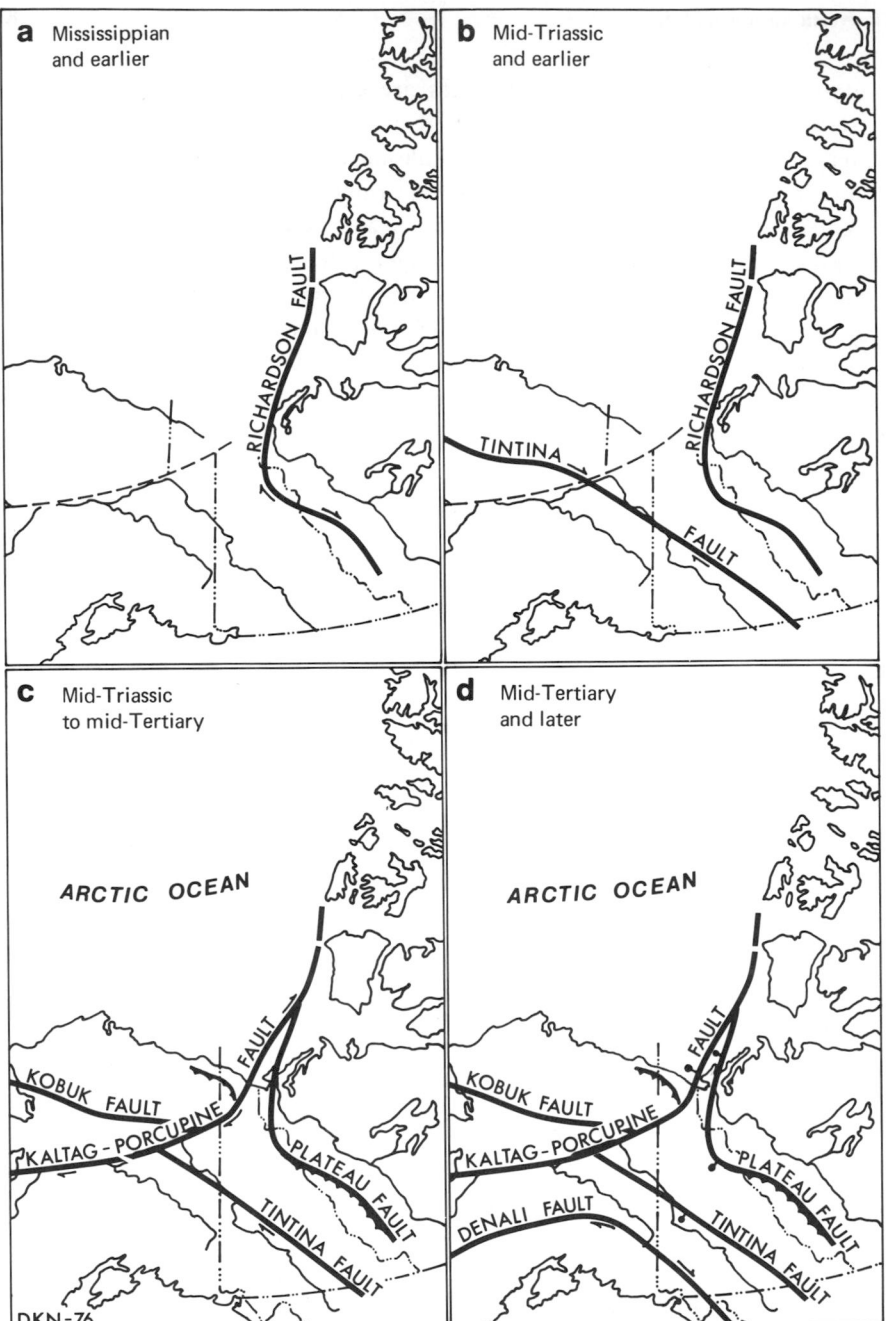

Fig. 13. Schematic portrayal of the spatial and approximate temporal relations among the major dextral strike-slip fault zones in the northern Cordillera from the Proterozoic to the Holocene. In each diagram arrows indicate the timing and sense of episodic strike slip and teeth the upthrown walls of associated reverse faults. Solid circles identify the downthrown walls for prominent, differential vertical motion from the mid-Tertiary onwards. Light dashed line in 13a and 13b identifies the incipient Kaltag–Porcupine Fault.

these is a Hadrynian grit unit which, according to Roddick (1967, p. 29), suggests a right-lateral offset of about 400 km.

Tempelman-Kluit (1976, p. 1343) suggested that, in central Yukon, movement on the fault occurred during three separate intervals: early Triassic, mid-Cretaceous, and Eocene or Oligocene. In addition, he indicated that a Pennsylvanian and Permian basalt, peridotite, chert, and limestone assemblage may document the first displacement along the fault. It may "mark the site of a closed Permian–Triassic rift in the continental crust, or it may have been a major transcurrent fault with important right-lateral movement" (Tempelman-Kluit, 1976, p. 1354, 1355).

Two hundred kilometers of mid-Cretaceous, right-lateral, strike-slip displacement is suggested by the truncation and separation of Lower Cretaceous rocks of the Kandik Thrust Belt from their possible continuation southwest of the fault in eastern Alaska (see King, P. B., 1969). Tempelman-Kluit (1976, p. 1343) recognized the difficulty of accommodating this motion along the supposed southward continuation of the Tintina Fault in the Rocky Mountain Trench. Vertical displacement of up to 800 m is documented in post-Paleocene time by the deformation of Upper Cretaceous and Lower Tertiary beds in the floor of the trench (Aho, 1959, p. 338).

The evidence for a major right-lateral, transcurrent fault in the Tintina Trench appears incontrovertible, that for 400 km of horizontal displacement is reasonably sound, but that for the time distribution of incremental displacement is uncertain. "Although the field evidence suggests that the Triassic displacement was largely vertical and that the mid-Cretaceous movement was most likely strike slip, the data permit the reverse interpretation" (Tempelman-Kluit, 1976, p. 1343). It appears, however, that strike-slip activity on Tintina Fault was confined largely to the period following deposition of the Hadrynian grit unit.

5. Kaltag–Porcupine Fault

The Kaltag–Porcupine Fault was first recognized by Gates and Gryc (1963, p. 276). Grantz (1966) made a more complete evaluation of its importance in the heirarchy of strike-slip faults in Alaska and named it the Kaltag Fault. Patton and Hoare (1968) discussed the feature at some length, indicated that 65 to 130 km of right-lateral offset may have occurred along it since Cretaceous time, and presented evidence to indicate that the fault has been active in the Holocene. Tailleur and Brosgé (1970, Fig. 12) applied the term "Kaltag–Yukon–Porcupine Fault" for the total feature in Alaska and indicated its northeastward extension across Yukon Flats to the vicinity of the International Boundary. D. K. Norris (1976a, p. 466) simplified the name to the Kaltag–Porcupine Fault and that name is used herein (Figs. 9 and 13).

The Kaltag–Porcupine Fault can be traced intermittently for approximately 1200 km from Norton Sound on the Bering Sea across Alaska and northern Yukon Territory to Beaufort Shelf on the southern rim of Canada Basin. At its southwestern extremity in Alaska, it splays (see King, P. B., 1969); one branch cuts diagonally across Koyukuk Basin into Norton Sound, and the other trends more southwesterly to skirt the basin and enter the continental shelf of Bering Sea a little farther south. Northeastward it extends across Yukon Flats Basin and enters Canada on the south flank of the Old Crow Granites (see Fig. 10). There, it forms the north boundary of Keele Platform, a part of the Aklavik Arch Complex. The thick, Lower Paleozoic dolomite and limestone succession comprising the platform is almost entirely absent north of the fault.

The Kaltag–Porcupine Fault is exposed in Canada on the Porcupine River approximately 14 km upstream from the village of Old Crow (Norris, D. K., 1976a, Fig. 97.6). At the river's edge, unnamed Lower Paleozoic dolomite and limestone are in juxtapositon across the fault with unnamed Upper Cretaceous sandstone. A short distance farther northeast, a string of highly sheared, basic intrusions occurs in the fault zone (Norris, D. K., 1974, p. 39), and at the headwaters of Driftwood River, the fault zone has a kink in it somewhat like that in the San Andreas Fault on the southwest flank of the Mojave Block. It merges with the north end of the Yukon Fault (Y), a major thrust and high-angle reverse fault extending from the Kandik Thrust Belt (Fig. 10), transecting the Aklavik Arch Complex and, together with the Kaltag–Porcupine (K–P), forming the principal component of the Rapid Fault Array (Fig. 9). Seaward the array comprises a fan-shaped family of nearly vertical faults striking northwest to northeast, cutting the Lower Tertiary and older sequence into a series of crudely pie-shaped blocks and lowering them systematically into the axial region of the Rapid Depression. There are no data to indicate that strike-slip displacement predominated there and it would appear that the horizontal component of motion elsewhere was attenuated through oblique-slip in the array both on land and seaward on Beaufort Shelf.

In Alaska, the best evidence for the timing of large-scale, right-lateral, strike-slip displacement along the Kaltag–Porcupine Fault probably is the lateral separation of mid-Cretaceous (Albian and Cenomanian) rocks on the east flank of the Yukon–Koyukuk Basin and the offset of gross geologic trends within the basin (Patton and Hoare, 1968, p. D149). Post-mid-Cretaceous displacement would appear to be necessary to account for these spatial relations. Patton and Hoare (1968, p. D152) further suggest that this motion probably postdates extrusion of Upper Cretaceous and Lower Tertiary(?) volcanic rocks in the eastern part of the basin. Moreover, Pliocene and Pleistocene basalt flows, which appear to straddle the westward extension of the fault on Norton Sound, show no significant lateral dislocation. They conclude that "the major lateral displacement of Mesozoic tectonic elements appears to have occurred in early and middle Tertiary time. Physiographic evidence, however, provides evidence that some movement has continued to the present" (Patton and Hoare, 1968, p. D153).

In Canada, the juxtaposition of Lower Paleozoic and Upper Cretaceous (Santonian) formations along Porcupine River supports the Alaskan evidence for late Cretaceous or younger movement in the Kaltag–Porcupine Fault Zone. The evidence is further substantiated by the presence of vertical coal measures of the Lower Tertiary Reindeer Formation in the Rapid Fault Array immediately west of the Mackenzie Delta (Norris, D. K., 1972b, p. 97).

C. Tectonic Synthesis

The geotectonic patterns discussed above show that three right-lateral, strike-slip fault zones played fundamental roles in the tectonic development of the Cordilleran Orogen north of latitude 60°N. These are the Richardson Fault Array, traceable for more than 1000 km from Beaufort Shelf southward to the core zone of the Mackenzie Foldbelt; the Tintina Fault, recognized for 1000 km between the 60th parallel of latitude and Yukon Flats in eastern Alaska; and the Kaltag–Porcupine Fault, traceable intermittently for more than 1200 km from Bering Sea across the northwestern tip of the North American plate to Beaufort Shelf. Their great length, linearity or curvilinearity, apparent continuity, and longevity suggest that they are deep and fundamental structures of the lithosphere, and that very large offsets can be expected along them.

The data on the life spans of each of these major shear zones suggest, moreover, that there was a general progression from older to younger faults from east to west, and that collectively the faults have undergone right-lateral, strike-slip motion since Proterozoic time (Fig. 13). It also appears that, in their later kinematic stages, vertical displacement may have predominated over horizontal slip.

The Richardson Fault Array, the most easterly of the shear zones, was an active bundle of right-lateral, strike-slip faults at least from the late Helikian to about the beginning of the Carboniferous, a time span of at least 600 m.y. (Fig. 13a). The Tintina Fault (and its northwestward continuation, the Kobuk Fault), on the other hand, does not appear to have been an active, strike-slip feature until the late Hadrynian; it may have had recurrent, right-lateral displacement along it until approximately the mid-Triassic, a span of about 400 m.y. (Fig. 13a).

The Kaltag–Porcupine Fault is postulated to have severed the Tintina–Kobuk Fault System (Freeland and Dietz, 1973, p. 385) between the mid-

Triassic and the mid-Cretaceous, and to have offset it right-laterally about 100 km prior to the deposition of the mid-Cretaceous clastic succession of the Yukon–Koyukuk Basin. Additional slip in the early or mid-Tertiary could account for the suggested 65- to 130-km offset of the basin, and for the net right-lateral separation of 200 km for the Tintina and Kobuk Faults (Fig. 13c; Norris, D. K., 1976b, p. 691).

This net horizontal displacement is assumed to have been dissipated among the several components of the Rapid Fault Array (Fig. 9). Beneath the Mackenzie Delta, the more easterly strands of the zone appear to merge with the Richardson Fault Array on the seaward flank of the Aklavik Arch Complex and to attenuate their displacement northeastward as a family of oblique-slip, down-to-basin faults on Beaufort Shelf between the Mackenzie Delta and Banks Island (Fig. 13d).

With the severing of the Tintina–Kobuk Fault System in the early or middle Mesozoic (Fig. 13c), that segment of the Kaltag–Porcupine Fault west of the intersection became the active continuation of the Tintina Fault. Further right-lateral motion on the Tintina, as suggested by Tempelman-Kluit (1976, p. 1343), therefore was accommodated on the new Kaltag–Tintina System. The first of the dextral transcurrent faults paralleling the curvature of the continental plate margin around the Gulf of Alaska was born. A second of this new series was the Denali Fault System (see Fig. 13d), with a possible right-lateral displacement of 300 km occurring in about mid-Tertiary time (Eisbacher, 1976, p. 1151). The Denali apparently lacks the long pre-Tertiary history of both the Tintina Fault and the Richardson Fault Array.

The protracted kinematic history and the great length of the Richardson Fault Array introduce pertinent questions concerning the southward continuation of the array within the Cordilleran Orogen and the role of the array in the early Paleozoic and Proterozoic evolution of the orogen. Absence of marked strike-slip activity since the late Devonian would cause the array to be masked by the cumulative blanket of younger sediments except where it has localized important later faulting, or where it has been uplifted regionally and exposed through differential erosion.

Well south of the project area (Fig. 2), the Plateau and Tsezotene Faults (Gabrielse et al., 1973) are on trend with the Richardson Fault Array, suggesting that they are reactivated structures, localized and oriented by the array. They are identified as thrusts that are presumed to have been active prior to late Cretaceous time (Gabrielse et al., 1973, p. 104 and 112). However, their association with the dextral Hayhook Fault and their position within regional arrays of right-hand, en echelon fold bundles (Norris, D. K., 1972a, Fig. 1) strongly suggest a basic, underlying, structural control, perhaps the southward continuation of the Richardson Fault Array. The writers suggest, moreover, that the pre-early Cambrian (probably dextral slip) faulting on the anomalous

Gundahoo Thrust (Taylor and Stott, 1973, p. 30) in northeastern British Columbia, in an equivalent structural position in the Cordilleran Orogen, may also be part of this ancestral fault array. As recognized by Taylor and Stott, "the fault was subsequently reactivated during Laramide deformation." Implicit in this reasoning is the concept that the Richardson Fault Array may be an ancient cratonic margin whose surface trace generally followed the curvature of the outer (eastern) flank of the orogen, and which possibly connected with a major Proterozoic (Aphebian?) strike-slip fault zone in the position of the Cenozoic, physiographic Rocky Mountain Trench. The Tintina Fault is a younger feature that either splayed from this ancestral shear zone or passed structurally west of it.

The complex pattern of curvilinear, nearly vertical faults trending northwest and constituting the structural grain of the Cordilleran Orogenic System (see Tipper, 1969; Eisbacher, 1972) may be an emulation on a grand scale of the braided pattern of strike-slip faults of the Richardson Fault Array in the core of Richardson Anticlinorium (Norris, D. K., 1975c). It is the cumulative effect of dextral shear at or near the boundary between the North American and Pacific Plates since the Proterozoic.

It is postulated that the pattern of faults in the northern Cordillera (Fig. 13) began with a relatively simple, curvilinear surface of shear failure in the early Proterozoic in the general position of the Rocky Mountain Trench and the Richardson Fault Array. Continued application of the fundamental drive resulted in the progressive deformation and locking of faults, with concomitant spasmodic generation of new slip surfaces, some of which disrupted old ones. The crust, therefore, was broken into an ever increasing number of blocks or slices that were free to move up (horsts), down (grabens), or laterally, depending upon whether they were in zones of compression, dilation, or dextral shear. As deformation continued, there resulted a growing network of diverging, converging, and intersecting curviplanar faults, with the throughgoing components being characterized by strike slip, and the shorter, connecting components by normal or reverse displacement. The fault network expanded westward through time away from the Cordilleran Foreland so that there was a general progression from older to younger blocks from the shelf, through the miogeocline and eugeocline in the direction of the ancestral Pacific Plate. As new faults were generated, they also became deformed and locked. Favorably oriented slip surfaces were reactivated so that the basic, westward progression in age from older to younger faults was continually camouflaged by renewed movement on some easterly faults. The net effect, however, has been the addition of successive tectonic belts to the North American continent as first suggested by Stoneley (1971, p. 623). P. B. King (1976, p. 685, 686) has identified the same phenomenon for the North American Cordillera near the 40th parallel. The entire Cordillera west of a line at or near the Rocky Mountain

Trench and Richardson Fault Array is allochthonous because of the wholesale, intermittent, differential translation of the Cordilleran Orogenic System dextrally since the Proterozoic.

The profound effects of this continual translation of successively younger belts on the paleogeography of the western margin of the North American plate is evident in the geological record. They are dramatically shown by the presence in southern Yukon Territory west of the Tintina Fault of thick Upper Triassic formations containing warm water coral faunas. Contemporary deposits east of the fault contain cooler-water pelagic faunas. The juxtaposition of these deposits can be explained by the northward movement of the western block in post-Triassic time (Tozer, 1970, p. 635). Moreover, paleomagnetic data from Jurassic rocks south of the Denali Fault in Alaska point to a much more southerly location for at least parts of Alaska (Packer and Stone, 1972, p. 25). The continental margin of the Alaska Peninsula would appear to have been approximately at the latitude of the Oregon–Washington continental margin in Jurassic time.

IV. RELATIONSHIP OF THE CANADA BASIN TO THE AMERICAN PLATE

A. Introduction

The relationship of the Canada Basin (Fig. 2) to the North American plate in terms of modern concepts of crustal development remains obscure. Permanent ice cover makes it difficult to obtain systematic, multiparameter geophysical, geological, and bathymetric data to the degree afforded by the remainder of the world's ocean basins. Although substantial amounts of data of various types have been obtained from ice platforms and nuclear submarines, much of it is equivocal and far from conclusive.

There have been hypotheses to explain continental margin geology in terms of the origin of the Arctic Ocean Basin, of which the Canada Basin forms a part. These fall into two main groups. One group considers that the Arctic Ocean Basin is a relatively young feature, formed by crustal subsidence and/or sea floor spreading since mid-Mesozoic time. This crustal subsidence, as proposed by Shatsky, Pushcharovsky, and Atlasov (in Churkin, 1973), implies an ancient landmass that supplied clastic sedimentary wedges to bordering regions of North America and Siberia before ultimately subsiding to form the Arctic Ocean Basin. The more conventional sea-floor spreading hypothesis, variations of which have been described by Carey (1955) and a host of followers, assumes that the floor of the Canada Basin is underlain by oceanic crust of Mesozoic and younger age. The basaltic nature of the crust is supported by aeromagnetic surveys, which illustrate high-frequency and high-amplitude anomalies over much of the region (King, E. R. *et al.*, 1966; Riddihough *et al.*, 1973). Bhattacharyya (1968) speculated that the marked increase in intensity and variation in direction of horizontal magnetic vectors beneath the Canada Basin was due to widespread folding and faulting. Moreover, a seismic refraction profile across the polar continental shelf of the Queen Elizabeth Islands supports a thinning crust toward the Canada Basin and thus, by inference, an oceanic crust beneath the Arctic Ocean Basin.

The second group of hypotheses describes the Arctic Ocean Basin as a relatively old feature dating at least from the Paleozoic. Within this group, two differing interpretations have been expressed. Meyerhoff (1973) described the Arctic Ocean Basin as a permanent feature, dating from late Proterozoic time, during which no plate motions are required to explain the orderly and systematic spatial and temporal relationships between upper Proterozoic to Lower Permian evaporite deposits and the observed continental geometry. The second interpretation (Churkin, 1973; Hall, 1973; Ostenso and Wold, 1973) describes the basin as an ancient feature, which since Paleozoic time has been opened and closed repeatedly. This view is supported by the apparent ease with which foldbelts are correlated around the edges of the basin (Churkin, 1973).

B. Geophysical Information

1. *Gravity*

The free-air gravity field beneath the Beaufort Sea obtained by Sobczak *et al.* (1973) has been reproduced in Fig. 3a. It shows several large positive curvilinear, elliptical anomalies that are part of a chain of such features extending from adjacent to the northern Queen Elizabeth Islands to Barrow, Alaska. Wold *et al.* (1970) interpreted them as representing the combined effects of a ridge in the crustal rocks, a thickening of uncompensated clastics beneath the continental slope and rise, and a rise in the crust–mantle interface as a result of thinning of the crust. Sobczak and Weber (1973, p. 525) attribute these anomalies to the transition from continental to oceanic crust; they can be explained best by a composite structure consisting of a sedimentary layer up to 10 km thick on the seaward side of the anomalies and a crust that thins by as much as 17 km on the landward side. More recently, Sobczak (1975*b*) analyzed density contrasts from borehole data in the Mackenzie Delta region and obtained a close fit with observed gravity profiles by modeling the effects of a large, uncompensated load of clastics beneath the continental margin. He found that such a load explains as much as 80 percent of the anomaly where the crestal value is in excess of 100 mgals. In such a case, "exceptionally deep sedimentary troughs associated with unusually high ridges at the crust–mantle boundary are demanded by the hypothesis" (Sobczak, 1975a, p. 381). Yorath and Norris (1975) largely supported the original interpretation of Wold *et al.* (1970) and suggested that there is a marginal ridge with a relief on the Proterozoic surface of about 4.5 km. They stated further that it appeared "unlikely that such a wedge of uncompensated sediments can fully or largely account for the margin anomaly belt" (Yorath and Norris, 1975, p. 601). Insofar as gravity modeling depends upon the choice of densities and resulting density contrasts between adjacent layers, the choice of 0.4 g/cm³ as the density contrast between consolidated sediments and "basement" by Wold *et al.* (1970) appears excessively high. The choice of 0.1 g/cm³ by Sobczak (1975b), however, is somewhat low.

In summary, the present authors suggest that the large, elliptical positive anomalies represent, in order of decreasing importance of effect, a large wedge of uncompensated clastics beneath the continental break, a rise in the crust–mantle interface consequent upon thinning of the crust, and a ridge adjacent to the continental–oceanic crust transition.

2. *Seismicity*

Seismicity in northern Canada and adjacent regions of the Canada Basin to date has not been related positively to specific faults because of the short time span over which relatively precise measurements have been made and because of the uncertainties in defining accurate locations and magnitudes of known earthquakes (Stevens, 1974, p. 403). Most earthquakes in Arctic Canada, however, have their foci within the crust (i.e., their depths are less than 40 km) and occur probably within plates. Some may be related to plate boundaries (Stevens, 1974). Limited observations suggest that seismic activity in northern Yukon Territory may be related to the Richardson Fault Array (Leblanc and Wetmiller, 1974, p. 1453).

Earthquake activity in the Arctic Islands is less than in northern Yukon Territory. Earthquakes that have occurred near Prince Patrick Island on the northwestern rim of the Sverdrup Basin may coincide with the north-trending, vertical faults of the Prince Patrick Uplift, an ancient feature that probably has been active intermittently from Paleozoic time (Tozer and Thorsteinsson, 1964).

Seismic activity in the Arctic Ocean Basin is concentrated along the Nansen Cordillera, the northward continuation of the Mid-Atlantic Ridge. The Alpha Cordillera, which bounds Canada Basin on the north, is believed to be an earlier (Tertiary) spreading axis (Ostenso and Wold, 1973). It is aseismic today. Within the Canada Basin, the few earthquakes that have been recorded and located appear too widely scattered to provide any trends. However, through an analysis of the dispersion characteristics of Rayleigh and Love waves generated by earthquakes in the Arctic Ocean, Laptev Sea, Norwegian Sea, and central Alaska, Hunkins (1963) concluded that the floor of the Arctic Ocean Basin, where greater than 2000 m in depth, was underlain by oceanic crust. This constitutes the first and only direct evidence of the nature of the crust in this region.

3. *Heat Flow*

Within the Arctic, no regional interpretations of heat flow patterns on land are yet possible and, with the exception of flux variations between shelves, abyssal plains, and ocean ridges, the same is true for oceanic areas. In the Canada Basin, Lachenbruch and Marshall (1966) determined that the heat flow is uniform at $1.41 \pm 4\%$ μcal/cm^2 per s over a distance of at least 75 km from the lower slopes of the Alpha Rise. This is close to the worldwide average. On the flank of the rise, the heat flow decreases to a minimum value of 0.77 units (1 unit = 1 μcal/cm^2 per s). Although the latter value is within the range of variation for normal crust, it is anomalously low. Lachenbruch and Marshall (1966, p. 1247) attributed the cause of the anomaly to a sharp lateral discontinuity in rock properties involving the entire crust, including an edge effect of lateral conductivity contrasts in crustal and subcrustal rocks. More recently, they stated that the values are "consistent with the view that the rise is a great accumulation of basalt such as might be expected in an extinct mid-ocean ridge" (Lachenbruch and Marshall, 1969, p. 308). However, the values may lend some support to the thesis of Herron *et al.* (1974) that the Alpha Cordillera represents a fossil subduction zone.

The continental shelf between Prince Patrick and Banks Islands has anomalously low heat flow values (Law *et al.*, 1965). The weighted mean value of 0.84 units is only 57 percent of the worldwide average for continental crust. In Crozier Channel and Kellett Strait, about 130 km farther to the northeast, however, a mean heat flow of 1.46 ± 0.16 units (Patterson and Law, 1966) is close to the worldwide average. Beneath the continental shelf northwest of Prince Patrick Island, moreover, the mean heat flow is 0.46 ± 0.08 units, about 30 percent of the world average. The measurements cited illustrate the lack of uniformity of heat flow in the region adjacent to the continental margin of Canada Basin.

4. *Magnetics*

No aeromagnetic data are available from within the report area. Ship-borne magnetic data were obtained in 1970 during the cruise of the CSS Hudson in the Beaufort Sea but the rapid, short term diurnal variations of the magnetic field in this region rendered such data unreliable.

Beyond the borders of the report area, however, extensive aeromagnetic surveys have been undertaken in the region of the Alpha Rise (Vogt and Ostenso, 1970; Ostenso and Wold, 1971) and the Queen Elizabeth Islands and adjacent shelf and ocean basin (Bhattacharyya, 1968; Riddihough *et al.*, 1973). Widely spaced data covering the Arctic Ocean between the North Pole and the North American continent were reported by E. R. King *et al.* (1966). Extensive aeromagnetic surveys of specific tectonic features are now being undertaken by the Office of Naval Research and the U.S. Geological Survey.

The aeromagnetic studies of E. R. King *et al.* (1966) illustrate that the Arctic Ocean Basin is divisible into several broad regions, the total fields of which range from magnetically featureless in some bordering areas, to 1000 gammas over the "Central Magnetic Zone," which corresponds to the Alpha Rise. They concluded that the Eurasian Basin on the northeast side of the Lomonosov Ridge is probably oceanic in nature and that the North American Basin, including the Alpha Rise, is formed by downdropped blocks of continental rocks, similar in magnetic character to the crystalline rocks of the Canadian Shield. Seismic reflection profiles of the Alpha Rise obtained from the Ice Island T-3 (Hall, 1970) show a rough basement topography similar to parts of the Mid-Atlantic Ridge. Both magnetic and gravity data suggest that the Alpha Rise is a fossil midocean ridge that became inactive in mid-Tertiary time (Vogt and Ostenso, 1970, p. 4935).

Bhattacharyya (1968), in reporting on aeromagnetic studies across the northwestern part of the Arctic Archipelago and adjacent continental shelf and Canada Basin, described an anomalous band of magnetic highs trending northeast, close to and parallel with the edge of the Shelf. He suggested that it was due to a rise in the basement underlying the continental shelf. Immediately to the north is a parallel band of magnetic lows, which he suggested may reflect a deep linear trough, a graben, or an oceanic trench. Riddihough *et al.* (1973) further speculated that these magnetic lows may be related to an earlier (pre-60 m.y.) period of sea-floor spreading. These observations and speculations led Yorath and Norris (1975) to the hypothesis that a spreading axis was located in the Canada Basin parallel to the northern Canadian continental margin before the inception of the Alpha Rise.

C. Geological Information

Data on the bedrock geology of the part of the southern rim of Canada Basin within the project area, presented earlier, established the lateral continuity of some formations on Beaufort Shelf and their correlation from the northern mainland to the Arctic Archipelago. Thus the circum-Arctic geosynclinal belt suggested by Churkin (1969, p. 549) doubtless existed. It served both as a center of marine sedimentation and an avenue for polar migration of faunas since at least the early Paleozoic (Churkin, 1973, p. 497). Gravity data suggest, moreover, that these formations thicken significantly along the outer edge of the continental shelf, but there is no information to document whether or not they are tectonically deformed. Thus the state of the clastic wedge at and near the interface between the Beaufort Shelf and the Canada Basin remains unknown and the geological history of this interface is purely speculative.

It is plausible, however, that the present curvilinear shape of the Beaufort Shelf edge was inherited from the ancestral south flank of the proto-Canada Basin. The reader will recall that the Rapid Fault Array flares out northward into a bundle of nearly vertical faults with strikes ranging from northwest to northeast (Fig. 9). Major components parallel both the northwest- and northeast-trending segments of the Beaufort Shelf and the present plate margin. If it can be assumed that the position and trend of some of the strike-slip faults have been controlled by fundamental shear zones at or near the margin of the North American plate, then the faults may help to establish the shape of the plate margin in the geological past. Available data indicate that the Richardson Fault Array is as old as the late Proterozoic and the Rapid Fault Array at least as old as the Mesozoic. It is postulated that the northeast-trending plate margin opposite the Arctic Archipelago may be as old as the Proterozoic, whereas that trending northwest parallel to the Romanzof Uplift may be somewhat younger. The present shape of the southern margin of the Canada Basin, therefore, may date from the Mesozoic Period.

As pointed out by Churkin (1973, p. 497), foldbelts are easily correlated around the edges of the Arctic Ocean Basin, but nor across it, suggesting that as a whole the basin is not a feature developed within a continental framework across which foldbelts of different trends can be matched. Rather, the basin represents the product of a series of openings and, perhaps, closings of ocean basins extending at least into early Paleozoic time.

V. CONCLUDING REMARKS

The North American plate from the Arctic Archipelago to the Romanzof Mountains embraces portions of the Interior Platform of mainland Canada and Banks Island, the adjoining Cordilleran Foldbelt, and Beaufort Shelf on the southern rim of Canada Basin. The rock succession there is in the form of an eastward- and generally northward-tapering sedimentary wedge, ranging in age from Proterozoic to Paleogene, and comprising two superposed, genetically and compositionally distinct, stratigraphic assemblages. The lower, Proterozoic to Lower Cretaceous miogeoclinal and eugeoclinal assemblage is the continuation of the relatively undeformed foreland sequence of the Interior Platform and Beaufort Shelf. At the time of deposition, the Canada Basin as we know it today may not have existed. The upper, Lower Cretaceous to

Paleogene exogeoclinal assemblage is the syn- and postorogenic suite of rocks derived from the deformed and uplifted regions farther into the Cordilleran Orogenic System (Norris, D. K., 1976b, p. 691). It was deposited along the southern margin of the developing Canada Basin.

Many unconformities within this sedimentary wedge attest to a long, regionally episodic history of tectonism extending from the Proterozoic to the Tertiary. Although commonly there are disconformities, locally the unconformities are spectacularly angular, as seen for example in and adjacent to the Richardson Anticlinorium, the Aklavik Arch Complex, and the Barn and Romanzof Uplifts. Of the plethora of tectonic episodes that affected the northern part of the Cordilleran Foldbelt and adjacent Beaufort Shelf, that in the late Cretaceous and early Tertiary appears to have been most widespread. It masks to varying degrees earlier deformations. Nevertheless this Laramide overprint did not destroy important evidence relating to preceding deformations, and here, unlike large segments of the Cordilleran Foldbelt to the south of the project area, the pre-Laramide tectonic history is more clearly revealed.

Four arrays or zones of right-lateral strike-slip faults played fundamental roles in the tectonic evolution of the Cordilleran Foldbelt north of latitude 60°N. The Richardson Fault Array, traceable for 1000 km from Beaufort Shelf southward to the core zone of the Mackenzie Arc, was demonstrably active as a system of dextral, strike-slip faults from the Proterozoic to the early Mississippian. The Tintina Fault Zone, also traceable for about 1000 km, is recognized from the 60th parallel of latitude to eastern Alaska. There it is postulated (Freeland and Dietz, 1973, Fig. 5; Norris, D. K., 1976b, p. 691) that its structural continuation is the Kobuk Fault, which continues westward across Alaska another 700 km to the continental shelf of Chukchi Sea. The Kaltag Fault Zone, extending more than 1200 km from Norton Sound in western Alaska to the Beaufort Shelf on the southern rim of Canada Basin, was active at least between mid-Cretaceous and mid-Tertiary time. It offsets the Tintina–Kobuk Fault Zone right laterally about 200 km, and in Canada it turns sharply seaward to dissipate this displacement as oblique slip among the several components of the Rapid Fault Array. Beneath Mackenzie Delta, the more easterly strands merge with the Richardson Fault Array on the seaward flank of the Aklavik Arch Complex and attenuate their displacements northeastward as a family of down-to-basin, oblique-slip faults on Beaufort Shelf. The Denali Fault, nested inside the Tintina–Kaltag Fault System, is a member of the family of major dextral, strike-slip faults paralleling the curvilinear margin of the North American plate in the Gulf of Alaska. It apparently lacks the long pre-Tertiary history of the three preceding fault zones.

Vertical components of displacement on the Richardson and Rapid Fault Arrays localized and controlled the timing, direction, shape, and extent of important, narrow depositional sites within the ancestral North American plate, the former the early and mid-Paleozoic Richardson Trough, and the latter, the late early Cretaceous flysch trough between Barn and Cache Creek Uplifts.

The many Phanerozoic rock units that extend from the Cordilleran to the Innuitian Foldbelts suggest that depositional basins were continuous between the two regions and that they had similar histories. There are no data, on the other hand, to indicate whether structures continued between the two foldbelts despite the fact that their tectonic histories are analogous. If there is a connection between the Lower Paleozoic mobile belts of northern Yukon Territory and the Arctic Islands, it lies now or lay at one time beneath the present Beaufort Sea. An unbroken, circum-Arctic Paleozoic mobile belt is, therefore, yet to be demonstrated for the North American plate.

The nature and geological history of the interface between the North American plate and the ocean floor beneath Canada Basin is speculative because of the paucity of geological and geophysical data available. The protracted history of the Richardson Fault Array as well as its great length suggest that it extended northeastward beyond Banks Island and may have exerted a fundamental structural control on the continental margin of southern Canada Basin adjacent to the Arctic Archipelago (Norris, D. K., 1974, p. 42). Similarly, some northwest-trending strands of the Rapid Fault Array may have controlled the structure of the continental margin opposite Romanzof Mountains. The position and shape of the southern Canada Basin between the Innuitian and Cordilleran Orgenic Systems, therefore, may date at least in part from Proterozoic time and identify a portion of the proto-Canada Basin, modified by plate interaction through the Phanerozoic.

ACKNOWLEDGMENTS

We are indebted to our many associates in industry and government with whom we have discussed geological problems of northwestern Canada and Alaska and who have helped us clarify our understanding of this most important part of the North American plate. We are particularly indebted to J. Wm. Kerr and E. R. W. Neale who critically read the manuscript and offered valuable commentary.

* The manuscript was approved for publication in May, 1977.

REFERENCES

Aho, A. E., 1959, Similar trenchlike lineaments in Yukon, *Can. Inst. Min. Met. Bull.*, v. 52, p. 337–338.

Aitken, J. D., Macqueen, R. W., and Usher, J. L., 1973, Reconnaissance studies of Proterozoic and Cambrian stratigraphy, lower Mackenzie River area, District of Mackenzie, *Geol. Surv. Can. Pap.* 73-9.

Baadsgaard, H., Folinsbee, R. E., and Lipson, J., 1961, Caledonian or Acadian granites of the northern Yukon Territory in: *Geology of the Arctic*, Raasch, G. O., ed., Toronto: University of Toronto Press, p. 458–465.

Balkwill, H. R., and Yorath, C. J., 1970, Brock River Map Area, District of Mackenzie (97D), *Geol. Surv. Can. Pap.* 70-32.

Bally, A. W., Gordy, P. L., and Stewart, G. A., 1966, Structure, seismic data, and orogenic evolution of southern Canadian Rocky Mountains, *Bull. Can. Petrol Geol.*, v. 14, p. 337–381.

Bamber, E. W., and Waterhouse, J. B., 1971, Carboniferous and Permian stratigraphy and paleontology, northern Yukon Territory, Canada, *Bull. Can. Petrol. Geol.*, v. 19, p. 29–250.

Bell, J. S., 1974, Late-Paleozoic orogeny in the northern Yukon, in: *Proceedings of the Symposium on the Geology of the Canadian Arctic*, Calgary: Canadian Society of Petroleum Geologists, p. 23–38.

Bhattacharyya, B. K., 1968, Analysis of aeromagnetic data over the Arctic Islands and continental shelf of Canada, *Geol. Surv. Can. Pap.* 68-44.

Bostock, H. S., 1970, Physiographic regions of Canada, *Geol. Surv. Can. Map* 1254A.

Brideaux, W. W., and McIntyre, D. J., 1975, Miospores and microplankton from Aptian–Albian rocks along Horton River, District of Mackenzie, *Geol. Surv. Can. Bull.* 252.

Brosgé, W. P., and Reiser, H. N., 1964, Preliminary map of the geology of the Chandalar Quadrangle, Alaska, *U.S. Geol. Surv. Misc. Geol. Inv. Map* I-375.

Carey, S. W., 1955, The orocline concept in geotectonics, part I, *Pap. Proc. R. Soc. Tasmania*, v. 89, p. 255–288.

Churkin, M., Jr., 1969, Paleozoic tectonic history of the Arctic Basin north of Alaska, *Science*, v. 165(3893), p. 549–555.

Churkin, M., Jr., 1973, Geologic concepts of Arctic Ocean Basin, in: *Arctic Geology*, Pitcher, M. G., ed., *Am. Assoc. Petrol. Geol. Mem.* 19, 485–499.

Churkin, M., Jr., 1975, Basement rocks of Barrow Arch, Alaska, and circum-Arctic Paleozoic mobile belt, *Bull. Am. Assoc. Petrol. Geol.*, v. 59, p. 451–456.

Cook, D. G., and Aitken, J. D., 1969, Erly Lake, District of Mackenzie (97A), *Geol. Surv. Can. Prelim. Ser. Map* 5-1969.

Cote, R. P., Rector, R. J., and Lerand, M. M., 1974, Gulf describes geology of Parsons Lake gas field, *Canadian Petroleum*, v. 15(4), p. 72–78.

Detterman, R. L., 1970, Sedimentary history of the Sadlerochit and Shublik Formations in northeastern Alaska, in: *Proceedings of the Geological Seminar on the North Slope of Alaska, Palo Alto, California, 1970*, Los Angeles: American Association of Petroleum Geologists, Pacific Section,p. O1–O13.

Douglas, R. J. W., Norris, D. K., Thorsteinsson, R., and Tozer, E. T., 1963, Geology and petroleum potentialities of northern Canada, *Geol. Surv. Can. Pap.* 63-31.

Dutro, J. T., Jr., 1970, Pre-Carboniferous carbonate rocks, northeastern Alaska, in: *Proceedings of the Geological Seminar on the North Slope of Alaska, Palo Alto, California, 1970*, Los Angeles: American Association of Petroleum Geologists Pacific Section, p. M1–M8.

Dutro, J. T., Jr., Brosgé, W. P., and Reiser, H. N., 1972, Significance of recently discovered Cambrian fossils and reinterpretation of Neruokpuk Formation, northeastern Alaska, *Bull. Am. Assoc. Petrol. Geol.*, v. 56, p. 808–815.

Dyke, L. D., 1971, Structural investigations in White Uplift, northern Yukon Territory, *Geol. Surv. Can. Pap.* 72-1A, p. 204–207.

Dyke, L. D., 1974, Structural investigations in Barn Mountains, northern Yukon Territory, *Geol. Surv. Can. Pap.* 74-1A, p. 303–308.

Dyke, L. D., 1975, Structural investigations in Campbell Uplift, District of Mackenzie, *Geol. Surv. Can. Pap.* 75-1A, p. 525–532.

Eisbacher, G. H., 1972, Major faults of Canadian Cordillera and southeast Alaska, in: *Faults, Frac-*

tures, Lineaments and Related Mineralization in the Canadian Cordillera [abs.], Vancouver: Geological Association of Canada, Cordilleran Section.

Eisbacher, G. H., 1976, Sedimentology of the Dezadeash flysch and its implications for strike-slip faulting along the Denali Fault, Yukon Territory and Alaska, *Can. J. Earth Sci.*, v. 13, p. 1495–1513.

Freeland, G. L., and Dietz, R. S., 1973, Rotation history of Alaskan tectonic blocks, *Tectonophysics*, v. 18, p. 379–389.

Gabrielse, H., Blusson, S. L., and Roddick, J. A., 1973, Geology of Flat River, Glacier Lake, and Wrigley Lake Map Areas, District of Mackenzie and Yukon Territory, *Geol. Surv. Can. Mem.* 366.

Gates, G. O., and Gryc, G., 1963, Structure and tectonic history of Alaska, in: *The Backbone of the Americas—Tectonic History from Pole to Pole, Am. Assoc. Petrol. Geol. Mem.* 2, p. 264–277.

Grantz, A., 1966, Strike-slip faults in Alaska, *U.S. Geol. Surv. Open-File Rep.* v. 267.

Green, L. H., 1972, Geology of Nash Creek, Larsen Creek, and Dawson Map Areas, Yukon Territory, *Geol. Surv. Can. Mem.* 364.

Hall, J. K., 1970, Arctic Ocean geophysical studies: The Alpha Cordillera and Mendeleyev Ridge, Lamont-Doherty Geol. Obs., Columbia University, I. R., No. 2.

Hall, J. K., 1973, Geophysical evidence for ancient sea-floor spreading from Alpha Cordillera and Mendeleyev Ridge, in: *Arctic Geology*, Pitcher, M. G., ed., *Am. Assoc. Petrol. Geol. Mem.* 19, p. 542–561.

Herron, E. M., Dewey, J. F., and Pitman, W. C. III, 1974, Plate tectonics model for the evolution of the Arctic, *Geology*, v. 2(8), p. 377–380.

Hills, L. V., and Fyles, J. G., 1973, The Beaufort Formation, Canadian Arctic Islands [abs.] in: *Program and Abstracts, Arctic Symposium*, Saskatoon, Saskatchewan: Canadian Society of Petroleum Geologists and the Geological Association of Canada.

Hofer, H., and Varga, W., 1972, Seismological experience in the Beaufort Sea, *Geophysics*, v. 37, p. 605–619.

Hornal, R. W., Sobczak, L. W., Burke, W. E. F., and Stephens, L. E., 1970, Preliminary results of gravity surveys over the Mackenzie Basin and Beaufort Sea, *Can. Dep. Energy Mines Resources, Earth Phys. Branch*, gravity maps Nos. 117, 118, 119.

Hunkins, K. 1963, Submarine structure of the Arctic Ocean from earthquake surface waves, in: *Proceedings of the Arctic Basin Symposium, Oct. 1962, Washington, D.C., Arctic Inst. North Am.*, p. 3–8.

Jeletzky, J. A., 1960, Uppermost Jurassic and cretaceous rocks, east flank of Richardson Mountains between Stony Creek and lower Donna River, Northwest Territories, *Geol. Surv. Can. Pap.* 59–14.

Jeletzky, J. A., 1961, Eastern slope, Richardson Mountains: Cretaceous and Tertiary structural history and regional significance, in: *Proc. First Int. Symp. Arctic Geol.*, Alberta Society of Petroleum Geologists, Toronto: University of Toronto Press, p. 532–583.

Jeletzky, J. A., 1962, Pre-Cretaceous Richardson Mountains Trough: Its place in the tectonic framework of Arctic Canada and its bearing on some geosynclinal concepts, *Trans. R. Soc. Can.*, v. 56(3), p. 55–84.

Jeletzky, J. A., 1967, Jurassic and (?) Triassic rocks of the eastern slope of Richardson Mountains northwestern District of Mackenzie, *Geol. Surv. Can. Pap.* 66-50.

Kay, M., 1945, North American geosynclines—their classification, *Geol. Soc. Am. Bull.* 56, p. 1172.

Kent, P. E., and Russell, W. A. C., 1961, Evaporite piercement structures in the northern Richardson Mountains, in: *Proc. First Int. Symp. Arctic Geol.*, American Association of Petroleum Geologists, p. 584–595.

King, E. R., Zietz, I., and Alldredge, L. R., 1966, Magnetic data on the structure of the central Arctic region, *Geol. Soc. Am. Bull.*, v. 77, p. 619–646.

King, P. B., compiler, 1969, *Tectonic Map of North America*, U.S. Geol. Surv. Map, scale 1:5,000,000, Washington, D.C.: U.S. Government Printing Office.

King, P. B., 1976, Cross section of North American Cordillera near the Fortieth Parallel [abs.], in: *25th International Geological Congress Abstracts*, v. 3, p. 685–686.

Klovan, J. E., and Embry III, A. F., 1971, Upper Devonian stratigraphy, northeastern Banks Island, N. W. T., *Bull. Can. Soc. Petrol. Geol.*, v. 19, p. 705–729.

Knipping, H. D., 1960, Late Paleozoic orogeny–north Yukon, A.A.P.G.–A.S.P.G. Regional Meeting, May 25–28, 1960, Banff, Alberta, Canada, Frontiers of Exploration in Canada, Preprint.

Lachenbruch, A. H., and Marshall, B. V., 1966, Heat flow through the Arctic Ocean floor: The Canada Basin–Alpha Rise boundary, *J. Geophys. Res.*, v. 71, p. 1223–1248.

Lachenbruch, A. H., and Marshall, B. V., 1969, Heat flow in the Arctic, *Arctic*, v. 22(3), p. 300–311.

Lathram, E. H., 1965, Preliminary geologic map of northern Alaska; *U.S. Geol. Surv. Open-File Rep.* 254.

Law, L. K., Paterson, W. S. B., and Whitham, K., 1965, Heat flow determination in the Canadian Arctic Archipelago, *Can. J. Earth Sci.*, v. 2(1), p. 59–71.

Lawrence, J. R., 1973, Old Crow Basin, in: *The Future Petroleum Provinces of Canada—Their Geology and Potential*, Can. Soc. Petrol. Geol. Mem. 1, p. 307–314.

Leblanc, G., and Wetmiller, R. J., 1974, An evaluation of seismological data available for the Yukon Territory and the Mackenzie Valley, *Can. J. Earth Sci.*, v. 11, p. 1435–1454.

Lenz, A. C., 1972, Ordovician to Devonian history of northern Yukon and adjacent District of Mackenzie, *Bull. Can. Petrol. Geol.*, v. 20(2), p. 321–361.

Lenz, A. C., and Perry, D. G., 1972, The Neruokpuk Formation of the Barn Mountains and Driftwood Hills, northern Yukon; its age and graptolite fauna, *Can. J. Earth Sci.*, v. 9, p. 1129–1138.

Lerand, M., 1973, Beaufort Sea, in: *The Future Petroleum Provinces of Canada—Their Geology and Potential*, Can. Soc. Petrol. Geol. Mem. 1, p. 315–386.

Mackay, J. R., 1963, The Mackenzie Delta area, N. W. T., *Can. Dept. Energy, Mines, Resources, Geogr. Br. Misc. Rep.* 23.

Macqueen, R. W., 1975, Lower and Middle Paleozoic sediments, northern Yukon Territory, *Geol. Surv. Can. Pap.* 75-1C, p. 291–301.

Martin, L. J., 1959, Stratigraphy and depositional tectonics of north Yukon–lower Mackenzie River area, Canada, *Bull. Am. Assoc. Petrol. Geol.*, v. 43(10), p. 2399–2455.

Meyerhoff, A. A., 1973, Origin of Arctic and North Atlantic Oceans, in: *Arctic Geology*, Pitcher, M. G., ed., *Am. Assoc. Petrol. Geol. Mem.* 19, p. 562–582.

Miall, A. D., 1974, Bedrock geology of Banks Island, District of Franklin, *Geol. Surv. Can. Pap.* 74-1A, p. 336–342.

Miall, A. D., 1975a, Stratigraphy of the Deminex CGDF FOC Amoco Orksut I-44 Well, *Geol. Surv. Can. Pap.* 75-1B, p. 257–259.

Miall, A. D., 1975b, Post-Paleozoic geology of Banks, Prince Patrick, and Eglinton Islands, Arctic Canada, *Can. Soc. Petrol. Geol. Mem.* 4, p. 557–587.

Miall, A. D., 1976, Proterozoic and Paleozoic geology of Banks Island, Arctic Canada, *Geol. Surv. Can. Bull.* 258.

Miall, A. D., 1977, Mesozoic and Tertiary geology of Banks Island, Arctic Canada: The history of an unstable craton margin, *Geol. Surv. Can. Mem.* 387.

Mountjoy, E. W., 1967a, Triassic stratigraphy of northern Yukon Territory, *Geol. Surv. Can. Pap.* 66-19.

Mountjoy, E. W., 1967b, Upper Cretaceous and Tertiary stratigraphy, northern Yukon Territory and northwestern District of Mackenzie, *Geol. Surv. Can. Pap.* 66-16.

Norford, B. S., 1964, Reconnaissance of the Ordovician and Silurian rocks of northern Yukon Territory, *Geol. Surv. Can. Pap.* 63-39.

Norris, A. W., 1968, Reconnaissance Devonian stratigraphy of northern Yukon Territory and northwestern District of Mackenzie, *Geol. Surv. Can. Pap.* 67-53.

Norris, A. W., 1971, Devonian biostratigraphy of northern Yukon Territory and adjacent District of Mackenzie, *Geol. Surv. Can. Pap.* 71-1A, p. 106.

Norris, D. K., 1970, Structural and stratigraphic studies, Blow River area, Yukon Territory, and western District of Mackenzie, *Geol. Surv. Can. Pap.* 70-1A, p. 230–235.

Norris, D. K., 1972*a*, En echelon folding in the northern Cordillera of Canada, *Bull. Can. Petrol. Geol.*, v. 20(3), p. 634–642.

Norris, D. K., 1972*b*, Structural and stratigraphic studies in the Tectonic Complex of northern Yukon Territory, north of Porcupine River, *Geol. Surv. Can. Pap.* 72-1B, p. 91–99.

Norris, D. K., 1973, Tectonic styles of northern Yukon Territory and northwestern District of Mackenzie, Canada, in: *Arctic Geology, Am. Assoc. Petrol. Geol. Mem.* 19, p. 23–40.

Norris, D. K., 1974, Structural geometry and geological history of the northern Canadian Cordillera, in: *Proceedings of the 1973 National Convention*, Calgary: Canadian Society of Exploration Geophysicists, p. 18–45.

Norris, D. K., 1975*a*, Geological maps of parts of Yukon Territory and Northwest Territories, *Geol. Surv. Can. Open-File Rep.* 279.

Norris, D. K., 1975*b*, Geological maps of part of Northwest Territories, *Geol. Surv. Can. Open-File Rep.* 302.

Norris, D. K., 1975*c*, Geological maps of parts of Northwest Territories and Yukon Territory, *Geol. Surv. Can. Open-File Rep.* 303.

Norris, D. K., 1976*a*, Structural and stratigraphic studies in the northern Canadian Cordillera, *Geol. Surv. Can. Pap.* 76-1A, p. 457–466.

Norris, D. K., 1976*b*, The North American Cordillera north and east of the Tintina Fault, in: *25th International Geological Congress, Sydney, Abstracts*, v. 3, p. 690–692.

Norris, D. K., and Bally, A. W., 1972, Coal, oil, gas, and industrial mineral deposits of the Interior Plains, Foothills, and Rocky Mountains of Alberta and British Columbia, in: *24th International Geological Congress, Montreal, Guidebook to Field Excursions*, p. A25–C25.

Norris, D. K., and Black, R. F., 1964, Paleomagnetic age of the pre-Ordovician rocks near Inuvik, Northwest Territories, *Geol. Surv. Can. Pap.* 64-2, p. 47–50.

Norris, D. K., and Hopkins, W. S. Jr., 1977, The geology of the Bonnet Plume Basin, Yukon Territory, *Geol. Surv. Can. Pap.* 76-10.

Norris, D. K., Price, R. A., and Mountjoy, E. W., 1963, Northern Yukon Territory and northwestern District of Mackenzie, *Geol. Surv. Can. Map* 10-1963 with marginal notes.

Ostenso, N. A., and Wold, R. J., 1971, Aeromagnetic survey of the Arctic Ocean: Techniques and interpretations, *Mar. Geophys. Res.* 1, p. 178–219.

Ostenso, N. A., and Wold, R. J., 1973, Aeromagnetic evidence for origin of Arctic Ocean Basin, in: *Arctic Geology*, Pitcher, M. G., ed., *Am. Assoc. Petrol. Geol. Mem.* 19, p. 506–516.

Packer, D. R., and Stone, D. B., 1972, An Alaskan Jurassic paleomagnetic pole and the Alaskan orocline, *Nature (London)*, v. 237, p. 25–26.

Paterson, W. S. B., and Law, L. K., 1966, Additional heat flow determinations in the area of Mould Bay, Arctic Canada, *Can. J. Earth Sci.*, v. 3(2), p. 237–246.

Patton, W. W., Jr., and Hoare, J. M., 1968, The Kaltag Fault, west-central Alaska, *U.S. Geol. Surv. Prof. Pap.* 600-D, p. D147–D153.

Payne, T. G., 1955, Mesozoic and Cenozoic tectonic elements of Alaska, *U.S. Geol. Surv. Misc. Geol. Inv. Map* I-84.

Reiser, H. N., 1970, Northeastern Brooks Range—a surface expression of the Prudhoe Bay section, in: *Proceedings of the Geological Seminar on the North Slope of Alaska, Palo Alto, California, 1970*, Los Angeles: American Association of Petroleum Geologists, Pacific Section, p. K1–K14.

Riddihough, R. P., Haines, G. V., and Hannaford, W., 1973, Regional magnetic anomalies of the Canadian Arctic, *Can. J. Earth Sci.* v. 10(2), p. 157–163.

Roddick, J. A., 1967, Tintina Trench, *J. Geol.*, v. 75(1), p. 23–33.

Sable, E. G., 1965, Geology of the Romanzof Mountains, Brooks Range, northeastern Alaska, University Microfilms Inc., Ann Arbor, Michigan.

Schatsky, N. S., and Bogdanoff, A. A., 1960, International tectonic map of Europe, Comm. Geol. Map of World, Subcommission for Tectonic Map of World, 21st Intern. Geol. Cong. Copenhagen.

Shearer, J. M., 1972, Geological structure of the Mackenzie Canyon area of the Beaufort Sea, *Geol. Surv. Can. Pap.* 72-1A, p. 179–180.

Sloss, L. L. Krumbein, W. C., and Dapples, E. C., 1949, Integrated facies analysis, in: *Sedimentary Facies in Geologic History*, *Geol. Soc. Am. Mem.* 39, p. 91–124.

Sobczak, L. W., 1975a, Gravity and deep structure of the continental margin of Banks Island and Mackenzie Delta, *Can. J. Earth Sci.*, v. 12(3), p. 378–394.

Sobczak, L. W., 1975b, Gravity anomalies and passive continental margins, Canada and Norway, in: *Canada's Continental Margins and Offshore Petroleum Exploration*, Yorath, C. J., Parker, E. R., and Glass, D. J., eds., *Can. Soc. Petrol. Geol. Mem.* 4, p. 743–761.

Sobczak, L. W., Stephens, L. E., Winter, P. J., and Hearty, D. B., 1973, Gravity measurements over the Beaufort Sea, Banks Island, and Mackenzie Delta, Gravity Map Series, Earth Physics Branch, No. 151.

Sobczak, L. W., and Weber, J. R., 1973, Crustal structure of Queen Elizabeth Islands and Polar Continental Margin, Canada, in: *Arctic Geology*, Pitcher, M. G., ed., *Am. Assoc. Petrol. Geol. Mem.* 19, p. 517–525.

Stevens, A. E., 1974, Seismicity of northern Canada, *Bull. Can. Petrol. Geol.*, v. 22, p. 387–404.

Stoneley, R., 1971, A note on the structural evolution of Alaska, *J. Geol. Soc. London*, v. 127, p. 623–628.

Tailleur, I. L., and Brosgé, W. P., 1970, Tectonic history of northern Alaska, in: *Proceedings of the Geological Seminar on the North Slope of Alaska, Palo Alto, California, 1970*, American Association of Petroleum Geologists, Pacific Section, p. E1–E20.

Taylor, G. C., and Stott, D. F., 1973, Tuchodi Lakes Map Area, British Columbia (94K), *Geol. Surv. Can. Mem.* 373.

Tempelman-Kluit, D. J., 1976, The Yukon Crystalline Terrane: Enigma in the Canadian Cordillera, *Geol. Soc. Am. Bull.*, v. 87, p. 1343–1357.

Thorsteinsson, R., and Tozer, E. T., 1960, Summary account of structural history of the Canadian Arctic Archipelago since Precambrian time, *Geol. Surv. Can. Pap.* 60-7.

Thorsteinsson, R., and Tozer, E. T., 1962, Banks, Victoria, and Stefansson Islands, Arctic Archipelago, *Geol. Surv. Can. Mem.* 330.

Tipper, H. W., 1969, Mesozoic and Cenozoic geology of the northeast part of Mount Waddington Map Area (92N), coast district, British Columbia, *Geol. Surv. Can. Pap.* 68-33.

Tozer, E. T., 1970, Marine Triassic faunas, *Geol. Surv. Can. Econ. Geol. Rep.*, v. 1, p. 633–640.

Tozer, E. T., and Thorsteinsson, R., 1964, Western Queen Elizabeth Islands, Arctic Archipelago, *Geol. Surv. Can. Mem.* 332.

Vogt, P. R., and Ostenso, N. A., 1970, Magnetic and gravity profiles across the Alpha Cordillera and their relation to Arctic sea-floor spreading, *J. Geophys. Res.*, v. 75, p. 4925–4937.

Wanless, R. K., Stevens, R. D., Lachance, G. R., and Delabio, R. N. D., 1974, Age determinations and geological studies, K–Ar isotopic ages, Report 12, *Geol. Surv. Can. Pap.* 74-2.

Wanless, R. K., Stevens, R. D., Lachance, G. R., and Rimsaite, R. Y. H., 1965, Age determinations and geologic studies, *Geol. Surv. Can. Pap.* 64-17, Part 1.

Wold, J., Woodzick, T. L., and Ostenso, N. A., 1970, Structure of the Beaufort Sea continental margin, *Geophysics*, v. 35(5), p. 849–861.

Yorath, C. J., 1973, Geology of the Beaufort–Mackenzie Basin and eastern northern Interior Plains, *Am. Assoc. Petrol. Geol. Mem.* 19, p. 41–47.

Yorath, C. J., Balkwill, H. R., and Klassen, R. W., 1969, Geology of the eastern part of the northern interior and arctic coastal plains, Northwest Territories, *Geol. Surv. Can. Pap.* 68-27.

Yorath, C. J., Balkwill, H. R., and Klassen, R. W., 1975, Franklin Bay and Malloch Hill Map Areas, District of Mackenzie, *Geol. Surv. Can. Pap.* 74-36.

Yorath, C. J., and Norris, D. K., 1975, The tectonic development of the southern Beaufort Sea and its relationship to the origin of the Arctic Ocean Basin, *Can. Soc. Petrol. Geol. Mem.* 4, p. 589–611.

Young, F. G., 1974a, Cretaceous stratigraphic displacements across Blow Fault Zone, northern Yukon Territory, *Geol. Surv. Can. Pap.* 74-1B, p. 291–296.

Young, F. G., 1974b, Mesozoic epicontinental, flyschoid, and molassoid depositional phases of

Yukon's north slope, in: *Proceedings of the Symposium on the Geology of the Canadian Arctic,* Calgary: Canadian Society Petroleum Geologists, p. 181–202.

Young, F. G., 1975, Upper Cretaceous stratigraphy, Yukon Coastal Plain and northwestern Mackenzie Delta, *Geol. Surv. Can. Bull.* 249.

Ziegler, P. A., 1969, *The Development of Sedimentary Basins in Western and Arctic Canada,* Calgary: Alberta Society of Petroleum Geologists.

Chapter 4

EVOLUTION OF THE CANADIAN ARCTIC ISLANDS: A TRANSITION BETWEEN THE ATLANTIC AND ARCTIC OCEANS

J. Wm. Kerr
J. Wm. Kerr and Associates
Calgary, Canada
(Formerly Geological Survey of Canada, Calgary, Canada)

I. INTRODUCTION

A. Geographic Limits of Study

The Canadian Arctic Islands and nearby mainland areas (Fig. 1) form a physiographic transition between the Atlantic and Arctic Oceans. Thus, the margins of two ocean basins are involved in this study.

The Canadian Arctic Islands are bordered on the southeast by Baffin Bay and by the Labrador Sea, which together form a major arm of the Atlantic Ocean. The relationship of both these features to the Arctic Islands is discussed.

The Arctic Islands are bordered on the northwest by the Arctic Ocean, where the study area partly overlaps with the northern mainland and Beaufort Shelf, treated in this volume by Norris and Yorath. On the northeast the study connects with northern Greenland, which is treated in this volume by Dawes and Peel.

B. Objectives and Format

This paper outlines geological relationships between the Canadian Arctic Islands and the two oceanic areas that border them on the southeast and northwest, which are, respectively, parts of the Atlantic and Arctic Oceans. An attempt is made to show how these three regions evolved. The Canadian Arctic (Fig. 1) is unique among lands surrounding the Arctic Ocean Basin because it is a fragmented and drowned part of a continent that now forms a shallow transitional connection between two deep ocean basins, both of which probably have an origin in rifting. Evolution of the region extended from Precambrian to Holocene, culminating in the fragmentation of a continent and formation of ocean basins.

This paper concentrates on the evolution and on the interrelationship of the main geological features. Details and documentation are sometimes not given, but appropriate sources are cited. Much documentation is contained in earlier syntheses (Douglas *et al.*, 1963; Thorsteinsson and Tozer, 1970; Trettin *et al.*, 1972; Kerr, 1977*a*; Balkwill, 1978). The plate tectonic history of the Canadian Arctic is summarized elsewhere by Kerr (1980*b*).

Evolution of the Canadian Arctic Islands occurred in two main phases that overlapped somewhat in time. First was a *constructional phase, the evolution of the continent as a substructure.* During this phase the continent developed by formation of a crystalline crust, and later by development of large depositional basins upon it. By the end of this phase almost the entire area shown by Fig. 1 was underlain by continental crust of approximately normal thickness (37 km). This crust was the substructure beside which and within which the ocean basins and their branches were to form later. The internal structure of that crust in large measure controlled the shapes and history of the sedimentary basins, and the present seaways.

Second was the *fragmentation phase, the evolution of the ocean basins and their branches within the continent.* This period extends from the earliest signs of continental breakup through the time of ocean formation to the present day.

C. Relationship of Sequences to Major Geological Provinces

Major geologic provinces in northern Canada are shown in plan view (Fig. 2), and in cross section (Fig. 3). They make up eight stratigraphic–structural sequences.

Sequence 1 is the Precambrian crystalline basement, a gneissic crust that lies on the mantle. It is the foundation of the continent, upon which younger tectonic units lie, and where exposed at the surface forms the Canadian Shield (Douglas, 1970).

The history of the Arctic region was dominated by the Innuitian Mobile Belt. This is a continental margin-type mobile belt, made up of sedimentary sequences that are designated by numbers 2 to 8 (Fig. 3). The temporal and spatial relationships of these sequences are shown in a tectonic time diagram

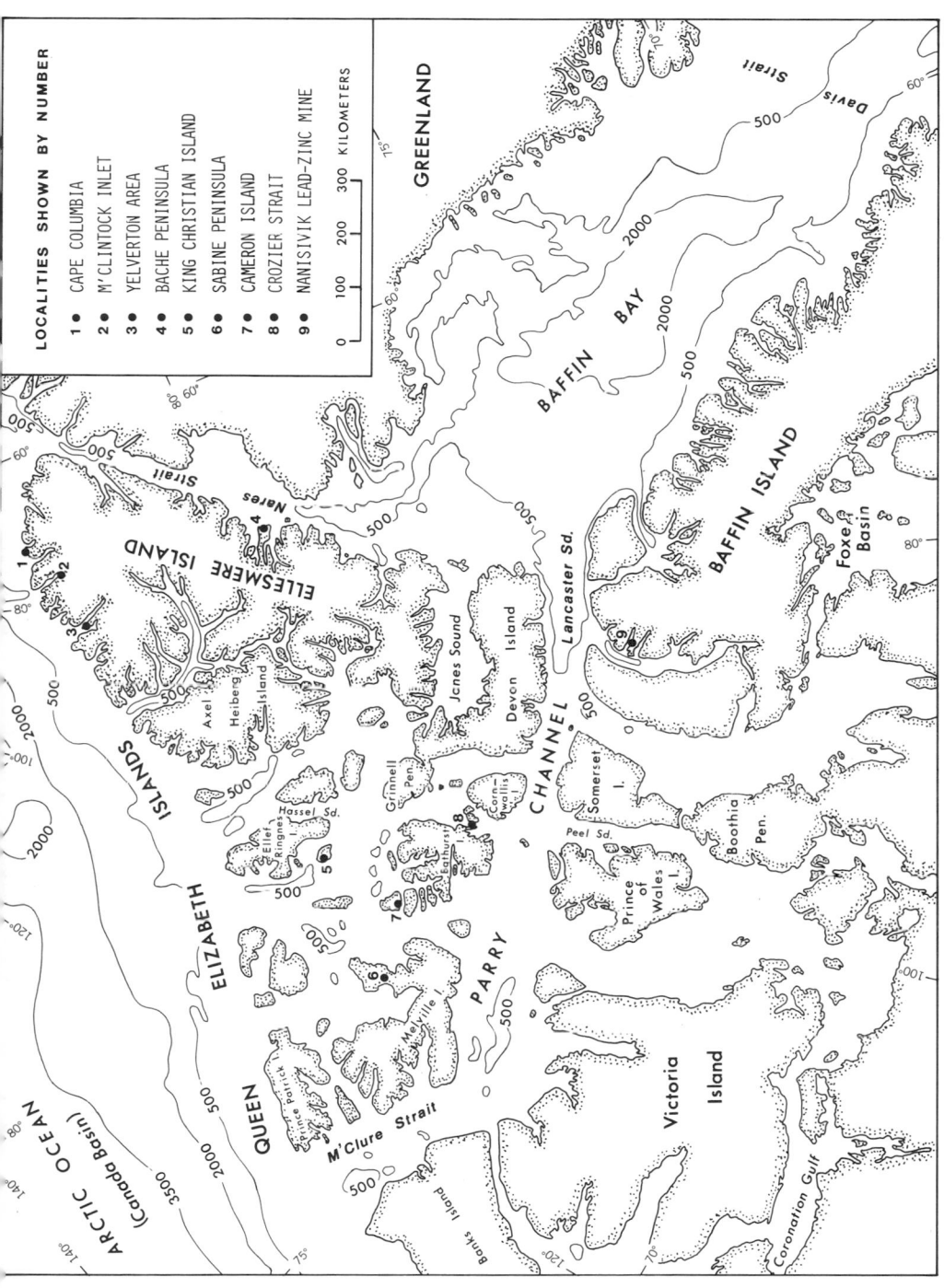

Fig. 1. Index map of the Canadian Arctic Islands showing localities referred to in the text. Bathymetry is in meters.

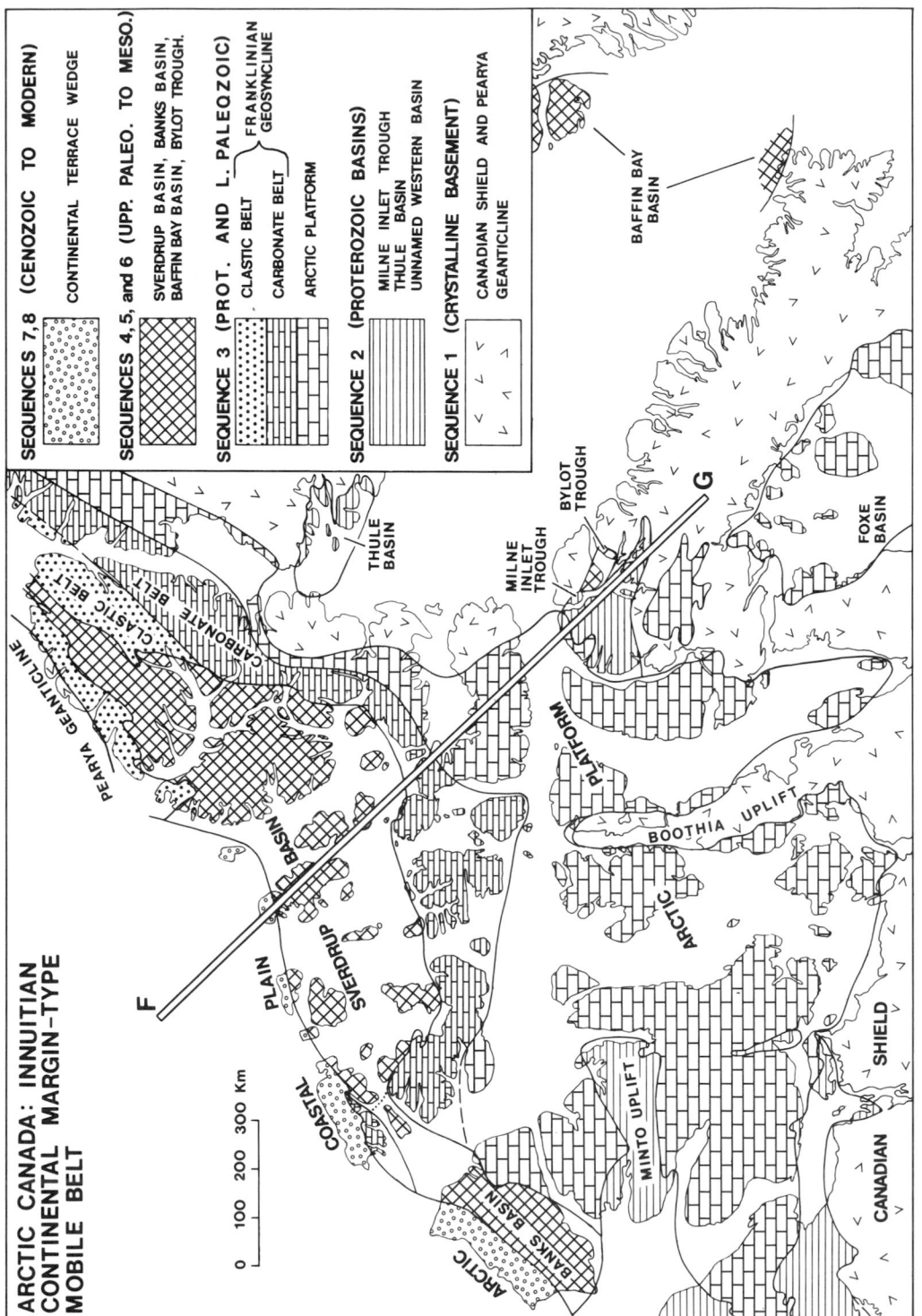

Fig. 2. Stratigraphic sequences and geological provinces of the Canadian Arctic Islands. FG is location of cross section (Fig. 3) and tectonic time diagram

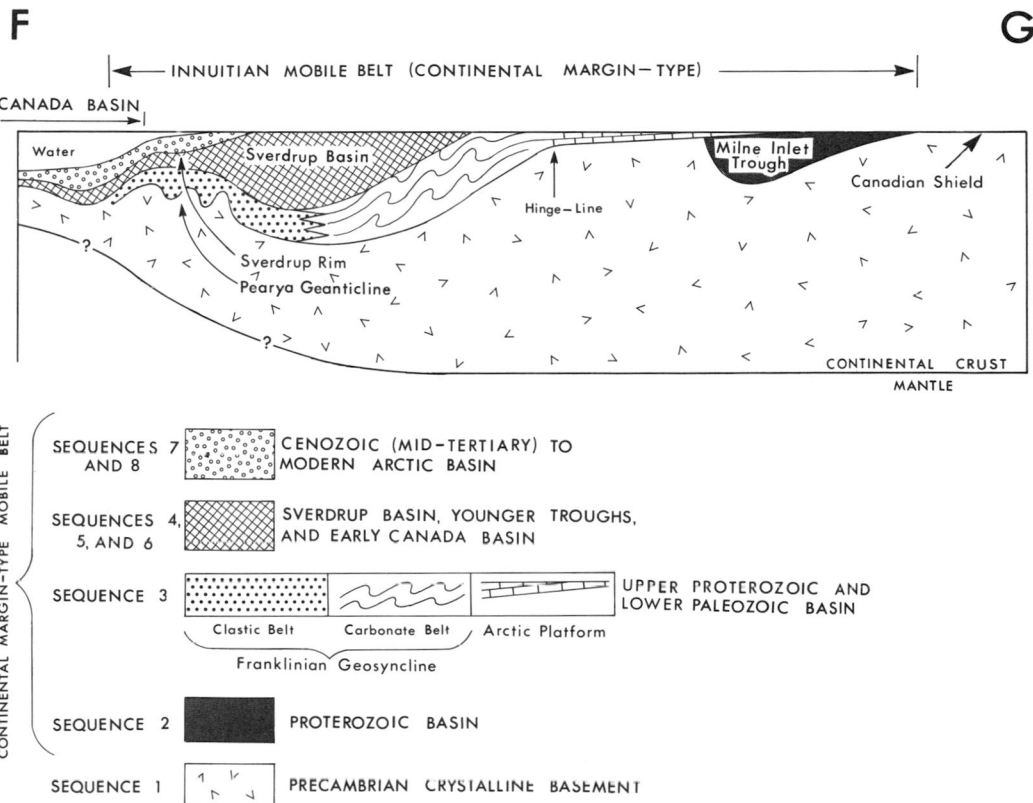

Fig. 3.   Cross section of the Innuitian Mobile Belt in the Canadian Arctic Islands. Location is shown in Fig. 2. Age relationships of sequences are shown in Fig. 4.

(Fig. 4). They are separated from each other by widespread angular unconformities, each related to a tectonic event.

Sequence 2 is Proterozoic rock, remnants of a former much more extensive basin. This basin was fragmented and its parts became isolated from each other before deposition of Sequence 3. Sequence 3 is a broad depositional basin of Proterozoic to Devonian age that consists of the Arctic Platform and the Franklinian Geosyncline (Figs. 2 and 3). The geosyncline includes a clastic belt in the northwest (formerly called the eugeosyncline) and a miogeocline on the south and east (formerly called the miogeosyncline).

The Pearya Geanticline (Trettin et al., 1972) has a core of gneissic rocks that were intruded, uplifted, and eroded at various times. It was primarily a positive feature and a major source of sediments, although sediments also were deposited upon the geanticline at times during the deposition of Sequence 3. These sediments are assigned to the clastic belt of the geosyncline. The geanticline may possibly be an extension of the shield, connected with it beneath the geosyncline (Fig. 3).

Sequences 4, 5, and 6 were deposited in several basins, the best known and best exposed being the Sverdrup Basin (Thorsteinsson and Tozer, 1970; Balkwill, 1978). The Sverdrup Basin is a successor basin of late Paleozoic and Mesozoic age that is unconformable on the deformed Franklinian Geosyncline and Pearya Geanticline. The southeastern limit of thick sedimentation in the basin was approximately at the southeastern limit of exposure (Fig. 3). On the northwest, the Sverdrup Basin was bordered throughout its history by the recurrently positive Sverdrup Rim, which lies above the southwestward extension of the Pearya Geanticline. Rocks of Sequences 4, 5, and 6 that are coeval with those of the Sverdrup Basin probably thin northwestward onto the Sverdrup Rim, but may thicken from there northward into the sedimentary wedge beneath the Canada Basin. Sequences 5 and 6 are present in the Baffin Bay Basin and are exposed on its margins (Fig. 2).

Sequences 7 and 8 include Cenozoic (mid-Tertiary) and younger rocks (Fig. 3). These sequences occur as thick bodies of sediment in the Canada Basin and Baffin Bay, and as thinner columns in the channels between the

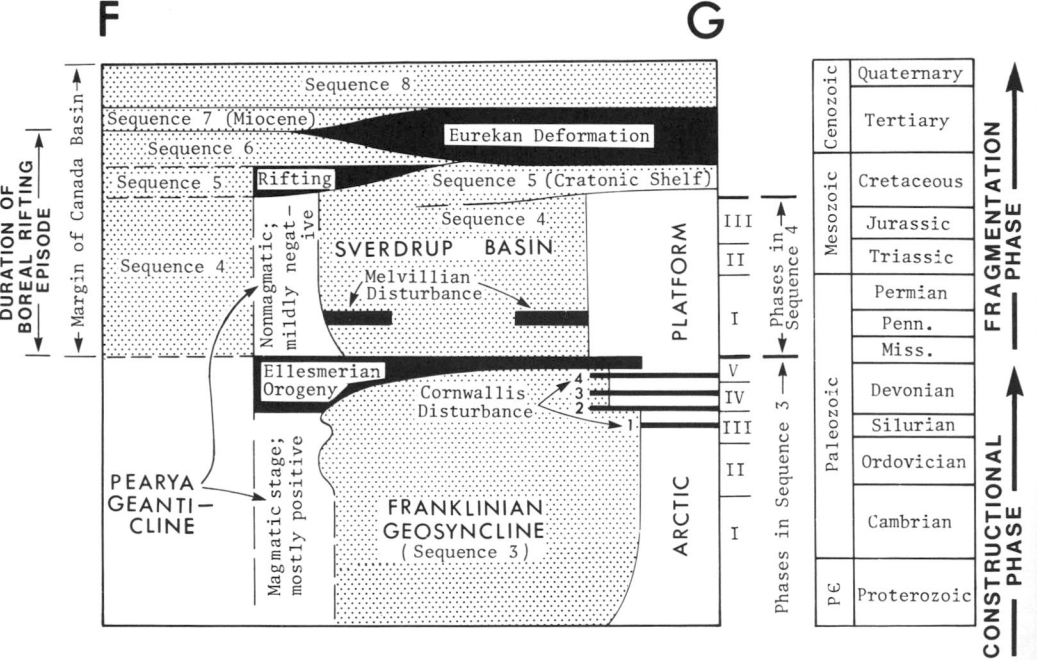

Fig. 4.   Tectonic time diagram for the Innuitian Mobile Belt. The stippled area shows a region where there was continuous or nearly continuous sedimentation. The solid black pattern indicates tectonic events. In the Ellesmerian Orogeny, Cornwallis Disturbance, and Melvillian Disturbance, erosion dominated, so widespread unconformities mark these events. The Boreal Rifting Episode and the Eurekan Deformation were plate tectonic events, with formation of broad basins and wider sedimentation. This figure resembles a cross section, with a location similar to that of Fig. 3.

Arctic Islands. Northwest of the Arctic Islands these sequences form a continental terrace wedge that is exposed in the Arctic Coastal Plain (Fig. 2) and thickens northwestward beneath the continental shelf and continental slope (Fig. 3). They form a more confined but thick wedge of sediments in Baffin Bay, but there the rocks of Sequences 7 and 8 are beneath the water.

II. CONSTRUCTIONAL PHASE: EVOLUTION OF THE CONTINENT AS A SUBSTRUCTURE

A. Sequence 1: Precambrian Crystalline Basement

The Precambrian crystalline basement forms the foundation of the North American continent and rests on the underlying mantle (Figs. 2 and 3). It crops out as the Canadian Shield, and is widely exposed in the southeastern part of the study area, including much of Greenland. The Canadian Shield is subdivided into structural provinces (Stockwell, 1964; Stockwell et al., 1970), based mainly on differences in structural trend, style of folding, and radiometric age.

A narrow belt of gneissic rocks on northwestern Ellesmere Island has metamorphic ages of 800 m.y. to 1000 m.y., indicating an intense metamorphism of about Grenvillian age (Sinha and Frisch, 1975, 1976). It is part of the Pearya Geanticline (Trettin et al., 1972), and may connect with the shield beneath the sedimentary sequences (Fig. 3). The thickness of the crust in the Arctic Islands is about 37 km (Sobczak and Weber, 1973), which is approximately normal for continental crust.

B. Innuitian Mobile Belt

The history of the Arctic continental margin of Canada from Ellesmere Island to Banks Island has been dominated by the Innuitian Mobile Belt from late Proterozoic time to the present (Figs. 2 and 3). This is a continental margin-type mobile belt that trends northeastward and includes features extending from the Proterozoic basins northwestward to the modern basin that trends along the margin of the Arctic Ocean. Several major depositional basins that make up the mobile belt lie one above the other. They are a Proterozoic basin (Sequence 2), the Franklinian Geosyncline (Sequence 3), the Sverdrup Basin and related troughs (Sequences 4, 5, and 6), and a mid-Tertiary to modern basin (Sequences 7 and 8). The northern part of the continent was undergoing a long lasting constructional phase until the end of Sequence 3 (Fig. 4). From that time to the present day it has been in a fragmentation phase.

C. Sequence 2: Proterozoic Basins

These oldest sediments of the Innuitian Mobile Belt are Proterozoic (Helikian and possibly Hadrynian), and are exposed in several areas that are partly or completely disconnected from each other (Fig. 2). They were subjected to intermittent tectonic activity including dike intrusion and faulting. A remnant of these Sequence 2 rocks that occurs on northern Baffin Island is shown in cross section (Fig. 3). It is isolated on this line of cross section, but on its west side probably connects with other parts of an originally larger unnamed western basin. These rocks form the Milne Inlet Trough (Geldsetzer, 1974; Jackson and Davidson, 1975; Jackson et al., 1975; Jackson et al., 1978). This trough plunges northwestward and probably connects in the subsurface with equivalent rocks of an originally very large Proterozoic basin farther west. The Nanisivik Lead–Zinc Mine occurs in a carbonate formation in the Milne Inlet Trough (Olson, 1977).

An isolated outcrop area of Proterozoic rocks in northwestern Greenland (Fig. 2) is called the Thule Basin (Davies et al., 1963; Kerr, 1967b; Dawes, 1976). Tectonism occurred there during Proterozoic time, but much of the isolation of the Thule Basin apparently resulted from Cretaceous–Tertiary faulting.

Thick Proterozoic rocks of Sequence 2 occur in northern Somerset and Prince of Wales Islands (Fig. 2). They form a north-plunging crescent-shaped body wrapped around the core of the north-plunging Boothia Horst (Kerr, 1977a). At least two pre-Paleozoic tectonic events affected these rocks. Each appears to have involved uplift, erosion, and diabase intrusion. The partial isolation of these rocks from equivalent rocks to the east and west was mainly achieved before Paleozoic deposition.

On Victoria Island (Fig. 2) a thick inlier of Sequence 2 occurs in the Minto Uplift (Thorsteinsson and Tozer, 1962; Young and Jefferson, 1975). A broad belt of the large unnamed western basin making up Sequence 2 occurs on the Canadian mainland southwest of Victoria Island (Fig. 2), trending eastward, and dipping partly northwest (Baragar and Donaldson, 1973).

The rocks of Sequence 2 in the study area are everywhere very thick and contain unconformities. They also were intruded at one or more times by diabase dikes (Fahrig and Jones, 1969; Fahrig et al., 1971). Their present distribution suggests that they were deposited in a single basin that trended generally northeastward. The present partial isolation of parts of this sequence resulted from a combination of syndepositional, pre-Paleozoic, and younger tectonism. The several remnants exposed in the west may largely connect in the subsurface beneath the Lower Paleozoic cover as an unnamed western basin (Fig. 2).

D. Sequence 3: Late Proterozoic to Late Devonian

1. *Foreword*

The late Proterozoic to late Devonian history of the Innuitian Mobile Belt is mainly the history of the Franklinian Geosyncline, the Arctic Platform, and the magmatic stage of the Pearya Geanticline (Fig. 4). The geosyncline was an orthogeosyncline in the sense of Kay (1951). The geosyncline was linear and lay between the Precambrian Shield and a geanticline (Figs. 2 and 3). Its history included five depositional phases (Fig. 5) that correspond to the three phases described from the miogeocline of Ellesmere Island (Kerr, 1967*a*,

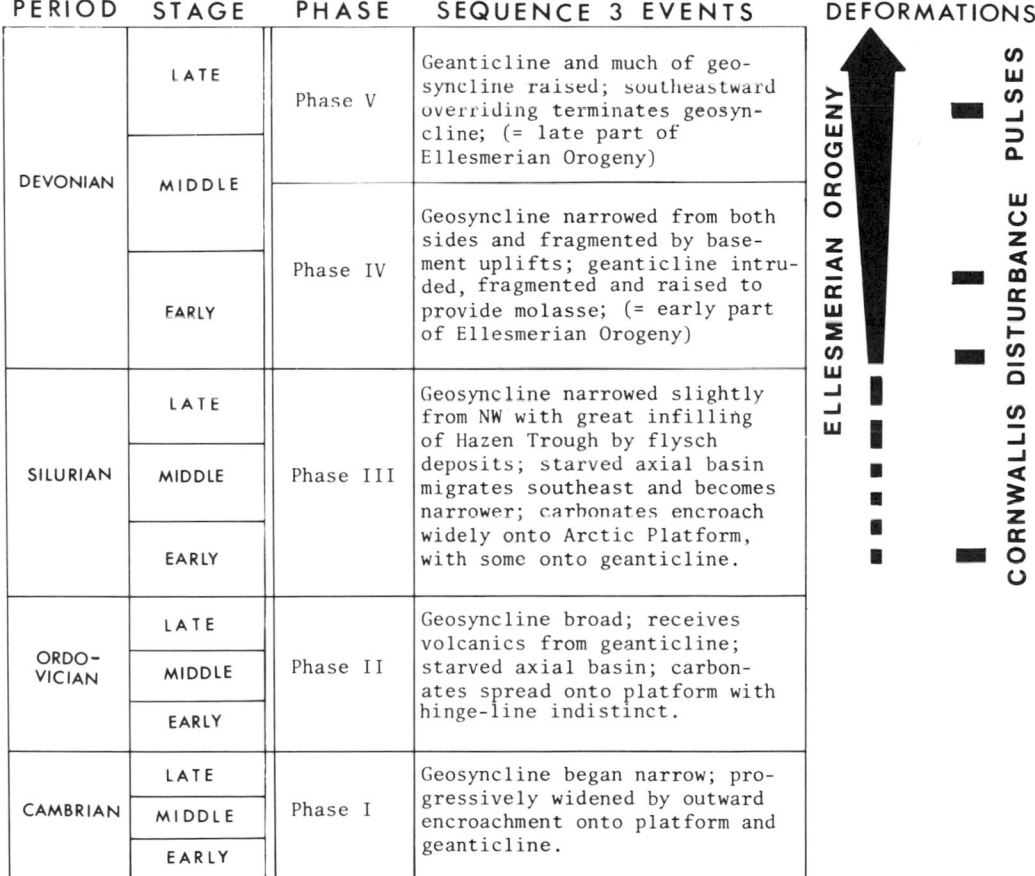

PERIOD	STAGE	PHASE	SEQUENCE 3 EVENTS	DEFORMATIONS
DEVONIAN	LATE	Phase V	Geanticline and much of geosyncline raised; southeastward overriding terminates geosyncline; (= late part of Ellesmerian Orogeny)	
DEVONIAN	MIDDLE	Phase IV	Geosyncline narrowed from both sides and fragmented by basement uplifts; geanticline intruded, fragmented and raised to provide molasse; (= early part of Ellesmerian Orogeny)	
DEVONIAN	EARLY	Phase IV		
SILURIAN	LATE	Phase III	Geosyncline narrowed slightly from NW with great infilling of Hazen Trough by flysch deposits; starved axial basin migrates southeast and becomes narrower; carbonates encroach widely onto Arctic Platform, with some onto geanticline.	
SILURIAN	MIDDLE	Phase III		
SILURIAN	EARLY	Phase III		
ORDO-VICIAN	LATE	Phase II	Geosyncline broad; receives volcanics from geanticline; starved axial basin; carbonates spread onto platform with hinge-line indistinct.	
ORDO-VICIAN	MIDDLE	Phase II		
ORDO-VICIAN	EARLY	Phase II		
CAMBRIAN	LATE	Phase I	Geosyncline began narrow; progressively widened by outward encroachment onto platform and geanticline.	
CAMBRIAN	MIDDLE	Phase I		
CAMBRIAN	EARLY	Phase I		

Fig. 5.   Phases in the history of Sequence 3 of the Innuitian Mobile Belt. Setting of this sequence is shown in Figs. 3 and 4; its evolution is shown in Figs. 7 and 9. Sequence 3 includes deposits of the Arctic Platform, Franklinian Geosyncline, and those on the Pearya Geanticline.

1968*a*, 1976*a*). The youngest of the original three is now itself divided into three phases, so that five are now recognized.

The stratigraphic section comprising Sequence 3 ranges in age from late Proterozoic to late late Devonian. It overlies older Proterozoic rocks unconformably, and is truncated at the top by a major unconformity related to the Ellesmerian Orogeny (Fig. 4). The sequence is not interrupted internally by major orogenic episodes, although some minor angular unconformities are present within it.

Sequence 3 was the most extensive sequence of the Innuitian Mobile Belt (Fig. 3). It formed an extensive, continuous depositional basin that at times covered the entire region (Fig. 1). The geosyncline in the Canadian Arctic first began to take shape in late Proterozoic or early Cambrian time. It began as a narrow basin and became progressively wider and deeper for more than 200 million years. It reached its maximum width and depth in late Ordovician to Early Silurian time. From then until its termination in late Devonian time, the geosyncline became progressively narrower and shallower. This resulted because uplifts adjacent to and within the geosyncline became active and filled it with sediments. Deposition ended in the geosyncline when the late Devonian to early Mississippian Ellesmerian Orogeny deformed and uplifted it.

The Franklinian Geosyncline was a wide basin with a thick sedimentary column. It was bounded on the southeast by the Arctic Platform, a stable area where a thin succession of rocks was deposited from time to time on the Precambrian Shield. Northwest of the geosyncline was the Pearya Geanticline, a linear positive feature with a crystalline Precambrian core (Fig. 2). At times during Sequence 3 the geanticline was stable and provided a narrow shelf on the northwest side of the geosyncline. At other times it was an active sedimentary source providing sediments to the geosyncline.

For some years the entire Lower Paleozoic mobile belt was called the Franklinian Geosyncline, and within it two divisions were recognized (Thorsteinsson and Tozer, 1960, 1970; Kerr, 1967*a*; Trettin, 1969*a*). On the southeast was a belt called the miogeosyncline, containing mainly carbonates. It was separated from the Arctic Platform by a hinge line across which there is marked change in thickness but little change in lithology (Fig. 3). Those parts of the mobile belt north and west of the miogeosyncline were called the eugeosyncline, following the terminology of Kay (1951). Trettin *et al.* (1972) suggested changes in this terminology, and these are employed in this report (Figs. 2 and 3). The name miogeocline replaces miogeosyncline, but the feature is unchanged. Four belts exist within the area of the former eugeosyncline as follows. The Pearya Geanticline is the partly exposed metamorphic belt on the extreme northwest; since it was mainly an area of uplift it is no longer considered part of the geosyncline. The remaining part of the former eugeosyncline is called the clastic belt of the geosyncline. It includes three divisions; from northwest to southeast these are a coastal plain, a shelf, and the Hazen Trough.

2. *Metamorphism*

Metamorphism of the Innuitian Mobile Belt was related to the magmatic stage of the Pearya Geanticline (Fig. 4). It was restricted to early Paleozoic and earlier time and therefore affected the rocks of Sequence 3. Metamorphism affected only the Pearya Geanticline and nearby parts of the clastic belt on northernmost Ellesmere and Axel Heiberg Islands (Figs. 2 and 3). Exposed gneisses of northern Ellesmere Island exhibit metamorphism that falls into two and possibly three episodes (Sinha and Frisch, 1975, 1976).

Major regional metamorphism about 1000 m.y. ago and about 800 m.y. ago may or may not be the same event. Both are Proterozoic, and one or both may be the metamorphic event reported by Trettin *et al.* (1972) as Cambrian or earlier. There was a subsequent retrograde metamorphism of the Cape Columbia complex (Fig. 1) between 500 and 600 m.y. ago (Sinha and Frisch, 1976) that may relate to Middle Ordovician or earlier cataclastic deformation and metamorphism (Frisch, 1974; Trettin *et al.*, 1972). Metamorphism associated with late Silurian–early Devonian deformation is based on K–Ar ages of 403 ± 17 and 389 ± 21 m.y. in the Cape Columbia region (Frisch, 1974). According to Trettin *et al.* (1972) these dates may represent heating associated with an intrusive episode rather than regional metamorphism. Regional metamorphism associated with the Ellesmerian Orogeny was related to intrusion and was tentatively interpreted as Middle Devonian (Trettin *et al.*, 1972).

According to Trettin *et al.* (1972), the metamorphic grade on northern Ellesmere Island generally increases to the northwest, away from the axis of the Franklinian Geosyncline and towards the core of the Pearya Geanticline (Fig. 2). Metamorphic processes may have been more or less continuous in the core of the Pearya Geanticline, and the metamorphic phases known on land may represent intervals during which the geanticline expanded into northern Ellesmere Island.

3. *Igneous Intrusion*

Devonian and older intrusions of the Innuitian Mobile Belt are restricted to northern Axel Heiberg and Ellesmere Islands, and were described by Trettin (1971*b*) and Trettin *et al.* (1972). They are restricted to the Pearya Geanticline (Figs. 3 and 6) and were emplaced during its magmatic stage (Fig. 4). A mafic–ultramafic body near M'Clintock Inlet (Fig. 1) was dated as 376 ± 16

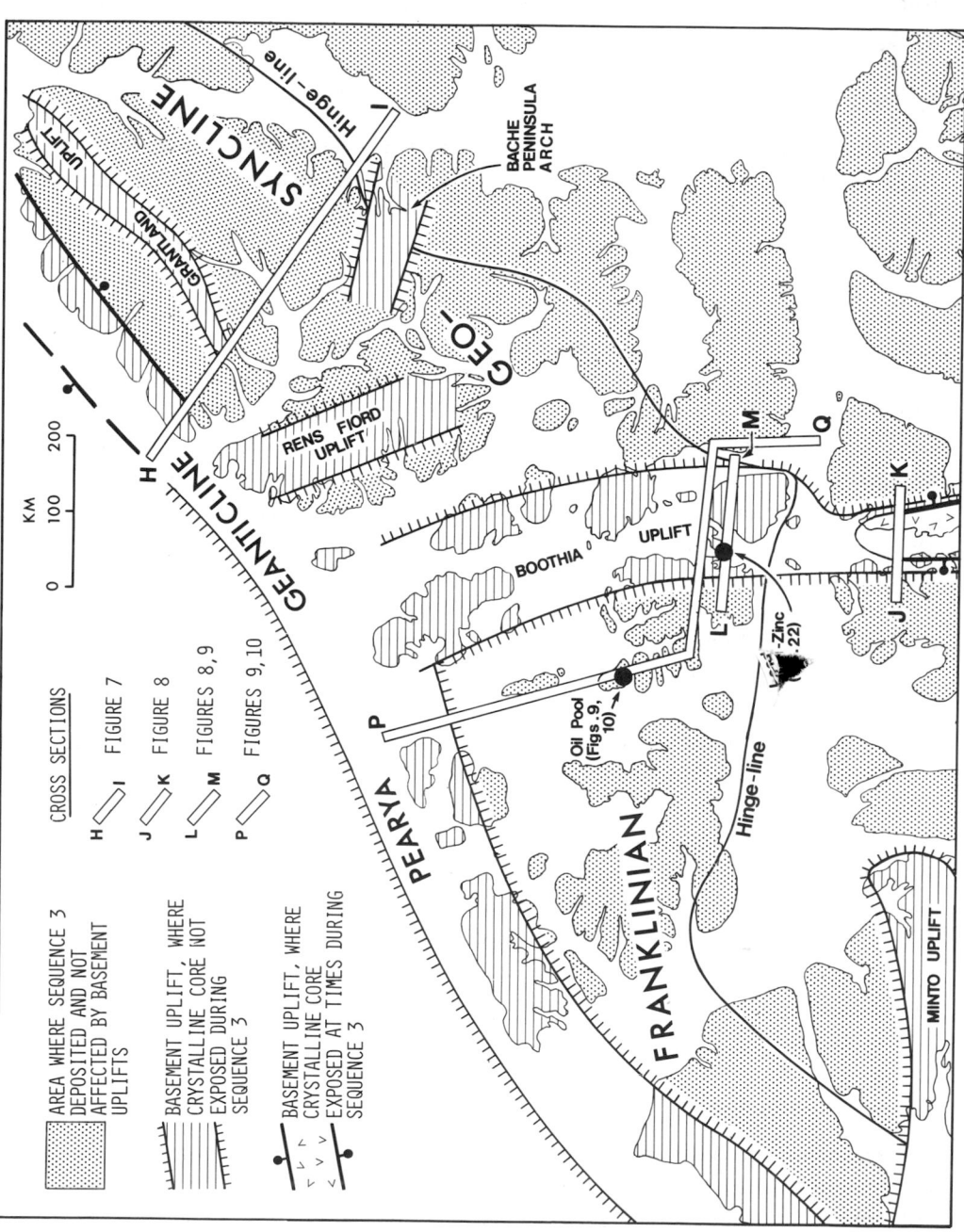

Fig. 6. Lower to middle Paleozoic tectonic elements of the Canadian Arctic. The basement highs shown affected Sequence 3, which

m.y., suggesting a late early Devonian event. Lower Paleozoic mafic dikes and sills in northwestern Ellesmere Island are post-Silurian, pre-Carboniferous, and probably were emplaced during an episode of crustal extension following the Ellesmerian Orogeny. Mafic sills and/or dikes on extreme northwestern Ellesmere Island may be related to Lower Devonian basic volcanism of northern Axel Heiberg Island.

Granitic plutons in northern Axel Heiberg and northern Ellesmere Islands (Trettin *et al.*, 1972) apparently are related to early and mid-Paleozoic orogenic activity. Some have been dated at 390 ± 20 m.y., or early Devonian (Frisch, 1974). A number of plutons were considered by Trettin *et al.* (1972) to be a genetic group of three assemblages. Others yield K-Ar age determinations of 345 ± 15 and 325 ± 14 m.y., and lie near the Devonian–Mississippian boundary (Trettin, 1971*b*). These are minimum values and so suggest intrusion in Devonian time, presumably during the folding and regional metamorphism of the Ellesmerian Orogeny (Fig. 4). Granitic intrusions on northern Axel Heiberg Island are post-kinematic (Trettin, 1969*a*, 1971*b*) and are younger than Lower or Middle Silurian strata, which they intrude (Trettin *et al.*, 1972). Radiogenic ages vary from 325 ± 25 to 360 ± 25 m.y. and probably indicate contemporaneity with the Ellesmerian Orogeny.

4. *Volcanism*

Lower Paleozoic volcanic rocks occur in a narrow belt along the present Arctic continental margin on Ellesmere and Axel Heiberg Islands, and there are three main ages (Trettin *et al.*, 1972; Trettin, 1976). Middle Ordovician volcanics are most widespread and occur as part of a volcanic island arc. Silurian volcanics are partly pyroclastic and brecciated, and they appear to be related to deformations that produced the Rens Fiord Uplift (Fig. 6). Lower Devonian (and/or younger) volcanism occurred on northern Axel Heiberg Island and probably was related to crustal extension that affected the Rens Fiord Uplift and produced the Svartevaeg Trough to the east of it.

Volcanic activity originated at different times and localities, and probably is mainly hidden beneath the continental shelf. Most volcanics are intermediate to somewhat silicic, whereas basic rocks are subordinate, and rhyolite is scarce or absent.

E. Sequence 3: Simple Evolution

Sequence 3 of the Innuitian Mobile Belt (Fig. 3) was dominated by the Pearya Geanticline, a major tectonic high that trends northeastward along the Arctic continental margin (Fig. 6). The sequence was also influenced by several smaller internal basement uplifts. A cross section through Ellesmere Island

(Fig. 7) shows the general or simple evolution of the mobile belt during deposition of Sequence 3. This is beyond the influence of any cross-trending internal basement uplifts that obstructed the history of the sequence (Fig. 6) and therefore is the simplest history of Sequence 3 that can be depicted. During the span of Sequence 3 the Franklinian Geosyncline formed, evolved, and terminated, with five phases being recognized in its history (Fig. 5).

1. *Phase I: Late Proterozoic (Hadrynian?) to Mid-Late Cambrian*

Phase I of the Franklinian Geosyncline (Figs. 5 and 7) shows a progressive pattern of sedimentary development (Kerr, 1967a). A thick column of Upper Proterozoic (Hadrynian?) mainly clastic sediments is present in the axis of the Franklinian Geosyncline of Ellesmere Island. These rocks are confined to the axis, suggesting that the geosyncline began as a narrow downwarp in Proterozoic time. Thickness patterns suggest that the Franklinian Geosyncline had taken on a linear form by late Proterozoic time, but was still very narrow. The geosyncline, however, had its characteristic broad but linear form by early early Cambrian time (Fig. 7), when great subsidence occurred and thick sediments accumulated. Thinner equivalent sedimentation occurred on the Arctic Platform to the southeast. The Pearya Geanticline may have been exposed in parts of Cambrian time, when it apparently was a sedimentary source, but this was followed by encroachment of the Grant Land Formation northwest into metamorphic basement (Trettin, 1976).

The Franklinian Geosyncline is interpreted to have formed initially as a downwarp on a continental crust (Fig. 3). A hinge line or flexure separated the geosyncline from the Arctic Platform and is located where the overall sedimentary thickness increases most abruptly. The hinge line was abrupt and rather narrowly defined during Phase I. It had a different location and nature during other phases of Sequence 3.

During Phase I the geosyncline had a general developmental pattern (Fig. 5). It may have begun initially in late Proterozoic time as a narrow downwarp of subaerially exposed continental crust. Then during the rest of Phase I it had a history of progressive widening by outward encroachment. By late early Cambrian time, rocks encroached outward onto the Arctic Platform at Bache Peninsula (Kerr, 1967a), and southeastern Devon Island (Kurtz et al., 1952). Clastic sediments were derived from both margins, the shield in the southeast and the exposed Pearya Geanticline in the northwest. Through time marine influence increased, as shown by the increasing volume of carbonate rock. By the end of Phase I in late Cambrian time, carbonate sedimentation dominated in the miogeocline and Arctic Platform, and may have been occurring throughout much or all of the southern Canadian Arctic. During Phase I in general, a thick sequence of rocks accumulated in the geosyncline and lesser thicknesses formed on the Pearya Geanticline and Arctic Platform.

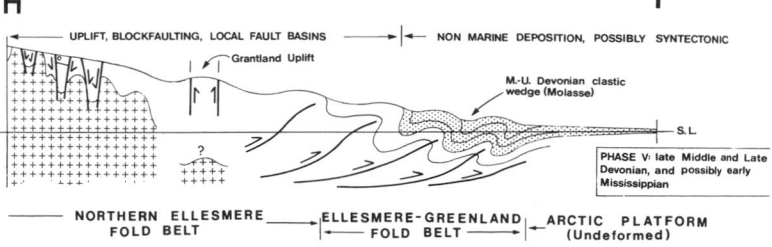

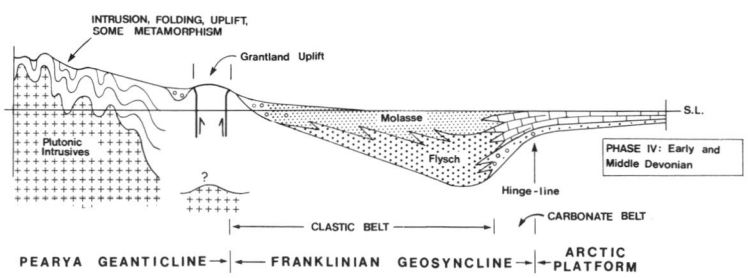

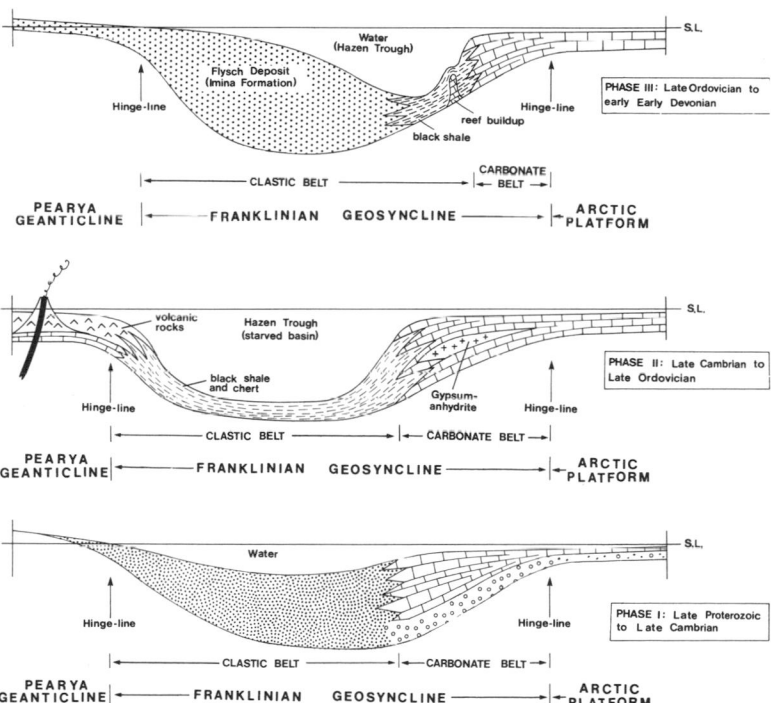

Fig. 7. Typical evolution of Sequence 3, showing the five phases in its development that are sum-marized also in Fig. 5. This is the simplest cross section of the geosyncline, located where that basin was not affected by internal cross-trending basement uplifts. Sequence 3 was part of the constructional phase of northern Canada (Fig. 4). Conventional lithologic symbols are used. The vertical scale is relative only, and varies from place to place. For location see Fig. 6.

In early Cambrian time central parts of the geosyncline were unstable tectonically as subsidence began and there was a great influx of clastic rocks from the southeast and the northwest. By late Cambrian time there was broader subsidence, with encroachment out of the geosyncline onto the former source areas (Fig. 7). Cambrian carbonates were deposited throughout the miogeocline and were spread as a sheet far to the south on the Arctic Platform including Boothia Peninsula (Miall and Kerr, 1977, 1980), northwestern Baffin Island (Trettin, 1975), and Victoria Island (Thorsteinsson and Tozer, 1962). Fine-grained clastic rocks that appear to be partly Upper Cambrian encroached unconformably onto metamorphic basement of the Pearya Geanticline (Trettin, 1976).

It was earlier thought that there was marine withdrawal and a disconformity at the end of Phase I (Kerr, 1967a). This unconformity is not present (Barnes, personal communication, 1975; Stuart Smith and Wennekers, 1977), so Phase I merged into Phase II during continuous sedimentation.

Phase I had a pattern of deepening and widening of the geosyncline, and it ended with a widespread marine incursion in which the hinge lines became less distinct. The progressive pattern of encroachment during Phase I is broadly similar to the pattern occurring in the Appalachian and Cordilleran miogeoclinal belts of North America.

2. *Phase II: Mid-Late Cambrian to Mid-Late Ordovician*

Phase II (Kerr, 1968a) was a long period of very widespread, continuous, and steady marine sedimentary accumulation in the Canadian Arctic (Figs. 5 and 7). A carbonate body that spans this phase occurs in the Franklinian Miogeocline and adjacent Arctic Platform. It extends from Victoria Island (Thorsteinsson and Tozer, 1962) to Somerset Island (Dixon, 1974; Miall and Kerr, 1977, 1980), northern Baffin Island (Trettin, 1975), eastern Ellesmere Island (Kerr, 1968a), and to northwest Greenland (Dawes, 1976). Sedimentation appears to have been continuous from Phase I to Phase II in the carbonate and clastic belt of the Franklinian Geosyncline. It may have been continuous also on the Pearya Geanticline, where Trettin (1976) reported fine-grained clastic rocks in the Cambro-Ordovician Grant Land Formation.

In Phase II the Franklinian Geosyncline reached its broadest extent, and marine deposition was most widespread. The hinge line was more poorly defined than at any other time (Fig. 7), although it had about the same location as earlier. The entire Canadian Arctic received marine sedimentation, at times, but subsidence and water depth were greatest in the geosyncline. The geosyncline was rather symmetrical during Phase II. The clastic belt of the geosyncline contained a starved axial basin, the Hazen Trough, with deep water and euxinic conditions, where black graptolitic shale and cherty limestone were deposited (Trettin *et al.*, 1972; Trettin 1973*a*, 1976, 1979). In parts of Ordovician time, sedimentation was about 2.5 to 4 times as slow in the Hazen Trough as it was in miogeoclinal parts of the geosyncline (Trettin, 1973*a*). Shallower water existed on both sides of the trough. On the southeast was a miogeocline, where a thick wedge of sediments was deposited (Kerr, 1968*a*). It is mainly carbonate, with three intervals of gypsum–anhydrite, one of which may be a sabhka deposit (Mossop, 1979). It contains only minor clastic rock.

The northwestern part of the clastic belt and the adjacent Pearya Geanticline contain a great variety of rock types and rapid facies changes (Trettin *et al.*, 1972; Trettin, 1973*a,b*), including clastic rocks, volcanic rocks, and carbonates. This is due to the intermittent activity of the Pearya Geanticline. At times during Phase II the geanticline and nearby parts of the clastic belt were stable and formed a shelf where carbonate rocks were deposited. At other times the geanticline was an active volcanic arc depositing thick volcanic rocks in the nearby clastic belt (Fig. 7).

3. *Phase III: Late Ordovician to Early Early Devonian*

Phase III (Figs. 5 and 7) is characterized by a large variety of rock types that show great variations in thickness and lithology. This phase was marked by a southeastward expansion of the Hazen Trough at the expense of the miogeocline. The carbonate belt was narrow, with thick sequences of rocks being deposited (Kerr, 1976*a*; Morrow and Kerr, 1977). It appears that the flexure between the Arctic Platform and the geosyncline on Ellesmere Island lay within the carbonate belt and was rather broad, with a gradual thickness change across it.

The Hazen Trough was the region of deepest subsidence during Phase III. In early parts of the phase, when water was very deep and sediment supply was low, graptolitic and shaly rocks were deposited there, the Cape Phillips Formation (Kerr, 1976*a*) and the partly equivalent upper member (cherty shale) of the Hazen Formation farther northwestward (Trettin, 1971*a*, 1973*a*, 1976). When the sediment supply increased because of uplift of the Pearya Geanticline, those rocks were succeeded by the Imina Formation, a flysch deposit that was dumped rapidly into the preexisting deep trough, and overlapped southeastward (Fig. 7).

In most places the shales of the Hazen Trough overlapped the carbonates of the miogeocline, indicating general and gradual expansion of the trough. In rare places the carbonate to shale facies change is very abrupt and occupied the same geographic location from late Ordovician to early Devonian time (Thorsteinsson, 1958; Thorsteinsson and Tozer, 1957; Thorsteinsson and Kerr, 1968; Kerr, 1976*a*).

A narrow reef occurred within the Hazen Trough, partly or completely isolated from the main carbonate shelf and enclosed by the Cape Phillips Shale. This reef extended at least intermittently and perhaps continuously from northern Melville Island (Tozer and Thorsteinsson, 1964) to Ellesmere Island (Kerr, 1976a) and to northern Greenland (Mayr, 1976a), a distance of about 1700 km.

The Imina Formation (Trettin, 1971a, 1973a, 1976) is a flyschlike deposit of turbidites that is as much as 2700 m thick (Fig. 7). It overlaps gradationally southeastward from the coastal plain and shelf to the Hazen Trough, so that its base is youngest in the miogeocline, where it grades in a complex way into equivalent formations. The flysch deposit largely filled up the trough that earlier had been a starved basin (Fig. 7). This flysch deposit was the precursor of the Ellesmerian Orogeny (Fig. 4).

The Hazen Trough, which is the axial zone of deep-water facies, had its maximum southeastward extent at the expense of the miogeocline during this phase, when shale deposition was widespread in the miogeocline. The carbonate belt of the geosyncline was very narrow.

Through most of Phase III the Pearya Geanticline was not strongly positive and a section at M'Clintock Inlet (Trettin, 1969b) shows the sequence. During an early part of this phase the geanticline was a carbonate shelf, in the middle of the phase it received flysch deposits, and in the latter part of the phase it again received carbonates. Lateral shifts of the northwestern margin of the Hazen Trough resulted from expansion and contraction of the Pearya Geanticline. The Imina Flysch was deposited on the geanticline when it contracted and carbonates were deposited there when it expanded. The Hazen Trough was deep during Phase III, with flysch deposits mainly in the northwest, and black shale farther southeast (Fig. 7). Subsidence probably exceeded sedimentation during an early part of Phase III, but later this trend reversed as uplift of the margins of the geosyncline began.

The closing part of Phase III was a narrowing of the geosyncline, an infilling, and a general shallowing that was influenced by the forthcoming Ellesmerian Orogeny (Fig. 7). The geosyncline was narrowed from the northwest by the gradually rising Pearya Geanticline, evidence of the raising being seen in the shallower facies in the late part of this phase in the clastic belt (Trettin et al., 1972; Trettin, 1976). Narrowing of the geosyncline from the southeast was due to an expansion and uplift of the Arctic Platform (Kerr, 1968b, 1977a). The end of Phase III is delineated by the unconformity produced by that uplifting event as the hinge line moved northwestward.

4. *Phase IV: Early Early Devonian to Middle Devonian*

During Phase IV there was a very different pattern of facies in the rocks of Sequence 3. This resulted from the developing Ellesmerian Orogeny, which gradually destroyed the geosyncline and terminated deposition of the sequence (Kerr, 1976a).

Deposition gradually became very restricted during this time of progressively more widespread emergence (Figs. 5 and 7), as the Franklinian Geosyncline was gradually filled in and the water-covered region was progressively reduced in size. This resulted largely from the Ellesmerian Orogeny, and associated increasing tectonic activity of the Pearya Geanticline (Trettin *et al.*, 1972; Trettin, 1978). Phase IV embraces an early part of the orogeny. At the same time that the Pearya Geanticline served as a source of sediments from the northwest, the Arctic Platform and Shield were also raised (Kerr, 1968b, 1977a) and became a source of sediments from the southeast.

A typical cross section of Phase IV from the Arctic Platform to the Hazen Trough (Fig. 7) involves rocks located far enough from linear cross-trending basement uplifts so as not to be affected by them.

Phase IV began when rapid uplift occurred at approximately the same time in the Pearya Geanticline and the Arctic Platform (Fig. 7), both of which contributed to narrowing the Franklinian Geosyncline. At the beginning of Phase IV, in early Devonian time, broad gentle uplift of the Arctic Platform caused it to expand generally. The zone of flexure between platform and geosyncline shifted basinward and remained there through Phase IV. As a result of this shift southern and eastern parts of the miogeocline were converted into Arctic Platform. This general expansion has been documented most clearly on central Ellesmere Island (Fig. 7), but probably affected the entire southeastern margin of the geosyncline (Kerr, 1968b). The base of the sequence in the southeast is an angular unconformity produced by uplift of the Arctic Platform, and that unconformity disappears northwestward within the geosyncline. The unconformity is succeeded on the platform and nearby parts of the miogeocline by synorogenic to postorogenic formations. These were deposited while uplift of the Arctic Platform occurred, and encroached onto the platform when uplift ceased (Fig. 7). These clastic rocks were succeeded by a narrow belt of thick carbonates in the miogeocline and equivalent thinner carbonates on the platform (Kerr, 1976a). The true starved basin of earlier times (Trettin *et al.*, 1972) had largely disappeared by uplift and infilling, and the geosyncline in this phase contains no true shales. Phase IV included marine and nonmarine rocks. It ended in the upper Middle Devonian with the beginning of deposition of a clastic wedge of largely nonmarine rocks including the Okse Bay and equivalent Formations (Kerr, 1974, 1976a; Embry and Klovan, 1976).

In Phase IV there was a marked increase in the uplift of the Pearya Geanticline, as it also broadened and became a much more active sedimentary source than in earlier phases. Throughout Phase IV the Ellesmerian Orogeny progressed and accelerated. This orogeny originated in and emanated from the

Pearya Geanticline. In this phase the geanticline continued to be a source of sediments to the geosyncline adjacent to it. In earlier parts of this phase a narrow Hazen Trough existed and received flysch deposits (Fig. 7). In later parts of Phase IV, as shallowing continued, those rocks were succeeded by molasse-type rocks. Northwestern parts of the clastic belt that earlier had been filled by flysch deposits may at this time have been raised and eroded, to be redeposited in the narrower basin to the southeast.

Plutonic intrusives were emplaced in the Pearya Geanticline and nearby clastic belt in early Devonian time (Trettin *et al.*, 1972; Frisch, 1974), and apparently produced the uplifting. There seems to have been an acceleration of uplift of the geanticline during this phase. It culminated with broadening that included uplift and erosion of nearby parts of the clastic belt of the geosyncline (Fig. 7).

A fragmentation of the geanticline and clastic belt apparently occurred during Phase IV as a result of the intrusion and uplift. At least one deeply subsiding basin developed, the Svartavaeg Trough (Trettin, 1969*a*) in which very thick Lower to Middle Devonian clastic and volcanic rocks were deposited.

The Grantland Uplift, which occurs within the clastic belt and trends along it, had more than 3 km of Ordovician and Silurian strata removed from it before being overlain by the Sverdrup Basin succession (Trettin *et al.*, 1972). The uplifting of that structure probably was contemporaneous with and a part of the Ellesmerian Orogeny, and may have been initiated in early Devonian time (Fig. 7). Because of its northeastward trend, the Grantland Uplift did not interfere with the general deformation of the geosyncline that was produced by the Ellesmerian Orogeny, but instead appears to have been a smoothly integrated part of that deformation. There may have been deposition in basins associated with the Grantland Uplift (Fig. 7), but the resulting rocks have subsequently been eroded.

5. *Phase V: Late Middle and Late Devonian, and Possibly Early Mississippian*

Phase V of Sequence 3 includes the last events in the history of the Franklinian Geosyncline (Fig. 7). It was intimately controlled by and coeval with the culmination of the Ellesmerian Orogeny.

The marine and partly nonmarine rocks of the earlier phase were succeeded by an upper Middle and Upper Devonian clastic wedge (Kerr, 1974, 1976*a*; Embry and Klovan, 1976), mainly nonmarine clastic rocks. These rocks were deposited in a narrow basin occupying the southeastern side of the miogeocline and nearby parts of the Arctic Platform. They were deposited during a time of high relief, and engulfed and covered the earlier marine rocks.

Beyond the influence of cross-trending linear basement uplifts (Fig. 7), Phase V has a simple and progressive history. The Pearya Geanticline, which had begun its rapid rise during Phase IV, continued rising during Phase V, and broadened farther. Fine grained clastic rocks that were derived from it were spread progressively farther southward and eastward. Because of the general uplift they were largely nonmarine. They are now preserved widely in the region of the former carbonate belt. Deposition of these sediments may have been partly contemporaneous with deformation of that belt. The huge clastic wedge of Middle to Upper Devonian rocks that covered the Franklinian Geosyncline and adjacent parts of the Arctic Platform constitutes Phase V. This writer considers that the source was largely the Pearya Geanticline and the exposed and eroded northwestern parts of the clastic belt of the geosyncline itself. Embry and Klovan (1976) suggest that the source was mainly uplifted regions of northeastern and eastern Greenland.

It appears that those parts of the Franklinian Geosyncline that were not near internal basement uplifts had a very similar history during Phase V. This applied along the full length of the geosyncline, from Ellesmere Island (Fig. 7), to Cornwallis Island (Thorsteinsson and Kerr, 1968; Thorsteinsson, 1973), to Bathurst Island (Kerr, 1974), and on to Banks Island (Miall, 1976a,b).

F. Pearya Geanticline

The Pearya Geanticline (Figs. 2, 6, and 7) is a linear tectonic high that trends northeastward along the Canadian Arctic continental margin. It is exposed today only on northwest Ellesmere Island and northern Axel Heiberg Island. The term Pearya Geanticline was introduced originally by Schuchert (1923), for an exposed borderland of crystalline rocks that he presumed to be of Precambrian age. The term fell into disuse when Thorsteinsson and Tozer (1960) applied the term eugeosyncline to a belt that included the crystalline rocks as well as the nearby Lower Paleozoic clastics and volcanics. At that time the crystalline rocks were interpreted as Lower Paleozoic intrusions. The term Pearya Geanticline was reintroduced by Trettin et al. (1972), who recognized that the tectonic belt that included northwestern Ellesmere Island and Axel Heiberg Island had been mainly positive, although it had been the site of sedimentation at times. It is now known that there is an exposed core of gneissic Precambrian rocks on northern Ellesmere Island that was metamorphosed about 800 to 1000 m.y. ago, and therefore is of about Grenvillian age (Sinha and Frisch, 1975, 1976). This supports the use of the term geanticline.

The term Pearya Geanticline now applies to a broad mainly positive belt that includes a Precambrian crystalline core, the mainly clastic Lower Paleozoic rocks that were deposited upon it, and the Lower Paleozoic plutonic rocks that were intruded into it. The Precambrian age of the core suggests that the geanticline and the shield are connected beneath the Franklinian Geosyncline (Fig. 3).

The Pearya Geanticline and the Franklinian Geosyncline are difficult to delineate precisely. Their histories were so long and complex that their mutual boundary varied in location with time. The geanticline (Fig. 6) was mainly a positive feature and source area, but at times it received thin deposition and at those times was part of the clastic belt of the geosyncline (Fig. 2). Northwestern parts of the geosyncline also behaved from time to time as a source area. Thus the boundary between the geanticline and geosyncline is a broad and transitional zone between a source area and a deep basin (Fig. 7).

The Pearya Geanticline had a magmatic stage that spanned Sequence 3 of the Innuitian Mobile Belt (Fig. 4). Its main period of tectonic activity was the Ellesmerian Orogeny, which spanned Phases IV and V of Sequence 3 (Figs. 5 and 7). The formation of the geanticline may have been largely the result of plutonic intrusives. The geanticline was the main source of the clastic sediments that filled the Franklinian Geosyncline and spread onto the adjacent Arctic Platform. During periods of gentle expansion or uplift of the Pearya Geanticline shallow water rocks of the clastic belt of the Franklinian Geosyncline were deposited on it. In the periods of more extreme uplift the geanticline was eroded, and probably metamorphosed or intruded. In periods of contraction or subsidence increasingly deeper-water sediments encroached onto the geanticline.

The Pearya Geanticline was a moderate source of clastic and volcanic sediments during Phases I and II of the Franklinian Geosyncline (Sequence 3), and had carbonate, clastic, and volcanic rocks deposited on it from time to time (Figs. 5 and 7). In Phase II it was positive and volcanic while the adjacent Hazen Trough was deeply negative. A metamorphic event in the Pearya Geanticline is recorded by radiometric dates between 465 ± 19 and 389 ± 21 m.y., falling in an early Ordovician to early Devonian range (Frisch, 1974). Stratigraphic evidence suggests that this metamorphism was before the Middle Ordovician, because unmetamorphosed Middle Ordovician rocks are unconformable on the metamorphosed Cape Columbia Complex (Trettin et al., 1972). This metamorphic event therefore preceded the Ellesmerian Orogeny.

In Phase III the Pearya Geanticline became a substantial source of flysch sediments that began to fill the Hazen Trough rapidly. Periodic uplift was prevalent during this phase, but there were also times when it received sediments. The youngest widespread sedimentary unit known in the Pearya Geanticline in the region of the typical cross section (Fig. 7) is the Silurian Marvin Formation. Deposition of this formation occurred in the upper part of Phase III, and preceded the Ellesmerian Orogeny. In early Devonian time, at the beginning of Phase IV, the Pearya Geanticline began to be uplifted more rapidly and was intruded by plutonic rocks (Fig. 7). This began the Ellesmerian Orogeny, which originated in and emanated from the Pearya Geanticline.

The oldest plutonic rocks that were intruded into the Pearya Geanticline or adjacent parts of the clastic belt of the geosyncline and that are attributed to the Ellesmerian Orogeny have yielded dates of 390 ± 20 and 390 ± 18 m.y. They are of early Devonian age and were emplaced during Phase IV. They occur in two places and were associated with forceful emplacement (Frisch, 1974). The youngest plutonic rocks attributed to the Ellesmerian Orogeny may have been emplaced in the Pearya Geanticline in an early part of Phase V, in about late Middle Devonian time (Fig. 7). This includes a small high-level granitic pluton that yielded an age of 360 ± 25 m.y. (late Middle Devonian), with confidence limits ranging from late early Devonian to early Mississippian. It was considered to have cooled quickly and the age is approximately the age of emplacement. Plutons with younger radiogenic ages are known, but they were considered to reflect slow cooling or postintrusive events (Trettin *et al.*, 1972). From early through late Devonian time, spanning Phases IV and V (Fig. 7), the Pearya Geanticline was affected by the Ellesmerian Orogeny. It was uplifted, and most of it probably was exposed and subjected to erosion. The geanticline provided an influx of clastic sediments to the geosyncline. Fragmentation of the geanticline and clastic belt also occurred during the orogeny. This fragmentation produced the large northwest trending Svartevaeg Trough in northern Axel Heiberg Island, where thick Lower Devonian clastic and volcanic rocks were deposited (Trettin, 1969a). Folding and uplift of the geanticline probably continued through Middle Devonian time, but had been largely completed by late Devonian time, probably by Frasnian time. A unit in the Yelverton area of northern Ellesmere Island that is only slightly deformed (Mayr, 1976b) has yielded fossils of Middle or Upper Devonian age (late Givetian or Frasnian; D. C. McGregor, personal communication, 1978).

The metamorphic grade of the exposed gneissic core of the Pearya Geanticline indicates that between 8 and 30 km of uplift and erosion occurred there prior to deposition of Upper Paleozoic strata of the Sverdrup Basin (Trettin *et al.*, 1972). There probably was more than one phase of uplift of the geanticline. An important uplift occurred in early Devonain time in an early part of the Ellesmerian Orogeny, coinciding with Phase IV of the geosyncline, when plutonic intrusion and fragmentation occurred. Another important uplift probably occurred in the late Middle and Upper Devonian time, coinciding with the later part of the Ellesmerian Orogeny, during Phase V of the geosyncline when widespread clastic sedimentation occurred (Fig. 7).

G. Ellesmerian Orogeny, Unobstructed Pattern

The Ellesmerian Orogeny (Fig. 4) was a tectonic event that intruded and deformed the rocks of Sequence 3 of the Innuitian Mobile Belt. It terminated and destroyed the Franklinian Geosyncline and brought the deposition of Sequence 3 to a close. The Ellesmerian Orogeny overlapped in time with

Phases IV and V of Sequence 3. It originated and spread from the Pearya Geanticline. Its ultimate cause was the plutonic intrusion and uplift of the geanticline. The Ellesmerian Orogeny produced relatively simple effects where it was not obstructed by cross-trending basement uplifts (Fig. 7).

There were precursors of the Ellesmerian Orogeny, which were the development of a foredeep in the geosyncline during Phase II (Fig. 7) and the subsequent partial filling of the foredeep by the flysch deposits during Phase III. The Ellesmerian Orogeny *sensu stricto*, and as here considered, began in early Devonian time, when substantial uplift, intrusion, and fragmentation began in the geanticline.

In an early stage of the Ellesmerian Orogeny in early Devonian time, during Phase IV of Sequence 3, the Pearya Geanticline was raised substantially and began to be fragmented and intruded by igneous rocks. By late mid- to late Devonian time the geanticline had been raised much higher, and parts of the clastic belt of the geosyncline were raised with it. The marine basin that formerly occupied the carbonate belt or miogeocline was eliminated by this broad uplift. The slope produced was sufficient so that parts of the geosynclinal column were transported away from the geanticline. Folds and thrust faults developed within the moving mass, presumably driven by gravitiational forces. This resulted in southeasterly overriding in structures within the Franklinian Geosyncline on Ellesmere Island (Thorsteinsson and Kerr, 1972a,b; Kerr, 1967b, 1973b,c) and southerly overriding on Bathurst Island (Kerr, 1974). These involved décollement within the Proterozoic rocks of Phase I, as well as within evaporite units of Phase II.

The Ellesmerian Orogeny involved great uplift and fragmentation in the Pearya Geanticline during Sequence 3 (Fig. 6). At this time the Rens Fiord Uplift also was raised, and Lower Devonian rocks were deposited east of it in the Svartevaeg Trough of northern Axel Heiberg Island (Trettin, 1969a). The great folding event of the Ellesmerian Orogeny occurred somewhat later. Folding was largely completed in the Pearya Geanticline and nearby parts of the clastic belt of Ellesmere Island by late Devonian, probably by Frasnian time. Lesser deformation continued on northern Ellesmere Island, but ended before deposition of the mid-Mississippian (Visean) Emma Fiord Formation (Thorsteinsson and Tozer, 1970; Trettin, 1973b).

The Ellesmerian Orogeny was the final event or culmination that ended the Franklinian Geosyncline. It produced overriding that affected the entire geosyncline, but died out rapidly southeastward within the Arctic Platform (Fig. 7).

It appears that deformation by the Ellesmerian Orogeny did not begin in the miogeocline until later than in the geanticline and the clastic belt. There may have been some folding contemporaneous with the Middle to Upper Devonian clastic wedge of the miogeocline. The intense folding of the miogeocline, however, involved the clastic wedge in the deformation (Kerr, 1974, 1976a; Embry and Klovan, 1976), and rocks in that wedge are as young as latest Devonian (Famennian). Deformation of the miogeocline had been completed prior to deposition of the mid-Mississippian Emma Fiord Formation.

The Ellesmerian Orogeny deformed the entire Franklinian Geosyncline and Pearya Geanticline, and its general effects are shown in a simple cross section (Fig. 7). The Ellesmere–Greenland Fold Belt was formed at this time and extended from eastern Grinnell Peninsula through Ellesmere Island to northern Greenland. Folding was most intense in the northwest in and near the geanticline, and diminished in intensity to the southeast. The age and extent of the Ellesmerian Orogeny are depicted symbolically in a tectonic time diagram (Fig. 4).

At the end of the Ellesmerian Orogeny in latest Devonian to early Mississippian time the entire Canadian Arctic Archipelago region, which now includes islands and channels (Fig. 6), may have been exposed and undergoing erosion. The known extent of the geosyncline and geanticline are shown in Fig. 2. The region now covered by the Sverdrup Basin probably also was part of those deformed features (Trettin et al., 1972; Meneley et al., 1975; Balkwill, 1978). The Franklinian Geosyncline continued at least as far southwest as Banks Island (Miall, 1976b). Those parts of the Pearya Geanticline that now lie under water northwest of the Arctic Archipelago presumably were exposed after the Ellesmerian Orogeny as well. It is probable also that the region that now forms the Canada Basin of the Arctic Ocean was a land area at that time. If so then the Franklinian Geosyncline and Pearya Geanticline made up a linear mobile belt lying within a large continental area. The Sverdrup Basin and presumably the Canada Basin subsequently developed on this eroded land-scape on a widespread angular unconformity.

H. Basement Uplifts Affecting Sequence 3

1. *Foreword*

Sequence 3 had a rather simple and straightforward evolutionary development in parts not affected by cross-trending internal basement uplifts (Fig. 7). The complete development of Sequence 3, however, was much more complicated than that, because of basement uplifts that extended across the basin (Fig. 6), and interfered with its development. Internal basement uplifts that strike across the northeastward depositional trend of the Pearya Geanticline and Franklinian Geosyncline caused Sequence 3 to have a much more complicated history.

Basement uplifts complicated the history of Sequence 3 in two ways. (1) They had a history of periodic uplifts that locally interrupted depositon of

Sequence 3 both in the geosyncline and platform, and caused unconformities as well as affecting facies and thickness patterns. (2) The uplifts were large basement cored bodies or plugs that had been forced upward or intruded into Sequence 3. The basement cores were more rigid than the surrounding sedimentary rocks and created a local anisotropy. When the rocks of Sequence 3 were subsequenctly deformed by the Ellesmerian Orogeny, the structures that formed within that sequence in the region of the basement uplifts were drastically different from the structures elsewhere, because of that anisotropy. Thus the structure of the uplifts interfered with and partly controlled the deformation of nearby parts of the geosyncline during the Ellesmerian Orogeny. The nature of the basement uplifts will be described below (Figs. 6 and 8), and subsequently their effects on the evolution of Sequence 3 will be discussed (Figs. 9 and 10).

There are four major basement uplifts in the Canadian Arctic Islands in addition to the Pearya Geanticline (Fig. 6). Three of these affected the Franklinian Geosyncline. A fourth, the Minto Arch, affected the Arctic Platform, and may also have affected the geosyncline at its western end.

2. *Boothia Uplift*

The Boothia Uplift (Figs. 6, 8, and 9) is a north-trending north-plunging structure at least 650 km long that had a number of discrete pulses of vertical uplift (Kerr, 1977a; Miall and Gibling, 1978). A cross section of the Boothia Uplift (Fig. 8) shows that it formed by vertical upward movement, by basement faulting, of a block of the Precambrian Shield called the Boothia Horst. The raising of this crystalline horst carried the overlying sediments with it, and deformed them to form the Cornwallis Fold Belt, a broad anticlinorium controlled by the basement faulting. The horst rose by means of numerous faults that trend northward and are concentrated near the margins of the uplift. In the deepest structural levels that can be observed, these faults are vertical. At higher elevations, or higher structural levels, the faults splay outward to become high-angle reverse faults, or steep thrusts. At still higher levels they die out to become overturned folds, which in turn merge into asymmetric folds. Thus the amount of vertical displacement on each zone diminished and died out upward.

The Boothia Uplift is mushroom shaped in cross section (Figs. 8 and 9) because the faults and folds splay out at higher elevations. This presumably was due to rock expansion that resulted from unloading by erosion and the consequent reduction of geostatic pressure. Each level within the raised uplift was brought to a higher elevation, and its confining pressure was reduced by the erosion that occurred above it, hence the uplift expanded.

The Boothia Uplift had one or more pulses of uplift in late Proterozoic time, when parts of Sequence 2 were removed and remnants of that sequence

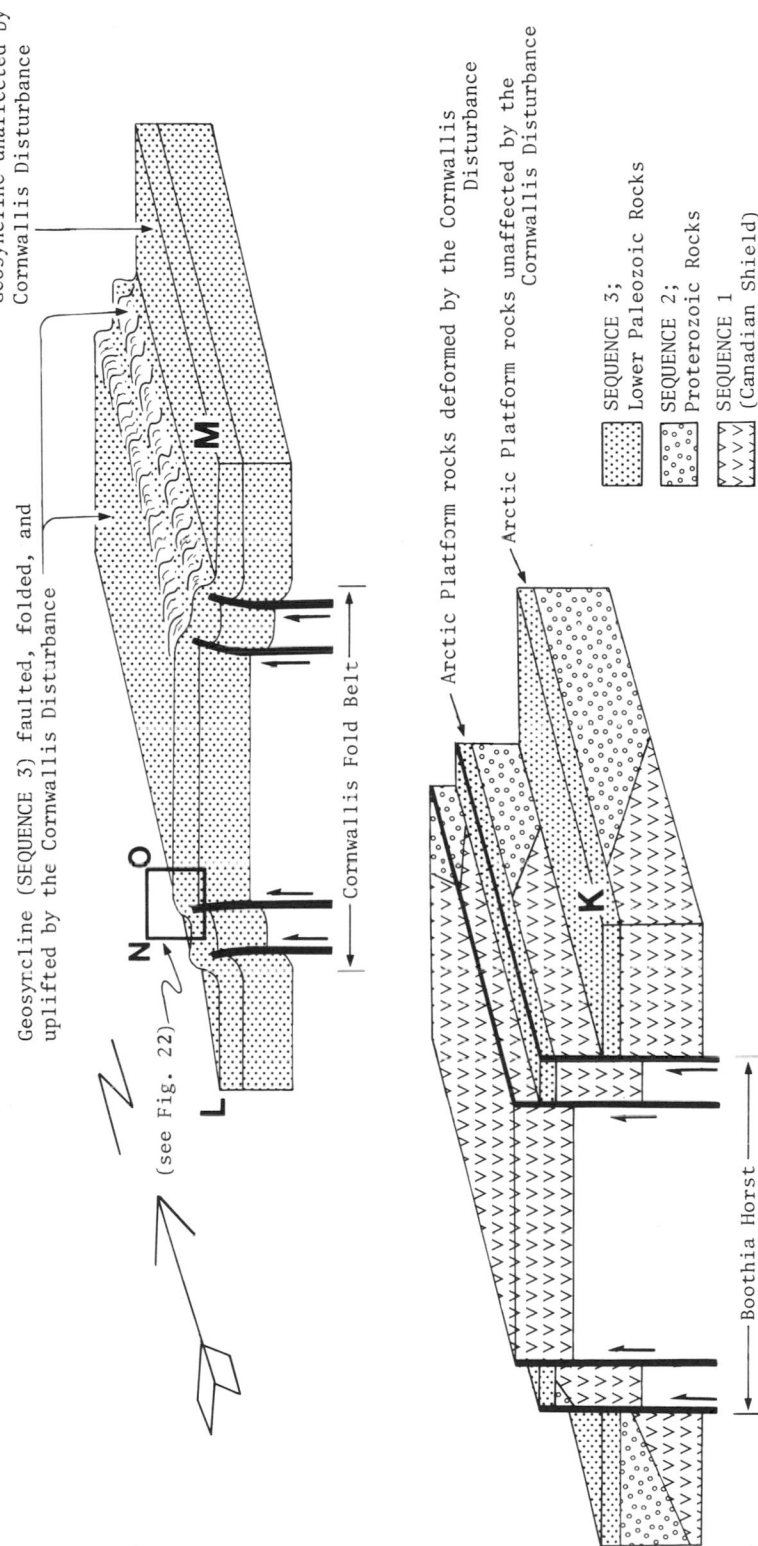

Geosyncline unaffected by Cornwallis Disturbance

Geosyncline (SEQUENCE 3) faulted, folded, and uplifted by the Cornwallis Disturbance

M

O

N

L

(see Fig. 22)

Cornwallis Fold Belt

Arctic Platform rocks deformed by the Cornwallis Disturbance

Arctic Platform rocks unaffected by the Cornwallis Disturbance

K

J

Boothia Horst

SEQUENCE 3;
Lower Paleozoic Rocks

SEQUENCE 2;
Proterozoic Rocks

SEQUENCE 1
(Canadian Shield)

Fig. 8. Formation of the Boothia Uplift by faulting and drape folding of the Cornwallis Disturbance. This shows the basic structure of the Boothia Uplift, which was the structure existing at the end of its formation, and before its modification (Kerr, 1977a). For locations of cross sections see Figs. 6 and 10. The Boothia Uplift includes the Boothia Horst and the Cornwallis Foldbelt.

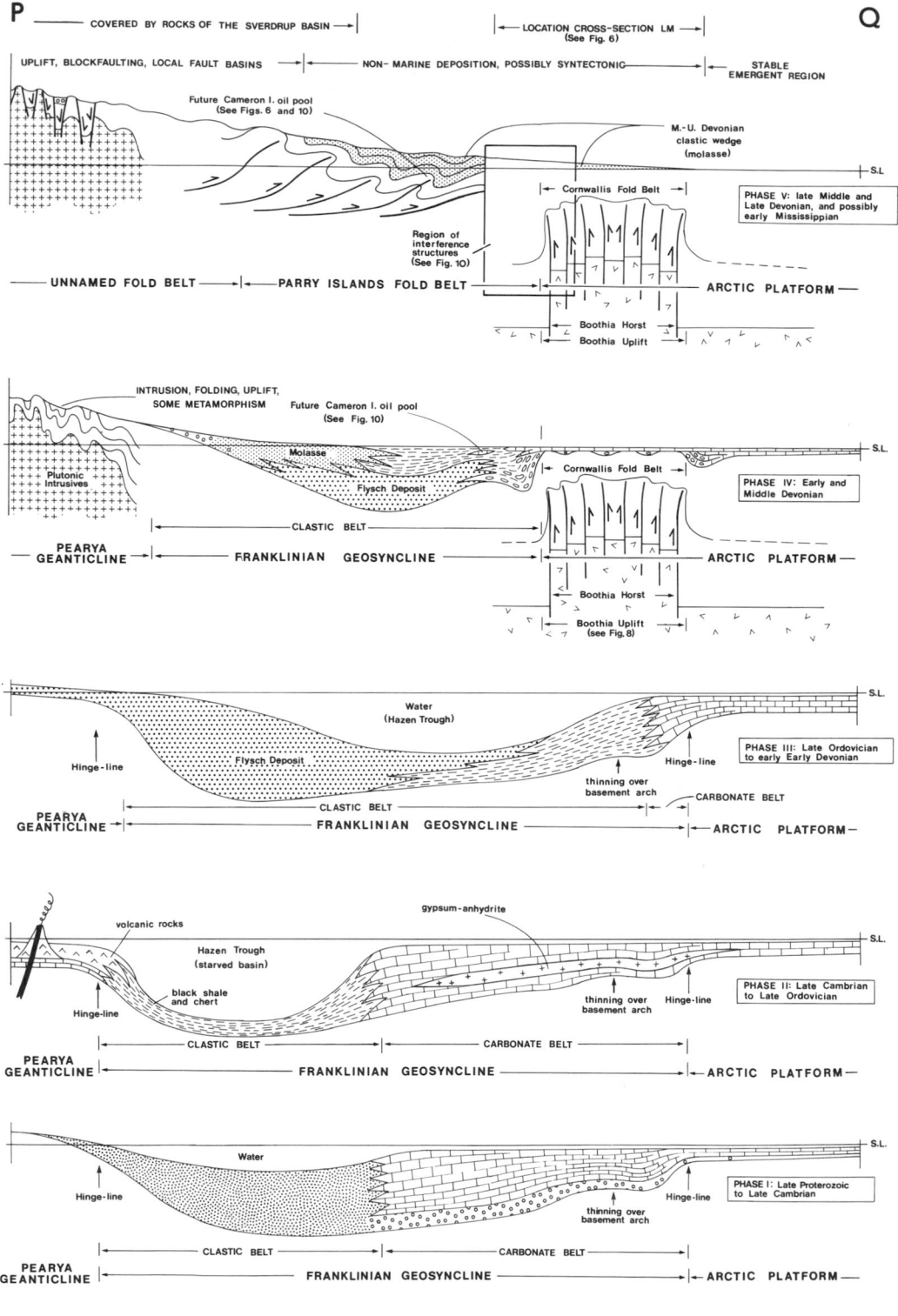

were left on either side. It was mildly positive in Cambrian to Silurian time during Phases I to III of Sequence 3 (Fig. 9). The main uplift, however, was the Cornwallis Disturbance in Silurian and Devonian time (Fig. 4), which included four pulses of uplift and erosion totaling at least 5300 m (Kerr, 1977a). In each pulse of the Cornwallis Disturbance the Boothia Uplift was raised and eroded, while deposition continued on either side.

The Boothia Uplift trends northward (Fig. 6), and this probably was largely controlled by the original northward trends of the gneisses in the shield that form its crystalline core. It also plunges northward so that the crystalline core is exposed in the south, and the folded sedimentary cover is exposed in the north (Fig. 8). The Boothia Uplift extends across the Arctic Platform and the exposed parts of the Franklinian Geosyncline. It probably extended completely across the geosyncline to connect with the Pearya Geanticline; however, this suggested northern extension is now covered by the Sverdrup Basin (Fig. 6).

The Cornwallis Disturbance occurred before the Ellesmerian Orogeny and was a deformation of much lesser intensity. The situation that existed at the end of the Cornwallis Disturbance is shown in a simplified way in Fig. 8, and is called the basic structure of the Boothia Uplift (Kerr, 1977a). This structure formed mainly in Devonian time during Phase IV of Sequence 3 (Fig. 5), while the Ellesmerian Orogeny was affecting the clastic belt, but before it affected the carbonate belt. The basic structure of the Boothia Uplift (Fig. 8) existed before the geosyncline was deformed by the Ellesmerian Orogeny (Fig. 4), so that it was modified by the orogeny.

3. Bache Peninsula Arch

The smallest and least active basement uplift in the Canadian Arctic Islands in is the Bache Peninsula Arch of Ellesmere Island (Fig. 6). It extends westward from the Arctic Platform across the miogeocline, and farther west is covered by the Sverdrup Basin.

This arch was first active in Ordovician and Silurian time (Kerr, 1968a), when it was a mildly positive swelling or uparching that subsided less than flanking basins. Its main activity was in early early Devonian time, when it was raised and eroded to become the source of red clastic sediments. At this time the part of the arch lying within the geosyncline was converted to part of

←

Fig. 9.   Atypical evolution of Sequence 3 showing the five phases in its development (cf. Fig. 7). This cross section of the geosyncline was obstructed by a cross-trending basement uplift (see Fig. 6). In Phases I, II, III, and IV, the Boothia Uplift influenced facies patterns. In Phase V the existence of the uplift as a structure interfered with the deformation produced by the Ellesmerian Orogeny (see Fig. 10). Conventional lithologic symbols are used. The vertical scale is relative only and varies from place to place. Where covered by the Sverdrup Basin the cross sections are speculative. There they are similar to corresponding parts of Fig. 7, and are compatible with available evidence.

the Arctic Platform. From then on the arch behaved as part of the Arctic Platform, with only thin sediments accumulating there.

Raising of the Bache Peninsula Arch did not produce any structural deformation that can now be observed at the surface. The extent of the arch has been determined by thickness and facies variations, and by the extent of the angular unconformity it produced. The mechanism of formation may have been similar to that of the Boothia Uplift (Fig. 8), but much weaker. It may have a mildly upfaulted basement core, with deformation dying out upward in the sedimentary column. Bache Peninsula Arch did not establish structural trends that were strong enough to influence later structural deformation now visible at the surface; however, such deformation may have occurred at depth.

4. *Rens Fiord Uplift*

The Rens Fiord Uplift (Fig. 6) is a fault-bounded structure exposed on northern Axel Heiberg Island (Trettin, 1969a; Trettin *et al.*, 1972). Its core is composed of Cambrian(?) and Ordovician rocks. The earliest know phase of uplift occurred in late Silurian to early Devonian time near the onset of Phase IV of Sequence 3 (Fig. 5). The Rens Fiord Uplift was then the source of a coarse red bed delta complex that prograded eastward into the Hazen Trough to form the Lower Devonian Stallworthy Formation, about 3000 m thick. Small plutonic intrusives in the core of the uplift are late Middle to Upper Devonian (360 ± 25 m.y.) and apparently were intimately related to a pulse of uplift. This uplifting was a part of the Ellesmerian Orogeny and the intrustions are postkinematic.

The main activity of the Rens Fiord Uplift occurred in one or more strongly upward pulses during which more than 3 km of rock were eroded from it on northern Axel Heiberg Island. These uplifts occurred between early Devonian time and mid-Mississippian time, within the span of the Ellesmerian Orogeny.

The Rens Fiord Uplift probably connected northward with or was a part of the Pearya Geanticline (Fig. 6). The activity of the two occurred at about the same time, as the rise of the Rens Fiord Uplift was approximately contemporaneous with the Ellesmerian Orogeny. The rise of both also was approximately contemporaneous with that of the Boothia Uplift. The Rens Fiord Uplift clearly was an important structure in Devonian time, extending south-southeastward from the Pearya Geanticline and reaching most of the way across the Franklinian Geosyncline. It apparently established a south-southeastward structural trend on Axel Heiberg Island at that time, and this trend has dominated later deformations there. The Rens Fiord Uplift may have been produced by a mechanism somewhat simlar to that of the Boothia Uplift (Fig. 8).

5. *Minto Arch*

The Minto Arch on Victoria Island (Thorsteinsson and Tozer, 1962, 1970) exposes a small area of the Precambrian Shield and a large area of Proterozoic sediments (Fig. 6). Structural trends within the arch are parallel to its outline. The arch is overlain unconformably by Cambrian and Ordovician sediments. In most places these dip away gently, but in places there is a faulted contact. The Minto Arch was active as a structural high at some time after deposition of the Ordovician sediments, but the time of this activity is unknown. It is very likely that the arch was raised in Devonian time, contemporaneous with the uplifting of similar basement highs elsewhere in the Canadian Arctic Islands, and its mechanism may also have been similar. The arch may be connected with the Pearya Geanticline but this is speculative.

6. *Summary of Basement Uplifts*

Phase IV of Sequence was dominated by the activity of basement uplifts. The Pearya Geanticline is a huge basement uplift (Figs. 4, 6, and 7) that was active throughout Sequence 3. It is overwhelmingly larger and more important than the other uplifts in the region (Fig. 6), and therefore has been treated separately in this report. The remaining uplifts, which are partly or entirely within the geosyncline, are much smaller and are fairly similar to each other in size though they vary in amount of vertical uplift, particularly along strike. Four of these basement uplifts trend crosswise to the Franklinian Geosyncline, and a fifth, the Grantland Uplift, trends along the geosyncline.

The several basement uplifts that were partly or entirely within the geosyncline had their main period of activity at about the same time, beginning in early Devonian. This coincided with the beginning of Phase IV of Sequence 3 (Fig. 5), with the raising of the Pearya Geanticline, the onset of the Ellesmerian Orogeny, *sensu stricto*, and with the general narrowing of the geosyncline by expansion of the Arctic Platform (Fig. 7). In Phase IV there was a great influx of clastic sediments and rapid infilling of the geosyncline. The sediments were derived partly from sources outside the geosyncline, the Pearya Geanticline and the Arctic Platform, and partly from uplifts within the geosyncline. The coincidence in time and the similarity of behavior to that of the Pearya Geanticline suggests that the raising of these uplifts was associated with and perhaps was a side effect of the Ellesmerian Orogeny, which was dominated by the raising of the Pearya Geanticline. The uplifts also apparently were related to the general narrowing of the geosyncline as well.

The ancient structural trends of the basement uplifts relative to the Pearya Geanticline, whether parallel, oblique, or perpendicular, were a very important influence on later sedimentation and deformation in the Canadian Arctic. The arrangement is exemplified by the two extreme cases. The Boothia Uplift trends at right angles to the Franklinian Geosyncline and Pearya Geanticline, and as such represents one extreme. The Grantland Uplift extends along the geosyncline and is parallel to the geanticline, and as such represents the other extreme. Other basement uplifts seem to be either nearly perpendicular, or quite parallel to the geosyncline.

I. Sequence 3: Obstructed Evolution

Several basement uplifts that trend crosswise to the Franklinian Geosyncline (Fig. 6) obstructed the normal evolution of Sequence 3. They first influenced facies patterns as the sequence was deposited and subsequently influenced structures as the sequence was terminated and deformed. Of these uplifts, the obstructing effects of the Boothia Uplift are best known (Figs. 9 and 10). This uplift affected the basin that formed Sequence 3 and caused local interference or obstructions in the general evolutionary pattern of that sequence. The other uplifts may have had similar but less marked obstructing influences.

The Boothia Uplift extends from the Arctic Platform across the Franklinian Geosyncline and probably connects with the Pearya Geanticline (Fig. 6). It influenced the evolutionary development of Sequence 3 in two ways. (1) Its influence on sedimentary thicknesses and facies of the sequence is shown by the cross section in Fig. 9. (2) Its influence on the structural deformation of the sequence by the younger Ellesmerian Orogeny is shown in plan view in Fig. 10.

In those parts of the line of cross section far from the Boothia Uplift (Fig. 9) the overall development of Sequence 3 was much like that which occurred elsewhere remote from uplifts (Fig. 7). In both cross sections the five phases of Sequence 3 can be readily summarized (Fig. 5) as follows: Phase I, origin and general widening; Phase II, starved basin; Phase III, activity of the geanticline begins to fill the starved basin with flysch deposits; Phase IV, substantial uplift as Ellesmerian Orogeny begins; Phase V, increasing uplift provides a clastic wedge, followed by southward overriding. The history was radically different only in the vicinity of the Boothia Uplift. Much of this cross section (Fig. 9) is entirely covered by rocks of the Sverdrup Basin (Fig. 6), but those parts of it that are exposed suggest that its history there was similar to the history of the sequence on Ellesmere Island (Fig. 7).

The effects of the Boothia Uplift on rocks of Phase I are uncertain because that part of the column is poorly exposed (Miall and Kerr, 1980) or known only from sporadic wells (Mayr, 1978). Thinning over the uplift is suspected. During Phase II the Franklinian Geosyncline appears to have developed regionally in its normal fashion. The Boothia Uplift may have been mildly positive then as well (Fig. 9).

Before Ellesmerian Orogeny   After Ellesmerian Orogeny

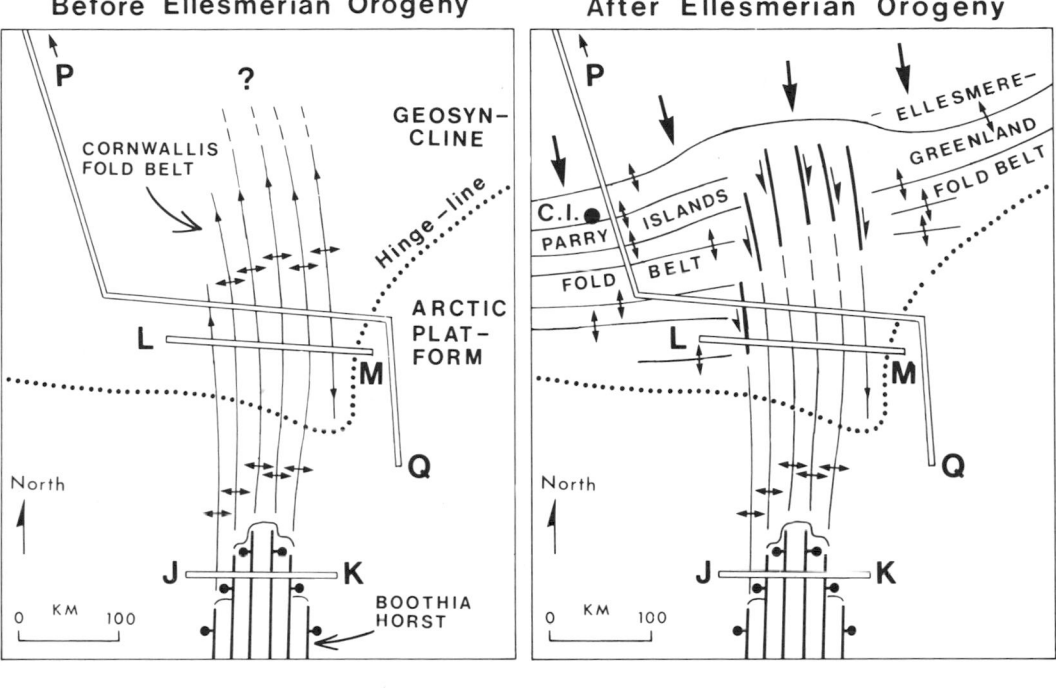

LEGEND

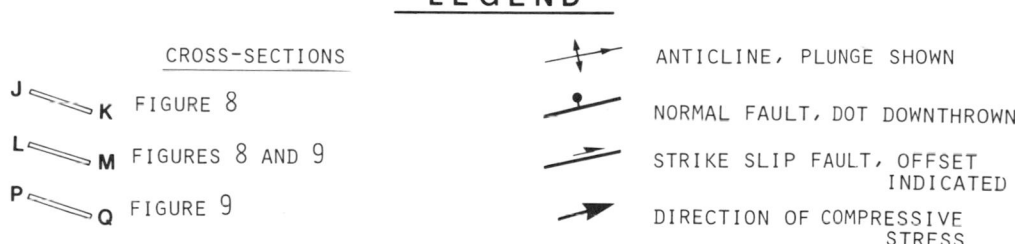

Fig. 10.   Maps showing how the Boothia Uplift interfered with the Ellesmerian Orogeny. Before: The basic structure of the Boothia Uplift was an anticlinorium (Cornwallis Fold Belt) draped over a crystalline core (Boothia Horst). After: Southward overriding of the Ellesmerian Orogeny produced new foldbelts in the previously unfolded parts of the geosyncline. Within the Cornwallis Fold Belt interference caused strike-slip faults that are left lateral on the west side and right lateral on the east side. The Cameron Island oil pool (Figs. 6 and 10) developed in one of the east–west-trending structures and is indicated by "C.I."

During Phase III in the development of Sequence 3, sedimentation was interrupted briefly by a minor uparching of the Boothia Uplift. Pulse 1 of the Cornwallis Disturbance occurred in early Silurian time (Figs. 4 and 5), when the Boothia Uplift was arched gently upward. About 450 m of sediments were eroded from its central part on western Grinnell Peninsula (Kerr, 1977a). An unconformity that developed on the southern part of the Boothia Uplift in early Silurian time (Miall and Kerr, 1980) probably is also due to Pulse 1.

This was a minor interruption in the sedimentation of the miogeocline and does not appear to have substantially altered sedimentary patterns there. Thus during Phase III of Sequence 3 the overall evolutionary pattern of the Franklinian Geosyncline continued, largely unaffected by the Boothia Uplift, which probably was a mild arch with an overall thinner sedimentary column.

The Boothia Uplift began to have a major influence on Sequence 3 during Phase IV. During Pulse 2 of the Cornwallis Disturbance (Fig. 4), the Boothia Uplift again was raised, and this time at least 4140 m of rocks were eroded (Fig. 9). From this time on the part of the geosyncline affected by the Boothia Uplift behaved instead as an enlarged part of the Arctic Platform and thereafter received only thin sediments. Thick sediments continued to be deposited on either side of the Boothia Uplift, however, within the region that continued as a geosyncline. Large amounts of clastic rock derived from the uplift were deposited in the nearby parts of the geosyncline.

Pulse 2 of the Cornwallis Disturbance occurred in early Devonian time (Fig. 4) It coincided with the beginning of Phase IV of the Franklinian Geosyncline and the onset of the Ellesmerian Orogeny in the Pearya Geanticline. It also coincided with the raising of the Rens Fiord Uplift (Trettin, 1969a) and the Bache Peninsula Arch (Kerr, 1968b, 1976a), and with the general northwestward expansion of the Arctic Platform by migration of the southeastern hinge line (Fig. 7). These events doubtless were related to and controlled by some broader but connected subcrustal mechanism.

After Pulse 2, sedimentation resumed on the Boothia Uplift (Fig. 9) as part of a widespread regional subsidence, and sedimentation continued there through much of the rest of Devonian time (Kerr, 1977a). The Boothia Uplift had a subsequent pulse during Phase IV, namely Pulse 3 in late early Devonian time. This was much weaker than Pulse 2, and a lesser thickness of rocks was eroded from the uplift.

The Boothia Uplift had a substantial effect on the sedimentary patterns of Sequence 3, by disrupting the normal pattern of sedimentation in a north-trending linear belt (Fig. 6). The effects on sedimentation during Phases I, II, and III were minor but during Phase IV they were substantial (Fig. 9).

During Phase IV the Franklinian Geosyncline was evolving regionally in its normal fashion (Fig. 7). The Pearya Geanticline was being raised and eroded, and the Ellesmerian Orogeny was progressing toward its culmination. In the region of the Boothia Uplift, there was a markedly different pattern (Fig. 9). During Phase IV several uplifts fragmented the geosyncline and contributed to its filling in the same way and at the same time as did the Pearya Geanticline. By analogy with the Rens Fiord Uplift, it is suggested that the Boothia Uplift was connected on the north with the Pearya Geanticline.

In Phase V of Sequence 3 the Pearya Geanticline continued to rise, and folds and thrusts were beginning to form in the geosyncline because of the Ellesmerian Orogeny. The Boothia Uplift was subjected to the last upward pulse of the Cornwallis Disturbance (Figs. 4 and 5) in late Devonian time (Kerr, 1977*a*). After that pulse the Boothia Uplift possessed its basic structure (Fig. 8), that of an anticlinorium cored by a crystalline horst (Fig. 9, Phase V; Fig. 10, left). In the line of cross section (Fig. 9) mass transport during Phase V, resulting from the Ellesmerian Orogeny, was directed southward away from the geanticline. In this part of the geosyncline the Boothia Uplift had a remarkable effect on the deformation of the Ellesmerian Orogeny and interfered with some of that mass transport (Fig. 10).

In the region of the Boothia Uplift the developmental history of the Franklinian Geosyncline and Arctic Platform was affected in two ways. First the Boothia Uplift influenced or modified regional facies patterns (Fig. 9). Secondly it influenced or obstructed the otherwise rather simple deformation produced by the Ellesmerian Orogeny (Fig. 10).

The basic structure of the Cornwallis Fold Belt (Fig. 8 and Fig. 9, Phase IV) was in existence prior to the overriding of the Ellesmerian Orogeny. In plan view at this time the basic structure (Fig. 10, left) was an anticlinorium, with a crystalline horst as a core, trending northward across an otherwise undeformed geosyncline. After the Ellesmerian Orogeny a very different situation existed (Fig. 10, right). In regions on either side of the Boothia Uplift, where there was no interference by a preexisting north-trending structure, the southward overriding of the Ellesmerian Orogeny deformed the Franklinian Geosyncline rather simply. This rather simple deformation formed the Parry Islands Fold Belt in the west and the Ellesmere–Greenland Foldbelt in the east. These belts contained anticlines that were subparallel to the Pearya Geanticline, from whence they were initiated, and subparallel to the depositional trend of the Franklinian Geosyncline. Both belts formed by the folding of an undeformed geosynclinal pile, and structures developed parallel to the trend of the geosyncline and overrode toward the Arctic Platform. The Ellesmerian Orogeny here involved a southward-directed décollement within one or more evaporite layers that lay within the rocks of Phase III, and the folding that this produced was a relatively surficial phenomenon (Kerr, 1974).

In and near the Boothia Uplift the rocks of Sequence 3 developed in quite a different situation during the Ellesmerian Orogeny than they did elsewhere (Figs. 9 and 10). The Boothia Uplift obstructed the rocks of the geosyncline that were being transported southward. A zone of interference developed within the Cornwallis Fold Belt, and that belt was deformed in an unusual way. The columns on either side were carried southward or southeastward by décollements. The strong north-trending structural grain that existed within the Cornwallis Fold Belt prevented a similar décollement from developing there.

Nevertheless the great force of the southward-moving mass was being imposed on the Cornwallis Fold Belt and deformed it in a more complicated way (Fig. 10). Vertical strike-slip faults developed within the Cornwallis Fold Belt, following approximately the axes of the preexisting anticlines. It is inferred that at depth these faults followed fundamental near-vertical normal faults in the crystalline basement that had controlled the anticlines of the Cornwallis Fold Belt (Fig. 8). The strike-slip faults in the western part of the Cornwallis Fold Belt were left lateral, while those in the eastern part of the fold belt were right lateral. The actual southward movement of material within the Cornwallis Fold Belt diminished from the edges toward the center.

The Rens Fiord Uplift (Fig. 6) may have interfered with the Ellesmerian Orogeny in a way similar to that of the Boothia Uplift. This cannot be determined because much of the Rens Fiord Uplift is covered by younger rocks. The Bache Peninsula Arch may have interfered with the Ellesmerian Orogeny to a small degree. There does not appear to have been any influence in the stratigraphic level that is presently exposed; however there may have been interference at greater depth. The relationship of the Minto Uplift to the effects of the Ellesmerian Orogeny is unknown.

The Franklinian Geosyncline, Pearya Geanticline, and Arctic Platform are large tectonic features with a tendency to develop regionally as shown in Fig. 7. Most cross sections drawn through those features between Banks Island (Miall, 1976a) and northern Greenland (Dawes, 1976) probably show a generally similar history. The basement uplifts (Fig. 6) are smaller tectonic features that produced local aberrations in the three larger features (Fig. 9). All these features, of course, developed in partial contemporaneity.

J. Continent as a Substructure Prior to Ocean Development

In latest Devonian time, at the end of the Ellesmerian Orogeny, all of northern Canada including the entire region of the Canadian Arctic Archipelago may have been above sea level and exposed (Fig. 7, Phase IV). Trending northeastward were the folded Franklinian Geosyncline and Pearya Geanticline, apparently both with high relief. Farther south was the exhumed Arctic Platform, where a thin sedimentary column covered much of the present Canadian Shield. The Shield itself probably was only locally exposed, in highs such as the Boothia Uplift, where there had been substantial Devonian erosion.

Little is known about the history of the water-covered region northwest of the Canadian Arctic Archipelago. The Pearya Geanticline probably was very broad and may have included much or all of the region that is now continental shelf and slope adjacent to the Canada Basin. The continent at this time also may have occupied adjacent parts of the Canada Basin. Recent investigations (Churkin, 1969; Vogt and Ostenso, 1970; Ostenso and Wold, 1973) have concluded that the Canada Basin is underlain by oceanic crust. It may however contain downfaulted continental rocks, because according to King *et al.* (1966), the rocks have continental characteristics.

The end of the Ellesmerian Orogeny in latest Devonian or early Mississippian time is an important turning point in the history of northern North America. It represents the end of a constructional phase in which stratigraphic Sequences 2 and 3 were deformed, and solidified in succession as part of a large continent. This primeval continent, largely above sea level, may have occupied the entire region shown by Fig. 1, including both land and sea areas of the present day.

III. FRAGMENTATION PHASE: EVOLUTION OF THE OCEAN BASINS AND BRANCHES WITHIN THE CONTINENT

A. Foreword

The fragmentation phase of northeastern North America began shortly after the Ellesmerian Orogeny in lastest Devonian or early Mississippian time (Fig. 4), and continued to the present day. This phase appears to have begun with an exposed, largely primeval continent that included the entire region shown by Fig. 2. The Sverdrup Basin and possibly the Canada Basin first developed as rifted and subsiding basins, unconformably upon part of this exposed continental landscape. Subsequently major rifting events produced the present configuration of lands and seas in the Canadian Arctic.

B. Boreal Rifting Episode

The Arctic and Atlantic Oceans are connected today via the Canadian Arctic by means of structurally controlled channels between them. Those structures resulted from rifting that emanated cratonward from the two oceans. Rifting appears to have begun first in the Canada Basin and to have occurred intermittently there during deposition of Sequences 4, 5, and 6 (Fig. 4). This rifting event, which may have originated in the Alpha Ridge, is here named the Boreal Rifting Episode. Side effects of this rifting reached some distance southeastward into the continent, where it apparently caused the Sverdrup Basin to form by crustal fracturing. The Boreal Rifting Episode was an unsuccessful attempt to connect the Arctic and Atlantic Oceans. It did not break fully through the continent to the southeast, and therefore was aborted.

C. Sequence 4: Mid-Mississippian (Visean) to Early Cretaceous (Pre-Isachsen Formation)

1. Foreword

The rocks of Sequence 4 of the Innuitian Mobile Belt (Figs. 3 and 4) were deposited in several basins (Fig. 11) the Sverdrup Basin, the Banks Basin, and presumably an open Arctic Ocean that became the Canada Basin They also were deposited at times on the Sverdrup Rim, an intermittently positive feature that overlay the Pearya Geanticline during its nonmagmatic stage.

The tectonic framework of Sequence 4 apparently was dominated by the events that took place in the Arctic Ocean. This basin appears to have been forming as a major depocenter during the time span of Sequence 4, by a rifting mechanism possibly centered on the Alpha Ridge (Fig. 11) and named here the Boreal Rifting Episode (Fig. 4). The Sverdrup Basin and Banks Basin are smaller depocenters, located on a primarily continental area. They formed in response to events in the Canada Basin, are marginal to it, and are separated from the major basin by the Sverdrup Rim and Storkerson Uplift (Fig. 11).

2. Sverdrup Basin

The evolution of the Sverdrup Basin (Fig. 11) has been summarized recently by Balkwill (1978), where many of the following data were presented. Stratigraphic nomenclature of the Sverdrup Basin has been discussed by Thorsteinsson and Tozer (1970), Plauchut (1971), Thorsteinsson (1974), and Henao-Londoño (1977), and subsidence rates were discussed by Sweeney (1977). The framework of the basin was discussed by Trettin et al. (1972), and its petroleum potential was discussed by Stuart Smith and Wennekers (1977).

The Sverdrup Basin contains mid-Mississippian (Visean) and younger rocks up to 13,000 m thick. It formed as a pericratonic basin along the northwest margin of Arctic Canada after the Ellesmerian Orogeny. It lies unconformably on folded rocks of the Franklinian Geosyncline and Pearya Geanticline (Fig. 3). Sequence 4 is mainly concordant in the axial part of Sverdrup Basin, but there are unconformities along the margins.

Sequence 4 includes the lower part of the column of the Sverdrup Basin (Fig. 4). Three phases are recognized within Sequence 4 there, each phase represented by a stratigraphic assemblage (Fig. 12). These are similar to the first three phases of the Sverdrup Basin as outlined by Balkwill (1978), with slight modification here in Phase III. These phases are (I) late Paleozoic (Carboniferous and Permian), when evaporites and marine muds were deposited in the axial region, and carbonates and mature sands were deposited on the margins; (II) early Mesozoic (Lower to mid-Upper Triassic), when great thicknesses of siltstone and shale accumulated in axial parts of the basin, with lesser

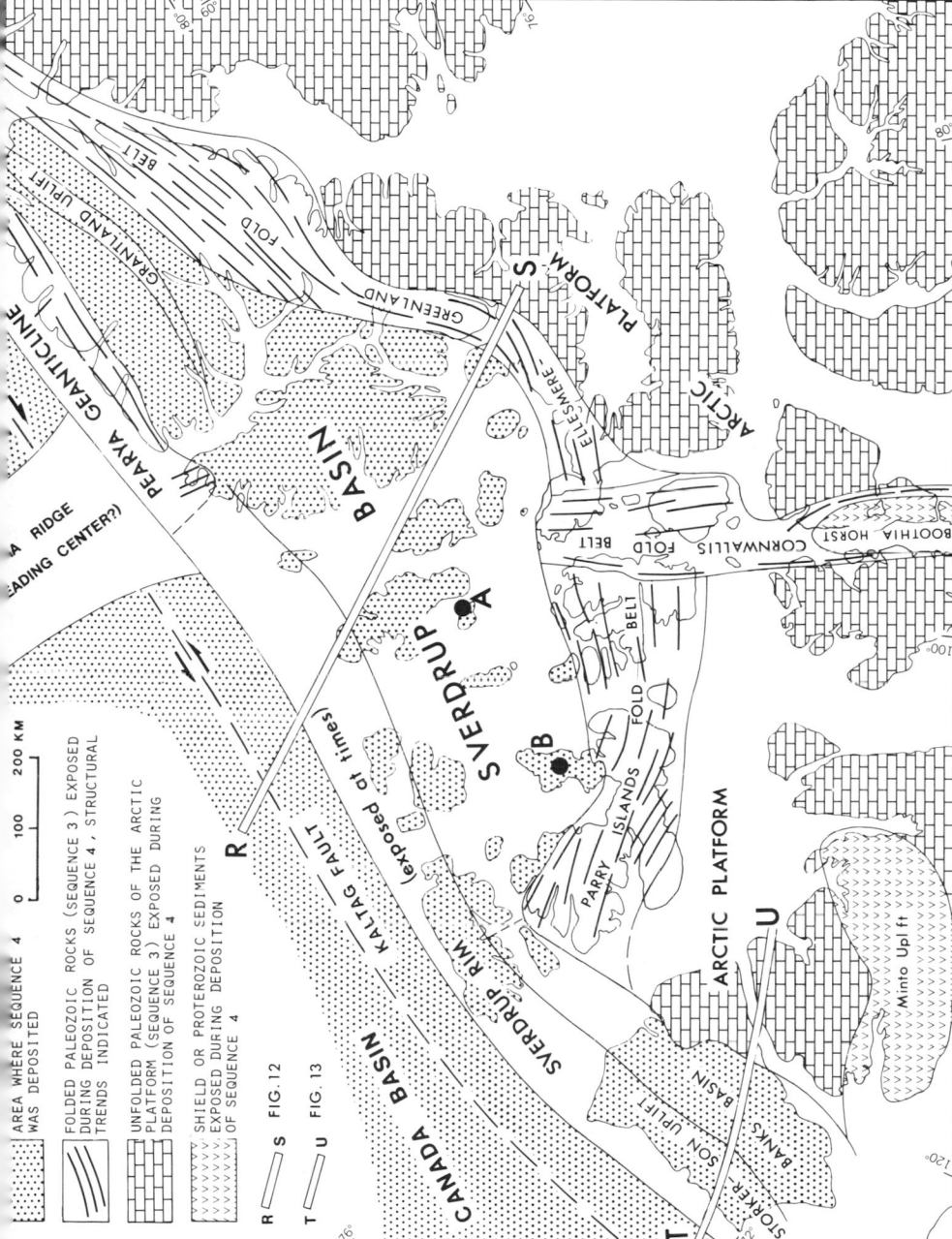

Fig. 11. Paleogeography and tectonic setting of Sequence 4 (Figs. 3, 4, and 12), which includes rocks extending from Mississippian to lower Lower Cretaceous (pre-Isachsen Formation) in Sverdrup Basin, Banks Basin, Sverdrup Rim, and presumably rocks along the margin of the ancestral Canada Basin. Sequence 4 began the fragmentation phase of northern Canada (Fig. 4). A, Ellef Ringnes gas fields region (Figs. 12 and 23); B, Sabine Peninsula gas fields region (Fig. 23).

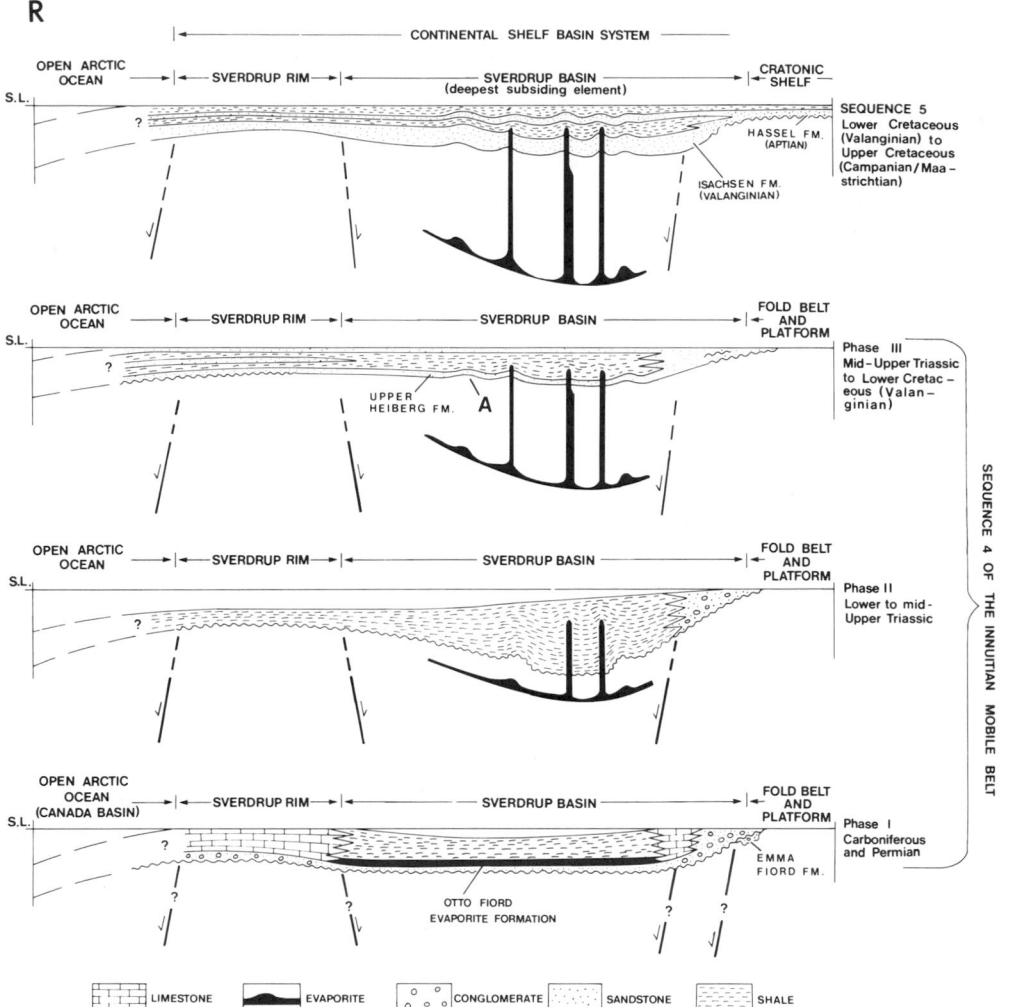

Fig. 12.   Cross sections showing the history of Sequences 4 and 5 of the Innuitian Mobile Belt. The history of these two sequences was dominated here by the Boreal Rifting Episode (Fig. 4). Sections are diagrammatic and scale is not shown. The location of these lines of section is shown in Fig. 11. Much of this diagram is adapted from Balkwill (1978). The Ellef Ringnes gas fields region is represented by the structure labeled "A" (see Fig. 11).

thicknesses of sand on the margin; (III) mid-Mesozoic (mid-Upper Triassic to Lower Cretaceous), when widespread terrigeneous clastic rocks accumulated slowly during transgressions and regressions.

The history of the three phases of Sequence 4 was one of continuous sedimentation with minor interruptions of deposition (Fig. 12). During Sequence 4 the Pearya Geanticline was in its nonmagmatic phase, and was mildly negative to positive (Fig. 4).

The southern and eastern margin of the Sverdrup Basin is at the major basal angular unconformity, with the basin lying on folded older rocks and dipping regionally northwestward (Figs. 11 and 12). This margin was also the approximate location of the shoreline during much of Phases I through III of the Sverdrup Basin. On its north side the Sverdrup Basin dips southeastward regionally. On Ellesmere and Axel Heiberg Islands the Pearya Geanticline was exposed through most of Sverdrup Basin history, so the present northwestern boundary there was the approximate shoreline at most times. Farther southwestward the Sverdrup Basin was bordered by the Sverdrup Rim, which was exposed from time to time during Phases I through III (Meneley *et al.*, 1975; Balkwill, 1978). This rim is aligned with and presumably is underlain by the Pearya Geanticline. It contains rocks equivalent to much of the Sverdrup Basin, but they are thinner, presumably due to the positive tendency of the geanticline beneath. A central high region of thinner rocks extended across the Sverdrup Basin (Meneley *et al.*, 1975), making two depocenters. Balkwill (1978) designated the depocenter to the west the Barrow segment and the one to the east the Axel Heiberg segment.

Phase I: Mid-Mississippian to Late Permian. The Sverdrup Basin began to take on its form as a subsiding basin shortly after the Ellesmerian Orogeny ended (Figs. 4 and 11). The oldest rocks of the basin are the Middle Mississippian (Visean) Emma Fiord Formation, which is sporadically exposed on its northern and southern edges (Thorsteinsson and Tozer, 1970; Trettin, 1973*b*; Kerr, 1976*b*). The Emma Fiord Formation occurs in small, downwarped and downfaulted structural lows that formed sedimentary embayments on the fractured margins of the Sverdrup Basin (Fig. 12). On the southern margin at northern Grinnell Peninsula (Kerr, 1976*b*), relief developed gradually during Emma Fiord deposition by downwarping and faulting. The Emma Fiord Formation occurs on the northern margin of the Sverdrup Basin, but in the central part this level is deeply buried and unknown. There probably was a similar tectonic setting throughout the Sverdrup Basin, with nonmarine Mississippian rocks deposited sporadically throughout it in narrow basins.

Mid-Mississippian (Visean) time appears to represent a very early stage of crustal fracturing with local small subsiding basins. This may have been the beginning of the Boreal Rifting Episode (Fig. 4), an extensional episode that followed the compression and exposure by the Ellesmerian Orogeny. Initial fracturing in mid-Mississippian time in the Sverdrup Basin was a transition that graded into the later widespread fracturing of younger parts of Phase I of the Sverdrup Basin (Fig. 12).

During later parts of Phase I, the crustal fracturing event increased and the Sverdrup Basin began to subside greatly. A widespread conglomerate unit was deposited throughout the basin in latest Mississippian to early Pennsylvanian time. This unit is called the Canyon Fiord Formation in the southeast and the Borup Fiord Formation in the northwest (Thorsteinsson, 1974). It was deposited during a time when there was high relief and faulting was active throughout the Sverdrup Basin. The crustal fracturing or extension developed local highs and lows within the overall low that formed the Sverdrup Basin. The highs were horsts or anticlines that provided a source of clastic sediments. The lows were grabens or synclines that received those sediments.

The crustal fracturing event that began to form the Sverdrup Basin during Phase I is well displayed on northern Grinnell Peninsula (Kerr, 1976*b*), but probably affected the entire Sverdrup Basin in much the same way. In central parts of the Sverdrup Basin for example, renewed differential uplift of the Grantland Uplift occurred during Phase I (Nassichuk and Christie, 1969; Trettin *et al.*, 1972; Mayr, 1976*b*).

It appears that the great crustal fracturing and faulting that initially produced the Sverdrup Basin was gradually but generally replaced during Phase I by subsidence or sagging. Gradually in the later part of Phase I the basin took on the form of a broad sag between the craton and the Pearya Geanticline. Relief was reduced by erosion, the proportion of conglomerate in the column progressively diminished, and fine-grained marine sedimentation finally prevailed (Fig. 12).

By Pennsylvanian and Permian time marine deposition was more widespread. Encroachment out of the southern part of the basin produced overstepping by Pennsylvanian rocks. The Melvillian Disturbance (Fig. 4), produced local disconformities between Pennsylvanian and Permian rocks along the southern margin of the Sverdrup Basin, according to Thorsteinsson and Tozer (1970). At this time the Sverdrup Rim was breached (Meneley *et al.*, 1975; Balkwill, 1978), and deeper-water facies followed. The Melvillian Disturbance apparently represents stepped-up activity in a long lasting but otherwise gradual crustal fracturing event referred to here as the Boreal Rifting Episode.

Continued broader subsidence during Phase I downwarped the Sverdrup Basin regionally, and thick sediments accumulated. The basal conglomeratic rocks of the basin were mainly nonmarine. They grade up to and were succeeded by fine-grained marine rocks, mainly limestones along the margins and evaporites and shales along the axis (Fig. 12). Thorsteinsson (1974) distinguished several concentric facies belts in the Upper Carboniferous to Lower Permian rocks. There were clastic rocks around the margin, carbonates within this, and deep-water evaporites and shales in the middle. The axial facies during Phase I formed a thinner column than the marginal facies (Fig. 12). In the lower part of the axial facies there are thick evaporites. They consist of gypsum and anhydrite in outcrop (Thorsteinsson, 1974; Davies and Nassichuk, 1975; Wardlaw and Christie, 1975), but include large volumes of halite as well in the subsurface (Davies, 1975). The upper part of the axial facies during Phase I was mainly shale.

The Canada Basin apparently was in existence during Phase I of the Sverdrup Basin (Figs. 11 and 12). This suggestion is supported by Upper Paleozoic faunas, which have a circum-Arctic affinity (Nassichuk, 1975) and which appear to have entered the Sverdrup Basin from the northwest (Nassichuk and Davies, 1981). The Canada Basin may have been initiated at the same time as the Sverdrup Basin, and their origins probably were related. The crustal fracturing process that developed the Sverdrup Basin during Phase I (Fig. 12) may have been a marginal effect of the more intense crustal fracturing, or rifting, and plate separation that produced the Canada Basin. These deformations may have originated in an active spreading center on the site of the Alpha Ridge (Fig. 11).

Phase II: Early Triassic to Mid-Late Triassic (Norian). Phase II of the Sverdrup Basin (Fig. 12) includes Lower Triassic to mid-Upper Triassic (Norian) rocks (Balkwill, 1978). The oldest Triassic rocks in the basin are as old as any Triassic rocks known (Tozer, 1967). They lie on a widespread regional disconformity that marks the beginning of Phase II. This is a basin-wide disconformity resulting from marine regression and separates the Permian and Triassic systems (Thorsteinsson, 1974).

Phase II is represented by two correlative facies, a marginal and an axial basin facies (Fig. 12). The marginal facies occurs in the east and south and consists of sandstone and conglomeratic sandstone. It grades very abruptly basinward into a thick succession of marine siltstones and shales that were deposited as turbidites. Some conglomeratic sandstone is present along the northwestern rim, but little detritus was contributed from that direction. The basinal succession in Phase II is as thick as 4500 m, which is almost as thick as the remainder of the Mesozoic succession. Phase II differed from Phase I in that the axial facies was considerably thicker than the adjacent marginal facies (Fig. 12), the opposite to conditions during Phase I. Phase II also differed from Phase I in having no significant carbonates or evaporites. Phase II differed from later phases in having deeper water.

Phase II ended in mid-late Triassic (mid-Norian) time, with a shallowing, when an Upper Triassic sandstone began to prograde widely across the Sverdrup Basin. The main large diapirs of the Arctic Islands formed by halokinesis (migration due to loading), a long continuing process that began in this phase (Triassic time). By mid-late Triassic time the combined thickness of Triassic and Upper Paleozoic rocks that overlay the Carboniferous Otto Fiord evaporites in eastern parts of the Sverdrup Basin was about 6000 m. Halokinesis was initiated by this loading (Balkwill, 1978).

Diapirs in the Arctic Islands occur only in central parts of the Sverdrup

Basin, because they originated in the Otto Fiord Formation, an axial facies. The natural gas pools in central parts of the Sverdrup Basin are related to these diapirs and are discussed later. There are numerous evaporitic piercement domes in the Sverdrup Basin (Gould and de Mille, 1964; Thorsteinsson, 1974). They originated in the Otto Fiord Formation, an evaporite unit within the rocks of Phase I (Fig. 12). Some developed halokinetically, beginning probably during Phase II in late Triassic time (Balkwill, 1978). They had intermittent activity that continued during deposition of younger rocks (Fig. 12). Other domes were generated later, during Tertiary folding and faulting. Evaporite has reached the surface and is exposed today in many structures, and in others it has reached only part way to the surface.

Phase III: Mid-Late Triassic (Norian) to Early Early Cretaceous (Valanginian). Phase III of Sequence 4 is similar to the third phase (Middle Mesozoic) of the Sverdrup Basin as outlined by Balkwill (1978). A difference is that he included older parts of the Isachsen Formation and equivalent rocks at the top of the third phase, and the present paper places the entire Isachsen Formation in the lower part of Sequence 5 (Fig. 12).

The Upper Triassic, Jurassic, and lower Lower Cretaceous rocks that constitute Phase III of Sequence 4 are characterized by alternating sandstone and shale formations (Fig. 12). The shales are marine, while the sandstones are largely but not wholly nonmarine. Throughout nearly all of Phase III the craton southeast of the Sverdrup Basin was a sedimentary source made up of the older foldbelts and the Arctic Platform. As a result the rocks are predominantly nonmarine sands along the southern, southeastern, and southwestern edges of the Sverdrup Basin. Moreover, along that margin there are several unconformities with very slight angularity that disappear basinward into continuous successions. Phase III included several southerly derived blanket sands that were spread fully across the Sverdrup Basin, and only two are shown (Fig. 12). In certain axial parts of the Sverdrup Basin it appears that shales were deposited more or less continuously throughout Phase III. Some sand however was deposited on the Sverdrup Rim and was largely enclosed by these shales (Fig. 12).

Upper Mesozoic clastic sediments of the Sverdrup Basin (Phase III) were largely derived from the southeast, presumably the folded Franklinian Geosyncline, the Arctic Platform, and the Precambrian Shield. The lowest rocks of Phase III are the upper or nonmarine part of the Heiberg Formation, a widespread Upper Triassic to Lower Jurassic delta deposit that prograded across the Sverdrup Basin from the southeast and represented almost complete marine withdrawal from that basin. Fine-grained marine equivalents of this great column of sand must exist still farther northwest beneath the Arctic Coastal Plain and Continental Shelf, as part of a thick and long-lived continental terrace wedge. This implies a proto-Arctic Ocean northwest of the Sverdrup Basin by late Triassic time on the site of the Canada Basin. A subsequent Jurassic marine transgression deposited widespread marine shales almost fully across the basin. From late early Jurassic until latest Jurassic time (late Volgian) shallow conditions of low relief continued in the Sverdrup Basin, and thin marine and nonmarine shales and sands alternated. From latest Jurassic to early early Cretaceous (late Valanginian) time, sedimentation rates increased as deltas prograded into Sverdrup Basin from the west, southeast, and northeast. This culminated by early early Cretaceous (late Valanginian) when there was a basin-wide marine regression and widespread exposure. This regression resulted from the deltaic alluviation of a thick succession of sands, directed principally from the southeast (Roy, 1974; Balkwill, 1978). This delta progradation deposited the Isachsen Formation, which was partly contemporaneous with tectonism (Rahmani, 1977). This widespread Isachsen Formation of sands marked a fundamental change in sedimentary and tectonic patterns in the Canadian Arctic. The top of Phase III of Sequence 4 is drawn in this paper in Valanginian time at the base of the Isachsen Formation of the Sverdrup Basin, which marked the onset of a pulse of increased tectonism.

The lower part of the Sverdrup Basin, which makes up Phases I, II, and III of Sequence 4 of the Innuitian Mobile Belt, had a long history that began with crustal fracturing that subsequently merged into a broader subsidence. This interval formed the lower part of the Sverdrup Basin and ended with the beginning of deposition of the Isachsen Formation.

During the period of its history that makes up Sequence 4, late Paleozoic to early early Cretaceous (Valanginian), the Sverdrup Basin appears to have been a small basin, lying on the continent, adjacent to a larger developing oceanic basin, the Canada Basin (Fig. 12). The events that affected the Sverdrup Basin in this interval probably were secondary events, marginal to larger scale rifting events that occurred in the adjacent Canada Basin. It happens that rocks of the Sverdrup Basin are well exposed and well described (Balkwill, 1978), so that events that took place in the Canada Basin can be inferred from them. The Canada Basin had a long history of sporadic deformation, with the Sverdrup Basin region reflecting side effects, specifically crustal fracturing or rifting. It is probable that the Canada Basin of the Arctic Ocean began to form in late Paleozoic time, by the Boreal Rifting Episode, and has been in existence over since. There is strong evidence from faunas (Tozer, 1961) that it has been inexistence continuously from late Triassic time onwards.

Balkwill (1978) pointed out that the depth to which the Sverdrup Basin would have subsided if not loaded with sediments is about the same as the present depth of the Canada Basin, i.e., about 3500 m. The Sverdrup Basin appears to have formed by vertical tectonics, prior to and in an early stage of plate breakup, in a manner similar to that outlined by Falvey (1974) for basins on other rifted continental margins.

3. *Banks Basin*

The Banks Island area lies southwest of the Sverdrup Basin (Fig. 11), and was part of an unstable cratonic margin in Jurassic and later time (Miall, 1975). Banks Basin (Figs. 11 and 13) is a narrow basin that at times was con-continuous with or merged into the Sverdrup Basin. It was bordered on the northwest by the Storkerson Uplift, which is the southwest continuation of the Sverdrup Rim.

The rocks of Sequence 4 of the Innuitian Mobile Belt that are known in the Banks Basin are of late early Jurassic to early early Cretaceous age (Fig. 13). They are equivalent to rocks of Phase III of the Sverdrup Basin (Fig. 12). They are thick marine sands and shales. At this time Banks Basin was part of a narrow marine trough that continued northeastward toward Sverdrup Basin, and southwestward toward the Beaufort Sea. This trough was bordered on the northwest by the exposed and positive Sverdrup Rim, which separated the subsiding Banks Basin from the deeper Canada Basin. This was a period of low to moderate tectonic activity, with relative uplift and subsidence, but apparently not strong faulting.

D. Sequence 5: Early Early Cretaceous (Valanginian) to Late Cretaceous (Campanian/Maastrichtian)

Sequence 5 begins with the Lower Cretaceous Isachsen Formation and includes all rocks up to the base of the Eureka Sound Formation. The sequence represents a marked change in sedimentary patterns from the earlier sequence (Figs. 4 and 12). The Isachsen Formation and equivalent rocks are a widespread wedge of nonmarine fine-grained sandstone that represents a marine regression and alluviation.

The Isachsen Formation is rather thick in the Sverdrup Basin (1400 m) and generally conformable. The formation is oldest in the basin, where it is Valanginian in age. It becomes younger where it encroaches onto the cratonic shelf (Fig. 12), and seems to be of mid-early Cretaceous age (about Aptian) there. The Isachsen Formation also is thinner on the margins of Sverdrup Basin (150 m) and in the wide cratonic shelf are where it encroached or overstepped southeastward, eastward, and southwestward, far beyond the long established margin of the basin. Prior to the Isachsen Formation when Sequence 4 was being deposited (Fig. 12), the Sverdrup Basin had, clearly definable limits. The regression and subsequent overstepping event that deposited the Isachsen Formation began Sequence 5, in which there was a

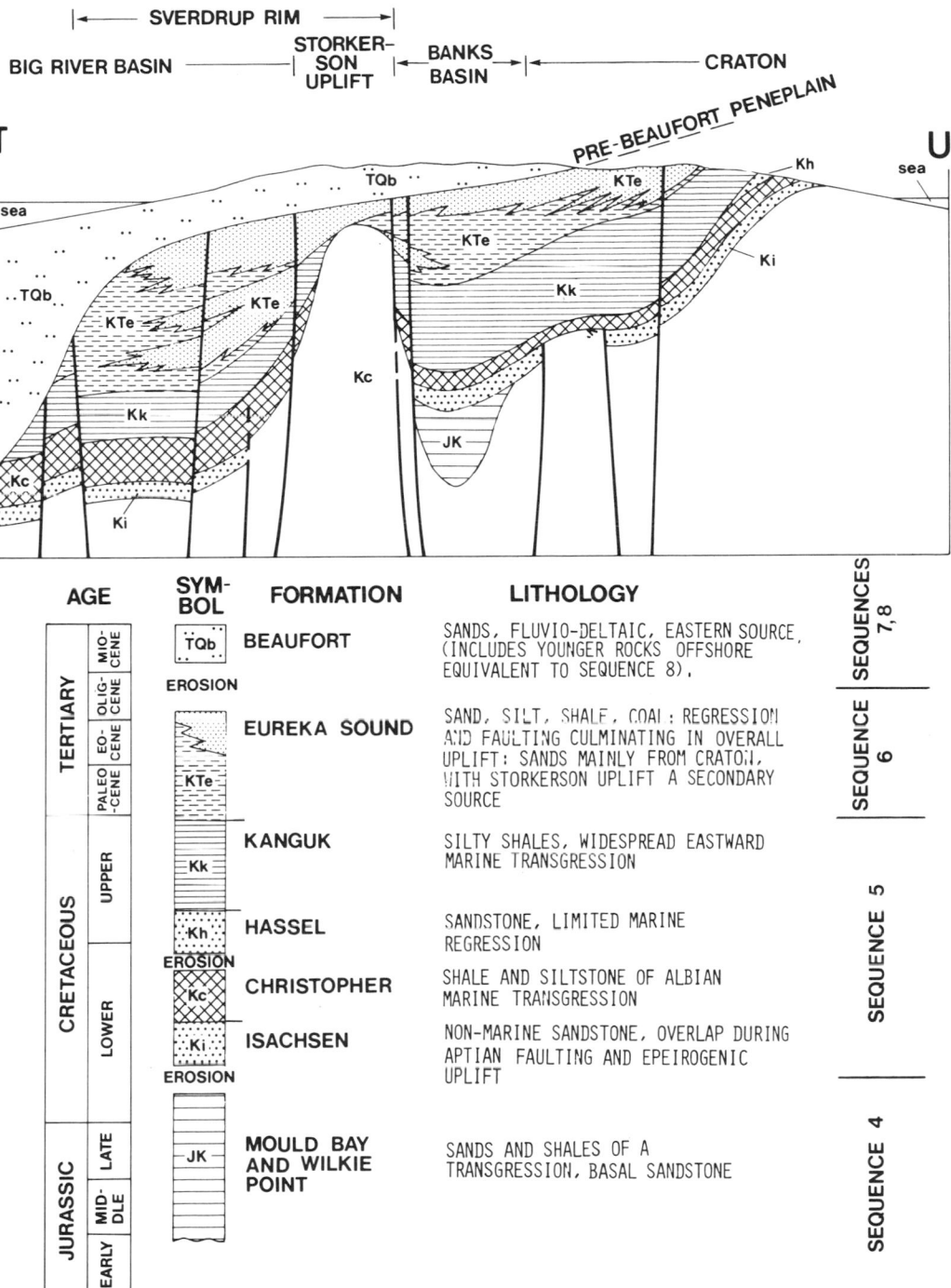

Fig. 13. Cross section showing the development of the margin of the Arctic Ocean Basin in the Banks Island area (after Miall, 1975). The approximate location is shown on Fig. 11.

much broader continental shelf basin system (Fig. 12). The Sverdrup Basin was the deepest subsiding element within that system. The region to the south and southeast, where Sequence 5 extended beyond the bounds of the Sverdrup Basin, was principally a cratonic shelf.

The Isachsen sands of the Sverdrup Basin were derived mainly from the craton southeast of the margin of the basin (Roy, 1974). The younger Isachsen sands of the cratonic shelf probably were derived from locally high parts of that shelf. The Isachsen Formation also occurs widely southeast of the normal bounds of the Sverdrup Basin on Banks Island (Fig. 13), where it is on an unstable cratonic margin (Miall, 1975). The rocks of the cratonic overlap assigned to the Isachsen Formation lapped out of the Sverdrup Basin to Bathurst Island (Kerr, 1974) and central Ellesmere Island (Thorsteinsson and Tozer, 1970), and extended far to the southeast.

Basal sands of Sequence 5 are present in the northern Baffin Island region (Jackson and Davidson, 1975; H. R. Balkwill, personal communication, 1978). They are younger there than the basal sands farther northwest, and equivalent to the Hassel Formation (Fig. 12). Similar basal sands are also present on western Greenland (Henderson et al., 1976). Basal sands of this sequence also diachronously overstepped onto the Anderson Plain of the northern Canadian mainland (Young et al., 1976). Encroachment onto the craton by the Isachsen, Hassel, and related basal formations was probably as a broad shallow epicontinental sea, rather than in deeply rifted basins. This overstepping in the Canadian Arctic was part of and synchronous with a worldwide marine transgression onto cratonic regions (Kent, 1976). It took place over much of the North American continent.

The Isachsen Formation of the Sverdrup Basin was deposited during an early early Cretaceous faulting event that began in Valanginian time (Rahmani, 1977). This faulting presumably was an extensional event that fractured the crust, subsided the Sverdrup Basin, and allowed an abnormally great thickness of the Isachsen Formation to accumulate in the depressed parts of the basin. The faults that produced this fracturing apparently did not extend upward far enough to cut Mesozoic rocks (Fig. 12). The Hassel Formation of the cratonic shelf represents a younger widespread transgression southeastward in middle early Cretaceous (about Aptian) time. While this transgression occurred onto the cratonic shelf, faulting and sedimentation continued within the Sverdrup Basin where mainly shale was deposited (Fig. 12). Fault bounded troughs apparently developed beneath the Sverdrup Basin, producing locally thick depocenters in the sags that formed above them. The pattern of these sags indicates that they deepened to the northwest, and they presumably connected with perhaps thicker rocks of the continental shelf and were tied to the Canada Basin. The troughs developed at the time of deposition of the Isachsen Formation, which is thickest in the troughs, but the formation later spread more widely through Sverdrup Basin. The pulse of faulting that occurred in the Sverdrup Basin and the Banks Island region during Isachsen deposition probably was a marginal effect of an enlarged pulse of the Boreal Rifting Episode in the nearby Canada Basin of the Arctic Ocean.

The post-Isachsen rocks of Sequence 5 are alternating shales and sands in the Sverdrup Basin. The rocks assigned to Sequence 5 in this paper have a total thickness of about 3000 m in the depoaxis of the Axel Heiberg segment of Sverdrup Basin (Balkwill, 1978). Rifting appears to have continued after Isachsen deposition, because the Christopher Formation of the Sverdrup Basin represents very rapid accumulation and presumably further subsidence. Post-Isachsen rocks of Sequence 5 include, in ascending order, the Christopher Formation (marine silty shale with basalt flows), the Hassel Formation (coaly sandstone), and the Kanguk Formation (acidic black shale). These rocks reflect, respectively, a subsidence and marine inundation, a marine withdrawal and alluviation, and a starved euxinic marine basin. A sequence of events similar to that which occurred in the rocks overlying the Isachsen Formation in the Sverdrup Basin (Plauchut, 1971) also occurred in Banks Basin (Miall, 1975; Plauchut and Jutard, 1976). The early Cretaceous overlap reached western Greenland (Henderson et al., 1976), where the succession is similar to that of Bylot Island. The pattern of rifting and cratonic overlap generally supports the view of Jeletzky (1978) that oscillations in Cretaceous sea levels in northern North America resulted from tectonic events there.

Sequence 5 ended with widespread marine withdrawal from Sverdrup Basin, Sverdrup Rim, the cratonic shelf, and Banks Basin. This withdrawal coincided with the onset of a new phase of tectonism that deposited the non-marine rocks of the Eureka Sound Formation. This tectonism was dominated by the Eurekan Deformation (Kerr, 1977a). It began in late Cretaceous (Campanian/Maastrichtian) time, and drastically changed sedimentary patterns. The top of Sequence 5 is at the onset of deposition of the Eureka Sound Formation, a mainly nonmarine sandstone. In places this contact is gradational but in others it is unconformable.

E. Connecting of Oceans by the Canadian Arctic Rift System

The Atlantic and Arctic Oceans are connected by a major seaway system that extends through the North American continent. This seaway makes up the famous Northwest Passage. It includes the Labrador Sea, Baffin Bay, Parry Channel, and other channels within and adjacent to the Canadian Arctic Islands (Fig. 1). This seaway was formed by geological events, and is still controlled by structures of the Canadian Arctic Rift System (Kerr, 1973a). The rift system is a northwest trending branch of the Mid-Atlantic Rift System, and formed by two major plate tectonic episodes (Kerr, 1980b).

The Canadian Arctic Rift System formed as a primeval continent broke up into plates and subplates. Plate activity took place in both the Atlantic and Arctic Oceans, and was propagated from these oceanic areas into the continent where the two rift deformations met. The Canadian Arctic Rift System formed as a result of fault propagation cratonward from both its ends at various times. The rifting episode that emanated southeastward from the Canada Basin was the Boreal Rifting Episode (Fig. 4), and began first. The episode that actively brought about the ultimate structural connection of the two oceans, however, was mainly the Eurekan Rifting Episode (Kerr, 1980*b*). That deformation propagated northwestward from the Atlantic Ocean, and occurred during deposition of Sequences 6 and 7 in the Canadian Arctic region.

F. Sequence 6: Late Cretaceous (Campanian/Maastrichtian) to Mid-Tertiary (Miocene)

Sequence 6 was deposited while two plate tectonic episodes were affecting the North American Arctic, the Eurekan Deformation (Kerr 1980*b*) and the Boreal Rifting Episode (Fig. 4). There were two main depocenters where thick sedimentary columns of Sequence 6 accumulated (Figs. 14 and 15). The largest depocenter was in the northwest. It includes a thick Cenozoic to modern continental terrace (Fig. 2) that extends along the margin of the Arctic Ocean. The history of that wedge was dominated by the history of the adjacent Canada Basin. The other includes the Baffin Bay Basin and the Labrador Sea (Fig. 16a). The Queen Elizabeth Islands subplate (Fig. 15) is a triangular-shaped block between those oceanic areas.

The two main depocenters are connected by rifted branches within the channels of the Canadian Arctic Archipelago. These channels form smaller local basins with lesser thicknesses of sediments (Fig. 14). Lancaster Sound, for example, projects westward from Baffin Bay and is fault controlled (Barrett, 1966; Jackson *et al.*, 1977). It contains the Lancaster Aulacogen (Kerr, 1980*a*), which has a westward-thinning extension of the thick sedimentation of the Baffin Bay Basin (Fig. 14). There also were several largely isolated, moderate-sized depocenters within the archipelago that are not confined to rifted channels. The central part of the Canadian Arctic Islands is the area least affected by rifting and forms a broad northeast-trending height of land (Figs. 14 and 15). Fault-controlled branches from the two oceanic areas project varying distances into this height of land.

A diagrammatic cross section through the Canadian Arctic Islands (Fig. 15) shows how that area may have existed at the end of Sequence 6, in early Tertiary time, in an intermediate stage of the Eurekan Deformation. It shows the main height of land, the Cornwall Arch, and the Sverdrup Rim.

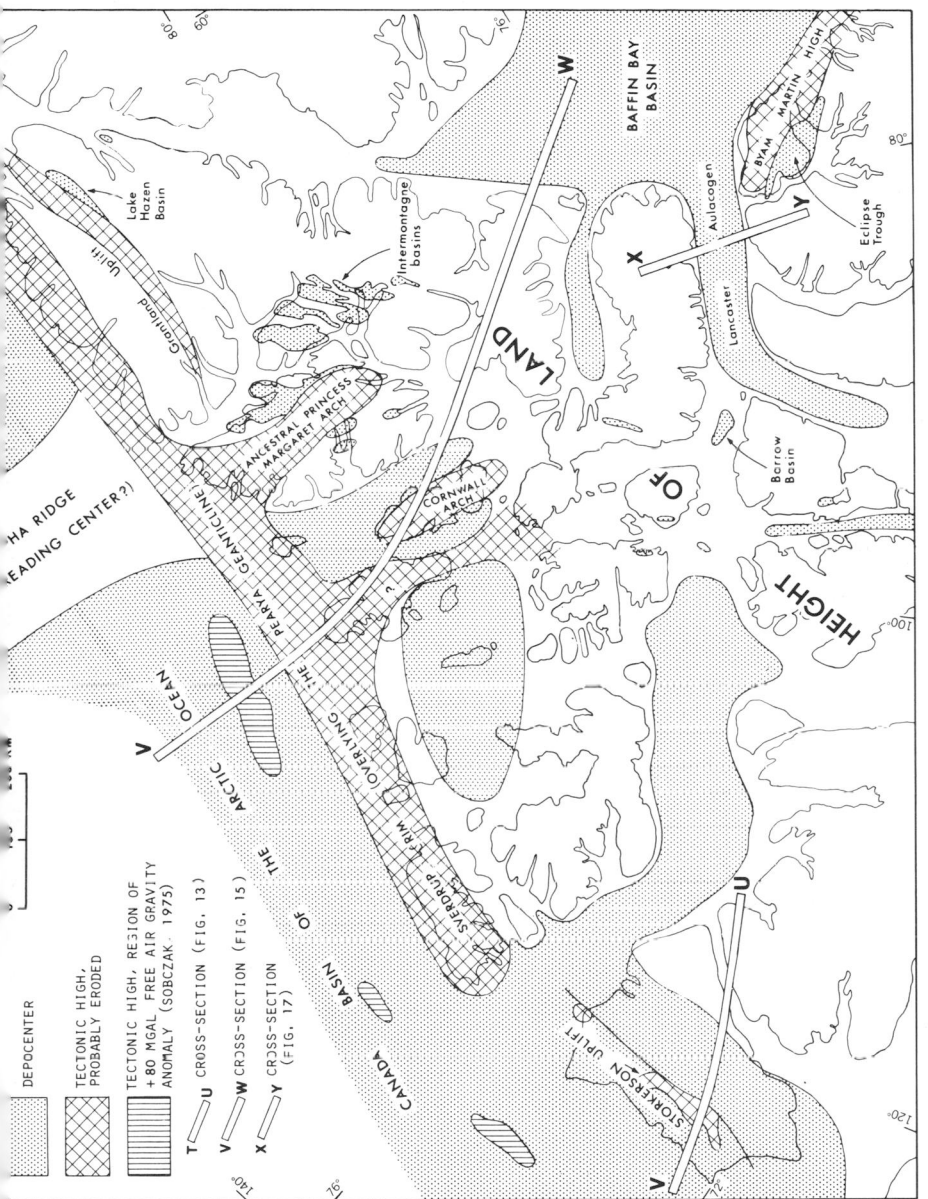

Fig. 14. Paleogeography of Sequence 6, the Eureka Sound Formation and equivalent rocks, which extends from late Cretaceous (Campanian–Maastrichtian) to early Tertiary (Paleocene–Eocene). This figure shows the transition that developed between the Atlantic and Arctic Oceans by means of the Canadian Arctic Rift System. The Canada Basin of the Arctic Ocean and the Baffin Bay Basin were actively forming rifted basins, separated by a broad height of land. These two main depocenters became connected by the rifted channels that separate the islands of the Canadian Arctic Archipelago. Finger-shaped projections of thinner rocks were deposited in parts of these channels.

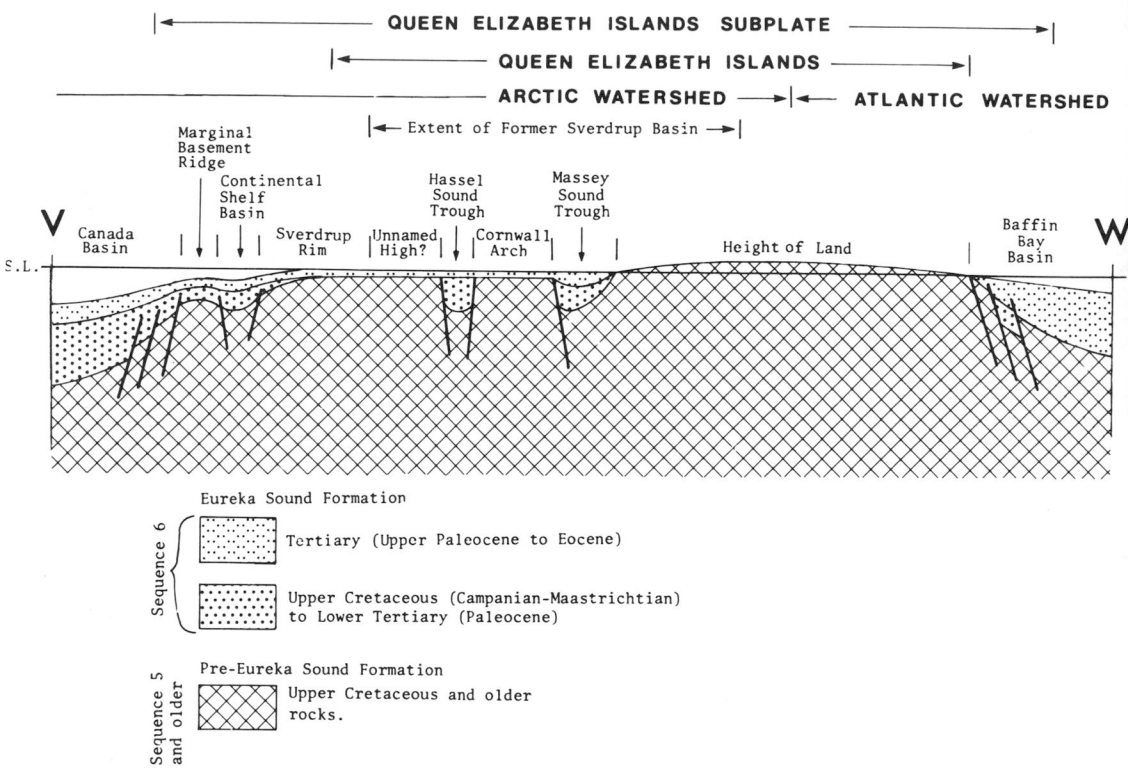

Fig. 15.   Generalized cross section from the Canada Basin to the Baffin Bay Basin reconstructed to about Eocene ti
showing Sequence 6 of the Innuitian Mobile Belt (see Figs. 3 and 4). The two basins had been actively forming as ri
oceanic basins, with the Queen Elizabeth Islands subplate being a continental transition between. The lower par
Sequence 6 was deposited mainly in the northwest, as substantial activity of the Boreal Rifting Episode continued (Fig.
early). The upper part of Sequence 6 was deposited more widely, when the Boreal Rifting Episode had nearly expired,
the Eurekan Rifting Episode was in an early stage and accelerating (Fig. 19, late). The time of the reconstruction is
prior to the climactic phase of the Eurekan Orogeny (Fig. 20), when Sequence 6 and older rocks were deformed by ex
sion southeast of the height of land and by compression in the northwest.

The Canada Basin of the Arctic Ocean apparently had been in existence
continuously since late Paleozoic or earlier time, and therefore/presumably
includes rocks older than Sequence 6. Thus the thick column along the Arctic
margin northwest of Sverdrup Rim (Figs. 14 and 15) probably includes Upper
Paleozoic, Triassic, Jurassic, and Cretaceous rocks in its lower part. Sequence
6 then presumably lies above these rocks along the margin of the Canada
Basin (Fig. 15), having accumulated contemporaneously with the Eurekan
Deformation (Fig. 4).

Sequence 6 apparently was deposited as plate movements expanded the
Canada Basin and the Baffin Bay Basin more or less stimultaneously. These
plate movements formed the basins within which Sequence 6 was deposited by

downfaulting them, but also raised adjacent areas to serve as sedimentary sources.

The Atlantic and Arctic Oceans became partly connected via channels within the Queen Elizabeth Islands. That island area is partly land and partly sea. It is therefore a structural and physiographic transition between a branch of the Atlantic Ocean (Baffin Bay) and the Arctic ocean.

G. Plate Reconstructions

1. *Foreword*

Sequence 6 in the North American Arctic was closely controlled by the plate tectonic processes that formed lands and seas, because the sequence was deposited while those tectonic events occurred. However, there are differences in interpretation of tectonics, which focus on two major problems. These problems are (1) the origin of the Canada Basin and its relationship to Alaska and Canada, and (2) the relationship of Greenland to the rest of North America. These questions are discussed below.

2. *Canada Basin*

The Canada Basin (Fig. 1) is a large deep part of the Arctic Ocean lying northwest of the Queen Elizabeth Islands and southwest of the Alpha Ridge. It has depths greater than 3500 m over a wide area. The edge of the continental shelf between the Canada Basin and the Queen Elizabeth Islands trends northeast, and is remarkably straight. It is a lineament controlled by the Kaltag Fault (Yorath and Norris, 1975).

Numerous contrasting views exist on the origin of the Canada Basin. They have been summarized by Churkin (1973) and by Sweeney and Haines (1978). On the one hand, Meyerhoff (1970, 1973) states that the Canada Basin and other features in the Arctic ocean have been in existence since late Proterozoic or earlier time. Standing in contrast are views that the Canada Basin is a relatively young feature. These latter ideas can be grouped into theories that suggest it is formed (1) by subsidence of continental crust, and (2) by rifting and lateral plate movements. Shatskiy (1935), Puscharovskiy (1960, 1978), and Atlasov (1964) concluded that the Canada Basin formed when the continental crust subsided.

A rifting origin for the Canada Basin was first proposed by Carey (1958). A rifting origin has been supported by numerous authors, who differ on its age and mechanism. Churkin (1960) suggested that the Canada Basin is at least as old as Paleozoic, Lambert (1974) that it developed in Carboniferous time, Tailleur (1973) that it formed in post-Triassic time, Herron *et al.* (1974) that it formed in Jurassic time, Sweeney *et al.* (1978) that it began to form in early Cretaceous time, and Vogt and Ostenso (1970) that it is no older than late Cretaceous. Geophysical work indicates that the Canada Basin is a true ocean basin with an abyssal plain floored by oceanic crust (Ostenso and Wold, 1973; Hall, 1973).

This paper concludes that the Canada Basin of the Arctic Ocean is a long persisting oceanic basin, formed by the Boreal Rifting Episode that was active intermittently between late Paleozoic (Mississippian) and Tertiary time (Fig. 4). It was suggested by Lambert (1973), Sweeney (1977), and Balkwill (1978) that extensional formation of the Canada Basin also may have produced extension in the Sverdrup Basin. That relationship is supported herein. The age of the Sverdrup Basin, which formed by rifting beginning in Carboniferous time, therefore is used here as a key to determining the age of the Canada Basin. The Canada Basin may have formed at the same time, and it is presumed, therefore, that the thick wedge of sediments northwest of the Arctic Islands along the margin of the Canada Basin (Sobczak and Weber, 1973) may include rocks of Sequence 4 as old as Carboniferous (Mississippian).

The Canada Basin seems to have been persistent and the lower part of the thick continental terrace wedge of sediments northwest of the Canadian Arctic Islands (Fig. 3) is tentatively assigned to Sequence 4. It is presumed to be equivalent to the lower part of the Sverdrup Basin. The Canadian Arctic region (Fig. 1) during the deposition of Sequence 4 (Figs. 4, 11, and 12) apparently had a history in which extensional structures were propagated southeastward from the Canada Basin, part way into the continental crust. This intermittent event, the Boreal Rifting Episode, apparently had a time span from late Paleozoic (Mississippian) to mid-Tertiary (Fig. 4). The active spreading center may have been the Alpha Ridge (Figs. 11 and 14). The boundary between great extension in the Canada Basin and lesser extension within the continent appears to be the Kaltag Fault, which Norris (1974) showed as extending along the continental margin. Yorath and Norris (1975) suggest that it had right-lateral displacement on the North American mainland. This author considers that it has been a transform fault, with intermittent pulses of both normal and strike-slip displacement between late Paleozoic and Cretaceous–Tertiary time. Adjacent to the Queen Elizabeth Islands its strike-slip displacement may have been left lateral (Fig. 11).

3. *Relationship of Greenland to the Rest of North America*

The history of Sequence 6 was particularly dependent on the relationship of Greenland to the rest of North America (Fig. 1). Theories to explain this relationship fall into three categories (Fig. 16). Most theories agree on the timing of tectonic events, but disagree on their mechanism.

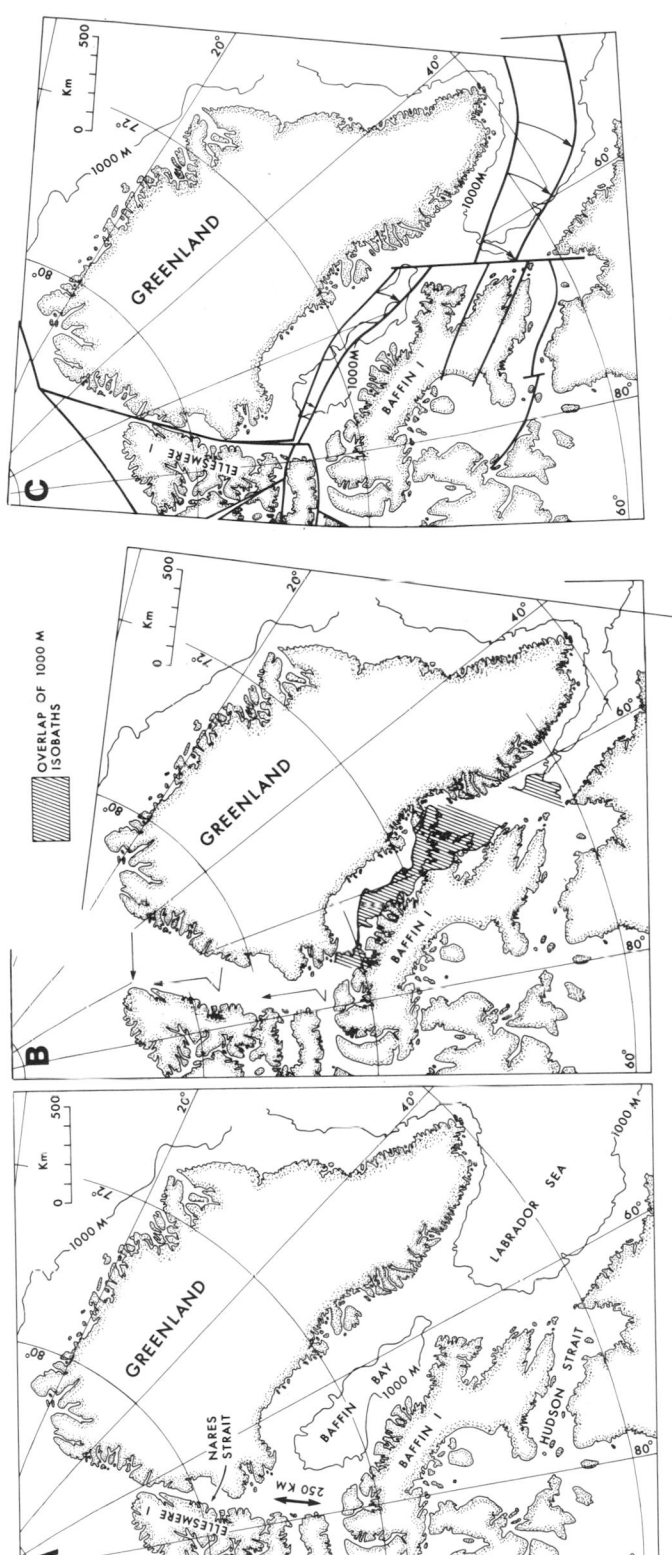

Fig. 16. Current theories of tectonics that explain the relationship of Greenland to the rest of North America. (A) Fixist theories (after Beloussov, 1970, and Meyerhoff, 1973), in which there has been no lateral movement of Greenland. (B) Conventional plate tectonic theories (moving continents or moving plates, after Wegener, 1924; Carey, 1958; Wilson, 1965b; Dewey, 1972; Herron et al., 1974; Srivastava, 1978; Newman and Falconer, 1978; and most modern-day workers), which consider that Greenland moved about 250 km along Nares Strait. The conventional reconstruction shown (after Srivastava, 1978) considers that Greenland first moved toward Ellesmere Island, closing up Nares Strait (horizontal arrows) and later had northward strike-slip displacement (vertical arrows). (C) Restoration according to the integrated theory of plate tectonics (Kerr, 1967b,c, and herein). Greenland and Canada rotated apart, with the lateral movement increasing to the southeast from Nares Strait to Baffin Bay to the Labrador Sea. The amount of restoration made here is shown by curving arrows. Much of the great movement apart in Labrador Sea is accounted for by minor rotational opening in southern Nares Strait. The 1000-m isobaths in Baffin Bay and Labrador Sea cannot be brought back together. The gap between them that appears at first to be missing continental material can be accounted for by a foundered remnant of continental crust.

(1) At one extreme are fixist theories and related oceanization theories (Fig. 16A). These both maintain that continents have always had their present geographic locations relative to each other. Shatskiy (1935) first suggested that large blocks of the continent foundered and have been oceanized where ocean now exists. This theory was modified by Beloussov (1968, 1970), who also maintains that certain marine areas formed when continental crust was down-faulted and oceanized to form oceanic crust. They are opposed to lateral drift of continents. Another fixist theory, which maintains that both continents and oceans are permanent and fixed is held by Meyerhoff (1970, 1973).

(2)   At the other extreme are theories that maintain that continents have moved great distances apart, and that their present continental shelf edges for-merly were adjacent. These suggest that Greenland drifted several hundred kilometers northward along a strike-slip fault in Nares Strait (Fig. 16B).

This lateral drift theory was first suggested by Wegener (1924), whose concept of continent drift resulted in large part from observations on Green-land. Later proponents of continental drift (Carey, 1958; Wilson, 1963a, 1965b) made little modification to Wegener's view concerning drift of Green-land though they have built upon it, and call it plate tectonics. Wilson (1963b) first suggested a Mid-Labrador Sea Ridge, and the existence of such a ridge was later confirmed (Drake et al., 1963). The Labrador Sea is also now known to have linear magnetic striping (Hood et al., 1967; Hood and Bower, 1975; Srivastava, 1978). Nearly all present-day workers who discuss tectonics of this area (Wilson, 1965b; Pitman and Talwani, 1972; Keen et al., 1972; Ross et al., 1973; Harland, 1973; Herron et al., 1974; Keen et al., 1974; Pelletier et al., 1975; Hyndman et al., 1973; Irving, 1977; Kristofferson and Talwani, 1977; Srivastava, 1978) suggest that the oceanic crust of Baffin Bay formed because Greenland drifted hundreds of kilometers laterally away from Baffin Island by a strike-slip fault in Nares Strait. They essentially conclude that the edges of the present continental shelves were once adjacent, which is the Bullard fit (Bullard et al., 1965).

The present writer refers to the group of reconstructions made or apparently favored by those papers as the *conventional plate tectonic theory*. This theory maintains that the continental shelf edges of Greenland and Baffin Island were formerly juxtaposed (Fig. 16B) and that Greenland drifted to the northeast along a left-lateral strike-slip fault in Nares Strait. Wilson called this the Wegener Transform Fault (1963a, 1965a) and indicated (1965b) that it had about 350 km of left-lateral motion. Recent workers who support this are Newman (1977) and Newman and Falconer (1978), who conclude that there was 250 km of strike-slip displacement, the amount shown in Figure 16B.

(3)   A third theory (Fig. 16C), intermediate between the extremes outlined above, was proposed over ten years ago by this writer (Kerr, 1967b,c). It sug-gests that there has been only a small to moderate amount of lateral movement

of the lands flanking Baffin Bay and the Labrador Sea. The oceanic areas between formed by a combination of (a) movement apart of plates (or drift) by rotation, and (b) subsidence and oceanization of a very large intervening segment of continental crust. The writer *suggests that the fixist theory and the conventional plate tectonic theory both are partly right.* This new proposal contains elements of and in fact reconciles both earlier theories. The reconstruction of western Greenland by Le Pichon *et al.* (1977) is somewhat similar to that of the present author (Kerr, 1967c, and herein), except that it involves somewhat more closure of the Labrador Sea.

The author's theory (Fig. 16C), which will be enlarged upon in future papers, integrates contrasting theories. It agrees with published conclusions that the Labrador Sea and Baffin Bay contain oceanic crust with linear magnetic striping. However, it disagrees with the conventional plate tectonic interpretation of how these submarine features formed. This new theory is a plate tectonic theory in that it considers that plates have moved apart. But it also is an oceanization theory, because it suggests that large parts of Baffin Bay and the Labrador Sea contain foundered and oceanized continental crust (Kerr, 1967c). The author suggests that existing plate tectonic theories and fixist theories have both gone too far and both need moderating.

4. *Baffin Bay*

The geology of Baffin Bay has been described in numerous pertinent papers over the years by Oliver *et al.* (1955), Pelletier (1966), Kerr (1967c) Hood and Bower (1975), Murray *et al.* (1970), Clarke and Upton (1971), Keen *et al.* (1972), McMillan (1973), Henderson (1973), Hyndman *et al.* (1973), Keen and Barrett (1973), Manchester and Clark (1973), Martin (1973), Ross (1973), Ross and Henderson (1973), Ross *et al.* (1973), Wallace (1973), Denham (1974), Wetmiller (1974), Beh (1975), Daae and Rutgers (1975), Ross and Falconer (1975), Jackson *et al.* (1977), and Srivastava (1978).

McMillan (1973) summarized the depth to magnetic basement along the Labrador Shelf, Baffin Shelf, and the area east of Devon Island. Along most of this coastline Precambrian Shield rocks crop out, generally in high sea cliffs. The depths to magnetic basement in the submarine areas indicate that there is a sedimentary wedge about 6000 m thick along the Baffin Island shelf. Thus Baffin Bay, whatever else it may be, is a deep depositional basin with a thick column of sediments. This column probably is largely assignable to Sequence 6 (Fig. 15).

The crust beneath central Baffin Bay was discussed by Keen *et al.* (1972), who showed that the Mohorovic discontinuity in most places is at a depth of about 10 km. They considered that the deep central region of Baffin Bay is underlain by oceanic crust. They found no conclusive evidence for a buried ridge beneath the bay, nor were magnetic lineations or fracture zones delineated at that time.

Pelletier *et al.* (1975) showed that there is a broad oceanic area in central Baffin Bay and suggested that the bay was formed by sea floor spreading. They designated an ocean–continent boundary on the basis of gravity and seismic refraction data, and superimposed this boundary on the 1000-m isobath of Baffin Bay. Their suggested boundary is located close to that obtained from aeromagnetic profiles (Hood and Bower, 1975), and they diverged no more than 50 km. Pelletier *et al.* (1975) suggested that there is a 40-km gap between oceanic and continental zones, which they suggest is occupied by non-magnetic material. They reported thick sediments in Baffin Bay.

Srivastava (1978) concluded from magnetic anomalies that seafloor spreading commenced in Baffin Bay in early Paleocene time (anomaly 24), and ceased during the early Oligocene (pre-anomaly 13). In Davis strait there is a great column of volcanic rocks (Park *et al.*, 1971), which also crops out on adjacent West Greenland (Munck and Noe-Nygaard, 1957; Rosenkrantz, 1970; Noe-Nygaard, 1974; Clarke and Pedersen, 1976) and on eastern Baffin Island (Clarke and Upton, 1971). Pelletier *et al.* (1975) reported that the main oceanic crustal layer in Baffin Bay is only 4 km thick, and omitting the sedimentary column, is significantly thinner than the normal thickness of 7 km for the crustal layer of most mature ocean basins. They followed Keen and Barrett (1972), who suggested that Baffin Bay has an oceanic crust overlain by a thick sequence of sediments.

Pelletier *et al.* (1975) suggested that geometrically the closure of Baffin Bay and Labrador Sea is effected in a reasonably satisfactory manner by two constructions that include rotation about a pole in Lancaster Sound, and translation along Nares Strait. Newman (1977) and Newman and Falconer (1978) support the suggestion that Greenland and Baffin Island have moved apart a great distance (Fig. 16B), because of their conclusion that there is 250 km of left-lateral displacement in Nares Strait. This follows essentially Wilson's (1965*b*) reconstruction, and the conventional theory of plate tectonics.

5. *Labrador Sea*

The geology of the Labrador Sea has been described in numerous pertinent papers by Kranke (1947), H. Holtedahl (1958), O. Holtedahl (1970), Drake *et al.* (1963), Wilson (1963*b*) Kerr (1967*c*), Fenwick *et al.* (1968), Grant and Manchester (1970), Mayhew *et al.* (1970), Le Pichon *et al.* (1971), Grant (1972, 1975), McMillan (1973), Austin (1973), Hyndman (1973), Hood and Bower (1975), Athavale and Sharma (1975), Beh (1975), McWhae and Michel (1975), Van der Linden (1975*a,b*), Van der Linden and Srivastava (1975), Kirstofferson and Talwani (1977), Wade *et al.* (1977), and Srivastava (1978).

The depth to magnetic basement was summarized by McMillan (1973), who showed that there is a thick wedge of sediments as much as 13000 m thick along the western margin of the Labrador Sea. He suggested that the western continental shelf is underlain at depth by fault blocks that formed by growth faults over which the sedimentary column is draped. The part of this column that was deposited during active tectonism is assignable to Sequence 6.

Srivastava (1978) concluded from magnetic anomalies that active sea floor spreading commenced in the southern Labrador Sea during the Campanian (anomaly 32) and in the northern Labrador Sea during the Maastrichtian (anomaly 28). The spreading ceased in the Labrador Sea as well as in Baffin Bay during the early Oligocene (pre-anomaly 13).

6. Nares Strait

A most crucial area to the interpretation of the origin of Baffin Bay and Labrador Sea is Nares Strait (Figs. 1 and 16), which separates Greenland from Ellesmere Island. There are two opposed views on the amount of displacement in Nares Strait.

(1)   The conventional plate tectonic theory requires several hundred kilometers of left-lateral strike-slip faulting in Nares Strait (Wilson, 1965b; Pitman and Talwani, 1972; Keen et al., 1972; Ross et al., 1973; Herron et al., 1974; Keen et al., 1974; Pelletier et al., 1975; Hyndman et al., 1973; Kristofferson and Talwani, 1977; Srivastava, 1978; Newman, 1977; Newman and Falconer, 1978). The most recent work (Newman, 1977; Newman and Falconer, 1978) uses a figure of 250 km, and this displacement was used in the conventional plate tectonic reconstruction (Fig. 16B).

(2)   The integrated theory of plate tectonics (Fig. 16C) of Kerr (1967b,c) interprets the strait as a submarine rift valley that behaved partly as a transform fault system, but concludes that there was very little strike-slip movement. Evidence for and against strike-slip movement in Nares Strait is presented below.

Recent suggestions that there has been great strike-slip movement in Nares Strait have resulted primarily from marine geophysics, and nearly all of this has been done outside of Nares Strait itself. The reasoning seems to be: (1) Baffin Bay contains oceanic crust; (2) oceanic crust forms by spreading apart of continental blocks; (3) the only zone that could have allowed Greenland to spread laterally from Baffin Island is along Nares Strait; (4) therefore Nares Strait contains a left-lateral strike-slip fault.

Early workers (Koch, 1929; Troelson, 1950) showed geological maps that included both sides of Nares Strait. From the evidence in those maps it was reasonable to suggest moderate displacement, but not hundreds of kilometers. Later work at Nares Strait (Kerr, 1967b) suggested that there was no more than several kilometers of strike-slip offset. Subsequently Dawes (1973) concluded that the present state of geological knowledge could not rule out displacement up to 250 km. Thus Keen *et al.* (1972) learned by personal communication with Dawes that a displacement of 150 to 200 km could not be disproved by the geology. They considered this to support their view (Keen *et al.*, 1972) that a reconstruction as in Figure 16B is the most reasonable. After considerably more field work along the entire Greenland coast bordering Nares Strait, Dawes presently considers (personal communication, 1978) that the geology suggests essentially no lateral displacement along Nares Strait, but that minor displacement in a left-lateral sense cannot be discounted. The present author holds to his original conclusion (Kerr, 1976*b*) that Nares Strait had only minor rotational opening from the south and minor strike-slip displacement. This strike-slip displacement, which is left lateral, could be as much as 25 km. An amount between 0 and 25 km probably cannot be resolved by comparison of facies belts.

The conventional plate tectonic reconstruction (Fig. 16B) recognizes that the amount of displacement is a matter of debate, but neglects the fact that the geology can put restraints on it. No reasonable interpretation of the geology of the two sides of Nares Strait can permit the left-lateral displacement required in reconstructions made by conventional plate tectonic theories (Fig. 16B), i.e., about 250 km. The geology places an upper limit of about 15 km on the displacement. That distance is the amount used in the present reconstruction (Fig. 16C).

7. *Summary of Greenland Relationships*

The three theories outlined above were applied to the Labrador Sea–Baffin Bay region where Sequence 6 is present, and the three maps (Fig. 16) represent reconstructions using the three methods. The theories of Beloussov (1970) and Meyeroff (1973) would maintain that the bordering lands have had fixed positions similar to those of the present (Fig. 16A). Alternatively, most field workers of those who have speculated on the origin of the Labrador Sea and Baffin Bay (Wilson, 1965*b*; Pitman and Talwani, 1972; Keen *et al.*, 1972; Pelletier *et al.*, 1975; Kristofferson and Talwani, 1977; Srivastava, 1978), favor the conventional plate tectonic theory (Fig. 16B). It indeed appears to be true that Labrador Sea and Baffin Bay formed by rifting and represent a branch of the rifted North Atlantic Ocean. The existence of rifting, along with evidence of marine geophysics, including magnetic striping, led to the further conclusion that Labrador Sea and Baffin Bay formed by great lateral movement apart of Greenland from the rest of North America. The lateral movement normally suggested in Labrador Sea is about 650 km if the edges of the present continental shelves were formerly joined and the displacement on Nares Strait is about 250 km (Fig. 16B).

The author's integrated theory of plate tectonics (Fig. 16C) suggests that Labrador Sea and Baffin Bay formed by a combination of lateral separation of plates by rotation, with foundering and oceanizing of a large remnant of continent between them (Kerr, 1967c). The restoration to an original predrift configuration (Fig. 16C) follows this reasoning and suggests that the lands bordering Baffin Bay and Labrador Sea moved apart by an amount that increases from northwestern Baffin Bay to southwestern Labrador Sea. The amount of restoration is measured by the movement back of the present 1000-m isobaths, which are approximately at the shelf edges. Those lines are not brought back completely together as was done by Bullard *et al.* (1965), and in other conventional reconstructions.

This integrated theory does not disagree with any of the geophysical observations or evidence from Labrador Sea or Baffin Bay, or with the geophysical interpretation of those observations. It disagrees only with the geological interpretation of those geophysical observations. This theory involves the lateral separation of plates, and in a reconstruction brings them back to their suggested original positions. However, it implies that the plates did not travel as far as conventional plate tectonic theory suggests.

The origin of the Baffin Bay–Labrador Sea area is intimately related to the origin of the Canadian Arctic Islands. The history of each area puts certain restrictions on the interpretation that can be allowed in the other area.

H. Eurekan Deformation

The Eurekan Deformation (Kerr, 1977a, 1980b) was a tectonic event that formed much of the southeastern part of the Canadian Arctic Rift System between late Cretaceous (Campanian/Maastrichtian) and mid-Tertiary (Miocene or possibly later) time (Fig. 4). The nature of this deformation is compatible with the reconstruction of Greenland made earlier (Kerr, 1967c) and outlined above (Fig. 16C).

The Eurekan Deformation includes two related and complementary structural phenomena. These are the Eurekan Rifting Episode and the Eurekan Orogeny, which are respectively the extensional and the compressional phases of a plate tectonic episode. The Eurekan Rifting Episode (Kerr, 1980b) originated in the southeast and advanced northwestward by propagating extension faults or rifts that resulted as Greenland and the rest of North America moved apart. The Eurekan Orogeny (Thorsteinsson and Tozer, 1970; Balkwill, 1978) was a compressional deformation located northwest of the area of rifting. The region in the southeast being subjected to extension was separated from the region to the northwest being subjected to compression, with an area of transition lying between them.

1. *Eurekan Rifting Episode*

The Eurekan Rifting Episode (Kerr, 1980*b*) was the rifting phase of the Eurekan Deformation. The rifting was solely responsible for producing most of the faults and fault-controlled channels in the southeastern part of the Canadian Arctic Rift System. It was responsible for intensifying the faults in the northwest part of the rift system that had been initiated by the Boreal Rifting Episode.

The Eurekan Rifting Episode began in the southeast in the area of the Labrador Sea and later affected Baffin Bay. It apparently formed those waterways by plate separation and foundering (Kerr, 1967*c*). By late Cretaceous time (?Campanian/Maastrichtian), as Greenland and the rest of North America rotated apart, the rifts or faults had been propagated northwestward into the Canadian Arctic Islands. Rifting in the Eurekan Episode was greatest at sea, but it also caused extension within the islands themselves.

Lancaster Aulacogen (Figs. 14 and 17) is a complex asymmetrical linear graben, and is the type of structure that was formed by extension during the Eurekan Rifting Episode. On the line of cross section (Fig. 17) the vertical displacement on the Parry Channel Fault during the rifting episode was about 8200 m, and the aulacogen contains 6000 m of unconsolidated to semiconsolidated Cretaceous to modern sediments. Displacement and thickness both increase eastward from that line toward Baffin Bay, and decrease westward. The aulacogen may contain rocks assignable to Sequences 5, 6, 7, and 8. Before the aulacogen developed the region had a rather flat topography, with Lower Paleozoic platform facies sediments (Sequence 3) exposed, presumably horizontal to gently northwest dipping. The Canadian Shield was the foundation of the continent, with remnants of a Proterozoic basin (Sequence 2) lying above it.

The aulacogen was dominated by the Parry Channel Fault, which extends along its north side and developed in the Eurekan Rifting Episode. It was first suggested by Wegener (1924), in his early work on continental drift, that Parry Channel was fault controlled. The Parry Channel Fault and other faults in the Lancaster Aulacogen are supported by various types of geophysical work (Gregory *et al.*, 1961; Barrett, 1966; Daae and Rutgers, 1975).

Sequence 5 may be present beneath Lancaster Sound in the lower part of the wedge-shaped sedimentary body of Cretaceous and younger sediments there (Fig. 17). It presumably is a remnant of a sheetlike deposit of the cratonic shelf (Fig. 12). Most of the semiconsolidated column in the aulacogen, however, may be part of Sequence 6, which was deposited as the aulacogen formed during the Eurekan Deformation. Faulted sediments overlain by undeformed sediments have been reported in Lancaster Sound (Jackson *et al.*,

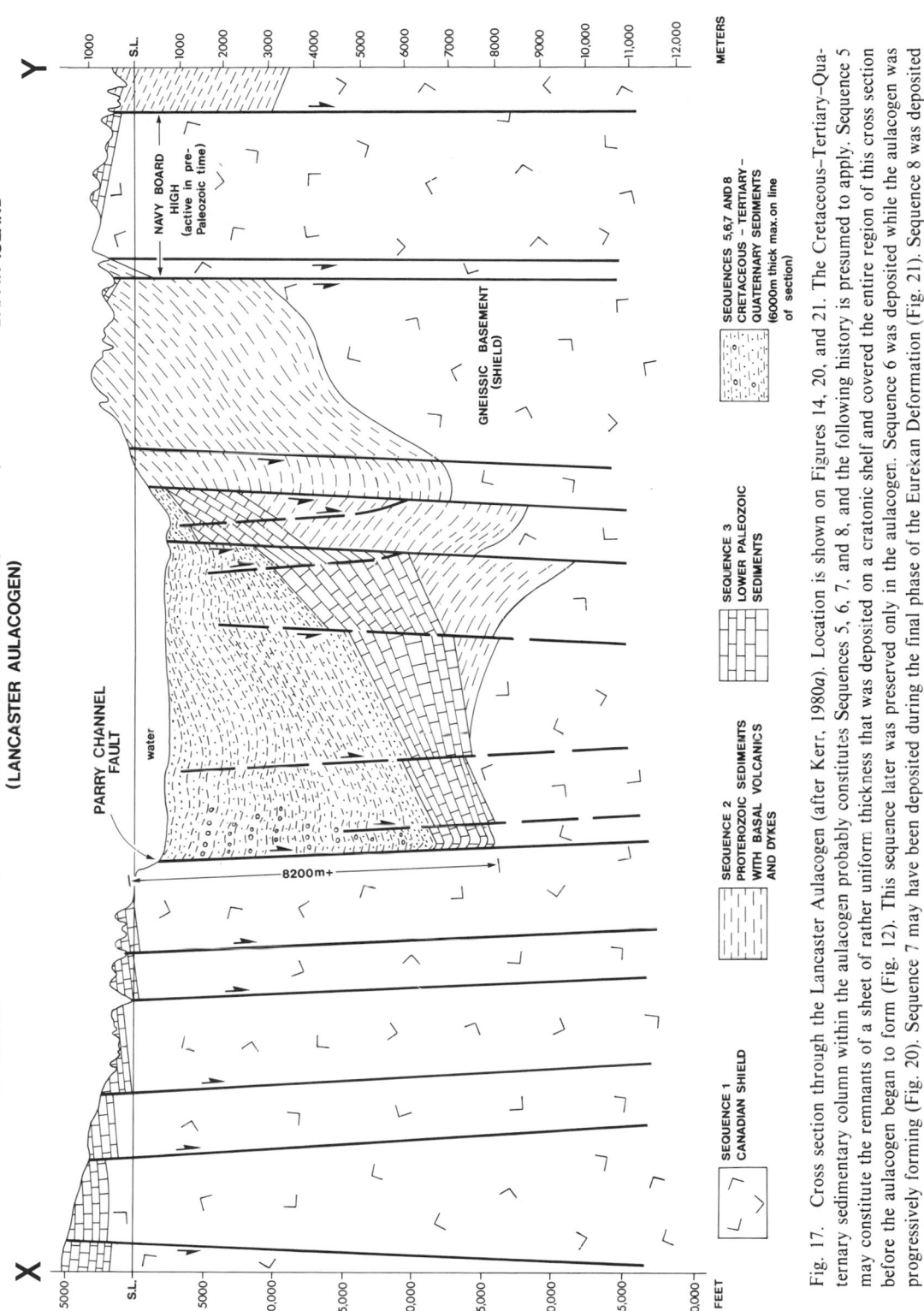

Fig. 17. Cross section through the Lancaster Aulacogen (after Kerr, 1980*a*). Location is shown on Figures 14, 20, and 21. The Cretaceous–Tertiary–Quaternary sedimentary column within the aulacogen probably constitutes Sequences 5, 6, 7, and 8, and the following history is presumed to apply. Sequence 5 may constitute the remnants of a sheet of rather uniform thickness that was deposited on a cratonic shelf and covered the entire region of this cross section before the aulacogen began to form (Fig. 12). This sequence later was preserved only in the aulacogen. Sequence 6 was deposited while the aulacogen was progressively forming (Fig. 20). Sequence 7 may have been deposited during the final phase of the Eurekan Deformation (Fig. 21). Sequence 8 was deposited after the period of active formation of the aulacogen.

1977), where the uppermost 2 km of the sedimentary section are not cut by major faults. The faults in Lancaster Aulacogen therefore originated at depth and were propagated upward. Some of these faults die out upward, so there are folds rather than faults near the surface, and the upper part of the column is not folded. Lancaster aulacogen (Figs. 14 and 17) is an indentation from the Baffin Bay Basin into the Canadian Arctic Islands. It is the type of structure formed in the major channels within the Canadian Arctic Archipelago by the Eurekan Rifting Episode.

2. *Eurekan Orogeny*

The Eurekan Orogeny (Thorsteinsson and Tozer, 1970; Balkwill, 1978) was a secondary phenomenon of the Eurekan Deformation (Kerr, 1980*b*). It developed in response to the Eurekan Rifting Episode, and was the complementary compressional phenomenon. Regionally the area of orogeny was apically opposed to the rifted area, with the two separated by a large area that constituted several transform pivots, by means of which extensional defomation was transformed into compression.

At the onset of the Eurekan Orogeny sedimentary and tectonic patterns changed rapidly in the Sverdrup Basin. That basin had undergone almost continuous subsidence and sediment accumulation for about 260 m.y., from late Paleozoic onward. In late Cretaceous (Campanian/Maastrichtian) time, the Eurekan Orogeny began to fragment and fold this basin, and Sequence 6 began to be deposited.

Balkwill (1978) recognized three phases of tectonism that he assigned to the Eurekan Orogeny: (1) regional uplift and erosion of the Sverdrup Rim and broad intrabasin arches in latest Cretaceous and early Tertiary; (2) compressive folding and faulting of the eastern part of Sverdrup Basin in the interval between middle Eocene and early Miocene; and (3) rejuvenated uplift of some arches in Miocene, and possibly Pliocene and later time. The first of these is herein assigned to a late phase of the Boreal Rifting Episode. The second is considered to be the Eurekan Orogeny, and the third is assigned to the final phase of the Eurekan Rifting Episode.

In the Sverdrup Basin the Eureka Sound Formation is a syntectonic deposit that was deposited during the first two events outlined above. The oldest part of the formation, Upper Cretaceous to Paleocene, was deposited adjacent to uplifted intrabasin arches (Fig. 15). A higher part (Paleocene) was more widespread and encroached onto those arches after there was a general collapse within the Sverdrup Basin. Still higher parts of the formation (Eocene and possibly Oligocene) were deposited contemporaneously with the compressive folding and faulting of the Eurekan Orogeny.

The Eureka Sound Formation (Fig. 15) is syntectonic, deposited during the first two of the above three phases. The height of land (Figs. 14 and 15) appears to have been a sedimentary source during most of Eureka Sound time, with detritus being shed northwestward and southeastward from it by rivers that fed into the interarch basins and graben-controlled channels.

I. Sequence 6: Penetration of the Continent by Rifting

The Canadian Arctic Rift System is a major feature that includes structures in the Labrador Sea, Baffin Bay, the Queen Elizabeth Islands, and that part of the Kaltag Fault lying northwest of the Queen Elizabeth Islands (Fig. 18). The Kaltag Fault is a boundary that separates the Canadian Arctic Rift System from other rifted structures farther northwest.

The Canadian Arctic Rift System severed the North American continent and connected the Atlantic and Arctic Oceans via the Canadian Arctic Islands. The connecting process involved a number of stages and is summarized in four sequential figures (Figs. 18–21). Two plate tectonic events that emanated from different oceans contributed to forming the Canadian Arctic Rift System. The initial penetration of the continent by the rift system was by southeastward propagation of extensional faults of the Boreal Rifting Episode during deposition of Sequences 4 and 5. This extension reached only into the central Queen Elizabeth Islands and was aborted there, apparently because of interfering structural trends. Most of the Canadian Arctic Rift System was formed by the Eurekan Rifting Episode, whose main activity occurred during deposition of Sequence 6.

Sequence 6 of the Innuitian Mobile Belt (Figs. 3 and 4) was deposited during a time when penetration of the North American continent by the Canadian Arctic Rift System increased greatly. Increased crustal fracturing occurred in the Sverdrup Basin, presumably due to an enlarged pulse of the Boreal Rifting Episode. The major event during deposition of Sequence 6, however, was the initiation and rapid development of the Eurekan Rifting Episode, which was propagated northwest from the Atlantic Ocean through Labrador Sea and Baffin Bay into the Queen Elizabeth Islands. The rift system advanced relatively quickly there and drastically altered depositional patterns (cf. Figs. 11 and 14); however, it was not able to break fully through the continent until somewhat after deposition of Sequence 6.

In early Cretaceous time, prior to the tectonic events that controlled deposition of Sequence 6 (Fig. 4), the Canadian Arctic was part of a rather stable continent (Fig. 18). Much of the present island and continent area was covered by very shallow seas, where thin sediments of Sequence 5 were accumulating (Fig. 12). The water probably was shallow on the Sverdrup Rim and to the southeast on the cratonic shelf, and was somewhat deeper in the Sverdrup Basin. The Canada Basin of the Arctic Ocean already existed as an

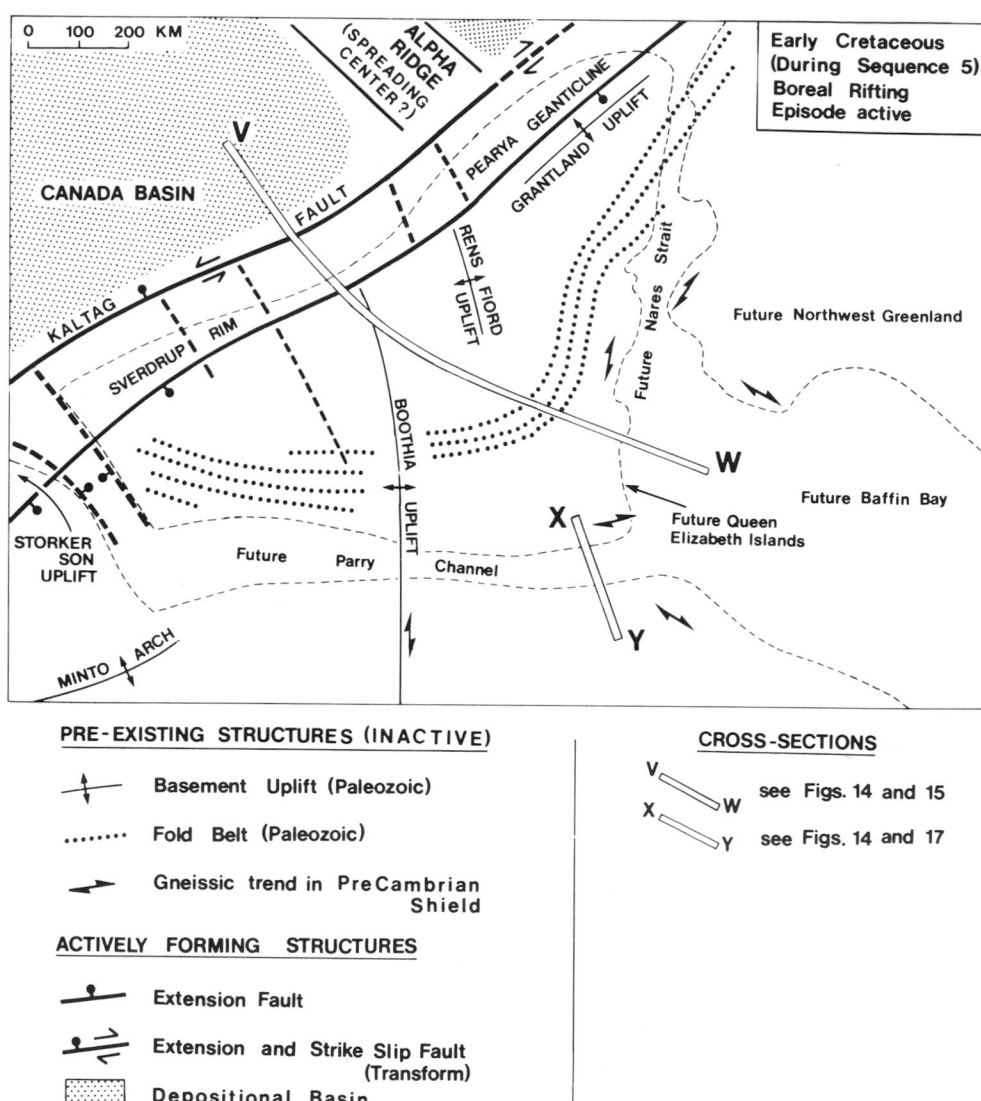

Fig. 18. Early formation of the Queen Elizabeth Islands by the Canadian Arctic Rift System. Dashed lines show outline of Queen Elizabeth Islands and main channels. This represents the structural situation existing in late Cretaceous time (pre-Eureka Sound Formation) that had been developed by the Boreal Rifting Episode (Fig. 4). The Canada Basin was being enlarged by the rifting episode. The continent was covered by an overlap of shallow seas that were depositing Sequence 5 (Fig. 12). The Pearya Geanticline and Sverdrup Rim were mildly positive. Trends of buried uplifts and gneisses influenced this and future rifting deformation.

oceanic basin at this time, lying northwest of the continent. It may have formed adjacent to a spreading center occupying the Alpha Ridge. Rifted branches that emanated from the Canada Basin presumably projected some distance southeastward into the Queen Elizabeth Islands. These had been formed earlier as a result of the Boreal Rifting Episode that was active during Sequences 4 and 5, and which continued to develop at this time (Fig. 18).

Prior to the Eurekan Deformation (Fig. 18) the present Arctic Islands and the continent to the south presumably were part of a continental crust of normal thickness (Fig. 3) that was made up of thick crystalline crust in its lower part (Sequence 1) and sedimentary basins in its upper part (Sequences 2, 3, 4, and 5). This crust contained within it at that time gross regional structural trends that had been actively forming in Paleozoic time, but probably were mainly inherited from an underlying Precambrian crystalline basement. Some of the structural trends that are known to have existed at the onset of rifting (Fig. 18) are those of the sedimentary cover or basement uplifts. It is inferred that there was an even stronger structural grain existing in the crystalline basement beneath those uplifts and having the same trends. The structural trends of the crystalline basement controlled the deformation that was produced by the Boreal Rifting Episode, as well as by the younger Eurekan Rifting Episode and the Eurekan Orogeny.

In late Cretaceous time (Fig. 19), the Canadian Arctic Islands region was subjected to two rifting events (Kerr, 1980b). The Boreal Rifting Episode emanating from the Canada Basin experienced enhanced activity that caused the Pearya Geanticline and its continuation southwest as the Sverdrup Rim to be actively raised (Fig. 19, early). According to Meneley et al. (1975) and Balkwill (1978), a few thousand meters of erosion occurred on the Sverdrup Rim in late Cretaceous time, prior to Eureka Sound deposition. Marine withdrawal from Sverdrup Basin proceeded southward at this time, prompted partly by the uplift and erosion along Sverdrup Rim. Two arches within the Sverdrup Basin were also raised and several thousand meters of erosion occurred on them (Fig. 15). Included were the Cornwall Arch (Balkwill, 1974), which may overlie a northward extension of the Boothia Uplift (Fig. 18), and the southward plunging ancestral Princess Margaret Arch (Balkwill et al.,1975), which overlies the Rens Fiord Uplift. That part of the Canadian Arctic Islands that included the Sverdrup Basin was being deformed into linear highs and lows. They were expressed at the surface as anticlines, but probably were fault blocks at depth, bounded by extension faults that were being propagated southeastward into the Canadian Arctic Islands from the Canada Basin. As this rifting deformation progressed and the structural highs within the Queen Elizabeth Islands were raised (Fig. 19, early), there was syntectonic deposition in the lows of thick Upper Cretaceous to Lower Tertiary sandstone of the Eureka Sound Forma-

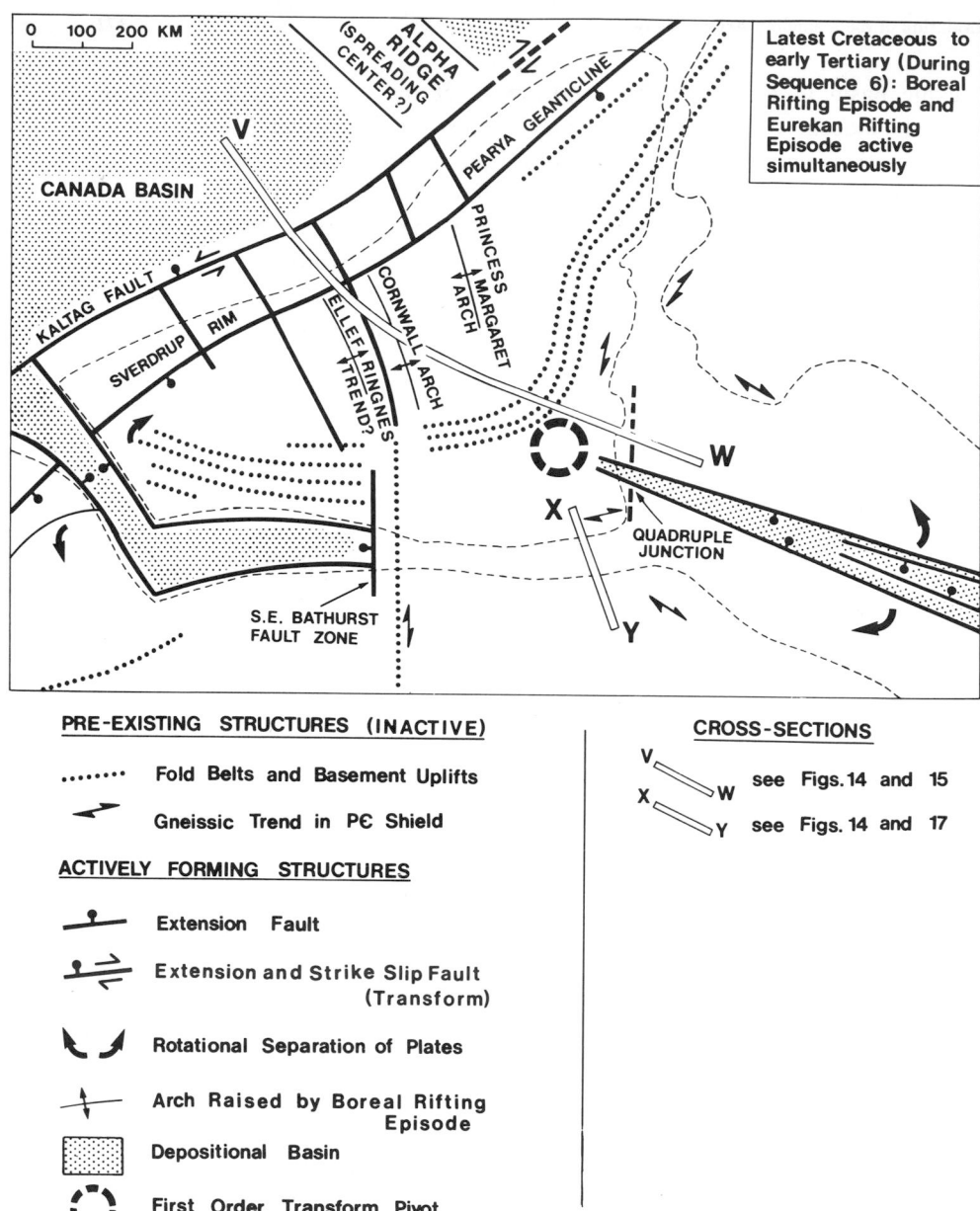

Fig. 19.   Simultaneous development of the Boreal Rifting Episode in the northwest and the Eurekan Rifting Episode in the southeast. Two time intervals are represented here. In late Cretaceous time, crustal fracturing emanating from the northwest caused the Pearya Geanticline, Sverdrup Rim, Storkerson Uplift, and arches within the Sverdrup Basin to be uplifted and eroded. Parry Channel probably opened from the west, terminated at the Southeast Bathurst Fault Zone within the Boothia Uplift. Simultaneously crustal fracturing also was being propagated to the area from the southeast. Later, in latest Cretaceous to early Tertiary time, further development of the Eurekan Rifting Episode began to neutralize the rifting effects of the Boreal Rifting Episode in the Sverdrup Basin. This resulted in widespread collapse of parts of the formerly uplifted arches, Sverdrup Rim, Storkerson Uplift, and the intrabasin arches, with sedimentary encroachment of Upper Paleocene to Eocene rocks onto each (Fig. 15).

tion. By late Paleocene time, the ancestral Princess Margaret Arch also was an active sedimentary source (Balkwill, 1978) that formed a major element of the height of land separating the Atlantic and Arctic watersheds.

The northeast trending Pearya Geanticline, which is a Precambrian-cored Lower Paleozoic structure, may have controlled the location of the younger Kaltag Fault, and thereby also the location of the margin of the Canada Basin (Fig. 19). Faults of the Boreal Rifting Episode were able to breach the Pearya Geanticline, but the distance they penetrated southeastward into the continent also appears to have been controlled by older structural trends farther southeast (Fig. 19).

Simultaneous with the Boreal Rifting Episode (Fig. 19, early) incipient rifts also were being propagated northwestward into the continental crust of the Canadian Arctic from a pull apart margin in the southeast. This was the Eurekan Rifting Episode (Kerr, 1980*b*). This rifting probably began to affect the Baffin Bay area by about latest Cretaceous to early Tertiary time, when tectonism also caused deposition of continental rocks of the Eureka Sound Formation in the Eclipse Trough of northern Baffin Island (Jackson and Davidson, 1975; H. R. Balkwill, personal communication, 1978). In this stage, in late Cretaceous time, there apparently was rifting in the southeast and northwest, with little or none in the region between.

Incipient rifting may have begun to form the future Baffin Bay Basin by Campanian/Maastrichtian time (Fig. 19, early). At approximately this time, separation apparently was further advanced in the Labrador Sea. The Labrador Sea had now reached the stage where magnetic lineations were forming in an oceanic crust. Srivastava (1978) reports that magnetic lineations formed in the southern Labrador Sea by Campanian time and in the northern Labrador Sea by Maastrichtian time.

By Paleocene time, faulting and spreading in Baffin Bay were sufficiently well developed for magnetic striping to begin to form (Srivastava, 1978). A pivotal area presumably existed at this time in the southeastern part of the Queen Elizabeth Islands, as a first-order transform pivot (Fig. 19, late). The separation within Baffin Bay created rotation, which occurred about the transform pivot area in the northwest. A four-armed junction of rifted zones, or a quadruple junction, may have begun to form at this time in northwestern Baffin Bay, and is still expressed in today's topography (Fig. 1). These arms formed Baffin Bay, Parry Channel, Nares Strait, and Jones Sound.

Increased separation in the southeast by Paleocene time may have caused the reduced rifting of the Boreal Rifting Episode in the northwest, and the widespread collapse of highs in the Sverdrup Basin. Upper parts of the sandy Eureka Sound Formation overstepped onto the flanks of the formerly high structures, onto the Sverdrup Rim in late Cretaceous to early Tertiary time,

(Balkwill, 1978), and onto the Cornwall Arch (Balkwill, 1974) in early Tertiary (Paleocene–Eocene) time (Fig. 15 and Fig. 19, late).

The Banks Island area was also fragmented by local uplifts in late Cretaceous time (Miall, 1975). This fragmentation may have extended eastward into Parry Channel as far as the Bathurst Island Fault Zone (Fig. 19), which formed in the late Cretaceous (Maastrichtian) time (Kerr, 1974).

Balkwill (1974, 1978) attributes the late Cretaceous–early Tertiary fragmentation of the Sverdrup Basin and elevation of intrabasin arches to a mechanism of crustal fracturing. This fracturing was most prevalent in structures on the northwest side of the study area near the northwest margin of the Arctic Islands, and diminished southeastward (Fig. 19). The present writer infers that these tectonic events stemmed from a larger event in the Canada Basin, perhaps a pulse of renewed rifting. Contemporaneous fracturing took place in the southeast and diminished northwestward. Thus, in late Cretaceous to early Tertiary time, rifting apparently took place simultaneously both northwest and southeast of the major height of land (Figs. 14 and 19).

It appears then that, in late Cretaceous time, Maastrichtian and possibly Campanian (Fig. 19), there were two episodes of extensional or rifting deformation being propagated toward the Queen Elizabeth Islands, one from the northwest and the other from the southeast (Kerr, 1980b).

In the northwest fragmentation by rifting in late Cretaceous time (Fig. 19, early) initiated marked uplift and erosion of Pearya Geanticline, Sverdrup Rim, and structures within the Sverdrup Basin (Balkwill, 1978), and raised the Storkerson Uplift (Miall, 1975). By latest Cretaceous to early Tertiary (Paleocene) time, rifting had become more advanced in the southeast, and ended in a large region in the northwest where there was then general collapse (Fig. 19, late). This led to widespread early Tertiary sedimentary overlap of some of the formerly uplifted structures by the Eureka Sound Formation (Fig. 15). The Pearya Geanticline remained high on Ellesmere Island, for the Lower Tertiary strata on Ellesmere Island become progressively finer and more marine southward, suggesting a northwerly source prograding southeastward (Bustin, 1977). This high may have been related to the proximity of a spreading center nearby to the north, occupying the Alpha Ridge.

The first-order transform pivot developed southeast of a prominent preexisting structural trend within the Queen Elizabeth Islands (Fig. 19, late). It appears that those trends impeded the northwestward advance of extensional faults of the Eurekan Rifting Episode, thereby controlling the location of the pivot. Subsequent rotation about that pivot produced the compression of the Eurekan Orogeny (Fig. 20).

The Eurekan Deformation reached its climactic phase (Fig. 20) in mid-Tertiary time (mid-Eocene to early Miocene). In this stage, spreading continued in Baffin Bay, and the rifts were propagated farther northwest. The

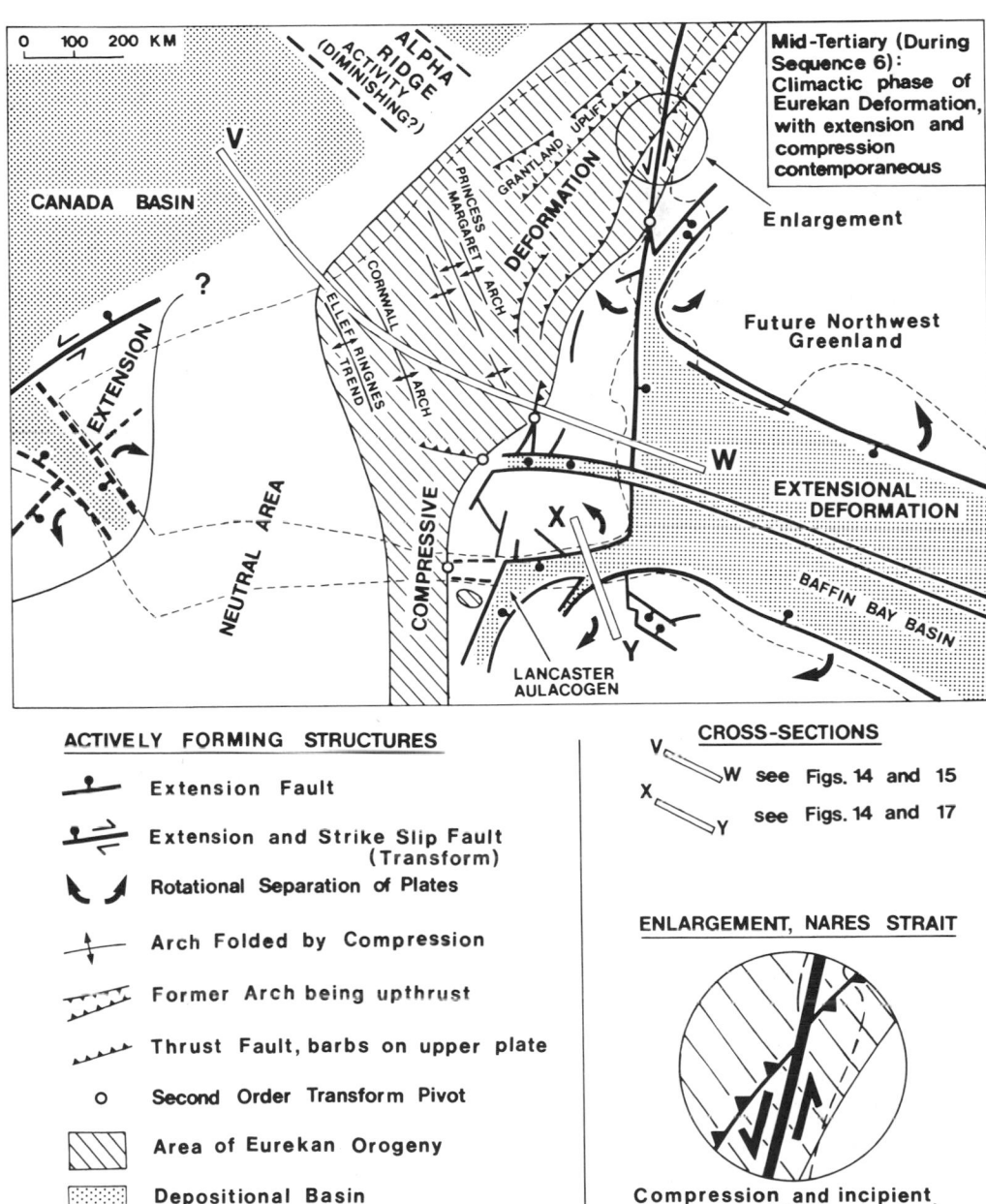

0   100   200 KM

Mid-Tertiary (During Sequence 6): Climactic phase of Eurekan Deformation, with extension and compression contemporaneous

CANADA BASIN

ALPHA RIDGE (ACTIVITY DIMINISHING?)

V

?

EXTENSION

NEUTRAL AREA

COMPRESSIVE

PRINCESS MARGARET ARCH

CORNWALL

ELLEF RINGNES TREND

ARCH

GRANTLAND UPLIFT

DEFORMATION

Enlargement

Future Northwest Greenland

W

X

Y

EXTENSIONAL DEFORMATION

BAFFIN BAY BASIN

LANCASTER AULACOGEN

ACTIVELY FORMING STRUCTURES

Extension Fault

Extension and Strike Slip Fault (Transform)

Rotational Separation of Plates

Arch Folded by Compression

Former Arch being upthrust

Thrust Fault, barbs on upper plate

o   Second Order Transform Pivot

Area of Eurekan Orogeny

Depositional Basin

CROSS-SECTIONS

V⊂──W see Figs. 14 and 15

X⊂──Y see Figs. 14 and 17

ENLARGEMENT, NARES STRAIT

Compression and incipient Strike Slip (Transpression)

Fig. 20. Climactic phase of the Eurekan Deformation, in mid-Tertiary time (between mid-Eocene and early Miocene), when extension in the southeast (Eurekan Rifting Episode) caused compressional deformation farther northwest (Eurekan Orogeny). Farther west, in a large neutral area, extension that earlier emanated from the Canada Basin was neutralized by the Eurekan Deformation. In the extreme west, in the Banks Island area, extension of the Boreal Rifting Episode continued uninterrupted, presumably because this was too far west to be affected by the Eurekan Deformation. This followed shortly after the situation depicted in Fig. 15.

advance of the rifts northwestward in the Queen Elizabeth Islands was impeded by the preexisting structural trends (Fig. 18), some of which were transverse to their paths and deflected their propagation (Fig. 20). One rift zone was propagated westward into Lancaster Sound to begin the formation of Lancaster Aulacogen (Figs. 14 and 17). It was then further deflected southwestward by the Boothia Uplift. Another rift zone was propagated northward into Nares Strait, but its advance also was impeded (Kerr, 1967b). Nares Strait appears to have had only minor strike-slip displacement at this time, perhaps 15 km (Fig. 20, enlargement). At this time Baffin Bay continued to develop as a rifted basin, with the blocks on its sides that were in the future to form Greenland and Baffin Island rotating apart. A region in the northwest was subjected to compressional deformation. This occurred because the blocks bordering Baffin Bay continued to rotate apart but the rifts could not propagate to the northwest, northeast, or west because of obstacles. This situation produced the Eurekan Orogeny (Fig. 20). As the two plates that now contain Greenland and Baffin Island moved apart slightly, a broad region of continental crust between them apparently foundered and was oceanized, according to Kerr (1967c).

The climactic phase of the Eurekan Deformation (Fig. 20) involved compressive folding and faulting in eastern parts of the Sverdrup Basin and in the underlying rocks, in mid-Eocene to early Miocene time, while extension continued in the Baffin Bay Basin. Thus, for a time, extension in the southeast (Eurekan Rifting Episode) apparently was coeval with compression in the northwest (Eurekan Orogeny). There was an area of transition between. This writer suggests that the extension in the southeast was the cause of the compressional deformation in the northwest. The effects of compressional deformation are most obvious in the Sverdrup Basin (Balkwill, 1978), because the existing rocks there were largely undeformed when compression began. Nevertheless compressional deformation of the Eurekan Orogeny also affected older rocks of the Franklinian Geosyncline and Arctic Platform that had been deformed previously (Kerr, 1967b; Kerr and Thorsteinsson, 1972). In those older rocks the effects of the Eurekan Orogeny are difficult to separate from earlier deformation. Eurekan deformation in the Sverdrup Basin (Fig. 20) was controlled in part by the regional stresses that were applied at the time, but also to a very large degree by the preexisting structural trends within and beneath the basin (cf. Figs. 18 and 20). The rifts propagating northwestward were obstructed on their advancing ends by ancient structural trends (Fig. 20), yet opening continued at their older ends. This combination of events, of which the obstruction was a vital part, produced the compressional deformation of the Eurekan Orogeny (Kerr, 1967b,c, 1980b).

The extensional deformation that had been propagating southeastward from the Canada Basin in late Cretaceous time (Fig. 19, early) apparently was later neutralized throughout a large region in latest Cretaceous to early Tertiary time by compressional effects that advanced northwest from the Eurekan Rifting Episode (Fig. 19, late). Subsequently (Fig. 20), this very widespread neutral state was soon replaced in a large part of the eastern Arctic by compressional deformation, and the neutral area became smaller and shifted westward.

The writer infers that the compressional deformation of the Eurekan Orogeny (Fig. 20) produced a general squeezing rather than a unidirectional overriding. It resulted from compression in the northwest that was apically opposed to the rotational opening of Baffin Bay and the Labrador Sea in the southeast.

The youngest deformation by the Eurekan Orogeny known on land is bracketed between Middle Eocene and Lower Miocene rocks on Axel Heiberg Island (Balkwill and Bustin, 1975). These are respectively Sequences 6 and 7 (Fig. 4).

Apparently the extensional deformation propagating to the northwest from Baffin Bay was stronger than that propagating to the southeast from the Canada Basin. The former compensated for and surpassed the latter in the broad area where compressional deformation occurred. The net effect in eastern parts of the Canadian Arctic Archipelago was that the compression resulting from the Eurekan Rifting Episode took over in the Sverdrup Basin area as the Eurekan Orogeny (Fig. 20). The central part of the Arctic Archipelago appears not to have been affected by the compressional deformation, which apparently diminished and died out toward the northwest.

All rocks younger than Maastrichtian (latest Cretaceous) are essentially unfolded along the Arctic Coastal Plain northwest of Ellef Ringnes Island (Meneley et al., 1975), and there may have been continuous sedimentation there through the time span of the Eurekan Orogeny. This therefore lay in the neutral area (Fig. 20). Parts of the Canada Basin northwest of the area affected by compressional deformation may also have been neutral (i.e., no deformation) through the time span of the Eurekan Orogeny. In certain parts of the western Arctic, however, extension apparently existed without interruption throughout the Eurekan Deformation (Fig. 20). In the Banks Island area there was differential uplift and faulting from late Maastrichtian to Eocene time, resulting in deltaic sedimentation without any interruptions by compressional deformation (Miall, 1975). The writer suggests that this faulting resulted from side effects of extensional processes that continued to operate in the Canada Basin through the time span of the Eurekan Orogeny. The Eurekan Orogeny was unable to interrupt and dominate the extensional faulting effects on Banks Island emanating from the Canada Basin (Fig. 20).

J. Sequence 7: (Miocene) Severing of the Continent by Rifting

Sequence 7 comprises the Miocene Beaufort Formation and equivalent rocks (Figs. 3 and 4). The sequence is exposed in three regions where the setting is very different. Deposition was partly contemporaneous with the final severing of the continent that was the last phase of the Eurekan Deformation (Fig. 19).

In the Arctic Coastal Plain (Fig. 2) Miocene sands of the Beaufort Formation are undeformed. The formation lies unconformably upon folded older rocks and dips northwestward (Thorsteinsson and Tozer, 1970; Hills *et al.*, 1974; Matthews, 1976), and apparently was not affected by tectonism of the Eurekan Deformation. It is part of a continental terrace wedge that thickens northwestward on northwest Ellef Ringnes Island. The base of this wedge is older than Miocene farther northwest in the subsurface of the Arctic Coastal Plain, where it includes unfolded Upper Cretaceous rocks at the base (Meneley *et al.*, 1975). Thus the Eurekan Orogeny did not affect that area with compressive deformation, and there may have been essentially continuous deposition in the offshore area from late Cretaceous (Eureka Sound) to Miocene (Beaufort) time as the Canada Basin subsided and grew.

Uplift, erosion, and faulting of the Eurekan Deformation continued in the eastern Arctic in Miocene and possibly later time. Miocene rocks occur in Axel Heiberg Island where they were faulted by penecontemporaneous uplift of the Princess Margaret Arch (Balkwill and Bustin, 1975). Miocene rocks also occur on northern Ellesmere Island (D. G. Wilson, 1976), where they are cut by northward-directed thrust faults. In both places Miocene rocks are faulted conglomerates and were considered to be equivalent to the Beaufort Formation. These late faulting events, which appear to have been contemporaneous, may have been part of the final pulse of the Eurekan Deformation that finally severed the continental crust (Fig. 21). The setting and relations of the three areas of Miocene rocks suggest (Kerr, 1980*b*) that this event was a continuation of the pattern shown earlier in the Eurekan Deformation, with extension in the southeast (Axel Heiberg Island), compression farther north (northern Ellesmere Island), and neutrality in the west (Arctic Coastal Plain). The final breaking through of rifts to reach the Arctic Ocean was by means of the eastern region, including Nares Submarine Rift Valley, more than through Parry Submarine Rift Valley (Fig. 21). The Beaufort Formation of the Arctic Coastal Plain apparently was not affected by post-Miocene faulting, but the Beaufort rocks of Axel Heiberg and Ellesmere Islands were. The breaking through in Parry Submarine Rift Valley may have involved very active extension faults in the east propagating west to connect with older extension faults that were inactive or only mildly active (Fig. 21).

Faulted Lower Miocene rocks in the eastern Queen Elizabeth Islands indi-

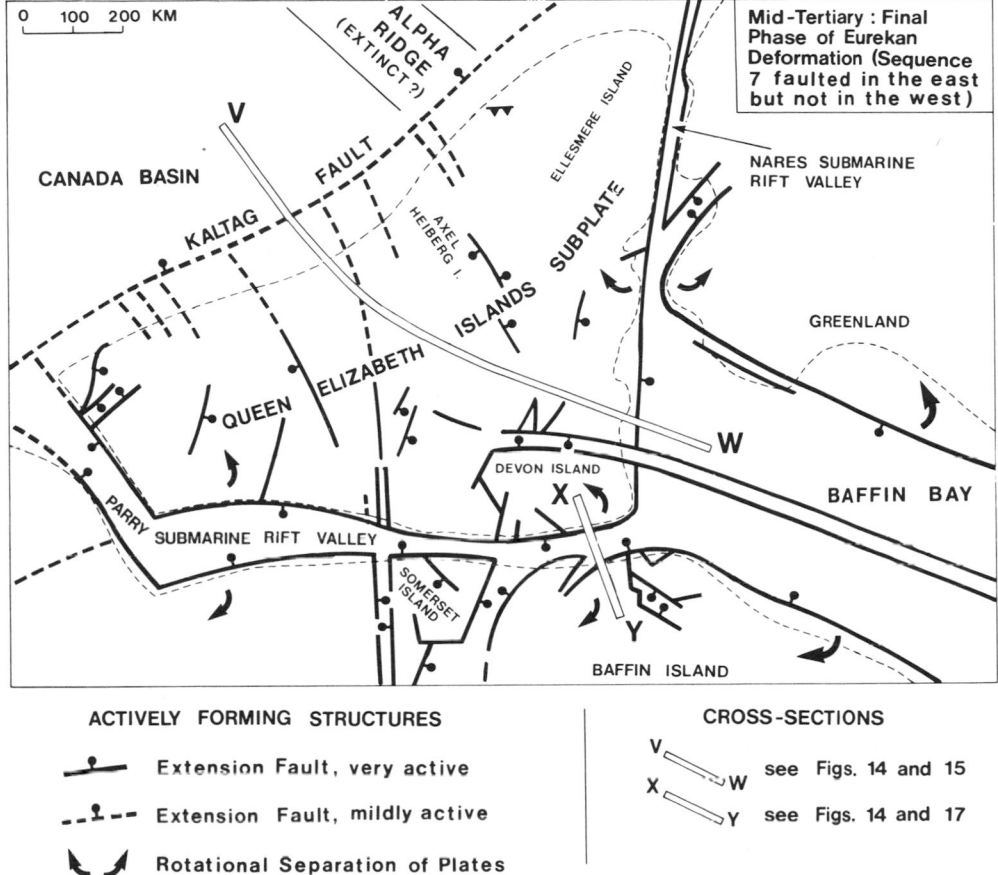

ACTIVELY FORMING STRUCTURES

●— Extension Fault, very active

--●-- Extension Fault, mildly active

↖ ↗ Rotational Separation of Plates

CROSS-SECTIONS

V, W   see Figs. 14 and 15

X, Y   see Figs. 14 and 17

Fig. 21. Final phase of the Eurekan Deformation (early Miocene or later), when the Queen Elizabeth Islands subplate became completely surrounded by rifted zones. In this event compressional faulting occurred only on northern Ellesmere Island. Extension faults then were able to break northward through the obstructions in the Nares Rift Valley and westward through obstructions in the Parry Rift Valley. For the first time a structural connection from the Atlantic Ocean to the Arctic Ocean was made through the North American continent. VW indicates location of the cross section shown in Fig. 15, and XY is the location of the cross section in Fig. 17.

cate that the Canadian Arctic Rift System was still active at that time. The pulse of activity in early Miocene or later time may have been its last. Because the rifting emanated from the southeast, it is probable that Baffin Bay, Nares Strait, and Parry Channel were also affected by that faulting event. It is likely that in those three waterways there will be a Miocene succession (Sequence 7) equivalent in age to the Beaufort Formation. If so it may be bounded by stratigraphic breaks. The lower break that separates it from Sequence 6 may result from the climactic event of the Eurekan Orogeny (Fig. 20). The upper break that separates it from Sequence 8 may result from the event that finally severed the continent (Fig. 20). The Eurekan Deformation ceased in mid-Tertiary (early Miocene or later) time, after extensional deformation breached the continent and reached the Arctic Ocean (Fig. 21). There apparently has been no strong tectonism since then in the study region (Fig. 1).

The final event in the Eurekan Deformation was widespread extensional faulting that broke through the Queen Elizabeth Islands (Fig. 21). The Nares Submarine Rift Valley (Kerr, 1967b), whose advance had been impeded previously, continued to develop at this time and broke the rest of the way northward through the continent to connect with the Arctic Ocean. A westward projection of the Lancaster Aulacogen (Kerr, 1980a) similarly broke through the Boothia Uplift and continued the rest of the way through the continent to form the Parry Submarine Rift Valley and reach the Arctic Ocean.

The Miocene rocks of Sequence 7 making up the Arctic Coastal Plain (Fig. 1) have not been faulted (Thorsteinsson and Tozer, 1970), so the final event of the Eurekan Deformation may not have affected the Arctic Coastal Plain. It appears that faults in the Parry Submarine Rift Valley broke westward through the Boothia Uplift at this time. In doing this they may have simply connected with existing older faults that had earlier formed the west end of that rift valley, but without substantially increasing the displacement on those western faults (Fig. 21).

With this final faulting event of the Eurekan Deformation a large triangular region became severed from the rest of the continent to become the completely fault bounded Queen Elizabeth Islands subplate (Fig. 15). This breaking through by extension faults (Fig. 21) produced the final structural connection between the Atlantic and Arctic Oceans. It may have occurred contemporaneously with the rejuvenated uplift and faulting of the Princess Margaret Arch on Axel Heiberg Island, which cuts Sequence 7, early Miocene rocks (Balkwill and Bustin, 1975; Balkwill, 1978). That is the youngest extensional event documented in the Queen Elizabeth Islands, and is interpreted as a side effect of the main extensional pulse that produced the final breakthrough of rifted structures to reach the Arctic Ocean. Thus the structural connection of the Atlantic and Arctic oceans apparently was achieved in early Miocene time or slightly later. The Alpha Ridge is not an active spreading center today. Its activity may have ceased when the continent was breached by the Eurekan Rifting Episode.

The severing of the Canadian Arctic by faults created the basic structural and physiographic configuration of the Arctic region (Figs. 1 and 21). Geological structure and tectonics controlled physiography on both very large and intermediate scales. The Canada Basin of the Arctic Ocean and the seaway forming Baffin Bay and the Labrador Sea were downfaulted as sedimentary basins, and are also deep physiographic basins. The channels within the Canadian Arctic Archipelago that connect the two main oceanic areas were downfaulted a lesser amount, and are now much shallower than the oceanic basins. The major channels bordering the Queen Elizabeth Islands contain rift valleys and are subplate boundaries. These rift valleys are structural projections of the oceanic areas into the continent. The Queen Elizabeth Islands subplate also is made up of a number of lower-order subplates that are separated by faults (Fig. 21). The faults controlled the shapes of the subplates, which in turn determined the shapes of the present day islands (Fig. 1).

The present study concludes that the Canadian Arctic Rift System formed the Labrador Sea, Baffin Bay, and the Queen Elizabeth Islands between late Cretaceous (Campanian/Maastrichtian) and mid-Tertiary (early Miocene or later) time. The main rifting was part of the Eurekan Deformation, which was propagated northwestward from the Atlantic Ocean. It met with, overpowered, and aborted the Boreal Rifting Episode, which was of lesser strength and that simultaneously was being propagated southeastward from the Canada Basin of the Arctic Ocean. The separation of Greenland from the rest of North America apparently occurred by a combination of rotation apart of subplates and foundering of continental blocks between. This occurred because of propagating faults that advanced northwestward. The nature and timing of events in the separation and rotation of Greenland are similar to those suggested in earlier papers (Kerr, 1967b,c). The timing conforms with that determined later form marine work in Labrador Sea and Baffin Bay (Pitman and Talwani, 1972; Srivastava, 1978). There appears to have been no closing of the Labrador Sea or Baffin Bay. Once they began to open they either continued to open or stopped, but did not close again.

The Haughton Astrobleme (Frisch and Thorsteinsson, 1978) is an impact crater of presumed meteoritic origin. Lake sediments of Miocene or possibly Pliocene age lie in the crater. They were deposited soon after impact, which apparently occurred in about Miocene time, perhaps contemporaneous with the Beaufort Formation and Sequence 7. By the time of impact that part of Devon Island was above sea level and the present peneplain of exposed Lower Paleozoic rocks had developed.

K. Sequence 8: Mid-Tertiary (Post-Miocene) to Present-Day Development

The Eurekan Deformation may have ceased rather suddenly after the early Miocene or later faulting event (Fig. 21), for it appears that the Canadian Arctic Rift System has been dormant since then. There is very little evidence of faulting since that event. Several long lineaments occur in the

Beaufort Formation of the Arctic Coastal Plain on Prince Patrick Island. These must be very recent faults because they create scarps in the modern soft-weathering landscape (Tozer and Thorsteinsson, 1964; Thorsteinsson and Tozer, 1970). These presumed faults show only a few feet of displacement and may represent minor readjustments on older structures.

Fault activity after the Eurekan Deformation may have been more prevalent in the marine area, where the major zones of weakness of the Canadian Arctic Rift System exist. This system could have been reactivated there by adjustments to more global plate motions. The region of the Queen Elizabeth Islands has been very stable tectonically or partly inextension since the final faulting of the Eurekan Deformation, that is from about mid-Miocene onward, or after about 20 m.y. ago. During this interval the present marine areas have probably changed shape very little. They probably also have been water covered continuously, except for near shore areas that are subjected to the vagaries of submergence and emergence due to glacial loading, isostasy, or eustatic seal level changes. There probably is no major post-Miocene stratigraphic break in the sedimentary sequence in the main channels. Whether deposition or erosion occurred in various parts of these basins since the end on the Eurekan Deformation probably depended on currents, sediment supply, and other factors, but probably not on active tectonics. It seems likely therefore that the sedimentary column extending from the end of the Eurekan Deformation (post-early Miocene) will not have been interrupted by major tectonic events, and therefore should make-up a single sequence at sea (Sequence 8, Figs. 3 and 4). This sequence also will include deposits related to Pleistocene glacial events.

It may be impossible to separate Sequences 7 and 8 in much of the western Arctic, beneath and northwest of the present Arctic Coastal Plain (Figs. 2 and 4), because the Beaufort Formation there apparently was not deformed by the Eurekan Deformation (Fig. 21). That area may be like Banks Island (Miall, 1975), Where the Middle to Upper Miocene Beaufort Formation is part of a major deltaic assemblage that has continued to accumulate to the present day (Fig. 13).

The Canadian Arctic Rift System is now a nearly dormant structure within the North American plate, as there is little seismic activity on it (Basham et al., 1977). There are numerous earthquakes, but their patterns indicate that tectonic forces characteristic of plate margins are not acting directly within the Canadian Arctic today. The present seismic activity may be mainly an expression of readjustment of existing structures, mainly the Canadian Arctic Rift System, to a regional stress field. Little plate tectonic activity is taking place now in northern Canada (Fig. 1), because its internal rift system is dormant and the entire region is travelling as a part of the North

American plate. the overall rotation of that plate may be localizing any seismic activity within the study region along dormant subplate boundaries.

From the end of the Eurekan Deformation in mid-Tertiary (probably Miocene) time to the present, the Canadian Arctic (Fig. 1) has been subjected to erosion in some places and deposition in others, but not to strong tectonism. Sequence 8 was deposited in this time span and is still accumulating. The present day physiography was mainly achieved by the end of the Eurekan Deformation. Since then it has only been modified further by the sculpturing and infilling of erosional processes. The present physiography therefore resulted largely from the interplay of two things. It was dominated by the geological structures that had been produced by the Eurekan Deformation and by other similar structural events to the northwest. It was also modified by erosion in the period following the Eurekan Deformation. Structure exerted vastly greater control on regional physiography than did erosion.

Structural control of physiography applies mostly to larger features, e.g., the main linear margins. The Canada Basin is topographically low because it appears to have been largely down faulted relative to the Arctic Islands. The Queen Elizabeth Islands subplate (Fig. 21), forms a triangular island group (Fig. 1), because it is rift bounded. It is bordered on the south and east by submarine rift valleys. On the northwest the island by group is bordered by the Kaltag Fault (Norris, 1974; Yorath and Norris, 1975; Norris and Yorath, this volume), which apparently controlled the continental margin there.

Baffin Bay and the Labrador Sea apparently were largely downfaulted to become deep waterways. Simultaneous with and related to that downfaulting, bordering lands, Greenland, Baffin Island, and Labrador, were raised, with high coastal mountain ranges developing along them. The main marine indentations into the Queen Elizabeth Islands group, such as Jones Sound, Hassel Sound, and others, appear to be grabens. The islands may have risen as horsts in conjunction with the formation of those grabens and a part of the same process.

The structural control of physiography applies also to structures of intermediate size, that are large enough to be shown on regional maps. Cornwallis Island, for example, is anticlinal (Thorsteinsson and Kerr, 1968), as are Cornwall and Amund Ringnes Islands (Balkwill, 1974, 1978). Ellef Ringnes Island may also be grossly anticlinal (Stott, 1968), with a structural low forming the channel to the east. Somerset Island (Fig. 21), is basically a horst, being almost completely surrounded by normal faults (Kerr and deVries, 1977).

A major height of land trending northeastward through the Canadian Arctic, which still exists, was produced largely by tectonic events during the Eurekan Deformation (Figs. 14 and 15). It is a mainly continental area lying between oceanic basins. The height of land was broken by narrow faulted channels during the Eurekan Deformation, when parts of it were deeply depressed to allow the oceans on either side to connect. The height of land has persisted since the Eurekan Deformation, and the narrow downdropped structures may have remained below sea level continuously since then. Since the Eurekan Deformation, the height of land has been modified by surficial processes operating both above and below sea level.

Fortier and Morley (1956) first suggested that the physiography of the Canadian Arctic Islands was produced by a combination of early Tertiary tectonic events and fluvial erosion, with the present marine channels being a structurally controlled former subaerial drainage system. That concept was enlarged by Pelletier (1966), Trettin et al. (1972), and Kerr (1977a, 1980b). The writer considers that the original concept of Fortier and Morley (1956) remains largely correct. He suggests, however, that the main channels may have been below sea level from an early stage of their formation, and that they were controlled only to a small degree by subaerial erosion. This applies particularly to the deeper or downstream parts of the channels, which were submerged in an early stage of tectonism, and may have remained below sea level ever since.

The Innuitian Ice Sheet (Blake, 1970, 1977) covered much or all of the Queen Elizabeth Islands region in Pleistocene time. This contributed to the sculpturing of the region by preferentially scouring the valleys on land, and perhaps also the shallow submarine areas, including the fiords and the shallower marine channels. Glaciation was followed by widespread post-glacial rebound. The amount of rebound appears to have been greater in the central Queen Elizabeth Islands than farther east, for the upper marine limit ascends westward on southern Ellesmere Island (Blake, 1970, 1975). It similarly increased westward on Devon Island (Kerr, 1977d).

With the retreat and melting of the glaciers, the Canadian Arctic Islands have emerged, such that there are marine raised beaches around many of the islands. The rate of post-glacial uplift in the Innuitian region was on the order of 1 to 2 mm per hundred years during the last six hundred years (Andrews, 1970), but this varies greatly from place to place. The rifting process that produced the Arctic Islands is now nearly dormant, the earthquake activity in the region is mild, and probably is limited to adjustments on existing structures (Basham et al., 1977).

L. Mineral and Petroleum Deposits

1. Cornwallis Lead–Zinc District

The Cornwallis Lead–Zinc District is located within Lower Paleozoic rock in the northern part of the Cornwallis Fold Belt (Figs. 6 and 22). The

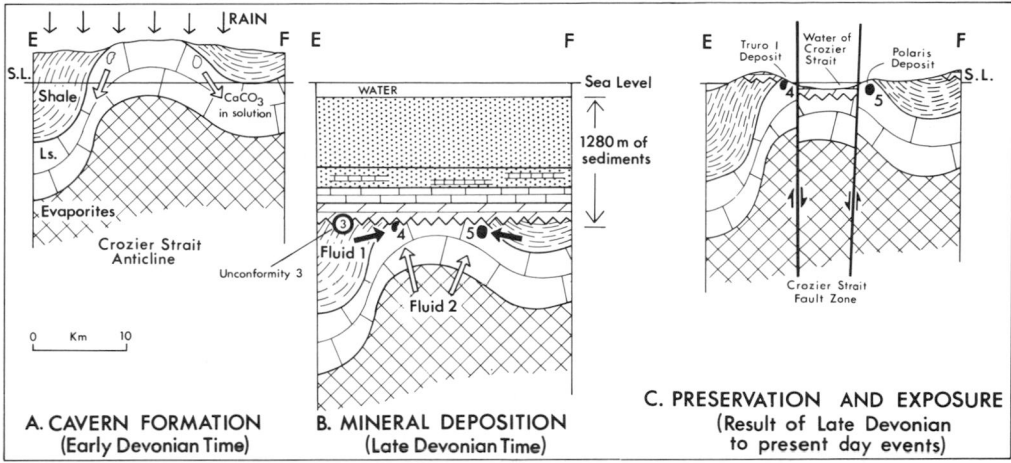

Fig. 22. Sequence of events that formed typical deposits of the Cornwallis Lead–Zinc District. For location of these cross sections see Figs. 6, 8, and 10. This is an enlargement of the area outlined in Fig. 8. (A) Cavern formation took place by karstification in an erosional interval during the Cornwallis Disturbance. (B) Mineral deposition took place in the caverns by precipitation from formation fluids, apparently while the caverns were buried during the interval between Pulses 3 and 4 of the Cornwallis Disturbance. (C) Preservation and exposure resulted from late Devonian and later deformation and erosion, particularly the Eurekan Rifting Episode, which formed the Crozier Strait Fault Zone.

lead–zinc deposits of that district all have a similar setting (Kerr, 1977*b*,*c*) and appear to have formed by a common mechanism. They were controlled by the tectonic history of the Boothia Uplift, having formed in the the upper or sedimentary level of that uplift (Fig. 8).

The lead–zinc deposits characteristically occur in an Ordovician carbonate formation, in anticlines that were formed by the Cornwallis Disturbance. They are of Mississippi Valley-type, apparently having formed in caverns at low temperatures (as high as 105°C), by precipitation from formational fluids.

An early stage in the formation of the lead–zinc deposits was cavern formation (Fig. 22A). These caverns formed in the Ordovician Thumb Mountain Formation, and preferentially in the upper fossiliferous part of that formation. That formation normally is limestone, but is dolomitized in the vicintiy of the lead–zinc deposits. The caverns were presumed by Kerr (1977*b*) to have formed in early Devonian time, by karst solution during Pulse 3 of the Cornwallis Disturbance. Callahan (1978, 1979) suggested that they may have formed in an earlier erosional event, but this was considered to be unlikely by Kerr (1978, 1979). Whichever timing is correct, it appears clear that the deposits were controlled by the tectonic history of the basement uplift within the geosyncline (Figs. 8 and 9).

Mineralization occurred in the caverns, probably in late Devonian time, apparently by precipitation when two appropriate fluids met (Fig. 22B). One of the fluids may have originated in a nearby black shale, obtaining one of more of the mineral components there, most probably the sulphur. The other essential components, metal ions, probably were carried in a second fluid that ascended from beneath. This second fluid may have originated in the deep crystalline basement of the Boothia Horst, which was being uplifted from time to time.

Various tectonic and erosional events that occurred subsequent to mineralization contributed to exposing the deposits. The Eurekan Rifting Episode was a major faulting event that formed the main channels in the Canadian Arctic, and occurred long after mineralization. It produced the Crozier Strait Fault Zone (Kerr and Ruffman, 1979) and was responsible for exposing the deposits on either side of Crozier Strait (Fig. 22C). This fault zone followed the structural trends of the Boothia Uplift that had been established during the Cornwallis Disturbance.

2. *Oil and Gas Discoveries*

Oil and gas discoveries in the Arctic Islands have been summarized by Drummond (1973), Meneley (1976), and Stuart Smith and Wennekers (1977). Source rock and maturation studies have been carried out by Snowdon and Roy (1975), Baker *et al.* (1975), and Powell (1978).

On Cameron Island (Figs. 1 and 6), live oil was discovered in two wells in Middle Devonian carbonate rocks that are part of a large isolated reef buildup. A subsequent follow-up well has been abandoned (Stuart Smith and Wennekers, 1977), suggesting that this may not be an important discovery. The location of the reef buildup is shown in Fig. 9, but its present setting is much more complicated because of subseqent deformation.

Exploration to 1977 had discovered 15 trillion cubic feet of gas in seven gas fields in the Western Sverdrup Basin, according to Meneley (1976). The fields are in two areas and appear to be in approximately the same stratigraphic level in a Triassic–Jurassic sandstone body. One area is in the axial part of the Sverdrup Basin (Fig. 11) in the region of Ellef Ringnes Island and King Christian Island. These are related to diapiric structures (Figs. 12 and 23) and are in a sandstone unit that overlies and was raised by those evaporite diapirs. These gas-bearing structures are ovate anticlines, with cores of evaporites which are not diapiric (Balkwill and Roy, 1977). The second area of gas fields is along the southern margin of the Sverdrup Basin on Melville Island (Fig. 11) and the nearby offshore region. These are the Sabine Peninsula fields (Fig. 23). They occur beneath a gentle regional angular unconformity (Meneley, 1976).

The size, shape, and mechanism of entrapment of gas were very different in the two regions (Meneley, 1976). The pools in the Ellef Ringnes Island

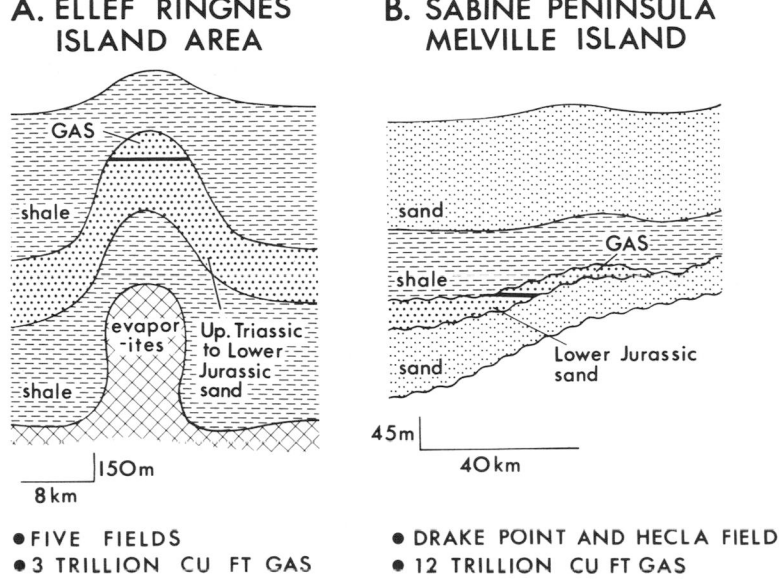

A. ELLEF RINGNES
ISLAND AREA

B. SABINE PENINSULA
MELVILLE ISLAND

- FIVE FIELDS
- 3 TRILLION CU FT GAS
- HIGH AMPLITUDE FOLDS
- THICK RESERVOIRS

- DRAKE POINT AND HECLA FIELDS
- 12 TRILLION CU FT GAS
- LOW AMPLITUDE FOLDS
- THIN RESERVOIRS

Fig. 23.   Generalized geology of gas discoveries in the Sverdrup Basin, summarized from Meneley (1976). Locations of these fields are shown in Fig. 11.

region have on the order of 150 m of gas, and are about 8 km in average breadth. In the Sabine Peninsula area of Melville Island the pools are thinner, being on the order of 45 m thick, but are much wider, being on order of 40 km across. About 80% of the gas reserves reported occur in the Sabine Peninsula fields (Fig. 23).

IV. SUMMARY AND DISCUSSION

The Canadian Arctic is a transition between two ocean basins, the Atlantic Ocean in the southeast and the Arctic Ocean in the northwest. The region of this transition that is most like a true continent geologically also is highest topographically. This is a northeast trending height of land that formed in Cretaceous–Tertiary time (Figs. 14 and 15) and persisted to the present, being expressed now by shallow bathymetry (Fig. 1). The height of land is the region that was least fragmented by the ocean forming plate tectonic process.

The Canadian Arctic is transitional first in a structural sense, with marginal parts of the Arctic being most oceanic in character and central parts of the Queen Elizabeth Islands subplate (Fig. 15) most continental. Labrador Sea is rather like a deep oceanic basin, with a crustal structure that resembles the oceanic crust of a major ocean. It had moderate opening or separation of the two sides, also resembling a rifted ocean. Taking a step from there northwestward into Baffin Bay the crust is oceanic, but perhaps quasioceanic, and even less like the deep abyssal parts of a true ocean basin than is the crust of Labrador Sea. Magnetic lineations are poorly developed in Baffin Bay (Srivastava, 1978). It has very thick sediments (Pelletier et al., 1975) and at least parts in the northwest have a crust that is intermediate between continental and oceanic crust (Wetmiller, 1974). In the region still farther northwestward, the Canadian Arctic Archipelago has a crust that is largely continental in character, with a thickness of about 37 km (Sobczak and Weber, 1973), but it has deep grabens and fissures in it as a result of crustal fracturing and extension.

A somewhat similar progression occurs from the Canada Basin of the Arctic Ocean southeastward to the central part of the Queen Elizabeth Islands subplate. The Canada Basin is a deep oceanic basin with a large region deeper than 3500 m. The crust beneath it is of oceanic or quasioceanic character. It has certain geophysical attributes of continental crust (King et al., 1966), but is far below sea level, approaching the depth of an abyssal plain. Farther southeastward is a continental slope and shelf, with a thick wedge of sediments that forms a continental terrace wedge (Fig. 3). Still farther southeastward are the Queen Elizabeth Islands, still more continental, and with a crust of about 37 km thickness.

The Canadian Arctic is transitional between the Atlantic and Arctic Oceans in a physiographic sense also. A central height of land in the Arctic Archipelago trends northeastward, expressed in topography and bathymetry. This is a part of the larger Queen Elizabeth Islands, parts of which are above, and parts below sea level as relatively shallow channels between those islands. The physiographic transition exists also on either side. Southeast of the Queen Elizabeth Islands there are progressively deeper basins, first Baffin Bay, then Labrador Sea, and finally the Atlantic Ocean proper. The physiographic transition also occurs northwestward from the Queen Elizabeth Islands toward the Canada Basin of the Arctic Ocean. The physiographic transition was produced by the structural transition.

The Canadian Arctic Islands region (Fig. 1) lies between two immature oceans, and it apparently resulted from side effects of the processes that developed those two oceans. Both oceans resulted from breaks in a primeval continental crust. Both were controlled in important ways by the structural trends in the Precambrian crystalline shield of the continent within which and beside which they formed. The study region (Fig. 1) is that region where the two ocean-forming events impinged on each other. Processes were different in nature and timing in the two oceanic areas, and this resulted in very different types of continental margins. At times these two events were in conflict with each other, producing a plate tectonic contest (Kerr, 1980*b*).

The Arctic Ocean margin of the Canadian Arctic Islands began its development very early. The trend of the margin can be traced back to the beginning of the Innuitian (continental margin-type) Mobile Belt (Fig. 3). That belt began to form along a northeast trend from the Banks Island region to northern Greenland, in late Precambrian time. It may have followed structural trends already existing in the older crystalline rocks beneath that make up Sequence 1. The Innuitian Mobile Belt developed through several stages, each depositing a huge stratigraphic sequence. These sequences lie one above the other and are imbricated, the oldest being in the southeast and the youngest in the northwest. Sequence 2 of Proterozoic age, and Sequence 3, the late Proterozoic to late Devonian Franklinian Geosyncline, have northeast trends. They either established this trend during deposition, or more likely reinforced an older one. The northeast structural grain is now followed by the Arctic Ocean-facing margin of the Canadian Arctic Islands. The history up to the end of Sequence 3 was the constructional phase of the continent (Fig. 4). It involved a linear geosyncline that may have lain within and upon a continental crust. It apparently did not involve the formation of a true ocean.

The fragmentation phase of the continent began in late Paleozoic time, probably Mississippian (Fig. 4), and has continued to the present. It formed the Canada Basin, Baffin Bay, and the other marine areas within the North American continent (Fig. 1). The Canada Basin of the Arctic Ocean may have begun to form during Sequence 4 by an unnamed rifting episode centered on the Alpha Ridge that began shortly after the Ellesmerian Orogeny. The Arctic facing margin of the study area apparently was guided by the Pearya Geanticline. The Canada Basin appears to have formed by rifting, and perhaps also foundering of continental crust. Lesser rifting marginal to it extended into the continent and apparently produced the Sverdrup Basin. A major fault zone lies between. There may have been great downdropping to the northwest of this fault and lesser downdropping on the southeast. Therefore it must also have had a transform component, presumably left lateral.

In early Cretaceous time, as Sequence 5 was deposited, a new pulse of activity in the Canada Basin caused accelerated rifting that extended farther southeastward across the main fault zone and into the continent. This rifting event faulted the northwest marginal parts of the Canadian Arctic Islands, simultaneous with a great continentwide overlap of shallow seas southward and southeastward onto a cratonic shelf.

In late Cretaceous time, the Canadian Arctic Rift System began to form more rapidly, mainly by activity in the southeast. The rift system is a branch that extended northwestward from the Mid-Atlantic Ridge through the

Canadian Arctic, and separated Greenland from the rest of North America. Major rifting apprently was propagated northwestward by extension faults that produced rotation about a pivotal area in the northwest. For a time the two rift systems were both active, propagating toward each other. The extension that was being propagated to the northwest was the strongest. It first nullified the other rifting, subsequently caused compressional deformation, and later managed to break through the Canadian Arctic Islands to reach the Arctic Ocean. Active rifting connected the Atlantic and Arctic Oceans in mid-Tertiary time, about early Miocene or later, and then stopped apparently as suddenly as it began. This suggests that the process of connecting the two oceans structurally ended the need for any further major rifting activity.

The Canadian Arctic Rift System is a major rift traversing a continent that became dormant or frozen in a very early stage of its development. It contains a well-exposed geologic record of the rocks from Precambrian to Recent time. These characteristics make it an excellent laboratory for the study of certain tectonic processes.

ACKNOWLEDGMENTS

The author is very grateful to the Geological Survey of Canada for the support provided in this research. The work benefited greatly by generous advice from M. G. Audley-Charles, H. R. Balkwill, A. W. Bally, R. K. H. Falconer, N. J. McMillan, W. W. Nassichuk, E. R. W. Neale, D. K. Norris, R. Thorsteinsson, and H. P. Trettin. The final manuscript was largely prepared while I was a visting scholar at Cambridge University. I wish to thank W. B. Harland and H. B. Whittington for generous assistance while there. The suggestions and counsel of helpful collegues has done much to improve the manuscript, but needless to say they are not responsible for the statements, for which I assume sole responsibility.

REFERENCES

Andrews, J. T., 1970, A geomorphological study of post-glacial uplift with particular reference to Arctic Canada, *Inst. British Geogr. Spec. Publ.* 2.

Athavale, R. N., and Sharma, P. V., 1975, Paleomagnetic results on Early Tertiary lava flows from West Greenland and their bearing on the Evolution of the Baffin Bay–Labrador Sea Region, *Can. J. Earth Sci.*, v. 12(1), p. 1–18.

Atlasov, I. P., ed., 1964, Tectonic map of the Arctic and Subarctic, Leningrad: Sci. Res. Inst. of Geology of Arctic, Scale 1:5,000,000.

Austin, G. H., 1973, Regional geology of Eastern Canada, *Am. Assoc. Petrol. Geol. Bull.*, v. 57, p. 1250–1275.

Baker, D. A., Illich, H. A., Martin, S. J., and Landin, R. R., 1975, Hydrocarbon source potential of sediments in the Sverdrup Basin, in: *Canada's Continental Margins and Offshore Petro-*

leum Exploration, Yorath, D. J., Parker, E. R., and Glass, D. J., eds., *Can. Soc. Petrol. Geol. Mem*. 4, p. 545–556.

Balkwill, H. R., 1974, Structure and tectonics of Cornwall Arch, Amund Ringnes and Cornwall Islands, Arctic Archipelago, in: *Geology of the Canadian Arctic*, Aitken, J. D., and Glass, D. J., eds., Proc. Symp. Geol. Canadian Arctic, Saskatoon, May 1973, Geol. Assoc. Can.–Can. Soc. Petrol. Geol., p. 39–42.

Balkwill, H. R., 1978, Evolution of Sverdrup Basin, Arctic Canada, *Am. Assoc. Petrol. Geol. Bull.*, v. 62, p. 1004–1028.

Balkwill, H. R. and Bustin, R. M., 1975, Stratigraphic and structural studies, central Ellesmere Island, and eastern Axel Heiberg Island, *Geol. Surv. Can. Pap*. 75-1A, p. 513–517.

Balkwill, H. R., and Roy, K. J., 1977, Geology of King Christian Island, District of Franklin, *Geol. Surv. Can. Mem.* 386.

Balkwill, H. R., Bustin, R. M., and Hopkins, W. S., Jr., 1975, Eureka Sound Formation at Flat Sound, Axel Heiberg Island, and chronology of the Eurekan Orogeny, *Geol. Surv. Can. Pap*. 75-1B, p. 205–207.

Baragar, W. R. A., and Donaldson, J. A., 1973, Coppermine and Dismal Lake map areas, *Geol. Surv. Can. Pap*. 71-39.

Barrett, D. L., 1966, Lancaster Sound shipborne magnetometer survey, *Can. J. Earth Sci.*, v. 3, p. 223–235.

Basham, P. W., Forsyth, D. A., and Wetmiller, R. J., 1977, The seismicity of Northern Canada, *Can. J. Earth Sci.*, v. 14(7), p. 1646–1667.

Beh, R. L., 1975, Evolution and geology of Western Baffin Bay and Davis Strait, Canada, in: *Canada's Continental Margins and Offshore Petroleum Exploration*, Yorath, C. J., Parker, E. R., and Glass, D. J., eds., *Can. Soc. Petrol. Geol. Mem.* 4, p. 453–476.

Beloussov, V. V., 1968, Some general aspects of development of the tectonosphere, 23rd Intern. Geol. Congr., Prague, 1968, 1, p. 9–17.

Beloussov, V. V., 1970, Against the hypothesis of ocean floor spreading, *Tectonophysics*, v. 9(6), p. 489–511.

Blake, W., Jr., 1970, Studies of glacial history in Arctic Canada I. Pumice, radiocarbon dates, and differential postglacial uplift in the eastern Queen Elizabeth Islands, *Can. J. Earth Sci.*, v. 7, p. 634–664.

Blake, W., Jr., 1975, Radiocarbon age determinations and postglacial emergence at Cape Storm, southern Ellesmere Island, Arctic Canada, *Geograf. Annal. Ser. A*, v. 57(1–2), p. 1–71.

Blake, W., Jr., 1977, Glacial sculpture along the east-central coast of Ellesmere Island, Arctic Archipelago, *Geol. Surv. Can. Pap*. 77-1C, p. 107–115.

Bullard, Sir Edward, Everett, J. E., and Smith, A. G., 1965, The fit of continents around the Atlantic, *Phil. Trans. R. Soc. London Ser. A*, v. 258, p. 41–51.

Bustin, R. M., 1977, The Eureka Sound and Beaufort Formations, Axel Heiberg and west-central Ellesmere Islands, District of Franklin, M. S. Thesis, University of Calgary, 208 pp.

Callahan, W. H., 1978, Cornwallis Lead–Zinc District; Mississippi Valley-type deposits controlled by stratigraphy and tectonics: Discussion, *Can. J. Earth Sci.*, v. 15, p. 459–60.

Callahan, W. H., 1979, Cornwallis Lead–Zinc District; Mississippi Valley-type deposits controlled by stratigraphy and tectonics: Discussion II, *Can. J. Earth Sci.*, v. 16 (3, Part 1), p. 614–615.

Carey, S. W., 1958, A tectonic approach to continental drift, in: *Continental Drift—A Symposium*, Carey, S. W., ed., Hobart: University of Tasmania, p. 117–355.

Churkin, M. J., 1969, Paleozoic tectonic history of the Arctic Basin north of Alaska, *Science*, v. 165, p. 549–555.

Churkin, M. J., 1973, Geologic concepts of Arctic Ocean Basin, in: *Arctic Geology*, Pitcher, M. G., ed., *Am. Assoc. Petrol. Geol. Mem.* 19, p. 485–499.

Clarke, D. B., and Pedersen, A. K., 1976, Tertiary volcanic provinces of West Greenland, in: *Geology of Greenland*, Escher, A., and Watt, W. S., eds., Copenhagen: The Geological Survey of Greenland, p. 364–385.

Clarke, D. B., and Upton, B. G. J., 1971, Tertiary basalts of Baffin Island: Field relations and tectonic setting, *Can. J. Earth Sci.*, v. 8, p. 248–258.

Daae, H. D., and Rutgers, A. T. C., 1975, Geological history of the Northwest Passage, *Bull. Can. Petrol. Geol.*, v. 23, pp. 84–108.

Davies, G. R., 1975, Hoodoo L-41: Diapiric halite facies of the Otto Fiord Formation in the Sverdrup Basin, Arctic Archipelago, *Geol. Surv. Can. Pap.* 75-1C, p. 23–29.

Davies, G. R., and Nassichuk, W. W., 1975, Subaqueous evaporites of the Carboniferous Otto Fiord Formation, Canadian Arctic Archipelago: A summary, *Geology*, v. 3(5), p. 273–278.

Davies, W. E., Krinsley, D. B., and Nicol, A. H., 1963, Geology of the North Star Bugt area, northwest Greenland, *Medd. Groenl.*, v. 162(12), 68 p.

Dawes, P. R., 1973, The North Greenland Foldbelt: A clue to the history of the Arctic Ocean Basin and the Nares Strait lineament, in: *Implications of Continental Drift to the Earth Sciences*, v. 2, Tarling, D. H., and Runcorn, S. K., eds., London: Academic Press, p. 925–947.

Dawes, P. R., 1976, Precambrian to Tertiary of Northern Greenland, in: *Geology of Greenland*, Escher, A., and Watt, W. S., eds., Copenhagen: The Geological Survey of Greenland, p. 247–303.

Denham, L. R., 1974, Offshore geology of northern West Greenland (69° to 75°N), *Geol. Surv. Greenland Rep.*, v. 63.

Dewey, J. F., 1972, Plate tectonics, *Sci. Am.*, v. 226(5), p. 56–68.

Dixon, J., 1974, Stratigraphy and sedimentary history of early Paleozoic rocks from Prince of Wales and Somerset Islands, N.W.T., in: *Geology of the Canadian Arctic*, Aitken, J. D., and Glass, D. J., eds., Calgary: Geological Association of Canada–Canadian Society of Petroleum Geology, p. 127–142.

Douglas, R. J. W., ed., 1970, Geology and economic minerals of Canada, *Geol. Surv. Can. Econ. Geol. Rep.*, v. 1.

Douglas, R. J. W., Norris, D. K., Thorsteinsson, R., and Tozer, E. T., 1963, Geology and petroleum potentialities of Northern Canada, *Geol. Surv. Can. Pap.* 63-31.

Drake, C. L., Campbell, N. L., Sander, G., and Nafe, N. E., 1963, A mid-Labrador sea ridge, *Nature (London)*, v. 200, p. 1085–1086.

Drummond, K. J., 1973, Canadian Arctic Islands, in: *Future Petroleum Provinces of Canada—Their Geology and Potential*, McCrossan, R. G., ed., *Can. Soc. Petrol. Geol. Mem.* 1, p. 443–472.

Embry, A. F., and Klovan, J. E., 1976, The Middle–Upper Devonian clastic wedge of the Franklinian Geosyncline, *Bull. Can. Petrol. Geol.*, v. 24(4), p. 485–639.

Fahrig, W. F., and Jones, D. L., 1969, Paleomagnetic evidence for the extent of the Mackenzie igneous events, *Can. J. Earth Sci.*, v. 6, pp. 679–688.

Fahrig, W. F., Irving, E., and Jackson, G. D., 1971, Paleomagnetism of the Franklin diabases: *Can. J. Earth Sci.*, v. 8(4), pp. 455–467.

Falvey, D. A., 1974, The development of continental margins in plate tectonic theory, *Austral. Petrol. Explor. Assoc. Jour.*, v. 14, pt. 1, p. 95–106.

Fenwick, D. K. B., Keen, M. J., Keen, C., and Lambert, A., 1968, Geophysical studies of the continental margin northeast of Newfoundland, *Can. J. Earth Sci.*, v. 5, p. 483–500.

Fortier, Y. O., and Morley, L. W., 1956, Geological unity of the Arctic Islands, *Trans. R. Soc. Can. Sect.* 4, v. 50, p. 3–12 (special volume).

Frisch, T. O., 1974, Metamorphic and plutonic rocks of northernmost Ellesmere Island, Canadian Arctic Archipelago, *Geol. Surv. Can. Bull.* 229.

Frisch, T. O., and Thorsteinsson, R., 1978, Haughton Astrobleme: A Mid-Cenozoic impact crater, Devon Island, Canadian Arctic Archipelago, *Arctic*, v. 31, pt. 2, p. 108–124.

Geldsetzer, H., 1974, The tectono-sedimentary development of an algaldominated Helikian succession on Northern Baffin Island, N.W.T., in: *Canadian Arctic Geology*, Aitken, J. D., and Glass, D. J., eds., Proc. Symp. Geol. Canadian Arctic, Saskatoon, May 1973, Geological Association of Canada—Canadian Society of Petroleum Geologists, p. 99–126.

Gould, D. R., and de Mille, G., 1964, Piercement structures in the Arctic Islands, *Bull. Can. Petrol. Geol.*, v. 12, p. 719–753.

Grant, A. C., 1972, The continental margin off Labrador and eastern Newfoundland—morphology and geology, *Can. J. Earth Sci.*, v. 19(11), p. 1394–1430.

Grant, A. C., 1975, Structural modes of the western margin of the Labrador Sea, *Geol. Surv. Can. Pap.* 74-30, p. 217–231.

Grant, A. C., and Manchester, K. S., 1970, Geophysical investigations in the Ungava Bay–Hudson Strait region of northern Canada, *Can. J. Earth Sci.*, v. 7(4), p. 1062–1076.

Gregory, A. F., Bower, M. E., and Morley, L. W., 1961, Geological interpretation of aerial magnetic and radiometric profiles, Arctic Archipelago, Northwest Territories, *Geol. Surv. Can. Bull.*, v. 73.

Hall, J. K., 1973, Geophysical evidence for ancient sea-floor spreading from Alpha Cordillera and Mendeleyev Ridge, in: *Arctic Geology*, Pitcher, M. G., ed., *Am. Assoc. Petrol. Geol. Mem.* 19, p. 542–561.

Harland, W. B., 1973, Tectonic Evolution of the Barents Shelf and Related Plates, in: *Arctic Geology*, Pitcher, M. G., ed., *Am. Assoc. Petrol. Geol. Mem.* 19, p. 599–608.

Henao-Londoño, D., 1977, Correlation of the producing formations in the Sverdrup Basin, *Bull. Can. Petrol. Geol.*, v. 25(5), p. 969–980.

Henderson, G., 1973, The geological setting of the West Greenland Basin in the Baffin Bay Region, *Geol. Surv. Can. Pap.* 71-23, p. 521–544.

Henderson, G., Rosenkrantz, A., and Schiener, E. J., 1976, Cretaceous–Tertiary sedimentary rocks of West Greenland, in: *Geology of Greenland*, Escher, A., and Watt, W. S., eds., Copenhagen: The Geological Survey of Greenland, p. 340–363.

Herron, E. M., Dewey, J. F., and Pitman, W. C. III, 1974, Plate tectonic model for the evolution of the Arctic, *Geology*, v. 2(8), p. 377–380.

Hills, L. V., Klovan, J. E., and Sweet, A. R., 1974, *Juglans eocinerea* n. sp., Beaufort Formation (Tertiary), southwestern Banks Island, Arctic Canada, *Can. J. Bot.*, v. 48(3), p. 457–464.

Holtedahl, H., 1958, Some remarks on geomorphology on continental shelves off Norway, Labrador, and Southeast Alaska, *J. Geol.*, v. 66, p. 461–471.

Holtedahl, O., 1970, On the morphology of the West Greenland shelf with general remarks on the "marginal channel" problem, *Mar. Geol.*, v. 8, p. 155–171.

Hood, P. J., and Bower, M. E., 1975, Aeromagnetic reconnaissance of Davis Strait and adjacent areas, in: *Canada's Continental Margins and Offshore Petroleum Exploration*, Yorath, C. J., Parker, E. R., and Glass, D. J., eds., *Can. Soc. Petrol. Geol. Mem.* 4, p. 433–451.

Hood, P. J., Sawatzky, P., and Bower, M. E., 1967, Progress report on low-level aeromagnetic profiles over the Labrador Sea, Baffin Bay, and across the North Atlantic Ocean, *Geol. Surv. Can. Pap.* 66-58.

Hyndman, R. D., 1973, Evolution of the Labrador Sea, *Can. J. Earth Sci.*, v. 10, p. 637–644.

Hyndman, R. D., Clarke, D. B., Hume, H., Johnson, J., Keen, M. J., Park, I., and Pye, G., 1973, Geophysical and geological studies in Baffin Bay and the Labrador Sea, *Geol. Surv. Can. Pap.* 71-23, p. 621–631.

Irving, E., 1977, Drift of the major continental blocks since the Devonian, *Nature (London)*, v. 270, p. 304–309.

Jackson, G. D., and Davidson, A., 1975, Bylot Island map area, District of Franklin, *Geol. Surv. Can. Pap.* 74-29.

Jackson, G. D., Davidson, A., and Morgan, W. C., 1975, Geology of the Pond Inlet map area, Baffin Island, District of Franklin, *Geol. Surv. Can. Pap.* 74-25.

Jackson, G. D., Iannelli, T. R., Narbonne, G. M., and Wallace, P. J., 1978, Upper Proterozoic sedimentary and volcanic rocks of northwestern Baffin Island, *Geol. Surv. Can. Pap.* 78-14.

Jackson, H. R., Keen, C. E., and Barrett, D. L., 1977, Geophysical studies on the eastern continental margin of Baffin Bay and in Lancaster Sound, *Can. J. Earth Sci.*, v. 14(9), p. 1991–2001.

Jeletzky, J. A., 1978, Causes of Cretaceous oscillations of sea level in western and Arctic Canada and some general tectonic implications, *Geol. Surv. Can. Pap.* 77-18.

Kay, M., 1951, North American geosynclines, *Geol. Soc. Am. Mem.* 48, 143 p.

Keen, C. E., and Barrett, D. L., 1972, Seismic refraction studies in Baffin Bay: An example of a developing ocean basin, *Geophys. J. R. Astron. Soc.*, v. 30(3), p. 253–272.

Keen, C. E., and Barrett, D. L., 1973, Structural characteristics of some sedimentary basins in Northern Baffin Bay, *Can. J. Earth Sci.*, v. 10(8), p. 1267–1278.

Keen, C. E., Barrett, D. L., Manchester, K. S., and Ross, D. I., 1972, Geophysical studies in Baffin Bay and some tectonic implications, *Can. J. Earth Sci.*, v. 9(3), p. 239–256.

Keen, C. E., Keen, M. J., Ross, D. I., and Lack, M., 1974, Baffin Bay: Small ocean basin formed by sea-floor spreading, *Am. Assoc. Petrol. Geol. Bull.*, v. 58, p. 1089–1108.

Kent, P. E., 1976, Major synchronous events in continental shelves; in: *Sedimentary Basins of Continental Margins and Cratons*, Bott, M. H. P., ed., *Tectonophysics*, v. 36, p. 87–91.

Kerr, J. Wm., 1967a, Stratigraphy of central and eastern Ellesmere Island, Arctic Canada, Part I, Proterozoic and Cambrian, *Geol. Surv. Can. Pap.* 67-27, pt. I.

Kerr, J. Wm., 1967b, Nares Submarine Rift Valley and the relative rotation of North Greenland, *Bull. Can. Petrol. Geol.*, v. 15(4), p. 483–520.

Kerr, J. Wm., 1967c, A submerged continental remnant beneath the Labrador Sea, *Earth Planet. Sci. Lett.*, v. 2(4), p. 283–289.

Kerr, J. Wm., 1968a, Stratigraphy of central and eastern Ellesmere Island, Arctic Canada, Part II: Ordovician, *Geol. Surv. Can. Pap.* 67-27, pt. I.

Kerr, J. Wm., 1968b, Devonian of the Franklinian Miogeosyncline and adjacent Central Stable Region, Arctic Canada, in: Proceedings of the International Symposium on the Devonian System, Oswald, D. H., ed., Calgary: Alberta Society of Petroleum Geologists, Alberta, v. 1, p. 677–692.

Kerr, J. Wm., 1973a, Canadian Arctic Rift System–A Summary [abs.] in: *Arctic Geology*, Pitcher, M. G., ed., *Am. Assoc. Petrol. Geol. Mem.* 19, p. 587.

Kerr, J. Wm., 1973b, Geology, Sawyer Bay, District of Franklin, *Geol. Surv. Can. Map* 1357A.

Kerr, J. Wm., 1973c, Geology, Dobbin Bay, District of Franklin, *Geol. Surv. Can. Map* 1358A.

Kerr, J. Wm., 1974, Geology of Bathurst Island Group and Byam Martin Island, Arctic Canada (Operation Bathurst Island), *Geol. Surv. Can. Mem.* 378.

Kerr, J. Wm., 1976a, Stratigraphy of central and eastern Ellesmere Island, Arctic Canada, Part III. Upper Ordovician (Richmondian), Silurian, and Devonian, *Geol. Surv. Can. Bull.* 260.

Kerr, J. Wm., 1976b, Geology of outstanding Arctic aerial photographs. 3. Margin of Sverdrup Basin, Lyall River, Devon Island, *Can. Soc. Petrol. Geol.*, v. 24(2), p. 139–153.

Kerr, J. Wm., 1977a, Cornwallis Fold Belt and the mechanism of basement uplift, *Can. J. Earth Sci.*, v. 14(6), p. 1374–1401.

Kerr, J. Wm., 1977b, Cornwallis Lead–Zinc District—Mississippi Valley-type deposits controlled by stratigraphy and tectonics, *Can. J. Earth Sci.*, v. 14(6), p. 1402–1426.

Kerr, J. Wm., 1977c, Four mineralization controls established for Arctic's Cornwallis Lead–Zinc District, *Northern Miner*, November 24, 1977, p. B16–B17.

Kerr, J. Wm., 1977d, An unusual sea stack at high elevation on northwest Devon Island, *Geol. Surv. Can. Pap.* 77-1C, p. 79–80.

Kerr, J. Wm., 1978, Cornwallis Lead–Zinc District; Mississippi Valley-type deposits controlled by stratigraphy and tectonics: Reply, *Can. J. Earth Sci.*, v. 15, p. 460.

Kerr, J. Wm., 1979, Cornwallis Lead–Zinc District—Mississippi Valley-type deposits controlled by stratigraphy and tectonics: Reply II, *Can. J. Earth Sci.*, v. 16, p. 615–617.

Kerr, J. Wm., 1980a, Structural framework of Lancaster Aulacogen, Arctic Canada, *Geol. Surv. Can. Bull.* 319 (in press).

Kerr, J. Wm., 1980b, A plate tectonic contest in Arctic Canada, in: *Crust of the Earth and Its Mineral Deposits*, Strangway, D. W., ed., *Geol. Assoc. Can. Spec. Pap.* (in press).

Kerr, J. Wm., and de Vries, C. D. S., 1977, Structural geology of Somerset Island and Boothia Peninsula, *Geol. Surv. Can. Pap.* 77-1A, p. 107–111.

Kerr, J. Wm., and Ruffman, A. R., 1979, The Crozier Strait Fault Zone, Arctic Archipelago, Northwest Territories, Canada, *Bull. Can. Petrol. Geol.*, v. 27(1), p. 39–52.

Kerr, J. Wm., and Thorsteinsson, R., 1972, Geology, Baumann Fiord, District of Franklin, *Geol. Surv. Can. Map* 1312A.

King, E. R., Zietz, I., and Alldredge, L. R., 1966, Magnetic data on the structure of the central Arctic region, *Bull. Geol. Soc. Am.*, v. 77, p. 619–646.

Koch, L., 1929, Stratigraphy of Greenland, *Medd. Groenl.*, v. 73(2).

Kranck, E. H., 1947, Indications of movements of the earth-crust along the coast of Newfoundland–Labrador, *Bull. Comm. Geol. Finl.*, v. 140, p. 89–96.

Kristofferson, Y., and Talwani, M., 1977, Extinct triple junction south of Greenland and the Tertiary motion of Greenland, relative to North America, *Geol. Soc. Am. Bull.*, v. 88, p. 1037–1049.

Kurtz, V. E., McNair, A. H., and Wales, D. B., 1952, Stratigraphy of the Dundas Harbour area, Devon Island, *Am. J. Sci.*, v. 250, p. 636–655.

Lambert, R. St. J., 1974, Global tectonics and the Canadian Arctic Continental Shelf, in: *Geology of the Canadian Arctic*, Aitken, J. D., and Glass, D. J., eds., Proc. Symp. on Geology of Canadian Arctic, Saskatoon, May 1973, *Geol. Assoc. Can.- Can. Soc. Petrol. Geol.*, p. 5–22.

Le Pichon, A., Hyndman, R. D., and Pautot, G., 1971, Geophysical study of the opening of the Labrador Sea, *J. Geophys. Res.*, v. 76, p. 4724–4743.

Le Pichon, X., Sibuet, J. C., and Francheteau, J., 1977, The fit of the continents around the North Atlantic Ocean, *Tectonophysics*, v. 38, p. 169–209.

Manchester, K. S., and Clarke, D. B., 1973, Geologic structure of Baffin Bay and Davis Strait as determined by geophysical techniques, in: *Arctic Geology*, Pitcher, M. G., ed., *Am. Assoc. Petrol. Geol. Mem.* 19, p. 536–541.

Martin, R., 1973, Cretaceous–Early Tertiary rift basin of Baffin Bay—Continental drift without sea-floor spreading, in: *Arctic Geology*, Pitcher, M. G., cd., *Am. Assoc. Petrol. Geol. Mem.* 19, p. 500–505.

Matthews, J. V., Jr., 1976, Insect fossils from the Beaufort Formation: geological and biological significance, in: *Geol. Surv. Can. Pap.* 77-1B, p. 217–227.

Mayhew, M. A., Drake, C. L., and Nafe, J. E., 1970, Marine geophysical measurements on the continental margins of the Labrador Sea, *Can. J. Earth Sci.*, v. 7(2), pt. I, p. 199–214.

Mayr, U., 1976a, Middle Silurian reefs in Southern Peary Land, North Greenland, *Bull. Can. Petrol. Geol.*, v. 24, p. 440–449.

Mayr, U., 1976b, Upper Paleozoic succession in the Yelverton area, Northern Ellesmere Island, District of Franklin, *Geol. Surv. Can. Pap.* 76-1A, p. 445–448.

Mayr, U., 1978, Stratigraphy and correlation of Lower Paleozoic formations, subsurface of Cornwallis, Devon, Somerset and Russell Islands, Canadian Arctic Archipelago, *Geol. Surv. Can. Bull.*, v. 276.

McMillan, N. J., 1973, Shelves of Labrador Sea and Baffin Bay, Canada, in: *Future Petroleum Provinces of Canada—Their Geology and Potential*, McCrossan, R. G., ed., *Can. Soc. Petrol. Geol. Mem.* 1, p. 473–517.

McWhae, J. R. H., and Michel, W. F. E., 1975, Stratigraphy of Bjarni H-81 and Leif M-48, Labrador Shelf, *Bull. Can. Petrol. Geol.*, v. 23(3), p. 361–382.

Meneley, R. A., 1976, *Exploration Prospects in the Canadian Arctic Islands*, presented to the Canadian Society of Exploration Geophysicists, May 27, 1976, Calgary, Alberta: Panarctic Oils Ltd.

Meneley, R. A., Henao, D., and Merritt, R. K., 1975, The northwest margin of the Sverdrup Basin, in: *Canada's Continental Margins and Offshore Petroleum Exploration*, Yorath, C. J., Parker, E. R., and Glass, D. J., eds., *Can. Soc. Petrol. Geol. Mem.* 4, p. 531–544.

Meyerhoff, A. A., 1970, Continental drift. II. High latitude evaporite deposits and geologic history of Arctic and North Atlantic Ocean, *J. Geol.*, v. 78(4), p. 406–444.

Meyerhoff, A. A., 1973, Origin of Arctic and North Atlantic Oceans, in: *Arctic Geology*, Pitcher, M. G., ed., *Am. Assoc. Petrol. Geol. Mem.* 19, p. 562–582.

Miall, A. D., 1975, Post-Paleozoic geology of Banks, Prince Patrick, and Eglinton Islands, Arctic Canada, in: *Canada's Continental Margins and Offshore Petroleum Exploration*, Yorath, C. J., Parker, E. R., and Glass, D. J., eds., *Can. Soc. Petrol. Geol. Mem.* 4, p. 557–587.

Miall, A. D., 1976a, Proterozoic and Paleozoic geology of Banks Island, Arctic Canada, *Geol. Surv. Can. Bull.* 258.

Miall, A. D., 1976b, Devonian geology of Banks Island, Arctic Canada, and its bearing on the tectonic development of the circum-Arctic region, *Geol. Soc. Am. Bull.*, v. 87, p. 1599–1608.

Miall, A. D., and Gibling, M. R., 1978, The Siluro-Devonian clastic wedge of Somerset Island, Arctic Canada, and some regional paleogeographic implications, *Sediment. Geol.*, v. 21, p. 85–127.

Miall, A. D., and Kerr, J. Wm., 1977, Phanerozoic stratigraphy and sedimentology of Somerset Island and northeastern Boothia Peninsula, in: *Geol. Surv. Can. Pap.* 77-1A, p. 99–106.

Miall, A. D., and Kerr, J. Wm., 1980, Cambrian to Upper Silurian stratigraphy, Somerset Island and northeastern Boothia Peninsula, District of Franklin, N.W.T., *Geol. Surv. Can. Bull.* 315.

Morrow, D. W., and Kerr, J. Wm., 1977, Stratigraphy and sedimentology of Lower Paleozoic formations near Prince Alfred Bay, Devon Island, *Geol. Surv. Can. Bull.* 254.

Mossop, G. D., 1979, The Ordovician Baumann Fiord Formation evaporites of Ellesmere Island, Arctic Canada, *Geol. Surv. Can. Bull.* 298.

Munck, S., and Noe-Nygaard, A., 1957, Age determination of the various stages of the Tertiary volcanism in the West Greenland Basalt Province: 20th Int. Geol. Congr., Mexico City, 1956, Section I, Book 1, p. 247–256.

Murray, J. W., Libby, W. G., and Chase, R. L., 1970, Baffin continental shelf potential oil and gas source, *Oilweek*, May 11, p. 50–54.

Nassichuk, W. W., 1975, Carboniferous ammonoids and stratigraphy in the Canadian Arctic Archipelago, *Geol. Surv. Can. Bull.* 237.

Nassichuk, W. W., and Christie, R. L., 1969, Upper Paleozoic and Mesozoic stratigraphy in the Yelverton Pass Region, Ellesmere Island, District of Franklin, *Geol. Surv. Can. Pap.* 68-31.

Nassichuk, W. W., and Davies, G. R., 1981, Stratigraphy, biochronology, and sedimentology of the Otto Fiord Formation—a major Mississippian–Pennsylvanian evaporite of subaqueous origin in the Canadian Arctic Archipelago, *Geol. Surv. Can. Bull.* 286.

Newman, P. H., 1977, The offshore and onshore geophysics and geology of the Nares Strait Region: Its tectonic history and significance in regional tectonics, M.S. Dissertation, Dalhousie University, Halifax, N.S.

Newman, P. H., and Falconer, R. K. H., 1978, Program of Joint Annual Meeting of Geol. Assoc. Can., Min. Assoc. Can., Geol. Soc. Am., Toronto, Canada, October 23–26, 1968, p. 463.

Noe-Nygaard, A., 1974, Cenozoic to recent volcanism in and around the North Atlantic Basin, in: *The Ocean Basins and Margins*, v. 2, Nairn, A. E. M., and Stehli, F. G., eds., New York: Plenum Press, p. 391–443.

Norris, D. K., 1974, Structural geometry and geological history of the northern Canadian Cordillera, in: Proceedings of the 1973 National Convention, Wren, A. E., and Cruz, R. B., eds., Calgary: Canadian Society of Exploration Geophysicists, p. 18–45.

Oliver, J., Ewing, M., and Press, F., 1955, Crustal structure of the Arctic regions from the Lg phase, *Bull. Geol. Soc. Am.*, v. 66, p. 1063–1974.

Olson, R. A., 1977, Geology and genesis of Zinc–Lead deposits within a late Proterozoic dolomite, northern Baffin Island, N.W.T., Ph.D. Dissertation, University of British Columbia, Vancouver, B.C.

Ostenso, N. A., and Wold, R. J., 1973, Aeromagnetic evidence for origin of Arctic Ocean Basin, in: *Arctic Geology*, Pitcher, M.G., ed., *Am. Assoc. Petrol. Geol. Mem.* 19, p. 506–516.

Park, K., Clarke, D. G., Johnson, J., and Keen, M. J., 1971, Seaward extension of the West Greenland Tertiary volcanic province, *Earth Planet. Sci. Lett.*, v. 10, p. 235–238.

Pelletier, B. R., 1966, Development of submarine physiography in the Canadian Arctic and its relation to crustal movement, in: *Continental Drift*, Garland, G. D., ed., Toronto: University of Toronto Press, p. 77–101.

Pelletier, B. R., Ross, D. I., Keen, C. E., and Keen, M. J., 1975, Geology and geophysics of Baffin Bay, *Geol. Surv. Can. Pap.* 74-30, v. 2: p. 247–258.

Pitman, W. C. III, and Talwani, M., 1972, Sea-floor spreading in the North Atlantic, *Bull. Geol. Soc. Am.*, v. 83, p. 619–646.

Plauchut, B. P., 1971, Geology of the Sverdrup Basin, *Bull. Can. Petrol. Geol.*, v. 19(3), p. 659–679.

Plauchut, B. R., and Jutard, G. G., 1976, Cretaceous and Tertiary stratigraphy, Banks and Eglinton Islands, and Anderson Plain (N.W.T.), *Bull. Can. Petrol. Geol.*, v. 24(3), p. 321–371.

Powell, T. G., 1978, An assessment of the hydrocarbon source rock potential of the Canadian Arctic Islands, *Geol. Surv. Can. Pap.* 78-12.

Puscharovskiy, Yu. M., 1960, Some general problems of the tectonics of the Arctic, *Dokl. Akad. Nauk SSSR, Ser. Geol.*, no. 9, p. 15–28.

Puscharovskiy, Yu. M., 1978, Tectonic movements in the oceans, *Geotectonics (USSR)*, v. 12(1), p. 1–9.

Rahmani, R. A., 1977, Fault control on sedimentation of Isachsen Formation in Sverdrup Basin, *Geol. Surv. Can. Pap.* 77-1B, p. 157–161.

Rosenkrantz, A., 1970, Marine Upper Cretaceous and lowermost Tertiary deposits in West Greenland, *Medd. Dansk Geol. Foren.*, v. 19(4), p. 406–453.

Ross, D. I., 1973, Free air and simple bouguer gravity maps of Baffin Bay and adjacent continental margins, *Geol. Surv. Can. Pap.* 73-37.

Ross, D. I., and Falconer, R. K. H., 1975, Geological studies of Baffin Bay, Davis Strait, and adjacent continental margins, *Geol. Surv. Can. Pap.* 75-1, Part A, p. 181–183.

Ross, D. I., and Henderson, G., 1973, New geophysical data on the continental shelf of central and northern West Greenland, *Can. J. Earth Sci.*, v. 10(4), p. 485–497.

Ross, D. I., Keen, C. E., Barrett, D. L., and Manchester, K. S., 1973, Geophysical studies on the structure of Baffin Bay, *Geol. Surv. Can. Pap.* 71-23, p. 633–638.

Roy, K. J., 1974, Transport directions in the Isachsen Formation (Lower Cretaceous), Sverdrup Islands, District of Franklin, *Geol. Surv. Can. Pap.* 74-1, Part A, p. 351–353.

Schuchert, C., 1923, Sites and nature of North American geosynclines, *Bull. Geol. Soc. Am.*, v. 34, p. 151–229.

Shatskiy, N. S., 1935, On the tectonics of the Arctic, in: *Geology of Mineral Resources of Northern U.S.S.R.*, 1st Geol. Inv. Conf., Moscow, v. 1, p. 149–168.

Shatskiy, N. W., 1946, Basic features of the structures and development of the East European Platform—Comparative tectonics of ancient platforms, *Akad. Nauk. SSR*, v. 12(1), p. 5–62.

Sinha, A. K., and Frisch, T. O., 1975, Whole-rock Rb-Sr ages of metamorphic rocks from northern Ellesmere Island, Canadian Arctic Archipelago. I. The gneiss terrain between Ayles Fiord and Yelverton Inlet, *Can. J. Earth Sci.*, v. 12, p. 90–94.

Sinha, A. K., and Frisch, T. O., 1976, Whole-rock Rb-Sr and zircon U-Pb ages of metamorphic rocks from northern Ellesmere Island, Canadian Arctic Archipelago. II. The Cape Columbia Complex, *Can. J. Earth Sci.*, v. 13(6), p. 774–780.

Snowdon, L. R., and Roy, K. J., 1975, Regional metamorphism in the Mesozoic strata of the Sverdrup Basin, *Bull. Can. Petrol. Geol.*, v. 23(1), p. 131–148.

Sobczak, L. W., and Weber, J. R., 1973, Crustal structure of Queen Elizabeth Islands and polar continental margin, Canada, in: *Arctic Geology*, Pitcher, M. G., ed., *Am. Assoc. Petrol. Geol. Mem.* 19, p. 517–525.

Srivastava, S. P., 1978, Evolution of the Labrador Sea and its bearing on the early evolution of the North Atlantic, *Geophys. J. R. Astron. Soc.*, v. 52(2), p. 313–357.

Stockwell, C. H., 1964, Fourth report on structural provinces, orogenies, and time—classification of rocks of the Canadian Precambrian Shield, *Geol. Surv. Can. Pap.* 64-17 (Part II).

Stockwell, C. H., McGlynn, J. C., Emslie, R. F., Sanford, B. V., Norris, A. W., Donaldson, J. A., Fahrig, W. F., and Currie, K. L., 1970, Geology of the Canadian Shield, *Geol. Surv. Can. Econ. Geol. Rep.*, v. 1, p. 45–150.

Stott, D. F., 1968, Ellef Ringnes Island, Canadian Arctic Archipelago, *Geol. Surv. Can. Pap.* 68-16.

Stuart Smith, J. H., and Wennekers, J. H. N., 1977, Geology and hydrocarbon discoveries of the Canadian Arctic Islands, *Bull. Am. Assoc. Petrol. Geol.*, v. 61(1), p. 1–27.

Sweeney, J. F., 1977, Subsidence rates of the Sverdrup Basin, Canadian Arctic Islands, *Geol. Surv. Am. Bull.*, v. 88, p. 41–48.

Sweeney, J. F., and Haines, G. V., 1978, Arctic geophysical review—an introduction, in: *Arctic Geophysical Review*, Sweeney, J. F., ed., *Can. Dep. Energy Mines Resources*, v. 45(4), p. 1–6.

Sweeney, J. F., Irving, E., and Geuer, J. W., 1978, Evolution of the Arctic Basin, in: *Arctic Geophysical Review*, Sweeney, J. F., ed., *Can. Dep. Energy Mines Resources*, v. 45(4), p. 91–100.

Tailleur, I. L., 1973, Probable rift origin of Canada Basin, Arctic Ocean, in: *Arctic Geology*, Pitcher, M. G., ed., *Am. Assoc. Petrol. Geol. Mem.* 19, p. 526–535.

Thorsteinsson, R., 1958, Cornwallis and Little Cornwallis Islands, District of Franklin, Northwest Territories, *Geol. Surv. Can. Mem.* 294.

Thorsteinsson, R., 1973, Prince Alfred Bay (59B), Resolute (58F), Baillie Hamilton Island (58G), Lowther Island (68E), and McDougall Sound (68H) map areas, Arctic Islands, *Geol. Surv. Can. Open-File Rep.* 139.

Thorsteinsson, R., 1974, Carboniferous and Permian stratigraphy of Axel Heiberg Island and western Ellesmere Island, Canadian Arctic Archipelago, *Geol. Surv. Can. Bull.* 224.

Thorsteinsson, R., and Kerr, J. Wm., 1968, Cornwallis Island and adjacent smaller islands, Canadian Arctic Archipelago, *Geol. Surv. Can. Pap.* 67-64.

Thorsteinsson, R., and Kerr, J. Wm., 1972a, Geology, Eureka Sound south, District of Franklin, *Geol. Surv. Can. Map* 1300.

Thorsteinsson, R., and Kerr, J. Wm., 1972b, Geology, Cañon Fiord, District of Franklin, *Geol. Surv. Can. Map* 1308A.

Thorsteinsson, R., and Tozer, E. T., 1957, Geological investigations in Ellesmere and Axel Heiberg Islands, 1956, *Arctic*, v. 10, p. 2–31.

Thorsteinsson, R., and Tozer, E. T., 1960, Summary account of structural history of the Canadian Arctic Archipelago since Precambrian time, *Geol. Surv. Can. Pap.* 60-7.

Thorsteinsson, R., and Tozer, E. T., 1962, Banks, Victoria, and Stefansson Islands, Arctic Archipelago, *Geol. Surv. Can. Mem.* 330.

Thorsteinsson, R., and Tozer, E. T., 1970, Geology of the Arctic Archipelago, *Geol. Surv. Can. Econ. Geol. Ret.* p. 548–590.

Tozer, E. T., 1961, Triassic stratigraphy and faunas, Queen Elizabeth Islands, Arctic Archipelago, *Geol. Surv. Can. Mem.* 316.

Tozer, E. T., 1967, A standard for Triassic time, *Geol. Surv. Can. Bull.* 156.

Tozer, E. T., and Thorsteinsson, R., 1964, Western Queen Elizabeth Islands, Arctic Archipelago, *Geol. Surv. Can. Mem.* 332.

Trettin, H. P., 1969a, Pre-Mississippian geology of northern Axel Heiberg and northwestern Ellesmere Island, Arctic Archipelago, *Geol. Surv. Can. Bull.* 171.

Trettin, H. P., 1969b, Geology of Ordovician to Pennsylvanian rocks, M'Clintock Inlet, north coast of Ellesmere Island, Canadian Arctic Archipelago, *Geol. Surv. Can. Bull.* 183.

Trettin, H. P., 1971a, Geology of Lower Paleozoic formations, Hazen Plateau and Southern Grant Land Mountains, Ellesmere Island, Arctic Archipelago, *Geol. Surv. Can. Bull.* 203.

Trettin, H. P., 1971b, Reconnaissance of Lower Paleozoic geology, Phillips Inlet region north coast of Ellesmere Island, District of Franklin, *Geol. Surv. Can. Pap.* 71-12.

Trettin, H. P., 1973a, Early Paleozoic evolution of the northern parts of the Canadian Arctic Archipelago, in: *Arctic Geology*, Pitcher, M. G., ed., *Am. Assoc. Petrol. Geol. Mem.* 19, p. 57–75.

Trettin, H. P., 1973b, Preliminary draft of 1 : 1,000,000 geological atlas sheets, Eureka Sound and Robeson Channel areas, Canadian Arctic Islands (NTS 560, 340, and Canadian part of 120), *Geol. Surv. Can. Open-File Rep.* 174.

Trettin, H. P., 1975, Investigations of Lower Paleozoic geology, Foxe Basin, northeastern Melville Peninsula, and parts of northwestern and central Baffin Island, *Geol. Surv. Can. Bull.* 251.

Trettin, H. P., 1976, Reconnaissance of Lower Paleozoic geology, Agassiz Ice Cap to Yelverton Bay, northern Ellesmere Island, *Geol. Surv. Can. Pap.* 76-1A, p. 431–444.

Trettin, H. P., 1978, Investigations of Devonian clastic formations, west-central Ellesmere Island, *Geol. Surv. Can. Bull.* 302.

Trettin, H. P., 1979, Middle Ordovician to Lower Devonian deep-water succession at southeastern margin of Hazen Trough, Cañon Fiord, Ellesmere Island, *Geol. Surv. Can. Bull.* 272.

Trettin, H. P., Frisch, T. O., Sobczak, L. W., Weber, J. R., Niblett, E. R., Law, L. K., DeLaurier, J. M., and Whitham, K., 1972, The Innuitian Province, in: *Variations in Tectonic Styles in Canada*, Price, R. A., and Douglas, R. J. W., eds., *Geol. Assoc. Can. Spec. Pap.* 11, p. 83–179.

Troelson, J. C., 1950, Contributions to the geology of northwest Greenland, Ellesmere Island, and Axel Heiberg Island, *Medd. Groenl.*, v. 149(7), p. 1–85.

Van der Linden, W. J. M., 1975a, Mesozoic and Cenozoic opening of the Labrador sea, the north Atlantic and the Bay of Biscay, *Nature (London)*, v. 253, p. 320–324.

Van der Linden, W. J. M., 1975b, Crustal attenuation and sea-floor spreading in the Labrador Sea, *Earth Planet. Sci. Lett.*, v. 27, p. 409–423.

Van der Linden, W. J. M., and Srivastava, S. P., 1975, The crustal structure of the continental margin off central Labrador, *Geol. Surv. Can. Pap.* 74-30, p. 233–245.

Vogt, P. R., and Ostenso, N. A., 1970, Magnetic and gravity profiles across the Alpha Cordillera and their relation to Arctic sea-floor spreading, *J. Geophys. Res.*, v. 75, p. 4925–4937.

Wade, J. A., Grant, A. C., Sanford, B. V., and Barss, M. S., 1977, Basement structure—Eastern Canada and adjacent areas, 1:2,000,000 scale, *Geol. Surv. Can. Map* 1400A.

Wallace, F. K., 1973, Geology of the Davis Strait Bathymetric Sill and associated sediments, offshore Baffin Island, Canada, in: *Canadian Arctic Geology*, Aitken J. D., and Glass, D. J. eds., Proc. Symp. Geol. Canadian Arctic, *Geol. Assoc. Can.–Can. Soc. Petrol. Geol. Mem.* 1, p. 81–97.

Wardlaw, N. C., and Christie, D. L., 1975, Sulphates of submarine origin in Pennsylvanian Otto Fiord Formation of Canadian Arctic, *Bull. Can. Petrol. Geol.*, v. 23, p. 149–171.

Wegener, A., 1924, *The Origin of Continents and Oceans*, translated by J. Biram, New York: Dover Publications (1966).

Wetmiller, R. J., 1974, Crustal structure of Baffin Bay from the earthquake-generated Lg phase, *Can. J. Earth Sci.*, v. 11(1), p. 123–130.

Wilson, D. G., 1976, Eureka Sound and Beaufort Formations, Yelverton Bay, Ellesmere Island, District of Franklin, *Geol. Surv. Can. Pap.* 76-1A, p. 453–456.

Wilson, J. T., 1963a, Continental drift, *Sci. Am.*, v. 208(4), p. 86–100.

Wilson, J. T., 1963b, Hypothesis of earth's behaviour, *Nature (London)*, v. 198, p. 925–929.

Wilson, J. T., 1965a, A new class of faults and their bearing on continental drift, *Nature (London)*, v. 207, p. 343–347.

Wilson, J. T., 1965b, Evidence from ocean islands suggesting movement in the Earth, in: Blackett, P. M. S., Bullard, E., and Runcorn, S. K., eds., p. 145–167.

Yorath, C. J., and Norris, D. K., 1975, The tectonic development of the southern Beaufort Sea and its relationship to the origin of the Arctic Ocean Basin, in: *Canada's Continental Margins and Offshore Petroleum Exploration*, Yorath, C. J., Parker, F. R., and Glass D. J., eds., Can. Soc. Petrol. Geol. Mem. 4, p. 589–611.

Young, G. M., and Jefferson, C. W., 1975, Late Precambrian shallow water deposits, Banks and Victoria Islands, Arctic Archipelago, *Can. J. Earth Sci.*, v. 12(10), p. 1734–1748.

Young, F. G., Myhr, D. W., and Yorath, C. J., 1976, Geology of the Beaufort–Mackenzie Basin, *Geol. Surv. Can. Pap.* 76-11.

Chapter 5

THE NORTHERN MARGIN OF GREENLAND FROM BAFFIN BAY TO THE GREENLAND SEA

Peter R. Dawes and John S. Peel

Geological Survey of Greenland
Copenhagen, Denmark

I. INTRODUCTION

Greenland, with a surface area of about 2,186,000 km², is the largest island in the world. It is composed of a central depression, now totally filled by the Inland Ice, and a mountainous coastal "ice-free" zone forming 20% of the total area.

The coastal zone reveals a long and complex geological history that has involved various Precambrian and Phanerozoic cycles of sedimentation, volcanism, and orogenesis stretching back to the earliest recognizable episodes in Earth's history. Precambrian crystalline rocks form a stable cratonic block (the Greenland Shield) around which sediments and volcanic rocks have accumulated from late Precambrian to the present day (Fig. 1).

The major rock divisions in Greenland correlate well with those in neighboring landmasses—North America to the west, Europe to the east. Greenland is bounded on all sides by seaways that are generally described as fundamental

The following Danish words appear throughout this chapter as part of official Greenland place names: *fjeld(e)*, mountain(s); *gletscher*, glacier; *isblink*, icecap; *kap*, cape; *ø*, island (plural *øer*, islands). Exceptions to the official nomenclature have been made in two instances, for which support exists in the literature: *Melville Bay* is used for Melville Bugt and *Wandel Sea* is used for Wandel Hav.

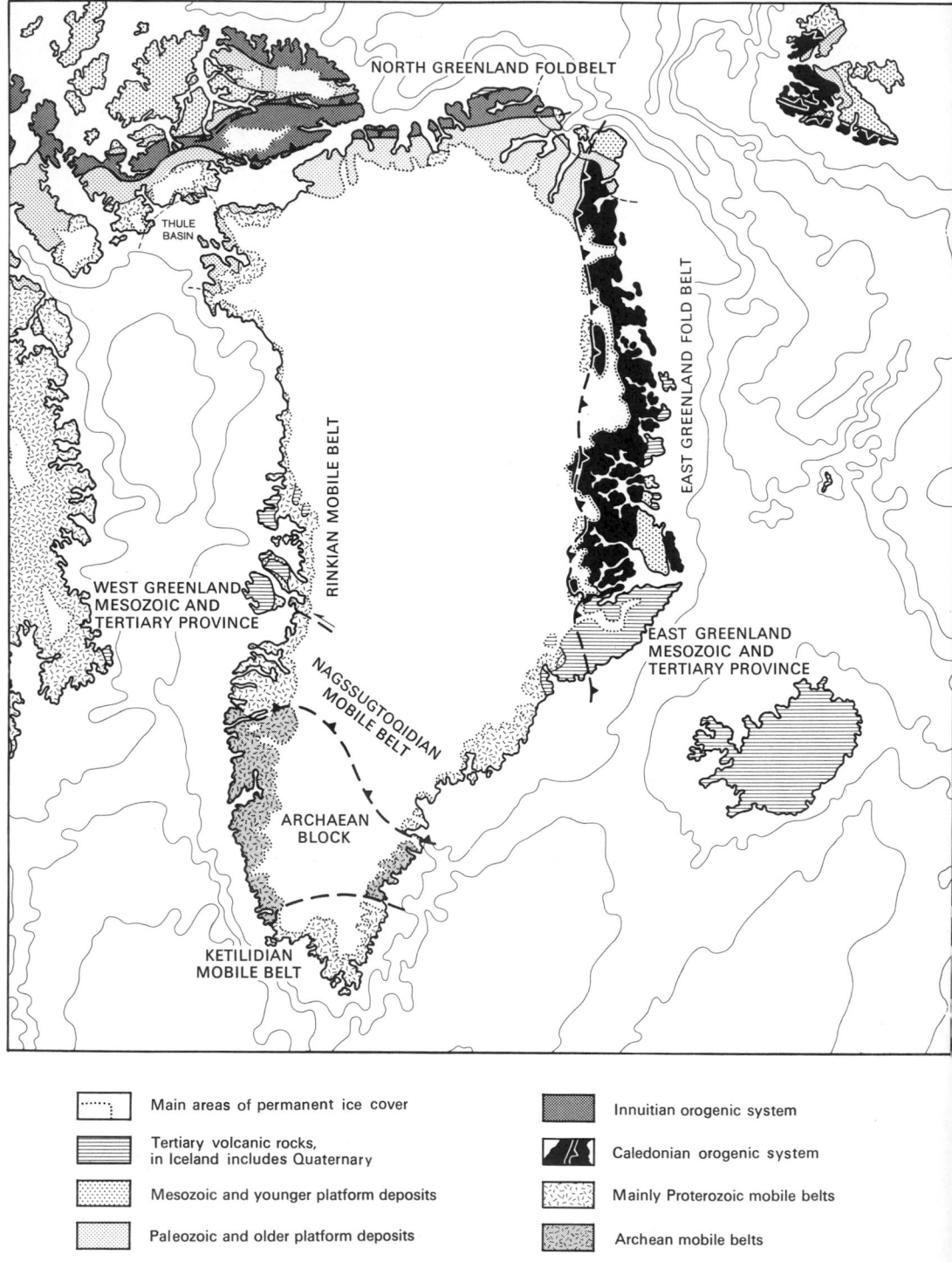

Fig. 1. Main geological provinces of Greenland and neighboring lands of the North Atlantic–Arctic region. (Modified from *Geology of Greenland*, 1976, Geological Survey of Greenland.)

tectonic structures of the crust and thus the island holds a key position in understanding the North Atlantic–Arctic Ocean system.

This chapter is concerned with the northern part of Greenland bordering the Arctic Ocean (Fig. 2). A previous chapter in Volume 2 of this series (Birkelund *et al.*, 1974) has described the southern and largest part of Greenland, including the East Greenland Caledonian Foldbelt. The northern part of

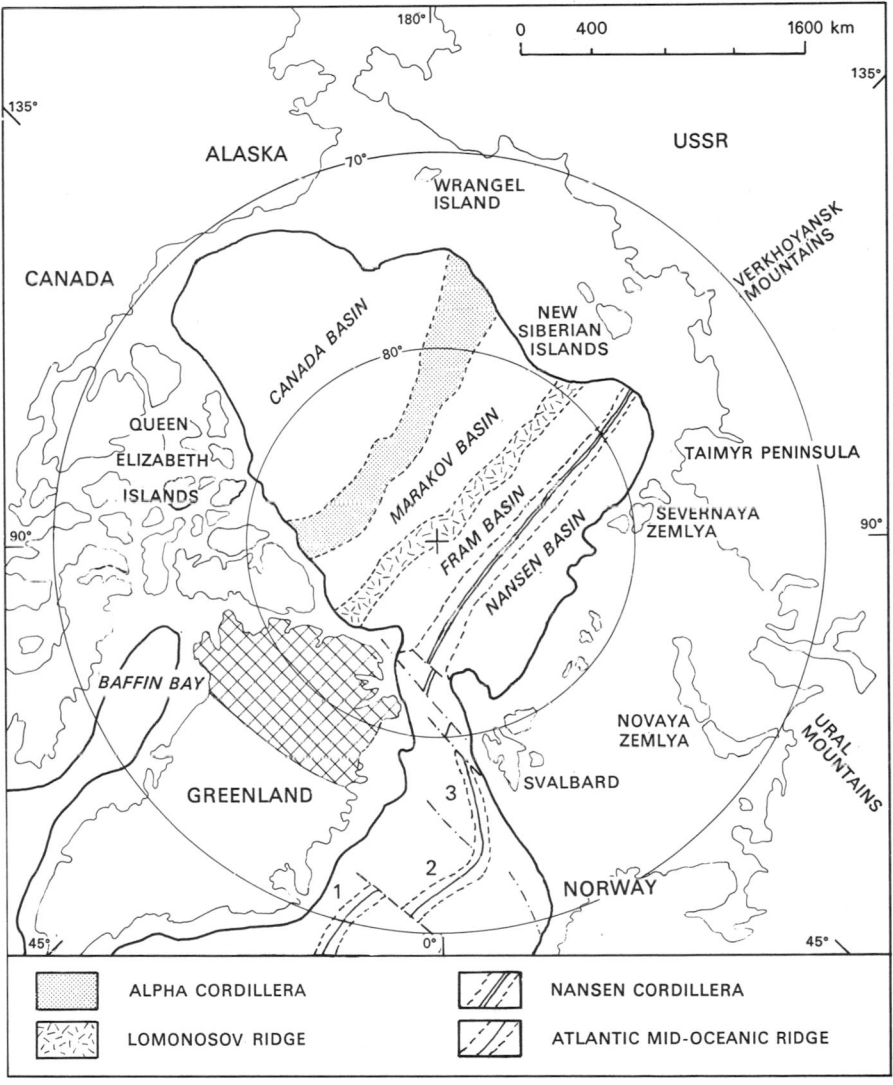

Fig. 2. Greenland's setting in the Arctic Ocean Basin. Cross-hatching covers the onshore region treated in this paper. Heavy full line represents boundary of the continental shelf. (1) Kolbeinsey Ridge; (2) Mohns Ridge; (3) Knipovich Ridge. (Modified from Dawes, 1973.)

this foldbelt is described here since it projects into the northern part of the Greenland Sea to become an important structural element of the Arctic Ocean margin. Some reference to the southerly continuation of the foldbelt into North-East and East Greenland, the main region dealt with by Birkelund *et al.* (1974), is consequently unavoidable.

A review of the bedrock of northern Greenland has recently been published (Dawes, 1976*a*).

II. REGIONAL GEOLOGICAL OUTLINE

A. Physiography

A large part of northern Greenland is covered by ice, in the form of glaciers and the northern extremity of the Inland Ice, or by isolated ice caps. The "ice-free" coastal zone has a total area of about 125,000 km² and in Peary Land is about 250 km wide. Plateaulike topography typical of the southern platform areas contrasts with the rugged and mountainous terrain of the Greenland Shield and the orogenic belt terrain of the north and east coasts. The highest peak is in northern Peary Land at about 2000 m.

B. Geological Provinces and Subprovinces

Northern Greenland is composed of four main geological provinces (Fig. 3): (1) the crystalline rocks of the Greenland Shield, mainly exposed in the south adjacent to the Inland Ice; (2) an extensive platform region—the North Greenland Platform—that stretches from the west to the east coast and is composed of homoclinal Proterozoic and Lower Paleozoic strata; (3) an east–west-trending orogenic belt of mainly Paleozoic rocks in the north—the North Greenland Foldbelt; and (4) a north–south-trending orogenic belt of Precambrian and Paleozoic rocks on the eastern coast—the East Greenland Caledonian Foldbelt.

Two subprovinces are recognized; (1) the Wandel Sea Basin of late Paleozoic to Tertiary rocks overlies the junction between the two Paleozoic orogenic belts in eastern North Greenland, and (2) the Kap Washington Group of late Cretaceous–early Tertiary bedded volcanics occurs along the northern coast forming a local boundary to the North Greenland Foldbelt.

C. Sedimentary and Structural Model

The central part of northern Greenland is occupied by the northern margin of the Greenland Shield, perhaps with a partial sedimentary cover. This largely ice-covered region forms a stable core flanked on the seaward side by substantial accumulations of sedimentary rock. These rocks outline three sedimentary basins: the Thule Basin in the Baffin Bay region to the west; the

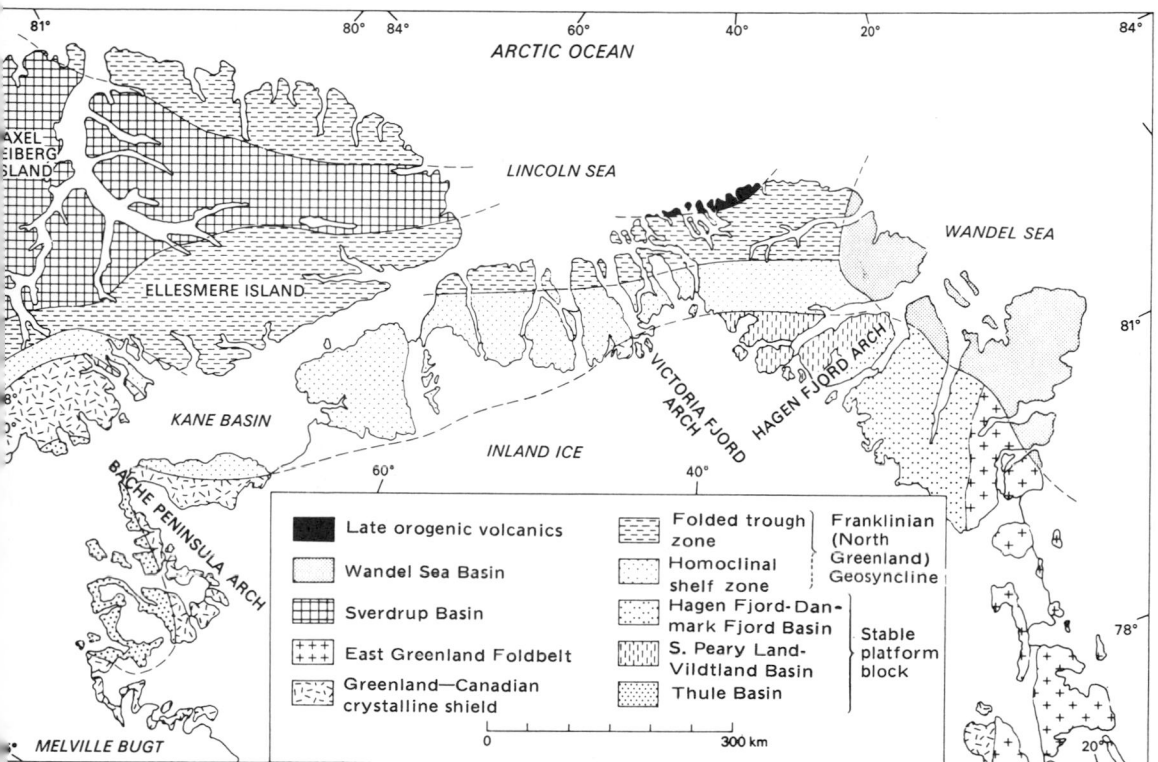

Fig. 3.   Main structural-stratigraphical units of northern Greenland and adjacent northern Arctic Canada. From Dawes (1976a).

North Greenland Geosyncline in the north, facing the Arctic Ocean; and the East Greenland Geosyncline on the east, facing the Greenland Sea. The Thule Basin is now a large graben and contains a faulted, unmetamorphosed Proterozoic sequence; the North and East Greenland Geosynclines are the sites of the North and East Greenland Foldbelts.

The North Greenland Platform, the dominant geological province of northern Greenland, occupies about 40% of the exposed land area. The platform performs a dual tectonic role, constituting a backland or hinterland to the North Greenland Foldbelt and, in the east, a foreland to the East Greenland Foldbelt. The Thule Basin, with its cratonic aspect, is considered as the western part of the platform.

The Wandel Sea Basin, containing a sequence of mainly marine platform sediments of late Paleozoic, Mesozoic, and Tertiary age, is deformed into a tectonic belt having a dominant northwest–southeast structural trend. The localization of regional subsidence in the Wandel Sea region may be controlled in some way by structural "weakness" caused by the junction of the North and East Greenland Foldbelts.

The crystalline shield dips shallowly north under the platform cover, but it must then plunge steeply to accommodate the thick geosynclinal sequence of the North Greenland Foldbelt. No Precambrian basement inliers outcrop in the foldbelt, nor are any reworked basement or tectonic slices suspected in the metamorphic complex of the northern coast. Likewise, there is a general eastward depression of the craton under the strata of the East Greenland Geosyncline, but in this case tectonic inliers of the craton are exposed within the foldbelt.

Certain magnetic and gravity anomalies in that part of the craton covered by the Inland Ice (see Haines *et al.*, 1970; Wilcox *et al.*, 1975; Coles *et al.*, 1976) may represent the presence of intracratonic sedimentary basins. Of particular interest in this respect is a large arc-shaped negative Bouguer anomaly parallel to the northern and eastern coasts that may represent a "sediment-filled structural depression" on the western edge of the East Greenland Foldbelt (Wilcox *et al.*, 1975). The widespread occurrence of quartzite erratics with *Scolithus* in East Greenland (Haller, 1971) is direct evidence of sedimentary areas under the Inland Ice in this region.

The sedimentary pattern of the North Greenland Platform and the North Greenland Geosyncline is an expression of the cratonic model outlined above. Thus, in the south an extensive stable platform of mainly carbonate rocks, characterized by a shallow northerly dip, passes to the north across a shallow shelf margin into a relatively thicker, mainly argillaceous and arenaceous sequence in the geosyncline. Proterozoic to Cambrian, mainly clastic strata overlying the crystalline basement in the south are succeeded northwards by a zone of Cambrian to Silurian carbonates. To the north an east–west linear Silurian reef mound complex contains extensive forereef and backreef associations. Further north a foreslope to basinal facies of argillaceous limestone with calcareous and graptolitic muds and cherts is characterized by allochthonous carbonate breccias. This sequence passes northwards into the more arenaceous deposits of the trough. These trough deposits extend into the Devonian and are characterized by abundant current structures indicating a steepening of the slope. The upper part of the trough sequence is a thick flysch unit of northerly provenance that encroaches southwards onto the shelf zone, surrounding and overlying the reef accumulations.

III. THE PRECAMBRIAN SHIELD

A. Introduction

The Precambrian shield forms the major part of the land surface of northern Greenland though much of this is now covered by the Inland Ice.

Thus, in terms of present-day exposure, the shield rocks make up only about 10% of the ice-free area.

Precambrian crystalline rocks outcrop in three areas: North-West Greenland between Melville Bay and the Humboldt Gletscher (75° to 79°30′N); central North Greenland in southern Wulff Land; and within the Caledonian Foldbelt in eastern North and North-East Greenland.

B. North-West Greenland

Precambrian crystalline rocks form the rugged and mountainous Melville Bay–Thule region and to the north, the lower, undulating peneplained surface of Inglefield Land. Toward the outer coast the rocks are unconformably overlain by the middle and late Proterozoic Thule Group, while Cambrian rocks are also present in Inglefield Land. The crystalline rocks have not been affected by major metamorphism and thermal episodes since the Hudsonian Orogeny in early Proterozoic time.

Much of the Melville Bay–Thule region (Fig. 4) is underlain by medium- to high-grade gneisses and granites and intimately associated metasedimentary and metavolcanic belts. Several of these supracrustal belts are associated with banded ironstone occurrences. Folded, in places migmatized, amphibolite dikes occur. One area of anorthosite and minor leucogabbro (at least 180 km²) in contact with supracrustal pelitic gneiss may form part of the basement to a thick psammite and amphibolite sequence. To the north in Inglefield Land, marble and pelite with calc-silicate rocks (Etah Group) form highly deformed belts up to 2 km wide. Farther north, towards the Humboldt Gletscher, granulite facies gneisses are common. Probably all of these rocks predate gabbroic to granitoid igneous rocks suites (Etah and Kap York metaigneous complexes) that form large areas and have suffered severe metamorphism and deformation. Younger gneisses, derived from these igneous rocks, can be recognized, although their regional extent is unknown.

Swarms of undeformed dolerite dikes cut all the above-mentioned rock types.

Regional foliation trends are generally easterly to northeasterly. Much of the basement has undergone polyphase deformation and metamorphism producing complex fold patterns. Uncertainty still surrounds the structural and age relationships of some units (Dawes, 1976a).

K–Ar mineral dates on gneisses, schists, and on metasedimentary and metaigneous rocks show a range of 1881 to 1610 m.y. (Larsen and Dawes, 1974), suggesting a regional thermal event in Hudsonian time. A K–Ar date of 1563 m.y. (Dawes et al., 1973) on a cross-cutting basic dike suggests that the regional thermal front had withdrawn by about 1500 m.y.

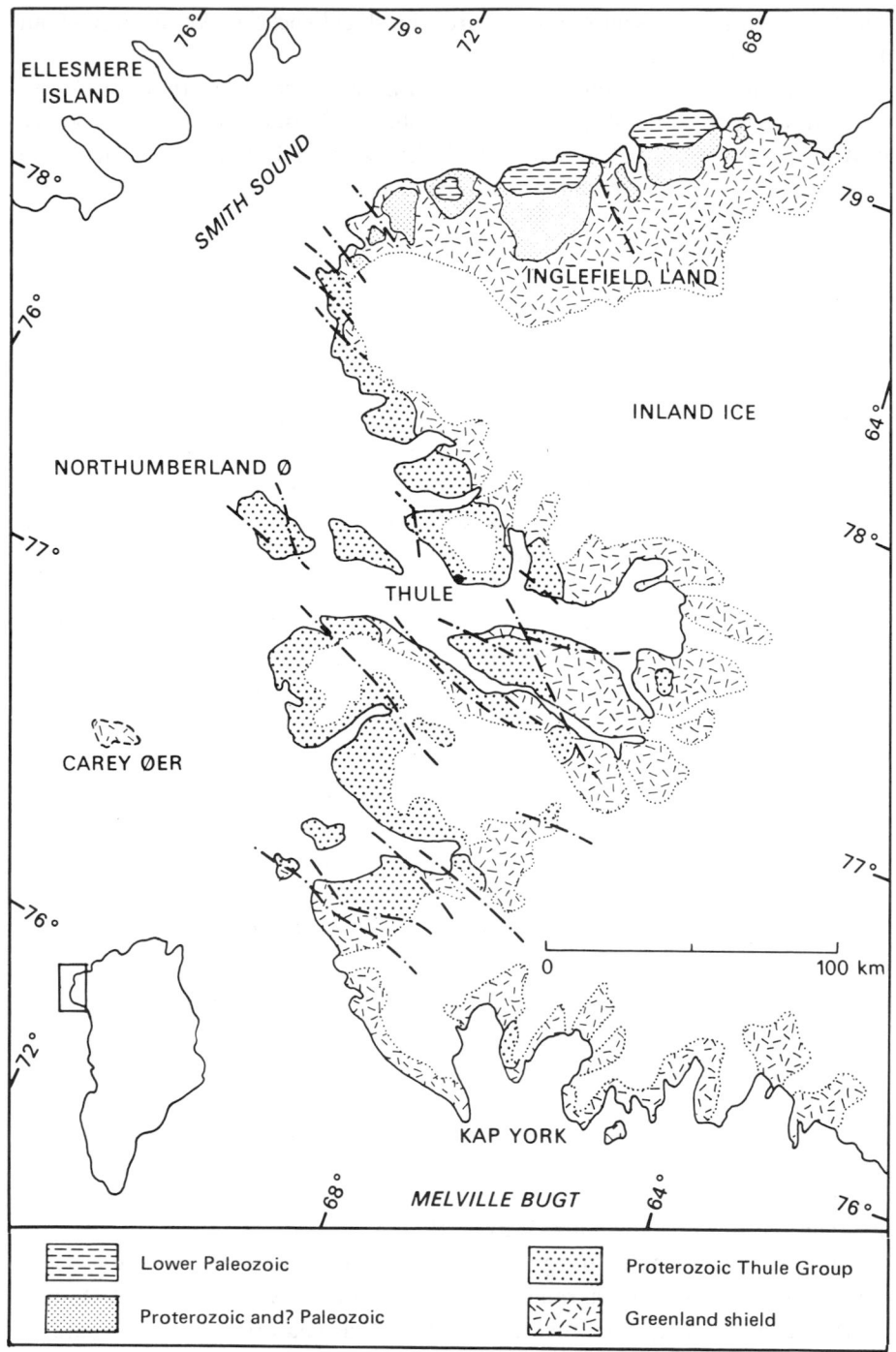

Fig. 4.   Geological sketch map of North-West Greenland.

Rb-Sr whole-rock analyses of Etah metaigneous rocks define an isochron at 1960 m.y., corresponding to the main metamorphism of the complex, while initial strontium ratios suggest that much older material is involved. Since at least one age of supracrustal rocks (Etah Group) can be shown to predate this complex, the presence in North-West Greenland of considerable amounts of Archean material in the crust is strongly suggested.

C. Central North Greenland

In northeastern Wulff Land and on various nunataks at the head of Victoria Fjord, small exposures (~200 km²) of crystalline rocks occur (Fig. 3). They form the core of the Victoria Fjord Arch of Dawes and Soper (1973). The arch is overlain by a thin succession of platform sediments of uncertain age (? late Proterozoic) followed by Cambrian to Silurian beds.

The crystalline complex is composed of a variety of amphibolite facies gneisses and schists, often garnet-bearing, with some granites and amphibolites that are cut by minor acid intrusions. The complex generally has a flat-lying foliation and it is cut by undeformed basic dikes of unknown composition and age.

The age of the Victoria Fjord Arch is uncertain. In the absence of isotopic dating the rocks are generally regarded as autochthonous stable shield, perhaps of the same age as the rocks of North-West Greenland. However, since the precise ages of the cross-cutting basic dikes and the overlying basal sediments are unknown, this cannot be unequivocally established. Plutonic activity, in say the period 1000 m.y. to 1600 m.y., may have affected this part of the crust.

D. Eastern North Greenland and North-East Greenland

Precambrian shield rocks are exposed in the Caledonian Foldbelt both within the orogenic zone and on the stable foreland (Fig. 5; for comprehensive accounts, see Haller, 1970, 1971; Henriksen and Higgins, 1976).

In Dronning Louise Land (77°N) gneisses, schists, granite, and some quartzites (the "western gneisses" of Peacock, 1956, 1958) are generally accepted as forming the eastern edge of the autochthonous shield, west of the orogenic belt. They are overlain by clastic deposits of the Trekant Series that is regarded as late Proterozoic (pre-Carolinidian) in age.

Within the orogenic zone, north of 76°N, gneiss, migmatites, and supracrustal rocks form several tectonic blocks and thrust sheets that may have been reworked during the Carolinidian and/or Caledonian Orogenies. The extreme northern part of the foldbelt (north of 78°N) remains little investigated and the extent of tectonic blocks or inliers of Precambrian shield, as distinct from younger crystalline complexes, is not known.

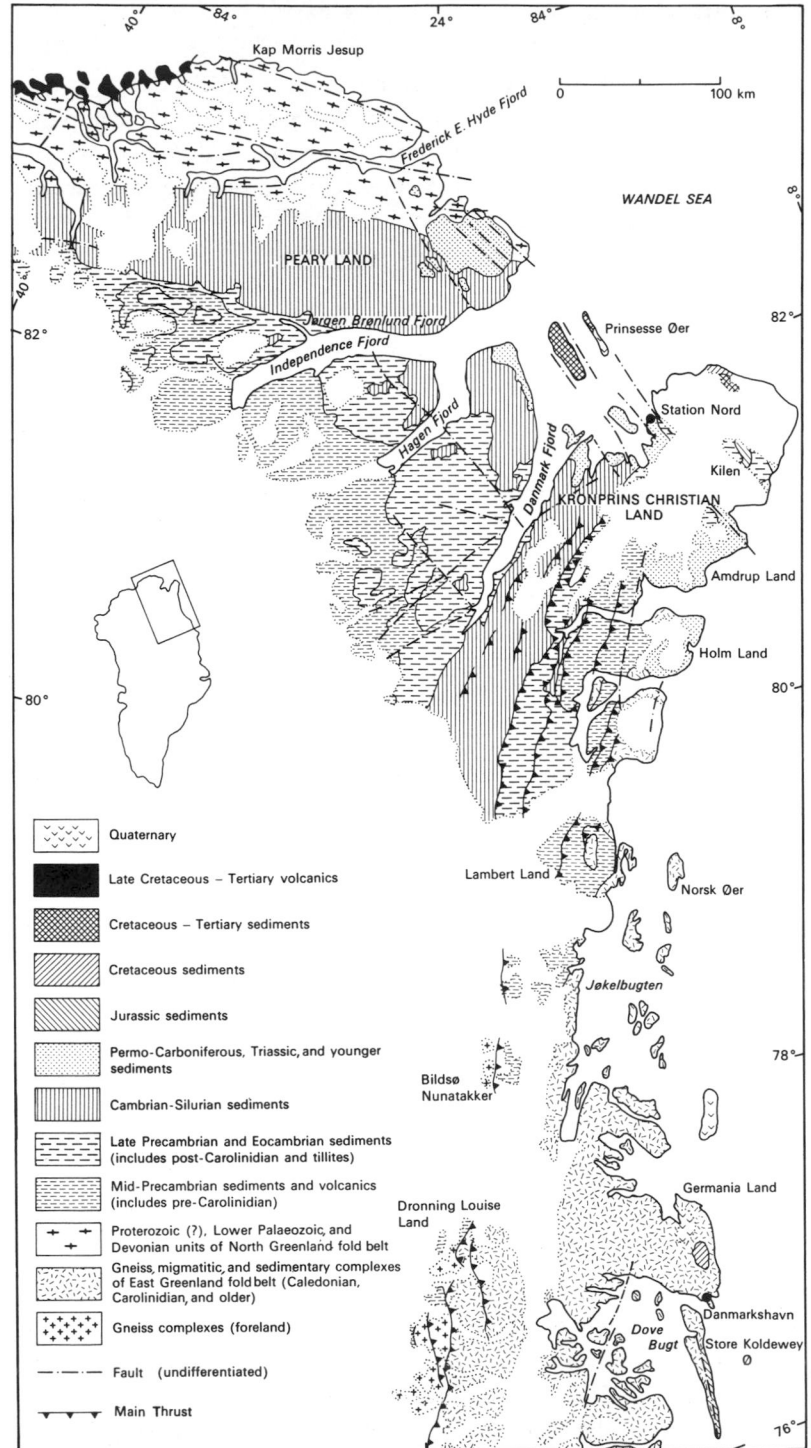

Fig. 5. Geological sketch map of North-East and eastern North Greenland. (Simplified, and with additions, from the *Tectonic/Geological Map of Greenland*, 1:2500000, 1970, Geological Survey of Greenland.)

IV. THE NORTH GREENLAND PLATFORM

A. Introduction

The crystalline basement of the stable platform is unconformably overlain by a sequence of Middle Proterozoic to Lower Paleozoic sediments that form a broad belt across North Greenland. The platform sequence is characterized by a shallow northerly dip so that, toward the north, Proterozoic, Cambrian, Ordovician, and Silurian strata outcrop as successive east–west-trending belts.

Several sedimentary basins can be delimited on the basis of tectonically high areas, the so-called arches (Fig. 3), and are particularly conspicuous during the Proterozoic. Clastic sediments and dolomites dominate the Proterozoic and earliest Phanerozoic, but a thick sequence of variable carbonates was deposited from the late Lower Cambrian through the Silurian. To the north these platform carbonates pass into equivalent clastic deposits in the North Greenland Foldbelt (Fig. 6).

B. Proterozoic

The Bache Peninsula Arch (Kerr, 1967a) separates the thick Proterozoic sequence of the Thule Basin to the south from a northerly dipping, relatively thin Proterozoic sequence extending east towards the Victoria Fjord Arch. Crystalline basement forms the exposed core of both arches. To the east of the Victoria Fjord Arch the Proterozoic succession thickens in the Peary Land–Vildtland Basin, which is itself bounded to the east by the most weakly defined of the three arches, the Hagen Fjord Arch. Sediments in the Hagen Fjord–Danmark Fjord Basin were deposited on the margins of the East Greenland Geosyncline.

Late Proterozoic tillite-bearing rocks, comparable to the Varangian Sequence of Scandinavia, are prominent in the Peary Land–Vildtland Basin.

1. *Thule Basin*

More than 4500 m of horizontal and shallowly dipping, variously colored, but commonly red clastics, shales, and carbonates comprising the Thule Group are preserved in a series of downfaulted blocks within the crystalline basement of North-West Greenland. The succession is composed of a lower sequence of sandstones, quartzites, and quartz-pebble conglomerates with minor shale and dolomite and with basaltic flows and sills, a middle sequence of shales, siltstones, and fine-grained sandstones with minor dolomite and with basaltic sills, and an upper sequence of cyclic alternating dolomites, limestones, red siltstones, and sandstones, with evaporites. Stromatolites and wormlike traces are the only recorded macrofossils (Dawes and Bromley, 1975).

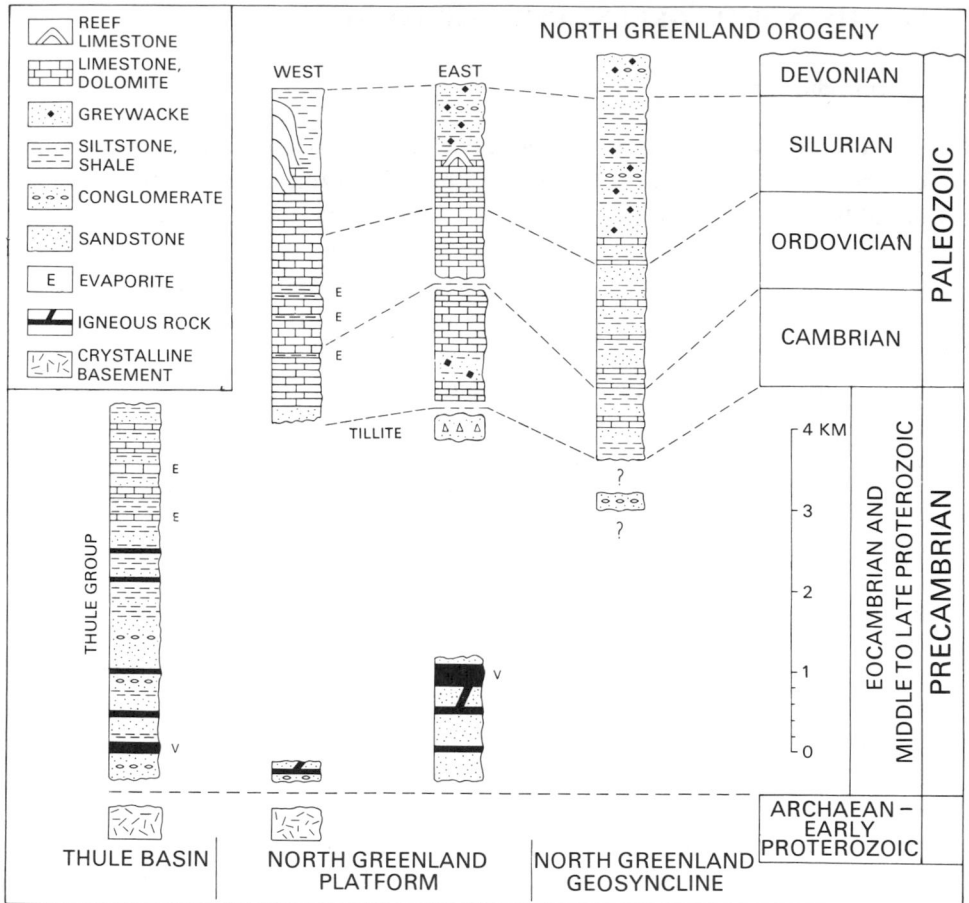

Fig. 6. Summary chart of Precambrian to Devonian stratigraphy of North-West and North Greenland (cf. Fig. 7). V denotes extrusive basic volcanics.

Radiometric ages from dolerite sills and dikes demonstrate that the lower sequence is at least 1200 m.y. old, while a date of 676 ± 25 m.y. has been obtained on a member of the cross-cutting WNW–ESE dike swarm that postdates the youngest sequence (Dawes *et al.*, 1973). Provisional microfossil (acritarch) identifications suggest a Riphean age for the lower and possibly middle sequences, and a Vendian age for the upper sequence (Gonzalo Vidal, personal communication).

2. *North Greenland Homocline*

Thin sequences of northerly dipping Proterozoic sediments occupy stratigraphic levels between the crystalline basement and overlying Lower Cambrian strata in Inglefield Land and the Victoria Fjord Arch area.

In southwestern Inglefield Land a 140- to 300-m sequence of ferruginous sandstones, shales, and some dolomites of the Rensselaer Bay Formation (Koch, 1933; Troelsen, 1950a; Cowie, 1961a) is intruded by dolerite sills of middle Proterozoic age (Dawes *et al.*, 1973). Overlying unfossiliferous dolomites have previously been grouped together with the formation, but recent field work has demonstrated the presence, especially in more northerly outcrops, of a much younger sandstone unit with *Cruziana* and *Scolithus* occurring between the Rensselaer Bay Formation and the dolomites. Limestones succeeding the dolomites yield Lower Cambrian fossils.

Supposed Proterozoic strata exposed in the Victoria Fjord Arch are thin, less than a few tens of meters thick. A basal breccia is followed by dolomite and sandstone units and this sequence is overlain by fossiliferous Lower Cambrian. The sequence is apparently not cut by basic dikes visible in the surrounding crystalline basement.

3. Peary Land–Vildtland Basin

The base of the sequence of Proterozoic and younger deposits in the Peary Land–Vildtland Basin (Koch, 1929a; Troelsen, 1949, 1956; Jepsen, 1971) is not exposed. The strata thin westwards, toward the Victoria Fjord Arch, and toward the Inland Ice to the south. However, stratigraphic units can still be recognized over the relatively weak Hagen Fjord Arch to the east.

A basal sequence of sandstone, arkose, and basic volcanics (Inuiteq Sø Formation, 1 km+ thick) is cut by dikes and sills indicating a middle Proterozoic age of at least 1000 m.y. for the deposits (Henriksen and Jepsen, 1970). Overlying tillites and associated deposits (Morænesø Formation, Jepsen, 1971, 0–130 m) have been correlated with the Varangian of Scandinavia (Troelsen, 1956). Succeeding dolomites (Portfjeld Formation, 200 m) and a turbiditic sequence (Buen Formation, 425 m) are considered early Cambrian in age although diagnostic Lower Cambrian macrofossils are first known from the middle of the latter unit (V. Poulsen, 1974).

4. Hagen Fjord–Danmark Fjord Basin

A comparable, but presumably thicker, sequence to that described from the Peary Land–Vildtland Basin is present to the southeast of the Hagen Fjord Arch (Fig. 7). The Middle Proterozoic sandstone–basalt sequence passes into the Carolinidian Geosyncline to the southeast (Haller, 1961). Late Proterozoic–early Phanerozoic deposits, including tillite horizons, form part of the present Caledonian Foreland and occur in nappes and thrust slices in the foldbelt (Fränkl, 1954; Haller, 1971).

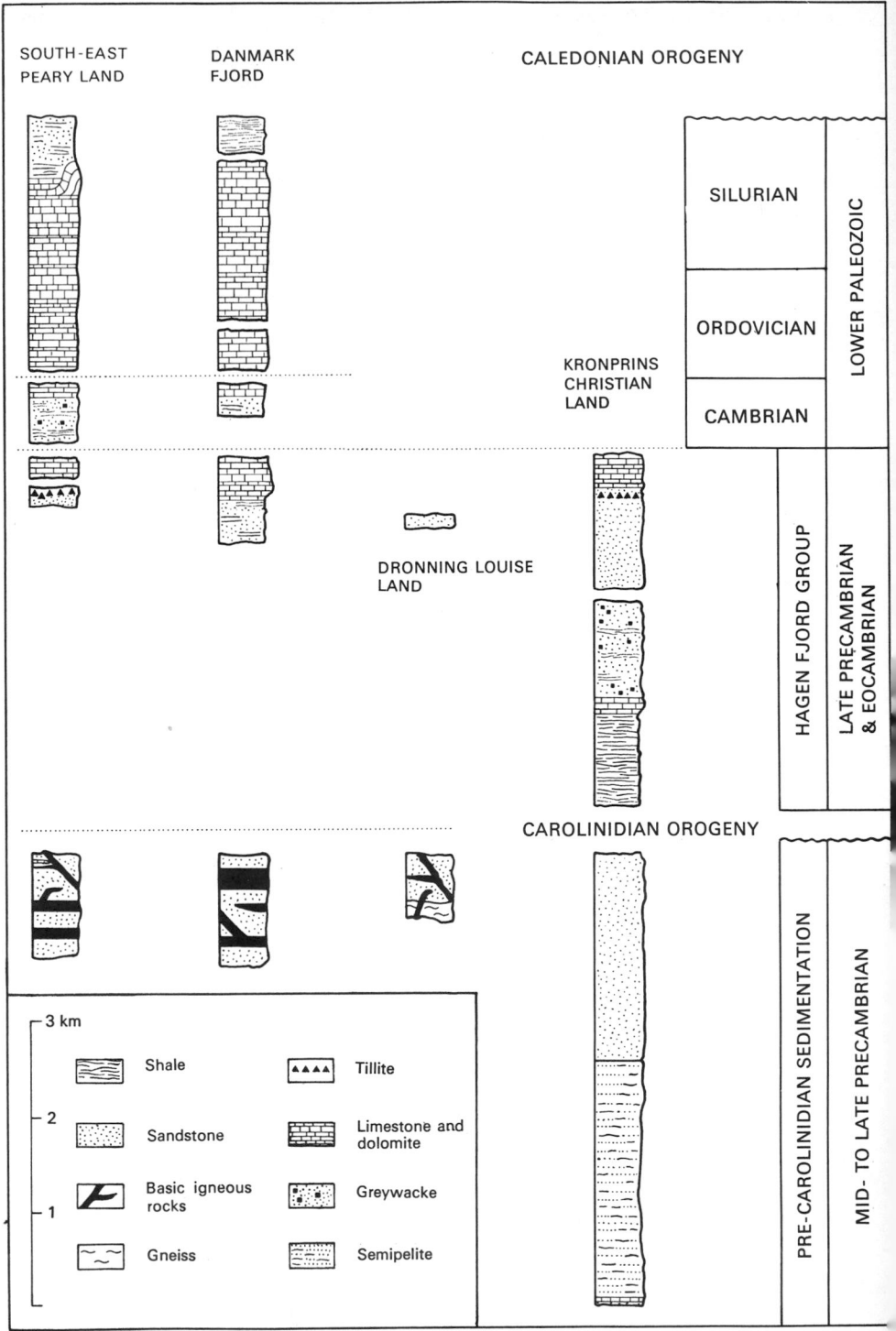

Fig. 7. Summary chart of Precambrian and Lower Paleozoic stratigraphy in eastern North and Nort[h] East Greenland. (From Birkelund *et al.*, 1974, with additions.)

C. Lower Paleozoic

Flat-lying strata of Lower Cambrian to Upper Silurian age overlie the essentially Proterozoic sedimentary sequences throughout North Greenland. However, the boundary between the Proterozoic and the Phanerozoic lies in a poorly known clastic and dolomite sequence. Late Lower Cambrian faunas are known from Inglefield Land in the west to Peary Land in the east (C. Poulsen, 1927, 1958; V. Poulsen, 1964, 1974; Peel and Christie, 1975). Underlying sediments may include deposits of Tommotian age (early Lower Cambrian) but fossil evidence is not available.

The youngest known Lower Paleozoic strata in the platform are of Upper Silurian (Ludlow) age, although graptolites and vertebrates from the foldbelt sequence to the north of the platform indicate Pridoli and Lower Devonian strata (Berry *et al.*, 1974; Bendix-Almgreen and Peel, 1974).

1. *Cambrian*

The most complete Cambrian sequence in northern Greenland occurs in the Inglefield Land–Washington Land area where fossiliferous Lower, Middle, and Upper Cambrian are present. Lower to Upper Cambrian strata also occur in western Peary Land, but only the Lower Cambrian is demonstrated in more easterly outcrops.

In Inglefield Land a sandstone unit with *Cruziana* and *Scolithus* overlies sandstones with middle Proterozoic dolerite intrusions and itself grades up into Lower and Middle Cambrian dolomites and limestones (100–150 m) with rich faunas (C. Poulsen, 1927, 1958; V. Poulsen, 1964). The sandstone unit with *Cruziana* and *Scolithus* is also present in southern Washington Land and may be equivalent to the Sverdrup Member of the Rensselaer Bay Formation of Bache Peninsula (Christie, 1967; Kerr, 1967b). However, strata in Inglefield Land included within the Rensselaer Bay Formation and previously correlated with the Sverdrup Member in Bache Peninsula are intruded by dolerite sills of Middle Proterozoic age (Dawes *et al.*, 1973). Overlying dolomites of Lower and Middle Cambrian age (Henriksen and Peel, 1976) precede limestones with intraformational conglomerates, thin dolomitic and anhydritic horizons yielding Middle and Upper (Dresbachian–Trempeleauian) Cambrian trilobites, followed by Lower Ordovician faunas without obvious stratigraphic break (Dawes, 1976a; Henriksen and Peel, 1976).

Cambrian rocks occur on the west side of the Victoria Fjord Arch but it is not known if other than the Lower Cambrian is represented (Dawes, 1976a).

Late Proterozoic–early Cambrian(?) dolomites in southern Peary Land are followed by fossiliferous Lower Cambrian clastics (425 m) and overlying carbonates (Troelsen, 1949; Jepsen, 1971; Peel and Christie, 1975). Middle and

Upper Cambrian limestones occur in western outcrops (Dawes, 1976b) but are absent from the east where there is a major unconformity between the Lower Cambrian and Lower Ordovician. A plausible lithological correlation can be made between unfossiliferous Cambrian outcrops in Kronprins Christian Land and southern Peary Land (Cowie, 1961b).

2. *Ordovician*

Shallow-water limestones and dolomites of the Cass Fjord Formation (400 m), with intraformational conglomerates and thin anhydritic horizons, continue without obvious stratigraphic break through the Middle and Upper Cambrian into the Lower Ordovician in Washington Land (Henriksen and Peel, 1976). The succeeding Lower Ordovician and early Middle(?) Ordovician strata consist of more than 750 m of alternating fossiliferous limestone units and pale anhydritic shales, both with thin intraformational conglomerates and dolomites. Comparable anhydritic horizons are well developed at approximately equivalent horizons in neighboring Arctic Canada. Overlying dark and more massive limestones (ca. 250 m) contain rich faunas of Middle and Upper Ordovician age (Troedsson, 1926, 1928; Koch, 1929b; Troelsen, 1950a).

In central southern Peary Land 320 m of pale and dark dolomites of late Lower Ordovician (Upper Canadian) and possibly younger age (Wandel Valley Formation) unconformably overlie the Lower Cambrian Brønlund Fjord Formation (Troelsen, 1949; Peel and Christie, 1975). Middle and Upper Cambrian strata intervene in more westerly areas of Peary Land. Succeeding dark mottled limestones (430 m), locally rich in chert, dominate the Middle and Upper Ordovician, as in Washington Land, although 150 m of dolomites of uncertain age straddle the boundary with the apparently conformably overlying Silurian.

A comparable Ordovician sequence is present in Kronprins Christian Land to the southeast, although subject to a different stratigraphic terminology (Adams and Cowie, 1953).

3. *Silurian*

Massive Silurian carbonates of the southern platform pass northwards into a prominent, generally east–west, belt of carbonate mounds, with associated intermound and basinal facies, which in turn gives way to the clastic sediments of the North Greenland Geosyncline. In Washington Land, and the area to the west of the Victoria Fjord Arch, the platform carbonates extend throughout the Silurian and reef mounds form complexes hundreds of meters thick and tens of kilometers in horizontal extent (Dawes, 1971, 1976a; Norford, 1972). In Peary Land, to the east of the arch, Lower Silurian platform carbonates are followed by Middle Silurian shales and Upper Silurian turbidites containing some chert-pebble conglomerates in the uppermost part. The mounds in these eastern outcrops are more isolated and less fully exhumed than their westerly counterparts (Peel and Christie, 1975; Mayr, 1976). A somewhat comparable sequence is present in Kronprins Christian Land to the southeast, although carbonate mounds have not been reported (Cowie, 1961b).

V. THE NORTH GREENLAND FOLDBELT

A. Introduction

The northern coast of Greenland is dominated by the mountains of an orogenic zone—the North Greenland Foldbelt. This belt of highly deformed and metamorphic, mainly Lower Paleozoic rocks stretches for 600 km from the Robeson Channel across to the Wandel Sea (Fig. 3). It is the eastern extension of the Innuitian Orogenic System of Arctic Canada (Fortier et al., 1954).

The North Greenland Foldbelt is characterized by a strong east–west-trending tectonic grain, parallel to the coast and continental margin. The belt shows its widest development, about 100 km, in Peary Land; in the west it is only 25 km wide. The present outcrop must represent only the southern margin of the entire orogenic zone, but how far the belt underlies the continental shelf is uncertain. Part, perhaps much, of the zone may have been removed or obliterated during later tectogenesis of the Arctic Ocean Basin (Dawes, 1973).

B. Geosynclinal Sedimentation

A major sedimentary basin, the North Greenland (Franklinian) Geosyncline, began its development on the northern edge of the Precambrian stable platform in late Proterozoic time. This basin evolved as a rapidly subsiding east–west-trending trough during the Lower Paleozoic with sedimentation continuing into the Devonian. The trough had a southerly provenance in the early phase of sedimentation. However, by late Ordovician time this source was added to from a northerly borderland and in the Silurian and Devonian large clastic sequences of flysch aspect accumulated.

Stratigraphical detail is known from only a few areas and the thickness, lithology, and age of many sections are uncertain. The succession, as currently known, is at least 5 km thick and appears to be entirely marine (Fig. 6). Stratigraphical units can be traced along the strike of the foldbelt, and despite the metamorphic and complexly deformed nature of the northernmost sections, and the general paucity of fossils, some correlation throughout the foldbelt and with adjacent Canada can be suggested (Fig. 8).

The succession is characterized by a predominance of clastic rocks,

	NE Ellesmere Island (Trettin, 1971)	Hall Land - Wulff Land (Dawes, 1971)	Northern Peary Land (Dawes & Soper, 1973): Frederick E. Hyde Fjord	Northern Peary Land: Roosevelt Fjelde, Group	Northern Peary Land: Roosevelt Fjelde, Formation
DEV.	Cape Rawson Group (Kerr, 1967a)		?	?	
SILURIAN	Cape Rawson Group (Kerr, 1967a): Marvin Formation (limestone)	Clastic unit (greywacke, flysch facies, some limestone)	Clastic unit (greywacke, flysch facies, some limestone)	Sydgletscher	Sydgletscher Sandstone Upper Sydgletscher Shales Sydgletscher Quartzite Lower Sydgletscher Shales Nysne Gletscher Mudstones
SILURIAN – ORDOVICIAN	Cape Rawson Group (Kerr, 1967a): Imina Formation (greywacke, siltstone, shale)	Limestone unit	- - - - - - - - - -	Polkorridoren	Polkorridoren Psammite Rusty, green quartz phyllite
ORDOVICIAN	Hazen Formation (limestone, shale, chert, breccia)	Limestone, shale, breccia, chert	Sandstone, siltstone, chert, shale, limestone, breccia	Paradisfjeld	Yellow limestone Green calcareous phyllite Dark grey limestone Graphitic and calcareous phyllite
CAMBRIAN	Grant Land Formation (clastic rocks, limestone)	?	?	- - - - - - - - - -	- - - - - - - - - -
CAMBRIAN	- - - - - - - - - -	Limestone, shale, dolomite, sandstone	Sandstone, shale		Ulvebakkerne Marble Kap Morris Jesup Quartz Phyllite
CAMBRIAN			Schley Fjord Formation	?	- - - - - - - - - -

Fig. 8. Correlation scheme of Lower Paleozoic geosynclinal strata across North Greenland to adjacent Ca (From Dawes, 1976*a*.)

particularly in the upper part, but with some important carbonate, breccia, chert, and argillaceous beds. No volcanic units have been encountered. At the present-day exposure level Proterozoic and Cambrian strata are apparently subordinate to Ordovician and Silurian rocks. However, nowhere has the base of the succession been seen.

The Proterozoic and Cambrian strata so far recognized are mainly sandstone, quartzite, grit, conglomerate, and shale with some limestone beds, which are in places stromatolitic. Proterozoic strata are known from two areas only; northern Wulff Land and eastern Peary Land. In the latter area at the edge of the foldbelt, light-colored quartzites containing dark basaltic intrusions and (?)flows correlate with the Proterozoic sequence known from the platform to the south (Dawes, 1976*a*).

Lower and Middle Cambrian rocks have so far been recognized. A Middle Cambrian trilobite fauna from the western part of the foldbelt is referred to the Atlantic province (V. Poulsen, 1969), while dark shales from the eastern part in Peary Land contain *Olenellus*, cf. *O. svalbardensis* (V. Poulsen, 1974), suggesting free faunal interchange with Svalbard in at least early Cambrian time.

The Ordovician is composed of a mixed shale–carbonate–sandstone sequence containing predominant units of dark silicified shale and siltstone, chert, bituminous limestone, and dolomitic rocks, some of which display spectacular allochthonous breccias derived from the carbonate platform edge (Dawes, 1976a). Graptolites from dark shales of the *Tetragraptus approximatus* and *Isograptus caduceus* zones in Peary Land have been referred to the Pacific Lower Ordovician fauna (Bjerreskov and Poulsen, 1973).

The Silurian and Devonian succession is dominated by a thick and monotonous, sparsely fossiliferous flysch sequence of calcareous greywacke, sandstone, siltstone, and shale that forms an extensive, more or less continuous tract across the foldbelt. The greywackes display abundant sedimentary structures and graded units indicating origin by turbidity currents. Graptolitic shale, limestone, and carbonate conglomerate with some megabreccia beds also occur. Thick units of conglomerate with grit are prominent in the upper part of the succession. These conglomerates are composed of subangular to well-rounded, multicolored chert pebbles, in places with radiolarian remains, and with occasional quartzite and crystalline clasts, set in a calcareous greywacke matrix. These rudaceous beds point to the presence of a proximal northern borderland as their source during at least the later development of the geosyncline (Fig. 9; Dawes, 1973).

During the Silurian the axis of the geosynclinal trough shifted gradually southwards and Ordovician platform carbonates were overlain by Silurian clastic beds. This expansion of the trough continued into at least late Silurian and early Devonian time when large areas of flysch encroached southwards over the Silurian carbonate platform, stifling the reef development of the shelf.

The restricted Silurian faunas known from the trough show forms in common with the better known Silurian assemblages of the platform. However, of particular significance is the presence of the brachiopod *Pseudoconchidium* from Peary Land (A. J. Boucot, in Dawes, 1976a), a genus known previously only from the northern Ural mountains of Russia.

The youngest strata so far recognized in the foldbelt outcrop in Hall Land. These are considered to be of early Devonian (Gedinnian) age on the basis of a graptolite compared to *Monograptus aequabilis* (Berry et al., 1974) and a vertebrate assemblage similar to the fauna of the Peel Sound Formation of Arctic Canada (Bendix-Almgreen and Peel, 1974). It would be coincidental if these fossiliferous horizons mark the precise top of the geosynclinal accumulation and it is thought most likely that younger Devonian beds were deposited in northern Greenland.

C. Structure and Metamorphism

The North Greenland Foldbelt is characterized by a tectonic and metamorphic pattern that is asymmetric. The intensity of metamorphism and deformation increases progressively northwards from the nonmetamorphic platform to the north coast where high-grade mineral assemblages occur in schist lithologies. Structural styles within the foldbelt are notably varied although in a regional sense the dominant direction of tectonic transport is northwards towards the Arctic Ocean and thus the Greenland Platform is considered as a hinterland. The northerly overturning of fold structures responsible for this northerly facing character of the foldbelt is particularly evident in northern Peary Land where stratigraphic and structural elements have a variable southerly dip. In the extreme north, on the border of the Arctic Ocean, the overturning has been severe and in many places shallow southerly dipping strata exist (Fig. 10; Dawes, 1976a, Fig. 261).

The southern boundary of the foldbelt is autochthonous; no regional dislocation planes separate the fold zone proper from the hinterland, as is the case with the East Greenland Foldbelt described later. In northwestern Peary Land the local northern boundary of the foldbelt is at the Kap Cannon Thrust (Dawes and Soper, 1973) along which the metamorphic folded strata have been pushed northwards over an essentially nonmetamorphic late Cretaceous–early Tertiary volcanic suite—the Kap Washington Group.

The metamorphic and structural history of the foldbelt is long and complex and as yet not fully understood. As currently known, the progressive increase of orogenic effects across the foldbelt allows subdivision, in the widest part of the foldbelt in Peary Land, into five tectonic–metamorphic zones (Dawes and Soper, 1973; see Fig. 11).

The highest grade, and most severely deformed, rocks underlie the extreme outer coast of Peary Land. These are garnet-bearing amphibolite facies schistose rocks that in places contain staurolite, andalusite, cordierite, sillimanite, and amphibole. The progressive change from undeformed, nonmetamorphic, fossiliferous platform strata to these high-grade schists takes place through lower grade zones composed of slate, phyllite, and marble, characterized by muscovite, chlorite, and biotite.

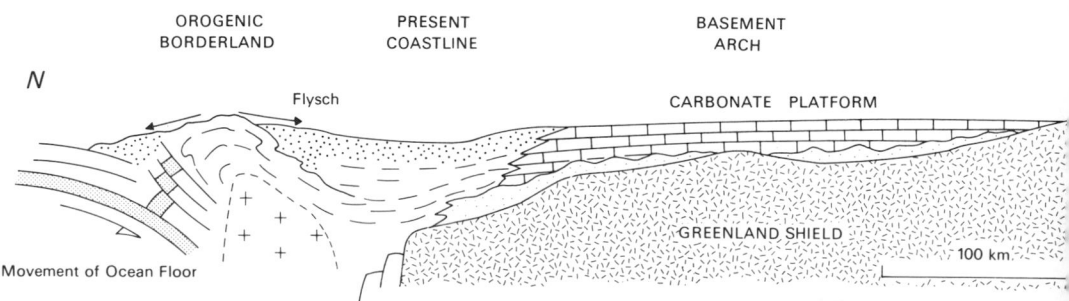

Fig. 9. Development of the North Greenland (Franklinian) Geosyncline in early Devonian time with the depositi flysch from a northern borderland.

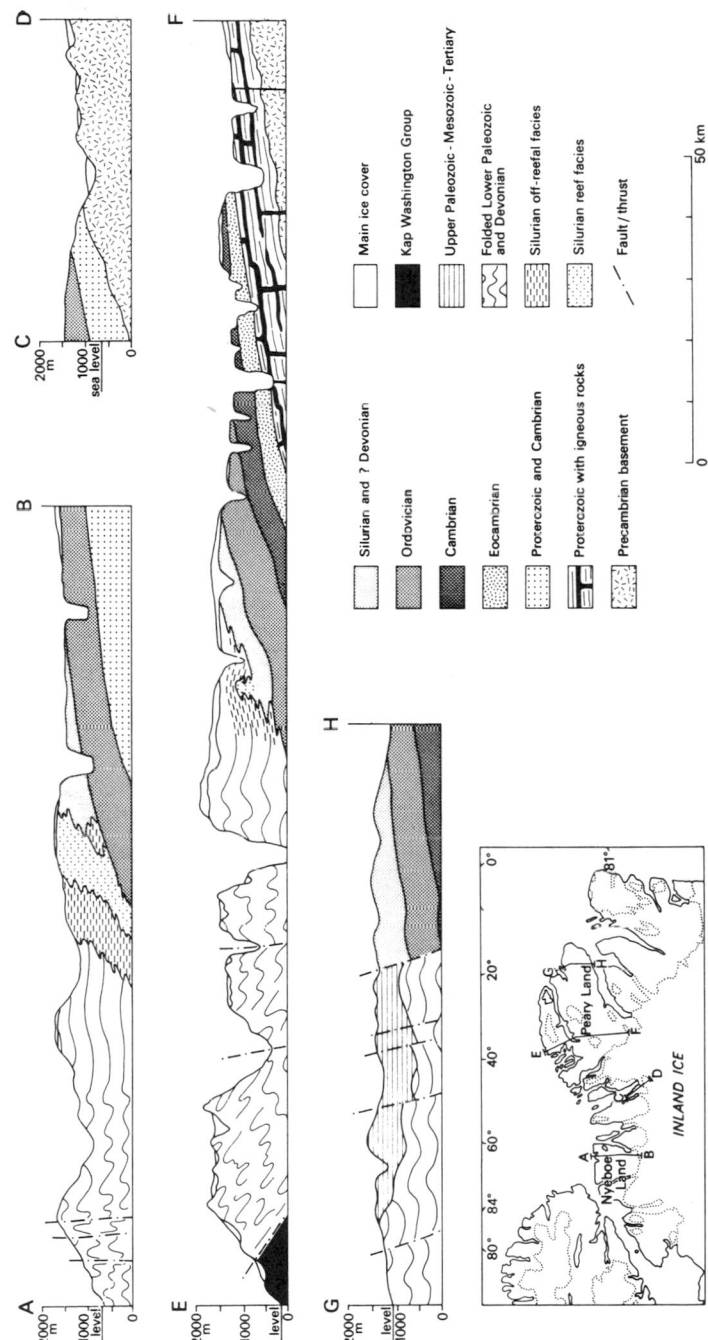

Fig. 10.   Simplified geological cross sections through the North Greenland Foldbelt and Platform. (From Dawes, 1976a.)

Fold structures and associated fabrics can be referred to three main episodes of tectogenesis (Dawes and Soper, 1973). Large-scale folds of the two earliest episodes are essentially coaxial and trend east–west, determining the disposition of the stratigraphical units. Major variance from the dominant east–west structural grain is evident in southern Peary Land where some northwest to southeast tectonic trends occur.

The margin of the foldbelt in Peary Land is well exposed. There, south of Frederick E. Hyde Fjord, the platform (which displays flexuring and block faulting) passes northwards into an area of open, symmetrical folds that at the foldbelt margin pass into large-scale, south-verging structures (Dawes, 1971). This suggests some southerly directed tectonic transport. However, to the north these folds appear to be superseded by "second" folds that take on a conspicuous north-verging style. Further northwards, with increasing metamorphic grade and a progressive destruction of sedimentary textures, the "second" folds develop a strong axial plane fabric that in northern Peary Land becomes a thoroughly penetrative schistosity and the dominant fabric of all rock types. This schistosity is deformed by "third" folds that locally develop southerly inclined schistosity.

Several dislocations of various trend, type, and age traverse the foldbelt. Some were generated during the Paleozoic, others are Tertiary in age (Haller and Kulp, 1962). Most significant are low to high-angle faults and thrusts, subparallel to stratigraphic units that have variably inclined planes indicating tectonic movements both to the north and south. Northern Peary Land has been likened to a horst block flanked by the post-orogenic WNW-ESE-trending Harder Fjord and Kap Bridgman faults (Haller and Kulp, 1962). However, such major dislocations are tantalizingly parallel to important offshore fractures (Spitsbergen Fracture Zone) and some transcurrent movements within Peary Land, at least in late Phanerozoic time, are to be expected in view of the sea-floor spreading history in the Greenland Sea (Dawes, 1973).

One such dislocation, the Harder Fjord Fault, is a major fault zone composed of several fault and crush planes. This zone, in one area north of Frederick E. Hyde Fjord, contains both acidic and basic igneous rocks of as yet unknown age and extent, while certain basic dikes of Mesozoic age (?Cretaceous) are aligned in the fault zone. Both the igneous rocks and to a lesser extent the dikes have suffered postintrusion fault crushing. This fault zone probably has a long and complex tectonic history. Recent solfatara activity along the eastern part of the fault zone (see later, p. 254), suggests the continuing activation of this disclocation and some transcurrent movements within Peary Land must be considered likely in view of the sea-floor spreading history in the Greenland Sea (Dawes, 1973).

The low-lying Kap Cannon Thrust in northern Peary Land is oblique to the metamorphic zones (Fig. 11) and indicates northerly directed Tertiary movements.

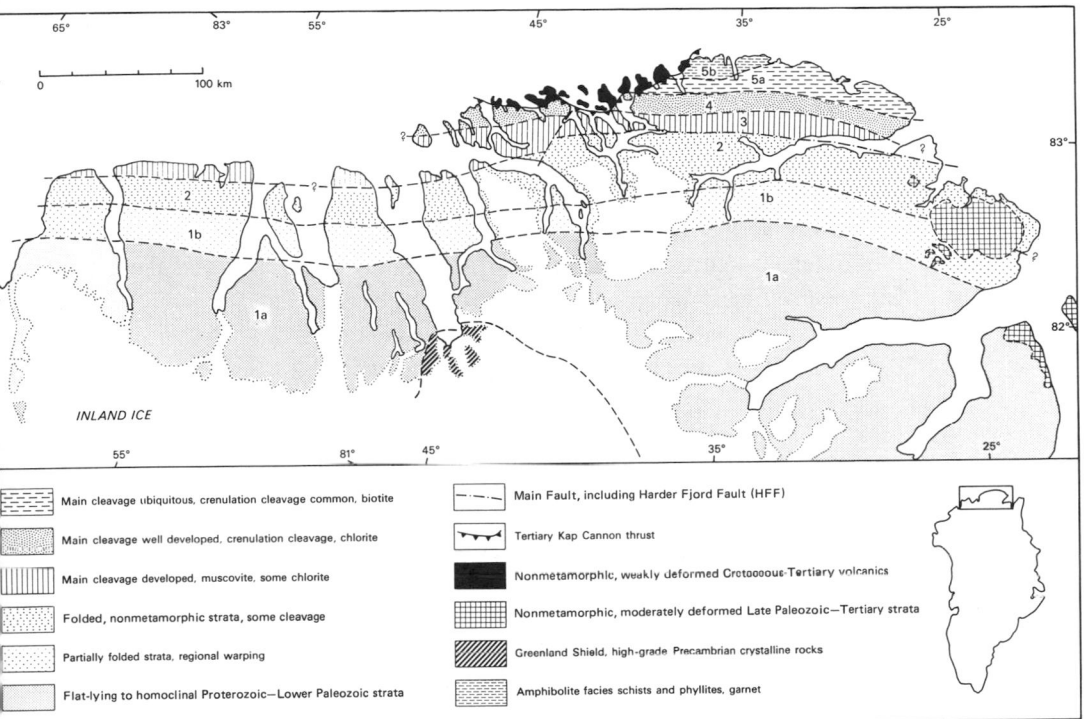

g. 11.   Tectonic–metamorphic units of the North Greenland foldbelt and Platform. (Redrawn after Dawes, 1976a.)

D. Age of Orogenesis

The North Greenland Foldbelt has been an unstable zone of the crust throughout Phanerozoic time and it is the product of both Paleozoic and late Phanerozoic orogenesis. It is still unclear to what extent the present structural and metamorphic character of the foldbelt is due to the late Phanerozoic orogenesis (Dawes and Soper, 1973). A particularly important consideration is to what extent the Tertiary northerly directed tectonism (e.g., Kap Cannon thrusting) has accentuated or modified the Paleozoic tectonic pattern.

Only in the southernmost part of the foldbelt in eastern Peary Land where deformation and metamorphic effects are weakest do cover rocks of known age occur (Fig. 11). Nevertheless, stratigraphic relationships closely resemble those in Arctic Canada between the folded strata of the Franklinian Geosyncline and the less severely deformed rocks of the Sverdrup Basin. Consequently, the base of the Wandel Sea Basin is taken as a minimum limit of the Paleozoic diastrophism of the foldbelt.

The basal rocks of the Wandel Sea Basin in eastern Peary Land are at least as old as Upper Carboniferous (late Pennsylvanian; possibly late Virgilian), but to the south, the basin contains Lower Carboniferous (Dinantian) beds. Thus the age of the main orogenesis is placed in the interval from early Devonian (Gedinnian) to early Carboniferous (Dawes, 1976a).

Tertiary earth movements within the foldbelt are indicated by regional faulting, thrusting and slight folding of the Kap Washington Group, slight deformation and metamorphism of late Cretaceous basic dikes, and mylonitization and greenschist metamorphism in connection with the Kap Cannon Thrust (Dawes and Soper, 1973). Moreover, Tertiary folds and faults, and probably older structures as well, affect the Wandel Sea Basin and these are superimposed on the Paleozoic tectonic pattern of eastern Peary Land.

K-Ar isotopic dates on Paleozoic metasediments from northern Peary Land show a range from 84 to 42 m.y., giving evidence of significant regional reactivation of the orogen in late Mesozoic and (?)Tertiary time (Dawes and Soper, 1971).

VI. THE EAST GREENLAND FOLDBELT

A. Introduction

The eastern coast of Greenland, north of 70°N, is dominated by the East Greenland Caledonian Foldbelt, a north–south-trending orogenic zone parallel to the coast and continental margin (Fig. 1). The foldbelt outcrops for a distance of 1500 km, forming the western part of the North Atlantic Caledonian Orogenic System. The most comprehensive accounts of the foldbelt are by Haller (1970, 1971). Important results of recent field work in the region between 70° and 74°N have been included in the accounts of Higgins (1976) and Henriksen and Higgins (1976).

The foldbelt is characterized throughout most of its extent by metamorphic crystalline complexes that show various age relationships to Precambrian and Lower Paleozoic sedimentary sequences. Archean, Proterozoic, and Paleozoic (Caledonian) orogenic and plutonic activity can be recognized, the Caledonian activity being traceable throughout the belt. The western margin of the belt is everywhere characterized by thrusts that indicate a main sense of tectonic transport towards the west.

B. Southern Part, 70° to 76°N

Precambrian crystalline basement composed of a variety of gneisses, granites, and high-grade supracrustal sequences makes up substantial areas of the foldbelt south of 76°N. Substantial evidence for widespread Archean and Proterozoic orogenesis now exists (Higgins, 1976). An extensive geosynclinal sequence, reaching a cumulative thickness of 17 km, is composed of the late

Precambrian Eleonore Bay Group, a tillite sequence and a Lower Paleozoic succession that includes strata of Cambrian to at least Middle Ordovician (Mohawkian) age. Caledonian orogenesis included folding, thrusting, and nappe development with associated regional metamorphism, migmatization, and reactivation, as well as late and postkinematic granite emplacement. Late and post-Caledonian (Devonian to Lower Permian) continental clastic strata followed by late Permian and Mesozoic, mainly marine beds cover parts of the eroded foldbelt. In the south Tertiary lavas overlie and completely conceal the belt to the south of Scoresby Sund.

C. Northern Part, 76° to 82°N

The northern part of the foldbelt is not known to the same degree as its southern counterpart and in many areas few or no ground observations exist. Most information is from the work of Fränkl (1954, 1955), Peacock (1956, 1958) and Haller (1956, 1961, 1970, 1971).

Like the southern part, the belt north of 76°N shows evidence of at least three orogenic cycles. The Caledonian fold pattern is superimposed on an earlier mountain chain, the Carolinides (Haller, 1961), thought to represent (?)Middle Proterozoic tectogenesis and associated migmatization with, in places, reactivation and reworking of earlier basement rocks. The distinction between Carolinidian and Caledonian structures, as well as between Carolinidian and older metamorphic complexes, is complicated and in some areas only sketchily known.

The oldest sedimentary sequences in the foldbelt can be separated into two ages by their relationship to the Carolinidian earth movements, viz., pre- and post-Carolinidian. The strata are preserved in different tectonic zones of the foldbelt and they have been correlated with the thinner and better known sequences of the foreland (Fig. 7).

In Kronprins Christian Land the foldbelt can be divided up into four, north–south-trending tectonic zones (Fig. 12):

(1) The autochthonous foreland—folded in the east, homoclinal in the west—is composed of Proterozoic and Lower Paleozoic strata. This zone represents the eastern part of the North Greenland Platform, described earlier.

(2) A main zone of westwards-facing Caledonian nappes that are composed, evidently entirely, of post-Carolinidian sediments.

(3) A central massif, the Prinsesse Caroline Mathilde Alper, that is composed of autochthonous blocks and parautochthonous thrust wedges of deformed pre-Carolinidian sediments.

(4) An eastern zone of gneisses and other crystalline rocks of Archean or younger age that have a partial cover of Upper Paleozoic platform strata of the Wandel Sea Basin.

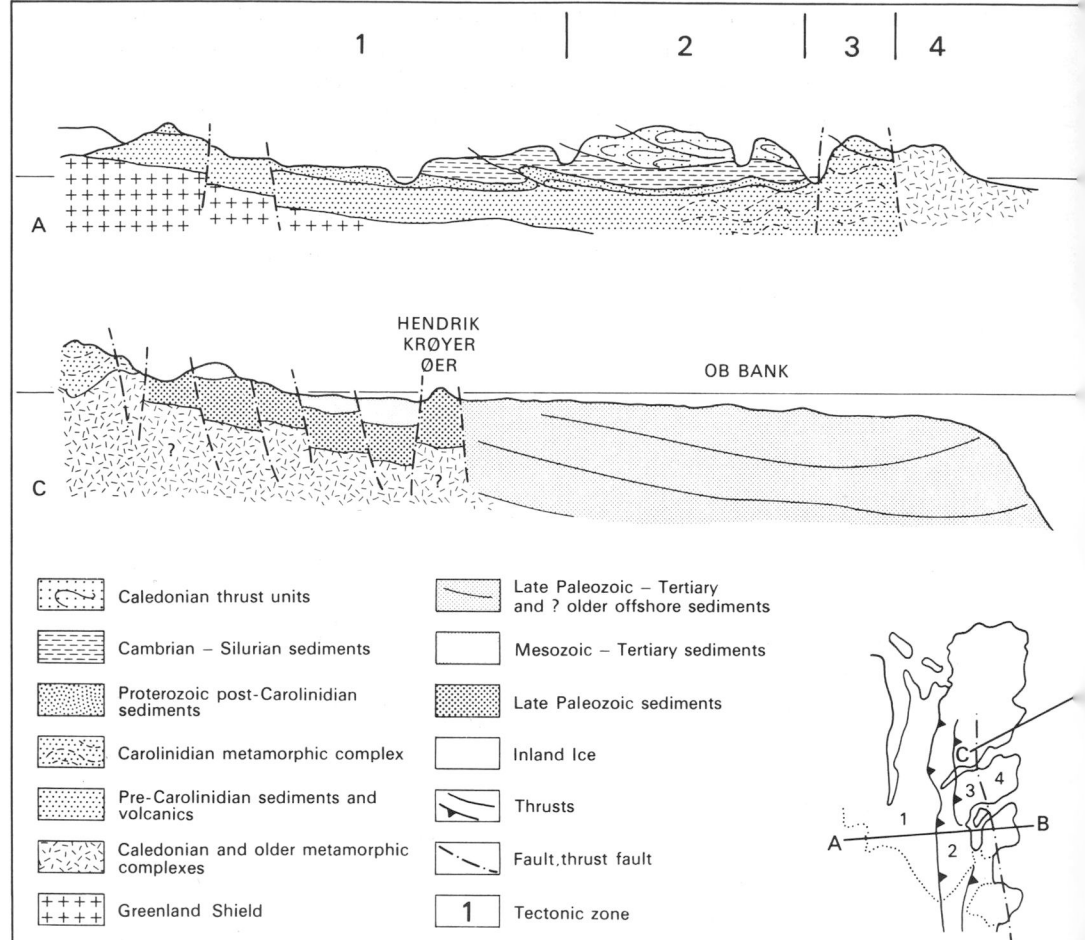

Fig. 12. Cross sections across the northern part of the East Greenland Foldbelt in Kronprins Christian Land and continental shelf.

1. *Pre-Carolinidian Sedimentation*

Sedimentation in this Proterozoic geosyncline resulted in a thick and variable sequence that, according to Haller (1971), comprises three distinct lithofacies. The composite sequence is quartzite–limestone (200–300 m) overlain by semipelites (2000–3000 m) and psammites (3000 m). The lower two facies compose the main geosynclinal accumulation, while the psammitic facies has a much wider distribution and respresents in part nonmarine depositional conditions. In both autochthonous and allochthonous regions of the foldbelt, igneous rocks, predominantly basic sills and dikes, are ubiquitous, and it is mainly by means of these igneous rocks that strata of the foreland (as far west as Peary Land) have been correlated with pre-Carolinidian phase of sedimentation. The igneous material is the result of two main periods of magmatism, one prior and one subsequent to the Carolinidian Orogeny. In at least the foreland the pre-Carolinidian sediments contain in addition to intrusions, thick units of bedded lavas and tuffs (Dawes, 1976a).

2. Post-Carolinidian Sedimentation

Following the Carolinidian diastrophism, the deformed and uplifted sediments and igneous rocks suffered strong denudation. Regional subsidence and transgression by the sea followed and a north–south-trending regional depression, the Caledonian Geosyncline, developed on the eroded roots of the Carolinidian mountain chain.

The post-Carolinidian Precambrian sequence—the Hagen Fjord Group, with a type area in the foreland—is preserved as nappes and in thrust sheets in the foldbelt, where a composite section 5 km thick is suggested. Fränkl (1954) envisages the original depositional trough to have been situated to the east of the Prinsesse Caroline Mathilde Alper but Haller (1971) believes the miogeosynclinal site to have been somewhere off the present-day coast.

The succession is composed of phyllite and greywacke with some marble, overlain by sandstone with shale and conglomerate. Overlying this is the Ulvebjerg Sandstone (20–35 m thick) considered by Fränkl (1954) to include a thin tillite bed (1–2 m) and a 400 m succession of limestone and dolomite containing stromatolites. The upper part resembles the base of the Hagen Fjord Group of the foreland (Fig. 7) and common formational names for the two successions have been used without hesitation by Adams and Cowie (1953) and Fränkl (1954, 1955).

No proven Lower Paleozoic rocks occur in the allochthonous cover but their eventual discovery seems probable. However, Lower Paleozoic rocks form a thick succession of mainly limestone and dolomite with sandstone and shale at the top on the autochthonous foreland and in several windows within the foldbelt. On the foreland this sequence reaches more than 3 km in thickness (Adams and Cowie, 1953).

3. Structure and Metamorphism

The tectonic style of the foldbelt north of 76°N is a product of both Carolinidian and Caledonian earth movements, evidence of which, according to Haller (1961, 1970, 1971), can be traced continuously from the south into Kronprins Christian Land.

Carolinidian Orogenesis. The trend of the Carolinidian folds varies considerably and this may be due to Caledonian overprinting. In the north, structures are NE–SW to N–S trending; in the south they have a more

NW–SE trend. Tectonism in the north (Kronprins Christian Land) and west (Dronning Louise Land) appears to be of superficial nature with simple fold style accompanied by low metamorphic grade. To the south, metamorphic grade is much higher and Carolinidian migmatization and mobilization features are evidently common. Haller (1971) regards the region between 76° and 78°N as the deep-seated central zone of the Carolinidian Foldbelt. However, recent isotopic work on gneisses around Danmarkshavn (77°N) giving ages of ca. 3000 m.y. (Steiger *et al.*, 1976), suggests some, perhaps large, areas of pre-Carolinidian basement rocks have survived subseqent reworking.

Caledonian Orogenesis. The Caledonian orogeny in the northern part of the foldbelt resulted in superficial structures and an alpine tectonic style characterized by a marginal zone of westward-facing thrusts and nappes. Fränkl (1955) divides the nappe zone into an eastern group of thrust sheets having contact to a root zone to the east of Prinsesse Caroline Mathilde Alper, and a western group of gliding nappes. The nappe zone overrides the autochthonous foreland and various windows reveal a minimum tectonic transport of 40 km. Haller (1970) envisages displacement far exceeding this, perhaps well over 100 km, from a root zone somewhere off the present-day coast. Caledonian folds affect the foreland up to 25 km from the main nappe zone, and folds, faults, and thrusts have reactivated the Carolinidian block to the east.

Kronprins Christian Land represents a high crustal level of the foldbelt. The late Proterozoic and Lower Paleozoic strata vary from unmetamorphosed rocks to low-grade metamorphics such as phyllite and slate, but no migmatites or plutons occur. Towards the south, metamorphic grade increases and Caledonian migmatites and gneiss are exposed (Haller, 1956, 1970).

The general trend of structural elements is N–S to NNE, but in Kronprins Christian Land structures show a broad flexure towards the east. Thus, in the northernmost exposures (south of Prinsesse Dagmars Ø) where some isoclinal folds are conspicuous on aerial photographs, the main structural trend is to the NE. Interpretation of structures from aerial photographs is difficult near the margins of the Flade Isblink because of heavy moraine cover and due to outcrops of Wandel Sea Basin strata. Thus, on the seminunatak Kilen, bordering the outer coast, NE- to ENE-trending structures may indicate a further flexuring of the foldbelt eastwards, although the presence of deformed Jurassic and other strata make it difficult to ascertain whether this trend is purely of Caledonian origin.

4. *Age of Orogenesis*

There is a growing body of evidence indicating that significant and widespread middle to late Proterozoic earth movements affected the east coast of Greenland. In the extreme southern part of the foldbelt isotopic ages on migmatites and granites fall between 900 and 1200 m.y., indicating an important orogenic episode that is considered to have affected regions much farther to the north (Higgins, 1976; Henriksen and Higgins, 1976). This may correspond in time with the Carolinidian orogenic activity known from the region north of 76°N.

The age of the Carolinidian diastrophism is uncertain (Steiger *et al.*, 1976). Isotopic evidence is restricted to the age of basic intrusions cutting platform strata of supposed pre-Carolinidian age. A ^{39}Ar-^{40}Ar age of 988 m.y. from a sill in Peary Land could reflect the age of the Carolinidian Orogeny, but is accepted as a minimum age for the sediments and a maximum age for the orogeny (Henriksen and Jepsen, 1970).

On the other hand, the age of the Caledonian Orogeny is relatively well fixed both by stratigraphic and isotopic evidence. Haller (1970, 1971) has divided the orogenic development into three main episodes. The main orogeny (Silurian, 420–400 m.y.), launched by deep-seated mobility, involved thrusting, folding, and nappe genesis, with associated regional metamorphism, migmatization, and reactivation, and was followed by "late Caledonian spasms" that affected restricted areas and included folding, intensive reactivation, and plutonic rock emplacement (Devonian, 400–350 m.y.). The "minor succeeding episodes," including warping, folding, and block faulting, occurred in the Carboniferous and lasted into Permian time (350–270 m.y.).

However, within this general time framework, it is clear now that the main Caledonian earth movements producing the dominant tectonic style of the foldbelt, with the development of westward-facing thrusts, is diachronous. The main thrusting in the northern part coincides, at least in part, with late orogenic activity in the south.

South of 76°N, the youngest pre-Caledonian strata belong to the Heim Bjerge Formation of late Middle Ordovician (Mohawkian), possibly Upper Ordovician (Cincinnatian) age (Cowie and Adams, 1957; Cowie, 1961*b*). Younger rocks may yet be encountered. The oldest post-Caledonian strata are late Middle Devonian (early Givetian) molasse dated by a vertebrate fauna. Hence, the main orogenic activity occurred in the interval limited by this stratigraphic data. Isotopic ages show a range of 470–370 m.y., with a concentration between 425 and 405 m.y. (Henriksen and Higgins, 1976).

In the northern part of the foldbelt, the age of the diastrophism is later, although based on less stringent evidence. The youngest pre-Caledonian strata affected by the strong westerly thrusting are the clastic rocks of the Profilfjeldet Formation that on the basis of a brachiopod and ostracod fauna are at least as young as late Silurian (Dawes, 1976*a*). The upper part of the succession both on the foreland (Cowie, 1961*b*) and within the foldbelt (Fränkl, 1955) remains little investigated and the age of the uppermost beds is unknown. The oldest "post-Caledonian" strata are the basal beds of the Wandel Sea Basin, which on floral evidence are suggested to be of Lower Carboniferous (Dinantian) age. Thus, the main diastrophism occurred in the interval between the late Silurian and early Carboniferous. No isotopic age dates are available from Kronprins Christian Land. However, a K-Ar date of 405 m.y. on a pegmatite from Lambert Land (79°20'N) (Haller and Kulp, 1962) suggests some plutonic activity of late Silurian to early Devonian age at that latitude.

Rb-Sr, K-Ar, and U-Th-Pb mineral ages between 320 and 380 m.y. on gneiss from Danmarkshavn at 77°N are remarked on by Steiger *et al.* (1976) to be "surprisingly . . . younger than one would expect for a Caledonian event." On the contrary, the present authors regard such isotopic ages as confirmatory for the diachronous orogenic regime of the Caledonian Foldbelt as outlined here.

VII. LATE CRETACEOUS–EARLY TERTIARY ACID VOLCANIC SUITE

A suite of mainly bedded extrusives, with some sediments and intrusive rocks, the Kap Washington Group, outcrops in northwestern Peary Land on islands and peninsulas bordering the Arctic Ocean. The rock suite structurally underlies the Paleozoic metasediments of the North Greenland Foldbelt, which have been thrust northwards on the southerly inclined Kap Cannon Thrust. The igneous rocks occupy more than 600 km^2 on land and an unknown area under the Arctic Ocean. A distinct positive magnetic anomaly over the continental shelf immediately offshore could depict the submarine extension of the group (Dawes, 1973; see below).

The suite is currently known from the easternmost part of the outcrop area (Fig. 11) where it is composed of a succession, at least 1500 m thick, of dominantly bedded lavas and agglomeratic tuffs, with some ignimbritic rock types. Many of the effusives are rhyolitic with SiO_2 exceeding 70% and with total alkalis ($Na_2O + K_2O$) between 7.5 and 9%, but basic lavas occur (SiO_2 between 44 and 50%). Interbedded sediments, arkose with conglomerate, and carbonate and calcareous shale, have been recognized in one area, while some granophyre and minor basic intrusions also exist.

The volcanic and sedimentary pile generally has a gentle southerly dip, and the rocks are nonmetamorphic, in strong contrast to the overlying metasediments. However, some folding affects the sequence and in places a weak cleavage occurs. In the vicinity of the Kap Cannon Thrust mylonites have developed.

The Cretaceous–Tertiary age assignment suggested for the volcanics on the basis of K-Ar whole-rock ages (Dawes and Soper, 1971) has been confirmed by recent Rb-Sr work (Ole Larsen, personal communication). Using a decay constant of 1.39×10^{-11}, the Rb-Sr data from five of the freshest acidic lavas available define an isochron corresponding to an age of 65 m.y., i.e., late Maestrichtian or early Paleocene. The Kap Washington Group is deformed by the Kap Cannon Thrust that is regarded as mid-Tertiary in age.

VIII. THE WANDEL SEA BASIN

A. Introduction

A series of slightly deformed Carboniferous to Tertiary strata, deposited in the Wandel Sea Basin forms scattered outliers in eastern North Greenland and probably underlies extensive areas of the Wandel Sea (Dawes and Soper, 1973). A major angular unconformity separates the mainly shallowly dipping strata of the Wandel Sea Basin from underlying, folded Lower Paleozoic (and Devonian?) sediments of the North and East Greenland Foldbelts and the flat-lying sequences of the stable platform.

A composite section of the Wandel Sea Basin, as currently known, comprises at least 3000 m of strata, although only scattered stratigraphic information is available (Fig. 13). The sequence commences with Lower Carboniferous terrestrial sandstones, conglomerates, and shales, presently known only in the more southerly outcrops in Kronprins Christian Land, passing up into a sequence of marine Upper Carboniferous and Permian limestones, shales, and sandstones. A thick Triassic sequence of mainly sandstones and shales in Peary Land is followed by Jurassic strata of unknown extent. Late Cretaceous and Tertiary plant-bearing sandstones and shales with thin coals form the upper-most units.

B. Carboniferous and Permian

Carboniferous and Permian strata occur in southeast Kronprins Christian Land (1300 m), in eastern Peary Land (700 m+), and in small exposures around Station Nord in northwestern Kronprins Christian Land.

1. *Kronprins Christian Land*

The sections in Amdrup Land and Holm Land are relatively well known. Grönwall (1916) divided the sequence into "Terrestrial; Lower Marine, and Upper Marine Groups" (Fig. 14).

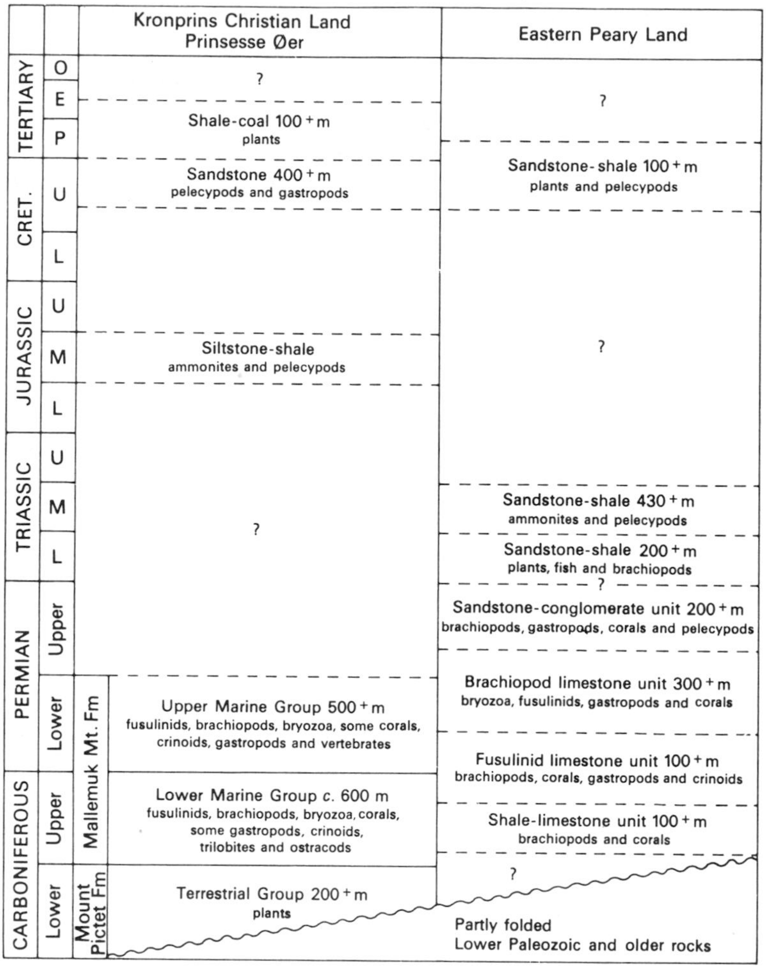

		Kronprins Christian Land Prinsesse Øer	Eastern Peary Land
TERTIARY	O	?	?
	E		
	P	Shale-coal 100 + m plants	
CRET.	U	Sandstone 400 + m pelecypods and gastropods	Sandstone-shale 100 + m plants and pelecypods
	L		
JURASSIC	U		?
	M	Siltstone-shale ammonites and pelecypods	
	L		
TRIASSIC	U		
	M	?	Sandstone-shale 430 + m ammonites and pelecypods
	L		Sandstone-shale 200 + m plants, fish and brachiopods
			?
PERMIAN	Upper		Sandstone-conglomerate unit 200 + m brachiopods, gastropods, corals and pelecypods
	Lower (Mallemuk Mt. Fm)	Upper Marine Group 500 + m fusulinids, brachiopods, bryozoa, some corals, crinoids, gastropods and vertebrates	Brachiopod limestone unit 300 + m bryozoa, fusulinids, gastropods and corals
			Fusulinid limestone unit 100 + m brachiopods, corals, gastropods and crinoids
CARBONIFEROUS	Upper (Mallemuk Mt. Fm)	Lower Marine Group c. 600 m fusulinids, brachiopods, bryozoa, corals, some gastropods, crinoids, trilobites and ostracods	Shale-limestone unit 100 + m brachiopods and corals
	Lower (Mount Pictet Fm)	Terrestrial Group 200 + m plants	?
			Partly folded Lower Paleozoic and older rocks

Fig. 13. Composite stratigraphical scheme of Upper Paleozoic, Mesozoic, and Tertiary strata of the Wandel Sea Basin. (From Dawes, 1976a.)

Rocks of the Terrestrial Group occur at the base of the sequence in Holm Land. Dark, plant-bearing sandstones, conglomerate, and some shales with thin seams (1 m) of poor quality coal form sequences up to 200 m thick, although Grönwall (1916) commented that a considerably greater thickness may be present. Nathorst (1911) compared the flora to that in occurrences in Spitsbergen and suggested a Lower Carboniferous (Dinantian) age.

The Lower Marine Group (600 m) disconformably overlies the Terrestrial Group, a 10-m-thick basal conglomerate being followed by an alternation of limestones, sandstones, shales, and some marls. A 35-m-thick gypsum–

anhydrite bed occurs in the middle of the unit in Amdrup Land. In Holm Land a conglomerate with quartzite and igneous pebbles occurs at the top. The fauna of the Lower Marine Group is of Upper Carboniferous age, with fusulinids comparable to the Soviet Moscovian (Dunbar, 1962; Ross and Dunbar, 1962).

The Upper Marine Group (500 m in Amdrup Land) is characterized by alternating dolomites and limestones, with an increasing arenaceous content in

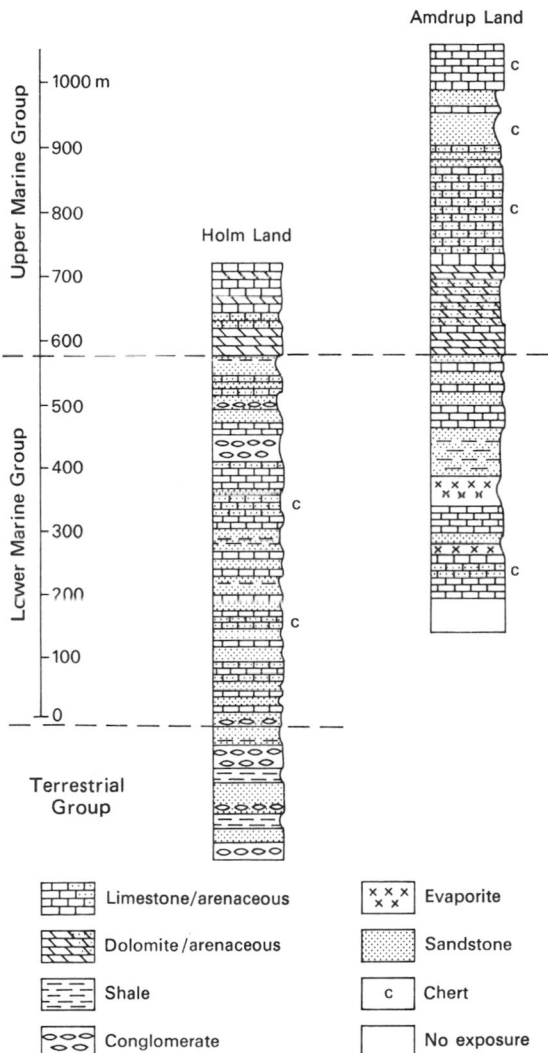

Fig. 14.   Permo-Carboniferous stratigraphical sections in the southern part of the Wandel Sea Basin, Kronprins Christian Land (see Fig. 13). (From Dawes, 1976a.)

the middle and upper parts. Bendix-Almgreen (1975) concluded that the higher beds are of latest Lower Permian age, an older Sakmarian age suggested by Dunbar *et al.* (1962) being discounted.

Thin-bedded arenaceous and silicified limestones around Station Nord are poorly known stratigraphically. Bryozoan mounds are conspicuous and contain a rich fauna, probably of Lower Permian age (C. Poulsen, in Dawes, 1971).

2. *Peary Land*

At least 700 m (possibly as much as 1 km) of Carboniferous and Permian strata are present in eastern Peary Land, although the sequence is much less well known than that in Kronprins Christian Land to the south (Dawes, 1971).

The oldest strata so far dated are from Hellefiskefjord, where at least 400 m of fossiliferous limestone is divided into a lower unit of fusulinid-rich limestone, violet shales, and thin marls and limestones, and an upper unit of buff-colored, brachiopod-rich limestones. Troelsen (1950*b*) noted *Triticites* from the fusulinid limestones, indicating an Upper Carboniferous (Upper Pennsylvanian) age. The identification has been confirmed by Allen A. Petryk (personal communication) who also noted *Pseudofusulinella* and *Schubertella*, possibly suggesting a late Virgilian age. The fusulinids are cosmopolitan, of the mid-continent Andean realm, with the exception of *Pseudofusulinella*, which appears to be of the Eurasian–Arctic realm. The fauna of the brachiopod-rich limestones has a general similarity to faunas from the late Lower Permian of East Greenland and Spitsbergen (F. G. Stehli, in Dawes, 1976*a*).

At least 200 m of clastic rocks at the top of the sequence are exposed in the area to the northwest of Mudderbugten. The yellow to red weathering sandstones, grits, conglomerates, and shales, with minor limestones, yield a brachiopod dominated fauna of Upper Permian aspect (W. E. Davies, in Dawes, 1971).

C. Triassic

Triassic rocks are presently known only from eastern Peary Land. At least 630 m of Lower and Middle Triassic strata have a probable faulted relationship to the older strata. A lower recessive unit (at least 200 m thick) is composed mainly of silty and sandy shales, with some cross-bedded sandstones. An upper, conformable unit (at least 400 m thick) is composed of gray- to yellow-weathering sandstones and some shales.

The fauna of the lower unit shows a relationship with the Lower Triassic of Spitsbergen and northern Russia. E. Nielsen (in Kummel, 1953) compared vertebrates to the Scythian fish horizon of Spitsbergen. The lower 150 m of the upper unit contain ammonoids of Lower Anisian (early Middle Triassic) age

(Kummel, 1953), indicating a relationship to strata in Siberia, Spitsbergen, and Cordilleran and Arctic North America. The occurrence of younger Triassic strata in the upper part of the section is considered probable on the basis of the comparison to the sequence in Spitsbergen.

D. Jurassic

At the present time Jurassic strata have only been recognized at Kilen, eastern Kronprins Christian Land, where a series of thinly bedded siltstones and dark shales of unknown thickness has yielded a mollusc fauna of Middle Jurassic (Bathonian) age (Dawes, 1976a).

E. Cretaceous—Tertiary

Poorly known strata of Cretaceous–Tertiary age are widely distributed in the Wandel Sea Basin. In Kronprins Christian Land, the lower half of a gently dipping sequence, at least 600 m thick, consists of dark sandstones and shales apparently of Cretaceous age. Higher strata are probably referable to the Tertiary. A similar, but thinner and gently folded sequence, containing some thin coals and coaly shales, is present in Peary Land (Troelsen, 1950b).

Tertiary fine-grained sandstones and shales in the Prinsesse Øer contain thin coals and have yielded Paleocene plants suggestive of a correlation with the Eureka Sound Formation of the Canadian Arctic (Dawes, 1976a).

F. Structure

The Wandel Sea Basin has been deformed into a tectonic belt, characterized by a dominant NW–SE trend (Fig. 5). Large- to medium-scale open folds with gentle dips characterize the belt, but locally steeply inclined folded strata occur. At least five major northwest-trending faults traverse the belt and form a series of tilted, possibly laterally displaced, blocks dipping generally to the southwest. This faulting has controlled much of the present topography and is well illustrated by the NW–SE-elongated shape of the islands in the southern part of the Wandel Sea—the Prinsesse Øer. Other directions of faults and folds are present in the basin but the relationships of these to stratigraphical units of known age is not everywhere evident (Dawes, 1976a). Furthermore, little is known about the nature of the fault movements and the parts played by normal, reversed, or transcurrent displacement.

Important northwest fault movements affect strata as young as Paleocene age; whether these major faults have an earlier history is not known. A regional unconformity below Cretaceous–Tertiary strata indicates important tectogenesis in late Mesozoic time, but the relationship of this movement (represented by both faulting and folding) to the Jurassic (Bathonian) outcrops is unknown. It may be that the Cretaceous–early Tertiary volcanic activity known from northern Peary Land (Kap Washington Group) correlates with this late Mesozoic hiatus.

IX. GEOLOGY AND PHYSIOGRAPHY OF THE CONTINENTAL MARGIN

A. Introduction

Relatively little information is available on the physiography and composition of the Greenland continental margin bordering the Arctic Ocean when comparison is made to the continental margin elsewhere in Greenland or at comparable latitudes elsewhere in the Arctic, for example, the Barents Shelf. From approximately 80°N in the west to about 76°N in the east, the coast is icelocked throughout the year. Thus, access is difficult and in contrast to regions farther south, few seaborne surveys have been undertaken on the continental shelf north of these latitudes.

Nevertheless, various airborne, iceborne, and submarine traverses have produced bathymetrical and geophysical data, although many areas remain essentially uninvestigated. A number of regional aeromagnetic surveys of the Arctic include northern Greenland (e.g., King et al., 1966; Haines et al., 1970; Ostenso and Wold, 1971) and some recent systematic low-level flights have been specifically undertaken off the northern Greenland coast (R. H. Feden, personal communication). Regional gravity maps compiled from data predicted by correlation methods are now available (e. g., Wilcox et al., 1975), but more detailed data in the form of Bouguer anomaly maps compiled from measurements at ground stations are restricted to the Lincoln Sea region (Sobczak and Stephens, 1974). A single traverse of seismic subbottom profiling carried out from the drifting ice island Arlis II (Ostenso and Pew, 1968) is the only seismic work known to the authors from the coast of eastern North Greenland.

Considerably more bathymetric and geophysical data are available to the south in the Baffin Bay–Nares Strait region, where shipborne investigations are possible (e.g., Pelletier, 1966; Keen, M. J., et al., 1972; Keen, C. E., et al., 1972; Keen and Barrett, 1973; Ross and Falconer, 1975). Dredging has been undertaken in this region, the most northerly samples being from the Kane Basin at latitude 80°N (J. H. Kravitz, personal communication).

The continental margin of northern Greenland shows marked variations in character and width (Fig. 15). For convenience the margin is described in

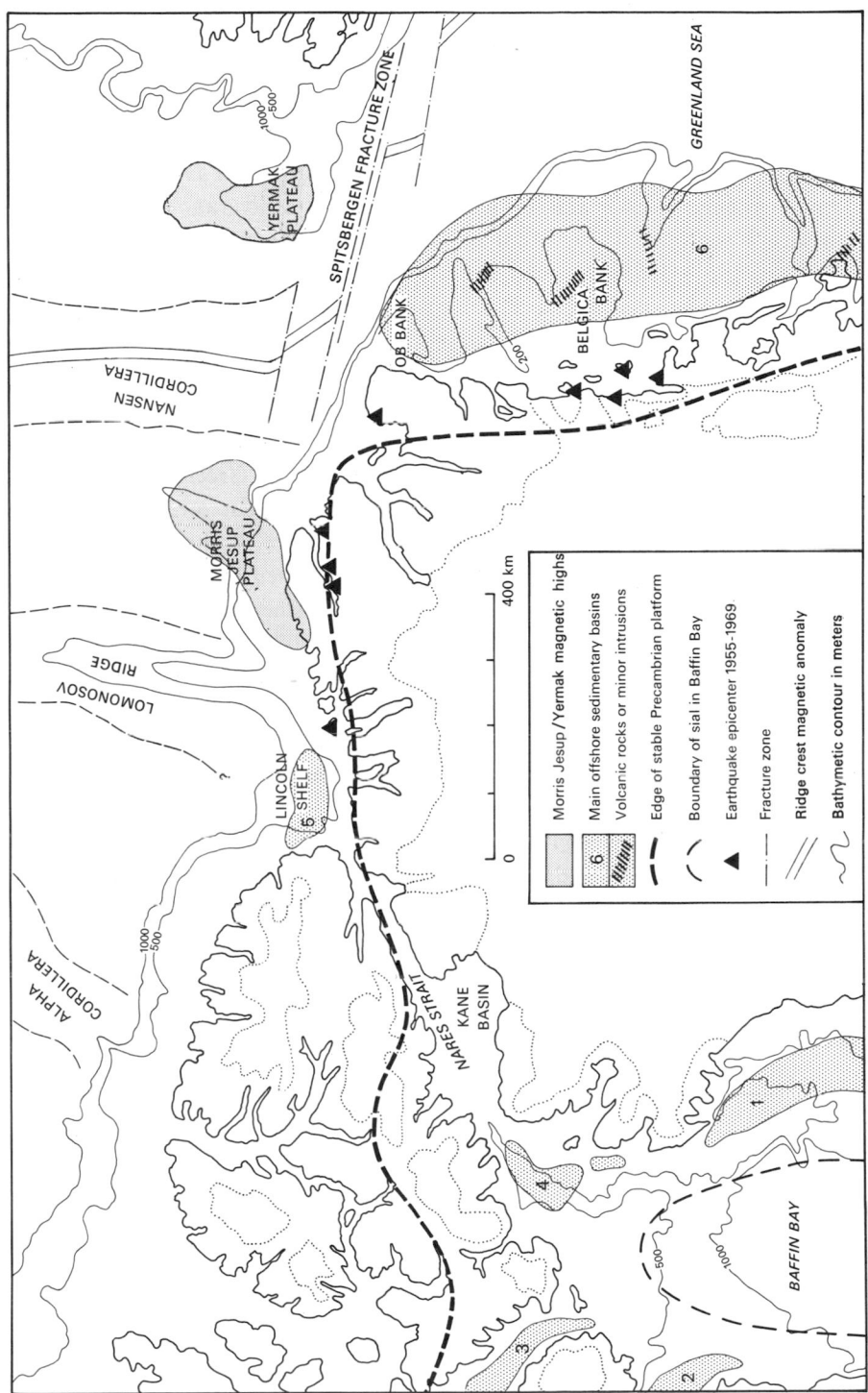

Fig. 15. Some major geological, geophysical, and physiographical features of the northern margin of Greenland. (Bathymetry from Heezen and Tharp, 1975; other data compiled from sources referred to in the text and from P. R. Vogt, personal communication.)

three parts, viz., bordering the Greenland Sea, the Eurasian Basin, and Baffin Bay–Nares Strait.

B.   Greenland Sea

Along the northern part of the east coast, the continental shelf is typically broad and irregular, with a well-defined shelf break at an average of 300 to 400 m and with a continental slope that is relatively steep, between 1:15 and 1:40 (see Volkov, 1961; Johnson and Eckhoff, 1966; Johnson and Heezen, 1967; Johnson, 1975). The shelf is characterized by a rough topography, intersected by a number of marked troughs or canyons, mainly at right angles or striking northwest (but some parallel) to the coast. Smooth areas are locally common, as well as a number of shoals less than 100 m deep.

The widest part of the shelf, the Belgica Bank, is about 300 km wide and located somewhat to the north of the western end of the Greenland Fracture Zone at latitude 78°N. To the north the shelf gradually narrows towards the Ob Bank at 81°N and farther to the north at the tip of Kronprins Christian Land it is reduced to 20 km in width with a distinct shelf break at about 250 m. At this latitude the continental slope is markedly steep and forms a linear, northwest-trending feature that persists for nearly 100 km into the Wandel Sea. The physiography of the continental margin off northern Kronprins Christian Land closely resembles the western side of the Yermak Plateau to the northwest of Svalbard, which is characterized by a steep slope that reaches the depths of the Lena Trough (>3500 m) and the site of the Spitsbergen Fracture Zone.

The wide continental shelf in the northern Greenland Sea is the site of thick sedimentary accumulations. Interpretation of aeromagnetic surveys from a number of sources indicates a large basin parallel to the coast between 72°N and 81°N, the center of which has a depth to magnetic basement of some 10 km (Henderson, 1976; see Fig. 15). The age of this offshore sedimentary sequence is unknown, but a wide age range of sedimentary rocks (Proterozoic, Paleozoic, and Mesozoic–Tertiary) occurs on land in the East Greenland Fold-belt. North of 78°N the late Paleozoic to Tertiary platform strata of the Wandel Sea Basin form the outer coast and the low-lying islands, including Hendrik Kröyer Øer some 40 km off the coast. It would appear that such rocks also cover much, if not all, of the inner shelf and probably make up a substantial part of the thick sedimentary accumulations farther eastwards.

Various igneous rocks are also indicated by the aeromagnetic data in the offshore sedimentary accumulation (Fig. 15). Intrusions and volcanic rocks are not known in the Wandel Sea Basin as exposed on land. However, in the Proterozoic sedimentary section, both within the foldbelt and in the foreland sequence, minor intrusions and volcanic rocks are ubiquitous.

Little is known about the geological structure of the shelf north of 78°N. Many of the offshore troughs are probably fault controlled. Both WNW- to NW-trending dislocations and dislocactions parallel to the coast are common on land. Nevertheless, many U-shaped depressions, for example those plotted at about 76°N and 77°N by Ostenso and Pew (1968) and by others farther north off the coast of Kronprins Christian Land, resemble glacial channels, suggesting a former advance of the Inland Ice onto the shelf. A major fault, with an assumed westward-dipping fault plane of about 20°, has been recognized at a depth of 2865 m in the Wandel Sea to the north of Kronprins Christian Land in the vicinity of 82°29'N and 11°41'W. Ostenso and Pew (1968) correlate this with the thrust front of the East Greenland Foldbelt.

C. Eurasian Basin

The Greenland continental margin of the Eurasian Basin is of irregular physiography and shows some variations in width. Although perhaps normal by world standards, it is notably the narrowest margin of the Arctic Basin and is conspicuously narrow when compared to the vast shelves on the Eurasian side of the ocean (Fig. 2). The continental shelf extends down to greater depths and generally has a less well-defined break than that in the Greenland Sea. The shelf is composed of two main features, the Morris Jesup Plateau (or Rise) in the east and the shallower Lincoln Shelf at the northern end of the Nares Strait.

The Morris Jesup Plateau, off Peary Land, forms a triangular segment approaching 200 km in width that projects northwards into that part of the Eurasian Basin bounded by the continental Lomonosov Ridge and the oceanic Nansen (Gakkel) Cordillera. From the scarce data available it seems that the plateau is developed as two benches that are separated by an east–west submarine trough. The part nearest the shore is about 60 km wide with a surface at between 200 and 400 m. The outer part has a much more irregular surface between 600 and 800 m deep, on the northern side of which there is a wide continental rise that falls away to the 4000-m-deep abyssal plain. The trough transecting the Morris Jesup Plateau is on strike with the main Spitsbergen Fracture Zone and is probably its continuation.

Westwards the shelf gradually narrows into the Lincoln Shelf, which, between western Peary Land and the Robeson Channel, seems also to be composed of two parts. The landward part, at an average depth of about 300 m, narrows westwards so that off Nyeboe Land it is less than 30 km wide. The seaward part, forming the main area of the Lincoln Shelf, has a more irregular topography and has a depth of about 600 m.

The precise nature of the continental break and slope of the Lincoln Shelf is not known. Moreover there is still uncertainty about the location of the

Lomonosov Ridge with respect to the Canada–Greenland block. Many sources indicate a termination of the ridge at the continental slope off northern Ellesmere Island, to the west of the Lincoln Shelf. However, a recent bathymetric map of the Arctic Basin (Heezen and Tharp, 1975) portrays the ridge between the Morris Jesup Plateau and the Lincoln Shelf, reaching the coast of western Peary Land.

The continental shelf off northern Greenland is assumed to be underlain by at least two main geological provinces; the Paleozoic and (?) older sediments of the North Greenland Geosyncline and the late Cretaceous–early Tertiary volcanics and sediments of the Kap Washington Group (Dawes, 1973). The mainly flat or nearly flat magnetic profiles described from the Paleozoic Franklinian Geosyncline in Ellesmere Island and northern Greenland suggest that the thick sedimentary section continues out under the adjacent continental shelf (King *et al.*, 1966). Nevertheless, various magnetic anomaly maps (e.g. Haines *et al.*, 1970; Ostenso and Wold, 1971; Coles *et al.*, 1976) show a large NE–SW-elongated, positive magnetic anomaly over the Morris Jesup Plateau, with a center at approximately 84°30′N, 26°W. This anomaly impinges on the Peary Land coast at the site of the volcanic province, the Kap Washington Group, with which it clearly invites correlation. However, on the basis of the available magnetic data alone it is impossible to establish the magnetic source as a surface or near surface body of high magnetization rather than a deep body of larger volume but lower magnetization (G. V. Haines, personal communication). Also, preliminary magnetic intensity measurements indicate that few of the Kap Washington lavas have suitably high values of magnetic intensity (N. A. Ostenso, personal communication). According to Feden *et al.* (1974), twin "high-amplitude, long-wavelength anomalies" occur over the Yermak Plateau and in an area that corresponds with eastern Morris Jesup Plateau. These anomalies flank the ridge crest of the Nansen Cordillera and they presumably represent parts of a single igneous province. In the absence of detailed geophysical information and of dredged material, comparisons remain uncertain but the correlation of the Kap Washington Group with the rocks forming the Kap Morris Jesup–Yermak Plateau province is hereby suggested.

In contrast, the Lincoln Shelf is characterized by low magnetic relief, suggesting the presence of a thick sedimentary section that is interpreted as the Paleozoic fill of the Franklinian Geosyncline. In addition, the gravity field of the Lincoln Shelf is markedly uneven, varying from +90 mgal in the northern part to a minimum of −30 mgal (Sobczak and Stephens, 1974). Primary gravity trends are parallel to the structural trend of the North Greenland–Ellesmere Island Foldbelt, suggesting the presence of widespread Franklinian sediments offshore (Fig. 16). The main anomalies are a prominent relative low of about −30 mgal (the Lincoln Sea Low), elongated parallel to

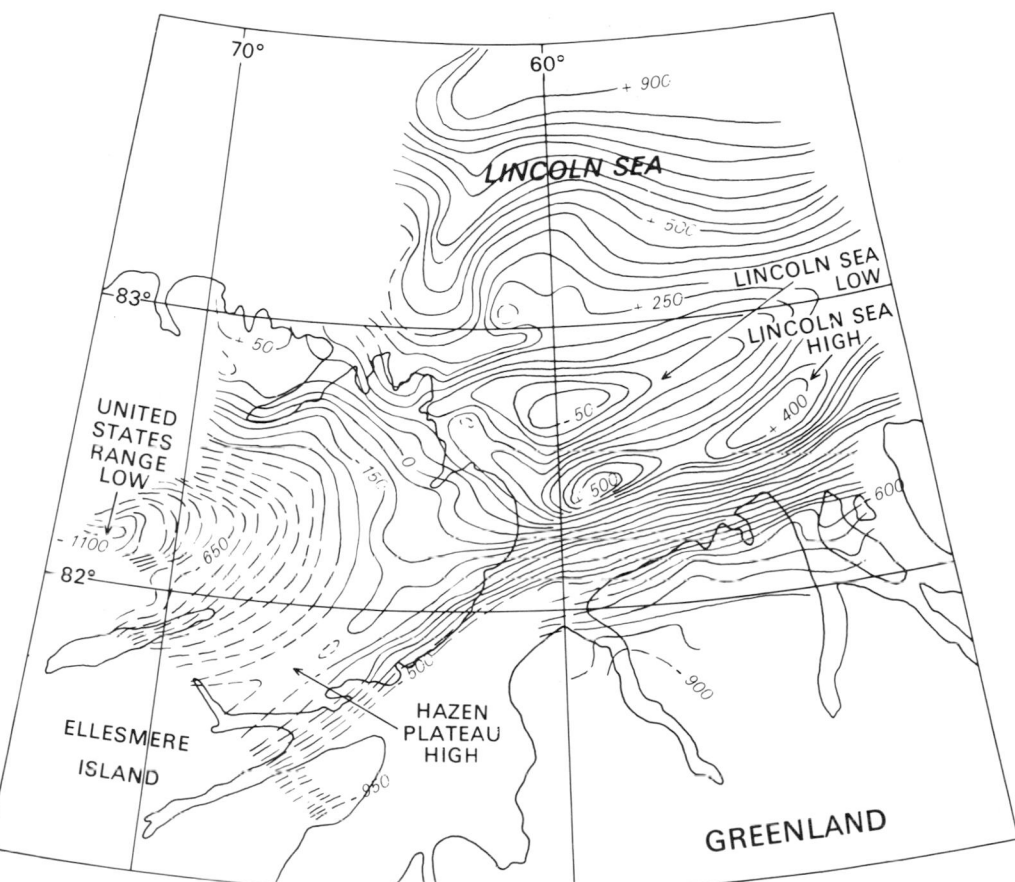

Fig. 16.   Bouguer anomaly map over the Lincoln Sea and adjacent Greenland and Ellesmere Island. Gravity contours at 5-mgal intervals. Redrawn from Sobczak and Stephens (1974).

the coast and covering an area exceeding 5500 km², and a relative high gravity anomaly (Lincoln Sea High), on the landward side.

Sobczak and Stephens (1974) correlate the Lincoln Sea Low with low density sediments having a mean density contrast of −0.33 g/cm³ and suggest the presence of a "young" basin containing at least 2 km of sediments. This may represent a northeastern extension of the Sverdrup Basin outcrops of the Lake Hazen region. The Lincoln Sea High, a twin anomaly with a maximum of +50 mgal, is interpreted as representing the edge of the stable craton, correlating well with the mantle–crust boundary proposed by Riddihough et al. (1973) from aeromagnetic data. This gravity high coincides with the edge of the inner part of the Lincoln Shelf.

D. Baffin Bay—Nares Strait

The Nares Strait is traceable as an NE–SW-trending depression for over 600 km from northern Baffin Bay into the Lincoln Sea (Fig. 15). The eastern boundary of this depression corresponds to the edge of the inner part of the Lincoln Shelf as described above. The depression is generally steep sided and in its central part between Smith Sound and the Robeson Channel it has a depth of between 400 and 450 m. Shallower areas make up the Hall Basin in the north and the larger Kane Basin. The latter has an average depth of about 200 m with some areas in the central part less than 100 m in depth (Pelletier, 1966). At the southern termination of the Nares Strait in Smith Sound the bottom topography is irregular, with the deeper part of the Channel (500–800 m) nearer the Greenland side than the Canadian side (M. J. Keen, et al., 1971).

In northern Baffin Bay the continental shelf is well developed and on the Greenland side approaches 200 km in width. It has an irregular surface topography but with a general bathymetric level of about 400 m. The continental break is well defined at about 550 to 600 m and the continental slope drops steeply down to reach depths of 2000 m in the center of the bay. The main bathymetrical feature is an extensive trough up to 1200 m deep (Melville Bay Graben, see below) that runs parallel to the coast from latitude 74° to 78°N. This is traversed by several troughs at right angles to the coast.

Geophysical data suggest that the northern Baffin Bay–Nares Strait seaway is floored by continental rocks, and that oceanic crust exists in the central and deeper parts of Baffin Bay to the south. Nares Strait cuts through the outer part of the Canadian–Greenland craton and across the strike of the platform and geosynclinal cover on its northern border. Various magnetic and seismic anomalies over the land areas of Greenland and Ellesmere Island (e.g., Riddihough et al., 1973; Sobczak and Stephens, 1974) show little or no change in intensity or trend over the seaway, suggesting no radical difference in the rocks flooring the strait.

In the northern Baffin Bay and Smith Sound region, shipborne magnetometer, aeromagnetic, and seismic data from a variety of sources (e.g., C. E. Keen, et al., 1972; M. J. Keen et al., 1972; Keen and Barrett, 1973; Hood and Bower, 1975a,b) indicate the presence of a number of offshore faulted basins that contain thick sedimentary sequences. Of these, the Melville Bay Graben on the Greenland Shelf is a north–south-trending trough, 400 km in length and between 50 and 75 km wide. The published data suggest that faulted and folded sediments, reaching over 10 km in thickness in the deepest part, occur in the graben and that a comparable thickness of sediment underlies northern Baffin Bay to the west of the graben. In the southern part there is evidence that a widespread unconformity separates an upper 2 km of mainly flat-lying strata from a lower sequence of variably folded beds.

Other basins with thick sedimentary sections occur on the Canadian Shelf, e.g., the Baffin Shelf Graben, the Lancaster Sound Graben, and Jones Sound Basin.

An extensive sedimentary basin, common to both the Greenland and Canadian shelf, occupies the southern part of Smith Sound between 76°45'N and 78°N. Aeromagnetic and seismic data delimit an arcuate U-shaped basin, some 50 km wide, that trends northwest between Carey Øer and Northumberland Ø and turns northwards along the shelf of southeast Ellesmere Island (Fig. 15; Hood and Bower, 1975a; Ross and Falconer, 1975). This basin is characterized by a linear belt of smooth, regular magnetic anomalies bounded by positive high-frequency magnetic anomalies indicative of shallow basement. The northern part of Smith Sound (78°N to 78°40'N) shows an east–west-orientated bottom topography and corresponding magnetic trends at a high angle to the strike of the seaway (M. J. Keen et al., 1971), suggesting very little sediment cover above crystalline basement. A basement high striking north from the Carey Øer may divide the sedimentary basin into two unequal parts. Depth determinations show a depth to crystalline basement of 20 km in the east and about 10 km to the west at longitude 76°W. The sediments are flat-lying to shallow dipping and broken in several places by normal faults.

The age of the sediments in the various basins in the Baffin Bay–Nares Strait region is unknown. In view of the age of offshore deposits established from dredging and bore holes farther south in the Atlantic, as well as the supposed inherent connection with the continental break up of Laurasia in late Phanerozoic time, it is generally held that the bulk of the sediment thickness is of Mesozoic and Cenozoic age. Also seismic velocities in many places are consistent with such an age. Some suggestions have been offered that the successions in Melville Bay and Lancaster Sound could contain Paleozoic strata (M. J. Keen et al., 1972; Henderson, 1973; Denham, 1974).

The present authors regard it as probable that strata older than Jurassic form important elements in the offshore sedimentary sequences in the northern Baffin Bay region and that strata as old as Proterozoic might be expected to be represented. This statement is based on four main lines of reasoning, in addition to consideration of the thickness of sediments involved:

(1) There is a growing body of evidence to suggest that the Davis Strait–Baffin Bay–Nares Strait lineament was in existence in some form in pre-Mesozoic time and thus a possible site of Paleozoic and older sediments. Evidence that the seaway follows an "old" line of crustal weakness is based on a variety of considerations including the trend of the Franklinian diabase dikes (Fahrig et al., 1973), the contrasted tectonic style of Paleozoic foldbelts in Greenland and Canada (Dawes and Soper, 1973), and comparison of global structural interactions (Burek, 1973).

(2) The fault-bounded Smith Sound Basin is on strike with prominent

onshore WNW- to NW-trending normal faults (Keen and Barrett, 1973). The Thule region is occupied by an asymmetrical composite graben filled by a least 4.5 km of relatively undeformed Proterozoic sediments and volcanics. Thule-type faulted Proterozoic outcrops also occur on the west side of Smith Sound in southeastern Ellesmere Island (Christie, 1972). Correlation of these Proterozoic successions, characterized on each side of Smith Sound by distinctive sedimentary and volcanic beds, and matching of thickness variations between onshore and offshore successions, suggest that Proterozoic strata could form part of the sediment thickness in the Smith Sound Basin.

(3)   The Melville Bay Graben strikes northwestwards towards the Kap York district suggesting a correlation with the onshore fault system of the Thule region (Kerr, 1967a; Henderson, 1973). Faults in the Thule region were active in Proterozoic time (Dawes, 1976a). Thus, if Kerr's and Henderson's correlation is correct, the Melville Bay Graben could also have had its initial origin at a similar time.

(4)   The West Greenland Shelf appears in many respects to be a mirror image of the Baffin Shelf, e.g., similarity in size, nature, and structure of continental margin, presence of coast-parallel grabens, coastal basalt provinces, etc. Offshore Ordovician carbonate rocks, assumed to be locally derived, have already been dredged on the Greenland Shelf between 60°N and 65°30'N (Johnson et al., 1975), while Ordovician limestone and shales on the Baffin Shelf are described as the isolated remnants of a formerly widespread platform terrain (Beh, 1975; Grant, 1975). The presence of the little publicized Ordovician limestone outliers at about 66°N within the Precambrian basement terrain of Greenland's west coast (V. Poulsen, 1966) provides evidence to show that Lower Paleozoic seas did indeed cover a large part of the Baffin Bay region.

X. REGIONAL CRUSTAL STRUCTURE AND RELATIONSHIPS

A. Introduction

The regional geological structure of northern Greenland has been known in outline for many years but it is only recently that data from the adjacent offshore regions have become available, and made possible elucidation of some of the relationships with the Arctic Ocean–North Atlantic Ocean System.

In the west, the Baffin Bay–Nares Strait Lineament, a narrow north-easterly trending seaway floored by continental rocks, separates Greenland from Arctic Canada. To the north and east the Greenland block is bounded by ocean basins floored by both continental and oceanic rocks that are the sites of presently active ocean ridges: the Nansen (Gakkel) Cordillera in the Arctic

Ocean Basin and the Mid-Atlantic Ridge in the North Atlantic. This latter ridge is traversed by several fracture zones. North of Iceland, the main segments of the ridge are, from south to north, the Kolbeinsey Ridge (also called the Iceland–Jan Mayen Ridge), the Mohns Ridge, and the Knipovich Ridge (Fig. 2).

Northern Greenland, lying to the west of this ridge system, not surprisingly shows evidence of late Phanerozoic crustal movements that are connected to the later evolution of the Arctic Ocean and North Atlantic Ocean System. The Proterozoic and Paleozoic earth movements, so conspicuous in northern Greenland, relate to earlier crustal interaction associated with the Laurentian–Eurasian boundary.

B. Deep Crustal Structure

The edge of the stable Precambrian craton in northern Greenland can be plotted with some certainty from the aeromagnetic and gravity data now available (Fig. 15). Terrain underlain by the Greenland Shield is typified by a complex magnetic pattern that is of the same type as that described by Morley *et al.* (1967) over corresponding parts of the Canadian Shield. The northern limit of this characteristic magnetic pattern is regarded as the edge of the stable craton (Riddihough *et al.*, 1973). The position of this margin is also defined in the Robeson Channel region by gravity data (Sobczak and Stephens, 1974).

The magnetic anomaly pattern over the craton in northern Greenland is characterized by a general north–south trend (Greenarctic Consortium, personal communication; Riddihough *et al.*, 1973). This is noticeably at right angles to the regional structural trends seen in the surface rocks across northernmost Greenland, but parallels the East Greenland Foldbelt and also the Victoria Fjord Arch—a basement "high" that projects northwards into the North Greenland Geosyncline. The full history of this arch is not yet clear but it was a positive feature in at least Proterozoic time, having a marked influence upon sedimentation (Dawes, 1976*a,b*). It is worth noting that the Victoria Fjord Arch coincides with the line of the Lomonosov Ridge as shown on Heezen and Tharp's (1975) bathymetric map of the Arctic Basin.

A crustal model for the Nyeboe Land region bordering the Robeson Channel suggests a standard continental crust about 32 km thick below Greenland with a northerly thinning to about 18 km below the Lincoln Sea (Sobczak and Stephens, in Trettin *et al.*, 1972). The prominent northeasterly trending Hazen Plateau and Lincoln Sea seismic "highs" (maximum anomaly $+50$ mgal) are separated by a steep horizontal gradient (maximum -3.68 mgal/km) from a parallel "low" (minimum anomaly -95 mgal) that strikes from Judge Daly Promontory in Ellesmere Island across to Greenland. Sobczak and Stephens (1974) interpret these data as indicating an abrupt step in the crust–mantle boundary off the coast of Nyeboe Land, with the gravity "low" corresponding to the thicker crust.

C. Summary of Crustal Movements

1. *Early Precambrian*

The close similarities in the early evolution of the Precambrian crystalline terrains bordering the Baffin Bay–Davis Strait (e.g., Bridgwater *et al.*, 1973) leave little doubt that the region once formed part of a single shield. In northern Greenland evidence for this unit is provided by the close correspondence of the Precambrian crystalline terrains on both sides of Smith Sound. Distinctive rock units such as the Etah Group and the Etah metaigneous complex (Dawes, 1972) have counterparts in the terrain mapped by Christie (1962*a,b*, 1972; T. O. Frisch, personal communication) in southeastern Ellesmere Island. Both regions appear to have been affected by the Hudsonian orogeny between 1800 and 1600 m.y. ago.

The eastern edge of the shield exposed within the East Greenland Foldbelt has not been investigated in detail north of 76°N. Nevertheless, the reappraisal of the crustal makeup of the foldbelt as containing widespread Precambrian elements agrees with the parallel reinterpretation of the Caledonian Orogen in Svalbard and Scandinavia. Since predrift reconstructions place western Norway in close proximity to East Greenland, there may be significance in the 3000 m.y. date for gneiss at Danmarkshavn at 77°N in Greenland and the 2800 m.y. age for gneisses at Langøy at 69°N on coastal Norway (Higgins, 1976).

2. *Late Precambrian*

Following the major uplift of the Greenland Shield after the Hudsonian orogenic episode, vast areas of Proterozoic sediments derived from the craton were deposited around its margins in both platform and geosynclinal environments. Volcanic activity of Helikian and Hadrynian age was of continental scale; sills, dikes, and effusives occur throughout northern Greenland (Fig. 17) and in the Arctic Islands of Canada.

In the west, within the stable craton, regional faulting probably accompanied the instability and subsidence to produce the Thule Basin—now a composite asymmetrical graben composed of a number of fault blocks. Regional faulting of Hadrynian age clearly affected a large part of the Baffin Bay–Smith Sound region, controlling the trends of igneous intrusions and initiating the formation of several graben structures in Greenland and Canada, and offshore in Baffin Bay and Smith Sound. These movements were of fundamental importance for shaping the structural pattern of this part of the Arctic with fault rejuvenation probably occurring throughout Phanerozoic time. Baf-

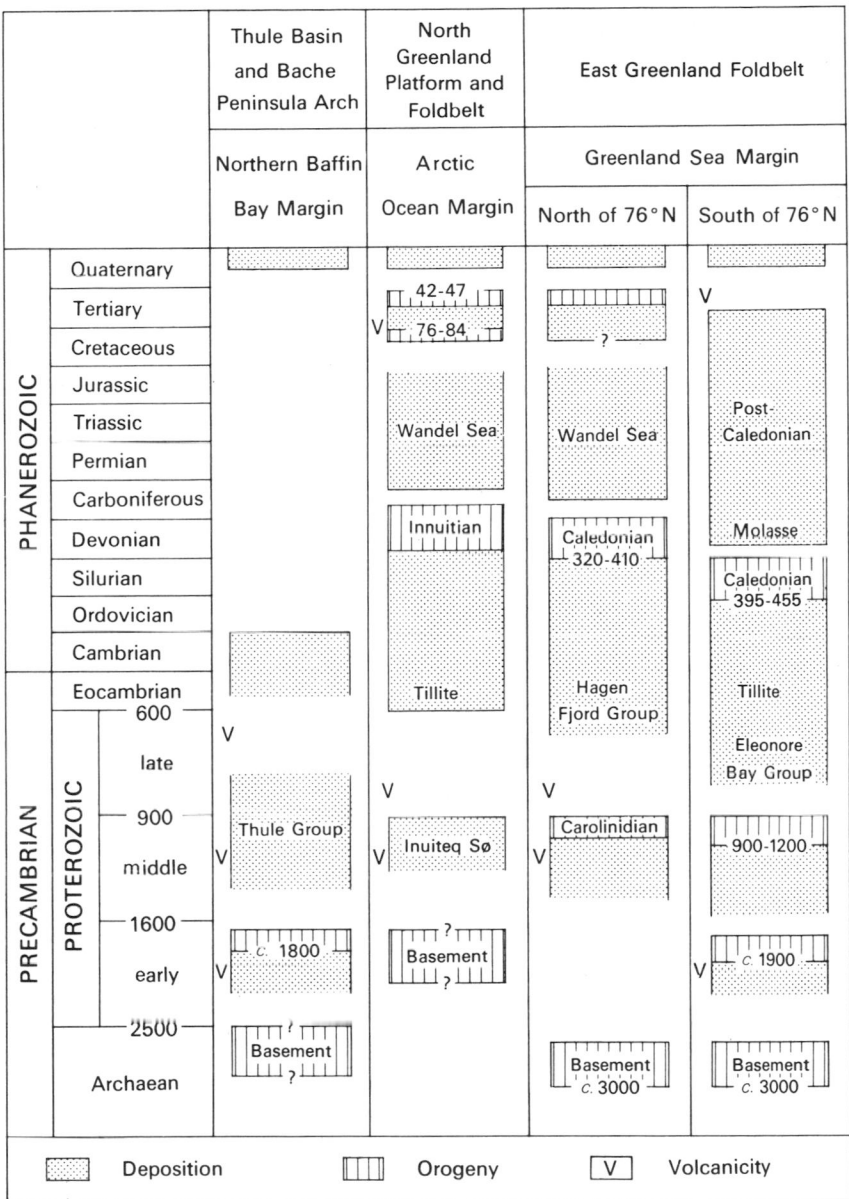

Fig. 17.   Chronological chart showing the main depositional, orogenic, and volcanic periods of the northern margin of Greenland.

fin Bay–Davis Strait (being parallel to important late Proterozoic fault trends) may indeed date back to a line of weakness established at this time (see Fahrig *et al.*, 1973).

In strong contrast, a major orogenic episode(s) in the east corresponding in time with the Grenville–Sveco–Norwegian orogenic events (900–1200 m.y.)

affected the eastern border of the craton and the thick Proterozoic geosynclinal accumulations. Isotopic evidence for migmatization, mobilization, and plutonic emplacement of this age to the south of 76°N in Greenland is abundant (Henriksen and Higgins, 1976); to the north, accurate dating of the presumably equivalent Carolinidian mountain belt is not yet available. The orogenic activity is seen as prolonged compression along the Eurasian–Laurentian plate boundary.

3. *Paleozoic*

Following the late Proterozoic and Lower Paleozoic subsidence along the northern and eastern rims of the Greenland Shield, the North Greenland (Franklinian) and the East Greenland (Caledonian) Geosynclines were deformed by mid-Paleozoic earth movements that show contrasted structural characterisitcs and timing of orogenic events. The most pertinent data are summarized in Table I.

The North Greenland Foldbelt is seen as a marginal compressional structure of the Paleo-Arctic ocean (Pelagus of Harland and Gayer, 1972) that involved the interaction of oceanic and continental crust. Bounded to the south by a stable platform that acted as a hinterland, the tectonic style of the foldbelt is regarded as the product of deep-seated southerly movements below the geosynclinal trough (Dawes and Soper, 1973). A model for such tectonic conditions could be a mid-Paleozoic subduction zone with an underthrusting of oceanic crust against the continental block, causing consumption and heat flow into the sedimentary prism above. This southerly movement may not have continued until the final stages of ocean contraction and the ultimate collision of Greenland with the corresponding Eurasian block. By at least late Carboniferous time the orogenic zone had been regionally uplifted as inferred by the unconformity below the Wandel Sea Basin.

The appearance in early Paleozoic time of a prominent borderland somewhere off the northern coast, from which the Silurian and Devonian clastic fill of the geosyncline could be derived, may have been generated by an early uplift phase of the orogen (Fig. 9). Evidence for an uplift is provided by the unconformities in the platform and miogeosynclinal sections in Greenland, and the strong indications of orogenic activity in the eugeosynclinal part of the orogen in northern Ellesmere Island as described by Trettin (1971). The borderland has been treated as the eastern extension of the Peary Geanticline (Dawes, 1973).

The East Greenland Foldbelt, as part of the circum-Atlantic Caledonian orogenic system linking Svalbard, Scandinavia, and the British Isles, is now generally agreed to be the result of contraction of an ocean (the Paleo-Atlantic or Iapetus of Harland and Gayer, 1972) causing collision of opposing

TABLE I

Summary of the Nature and Timing of the Mid-Paleozoic ("Caledonian") Orogeny in the North and East Greenland Foldbelts

	North Greenland Foldbelt	East Greenland Foldbelt	
		76°N to 82°N	70°N to 76°N
Sedimentary regime			
Age of youngest known geosynclinal strata	Lower Devonian (Gedinnian)	Upper Silurian (Ludlovian)	Middle Ordovician (Mohawkian)
Age of oldest known postorogenic strata	Upper Carboniferous (late Pennsylvanian)	Lower Carboniferous (Mississippian)	Middle Devonian (Givetian)
Tectonic regime			
Direction of main tectonic transport	North toward the Arctic Ocean	West away from the Atlantic Ocean	
Character of foldbelt margin	Autochthonous—gradual incomings of structural elements	Allochthonous—thrust and nappe front	
Tectonic status of platform	Hinterland	Foreland	

continental margins. There is however much disagreement as to the nature and contraction history of this ocean(s), of the significance of the marginal geosynclinal tracts, and indeed on the relative positions of Greenland and Europe and the age of initial separation. The different explanations and models based on lithosphere plate tectonics put forward to explain all aspects of the orogenic system and its relation to the Arctic Ocean are too numerous to mention here.

What is significant for our discussion is that the main E–W compression characterized by an intense deformation and suturing, causing the projection of accreted geosynclinal units onto the foreland, was a prolonged event that probably lasted for some 60 m.y., i.e., from the late Ordovician to latest Silurian or Devonian. Furthermore, the diastrophism in Greenland transgressed northwards as demonstrated by the contrasted stratigraphical record between segments of the foldbelt north and south of 76°N. The Lower Paleozoic platform and geosynclinal succession, and the faunal character (as currently known) of the segment north of 76°N, compares well with the North Greenland Foldbelt. Indeed, to such a degree that the two sedimentary tracts can be considered to form a continuous province around the northeastern corner of Greenland. The diastrophism that brought the geosynclinal deposition to a close in these two regions cannot have been of markedly different age. The orogenic pattern suggests that the E–W Caledonian compression was perhaps due more to a rotational collision causing a delayed impact of the region north of 76°N, allowing the development of a sedimentary regime with northern aspect.

4. *Mesozoic and Tertiary*

Following the uplift and denudation of the folded and metamorphosed Franklinian and Caledonian geosynclinal belts, late Paleozoic regional subsidence affected large parts of eastern North Greenland. A major Carboniferous transgression over this area heralded the beginning of the Wandel Sea Basin, in which subsidence more or less kept pace with sedimentation. Mesozoic and Tertiary earth movements affected this basin as well as the North Greenland Foldbelt and Platform. The movements are summarized in Fig. 18.

Orogenic activity was concentrated in the northeastern part of northern Greenland, i.e., adjacent to the Eurasian Basin and the Greenland Sea. The North Greenland Platform remained stable and suffered faulting with "normal" displacements of 1 km or more and, with, to the east of the Victoria Fjord Arch, injection of basic dikes.

The mid-Paleozoic deformation and metamorphic pattern of the North Greenland Foldbelt has been overprinted to a varying degree by Mesozoic and Tertiary regional metamorphism and deformation with the Kap Cannon

Age	North Greenland Platform	North Greenland Foldbelt	Wandel Sea Basin
Cenozoic — Quaternary	Minor volcanicity	Seismically active minor volcanicity	Seismically active
	Isostatic changes	Isostatic changes	Isostatic changes
Cenozoic — Tertiary	Faulting	Late Tertiary E–W faulting	Faulting
	Faulting	Middle Tertiary thrusting, folding, faulting, metamorphism	Post-Paleocene NW folding, faulting
Mesozoic	Late Cretaceous faulting, NW basic dikes	Kap Washington Group, rhyolitic volcanicity Regional reheating Late Cretaceous E–W; N–S, NE–SW basic dikes	Late Jurassic–Cretaceous uplift, folding, faulting

Fig. 18. Summary of late Mesozoic and Cenozoic tectonism and volcanism in the northeastern corner of Greenland, Peary Land to Kronprins Christian Land.

thrusting accentuating the northerly overturning of structural elements. The regional reheating of the orogen, apparent through the late Cretaceous radio-metric age dates (Dawes and Soper, 1971), was connected with the expulsion of the large volume of predominantly acidic magma at about the time of the Cretaceous–Tertiary boundary (Kap Washington Group).

According to most current ideas on Arctic Ocean genesis, the Eurasian Basin is a "new" ocean formed by expansion from the actively spreading Nansen Cordillera. Vogt and Ostenso (1970) consider an age of 40 m.y. (late Eocene) for the initiation of this spreading; Pitman and Talwani (1972) and Feden et al. (1974) conclude that it goes back to earliest Tertiary (Danian) time, i.e., 65–63 m.y. ago or at about anomaly 25 time. Although there is still disagreement among advocates of continental drift about the initial positions and movement paths of the Arctic lands, most models accept Wilson's (1963) suggestion that the nonoceanic Lomonosov Ridge represents a fragmented part of the northern Barents Shelf. The drift path of the ridge in response to spread-ing may have been sinistrally transcurrent in close juxtaposition to the continental margin of northern Greenland.

Nevertheless, sometime prior to the initiation of spreading beneath the Barents Shelf and the development of the Nansen axis, the Kap Washington rhyolitic volcanics were extruded. These volcanics, together with the late Cretaceous basic dike swarms that cut the North Greenland Foldbelt (Dawes, 1976a), are indicative of the initial fracturing and rifting of the continental crust and margin of northern Greenland. Delineation of the relationship between the acidic vulcanicity and oceanic expansion that followed it must wait until more geological and geophysical detail is available from the Eurasian Basin (see Feden *et al.*, 1974). Furthermore, modern data from the little publicized Cretaceous volcanic rocks of Franz Josef Land (e.g., Dibner, 1957) and the Cretaceous and Tertiary volcanics of Svalbard (e.g., Harland, 1973) are essential links in any geotectonic reconstruction of the Barents–Greenland continental margin.

A critical question in continental drift studies in the North Atlantic and Arctic region is Greenland's status as a rigid block or plate in relation to North America and Europe during the Mesozoic and Cenozoic. As a conseqence of the modern interpretation of magnetic and paleomagnetic data from the North Atlantic (e.g., Johnson and Heezen, 1967; Pitman and Talwani, 1972; Johnson and Vogt, 1973; Athavale and Sharma, 1975), Greenland is generally thought to have moved northwards relative to Arctic Canada, on one side, and Svalbard and Europe on the other. Furthermore, a model that has received wide acclaim assumes that prior to 63 m.y. ago (anomaly 25) Greenland was a part of the Eurasian plate, and that at about 47 m.y. ago (anomaly 19), Greenland became part of the North American plate, thus placing the independent movement of Greenland in the period between 63 and 47 m.y. (Pitman and Talwani, 1972). The timetable covering the movement events and the routes adopted by the moving parts is still very much in debate.

The transport of the North Greenland Foldbelt over the late Cretaceous–early Tertiary Kap Washington Group of volcanics and sediments along the Kap Cannon Thrust could well be explained as a marginal compressional feature due to a northerly motion of the Greenland block. This overthrusting and the associated overfolding and metamorphism are considered to be mid-Tertiary (Oligocene) events. K-Ar dates of 34.9 and 32.3 m.y. on lavas are interpreted as giving a maximum age for the thrusting (Dawes and Soper, 1971).

Likewise, the post-Paleocene deformation of the Wandel Sea Basin to produce a NW-trending foldbelt was probably a compressive regime at the leading plate margin, the northerly movement of Greenland taking place along the Spitsbergen Fracture Zone. The similarity between the Mesozoic–Tertiary platform sequences of Arctic Canada (Sverdrup Basin) and Svalbard has long been realized. The Eurekan and West Spitsbergen Orogenies, respectively, have been attributed (in both sea-floor spreading and rifting models of continental drift) to northerly motion of the Greenland block. Depending on which "predrift" reconstruction of Greenland and Svalbard is favored (and there are many), Svalbard may well have been near to, in close juxtaposition with, eastern Peary Land in Paleocene time.

Thus, the Tertiary motion of Greenland also must have affected the Barents block, resulting in collision against Svalbard in perhaps late Eocene–early Oligocene time. The West Spitsbergen orogeny has been described as resulting from dextral transcurrent movement relative to Greenland (Harland, 1965; Lowell, 1972; Birkenmajer, 1972). The time relationships between Tertiary deposition and horizontal and vertical movements in Svalbard are complex (Kellogg, 1975). Relationships are not known in the same detail in the Wandel Sea Basin, but compressional, tensional, and uplift events are seen as a consequence of the varying nature of the interaction of Greenland with the Barents block (Dawes, 1976a).

5. *Quaternary*

The migration of polar temperatures southward that was already taking place in the Mesozoic intensified in the Tertiary and culminated in the Quaternary ice age. Davies (1972) and Weidick (1972, 1976) have described the main aspects of Pleistocene and Holocene geology of northern Greenland and little will be added here. It is generally assumed that the late Tertiary and Quaternary was marked by increasing aridity and progressively colder interglacials and thus the lands of the Arctic Basin were probably dry throughout most of Quaternary time.

Apart from various areas of high ground in the Roosevelt Fjelde in northern Peary Land and perhaps some in northern Melville Bay and in Kronprins Christian Land, all of northern Greenland was once covered by the Inland Ice. Isostatic uplift occurred during the postglacial recession in the Holocene (10,000–6000 b.p.), resulting in renewed erosion, the deposition of various ice margin features, and the loading of detritus onto the continental shelves.

One of the most conspicuous oceanic features of the Arctic, the Spitsbergen Fracture Zone, is widely regarded as a great dextral transform fault that offsets the actively spreading Nansen Cordillera and the Knipovich Ridge. A glance at a bathymetric map of the Arctic indicates that this fracture zone forms the northern margin of the Greenland block along the coast of northern Peary Land and that this region is the closest land—less than 20 km—to the transform fault and to the Nansen Cordillera. Throughout the Quaternary this northeastern corner of Greenland must have been affected by ocean-floor generation and by the dextral dislocation. It is therefore hardly surprising that seismic activity, in the form of earthquake epicentres, has been reported on land in northern Kronprins Christian Land and eastern Peary Land (Fig. 15; Sykes, 1965; Vogt *et al.*, 1970).

The presence of WNW-trending faults in northern Peary Land on line with the Spitsbergen Fracture Zone has already been noted (Dawes, 1973), and although the age and geometry of the latest movements along the faults is still unclear, transcurrent displacements in Peary Land are to be expected. Furthermore, the presence of active volcanicity in eastern Peary Land (Troelsen, 1949), associated in some cases with E–W-trending faults, is highly significant. The volcanic activity takes the form of solfataras that, ideally, are small cones 2 to 3 m high and up to 30 m across, composed of sulphur, gypsum, copiapite, fibroferrite, and other iron compounds. The occurrences are situated in apparent linear fashion along sides of Jørgen Brønlund Fjord and Frederick E. Hyde Fjord. One main occurrence on the south coast of the latter fjord is situated on the E–W-trending Harder Fjord Fault.

Solfataras are generally regarded as characteristic of waning volcanicity. The Greenland examples are apparently not connected with lava and pyroclastic rocks and, thus, may mark the initiation rather than the cessation of volcanic period (Troelsen, 1949). Similar volcanic activity of Holocene age in the form of basic volcanoes and hot springs, aligned along fault lines, is known in Svalbard (Hoel, 1914; Gjelsvik, 1963).

D. Concluding Remarks

A better understanding of the relationships throughout geological time of the lands surrounding the Arctic Basin must be the main aim of this volume. With this aim in view, but without entering into detail or attempting discussion of the literature, some concluding remarks are given here on the geological links of the northern margin of Greenland. Various comments have already been given on the relationship to Svalbard, Scandinavia, and Canada in the foregoing "summary of crustal movements."

1. *The Problematic Northeast Corner of Greenland*

The age relationship of the North and East Greenland Foldbelts has important implications in regional orogenic syntheses. Tantalizingly, the junction of the foldbelts is not exposed on land in Greenland: its presumed site in the Wandel Sea is now an important late Paleozoic–Tertiary sedimentary basin.

From evidence presented earlier we consider that Proterozoic and Paleozoic platform and geosynclinal sedimentation was essentially continuous around the northeast corner of Greenland and that the main mid-Paleozoic orogenesis in North and North-East Greenland (Caledonian in the broader sense) was of comparable age, although of contrasting character. The Wandel Sea is the site of the junction of tectonic regimes created by the opposing crustal configuration of the early Arctic and Atlantic Ocean Basins.

The regional strike of the two Greenland foldbelts is towards the Nansen Cordillera and the Barents Shelf. If the Lomonosov Ridge is indeed a detached portion of the northern Barents Shelf, caused by the spreading along the Nansen Cordillera, then it is likely to contain sedimentary and structural elements in common with either or both of the foldbelts. The fact that the Lomonosov Ridge is a relatively narrow, presumably continuous belt of continental rocks flanked by oceanic crust suggests, in any case, that the ridge is a rigid, coherent structural entity, i.e., an orogenic belt. The early Russian ideas of connection between the North Greenland Foldbelt (Innuitian) and the orogenic belt(s) of northern Russia (Uralides) must gain support, irrespective of whether or not the Lomonosov Ridge represents the displaced line of such an Arctic Ocean orogenic system (for references and discussion see, e.g., Hamilton, 1970; Harland and Gayer, 1972; Dawes, 1973).

2. *Nares Strait Lineament*

The Nares Strait is a narrow linear seaway separating northern Greenland from Arctic Canada. It represents the only border of Greenland not floored by oceanic crust. While there is universal agreement about the unity of Canadian and Greenland geology across the strait, there is much disagreement about the nature and history of the Nares Strait Lineament. This disagreement centers upon whether or not transcurrent motion has taken place along the strait and if so, the age(s) and amount of displacement(s) involved. Early advocates of continental drift in the Arctic, e.g., F. B. Taylor, A. Wegener, A. L. du Toit, and later S. W. Carey and J. T. Wilson, focused attention on the prominent linearity of the strait and suggested substantial transcurrent or transform displacement (200–500 km).

The crystalline basement, platform, and geosynclinal units of northern Greenland have westerly counterparts in Ellesmere Island; there is no doubt that the Innuitian Orogenic System continues into Greenland. The correspondence of regional geology is so close that some formational names in both platform and foldbelt sequences are common to both Ellesmere Island and Greenland. Furthermore, the main Proterozoic–Paleozoic facies belts in the southern part of the Franklinian Geosyncline on both sides of the Nares Strait have an on-line regional strike. Of particular importance is the close similarity in stratigraphic detail between the Proterozoic Thule Group sediment–volcanic succession in the western part of the Thule Basin, in Greenland, and the sections in adjacent Ellesmere Island (Dawes, 1976c). Despite the unfortunate fact that stratigraphic boundaries, and such structural markers as the southern border of the mid-Paleozoic foldbelt, cut the strait with large angles of intersection, making simple and direct interpretation of their form as "offset" or "continuous" open to some discussion, large-scale strike-slip movements as envisaged by the early workers create many problems of correlation and essentially disrupt the regional trend of the facies belts.

The bedrock geology in the immediate strait area clearly places some restraint on overall transcurrent motion and we concur with the view expressed by Kerr (1967a) that in Proterozoic and early Paleozoic time Ellesmere Island and Greenland were situated not far removed from their present-day relative positions. Geophysical data from the immediate Nares Strait area are more difficult to interpret in terms of relative motions between Ellesmere Island and Greenland but no data suggesting any appreciable offset are known to the present authors. Thus the "predrift" fits that depict Ellesmere Island and Greenland widely displaced (e.g., the "Bullard fit" and many others) are not in congruence with the regional geology, unless a complicated movement history is envisaged.

Thus, since geophysical and paleomagnetic data from the Atlantic Ocean seem to demand a northerly movement of Greenland independent of the North American and European blocks—a movement also supported by the compressional effects along the northern margin of Greenland—Ellesmere Island must have remained a part of, and moved with, the Greenland block during the motion. Assuming that Greenland remained a rigid block not traversed by major transcurrent dislocations, it seems necessary to look elsewhere for the site of any substantial motion thay may be needed to accomplish the late Phanerozoic crustal breakup of the Arctic.

XI. APPENDIX

Since this chapter was prepared in 1976, geological exploration in northern Greenland has gathered momentum and there also have been important results from the offshore regions, viz., the Greenland Sea and the Eurasian Basin (e.g., Talwani and Eldholm, 1977; Vogt et al., 1979; Phillips et al., 1980).

Onshore operations, often in previously poorly known areas, have been undertaken by the Geological Survey of Greenland each year, culminating in 1978 with the commencement of a five-year systematic mapping project. Results are outlined in the Survey's Report series, e.g., numbers 85 (1977), 90 (1978), and 88, 91, 95 (1979). One of the results of this field effort will be a flood of stratigraphic names and data as new successions are investigated and as the existing nomenclature, much of it erected over 50 years ago, is reassessed and redefined.

In addition, several accounts have been published describing results of earlier field work. Report 82 (1977) describes Cambrian–Silurian platform stratigraphy in southern Peary Land and Report 93 (1979) presents some structural and stratigraphic details from the North Greenland Foldbelt. Bulletins 121 (1977) and 131 (1979) describe Lower Paleozoic faunas.

A. The North Greenland Platform

1. *Proterozoic*

The Proterozoic sandstone–basaltic rocks forming the base of the platform sequence between Peary Land and Kronprins Christian Land are now known to have a composite thickness of over 3 km. A topographical unconformity with relief of up to 70 m separates a lower sandstone formation (800 m thick) from an upper sandstone and minor siltstone formation (900 m thick) (Collinson, 1979). Basic intrusions occur within both formations. The extensive basaltic extrusives (mapped previously by Greenarctic Consortium) form flows up to 100 m thick and cap the intrusion-filled sandstones. The basalt succession has a maximum thickness of 1300 m (Jepsen and Kalsbeek, 1979).

The late Proterozoic sequence that unconformably overlies the basalt flows reaches a thickness of at least 1200 m. Within this, the tillite-bearing Morænesø Formation, restricted to southern Peary Land, is considered to be mainly fluvial or lacustrine in origin, although some beds are thought to be true glacial deposits (Clemmensen, 1979). Equivalent shallow marine strata outcrop over large areas to the southeast towards Kronprins Christian Land.

2. *Paleozoic*

The supposed early Cambrian Portfjeld Formation of shallow water dolomite with black cherts that overlies the late Proterozoic sequence is now known to thin markedly from southern Peary Land (300 m) southeastwards to Hagen Fjord (30 m) (O'Connor, 1979). The tracing of this formation between Peary Land and Kronprins Christian Land suggests important modifications to the established stratigraphic correlation (see also V. Poulsen, 1978).

The unconformity at the base of the Wandel Valley Formation (late Lower Ordovician) is now recognized as the principal structural feature within the Cambro-Ordovician platform sequence of southern Peary Land (Peel, 1979). An extensive (1100 m thick) Lower to Upper Cambrian, mainly carbonate sequence in western Peary Land represents a slope sequence with passage from carbonate platform deposition in the southeast to a more basinal sequence to the northwest. The sequence is unconformably overlain by the late Lower Ordovician carbonates, which progressively overstep to the southeast the Middle and Lower Cambrian. The unconformity may be equivalent to the Middle Canadian evaporites in Washington Land in western North Greenland.

The Silurian platform and shelf deposits have been studied in detail by Hurst (1979, 1980a,b) who recognizes a maximum of about 2 km of strata in the west (Washington Land) and about 1.5 km in the east (Peary Land). The Silurian in the west is a mosaic of basin, slope, and platform strata in which clastic and carbonate rocks interfinger and in which rapid facies changes exist. Buildups occur at several stratigraphical levels. In the east, the established succession of Lower Silurian carbonates, Middle Silurian shales, and Upper Silurian turbidites is replaced along strike in J. P. Koch Fjord in westernmost Peary Land by a totally different succession of shelly limestones and a different style of shelf–basin contact.

B. The North Greenland Foldbelt

Detailed mapping in the southernmost segment of the foldbelt in eastern Peary Land by Christie and Ineson (1979) has established the presence of Proterozoic, Cambro-Ordovician, and Silurian sequences correlatable with the better known units of the platform to the south. This is particularly important for the determination of the history of the shelf–basin contact. Structural analysis by Pedersen (1979) of the foldbelt margin in one part of Peary Land has revealed a series of southerly directed thrust sheets of presumed regional dimensions.

C. The Wandel Sea Basin

Systematic mapping of the northern part of the Wandel Sea Basin has revealed a composite sequence of sediments exceeding 3500 m in thickness that, as expected, shows an overall similarity to strata of comparable age in Svalbard (Håkansson, 1979). An important addition to the previously known sequence is a 200 m thick section of Upper Jurassic sandstones, conglomerates, and sandy shales that represent marine to limnic deposits. Faunal and microfloral identifications indicate that the strata span at least early Volgian to Rhyazanian.

The increased knowledge of the late Paleozoic–Mesozoic succession allows some firmer control of tectonic events. A slight angular unconformity between the Upper Permian and Triassic deposits suggests some minor movements of probable late Permian age, while the main late Mesozoic unconformity recognized previously is now known to predate the newly discovered Upper Jurassic beds.

Fault tectonics are complicated and there is evidence to suggest that the dominant NW-trending faults are Mesozoic or even older features strongly reactivated in the Tertiary. Furthermore, field evidence suggests that strike-slip movements on the NW faults amount to a net sinistral displacement of the fault system of at least 2.5 km, possibly considerably more (Håkansson, 1979).

ACKNOWLEDGMENTS

This paper is published with the permission of the Director of the Geological Survey of Greenland. A. K. Higgins and T. O. Frisch are thanked for critical reading of the text.

REFERENCES

Adams, P. J., and Cowie, J. W., 1953, A geological reconnaissance of the region around the inner part of Danmarks Fjord, Northeast Greenland, *Medd. Groenl.*, v. 111(7), 24 pp.

Athavale, R. N., and Sharma, P. V., 1975, Paleomagnetic results on Early Tertiary lava flows from West Greenland and their bearing on the evolution history of the Baffin Bay-Labrador Sea region, *Can. J. Earth Sci.*, v. 12, p. 1–18.

Beh, R. L., 1975, Evolution and geology of western Baffin Bay and Davis Strait, Canada, in: *Canada's Continental Margins and Offshore Petroleum Exploration*, Yorath, C. J., Parker, E. R., and Glass, D. J., eds., *Can. Soc. Petrol. Geol. Mem.* 4, pp. 453–476.

Bendix-Almgreen, S. E., 1975, Fossil fishes from the marine late Palaeozoic of Holm Land–Amdrup Land, North-East Greenland, *Medd. Groenl.*, v. 195(9), 38 pp.

Bendix-Almgreen, S. E., and Peel, J. S., 1974, Early Devonian vertebrates from Hall Land, North Greenland, *Rapp. Groenlands geol. Unders.* 65, p. 13–16.

Berry, W. B. N., Boucot, A. J., Dawes, P. R., and Peel, J. S., 1974, Late Silurian and early Devonian graptolites from North Greenland, *Rapp. Groenlands geol. Unders.* 65, p. 11–13.

Birkelund, T., Perch-Nielsen, K., Bridgwater, D., and Higgins, A. K., 1974, An outline of the geology of the Atlantic coast of Greenland, in: *The Ocean Basins and Margins*, v. 2, Nairn, A. E. M., and Stehli, F. G., eds., New York: Plenum Press, p. 125–159.

Birkenmajer, K., 1972, Tertiary history of Spitsbergen and continental drift, *Acta Geol. Pol.*, v. 22, p. 193–218.

Bridgwater, D., Escher, A., Jackson, G. D., Taylor, F. C., and Windley, B. F., 1973, Development of the Precambrian Shield in West Greenland, Labrador, and Baffin Island, in: *Arctic Geology*, Pitcher, M. G., ed., *Am. Assoc. Petrol. Geol. Mem.* 19, p. 99–116.

Bjerreskov, M., and Poulsen, V., 1973, Ordovician and Silurian faunas from northern Peary Land, North Greenland, *Rapp. Groenl. geol. Unders.* 55, p. 10–14.

Burek, P. J., 1973, Structural deduction of the initial age of the Atlantic rift systems, in: *Implications of Continental Drift to the Earth Sciences*, v. 2, Tarling, D. H., and Runcorn, S. K., eds., London: Academic Press, p. 815–830.

Christie, R. L., 1962a, Geology, southeast Ellesmere Island, District of Franklin (map with marginal notes), *Geol. Surv. Can. Map* 12-1962.

Christie, R. L., 1962b, Geology, Alexandra Fiord, Ellesmere Island, District of Franklin (map with marginal notes), *Geol. Surv. Can. Map* 9-1962.

Christie, R. L., 1967, Bache Peninsula, Ellesmere Island, Arctic Archipelago, *Mem. Geol. Surv. Can.* 347, 63 pp.

Christie, R. L., 1972, Central stable region, in: *The Canadian Arctic Islands and the Mackenzie Region*, Glass, D. J., ed., 24th Int. Geol. Congr., Montreal, Guidebook Field Excursion A66, p. 40–87.

Christie, R. L., and Ineson, J. R., 1979, Precambrian–Silurian geology of the G. B. Schley Fjord region, eastern Peary Land, North Greenland, *Rapp. Groenl. geol. Unders.* 88, p. 63–71.

Clemmensen, L. B., 1979, Notes on the palaeogeographic setting of the Eocambrian tillite-bearing sequence of southern Peary Land, North Greenland, *Rapp. Groenl. geol. Unders.* 88, p. 15–22.

Coles, R. L., Haines, G. V., Hannaford, W., 1976, Large scale magnetic anomalies over western Canada and the Arctic: a discussion, *Can. J. Earth Sci.*, v. 13, p. 790–802.

Collinson, J. D., 1979, The Proterozoic sandstones between Heilprin Land and Mylius-Erichsen Land, eastern North Greenland, *Rapp. Groenl. geol. Unders.* 88, p. 5–10.

Cowie, J. W., 1961a, Contribution to the geology of North Greenland, *Medd. Groenl.*, v. 164(3), 47 pp.

Cowie, J. W., 1961b, The Lower Palaeozoic geology of Greenland, in: *Geology of the Arctic*, v. 1, Raasch, G. O., ed., Toronto: Toronto University Press, p. 160–169.

Cowie, J. W., and Adams, P. J., 1957, The geology of the Cambro-Ordovician rocks of Central East Greenland, Part 1, Stratigraphy and structure, *Medd. Groenl.*, v. 153(1), 193 pp.

Davies, W. E., 1972, Landscape of northern Greenland, *Spec. Rep. Cold Reg. Engng. Lab.* 164, 55 pp.

Dawes, P. R., 1971, The North Greenland fold belt and environs, *Bull. Geol. Soc. Den.*, v. 20, p. 197–239.

Dawes, P. R., 1972: Precambrian crystalline rocks and younger sediments of the Thule district, North Greenland, *Rapp. Groenl. geol. Unders.* 45, p. 10–15.

Dawes, P. R., 1973, The North Greenland fold belt: a clue to the history of the Arctic Ocean Basin and the Nares Strait Lineament, in: *Implications of Continental Drift to the Earth Sciences*, v. 2, Tarling, D. H., and Runcorn, S. K., eds., London: Academic Press, p. 917–939.

Dawes, P. R., 1976a, Precambrian to Tertiary of northern Greenland, in: *Geology of Greenland*, Escher, A., and Watt, W. S., eds., Copenhagen: Geological Survey of Greenland, p. 248–303.

Dawes, P. R., 1976b, Reconnaissance of Eocambrian and Lower Palaeozoic strata in south-western Peary Land, North Greenland, *Rapp. Groenl. geol. Unders.* 80, p. 9–14.

Dawes, P. R., 1976c, 1:500,000 mapping of the Thule district, North-West Greenland, *Rapp. Groenl. geol. Unders.* 80, p. 23–28.

Dawes, P. R., and Bromley, R. G., 1975, Late Precambrian trace fossils from the Thule Group, western North Greenland, *Rapp. Groenl. geol. Unders.* 75, p. 38–42.

Dawes, P. R., and Soper, N. J., 1971, Significance of K/Ar age determinations from northern Peary Land, *Rapp. Groenl. geol. Unders.* 35, p. 60–62.

Dawes, P. R., and Soper, N. J., 1973, Pre-Quaternary history of North Greenland, in: *Arctic Geology*, Pitcher, M. G., ed., *Am. Assoc. Petrol. Geol. Mem.* 19, p. 117–134.

Dawes, P. R., Rex, D. C., and Jepsen, H. F., 1973, K/Ar whole rock ages of dolerites from the Thule district, western North Greenland, *Rapp. Groenl. geol. Unders.* 55, p. 61–66.

Denham, L. R., 1974, Offshore geology of northern West Greenland (69° to 75°N), *Rapp. Groenl. geol. Unders.* 63, 24 pp.

Dibner, V. D., 1957, The geological structure of Franz Josef Land, in: *Geologiia Sovetskoi Arktiki: Nauchno-Issled*, v. 81, Markov, F. G., and Nalivkin, D. V., eds., Leningrad: Institut Geologii Arktiki Trudy, p. 11–20 [in Russian].

Dunbar, C. O., 1962, Faunas and correlation of the late Paleozoic rocks of Northeast Greenland. III. Brachiopoda, *Medd. Groenl.*, v. 167(6), 14 pp.

Dunbar, C. O., Troelsen, J. [C.], Ross, C. [A.], Ross, J. P., and Norford, B. [S.], 1962, Faunas and correlation of the late Paleozoic rocks of Northeast Greenland. I. General discussion and summary, *Medd. Groenl.*, v. 167(4), 16 pp.

Fahrig, W. E., Irving, E., and Jackson, G. D., 1973, Test of nature and extent of continental drift as provided by study of Proterozoic dike swarms of Canadian Shield, in: *Arctic Geology*, Pitcher, M. G., ed., *Am. Assoc. Petrol. Geol. Mem.* 19, p. 583–586.

Feden, R. H., Fleming, H. S., Phillips, J. D., Massinghill, J. V., and Perry, R. K., 1974, Aeromagnetic study over the Eurasian Basin, Arctic Ocean, *Geol. Soc. Am. Abstr. Progr.*, v. 6(7), p. 730.

Fortier, Y. O., McNair, A. H., and Thorsteinsson, R., 1954, Geology and petroleum possibilities in Canadian Arctic Islands, *Bull. Am. Assoc. Petrol. Geol.* 38, p. 2075–2109.

Fränkl, E., 1954, Vorläufige Mitteilung über die Geologie von Kronprins Christians Land (NE-Grönland, zwischen 80–81°N und 19–23°W), *Medd. Groenl.*, v. 116(2), 85 pp.

Fränkl, E., 1955, Weitere Beiträge zur Geologie von Kronprins Christians Land (NE-Grönland, zwischen 80° und 80°30′N), *Medd. Groenl.*, v. 103(7), 35 pp.

Gjelsvik, T., 1963, Remarks on the structure and composition of the Sverrefjellet volcano, Bockfjorden, Vestspitsbergen, *Nor. Polarinst. Årbök*, 1962, p. 50–54.

Grant, A. C., 1975, Geophysical results from the continental margin off southern Baffin Island, in: *Canada's Continental Margins and Offshore Petroleum Exploration*, Yorath, C. J., Parker, E. R., and Glass, D. J., eds., *Can. Soc. Petrol. Geol. Mem.* 4, p. 411–431.

Grönwall, K. A., 1916, The marine Carboniferous of North-east Greenland and its brachiopod fauna, *Medd. Groenl.*, v. 43, p. 509–618.

Haines, G. V., Hannaford, W., and Serson, P. H., 1970, Magnetic anomaly maps of the Nordic countries and the Greenland and Norwegian Seas, *Publ. Dominion Observ. Ottawa*, v. 39, p. 119–149.

Håkansson, E., 1979, Carboniferous to Tertiary development of the Wandel Sea Basin, eastern North Greenland, *Rapp. Groenl. geol. Unders.* 88, p. 73–83.

Haller, J., 1956, Die Strukturelemente Ostgrönlands zwischen 74° und 78°N, *Medd. Groenl.*, v. 154(2), 27 pp.

Haller, J., 1961, The Carolinides: an orogenic belt of late Precambrian age in Northeast Greenland, in: *Geology of the Arctic*, v. 1, Raasch, G. O., ed., University of Toronto Press, p. 155–159.

Haller, J., 1970, Tectonic map of East Greenland (1 : 500,000). An account of tectonism, plutonism, and volcanism in East Greenland, *Medd. Groenl.*, v. 171(5), 286 pp.

Haller, J., 1971, *Geology of the East Greenland Caledonides*, London: Interscience Publishers, 413 pp.

Haller, J., and Kulp, J. L., 1962, Absolute age determinations in East Greenland, *Medd. Groenl.*, v. 171(1), 77 pp.

Hamilton, W., 1970, The Uralides and the motion of the Russian and Siberian platforms, *Bull. Geol. Soc. Am.*, v. 81, p. 2553–2576.

Harland, W. B., 1965, Tectonic evolution of the Arctic–North Atlantic region, *Phil. Trans. R. Soc. London*, v. 258, p. 59–75.

Harland, W. B., 1973, Mesozoic geology of Svalbard, in: *Arctic Geology*, Pitcher, M. G., ed., *Am. Assoc. Petrol. Geol. Mem.* 19, p. 135–148.

Harland, W. B., and Gayer, R. A., 1972, The Arctic Caledonides and earlier oceans, *Geol. Mag.*, v. 109, p. 289–384.

Heezen, B. C., and Tharp, M., 1975, Map of the Arctic region (bathymetry), Washington, D.C.: American Geographical Society.

Henderson, G., 1973, The geological setting of the West Greenland basin in the Baffin Bay region, *Geol. Surv. Can. Pap.* 71-23, p. 521–544.

Henderson, G., 1976, Petroleum Geology, in: *Geology of Greenland*, Escher, A., and Watt, W. S., eds., Copenhagen: Geological Survey of Greenland, p. 488–505.

Henriksen, N., and Jepsen, H. F., 1970, K/Ar age determinations on dolerites from southern Peary Land, North Greenland, *Rapp. Groenl. geol. Unders.* 28, p. 55–58.

Henriksen, N., and Higgins, A. K., 1976, East Greenland Caledonian fold belt, in: *Geology of Greenland*, Escher, A., and Watt, W. S., eds., Copenhagen: Geological Survey of Greenland, p. 180–246.

Henriksen, N., and Peel, J. S., 1976, Cambrian–Early Ordovician stratigraphy in south-western Washington Land, western North Greenland, *Rapp. Groenl. geol. Unders.* 80, p. 17–22.

Higgins, A. K., 1976, Pre-Caledonian metamorphic complexes within the southern part of the East Greenland Caledonides, *Q. J. Geol. Soc. London*, v. 132, p. 289–305.

Hoel, A., 1914, Nouvelles observations sur le district volcanique du Spitsberg du nord, *Vidensk. Selsk. Skr. Mat. Nat.*, v. 1(9), Kristiania.

Hood, P. J., and Bower, M. E., 1975a, Northern Baffin Bay: low-level aeromagnetic profiles obtained in 1974, *Geol. Surv. Can. Pap.* 75-1, Pt. A, p. 89–93.

Hood, P. [J.], and Bower, M. [E.], 1975b, Aeromagnetic reconnaissance of Davis Strait and adjacent areas, in: *Canada's Continental Margins and Offshore Petroleum Exploration*,

Yorath, C. J., Parker, E. R., and Glass, D. J., eds., *Can. Soc. Petrol. Geol. Mem.* 4, p. 433–451.

Hurst, J. M., 1979, Uppermost Ordovician and Silurian geology of north-west Peary Land, North Greenland, *Rapp. Groenl. geol. Unders.* 88, p. 41–49.

Hurst, J. M., 1980a, Paleogeographic and stratigraphic differentiation of Silurian carbonate buildups and biostromes of North Greenland, *Bull. Am. Assoc. Petrol. Geol.* v. 64, p. 527–548.

Hurst, J. M., 1980b, Silurian stratigraphy and facies distribution in Washington Land and western Hall Land, North Greenland, *Bull. Groenl. geol. Unders.* 138 (in press).

Jepsen, H. F., 1971, The Precambrian, Eocambrian and early Palaeozoic stratigraphy of the Jørgen Brønlund Fjord area, Peary Land, North Greenland, *Bull. Groenl. geol. Unders.* 96 [also *Medd. Groenl.*, v. 192(2)], 42 pp.

Jepsen, H. F., and Kalsbeek, F., 1979, Igneous rocks in the Proterozoic platform of eastern North Greenland, *Rapp. Groenl. Geol. Unders.*, v. 88, p. 11–14.

Johnson, G. L., 1975, The morphology and structure of the Norwegian–Greenland Sea, D.Phil. Thesis, University of Copenhagen, 157 pp., 95 figs.

Johnson, G. L., and Eckhoff, O. B., 1966, Bathymetry of the North Greenland Sea, *Deep Sea Res.*, v. 13, p. 1161–1173.

Johnson, G. L., and Heezen, B. C., 1967, The morphology and evolution of the Norwegian–Greenland Sea, *Deep Sea Res.*, v. 14, p. 755–771.

Johnson, G. L., and Vogt, P. R., 1973, Marine geology of the Atlantic Ocean north of the Arctic Circle, in: *Arctic Geology*, Pitcher, M. G., ed., *Am. Assoc. Petrol. Geol. Mem.* 19, p. 161–170.

Johnson, G. L., McMillan, N. J., Rasmussen, M., Campsie, J., and Dittmer, F., 1975, Sedimentary rocks dredged from the southwest Greenland continental margin, in: *Canada's Continental Margins and Offshore Petroleum Exploration*, Yorath, C. J., Parker, E. R., and Glass, D. J., eds., *Can. Soc. Petrol. Geol. Mem.* 4, p. 391–409.

Keen, C. E., and Barrett, D. L., 1973, Structural characteristics of some sedimentary basins in northern Baffin Bay, *Can. J. Earth Sci.*, v. 10, p. 1267–1278.

Keen, C. E., Barrett, D. L., Manchester, K. S., and Ross, D. I., 1972, Geophysical studies in Baffin Bay and some tectonic implications, *Can. J. Earth Sci.*, v. 9, p. 239–256.

Keen, M. J., Loncarevic, B. D., and Ewing, G. N., 1971, Continental margin of eastern Canada: Georges Bank to Kane Basin, in: *The Sea*, v. 4, Maxwell, A. E., ed., New York: Wiley-Interscience, p. 251–291.

Keen, M. J., Johnson, J., and Park, I., 1972, Geophysical and geological studies in eastern and northern Baffin Bay and Lancaster Sound, *Can. J. Earth Sci.*, v. 9, p. 689–708.

Kellogg, H. E., 1975, Tertiary stratigraphy and tectonism in Svalbard and continental drift, *Bull. Am. Assoc. Petrol. Geol.*, v. 59, p. 465–485.

Kerr, J. W., 1967a, Nares submarine rift valley and the relative rotation of North Greenland, *Bull. Can. Petrol. Geol.*, v. 15, p. 483–520.

Kerr, J. W., 1967b, Stratigraphy of central and eastern Ellesmere Island, Arctic Canada, Pt. 1, Proterozoic and Cambrian, *Geol. Surv. Can. Pap.* 67-27, 63 pp.

King, E. R., Zietz, I., and Alldredge, L. R., 1966, Magnetic data on the structure of the central Arctic region, *Bull. Geol. Soc. Am.*, v. 77, p. 619–646.

Koch, L., 1929a, Stratigraphy of Greenland, *Medd. Groenl.*, v. 73(2), p. 205–320.

Koch, L., 1929b, The geology of the south coast of Washington Land, *Medd. Groenl.*, v. 73(1), 39 pp.

Koch, L., 1933, The geology of Inglefield Land, *Medd. Groenl.*, v. 73(1), 38 pp.

Kummel, B., 1953, Middle Triassic ammonites from Peary Land, *Medd. Groenl.*, v. 127(1), 21 pp.

Larsen, O., and Dawes, P. R., 1974, K/Ar and Rb/Sr age determinations on Precambrian crystalline rocks in the Inglefield Land–Inglefield Bredning region, Thule district, western North Greenland, *Rapp. Groenl. geol. Unders.* 66, p. 4–8.

Lowell, J. D., 1972, Spitsbergen Tertiary orogenic belt and the Spitsbergen Fracture Zone, *Bull. Geol. Soc. Am.*, v. 83, p. 3091–3102.

Mayr, U., 1976, Middle Silurian reefs in southern Peary Land, North Greenland, *Bull. Can. Petrol. Geol.*, v. 24, p. 440–449.

Morley, L. W., McLaren, A. C., and Charbonneau, B. W., 1967, Magnetic anomaly map of Canada, *Geol. Surv. Can. Map* 1255A.

Nathorst, A. G., 1911, Contributions to the Carboniferous flora of North-eastern Greenland, *Medd. Groenl.*, v. 43, p. 337–346.

Norford, B. S., 1972, Silurian stratigraphic sections at Kap Tyson, Offley ∅ and Kap Schuchert, Northwestern Greenland, *Medd. Groenl.*, v. 195(2), 40 pp.

O'Connor, B., 1979, The Portfjeld Formation (?early Cambrian) of eastern North Greenland, *Rapp. Groenl. geol. Unders.* 88, p. 23–28.

Ostenso, N. A., and Pew, J. A., 1968, Sub-bottom seismic profile off the east coast of Greenland, in: *Arctic Drifting Stations*, Sater, J. E., coord., Washington: Arctic Institute of North America, p. 345–363.

Ostenso, N. A., and Wold, R. J., 1971, Aeromagnetic survey of the Arctic Ocean: Techniques and interpretations, *Mar. Geophys. Res.*, v. 1, p. 178–219.

Peacock, J. D., 1956, The geology of Dronning Louise Land, N.E. Greenland, *Medd. Groenl.*, v. 137(7), 38 pp.

Peacock, J. D., 1958, Some investigations into the geology and petrography of Dronning Louise Land, N.E. Greenland, *Medd. Groenl.*, v. 157(4), 139 pp.

Pedersen, S. A. S., 1979, Structural geology of central Peary Land, North Greenland, *Rapp. Groenl. geol. Unders.* 88, p. 55–62.

Peel, J. S., 1979, Cambrian–Middle Ordovician stratigraphy of the Adams Gletscher region, south-west Peary Land, North Greenland, *Rapp. Groenl. geol. Unders.* 88, p. 29–39.

Peel, J. S., and Christie, R. L., 1975, Lower Palaeozoic stratigraphy of southern Peary Land, eastern North Greenland, *Rapp. Groenl. geol. Unders.* 75, p. 21–25.

Pelletier, B. R., 1966, Development of submarine physiography in the Canadian Arctic and its relation to crustal movements, *R. Soc. Can. Spec. Publ.* 9, p. 77–101.

Phillips, J. D., Feden, R., Fleming, H. S., and Tapscott, C., 1980, Aeromagnetic studies of the Greenland/Norwegian Sea and Arctic Ocean, *Bull. Geol. Soc. Am.* (in preparation).

Pitman, W. C., and Talwani, M., 1972, Sea-floor spreading in the North Atlantic, *Bull. Geol. Soc. Am.*, v. 83, p. 619–646.

Poulsen, C., 1927, The Cambrian, Ozarkian, and Canadian faunas of Northwest Greenland, *Medd. Groenl.*, v. 70(1), p. 233–343.

Poulsen, C., 1958, Contribution to the palaeontology of the Lower Cambrian Wulff River Formation, *Medd. Groenl.*, v. 162(2), 25 pp.

Poulsen, V., 1964, Contribution to the Lower and Middle Cambrian paleontology and stratigraphy of Northwest Greenland, *Medd. Groenl.*, v. 164(6), 105 pp.

Poulsen, V., 1966, An occurrence of Lower Palaeozoic rocks within the Precambrian terrain near Sukkertoppen, *Rapp. Groenl. geol. Unders.* 11, p. 26.

Poulsen, V., 1969, An Atlantic Middle Cambrian fauna from North Greenland, *Lethaia*, v. 2, p. 1–14.

Poulsen, V., 1974, Olenellacean trilobites from eastern North Greenland, *Bull. Geol. Soc. Den.*, v. 23, p. 79–101.

Poulsen, V., 1978, The Precambrian–Cambrian boundary in parts of Scandinavia and Greenland, *Geol. Mag.*, v. 115, p. 131–136.

Riddihough, R. P., Haines, G. V., and Hannaford, W., 1973, Regional magnetic anomalies of the Canadian Arctic, *Can. J. Earth Sci.*, v. 10, p. 157–163.

Ross, C. A., and Dunbar, C. O., 1962, Faunas and correlation of the late Paleozoic rocks of northeast Greenland. II. Fusulinidae, *Medd. Groenl.*, v. 167(5), 55 pp.

Ross, D. I., and Falconer, R. K. H., 1975, Geological studies of Baffin Bay, Davis Strait, and adjacent continental margins, *Geol. Surv. Can. Pap.* 75-1, Pt. A, p. 181–183.

Sobczak, L. W., and Stephens, L. E., 1974, The gravity field of northeastern Ellesmere Island, part of northern Greenland and Lincoln Sea, *Can. Dept. Energy Mines Resources, Gravity Map Series* 114, 9 pp.

Steiger, R. H., Harnik-Šoptrajanova, G., Zimmermann, E., and Henriksen, N., 1976, Isotopic age and metamorphic history of the banded gneiss at Danmarkshavn, East Greenland, *Contrib. Mineral. Petrol.*, v. 57, p. 1–24.

Sykes, L. R., 1965, The seismicity of the Arctic, *Bull. Seism. Soc. Am.*, v. 55, p. 501–518.

Talwani, M., and Eldholm, O., 1977, Evolution of the Norwegian–Greenland Sea, *Bull. Geol. Soc. Am.*, v. 88, p. 969–999.

Trettin, H. P., 1971, Geology of Lower Paleozoic formations, Hazen Plateau and southern Grant Land mountains, Ellesmere Island, Arctic Archipelago, *Geol. Surv. Can. Bull.* 203, 134 pp.

Trettin, H. P., Frisch, T. O., Sobczak, L. W., Weber, J. R., Niblett, E. R., Law, L. K., DeLaurier, J. M., and Whitham, K., 1972, The Innuitian Province, in: *Variations in Tectonic Styles in Canada*, Price, R. A., and Douglas, R. J. W., eds., *Geol. Assoc. Can. Spec. Pap.* 11, p. 83–179.

Troedsson, G. T., 1926, On the Middle and Upper Ordovician faunas of northern Greenland. I. Cephalopods, *Medd. Groenl.*, v. 71, 157 pp.

Troedsson, G. T., 1928, On the Middle and Upper Ordovician faunas of northern Greenland. II, *Medd. Groenl.*, v. 72(1), 197 pp.

Troelsen, J. C., 1949, Contributions to the geology of the area round Jørgen Brønlunds Fjord, Peary Land, North Greenland, *Medd. Groenl.*, v. 149(2), 29 pp.

Troelsen, J. C., 1950*a*, Contributions to the geology of Northwest Greenland, Ellesmere Island, and Axel Heiberg Island, *Medd. Groenl.*, v. 149(7), 86 pp.

Troelsen, J. [C.], 1950*b*, Geology, in: *A Preliminary Account of the Danish Pearyland Expedition, 1948–9*, J. Arc. Inst. N. Amer., v. 3, p. 6–8.

Troelsen, J. C., 1956, The Cambrian of North Greenland and Ellesmere Island, 20th Int. Geol. Congr. Mexico City, Section 3, Book 1, p. 71–90.

Volkov, P., 1961, New exploration of the bottom topography in the Greenland Sea, *Morsk. Flot*, v. 3, p. 35–37 [in Russian].

Vogt, P. R., and Ostenso, N. A., 1970, Magnetic and gravity profiles across the Alpha Cordillera and their relation to Arctic sea-floor spreading, *J. Geophys. Res.*, v. 75, p. 4925–4937.

Vogt, P. R., Ostenso, N. A., and Johnson, G. L., 1970, Magnetic and bathymetric data bearing on sea-floor spreading north of Iceland, *J. Geophys. Res.*, v. 75, p. 903–920.

Vogt, P. R., Taylor, P. T., Kovacs, L. C., and Johnson, G. L., 1979, Detailed aeromagnetic investigation of the Arctic Basin, *J. Geophys. Res.*, v. 84, p. 1071–1089.

Weidick, A., 1972, Holocene shore lines and glacial stages in Greenland—an attempt at correlation, *Rapp. Groenl. geol. Unders.* 41, 39 pp.

Weidick, A., 1976, Glaciation and the Quaternary of Greenland, in: *Geology of Greenland*, Escher, A., and Watt, W. S., eds., Copenhagen: Geological Survey of Greenland, p. 430–458.

Wilcox, L. E., Voss, J. T., and Pals, P. F., 1975, Regional gravity and elevation maps of Greenland, Dept. of Defense, St. Louis Air Force Station DMAAC/TP 75-001, 8 pp., 5 maps.

Wilson, J. T., 1963, Continental drift, *Sci. Am.*, v. 208, p. 86–100.

Chapter 6

THE GEOLOGY OF SVALBARD, THE WESTERN PART OF THE BARENTS SEA, AND THE CONTINENTAL MARGIN OF SCANDINAVIA

K. Birkenmajer

Institute of Geology
Polish Academy of Sciences
Cracow, Poland
and Institute of Historical Geology and Paleontology
Copenhagen, Denmark

I. INTRODUCTION

The Svalbard Archipelago is located on the northwestern corner of the Barents Shelf (Fig. 1). Its position at the upturned edge of the Eurasian continental plate, opposite the Greenland continental block, and its long geological record, with evidence of major Precambrian, Paleozoic, and Cenozoic deformation, offer a potential for intercontinental correlation. Investigations of Svalbard have provided evidence of a long history of vertical and horizontal movement. These data supplemented with knowledge of the topography and structure of the neighboring sea floor helps us to reconstruct the evolution of the North Atlantic and Arctic Ocean Basins.

Svalbard is the collective name of a group of islands situated between 74° and 81° N latitude and between 10° and 35° E longitude, which have a total area of 62,000 km². Spitsbergen is the largest of the islands (39,000 km²), Nordaustlandet, Edgeøya, Barentsøya, Prins Karls Forland, Kong Karls Land, Kvitøya, and Bjørnøya (Bear Island) are successively smaller main islands, while there are hundreds of adjacent small islands and islets. The major islands

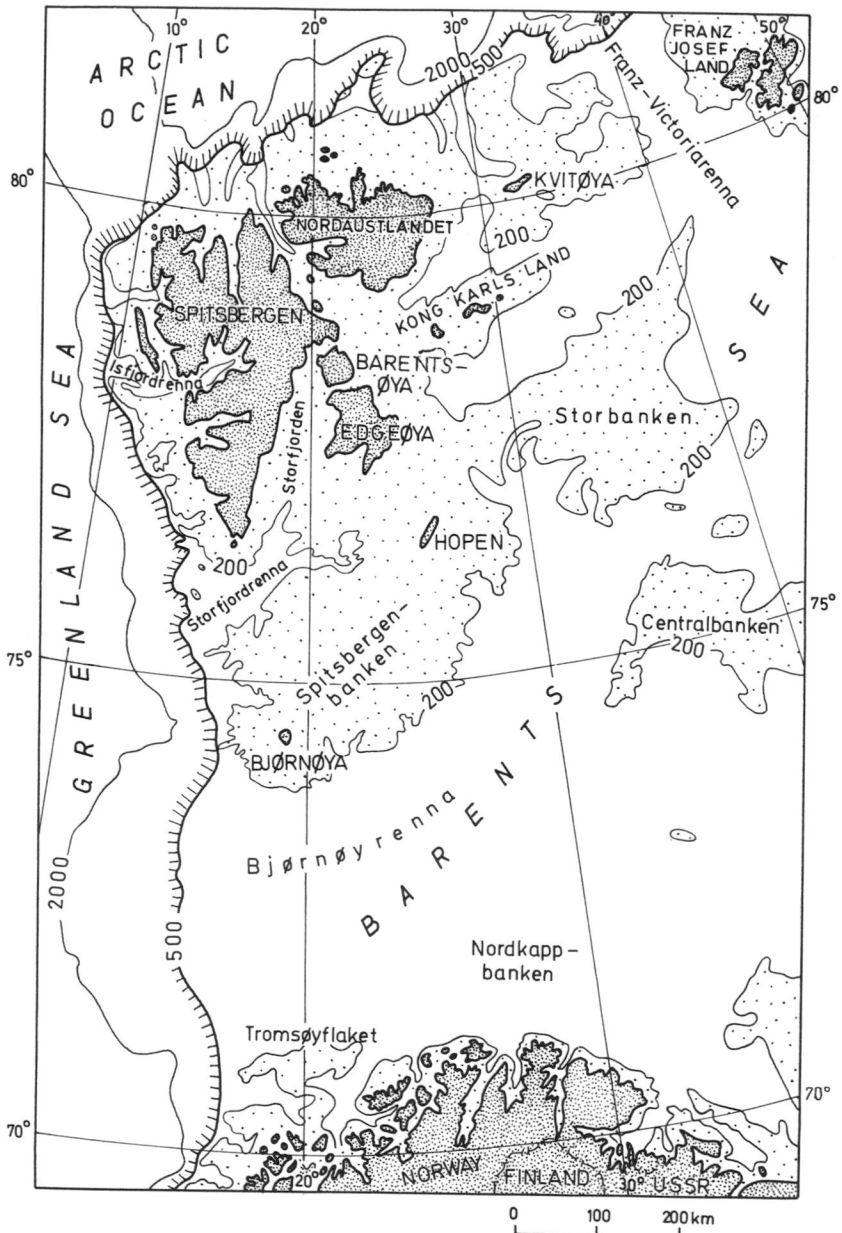

Fig. 1. Svalbard and the western Barents Shelf. Morphologic features of the sea bottom based mainly on Åm (1975a). Submarine contours in meters; barbs on 500 m isobath indicate margin of continental shelf.

are cut by long, deep fjords that expose excellent geological sections in coast profiles and on the steep mountain slopes. About two thirds of the land area is covered by glaciers and small ice caps, with the largest ice plateaus in Nordaustlandet and northeast Spitsbergen (Ny Friesland–Olav V Land).

A ragged mountain chain, with sharp peaks rising to 1000–1400 m, stretches along the west coast of Spitsbergen and gives the island its name. Flat-topped mountains and dissected plateaus 500–600 m high are characteristic for the southeastern part of the archipelago. The highest mountain in Svalbard is Newtontoppen (1717 m), which projects as a nunatak through the ice cap in northeast Spitsbergen.

The continental shelf north of Svalbard is approximately 70 km wide. It narrows to less than 18 km along the northwest tip of Spitsbergen close to the submarine Yermak Marginal Plateau (Fig. 22). Along the west coast of Spitsbergen, the shelf is from 25 to 45 km wide, its margin being very close to the Spitsbergen Fracture Zone and the Knipovitch Ridge (Fig. 22), two important ocean floor features of the northern Greenland Sea.

The margin of the Barents Shelf continues due south past Bjørnøya, and on approaching Norway, at about 70°N, it turns rapidly southwestward. The width of the Norwegian continental shelf is least (40–50 km) at 70°N, but increases to more than 200 km at 67°N on approaching the submarine Vøring Marginal Plateau.

II. CALEDONIAN FOLDBELT OF SVALBARD

The Caledonian Foldbelt of Svalbard crops out mainly in the west and north-central parts of its largest island, Spitsbergen, in Nordaustlandet and adjacent smaller islands (Fig. 2), and an isolated occurrence is known from Bjørnøya situated halfway between Spitsbergen and Norway, close to the margin of the Eurasian continental plate. The width of the belt (between Spitsbergen and Nordaustlandet) exceeds 400 km, the length (between Nordaustlandet and Bjørnøya) is about 700 km. To the north the belt abruptly terminates at the margin of the Barents Shelf, to the south it is separated from the Scandinavian Caledonides by a vast expanse of the Barents Shelf with its Cenozoic, Mesozoic, and possibly late Paleozoic platform cover. The East Greenland Caledonides are separated from those of Svalbard by the young oceanic crust of the Atlantic Ocean (Greenland Sea).

The Svalbard and Scandinavian Caledonian belts occur within the same Eurasian continental plate, but show obvious differences in the stratigraphic column, in the ages of major disturbances, and in the polarity trends of the geosyncline. The East Greenland and Svalbard Caledonian belts have much

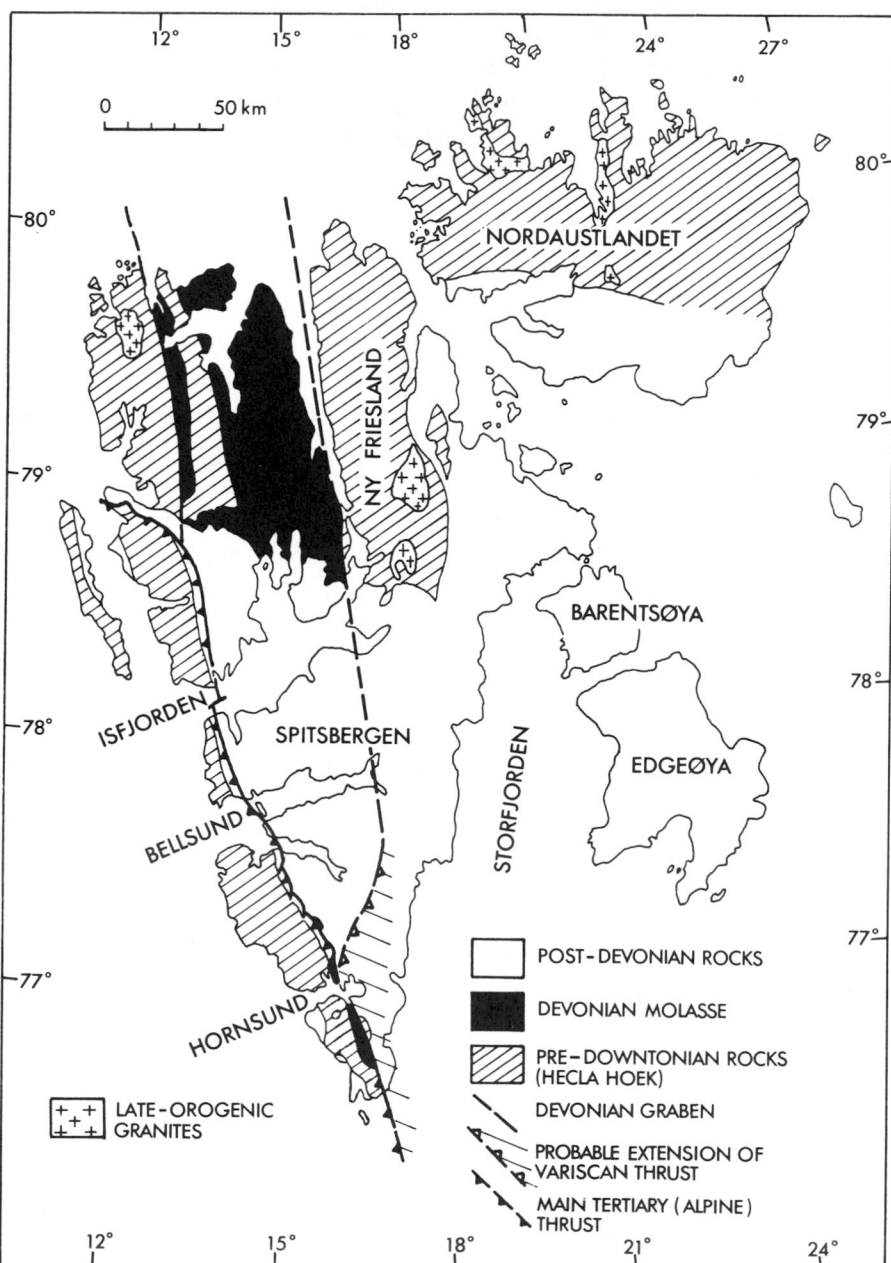

Fig. 2.   Major exposures of the Caledonian mountain system in Svalbard.

more in common, and seem to be parts of a western branch of the circum-Atlantic Caledonides. This becomes apparent on late Cretaceous palinspastic reconstructions based on the relative positions of the North America–Greenland plate versus the Eurasian plate (e.g., Harland, 1967, 1969, 1973*a*; Harland and Gayer, 1972; Birkenmajer, 1972*a*, 1975*b*) (Fig. 3).

A. Main Precambrian through Early Paleozoic Successions

The Caledonides of Svalbard comprise rocks of Precambrian through Middle Ordovician age, usually referred to as the *Hecla Hoek Succession*. The stratigraphic column includes between 15 and 20 km of metasediments and sediments usually with subordinate metavolcanics and volcanics, which formed in geosynclinal basins of westward polarity. Major unconformities help define these basins, especially in marginal parts of the geosynclinal trough.

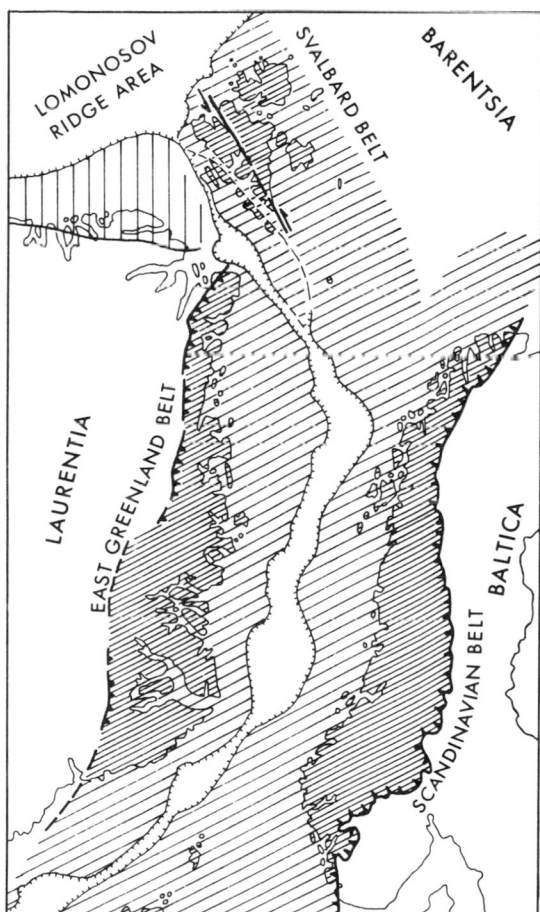

Fig. 3. Palinspastic reconstruction of the circum-Atlantic Caledonian mountain system (obliquely shaded) prior to continental drift in late Cretaceous time. Heavy dentated lines denote edges of Caledonian thrusts; vertical shading indicates Variscan mountain system; thin dentated lines denote margins of continental blocks. (After Birkenmajer, 1975*b*.)

Exposures along the west coast of Spitsbergen (between Sørkapp and Kongsfjorden, within the Tertiary thrust–fold belt), represent the western parts of the Hecla Hoek Geosyncline, while the central part is found in north-central Spitsbergen (Ny Friesland and Olav V Land) and the eastern parts may be seen in Nordaustlandet.

1. *South Spitsbergen*

The Hecla Hoek Succession on the west coast of Spitsbergen is best known from the area between Bellsund and Sørkapp (Major and Winsnes, 1955; Birkenmajer, 1958, 1959, 1960*a,b*, 1972*c*, 1975*b*; Birkenmajer and Narebski, 1960; Smulikowski, 1965, 1968; Winsnes, 1965; Hjelle, 1969). The thickness of the Precambrian through Lower Ordovician column totals between 15 and 17 km in the Hornsund area (Fig. 7), and the succession is characterized by several important unconformities (Figs. 4–6).

The *Isbjørnhamna Group* (max. 2850 m) includes garnet-mica schists (Skoddefjellet Fm. and Revdalen Fm.) intercalated in the middle part of the Group with gray marble (Ariekammen Fm.). An unconformity may occur below the overlying Eimfjellet Group.

The *Eimfjellet Group* (2500 to 4650 m) begins with light well-bedded quartzites with subordinate chloritic schists (Gulliksenfjellet Fm.), followed by alternating quartzite and schist with concordant amphibolite intercalations (Steinvikskardet Fm.). The upper part of the group is regionally differentiated into the Skålfjellet Subgroup on the east, and the Vimsodden Subgroup on the west. The Skålfjellet Subgroup (1400 m) includes layered amphibolites alternating with greenschists (Torbjørnsenfjellet Fm. and Brattegga Fm.), with a granitization zone (Gangpasset Fm.) in the middle. A tilloid intercalation occurs near the granitization zone. The Vimsodden Subgroup (3550 m) starts with phyllites and schists, intercalated near the base with quartzite and higher up with marble and concordant amphibolite layers (Nottinghambukta Fm.). Quartzite tilloids (subaqueous slump breccia type) and rhyolite conglomerate occur near the base, and albitic gneiss intercalations occur near the top of the Nottinghambukta Formation. The upper part of the Vimsodden Subgroup (Elveflya Fm.) consists of quartz schists, phyllite, and slate with three tilloid intercalations. Green quartzite and amphibolite–greenschist intercalations occur in its upper part in the eastern area.

The *Deilegga Group* (3500 m) is separated from the Eimfjellet Group by an unconformity. Its lower part (Tonedalen Fm.) begins with quartzite conglomerate followed by dolostone and by a thick green phyllite–slate complex; thin dolostone and gypsiferous shale appear in its middle part (Bergnova Fm.) and light quartzite intercalations in its upper part (Bergskardet Fm.). A major unconformity separates the Deilegga and Sofiebogen groups.

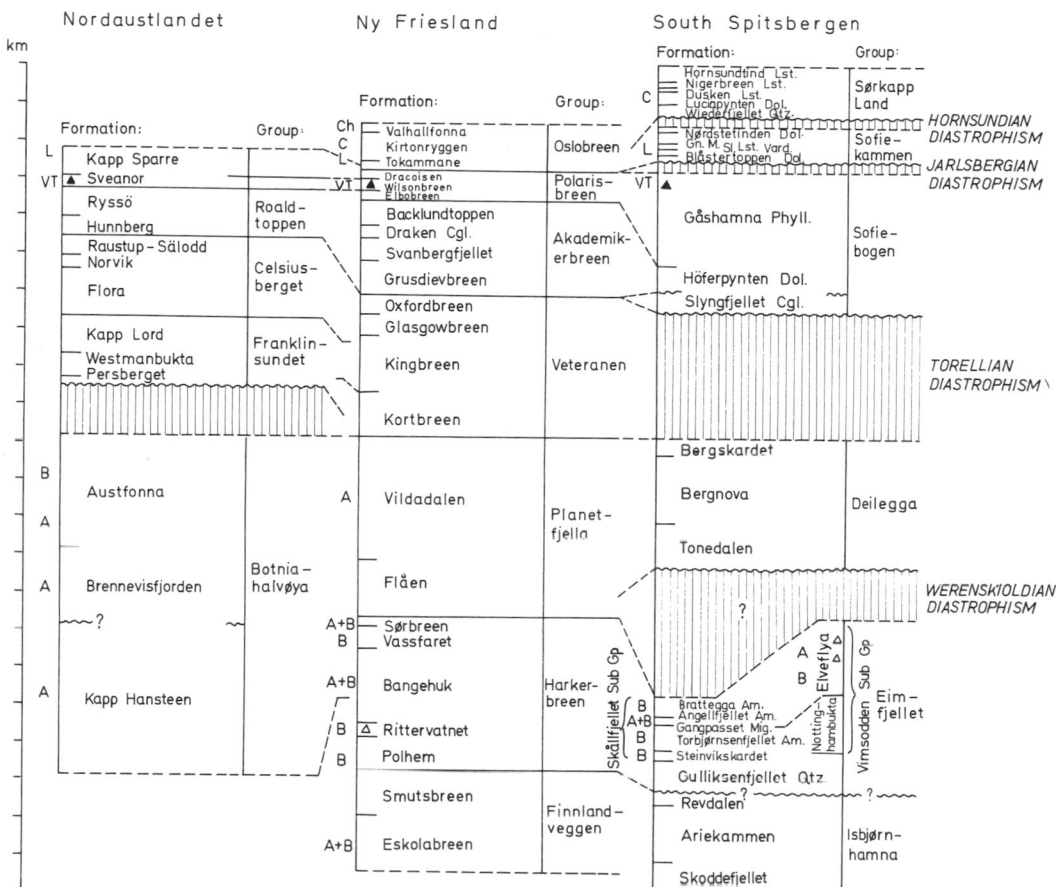

Fig. 4.   Correlation of the Hecla Hoek rocks in Svalbard (after Birkenmajer, 1975b). The thicknesses used are maximum values from Figs. 7–9. (A) Acid igneous activity; (B) basic igneous activity. Faunas: (Ch) Champlainian, (C) Canadian, (L) Lower Cambrian; (VT) Varangian tillites. Tillites marked by full triangles, metatilloids by open triangles. Wavy lines denote unconformities, vertical shading denotes erosional breaks.

The *Sofiebogen Group* (ca. 3700 m) begins with thick conglomerates (Slyngfjellet Fm.) resting unconformably upon various formations of the Deilegga Group or even upon the Vimsodden Subgroup. A minor unconformity is present between the conglomerate and the Höferpynten Formation, the latter consisting of limestone in the lower part followed by dolostone with cherts, sedimentary breccias and stromatolitic structures, and by oolitic-pisolitic dolostone. The dolostone passes upward into a thick phyllite–slate complex (Gåshamma Fm.) with subordinate quartzite, dolostone, and limestone intercalations. At Bellsund, Isfjorden (Kapp Linné), and St. Jonsfjorden, tillites have been found, presumably in the upper part of the formation.

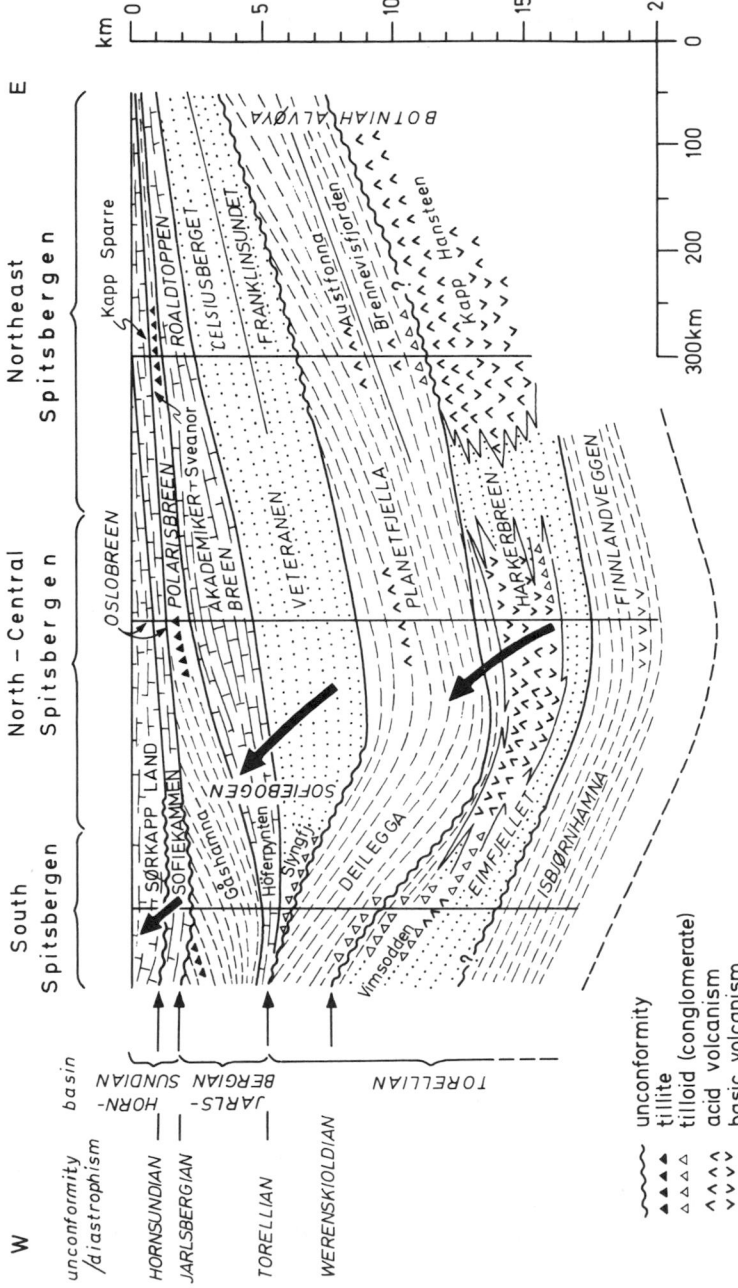

Fig. 5. Reconstruction of the Hecla Hoek geosynclinal trough in Svalbard (after Birkenmajer, 1975b). Width of the trough assumed to be twice that of the present Caledonian Foldbelt. Thick arrows denote migration of basin axis (polarity of geosyncline). The thicknesses used are maximum values from Figs. 7–9. West and east margins of geosyncline unknown. Northeast Spitsbergen = Nordaustlandet.

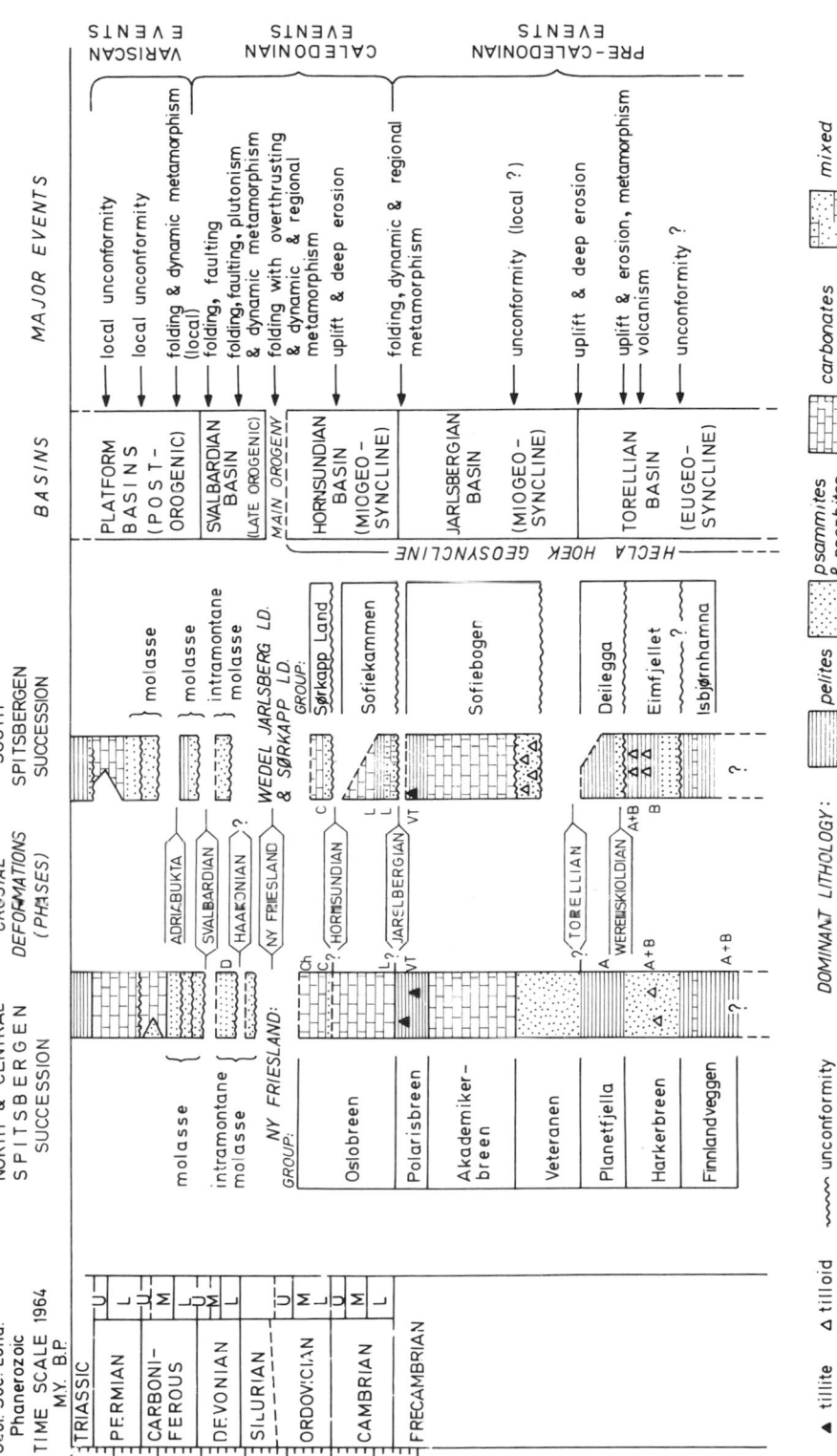

Fig. 6. Major pre-Caledonian, Caledonian, and Variscan events in Spitsbergen (after Birkenmajer, 1975b). (D) Downtonian fish. Other explanations as in Fig. 5.

Supergroup	Group		Formation	
Hornsund	Sørkapp Land		Hornsundtind Lst. { Tsjebysjovfjellet Lst. Mbr	400m
			Rasstupet Lst. Mbr	80m
			Nigerbreen Lst.	120m
			Dusken Lst.	100m
			Luciapynten Dol.	400m
			Wiederfjellet Quartzite	300m
	Sofiekammen		Nørdstetinden Dol.	150m
			Gnålberget Marble	250 – 300m
			Slaklidalen Lst.	10 – 120m
			Vardepiggen	130 – 215m
			Blåstertoppen Dol.	100 – 150m
	Sofiebogen		Gåshamna Phyllite	1,500–2,500m
			Höferpynten Dol.	120 – 710m
			Slyngfjellet Cgl.	10 – 500m
	Deilegga		Bergskardet	+500m
			Bergnova	1,800m
			Tonedalen	+1,200m
Torellbreen	Eimfjellet	Skål-fjellet Sub Gp.	Brattegga Amphibolite	300 – 500m
			Angellfjellet Amphibolite	0 – 200m
			Gangpasset Migmatite	0 – 100m
			Torbjørnsenfjellet Amphibolite	350 – 600m
		Vims- odden Sub Gp.	Steinvikskardet	100 – 250m	Elveflya 2,000m
			Gulliksenfjellet Quartzite	500 – 850m	Nottinghambukta 1,550m
	Isbjørnhamna		Revdalen	250 – 350m
			Ariekammen	500 – 1,500m
			Skoddefjellet	+1,000m

Fig. 7. Hecla Hoek Succession in south Spitsbergen, Hornsund area (revised scheme based on Major and Winsnes, 1955; Birkenmajer, 1958, 1959; Birkenmajer and Narebski, 1960). From Birkenmajer, 1975b.)

The *Sofiekammen Group* (935 m) begins with arenaceous–dolomitic rocks (Blåstertoppen Fm.) lying unconformably upon the Gåshamna Phyllite, and containing intercalations of sedimentary breccias, shale, and Olenellid-rich limestone. Together with the overlying Olenellid-rich dark shale with sedimentary breccia and limestone intercalations (Vardepiggen Fm.) these two formations represent the basal part of the Lower Cambrian. There follow limestones (Slaklidalen Fm.) with trilobites indicative of the upper part of the Lower Cambrian, overlain by unfossiliferous limestone and marble (Gnålberget Fm.) and finally by massive and platy dolostone (Nørdstetinden Fm.) with scarce *Lingulella*. A Middle or Middle–Upper Cambrian age is tentatively suggested for the upper two formations.

The *Sørkapp Land Group* (1400 m) is separated by an unconformity from the Sofiekammen Group. Its basal quartzite sandstone (Wiederfjellet Fm.) rests upon various formations of the latter group, often filling karstlike pockets in dolostone or limestone. Higher up there follows banded dolostone often with chert intercalations (Luciapynten Fm.) overlain by a thick sequence of limestones that are laminated in the lower part (Dusken Fm.), contain chert intercalations in the middle (Nigerbreen Fm.), and are massive in the upper part (Hornsundtind Fm.). The upper two formations have yielded Canadian fossils comparable with those of Beekmantown Group.

2. *North-Central Spitsbergen*

The Hecla Hoek Succession of Ny Friesland and Olav V Land (southeast of Ny Friesland) is representative of the north-central part of the Spitsbergen Caledonides (Harland and Wilson, 1956; Wilson, 1958, 1961; Harland, 1959, 1960, 1969; Gobbett and Wilson, 1960; Wilson and Harland 1964; Gayer and Wallis, 1966; Harland *et al.*, 1966; Vallance and Fortey, 1968; Fortey and Bruton, 1973; Fortey, 1974, 1975). Its thickness approximates 19 to 20 km (Fig. 8). Unlike south Spitsbergen, the succession here is claimed to be devoid of major unconformities, although Krasilshchikov (1973, Fig. 14) believes that a substantial break exists between the Harkerbreen and the Planetfjella Groups.

The *Finnlandveggen Group* (2700 m) is characterized in the lower part (Eskolabreen Fm.) by gneissose feldspathites associated with coarse-grained semipelites, sometimes alternating with amphibolites that are often of very coarse texture, suggesting a plutonic environment; the upper part (Smutsbreen Fm.) is characterized by semipelites, often garnetiferous, associated near the top with marbles.

The *Harkerbreen Group* (4100 m) begins with pale quartzites and quartz psammites containing rare bands of feldpathic psammite and concordant amphibolite bands (Polhem Fm.), followed by marble, quartzite, and psam-

Supergroup	Group	Formation	
Hinlopenstretet (= Upper Hecla Hoek)	Oslobreen	Valhallfonna	255m
		Kirtonryggen	+750m
		Tokammane	192m
	Polarisbreen	Dracoisen	280m
		Wilsonbreen	280m
		Elbobreen	270m
Lomfjorden (= Middle Hecla Hoek)	Akademiker – breen	Backlundtoppen	360–700m
		Draken Cgl.	25–300m
		Svanbergfjellet	100–625m
		Grusdievbreen	865m
	Veteranen	Oxfordbreen	550m
		Glasgowbreen	540m
		Kingbreen	1,500m
		Kortbreen	1,200m
Stubendorff – breen (= Lower Hecla Hoek)	Planetfjella	Vildadalen	3,250m
		Flåen	1,500m
	Harkerbreen	Sørbreen	+ 250m
		Vassfaret	600m
		Bangehuk	2,000m
		Rittervatnet	350m
		Polhem	900m
	Finnlandveggen	Smutsbreen	1,200m
		Eskolabreen	1,500m

Fig. 8.   Hecla Hoek Succession in north-central Spitsbergen—Ny Friesland and Olav V Land (after Harland *et al.*, 1966; Harland, 1960; Vallance and Fortey, 1968).

mite–semipelite with concordant amphibolites (Rittervatnet Fm.). Two metatilloid horizons occur in the upper part of the latter formation. There follow gneissose feldspathites with concordant amphibolites, with some psammites and with pegmatites (Bangehuk Fm.), and again semipelites and psammites with psammitic feldspathites and concordant amphibolites (Vassfaret Fm.). At the top of the group, quartzites, quartz psammites, and plagioclase psammites predominate over the amphibolites, which are sometimes associated with acid tuffs.

The *Planetfjella Group* (4750 m) consists of coarse-grained, well-laminated semipelitic to psammitic rocks in the lower part (Flåen Fm.) and of a repeating monotonous sequence of finely laminated psammitic–semipelitic, massive, pure quartzites and marbles in the upper part (Vildadalen Fm.).

The *Veteranen Group* (3790 m) begins with flaggy limestone with dolostone bands interbedded with dark shale and quartzose flags, followed by a uniform sequence of dark- or light-colored quartzite with subordinate darker shale bands (Kortbreen Fm.). There follow (a) quartzites and greywackes with phyllitic and quartzose shales, (b) black flaggy limestones interbedded with dark shale and phyllite, and (c) impure quartzites with subordinate greywacke and shale (a–c: Kingbreen Fm.). The highest part of the Group is formed by pale thickly bedded quartzites passing into a greywacke–quartzite complex (Glasgowbreen Fm.) and finally by quartzose shales and flags with subordinate quartzite and dolomitic shale (Oxfordbreen Fm.).

The *Akademikerbreen Group* (2490 m) consists predominantly of carbonates: (a) limestone, often dolomitic or silicified, sometimes with stromatolites (Grusdievbreen Fm.), (b) stromatolitic dolostone and cherty dolostone (Svanbergfjellet Fm.), (c) dolostone conglomerate and microconglomerate with cherts (Draken Fm.) and (d) dark oolitic–pisolitic, often silicified limestone and dolostone with stromatolites (Backlundtoppen Fm.).

The *Polarisbreen Group* (830 m) consists mainly of dark paper shales with black shaly limestone and siltstone intercalations at the bottom (Elbobreen Fm.) and with sandstone intercalations near the top (Dracoisen Fm.), with marine tillites in the middle (Wilsonbreen Fm.).

The *Oslobreen Group* (ca. 1400 m) begins with basal quartzite followed by dolostone and dolomitic limestone with an intraformational carbonate conglomerate (Tokammane Fm.). The presence of *Salterella* indicates the Lower Cambrian age of the carbonates. There follow limestones, often cherty, with stromatolites and intraformational carbonate conglomerate intercalations, and again more limestones and intraformational conglomerates (Kirtonryggen Fm.); the fossils indicate a Lower to Upper Canadian age for the formation. The highest part of the succession is built of black bituminous limestone and shale, often graptolitic, sometimes with chert bands, followed by bioclastic and algal limestone with subordinate shale (Valhallfonna Fm.). Rich assemblages of fossils indicate an Arenigian age for its lower part and the Whiterock stage (latest Arenigian–early Llanvirnian) for its upper part.

3. *Nordaustlandet*

The Hecla Hoek Succession of Nordaustlandet (Fig. 9) is representative for the northeastern part of the Svalbard Caledonides (Kulling, 1934; Sandford, 1950, 1956, 1963; Krasilshchikov, 1965, 1973; Krasilshchikov et al., 1965; Flood et al., 1969). Its thickness exceeds 15 km. There is a close resemblance of the higher part of the Nordaustlandet succession to that of Ny Friesland, but several major unconformities are present, possibly correlating with those of south Spitsbergen.

The *Botniahalvøya Group* (ca. 9000 m) begins with a huge acid volcanic complex ranging from volcanic breccias and agglomerates to tuffs and tuffaceous rocks, with quartz porphyry intrusions (Kapp Hansteen Fm.). It is followed by a monotonous sequence of quartzite, siltstone, and shale (Brennevisfjorden Fm.) with quartz porphyry plugs and dikes. A conglomerate at the base of the Brennevisfjorden Formation may suggest an unconformity. The upper part of the Group consists of quartzite and calcareous rocks intercalated

Supergroup	Group	Formation	
Murchison–fjorden		Kapp Sparre	700–800m
		Sveanor	200–300m
	Roaldtoppen	Ryssö	750m
		Hunnberg	500m
	Celsiusberget	Raustup – Sälodd	550m
		Norvik	340m
		Flora	1,250m
	Franklinsundet	Kapp Lord	c.1,000m
		Westmanbukta	625m
		Persberget	+150m
	Botniahalvøya	Austfonna (+Kapp Platen)	c.3,000m
		Brennevisfjorden	c.2,000m
		Kapp Hansteen	c.4,000m

Fig. 9.   Hecla Hoek Succession in Nordaustlandet (after Kulling, 1934; Flood *et al.*, 1969).

in a monotonous sequence of dark phyllites, with some intrusive porphyry and amphibolite layers (Austfonna Fm. and Kapp Platen Fm.).

The *Franklinsundet Group* (ca. 1780 m) rests unconformably upon the Botniahalvøya Group as is shown by a basal quartzite conglomerate overlain by light quartzite with subordinate dark shale (Persberget Fm.). There follow mudstones with some quartzite intercalations (Westmanbukta Fm.) and alternating mudstone, quartzite, and limestone beds (Kapp Lord Fm.).

The *Celsiusberget Group* (2140 m) begins with predominantly light-colored quartzose sandstones with subordinate siltstone and shale (Flora Fm.) followed by a shale–siltstone–sandstone complex (Norvik Fm.), and that in turn by multicolored shales with some quartzose sandstone and light dolostone (Raustup-Sälodd Fm.).

The *Roaldtoppen Group* (1250 m) is predominantly carbonate: dark limestone with chert concretions, conglomerate, and oolite dominate in the lower part (Hunnberg Fm.), while light-grey massive dolostone with stromatolitic structures, oolitic limestone, chert, and subordinate clayshale occur in the upper part (Ryssö Fm.).

The *Sveanor Formation* (200–300 m) consists of alternating variably colored shale, sandstone, marl, and arenaceous limestone, with tillites near the top.

The *Kapp Sparre Formation* (700–800 m) consists of calcareous-dolomitic and psammitic–pelitic rocks in the lower part, and of dolostones and limestones that often are oolitic in the upper part. The presence of *Obolus* and *Lingulella* indicates the Cambrian age of the formation.

4. *Bjørnøya*

Only the youngest units of the Hecla Hoek Succession crop out in Bjørnøya (Holtedahl, 1919; Horn and Orvin, 1928; Winsnes, 1965). The succession totals 1.2 km in thickness and consists of four units: (1) older dolostone with oolites and stromatolites—400 m (Russehamna Fm. of Krasilshchikov and Livshic, 1974), (2) slate–quartzite—175 m (Sørhamna Fm., *op. cit.*), (3) younger dolostone with a fossiliferous zone in its lower part—400 m, (4) *Tetradium* limestone—240 m (3–4—Ymerdalen Fm., *op. cit.*). The succession seems to correspond to Ny Friesland–Nordaustlandet successions rather than to that of south Spitsbergen. The older dolostone correlates with late Precambrian oolite-stromatolite carbonates of Spitsbergen and Nordaustlandet (Akademikerbreen Gp., Roaldtoppen Gp., and Höferpynten Fm.) and with the Porsanger Dolostone of north Norway. The succeeding slate–quartzite is thought to be contemporaneous with the sediments of the Varangian glacial period in Svalbard (Sveanor Fm., Polarisbreen Gp., Gåshamna Fm.) and its age equivalents in northern Norway. The two upper units yield fossils indicative of Canadian (3) and late Canadian to Whiterock ages (4). A hiatus corresponding to the Cambrian is thus to be expected between the units 2 and 3.

B. Stratigraphic Correlation of Rocks and Events

Correlation of the Precambrian succession of Svalbard is most difficult in the lowest units owing to strong metamorphic alteration of the sediments. The presence of acidic and basic igneous activity, tilloids and tillites, and major unconformities usually makes lithological correlation possible in the late Precambrian sequences at the group level, and less frequently at the formation level. Tillites, tilloids, and oolite–stromatolite beds serve as marker horizons (see, e.g., Harland, 1960, 1969; Birkenmajer and Narebski, 1960; Winsnes, 1965; Harland *et al.*, 1966; Flood *et al.*, 1969; Birkenmajer, 1972c, 1975b; Krasilshchikov, 1973). Precambrian radiometric dates are usually strongly overprinted by the main Caledonian (Ordovician) event. Fossils serve correlation purposes in the Cambrian and Ordovician strata but they are far from common. The correlation scheme used here (Figs. 4–6) is that proposed by Birkenmajer (1975b). It corresponds to a great extent to the latest correlation by Harland (1969), and differs in many respects from that proposed by Krasilshchikov (1973).

The basal Isbjørnhamna and Finnlandveggen Groups are correlated on common occurrence of garnet-mica schists with marble intercalations. The following volcanic–sedimentary complexes of the Eimfjellet and Harkerbreen Groups and the Kapp Hansteen Formation seem to correlate with one another on the common occurrence of concordant basic rocks (amphibolites) in the former two units, and acid volcanic rocks in the Kapp Hansteen Formation and in the Eimfjellet (Vimsodden Subgroup) and Harkerbreen Groups. Another criterion is the presence of metatilloids within the Harkerbreen (Rittervatnet metatilloids) and Eimfjellet Groups (Vimsodden metatilloids).

The Deilegga and Planetfjella Groups on the one hand and the Brenne-visfjorden and Austfonna Formations on the other have similar lithologies. Acid volcanism is present in the Planetfjella Group and the Brennevisfjorden and Austfonna Formations. The unconformity and basal conglomerate known from the bottom of the Deilegga Group may correlate with that below the Brennevisfjorden Formation.

The Veteranen, Franklinsundet, and Celsiusberget Groups of Ny Fries-land and Nordaustlandet are correlated on lithological grounds. Further south (Hornsund area) there is a long erosional break at the base of the Slyngfjellet Conglomerate. An important unconformity at the base of the latter may cor-relate with that at the base of the Franklinsundet Group.

The Roaldtoppen and Akademikerbreen Groups and the Höferpynten Formation are readily correlated by common lithologies (the presence of oolitic and stromatolitic rocks). The Varangian tillites of the Sveanor Forma-tion and Polarisbreen Group probably occupy a similar position in southwest Spitsbergen near the top of the Gåshamna Phyllite.

The Cambrian Sofiekammen Group correlates with the lower part of the Oslobreen Group and with the Kapp Sparre Formation. An important uncon-formity at the base of the Cambrian recognized in south Spitsbergen may also exist in other areas of Svalbard.

The Ordovician Sørkapp Land Group correlates with the upper part of the Oslobreen Group, the latter containing the youngest Ordovician strata so far known from Svalbard. A well-marked unconformity at the base of the Ordovician, recognized in south Spitsbergen, probably occurs also in Ny Fries-land.

C. Problem of Precambrian Basement

Sandford (1950, 1956) postulated that much of Nordaustlandet consists of Archean basement rocks unconformably underlying the Hecla Hoek. Com-parison with central-north Spitsbergen (Harland, 1961; Harland et al., 1966, 1974), and new evidence from Nordaustlandet (Flood et al., 1969) leave this postulate in doubt. "Basement has yet to be established" (Harland et al., 1966, p. 78).

Krasilshchikov (1973, Fig. 14) accepts a major break between the Lower–(?)Middle Proterozoic "crystalline basement" and the late Proterozoic succession in Nordaustlandet, Ny Friesland, and west Spitsbergen. The crystalline "basement," considered by him as older than 1900 m.y., includes the Finnlandveggen and Harkerbreen Groups of Ny Friesland, the Isbjørn-hamna and Eimfjellet Groups of south Spitsbergen, and the gneiss–granite complex of Nordaustlandet ("Archean" of Sandford).

A late Proterozoic (Riphean) age for the lower Hecla Hoek succession of

Svalbard has recently been advocated by Harland and Gayer (1972, Table 1). Harland *et al.* (1974) consider the base of the Finnlandveggen Group as corresponding to about 1000 m.y. (?). However, there is radiometric evidence from the Hecla Hoek Succession (see Gayer *et al.*, 1966; Krasilshchikov, 1973, Fig. 11) for much older events (see p. 282).

D. Precambrian Sedimentation and Igneous Activity—Torellian Basin

The lower part of the Hecla Hoek, comprising between 9 and 11.5 km of metasediments and metavolcanics (Botniahalvoya Group in Nordaustlandet 9 km; Finnlandveggen, Harkerbreen, and Planetfjella Groups in north-central Spitsbergen 11.5 km; Isbjørnhamna, Eimfjellet, and Deilegga groups in south Spitsbergen 9–11 km), shows the characteristic features of eugeosynclinal development. This geosyncline has been named the Torellian Basin from the succeeding Torellian diastrophism (Birkenmajer, 1975*b*). Basic volcanics of the Eimfjellet and Harkerbreen Groups could be considered to have been formed close to the spreading center of an oceanic rift. The occurrence of eclogites in equivalent rocks in northwest Spitsbergen (Gee, 1966) is consistent with this assumption. The thick pile of acidic effusives and intrusives in the Botniahalvoya Group of Nordaustlandet may indicate the presence of a continental-type crust below the eastern margin of the geosyncline, possibly even an island arc and subduction zone at the boundary of the Barents Craton (on the east) and the Torellian Geosyncline (on the west).

E. Werenskioldian Diastrophism

An important unconformity at the base of the Deilegga Group in south Spitsbergen separates highly metamorphosed Eimfjellet sediments and volcanics from much less metamorphosed Deilegga sediments. Metamorphism under conditions of amphibolite facies (staurolite-kyanite subfacies, e.g., Isbjørnhamna Group) and albite-epidote-amphibolite and greenschist facies (e.g., Eimfjellet Group), followed by some migmatization (granitization) in the central and western parts of the geosynclinal trough, associated with some folding and with subsequent uplift and erosion along eastern and western margins of the trough, are the main events suggested for the Werenskioldian diastrophism (Birkenmajer, 1975*b*).

No major unconformity between the Harkerbreen and Planetfjella Groups in Ny Friesland has been recognized by British authors (e.g., Harland, 1969; Harland *et al.*, 1974). In contrast, Krasilshchikov (1973) claims a major break between his "Atomfjella Series" (= Finnlandveggen and Harkerbreen Groups) and the "Mosselbukta Series" (= Planetfjella Group), corresponding to that between the Deilegga Group and the Eimfjellet Group in south Spitsbergen.

According to Krasilshchikov, folding, magmatism, and metamorphism (including ultrametamorphism) affected the "crystalline basement" in Svalbard during that mid-Proterozoic diastrophic epoch (1700 to 1900 m.y. ago).

Isotopic ages of the oldest Hecla Hoek rocks usually show overprinting by younger, late Precambrian and Caledonian metamorphism (Gayer *et al.*, 1966). However, some high pyroxene ages from eclogites of northwest Spitsbergen, 1393 to 1939 m.y., generally disregarded in the discussion on pre-Caledonian metamorphic events (e.g., Harland *et al.*, 1966; Harland, 1969), might well correspond with postulated Werenskioldian diastrophism and related metamorphism.

As the result of the Werenskioldian diastrophism the axis of the geosynclinal basin in Svalbard shifted westward (Fig. 5) as is indicated by the increasing thickness of the sediment pile in that direction (Planetfjella Gp.).

F. Torellian Diastrophism

Another important unconformity at the base of the Slyngfjellet Conglomerate and Franklinsundet Formation is best observed in south Spitsbergen (Birkenmajer, 1975*b*). In the marginal parts of the geosynclinal trough the Torellian diastrophism produced uplift and deep erosion, the axial part of the trough being seemingly unaffected by these events. Appearance of granite and contact metamorphic rock pebbles in the Slyngfjellet Conglomerate may be an indication of plutonism subsequent to Werenskioldian diastrophism. If correctly correlated with diastrophic events in East Greenland, the Werenskioldian and Torellian diastrophisms would correlate with the Carolinidian diastrophism.

G. Late Precambrian Sedimentation—Jarlsbergian Basin

The middle part of the Hecla Hoek, comprising between 1.5 and 7 km of slightly metamorphosed sediments (Roaldtoppen Gp. and Sveanor Fm. in Nordaustlandet, 1.5 km; Veteranen, Akademikerbreen, and Polarisbreen Groups in north-central Spitsbergen, 7.1 km; Sofiebogen Gp. in south Spitsbergen, 3.7 km), shows features of miogeosynclinal development. This geosynclinal trough has been referred to as the Jarlsbergian Basin, the name derived from the succeeding Jarlsbergian diastrophism (Birkenmajer, 1975*b*).

The basal psammite–pelite to psephite beds (Franklinsundet, Celsiusberget, and Veteranen Groups) are partly shallow marine (Veteranen Gp.) with current bedding and ripplemarks (Franklinsundet and Celsiusberget Groups), and partly molassic in character (Slyngfjellet Conglomerate). There follows a shallow-marine carbonate facies with stromatolites, oolites, and intraformational conglomerates (Roaldtoppen and Akademikerbreen Groups, Höferpynten Fm., and "lower dolostone" of Bjørnøya) passing upward into a monotonous, predominantly pelite–psammite shallow-marine sequence with tillite horizons near the top, corresponding to the Varangian glaciation (Sveanor Fm., Polarisbreen Gp., and Gåshamna Fm.). The thickest sediments occur in south Spitsbergen (Gåshamna Fm.: up to 2.5 km), indicating a further westward shift of the sedimentary axis (Fig. 5).

H. Jarlsbergian Diastrophism

An important unconformity at the base of the Cambrian Sofiekammen Group separating the Gåshamna Phyllite from almost unmetamorphosed Sofiekammen sediments in Wedel Jarlsberg Land, south Spitsbergen, is related to the Jarlsbergian disastrophism. Folding, dynamic and regional metamorphism of the whole Precambrian Hecla Hoek, and subsequent uplift and erosion are the main events of the Jarlsbergian diastrophism (Birkenmajer, 1975b). Evidence of the stratigraphic age of the metamorphism is found in the presence of sedimentary breccias containing phyllite fragments from the Gåshamna Formation in the Lower Cambrian Blåstertoppen Dolostone. Isotopic determinations from 529 to 636 m.y. of rocks older than Cambrian, from south, northwest, and northeast Spitsbergen (see Gayer et al., 1966; Gee and Hjelle, 1966; Harland et al., 1966; Harland, 1969, 1972; Gee, 1972), though possibly showing some Caledonian overprinting, suggest that the age of the Jarlsbergian diastrophism may lie at about 600 m.y., i.e., close to the boundary of Cambrian and Precambrian (Birkenmajer, 1975b). The "Raudfjorden Event" of Harland (1973a) in northwestern Spitsbergen, dated radiometrically at approximately 600 m.y., may correlate with the Jarlsbergian diastrophism.

I. Early Paleozoic Sedimentation—Hornsundian Basin

The upper part of the Hecla Hoek, comprising between 0.8 and 2.3 km of generally unmetamorphosed sediments (Kapp Sparre Fm. in Nordaustlandet, 0.8 km; Oslobreen Gp. in north-central Spitsbergen, 1.3 km; Sofiekammen and Sørkapp Land Groups in South Spitsbergen, 2.3 km), also shows features of miogeosynclinal development. This geosynclinal trough has been referred to as the Hornsundian Basin, the name derived from the well-marked Hornsundian unconformity between the Cambrian and Ordovician in south Spitsbergen (Birkenmajer, 1975b).

Apart from the basal psammitic rocks, best recognized in south Spitsbergen (the lower ones succeeding the Jarlsbergian unconformity and the upper ones lying above the Hornsundian unconformity), the sediments of the Hornsundian Basin are predominantly carbonates. They are thickest in south Spitsbergen (Fig. 5), with the subordinate shales (with Cambrian trilobites and

Ordovician graptolites) indicating deeper water sedimentation. The occurrence of algal structures, oolites, and sedimentary breccias in various parts of the succession indicate that shallow-marine conditions occasionally prevailed.

J. Hornsundian Unconformity

The Hornsundian unconformity was caused by uplift along the western margin of the basin and deep erosion of Cambrian strata. The succeeding Lower Ordovician Wiederfjellet Quartzite rests on various stratigraphic units of the Cambrian Sofiekammen Group, often filling karstlike holes in the underlying carbonates. The Hornsundian unconformity has not been recognized either in north-central or northeastern Svalbard. In Bjørnøya, a hiatus is to be expected between the slate–quartzite complex (correlated with the Varangian deposits) and the upper dolostone (correlated with lower Canadian). The hiatus could well be the result of the Hornsundian unconformity.

K. Main Caledonian (Ny Friesland) Orogeny

The main Caledonian orogeny, the Ny Friesland Orogeny (Harland, 1961), is responsible for folds, thrusts, and both dynamic and regional metamorphism within the Hecla Hoek Succession of Svalbard. Radiometric ages of (Gayer et al., 1966) 430 to 450 m.y. occur in all the main metamorphic areas of northern Spitsbergen and Nordaustlandet and suggest that the main metamorphic event occurred at 440–450 m.y. This would imply a late Ordovician orogeny (Gee, 1972). However, if the latest recorded metamorphic event of 419–436 m.y. is taken into account, a Silurian age for the culmination of tectogenesis would be appropriate (Harland et al., 1974).

The stratigraphic gap between the youngest strata of the Valhallfonna Formation (latest Arenigian–early Llanvirnian in age) (Fortey and Bruton, 1973) and the Downtonian (late Silurian) Siktefjellet Group of the Old Red Sandstone (Gee and Moody-Stuart, 1966) accounts for most of the Silurian.

The structure resulting from the Ny Friesland Orogeny is often complex, showing folds and thrust-folds generally indicating a west–east compression. Along the west coast of Spitsbergen, there has been much reorientation by Tertiary folding. In the Svalbard area as a whole, no attempt has yet been made to separate the Caledonian and pre-Caledonian structural patterns. Harland (1966) suggests a shortening of at least 200 km across the whole zone, but this seems to be a minimum value. Sinistral strike-slip movement (Harland, 1965, 1971, 1973a) may have been subsequent to the west–east compression, and a late stage of the Ny Friesland Orogeny could be represented by north–south-trending basic dikes (Harland et al., 1974).

L. Late Orogenic Plutonism

Late orogenic plutonism is known from the northwest, north-central, and northeast areas of Svalbard (Fig. 2). Granite plutons in the north-central area are known to postdate the main Caledonian tectogenesis. Isotopic determinations give an age of plutonism of between 420 and 340 m.y. (Gayer *et al.*, 1966) with a peak for migmatization and plutonism at 390 to 400 m.y., i.e., close to the Silurian–Devonian boundary. Lamprophyre dikes are sometimes associated with the plutons. Ore-bearing quartz-ankerite mineralization was possibly subsequent to granite emplacement but older than basal Old Red Sandstone sediments (Siegenian) in south Spitsbergen (Birkenmajer and Wojciechowski, 1964).

III. DEVONIAN MOLASSE OF SVALBARD AND PLATE MOTIONS

A. Late Orogenic Sedimentation—Svalbardian Basin and Haakonian Movements

The Ny Friesland Orogeny was followed by extensive denudation, most marked in the north where, prior to the Gedinnian, some 15 km of overburden was eroded from northwest Spitsbergen (Gee, 1972), and prior to the Tournaisian, nearly 20 km (Harland, 1969) was removed. The late orogenic Old Red Sandstone molasse was laid down in a narrow intramontane basin some 400 km long, possibly widening from south to north to about 70 km (Fig. 10). This graben has been interpreted as synsedimentary (Friend, 1967) or as the failed arm of a triple junction, the other arms causing the opening of an oceanic basin situated farther north (Birkenmajer, 1975*b*). An alternative explanation takes into account the lack of coarse sediments in the Lower Devonian along the eastern flank of the present graben and considers the faulting as postdating the Old Red Sandstone sedimentation (Harland, 1969; Friend and Moody-Stuart, 1972; Harland *et al.*, 1974).

Sediments started to accumulate in the north in the late Silurian or Downtonian (Siktefjellet Gp., 1.5 to 2 km) and they were folded and thrust prior to Gedinnian Red Bay Group sedimentation during the Haakonian movements (Gee, 1972). Harland's (1973*a*) "Forlandsundet Phase" of folding distinguished in Oscar II Land (northwest Spitsbergen) and Prins Karls Forland probably corresponds to the Haakonian movements of Gee (1972). The Gedinnian to Lower Frasnian succession is about 6 km thick to the north (Friend, 1961), while only about 1 km of sediments accumulated in the south during late Siegenian to Eifelian times (Birkenmajer, 1964).

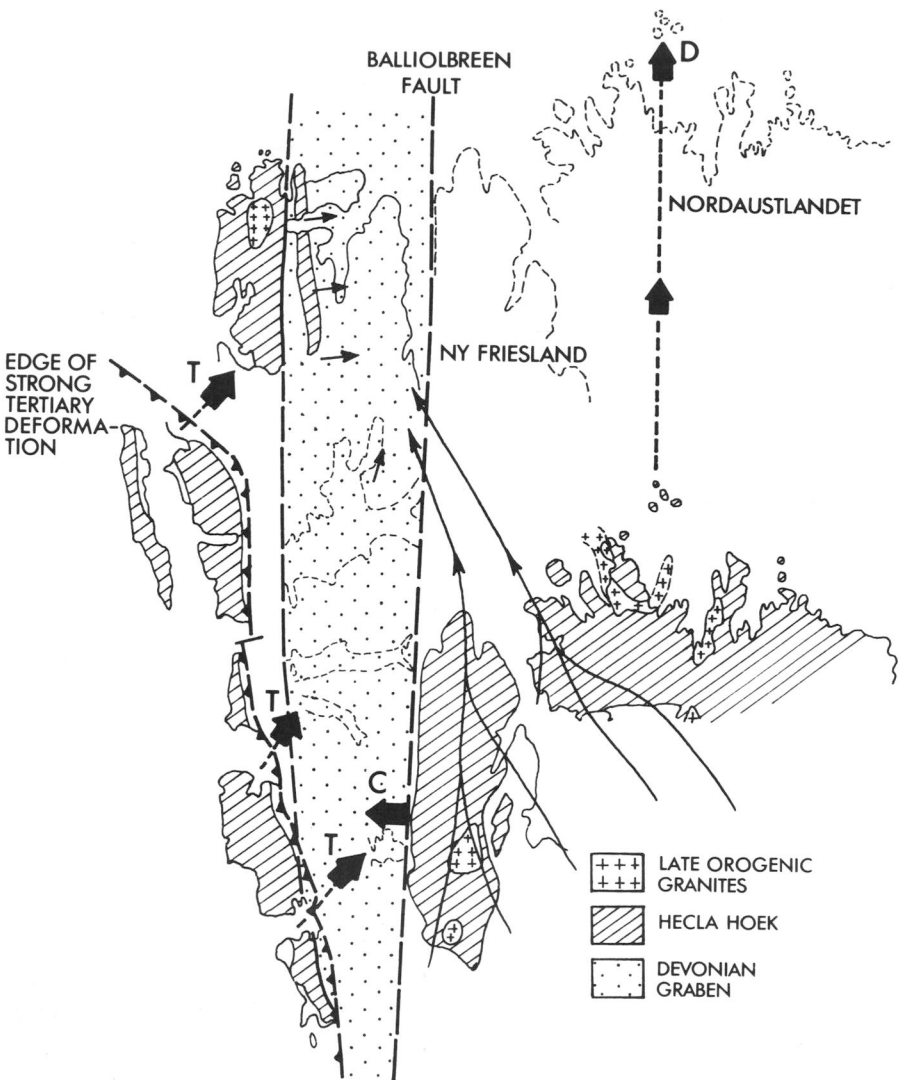

Fig. 10.   Palinspastic map of Spitsbergen and Nordaustlandet in Lower Devonian time before late Devonian left-lateral movement along the Balliolbreen Fault (D). Caledonian mountain belt of the west coast of Spitsbergen shown in a position prior to Tertiary diastrophism (T); southern termination of the Devonian graben shown in a position prior to mid-Carboniferous thrust (C). Small arrows indicate Lower Devonian river directions. (After Friend and Moody-Stuart, 1972, with modifications by Birkenmajer, 1975*b*.)

The lower part of the Old Red Sandstone sedimentation is predominantly fluviatile both to the north (Siktefjellet Gp., Red Bay Gp., and Wood Bay Fm.) and to the south (Lower Marietoppen Fm.). Two main sources of clastics have been recognized (Friend and Moody-Stuart, 1972): a major source located to the south southeast and south supplied material roughly along the

basin, while another source located to the northwest provided a lateral supply
from the western margin of the basin (Fig. 10). Most fluviatile to lacustrine
formations contain vertebrate and plant remains, and on occasion ostracods,
estheriids, and eurypterids. Bivalves and gastropods are less frequent and
appear mainly in supposedly brackish or shallow-marine sediments near the
top of the succession (Grey Hoek, Wijde Bay, and upper Marietoppen Formations). No typical marine indicators are known. The stratigraphic succession of
the Old Red Sandstone in Svalbard is shown in Fig. 11.

B. Svalbardian Movements

The late Devonian (post Upper Givetian–pre-Tournaisian) Svalbardian
movements (so named by Vogt, 1928) caused folding and faulting in the Old
Red Sandstone and subsequent transcurrent sinistral strike-slip movement
along the Balliolbreen Fault that forms the eastern flank of the present
Devonian graben in Spitsbergen (Harland, 1965, 1969; Friend and Moody-
Stuart, 1972; Harland et al., 1974) (Fig. 10), and possibly along other transcurrent faults as postulated by Harland (1971, 1972). Strike-slip movements along
these faults may have caused separation of the eastern plate (including central-

SERIES	STAGE	NORTH & CENTRAL SPITSBERGEN	SOUTH SPITSBERGEN (HORNSUND)
UPPER	FAMENNIAN	*Svalbardian movements*	
	FRASNIAN		
MIDDLE	GIVETIAN	Wijde Bay Fm. 500m	Mimer Valley Fm. 1,000m
	EIFELIAN		
	EMSIAN	Grey Hoek Fm. 1,000m	Marietoppen Fm. 1,100m
LOWER	SIEGENIAN	Wood Bay Fm. 2,900m	
	GEDINNIAN	Red Bay Gp. 0–2,000m	Ben Nevis Fm. / Fraenkelryggen Fm. / Andréebreen Sdst. Fm. / Red Bay Conglomerate
	DOWN-TONIAN	*Haakonian movements*	
UPPER		Siktefjellet Gp. 0–2,000m	

11.   Old Red Sandstone succession in Spitsbergen (after Friend, 1961; Birkenmajer, 1964; Friend et al., 1966, sim-
ed).

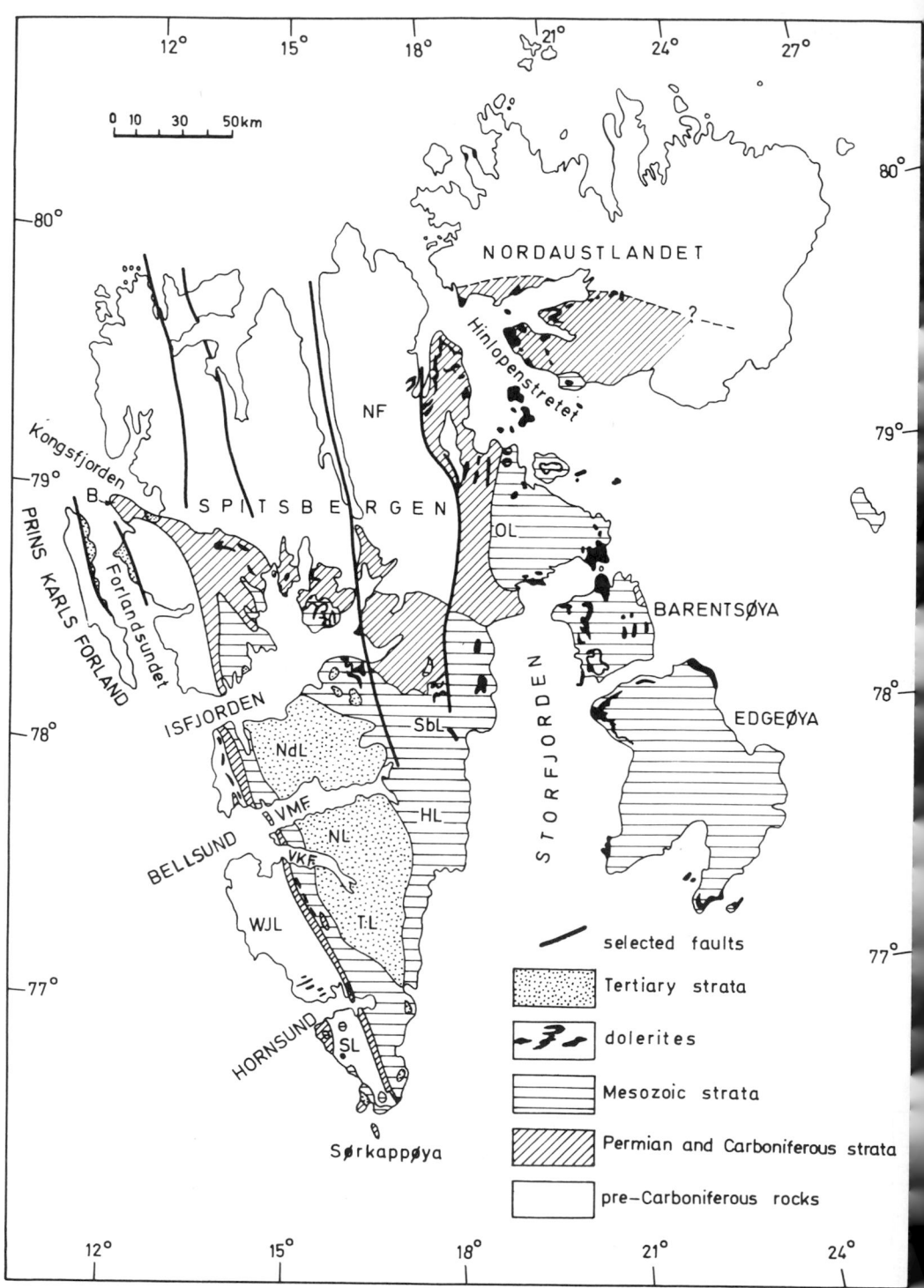

north and northeast Svalbard adjacent to the Barents craton) from the western Greenland plate, through a set of slices in an intermediate area now represented by Hecla Hoek rocks of northwest, west and, southwest Spitsbergen (Harland, 1971, 1972). Paleomagnetic investigation of the Wood Bay rocks (Lower Devonian) indicates pronounced thermal remagnetization of the sandstones along the border faults of the graben, which is attributable to the Upper Devonian Svalbardian transcurrence. No rotation of the archipelago resulted from these movements (Storetvedt, 1972). Palinspastic reconstruction of the Devonian sedimentary basin in Svalbard prior to the Svalbardian movements (Fig. 10) shows the north-central and northeast Svalbard Caledonides situated some 200–250 km farther south than at present (Friend and Moody-Stuart, 1972). Some even larger estimates, up to 1,000 km of displacement as discussed by Harland and Gayer (1972) and Harland *et al.* (1974), seem to be opposed by evidence from the development and succession of the Hecla Hoek rocks, which show obvious analogies in the whole area of Svalbard (see Figs. 4–6) and suggest a close connection.

IV. CARBONIFEROUS THROUGH TERTIARY PLATFORM BASINS OF SVALBARD

Beginning with the Carboniferous, the Svalbard area became part of the continental Barents–Kara platform (Fig. 12) (see Atlasov *et al.*, 1964; Gakkel and Dibner, 1967) and was subject to recurrent denudation and the accumulation of mainly terrigenous deposits (see reviews by Orvin, 1940; Frebold, 1951; Harland, 1961, 1969; Sokolov *et al.*, 1968; Livšic, 1973, 1974). Local postorogenic molasse basins developed during Lower Carboniferous to Lower Permian time, and some folding and thrusting took place along the Balliolbreen Fault during a mid-Carboniferous Adriabukta Phase. Permian to Tertiary time witnessed repeated transgressions and regressions of shallow seas, and the stratigraphic column shows numerous breaks in sedimentation caused by vertical movements, especially along the western border of the Main Spitsbergen Basin.

A. Carboniferous and Permian Sedimentation

Carboniferous and Permian rocks form the base of the post-Caledonian platform sequence of Svalbard. Their lithofacies reflect unstable land–shallow

Fig. 12.   Post-Caledonian platform cover in Svalbard (simplified from several sources). (B) Brøg-gerhalvøya; (HL) Heer Land; (NdL) Nordenskiöld Land; (NF) Ny Friesland; (NL) Nathorst Land; (OL) Olav V Land; (TL) Torell Land; (SbL) Sabine Land; (SL) Sorkapp Land; (VKF) Van Keulenfjorden; (VMF) Van Mijenfjorden; (WJL) Wedel Jarlsberg Land.

sea conditions of sedimentation and the changing relief of the basin borders due to vertical movements. The diversity of sediments, often very rich in marine fossils (brachiopods, bryozoans, corals, fusulinids, etc.), resulted in a proliferation of both formal and informal lithostratigraphic names, often of very restricted significance. Figure 13 shows a simplified standard for three main areas of Svalbard, i.e., central Spitsbergen, south Spitsbergen, and Bjørnøya. For more details the reader is referred to papers by Nathorst (1910), Horn and Orvin (1928), Gee et al. (1952), Dineley (1958), Forbes et al. (1958), Birkenmajer (1964), and Cutbill and Challinor (1965).

The Lower Carboniferous Billefjorden Group is characterized by freshwater, often coal-bearing sediments, which are predominantly sandstones, quartzites, and shales, with subordinate conglomerates. An angular unconformity is always recognizable at the base of the group over the Hecla Hoek or Old Red Sandstone rocks. The Lower Carboniferous postorogenic molasse deposits were laid down in relatively small, separate basins with tectonically

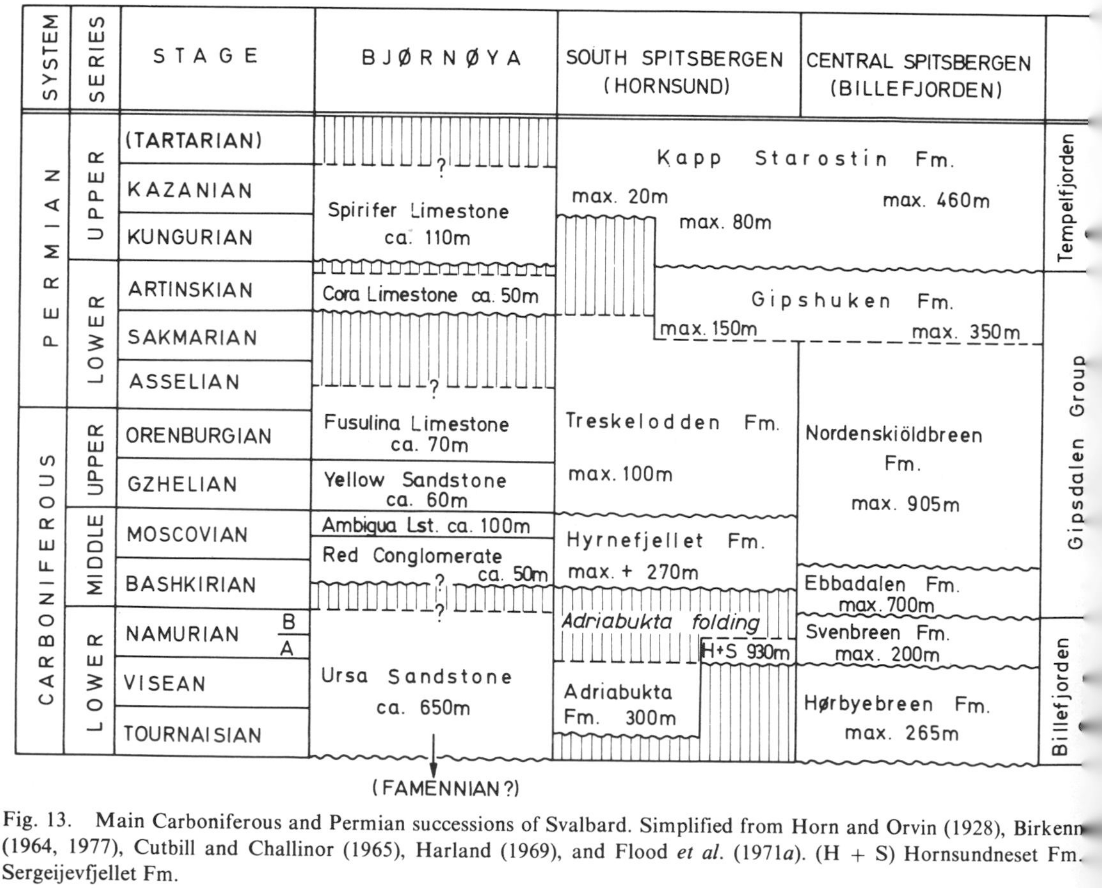

SYSTEM	SERIES	STAGE	BJØRNØYA	SOUTH SPITSBERGEN (HORNSUND)	CENTRAL SPITSBERGEN (BILLEFJORDEN)	
PERMIAN	UPPER	(TARTARIAN)		Kapp Starostin Fm.		Tempelfjorden
		KAZANIAN	Spirifer Limestone ca. 110m	max. 20m	max. 460m	
		KUNGURIAN		max. 80m		
	LOWER	ARTINSKIAN	Cora Limestone ca. 50m		Gipshuken Fm.	Gipsdalen Group
		SAKMARIAN		max. 150m	max. 350m	
		ASSELIAN				
CARBONIFEROUS	UPPER	ORENBURGIAN	Fusulina Limestone ca. 70m	Treskelodden Fm. max. 100m	Nordenskiöldbreen Fm. max. 905m	
		GZHELIAN	Yellow Sandstone ca. 60m			
	MIDDLE	MOSCOVIAN	Ambigua Lst. ca. 100m Red Conglomerate ca. 50m	Hyrnefjellet Fm. max. + 270m	Ebbadalen Fm. max. 700m	
		BASHKIRIAN				
	LOWER	NAMURIAN B / A	Ursa Sandstone ca. 650m	Adriabukta folding H+S 930m	Svenbreen Fm. max. 200m	Billefjorden
		VISEAN		Adriabukta Fm. 300m	Hørbyebreen Fm. max. 265m	
		TOURNAISIAN				

(FAMENNIAN ?)

Fig. 13. Main Carboniferous and Permian successions of Svalbard. Simplified from Horn and Orvin (1928), Birkenm (1964, 1977), Cutbill and Challinor (1965), Harland (1969), and Flood et al. (1971a). (H + S) Hornsundneset Fm. Sergeijevfjellet Fm.

active borders. The thickness of the continental deposits amounts to 300–400 m in the central depression of Spitsbergen, but grows to more than 900 m in some basins along the west coast of the main island (e.g., at Hornsund). Palynologic evidence shows that in Spitsbergen the base of the Billefjorden Group is either Tournaisian, Visean, or Namurian. In Bjørnøya the base of the sequence (lower part of the Ursa sandstone, 100–360 m) was thought to belong to the Famennian (Horn and Orvin, 1928), but this has not been confirmed by paleobotanical investigation (Playford, 1962–1963).

The *Gipsdalen Group* (Middle Carboniferous–Lower Permian) usually rests unconformably, sometimes with angular unconformity, upon the Lower Carboniferous or directly upon the Hecla Hoek rocks. The base of the group is characterized by a variety of clastic sediments ranging from conglomerates to siltstones and shales, and often recognizable as alluvial fan and floodplain deposits formed in an arid, warm climate, with red colors predominating (e.g., Hyrnefjellet Fm.). In the central part of the Main Spitsbergen Basin, gypsum and anhydrite (lower evaporate unit) associated with limestone and clastics formed close to the seashore, and were followed by widespread carbonate facies with a warm shallow-marine fauna (Nordenskiöldbreen Fm.). The southwestern margin of the basin was marked by the predominance of clastics with subordinate fossiliferous limestone intercalations often containing coral bioherms (Treskelodden Fm.).

The evaporate facies reappeared in the Lower Permian (Gipshuken Fm.) and is often followed by a break in deposition near the Lower–Upper Permian boundary. The total thickness of the Gipsdalen Group is usually 600 to 900 m.

The *Tempelfjorden Group* is entirely marine and is represented at the base mainly by fossiliferous limestone with some sandstone and conglomerate, and higher in the sequence by siliceous limestone, silty or arenaceous limestone, and shale. Brachiopods, bryozoans, and sponges are the main faunal elements of the group. The shallow marine basin was poorly aerated, as indicated by the predominance of dark colors and the bitumen content. The thickness of the group is from 200 to 460 m in the central part of the basin in Spitsbergen but diminishes to 5–12 m or less along its southwest margin.

The Carboniferous to Permian marine basin of Svalbard represents an epicontinental embayment that covered much of the Barents Shelf. Its western margin lay within the Caledonian Foldbelt of western Spitsbergen, which had become leveled with time. Another land bordered the bay on the north and northeast.

B. Variscan Movements

Numerous breaks in sedimentation and erosional (occasionally angular) unconformities (Fig. 13), together with the coarse clastic character of many

Carboniferous to Lower Permian formations, are evidence of tectonic instability in Svalbard, (e.g., Orvin, 1940a; Birkenmajer, 1964; Harland, 1969). The instability gradually diminished until during Upper Permian time stable platform conditions obtained.

There is no record of Variscan folding on a large scale, and the Carboniferous movements in Svalbard are regarded by Harland (1969) as adjustments at the conclusion of the Svalbardian (Upper Devonian) movements, representing resurgent late Caledonian activity. The only major tectonic deformation is seen along the eastern limb of the Devonian graben in south Spitsbergen (Hornsund), where Precambrian and Lower Carboniferous rocks have been thrust westward over Lower Carboniferous and Devonian sediments, then eroded and unconformably covered by Middle Carboniferous conglomerates (Fig. 21). This deformation, correlated with the Erzgebirge Phase (Birkenmajer, 1964), later named the Adriabukta Phase after the main area of deformation (Birkenmajer, 1975b), caused strong folding and slight dynamic metamorphism of Lower Carboniferous shales in a narrow zone corresponding to the Svalbardian Graben. The thrusting of the eastern limb of the graben to the west indicates that the transcurrent movement along the Balliolbreen Fault in Spitsbergen had ceased by Mid-Carboniferous times, or even earlier (i.e., before the Tournaisian). No major deformation attributable to Carboniferous–Permian time has been recognized along this fault in north-central Spitsbergen (Harland et al., 1974). Evidence of increased tectonic activity in that area may be found in the presence of north–south-striking lamprophyre dikes corresponding to an east–west extension, dated as Middle Carboniferous (Gayer et al., 1966; Harland et al., 1974). These dikes probably postdate the Adriabukta deformational phase.

A pronounced unconformity between the *Fusulina* limestone and the Cora limestone recognized in Bjørnøya has been assigned to the Asturic Phase by Orvin (1940). However, given the new stratigraphical position of these units (Fig. 13), it corresponds to a Lower Permian phase of uplift.

A hiatus recognizable at the base of the Upper Permian Kapp Starostin Formation may correspond to the Saalic Phase of uplift. At Hornsund (south Spitsbergen) it is expressed as a slight angular unconformity (Birkenmajer, 1964).

C. Triassic through Basal Jurassic Sedimentation

Stable platform conditions prevailed in Svalbard through most of Triassic and Jurassic time, with minor breaks caused by positive movements. The marine Triassic basin was, in general, a continuation and further extension of the Upper Permian basin. In the central part of the basin there is either a continuous lithological passage from the Permian to Triassic or a break of minor order. Longer breaks in sedimentation and the appearance of basal Triassic conglomerate directly upon pre-Upper Permian rocks mark the western and southwestern borders of the basin.

During the Griesbachian to Carnian time a shallow sea covered the continental platform and part of the continental slope in the western part of the Barents Shelf including most of the Svalbard Archipelago, as far south as Bjørnøya. A narrow bay branched off to the south as far as central East Greenland (Fig. 14A), while an open sea lay further north.

There is a variety of marine fossils, including ammonoids, bivalves, brachiopods, vertebrates, etc. (see, e.g., Frebold, 1929; Buchan *et al.*, 1965; Pchelina, 1965, 1967; Tozer and Parker, 1968; Korchinskaja, 1972) but, as a whole, the fossil content is poor compared with that of the Upper Permian. There is no clear indication of the cause of quick disappearance of the rich warm-water

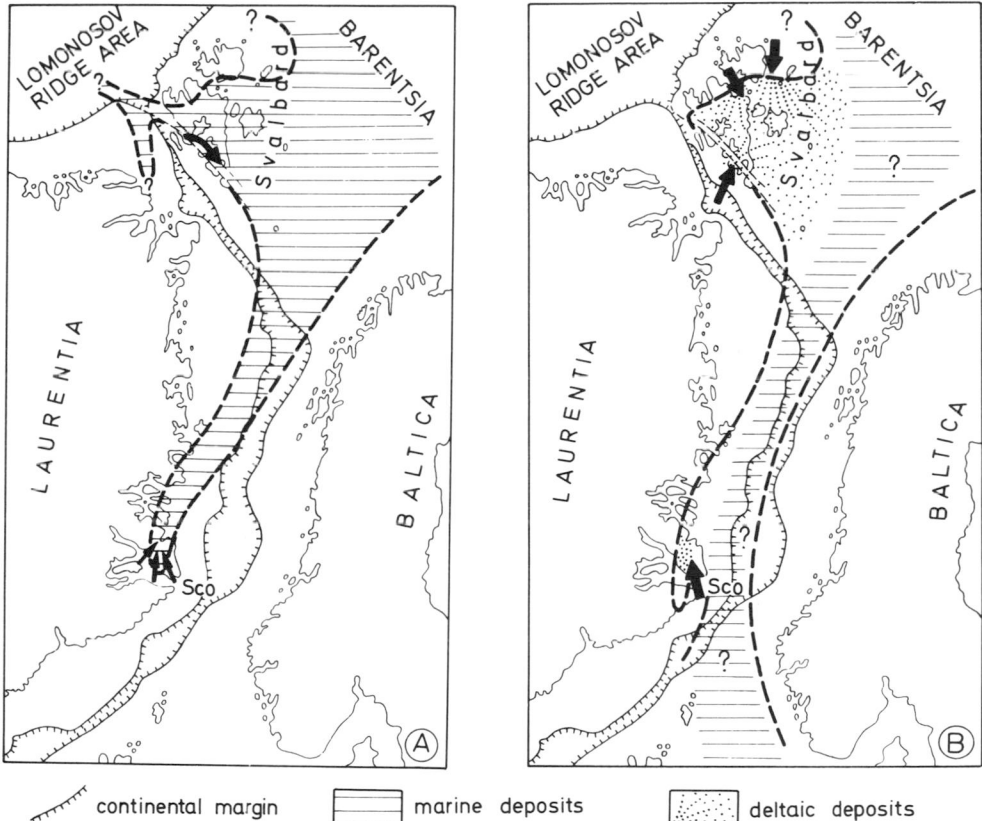

continental margin   marine deposits   deltaic deposits

Fig. 14. Palinspastic–paleogeographical reconstructions of the Lower Triassic (Griesbachian) (A) and Rhaetian–Hettangian (B) basins between Svalbard and Greenland. Predrift position of continental plates after Bullard *et al.* (1965), modified by Birkenmajer (1977). Arrows indicate transport direction in marine sediments (A) and direction of supply in deltaic deposits (B). (Sco) Scoresby Sund.

Permian fauna or of the paucity of fossils in the succeeding lowermost Triassic strata. The opening of a new seaway between the hitherto warm-water Upper Permian basin of Svalbard and a northern ocean (Fig. 14A) that introduced a cold current is a possible cause (Birkenmajer, 1977).

The subdivision of the Triassic–lowermost Jurassic deposits of Svalbard is shown in Fig. 15. The lower portion of the column up to the Carnian is represented by marine, predominantly dark-colored, often bituminous sediments: shale, siltstone, fine-grained sandstone, and usually impure limestone. Coarser clastics (conglomerate, sedimentary breccia) are uncommon but occur at the base of the succession along the western margin of the basin. Directional sedimentary structures and facies variations indicate that the basin axis lay between the mouths of Isfjorden and inner Hornsund, with a paleoslope to the SSW (Fig. 16). This is reflected in sediment thickness variation (Fig. 17), which shows the western platform, corresponding to a promontory of North Greenland, where sediments reached a thickness of 100–500 m, a main trough where the sediments reached a thickness of 500–800 m, and an eastern Barents platform with sediments 100–500 m thick.

A break in sedimentation caused by uplift along the western margin of the basin occurs at the base of the predominantly fresh-water sandstones, quartzites, and shales of the De Geerdalen Formation (Carnian–Hettangian). In the remaining areas of Svalbard, most authors accept a transition from the marine Tschermakfjellet Shale to the fresh-water to marine De Geerdalen Formation and its regional equivalents (Fig. 15). There is a marked shift of the sedimentary basin axis toward the east (northeast) at the boundary between the Carnian and Norian. The thickest deltaic-type sediments (460 m +) were laid down in eastern Svalbard. Western and northern sources of clastics were then active (Fig. 14B); the latter also could have been the source for the predominantly fluviatile Norian to Rhaetian sediments of Franz Josef Land (Dibner et al., 1962; Birkenmajer, 1977). Thin coal seams may be found in the predominantly fresh-water Upper Triassic—lowermost Jurassic sequence, and there is a good fossil flora (e.g., Nathorst, 1910; Vasilevskaja, 1972; Smith et al., 1975.

D. Jurassic through Lower Cretaceous Sedimentation

During the period of deltaic sedimentation in Svalbard at the Triassic–Jurassic boundary, the coastline receded southeastward, toward the central part of the Barents Shelf. Minor marine incursions recognizable in the predominantly fresh-water sequence of the De Geerdalen Formation and its equivalent units of eastern Svalbard probably came from that direction. Regression continued in the lowermost Jurassic, and it seems probable that during the Hettangian, land conditions were established in Svalbard. There

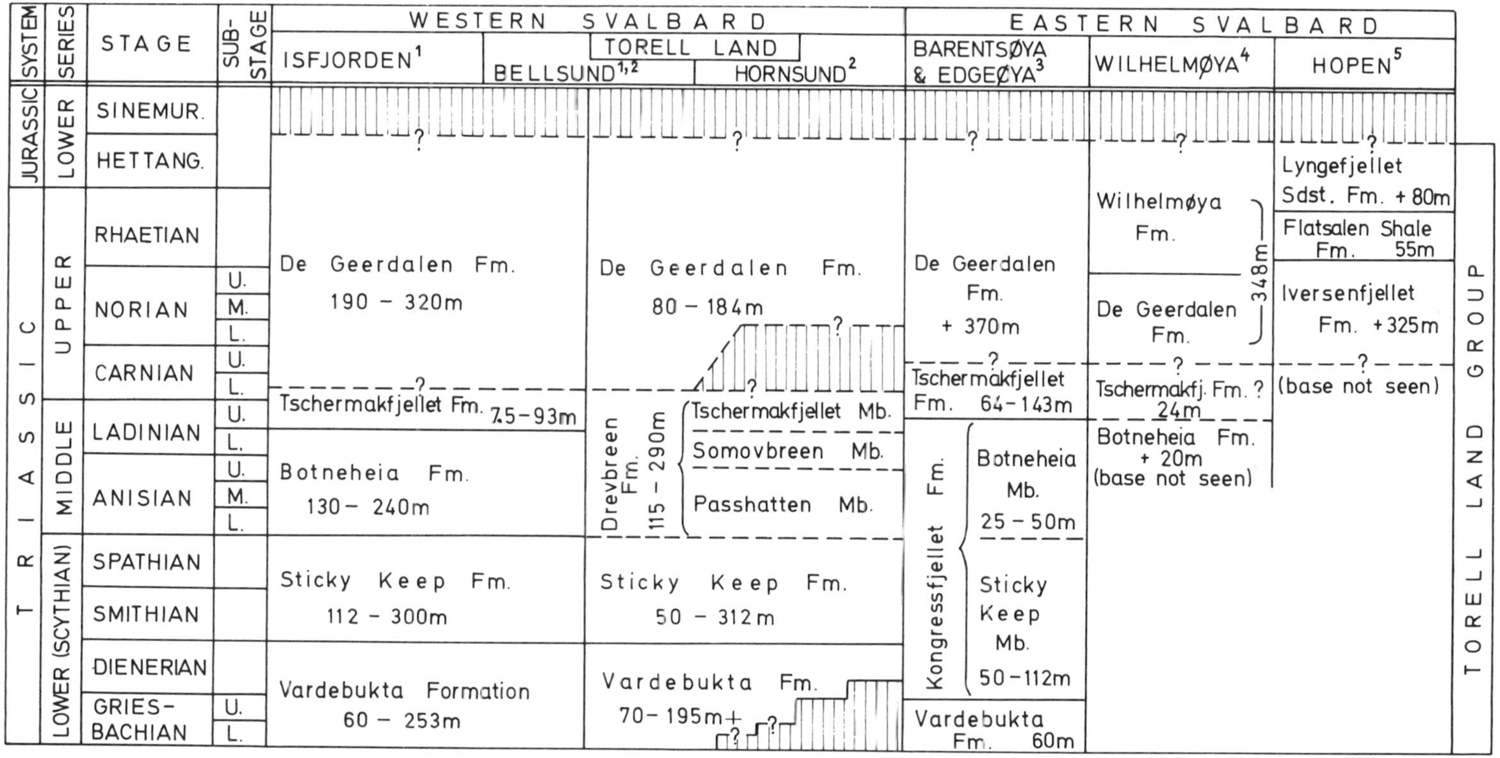

		WESTERN SVALBARD			EASTERN SVALBARD		
STAGE	SUB-STAGE	ISFJORDEN[1]	BELLSUND[1,2]	TORELL LAND HORNSUND[2]	BARENTSØYA & EDGEØYA[3]	WILHELMØYA[4]	HOPEN[5]
SINEMUR.							
HETTANG.							Lyngefjellet Sdst. Fm. +80m
RHAETIAN		De Geerdalen Fm. 190–320m	De Geerdalen Fm. 80–184m	De Geerdalen Fm.	De Geerdalen Fm. +370m	Wilhelmøya Fm.	Flatsalen Shale Fm. 55m
NORIAN	U. / M. / L.					De Geerdalen Fm. (348m)	Iversenfjellet Fm. +325m
CARNIAN	U. / L.						(base not seen)
LADINIAN	U.	Tschermakfjellet Fm. 75–93m	Drevbreen Fm. 115–290m (Tschermakfjellet Mb., Somovbreen Mb., Passhatten Mb.)	Tschermakfjellet Mb.	Tschermakfjellet Fm. 64–143m	Tschermakfj. Fm.? 24m	
ANISIAN	U. / M. / L.	Botneheia Fm. 130–240m			Botneheia Mb. 25–50m (Kongressfjellet Fm.)	Botneheia Fm. +20m (base not seen)	
SPATHIAN		Sticky Keep Fm. 112–300m	Sticky Keep Fm. 50–312m	Sticky Keep Fm.	Sticky Keep Mb. 50–112m (Kongressfjellet Fm.)		
SMITHIAN							
DIENERIAN		Vardebukta Formation 60–253m	Vardebukta Fm. 70–195m+	Vardebukta Fm.	Vardebukta Fm. 60m		
GRIES-BACHIAN	U. / L.						

TORELL LAND GROUP

Fig. 15. Triassic–lowermost Jurassic succession of Svalbard. (1) After Buchan et al. (1965), Tozer and Parker (1968), and Smith et al. (1975); (2) after Birkenmajer (1976); (3) after Flood et al. (1971b); (4) after Buchan et al. (1965), Worsley (1973), and Smith et al. (1975); (5) after Smith et al. (1975).

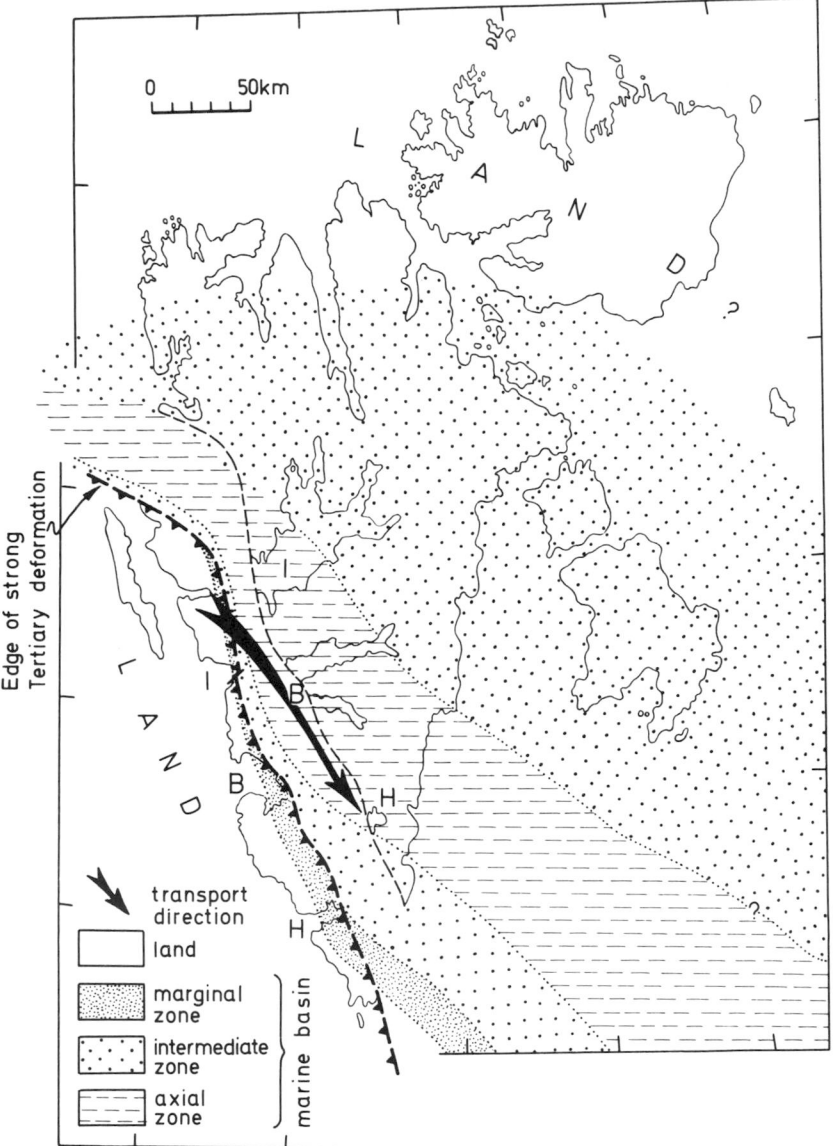

Fig. 16. Lower Scythian sea in Svalbard. Predrift palinspastic reconstruction after Birkenmajer (1977). (B) Bellsund; (H) Hornsund; (I) Isfjorden.

was probably a period of nondeposition during the Sinemurian (Fig. 18), followed by short incursions of the Pliensbachian (?), Toarcian, and Bajocian seas that deposited thin phosphatic layers (condensed sedimentation) with marine invertebrate shells (mainly ammonites and bivalves), which occur as secondary deposits in the Callovian Brentskardhaugen Bed at the base of the Janusfjellet Formation (*Aucella* or *Buchia* shale).

A lowland, which bordered the Jurassic sea on the west, supplied a small amount of predominantly clayey material during the sedimentation of the Janusfjellet Formation (Callovian–Hauterivian), which thus consists mostly of black shale with clay–ironstone intercalations and nodules, and subordinate, often ferruginous sandstone. The invertebrate fauna (ammonites, belemnites, bivalves, e.g., *Buchia*) shows a boreal character and close connections with epicontinental seas of the East European Platform (e.g., Frebold, 1930, 1951; Sokolov and Bodylevsky, 1931; Rozycki, 1959; Pchelina, 1965, 1967; Parker, 1967; Efremova, 1972; Birkenmajer, 1975a; Birkenmajer and Pugaczewska,

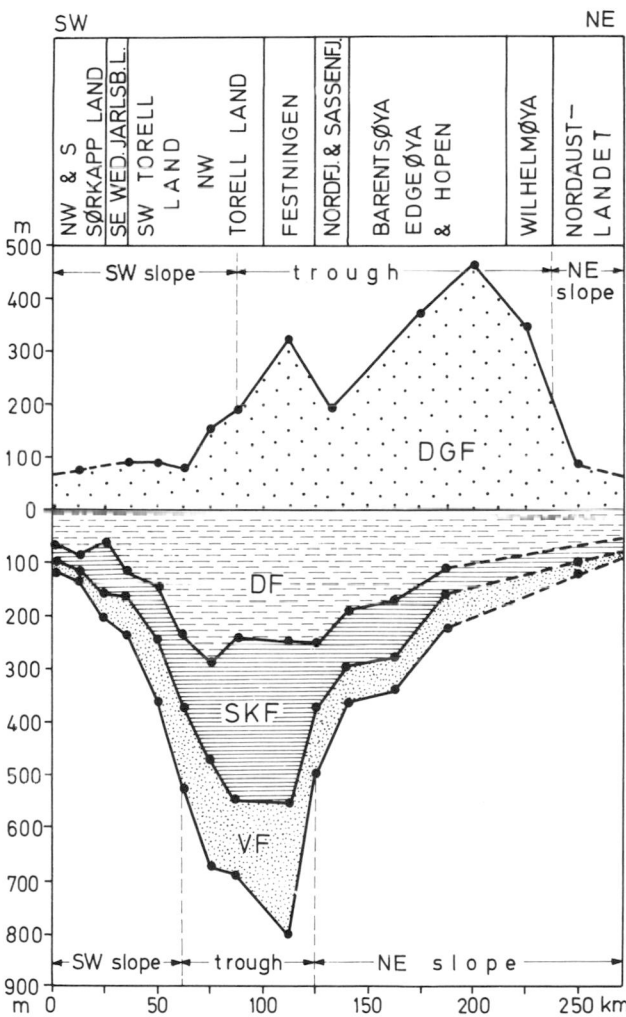

Fig. 17. Thickness variation of the Triassic–lowermost Jurassic sediments in Svalbard (after Birkenmajer, 1977). (DGF) De Geerdalen Formation and equivalents; (DF) Drevbreen Formation and equivalents; (SKF) Sticky Keep Formation; (VF) Vardebukta Formation (see Fig. 15).

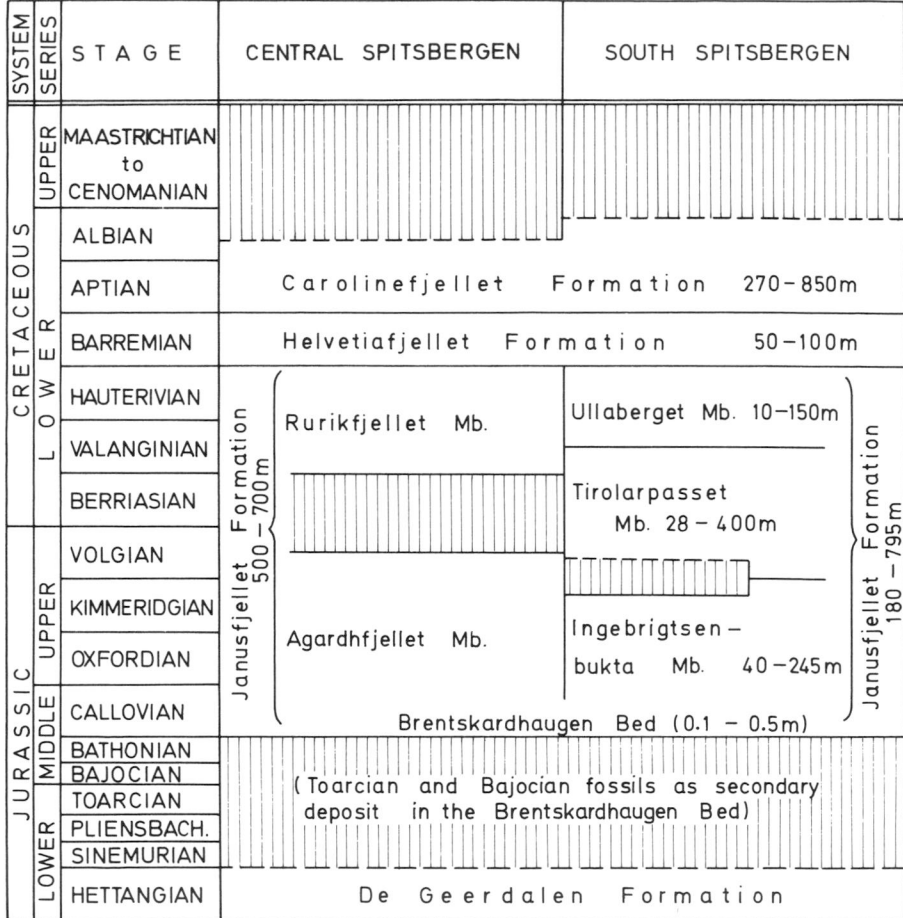

Fig. 18.   Jurassic and Cretaceous succession of Spitsbergen. After Rozycki (1959), Parker (1967), Nagy (1970), Major and Nagy (1972), and Birkenmajer (1975a).

1975). The extent of the Callovian to Hauterivian sediments in Svalbard is roughly the same as that of the Ladinian to Carnian sediments.

The monotonous sedimentation of the *Buchia* shale (Janusfjellet Formation) was interrupted first during the Upper Jurassic (Neocimmerian uplift) south of Van Keulenfjorden about the boundary of the Kimmeridgian and Volgian, then in inner Isfjorden near the boundary of the Volgian and Berriasian (Fig. 18). There was continuous marine sedimentation in southern Spitsbergen (Hornsund area). Local folding and faulting associated with minor basic intrusive activity (dolerite sills) between the deposition of the Agardhfjellet and Rurikfjellet Members (Fig. 18) is recognized in east-central Spitsbergen (Parker, 1966, 1967) and coincides with the southern prolongation of the Balliolbreen Fault (lineament), or the Billefjorden Fault Zone (see Harland *et al.*, 1974).

Another uplift, during the Barremian, was of regional dimension, and caused regression of the sea from the whole Svalbard area. Deltaic fans of cross-bedded sandstones supplied from several sources situated on the west, north, and southeast spread over Svalbard, and thin coal-bearing shale–sandstone sequences developed as floodplains between the fans (Helvetiafjellet Fm.). The sea reappeared in the Aptian, leaving a thick shale succession with several laminated sandstone horizons (Carolinefjellet Fm.) that may in part be Albian (Nagy, 1970). The Aptian–Albian sea was generally very shallow and deposition occurred close to wave base; however, no regressive deposits are recognized.

No Upper Cretaceous sediments have been found in Svalbard. There is evidence of warping and erosion, which is strongest in the north, and basal Tertiary (Danian–Montain) predominantly fresh-water strata lie upon different members of the Carolinefjellet Formation. This indicates a regional unconformity between Tertiary and Cretaceous rocks with a local angular discordance of up to 10 degrees (see Birkenmajer, 1972a), although both successions usually lie essentially parallel to each other.

E. Mesozoic Igneous Events

Basic volcanics of tholeiitic composition in the form of dolerite sills and dikes and basalt lava flows are regarded as a characteristic feature of platform development in Svalbard during the Mesozoic. The cliff-forming dolerite sills were intruded mainly into Carboniferous through Triassic sediments, while the dolerite dikes cut most of the rocks from the Precambrian through the Upper Jurassic (e.g., Backlund, 1907; Nathorst, 1910; Tyrrell and Sandford, 1933; Orvin, 1940; Birkenmajer and Morawski, 1960; Harland, 1961, 1973b).

Isotopic age determinations of the central Spitsbergen dolerites by Gayer et al. (1966) show a wide time distribution (110 to 149 m.y.). These are probably minimum ages, because the rocks examined were altered to some degree, and the K-Ar whole-rock method was used. A pre-Cenomanian but post-Middle Jurassic age seems probable. The structural evidence from east-central Spitsbergen indicates a Berriasian age for some of the intrusions that cut through the Agardhfjellet (Callovian to Volgian) but predate the Rurikfjellet (Valanginian to Hauterivian) deposition (Parker, 1966). The unconformity and hiatus between these two units could represent the time span for most if not all late Mesozoic intrusions in Svalbard (Harland, 1973b).

A younger stage of mafic volcanism in Svalbard is evidenced by capping lavas in Kong Karls Land which are post-Hauterivian, possibly Barremian or later. A Barremian "tuff conglomerate" in the upper part of the Helvetiafjellet Formation in Spitsbergen is regarded as correlative with this event (Harland, 1973b). In Franz Josef Land, basaltic lavas and tuffs alternate with

Hauterivian through Aptian and Albian(?) sediments, while dolerite sills and dikes cut through most of the Mesozoic succession, including the marine Lower Cenomanian (Dibner, 1962; Dibner et al., 1962).

It seems possible that basic volcanism started first in Spitsbergen (Berriasian or older), then shifted eastward to Kong Karls Land (Barremian or later) and Franz Josef Land (Hauterivian through Cenomanian), where it died out at the beginning of the Upper Cretaceous. Rafted dolerite pebbles occur as secondary deposits in Lower Tertiary strata of eastern Spitsbergen (Birkenmajer and Narebski, 1963; Birkenmajer et al., 1971).

F. Early Tertiary Sedimentation

The two major areas of Tertiary sediments in Svalbard are in the Main Spitsbergen Basin (Van Mijenfjorden Group) and in the Forlandsundet Graben (Forlandsundet Group). Smaller occurrences are known along the west coast of Spitsbergen south of Forlandsundet in Bellsund and in Sørkapp (Fig. 19). The main basin is 320 km long, widening from 10–15 km in the north to 90 km or more in the south. The Forlandsundet–Bellsund Basin is a minimum of 180 km long but only 10–20 km wide. Lack of Tertiary deposits on Spitsbergenbanken (between Skørkapp and Bjørnøya) suggests a closure of the Main Spitsbergen Basin in that direction (Edwards, 1975; Björlykke and Elverhøi, 1975; Ronnevik et al., 1975).

1. Main Spitsbergen Basin

The basal Tertiary strata in the Main Spitsbergen Basin are unconformable upon various members of the Aptian–Albian Carolinefjellet Formation to the south, and on Lower Triassic and Permian rocks to the north (Kongsfjorden). The stratigraphic succession (Fig. 20) generally follows Nathorst's (1910) subdivision into six units for which new lithostratigraphic names have been proposed by Major (1964; see also Harland, 1969; Birkenmajer, 1972a; Major and Nagy, 1972). Independent schemes have been proposed by Livšic (1967, 1974) and Vonderbank (1970).

Lower Paleocene (Danian–Montian) ages seem to be well established for the lower part of the succession in the Main Basin on mollusc and

$\longrightarrow$

Fig. 19.   Relation of Paleogene basins to the Alpine Foldbelt (A) and Alpine structural elements of Spitsbergen (B). Thick arrows denote direction of tectonic transport (after Birkenmajer, 1972b). (1) Paleogene sediments [(f) basin axis of the basal Firkanten Fm.; (b) basin axis for the successive Basilika Fm.]; (2) possible maximum extension of Paleogene basins; (3) structures produced by the Spitsbergenian Phase [(a) overthrusts and reverse faults; (b) foldbelt–major thrust zone and transitional zone; (c) normal faults; (d) zone of gentle folded foreland—transition zone]; (4) faults and fractures younger than the Spitsbergenian Phase; (5) centers of late Pleistocene/early Holocene volcanic activity. (B) Brøggerhalvøya; (K) Kopernikusfjellet; (Ke) Keilhaufjellet; (KL) Kapp Lyell; (M) Midterhuken.

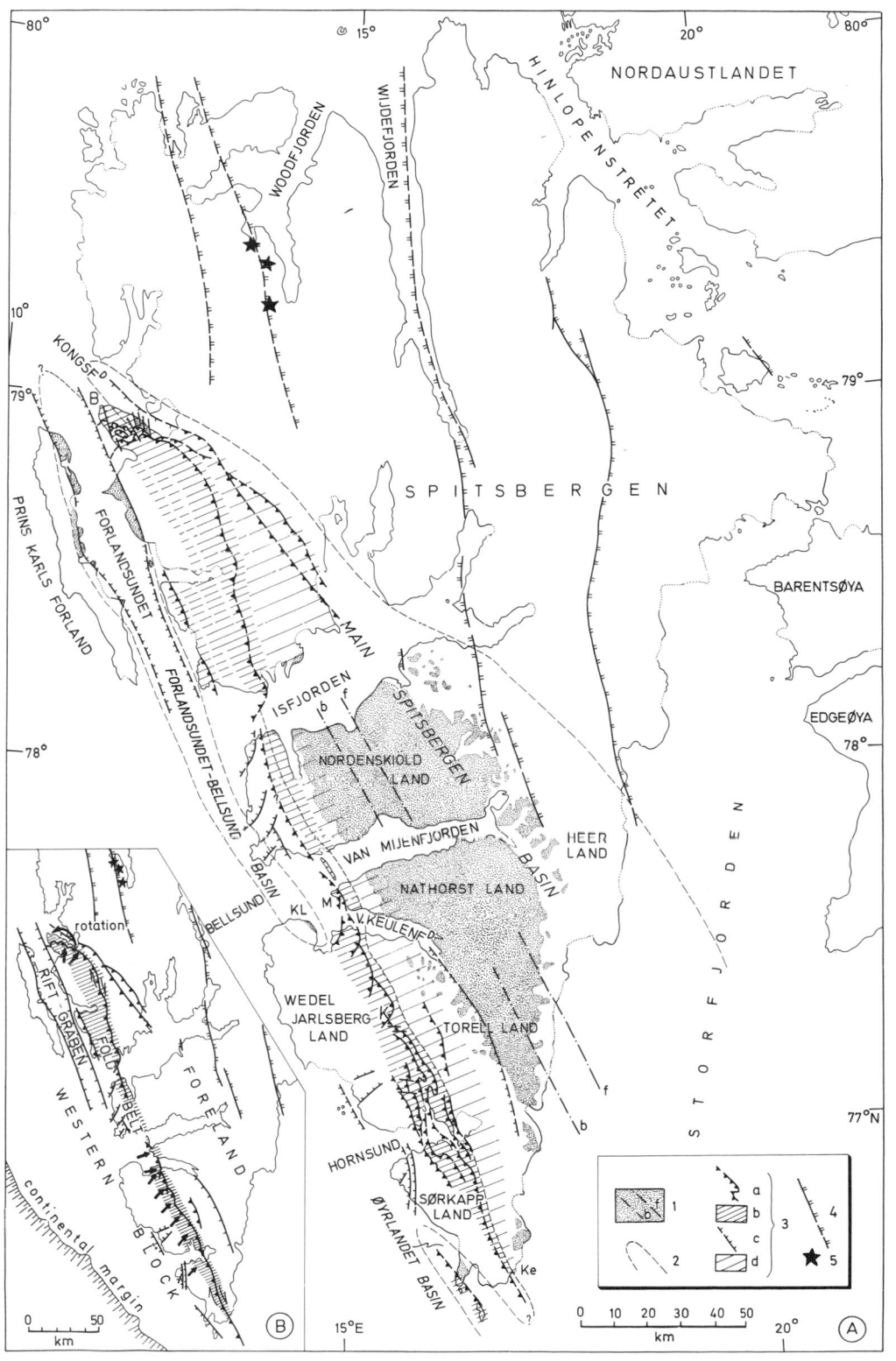

80°   15°   20°   80°

NORDAUSTLANDET

H I N L O P E N S T R E T E T

WOODFJORDEN
WIJDEFJORDEN

10°

KONGSF

79°   S P I T S B E R G E N   79°

PRINS KARLS FORLAND

FORLANDSUNDET

FORLANDSUNDET-BELLSUND

BARENTSØYA

MAIN

ISFJORDEN

SPITSBERGEN

EDGEØYA

78°   NORDENSKIOLD LAND   78°

BASIN

VAN MIJENFJORDEN   HEER LAND

BELLSUND   NATHORST LAND

KL M

V. KEULENF

ST ORFJORDEN

WEDEL JARLSBERG LAND

TOREL LAND

f

b

rotation

RIFT GRABEN

FOLD THRUST BELT

FORELAND

HORNSUND

77°N

SØRKAPP LAND

WESTERN BLOCK

continental margin

ØYRLANDET BASIN

Ke

0   50
km

B

15°E

	1
b	
	2

⊣⊢	a
▨	b
▨	c
▨	d

3

| ⊣⊢ | 4 |
| ★ | 5 |

0 10 20 30 40 50
km

20°   A

SUPPOSED AGE[1]	AGE by LIVŠIC (1974)	MAIN SPITSBERGEN BASIN Van Mijenfjorden Group[2]		FORLANDSUNDET GRABEN Forlandsundet Group[3]	
		MAJOR (1964); MAJOR & NAGY (1972)[4]	LIVŠIC (1967, 1973, 1974)	Forlandsundet (1967, 1973, 1974)	ATKINSON (1963)
LOWER EOCENE ?	OLIGO-CENE	Aspelintoppen Fm. 500–600m	Storvola Fm. 700 m +	Marchaislaguna Fm. 2,000m +	McVitie Formation ca 2,000 m and Sars Fm.
UPPER PALEOCENE ?	E O C E N E	Battfjellet Fm. 200–250m	Collinderodden Fm. 100–500m	Krokodillen Fm. 400m	
		Gilsonryggen Formation 125–300m	Frysjaodden Formation 200–400m	Reinhardpynten Fm. 210m / Sesshøgda Fm. 120–300m +	
		Sarkofagen Fm. 70–250m	Hollendardalen Fm. 130m	Selvågen Fm. 30–1000m +	Selvågen Cgl. Fm. 60–150m
			Grumantbyen Fm. 160–240m		
(DANIAN – MONTIAN) LOWER PALEOCENE	PALEOCENE	Basilika Fm. 50–220m	Colesbukta Fm. 20–350m		
		Firkanten Fm. 15–130m	Barentsburg Fm. 80–230m		

Fig. 20. Tertiary succession of Svalbard. (1) Age based on Ravn (1922), Rosenkrantz (1942), Manum (1960, 1962), and Vonderbank (1970); see also Birkenmajer (1972a). (2, 3) Group names after Harland (1969). (4) Thickness ranges from Nathorst (1910), Orvin (1940), Flood et al. (1971a), and Birkenmajer (1972a).

foraminiferal evidence (Ravn, 1922; Rosenkrantz, 1942; Vonderbank, 1970). The plant evidence is less conclusive stratigraphically, indicating a Paleocene to Eocene age (Manum, 1962). Upper Paleocene and Lower Eocene(?) ages are suggested for the youngest two formations on paleobotanical (Manum, 1962) and structural evidence (Harland, 1969, 1973a, 1975; Birkenmajer, 1972a), but much younger, Eocene to ?Miocene (Livšic, 1967) or Eocene to Oligocene (Livšic, 1973, 1974) ages have also been proposed.

Of the six Tertiary formations in the Main Basin, four (1st, 3rd, 5th, and 6th) consist predominantly of sandstone, the remaining two (2nd and 4th) of shale. The Firkanten Formation begins either with a basal conglomerate or with dark shale usually resting on weathered Lower Cretaceous rocks. Predominantly fresh-water light-colored sandstone with subordinate shale and conglomerate dominate the formation. Coal seams, mined at several sites, occur near the base, and shale intercalations with marine fossils appear higher in the succession.

The Basilika Formation consists of dark grey shale with infrequent marine fossils, often passing to sandy shale and fissile sandstone. The Sarkofagen Formation is represented mainly by homogeneous often glauconitic sandstone with poor marine fossils in the upper part. The Gilsonryggen Formation consists of black shale with marine fossils. Locally, a break in deposition and a basal conglomerate are recognized at the base of the shale. Thin tuff intercalations have been reported from the Firkanten, Basilika, and Gilsonryggen Formations.

Alternating black clayshale and grey-greenish flaggy sandstone with some marine faunas and with plant remains characterize the Battfjellet Formation. The nonmarine Aspelintoppen Formation consists predominantly of sandstone with subordinate siltstone, marl, and coal seams.

The Tertiary sequence in the Main Basin grows from 950 m at Isfjorden to 1500 m or more in Nathorst Land and still further south along the axis of the basin. There is a marked reduction in thickness of particular formations across the basin, both to the southwest and northeast. Low-profile borderlands that supplied clastics to the basin probably included northern, northeastern, and eastern Svalbard (north Spitsbergen, Nordaustlandet, Barentsøya, and Edgeøya) and an area along the west coast of Spitsbergen.

2. *Forlandsundet Graben*

In the Forlandsundet Graben, Tertiary sediments directly overlie the Precambrian metamorphic complex (Hecla Hoek). The Tertiary sequence, referred to as the Forlandsundet Group (Harland, 1969) has been subdivided by Atkinson (1963) into the Selvågen Conglomerate Formation at the base and the succeeding McVitie (correctly McVitiepynten) Formation on the west, and the Sars (correctly Sarsbukta) Formation on the east. Livšic (1967, 1973, 1974) subdivided the rocks overlying the Selvagen Conglomerate into four formations (Fig. 20).

The Forlandsundet Group is considered to be younger than Lower Paleocene (Orvin, 1940; Harland, 1961, 1969, 1973a; 1975; Atkinson, 1962, 1963; Livšic, 1967, 1973, 1974). However, its exact age remains uncertain. The microfauna from the Sarsbukta Formation gives an age of no greater than Upper Paleocene (Manum, 1960). Livšic (1973, 1974) accepts an Eocene (Upper Eocene) age for the lower four formations and suggests an Oligocene age for the fifth. His paleontological evidence is inconclusive.

The thickness of the Tertiary strata at Forlandsundet was estimated to be about 2100 m by Atkinson (1963), but Livšic (1973, Table 3; 1974, Table 2) gives a value closer to 4000 m. To the west, the sequence begins with the Selvågen Conglomerate, which includes boulders more than 1 m in diameter. The succeeding Sesshøgda Formation (and the Sarsbukta Fm. to the east of the graben) consists mainly of fresh-water sandstone, conglomerate, and shale with some thin coal seams. The succeeding Reinhardpynten Formation consists of siltstone with marine bivalves in the lower part. More shale and siltstone appear in the overlying Krokodillen Formation. The Marchaislaguna Formation consists of rhythmically alternating sandstones, siltstones, shales, and conglomerates with plant detritus. A slight unconformity appears at its base.

There is some reason to believe that the Forlandsundet sequence represents a sedimentary event postdating Tertiary sedimentation in the Main Spitsbergen Basin (Orvin, 1940; Harland, 1961, 1973a, 1975). The differences in lithology, and especially in the abundance of conglomerates, their thickness and the size of boulders, and the direct contact with strongly weathered Hecla Hoek rocks seem to favor an independent, fault-controlled basin with clastics possibly supplied from the north and northwest (Atkinson, 1963). The sediment corresponds to a diastrophic molasselike fill of an intramontane rift valley subsequent to folding of the Tertiary in the Main Basin (see Harland, 1961; Atkinson, 1962, 1963; Birkenmajer, 1972a).

V. TERTIARY FOLDING IN SVALBARD AND PLATE MOTIONS

During the Tertiary, Spitsbergen was subject to strong tectonic deformation. The Alpine structural pattern includes the zone of strongest deformation, the foldbelt, and the zones of less intense deformation, the foreland (central depression) and the hinterland (western block) (see Figs. 19 and 21).

The Alpine Foldbelt of Spitsbergen stretches nearly 300 km NNW–SSE from Brøggerhalvøya (Kongsfjorden) in the north to Keilhaufjellet (Sørkapp

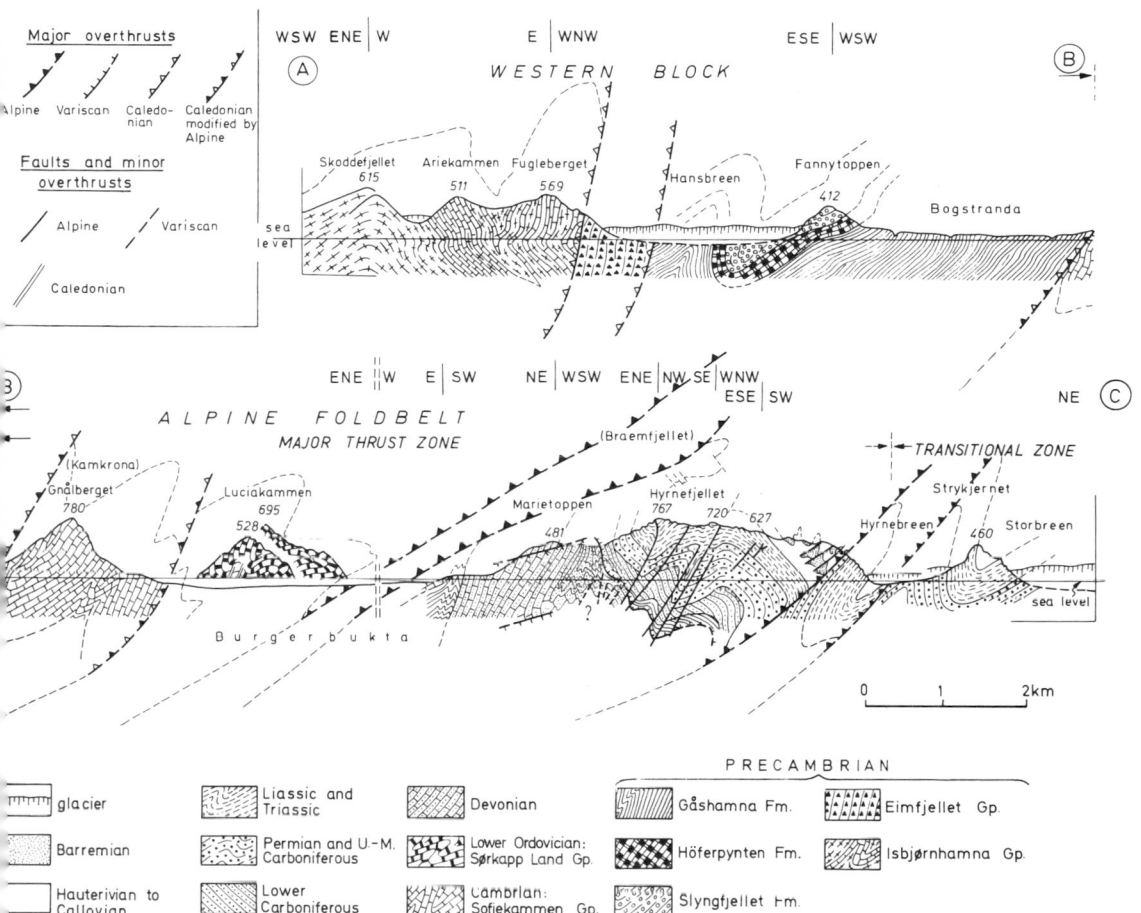

Fig. 21.   Caledonian, Variscan, and Alpine structures in south Spitsbergen: Hornsund (after Birkenmajer, 1972b).

Land) in the south. Deformation affected the western margin of the post-Caledonian platform of Svalbard, causing strong folding and low-angle thrusting in
a zone 10–20 km wide. The foldbelt consists of two structural units (Birkenmajer, 1972a,b): (1) the major thrust zone (i.e., the easternmost part of the
overthrust western block) and (2) the transitional zone (i.e., the western margin
of foreland). The major thrust zone has been studied in most detail at Brøggerhalvøya (Orvin, 1934; Barbaroux, 1966, 1967, 1968; Challinor, 1967; Harland and Horsfield, 1974) and in the area between Van Keulenfjorden and
Hornsund (Rozycki, 1959; Birkenmajer, 1960b, 1964, 1972a,b). At Brøggerhalvøya the thrusting was generally toward the north or northeast and
involved Precambrian to Paleogene rocks. Up to three (according to Challinor)
or five (according to Barbaroux) thrust sheets were formed which overrode one
another for distances of 1 to 4 km. Between Van Keulenfjorden and Hornsund

thrusting was toward the northeast (or ENE) and involved the Precambrian to Lower Cretaceous and Paleogene rocks. One or two thrust sheets were formed between Bellsund and Kopernikusfjellet (Torell Land), and up to five thrust sheets to the south of Kopernikusfjellet overrode one another for distances of up to 6 km.

Only the narrow western margin of the foreland (transitional zone: 1–1.5 km wide) was involved in intense tectonic deformation (small-scale thrusts, disharmonic folds, etc.). Further east, two to three major anticlines appear parallel to the zone of thrusting. Steeper and more complex close to the fold belt, they become increasingly gentle further away from it between Van Keulenfjorden and Hornsund (Rozycki, 1959; Birkenmajer, 1972b). Gentle undulations (depressions and swells) with axes parallel to the foldbelt are present in the axial part of the central depression between Isfjorden and Van Keulenfjorden (see Sokolov et al., 1968; Major and Nagy, 1972), while only a single brachyanticline appears in central Torell Land (Birkenmajer, 1972b) involving Paleogene and older strata. A tableland structure gently uptilted along the eastern flank of the central depression is characteristic of the foreland around Storfjorden.

The main Tertiary deformation phase in Spitsbergen has been called the "West Spitsbergen Orogeny" by Harland (1969; see also Harland and Horsfield, 1974) and the "Spitsbergenian Phase" by Birkenmajer (1972b). The age of folding is certainly post-Lower Paleocene. The upper age limit of the folding cannot as yet be determined with precision, and there is a serious difference of opinions concerning the age of the youngest Tertiary strata in the Main Basin (Fig. 20), which have been subjected to folding (see Harland, 1975). A pre-Oligocene date of folding is preferred by Harland (1969), Birkenmajer (1972a,b), and Harland and Horsfield (1974) as better corresponding to stages in the evolution of the North Atlantic Basin, but a post-Oligocene date has been proposed by Lowell (1972).

The Spitsbergenian Phase includes three stages of deformation (Birkenmajer, 1972a,b) corresponding in succession to (1) compression, (2) extension, and (3) compression. The first stage (1) involved a forceful push from the southwest of the western block, which was thrust over the southwest margin of the central depression of Spitsbergen. The folding proceeded in a shallow zone and no metamorphic alterations developed. Lowell (1972) assumed that the continental deposits at the top of the Tertiary sequence in the Main Basin were tilted while folding and thrusting occurred along the orogenic front to the west. Other authors (e.g., Harland, 1969, 1975; Birkenmajer, 1972a,b; Harland and Horsfield, 1974) consider the whole Van Mijenfjorden Group as predating the folding. The crustal shortening in the foldbelt due to thrusting amounted to about 10 km in the north (Brøggerhalvøya), growing to about 15 km in the south (between Bellsund and Sørkapp). Birkenmajer (1972a,b) assumed that the whole NNW-trending strip of the west coast of Spitsbergen between Kongsfjorden and Sørkapp, including the major thrust zone and its hinterland (western block and continental shelf to the west of it), had been translated to the NNW from a more southerly location, possibly some 30 or so kilometers, and sandwiched between the continental blocks of Svalbard and Greenland. At Kongsfjorden this zone collided with the rigid mass of northeast Spitsbergen, thus producing a counterclockwise rotation of the Alpine and Caledonian structures at Brøggerhalvøya. It seems that the contact of the major thrust zone with the western margin of the foreland is a dextral transcurrent fault.

In the second stage (2), subsequent to the main deformation and as a result of vertical arching of the western block under the west–east compression, a rift valley (graben) developed between Forlandsundet and Bellsund (possibly also at Øyrlandet in Sørkapp Land) west of the foldbelt, and was simultaneously filled with orogenic molasse (Forlandsundet Group). The termination of the Spitsbergenian Phase, the third stage (3), was marked by the formation of wrench faults in the foldbelt, and an analogous system of faults in the western block, reflecting horizontal stress from the west (Birkenmajer, 1972a,b).

The Tertiary foldbelt of Spitsbergen differs from other collision belts in many respects, one of which is the absence of geosynclinal characteristics in the folded late Paleozoic to Tertiary sediments which, on the contrary, are typical platform cover. The total lack of volcanism related to folding and the absence of a molasse-filled foredeep indicate that no major subduction of a downgoing lithospheric slab was associated with the Tertiary foldbelt of Spitsbergen.

The Tertiary compression in the North Atlantic area, which produced the Alpine Foldbelt of Spitsbergen and equivalent but much weaker structures in North Greenland (Kronprins Christians Land and Peary Land), and possibly some structures along the margin of the Barents Shelf between Bjørnøya and Norway (see Chapters 5 and 7), is considered to be the result of a dextral horizontal translation of the Greenland continental block with respect to the Svalbard–Barents continental block. Translation occurred along a geosuture known as the De Geer Line (shear zone) and its present expression— the Spitsbergen Fracture Zone. This idea was put forward by Harland (1965, 1967, 1969; see also Harland and Horsfield, 1974) and tested on a clay model by Lowell (1972) who was able to reproduce en echelon thrust-fold structures resembling in many respects those of the Spitsbergen belt. Harland (1969, 1973a) and Birkenmajer (1972a) correlated the stages of Tertiary deformation in Spitsbergen with the stages of opening and evolution of the North Atlantic Basin, but with inadequate dating of the events, much still remains conjectural. Three successive stages of Tertiary folding in Spitsbergen express changes in the stress regime in the zone of transcurrence between the moving continental plates, i.e., (1) transpression, (2) transtension, and (3) transpression (Harland's, 1971, terminology). All these events need to be better dated with stratigraphic, isotopic, and paleomagnetic evidence from the study of Svalbard and the adjoining Greenland sea floor.

VI. NEOGENE TO QUATERNARY EVENTS

The post-Spitsbergenian Phase tectonic events include the formation or rejuvenation of great north–south-striking normal faults, which cut through the Carboniferous-Paleogene platform cover, and the Devonian and Hecla Hoek rocks. These faults are roughly parallel to the present Spitsbergen Fracture Zone, which is a major feature of the ocean floor between Svalbard and Greenland (Fig. 22). The history of some of these faults, e.g., the Billefjorden Fault Zone (see Harland et al., 1974), goes back to the Upper Devonian. The post-Paleogene tension generated at the Spitsbergen Fracture Zone as the result of ocean-floor spreading may have been responsible for rejuvenation of these old fractures in Svalbard. The Spitsbergen Fracture Zone itself may have developed along one such preexisting fracture (De Geer Line). The deep fault line passing through Bockfjorden, northwest Spitsbergen, served as a feeder for basaltic central volcanoes active during late Pleistocene to early Holocene times (see Hoel and Holtedahl, 1911; Gjelsvik, 1963; Harland, 1969), and Recent hot springs are known from that area (Hoel and Holtedahl, 1911; Orvin, 1940). The occurrence of olivine and peridotite nodules in the basaltic (trachydoleritic) lava of the Sverrefjellet volcano, Bockfjorden, indicates that the Bockfjorden Fault may be a very deep fracture, perhaps reaching to the Mohorovičić discontinuity (Gjelsvik, 1963).

Plateau basalts (plagioclase basalts), which cap many mountains in northwest Spitsbergen (Hoel and Holtedahl, 1911), have been considered as possibly Tertiary (Harland, 1969, 1973b), extruded in connection with the opening of the Arctic–North Atlantic Ocean Basin. No radiometric data are available to support this assumption, while paleomagnetic investigation of the lavas by Sandal and Halvorsen (1973) suggests a Mesozoic, probably Cretaceous age for the basalts, with late Tertiary remagnetization.

Traces of a peneplain usually referred to the late Tertiary (e.g., Harland, 1961) cut through the Alpine Foldbelt and younger fault structures at altitudes near or above 500 m above sea level. There is no evidence of sediment deposition in Svalbard during the Neogene. The sediments, if they once existed, have been removed by intense late Tertiary and Quaternary denudation. Late Pleistocene and Holocene glacial deposits, and Holocene raised marine beaches, terraces, and cliffs, are characteristic features of the Svalbard archipelago and border the present zone of glaciation.

During the Pleistocene, the subsidence and uplift in Svalbard seem to have

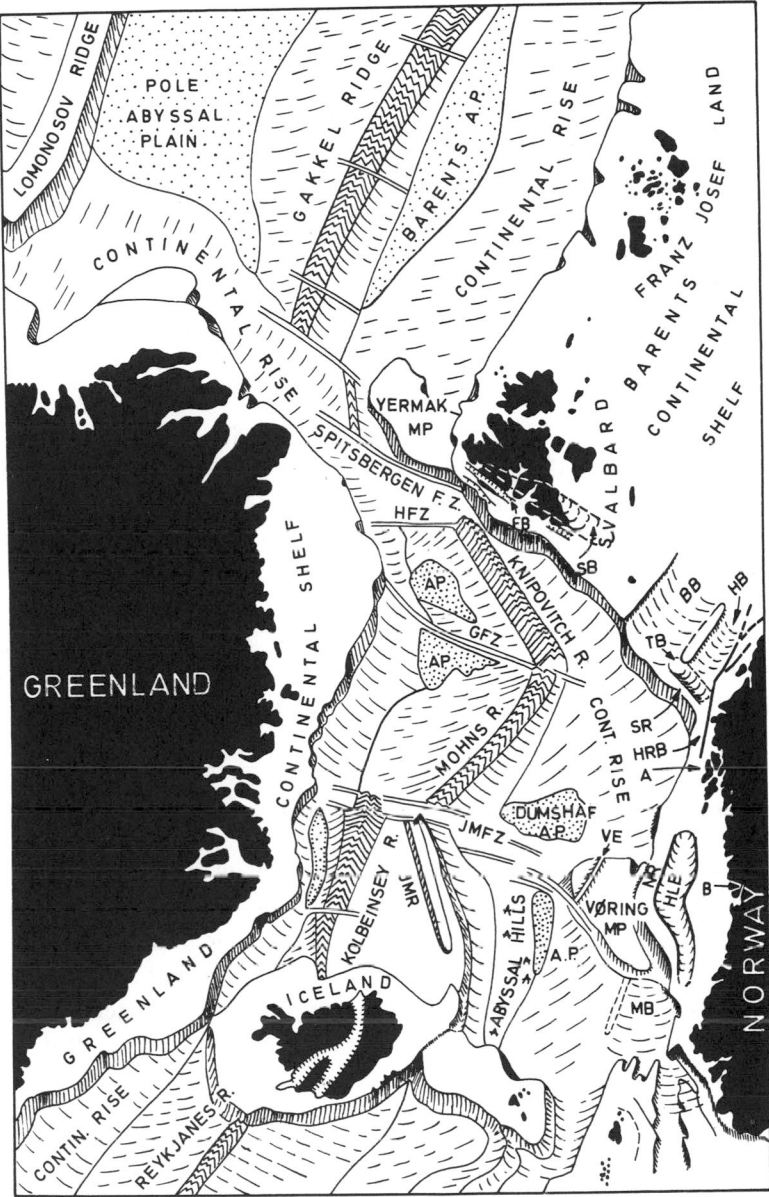

Fig. 22. Selected Cenozoic structural elements of the Barents–Norwegian continental shelf and adjoining ocean bottom. Compiled from Johnson and Heezen (1967), Johnson (1975), and Ronnevik *et al.* (1975). (A) Andøya; (AP) Abyssal Plain; (B) Beitstadfjorden; (BB) Bjørnøyrenna Basin; (FB) Forlandsundet Basin; (GFZ) Greenland Fracture Zone; (HB) Hammerfest Basin; (HFZ) Hovgaard Fracture Zone; (HLB) Helgeland Basin; (HRB) Harstad Basin; (JMFZ) Jan Mayen Fracture Zone; (JMR) Jan Mayen Ridge; (MB) Møre Basin; (MP) Marginal Plateau; (NR) Nordland Ridge; (SB) Main Spitsbergen Basin; (SR) Senja Ridge; (TB) Tromsø Basin; (VE) Vøring Escarpment.

been controlled by the growth and decay of an ice cap, either local or covering the whole Barents Shelf (see Corbel, 1965; Grosswald *et al.*, 1967; Schytt *et al.*, 1967). Vertical displacement of the archipelago during that time probably did not exceed 300–400 m, as may be inferred from the highest altitudes of raised marine strandlines in southern Spitsbergen, recognized by Werenskiold (1922) and considered to be Pleistocene (Birkenmajer, 1960c). The Holocene land uplift pattern based on radiocarbon-dated samples (shells, wood, whale bones) indicates maximum uplift in eastern Svalbard, apparently in the direction toward the center of the Barents Shelf, with isobases gradually decreasing toward the edge of the continental margin (see Grosswald *et al.*, 1967; Schytt *et al.*, 1967). The greatest uplift during the past 10,000 years occurred in eastern and central Svalbard, where it amounted to 60–90 m (Feyling-Hanssen and Olsson, 1959–1960; Feyling-Hanssen, 1965a), with lower values (10–40 m) along the west and southwest coast of Spitsbergen (Birkenmajer and Olsson, 1971) and in Nordaustlandet (Blake, 1961). Minor marine transgressions occurred in the coastal zone of Svalbard around 5000 and 3000 B.P. (Feyling-Hanssen, 1965b; Hyvärinen, 1969; Birkenmajer and Olsson, 1971).

VII. GEOLOGICAL FEATURES OF THE WESTERN BARENTS SEA AND THE CONTINENTAL MARGIN BETWEEN SVALBARD AND SCANDINAVIA

A. Structure of the Continental Margin and Slope

The northern and western continental margins of the Barents Shelf mark the nearly rectangular northwest termination of the Eurasian continental plate (Fig. 22). The northern margin is parallel to the Arctic mid-ocean ridge (Gakkel Ridge or Nansen Cordillera), and the western margin runs between Spitsbergen and Bjørnøya, subparallel to the Spitsbergen Fracture Zone and the Atlantic mid-ocean ridge (Knipovitch Ridge or Atka Ridge). The continental margin of Scandinavia between Lofoten and Trondheim runs subparallel to the Atlantic mid-ocean ridge (Mohns Ridge).

1. *Svalbard Continental Margin and Rise*

The continental shelf north, west, and south of Svalbard is both deeper and more irregular than normal continental shelves, and in this respect resembles the Greenland continental shelf (Johnson and Eckhoff, 1966; Johnson, 1975). North of Spitsbergen, at 13°E, the shelf break occurs at about 220 m and the shelf is approximately 70 km wide. Along the northwest tip of Spitsbergen, the shelf break occurs at approximately 55 m and the shelf narrows to less than 18 km. Along the west coast of Spitsbergen, the shelf margin lies between 200 and 220 m and the shelf broadens to 43 km between Magda-lenefjorden and Kongsfjorden (Kapp Mitra), but narrows again to about 27 km off Prins Karls Forland where the shelf break occurs slightly deeper than the 90 m isobath. South of Isfjorden, along the southwest coast of Spitsbergen, the shelf averages 35 km in width, the shelf margin lowers to 275–295 m and to 355 m at 76°N. West of Bjørnøya the shelf break rises again to about 240 m (Johnson and Eckhoff, 1966; Johnson, 1975).

North of Hinlopenstretet, west of Kongsfjorden and off Isfjorden (Isfjordrenna) the shelf is cut by deep canyons, possibly of glacial origin. Similar, the smaller, canyons occur south of Bellsund and south of Hornsund. A broad depression (Storfjordrenna) in the sea floor is found between Spitsbergen and Bjørnøya, and a flat depression (Bjørnøyrenna) occurs between Bjørnøya and Norway.

Aeromagnetic surveys of Svalbard (Åm, 1975a) indicate that to the north of Spitsbergen and Nordaustlandet, the shelf has a sediment cover (Devonian and younger?) of up to 4 km. Crystalline basement of the west coast of Spitsbergen, with some strongly magnetic zones, is recognizable as far south as Bjørnøya, and principal Tertiary fault structures such as the Billefjorden Fault Zone and the Forlandsundet Graben are traceable northward in the shelf area. A fault line parallel to the coast has been recognized immediately west of Prins Karls Foreland (itself a horst structure) by the aeromagnetic survey (Åm, 1975a), confirming an earlier interpretation by Holtedahl (1937) based on sub-marine topography.

The base of the continental slope west of Svalbard ranges between 1100 and 1650 m, deepening northward. The slope gradients are lower on the south and become steeper on the north where the Spitsbergen Fracture Zone lies at the base of the slope. The continental slope there descends by a series of small benches to depths greater than 3300 m, with the two largest benches, at 1010–1280 m and at 1535–1645 m, occurring between 79° and 80°N. These benches possibly reflect a step-fault structure of the northwestern margin of the continental plate.

At the northwest corner of Spitsbergen, close to the Yermak Marginal Plateau, the base of the continental slope occurs at only 400 m and is marked by steep escarpment with a 1:15 gradient (Johnson and Eckhoff, 1966; Johnson, 1975). Further to the east, between Svalbard and Franz Josef Land, the depths of continental slope are inadequately known.

The continental rise seems to be rather wide between Svalbard and Franz Josef Land where it descends to the Barents Abyssal Plain (known also as the Nansen Basin; see Gakkel and Dibner, 1967). It narrows toward the northwest where it terminates against the Gakkel Ridge and the Spitsbergen Fracture Zone.

The continental rise off the west margin of the Barents Shelf is well developed in the southern sector, between Bjørnøya and Norway, where it averages 160 km in width. The rise begins to narrow at 76°30′N, i.e., off the southern tip of Spitsbergen, and terminates at 77°30′N (i.e., opposite Bellsund), where the mid-ocean ridge (Knipovitch Ridge) intersects the continental block of Svalbard. The rift mountains of the ridge dam up the detritus supplied by Svalbard; thus the continental rise may have low enough gradients to be classified as a perched plain (Johnson, 1975).

2. *Yermak Marginal Plateau*

The Yermak Marginal Plateau (Dietz and Shumway, 1961; Johnson and Eckhoff, 1966; Johnson, 1975) lies off the northwestern corner of Spitsbergen and is separated by a steep escarpment (1:10 to 1:15) from the continental margin of Svalbard. The plateau extends for 235 km. from about 80°N to beyond 82°N and although slightly irregular, the width averages in excess of 125 km. The crest of the plateau lies at depths of less than 915 m, with prominent 490- and 570-m benches in its southern part. The marginal escarpment on the east side has a gradient of about 1:10 and descends to about 1830–2200 m (possibly structural bench). The northern slope of the plateau is described as gentle, gradually merging into the continental rise. The western margin of the plateau drops off rapidly to a poorly known basin at 3300–3500 m, while the southwestern portion of the plateau parallels the Spitsbergen Fracture Zone.

No information is so far available as to the geological structure of the Yermak Marginal Plateau. Its position just northwest of Svalbard in a corner between the Spitsbergen Fracture Zone and the Gakkel Ridge may indicate that the plateau is a downfaulted splinter of the Svalbard Shelf left in this position after the spreading Gakkel Ridge sliced the Lomonosov Ridge from the northern margin of the Barents–Kara Shelf. The plateau lies so close to Quaternary centers of volcanic activity in Bockfjorden, northwest Spitsbergen, that volcanic material of Quaternary or even late Tertiary age could be expected in the structure of the plateau (see Birkenmajer, 1972a).

3. *Norwegian Continental Margin and Rise*

The continental shelf along the west coast of Norway varies in width from about 60 km seawards of Stad (62°N) to 200 km off Helgeland (65–67°N), and to less than 60 km off Lofoten–Vesterålen (68–69°N). Off Norway, between Stad and Lofoten, the shelf area shows the presence of sediments exceeding 10 km in thickness, according to aeromagnetic investigations (Åm, 1975b). A very characteristic feature of the shelf is the presence of depressions parallel to the coastline. These marginal channels divide the shelf into inner and outer portions, separating seaward an uneven, rocky ground of submerged strandflat from the shoreward area of banks with loose glacial material (Holtedahl, 1960; Gakkel and Dibner, 1967; Johnson, 1975; Holtedahl and Bjerkli, 1975). The channels are believed to have been carved during periods of lowered sea level by fluvial and glacial erosion along the principal geological boundary between the hard metamorphic bedrock of the Caledonian mountain chain and less resistant, possibly Mesozoic, platform cover. Faulting, possibly in Tertiary time, indicated by faults cutting through the Oxfordian to Lower Cretaceous sediments at Andøya, Vesterålen Islands (see Ørvig, 1960; Dalland, 1975), took place along the strandflat border with relative uplift on the land side. The sea floor expressions of such faults are channels transverse to the banks. Profiles across the shelf and slope show blocks of hard bedrock rising stepwise to the shoreline. The sediment–basement interface dips towards the shelf edge with particular sedimentary (Mesozoic to Tertiary?) horizons dipping toward the ocean floor (Eldholm, 1970; Talwani and Eldholm, 1972; Johnson, 1975; Sellevoll, 1975).

The Norwegian continental slope is a steep escarpment with its base at the 2000 m isobath. Progradation of the slope is evident in some areas (Johnson, 1975).

Several structural elements have recently been discovered at the junction of the Barents Shelf and the Norwegian Shelf (Sundvor, 1975; Åm, 1975b; Ronnevik et al., 1975). A major sedimentary basin occurs between the Bjørnøya high (also known as Spitsbergenbanken or Svalbardbanken; Fig. 22) and the Norwegian mainland. This is a wedge of undisturbed or slightly disturbed, probably Tertiary, sediments of great thickness, which may indicate that the Barents Shelf was an area of erosion during the deposition of the wedge. At 72°N a narrow structural high running WSW–ENE divides this basin into the northern Bjørnøyrenna Basin and the southern Hammerfest Basin (Fig. 22). The older sediments increase in thickness over the high while the younger ones clearly thin over it. The structural high coincides with the magnetic high and may be the result of deep-seated plutonic activity.

The Hammerfest Basin is separated on the south from the Norwegian mainland by a SW–NE-trending fracture system that dies out on the east. On the west, the basin terminates against the N–S-trending Tromsø Basin, which is bounded offshore by the Senja Ridge. The older layers of the Tromsø Basin are downfaulted along eastern and western hinge lines, whereas the younger layers are deformed by epeirogenic subsidence. Diapirs, possibly consisting of Permian evaporites, have been recognized in the axial part of the Tromsø Basin.

The deeper sediments of the Senja Ridge are highly disturbed perhaps as a result of the Tertiary compression recognized in the Alpine Foldbelt of Spitsbergen. The fault-bounded N–S-trending Tromsø Basin may be an equivalent to the Tertiary Forlandsundet Graben in Spitsbergen. The

WSW–ENE-trending fault system within the main basin between Bjørnøya and the Norwegian mainland seems to be younger than the major Tertiary compression (transpression) as recognized in Spitsbergen. It may correspond to a stage of tension caused by sea-floor spreading along an axis situated close to the continental margin of Norway. The extinct Vøring rift axis (Talwani and Eldholm, 1972) with Upper Paleocene volcanic activity may have been the cause.

South of the Tromsø Basin and parallel to the coastline of Norway, the poorly known Harstad Basin stretches south as far as Lofoten. Still further south, the Lofoten–Møre–Shelf shows a structure (see Grønlie and Ramberg, 1971; Ronnevik et al., 1975; Åm, 1975b) that is characteristic of passive continental margins. On the landward side, the shelf sediments are bordered by early Paleozoic to Precambrian rocks of the Caledonian Foldbelt (see Strand and Kulling, 1972; Nicholson, 1974; Kvale, 1975). Two small inliers of Mesozoic rocks are known, the one at Andøya, Vesterålen Islands, with Upper Jurassic to Lower Cretaceous sediments (Ørvig, 1960; Dalland, 1975), and another in Beitstadfjorden, north of Trondheim, where erratic boulders of Middle Jurassic age have been found (Os Vigran, 1970; Oftedahl, 1972). Sandstone fragments, which may be of Upper Cretaceous age, have been dredged along a submarine canyon 20–40 km NNW of Andenes at Andøya (Manum, 1966).

The Nordland Ridge, situated at the edge of the continental margin, separates the outer Vøring Basin and its prolongation south of the Jan Mayen Fracture Zone, the Møre Basin, from the inner Helgeland Basin (Ronnevik et al., 1975). The Helgeland Basin is a shallow oval basin probably filled by Tertiary and older sediments. A core of bottom sediment taken at 63°52'N, 7°49'E contained palynomorphs and foraminifers indicating an Upper Cretaceous–Lower Tertiary age of the bedrock, possibly in situ (Bugge et al., 1975). The faulting within the Helgeland Basin seems to stop at a reflector believed to occur near the top of the Lower Cretaceous (Ronnevik et al., 1975).

4. Vøring Marginal Plateau

The Vøring Marginal Plateau is the most prominent feature of the Norwegian continental margin. Bounded on the east by the Nordland Ridge and on the south by the Jan Mayen Fracture Zone (Fig. 22), it has an areal extent of over 25,000 km², and the margin of its generally smooth surface lies close to the 900 m isobath. Seismic reflection profiles reveal at least 1.5 km of sediments, which may date back to the Paleozoic, overlying a continental basement (Johnson and Heezen, 1967; Johnson et al., 1968; Grønlie and Ramberg, 1971; Talwani and Eldholm, 1972; Johnson, 1975; Sellevoll, 1975). Faulting prior to the basal Tertiary reflector horizon delimites the eastern flank of the

Vøring Basin (Ronnevik *et al.*, 1975). Its western boundary seems to correspond to a buried structure 18 km wide parallel to the continental margin and plunging to the southeast (Johnson *et al.*, 1968). This structure has been interpreted as a buried escarpment (Vøring Escarpment), probably of volcanic origin, which may mark the original rift site for the Tertiary opening of the Norwegian Sea (Talwani and Eldholm, 1972). Some other "escarpments" seem to exist in the Vøring Marginal Plateau, indicating a rather complicated pattern of rifting, combined with vertical block movements (Sellevoll, 1975). Volcanic lutites with sands derived from submarine erosion of basalt were cored on the west slope of the plateau (66°21′N, 00°18′E). Their planktonic and benthonic microfauna indicates an Upper Paleocene age for the volcanic activity.

The plateau has a fairly gentle slope toward the ocean floor. The slope, which seems to be dominated by prograded sedimentary layers, has a base at depths from 3100 to 3500 m in the south.

The large size of the Vøring Plateau was a major obstacle in obtaining a good fit of the opposing continental margins of Scandinavia and Greenland in the predrift position of the continents (Birkenmajer, 1972a). The evidence provided by Talwani and Eldholm (1972) and Sellevoll (1975) seems to indicate that only a part of the plateau, landward of the buried escarpment, is continental, the remaining outward portion being oceanic in origin. However, the existence of "continental crust" overlain by basaltic lava west of the Vøring Escarpment cannot yet be ruled out (Sellevoll, 1975).

B. Relation to Mid-Ocean Ridges

The mid-ocean ridge of the Norwegian–Greenland Sea splits into three major segments (Fig. 22): the Kolbeinsey Ridge, the Mohns Ridge, and the Knipovitch Ridge. The Kolbeinsey Ridge strikes NNE from Iceland and is offset about 200 km to the east by the Jan Mayen Fracture Zone (transform fault). The Mohns Ridge strikes NE from Jan Mayen toward Bjørnøya and turns sharply at its intersection with the Greenland Fracture Zone, from whence it continues due north as the Knipovitch Ridge (known also as the Atka Ridge) as far as the southwestern border of the Yermak Plateau, where it is offset about 540 km to the northwest by the Spitsbergen Fracture Zone. Then it resumes its course as the Gakkel Ridge (known also as the Nansen Cordillera or Ridge) through the Arctic Ocean, parallel to the northern margin of the Barents–Kara Shelf (see Johnson and Eckhoff, 1966; Johnson and Heezen, 1967; Avery *et al.*, 1968; Vogt and Avery, 1974; Johnson, 1975).

The Mohns Ridge shows characteristics of an active mid-ocean ridge. It has a deep axial rift that apparently bifurcates at 4°W. A positive magnetic anomaly of about 1000 gamma is associated with the rift valley. The youth of the ridge is suggested by the lack of sediment along its crest. The ridge is separated from the Norwegian continental margin by a wide continental rise and by the Dumshaf Abyssal Plain. A 200-m-thick "turbidite" sequence covers about 1 km of pelagic sediments on this plain (Johnson et al., 1970).

The Knipovitch Ridge is inadequately known. Its crestal area is barely recognizable on magnetic anomalies, which do not show the characteristically high-amplitude Brunhes normal anomaly. A well-marked trench abruptly terminating at 78°30′N off Prins Karls Forland accompanies the ridge along its eastern flank (Bjørnøya Trench of Johnson, 1975), but its character is unknown. Johnson (op. cit.) considers the possibility that this trench could have been the site of destruction of sea floor within recent times and that the axis of the Knipovitch Ridge, either active or recently inactive, is located 50 km west of the trench. However, he does not exclude the possibility that the Bjørnøya Trench is an axial valley of the Knipovitch Ridge, and that its low magnetic signature is due to sediment fill.

The character of the Spitsbergen Fracture Zone (Johnson and Eckhoff, 1966) is even less well known than the Knipovitch Ridge. Both J. T. Wilson (1965) and Harland (1969) suggested that this zone, known also as the De Geer Line (shear zone), is a transform fault. Data from Johnson and Heezen (1967) fail to reveal any well-defined ridge and rift system; however, Vogt et al. (1970) reinterpreted the northern termination of the zone (between 80° and 85°N) as a very narrow ridge crest dissected by northwest-trending transform faults. Earthquake data as interpreted by Horsfield and Maton (1970) confirm the existence of a steeply dipping plane striking 310°. Tracings of four echograms across the Spitsbergen Fracture Zone reveal an elongated depression with depths in excess of 4390 m, lying between 79°10′ and 80°N. It shoals rapidly at both ends. The trench is flanked to the southwest by a ridge with minimum depths of less than 2380 m. Southwest of this ridge lies a major trough with depths greater than 3100 m (Johnson and Eckhoff, 1966; Johnson, 1975).

The Gakkel Ridge (Nansen Cordillera or Ridge), considered to be the active boundary of the Eurasian and Arctic plates in the Arctic Ocean, parallels the northern continental margin of Eurasia between Spitsbergen and Severnaya Zemlya (North Land)–Taimyr Peninsula. Sea-floor spreading along this ridge generated the adjoining basins known as the Barents Abyssal Plain (Nansen Basin) and the Pole Abyssal Plain (Amundsen Basin) in approximately the last 40 m.y. (see review by Harland, 1973a).

VIII. CONCLUDING REMARKS

The geological history of Svalbard, with its long record of sedimentary, igneous, and metamorphic events and vertical and horizontal movements, provides a valuable standard for tracing the complex evolution of a major plate boundary between Eurasia and Laurentia.

(1)   Prior to the main Caledonian diastrophism this boundary lay within a vast geosynclinal trough, the folded remains of which constitute the East Greenland and Svalbard Caledonian Foldbelts. During an early eugeosynclinal stage, between 9 and 11.5 km of sediments and volcanics were laid down in the trough in Svalbard (Torellian Basin), and there is evidence of diastrophism along the western and eastern margins of the trough on approaching the Laurentia and Barents cratons (Birkenmajer, 1975b). This stage corresponds with the proto-Iapetus stage of the early Atlantic (Harland and Gayer, 1972; Harland, 1973a) and with pre-Carolinidian sedimentation in the East Greenland Geosyncline (see Haller, 1971; Birkelund et al., 1974).

(2)   During the late Precambrian to early Paleozoic stages, two superimposed miogeosynclinal basins (Jarlsbergian and Hornsundian) developed in the Svalbard trough, with 1.5–7 and 0.8–2.3 km of sediments, respectively. These basins are defined by major unconformities at the base of the late Precambrian and Cambrian along the western margin of the trough (Birkenmajer, 1975b). Miogeosynclinal sedimentation prevailed during late Precambrian to Ordovician time in the East Greenland trough (Haller, 1971; Birkelund et al., 1974). This implies that the plate boundary was relocated eastward to the Scandinavian eugeosynclinal trough that, in its northeastern termination, wedged between the Barents and Baltic (Fennoscandian) cratons (see Harland and Gayer, 1972; Harland, 1973a; Birkenmajer, 1975b; Siedlecka, 1975). This stage would correspond with the Iapetus stage of the early Atlantic of Harland and Gayer (1972).

(3)   The rates of sedimentation for the Precambrian to Ordovician stages of Svalbard are difficult to calculate owing to the inadequate age determination of unit boundaries and to the unknown duration of sedimentary breaks in the succession. Average rates of vertical movement in meters per million years (m/m.y.) have been calculated by Harland et al. (1974): 60–200 m/m.y. for the eugeosynclinal stage (Stubendorff Supergroup), and 76–200 m/m.y. (Veteranen Gp.), 20 m/m.y. (Akedemikerbreen Gp.), 8 m/m.y. (Polarisbreen Gp.), and 12.5 m/m.y. (Oslobreen Gp.) for the miogeosynclinal stages.

(4)   The main Caledonian Orogeny in Svalbard, East Greenland, and Scandinavia brought the early Atlantic geosyncline to a close. In Svalbard, the main metamorphic event at about 440–450 m.y. suggests a late Ordovician age for the Ny Friesland Orogeny (Gee, 1972) or a Silurian age if the latest recorded metamorphic event of 419–436 m.y. is taken into account (Harland et al., 1974). The orogeny was followed by migmatization and plutonism at about 390–400 m.y. (i.e., near the Silurian–Devonian boundary) and by local folding at about the same time (Gayer et al., 1966; Gee, 1972). In East Greenland, the main orogeny has been determined as Silurian (420–400 m.y.) with subsequent

late Caledonian spasms extending into the Devonian (400–350 m.y.; see Haller, 1971; Birkelund *et al.*, 1974).

In Scandinavia, three of the seven Caledonian disturbances distinguished seem to be of major importance (Kvale, 1975): (1) Lower Ordovician Trondheim Phase at about 490 m.y. (intrusion and metamorphism), (2) Upper Ordovician(?) Ekne Phase, about 445–435 m.y. (intrusion and metamorphism), (3) Upper Silurian Ardennian Phase, about 415–405 m.y., with large-scale horizontal displacement and subsequent major uplift.

Geosynclinal sedimentation continued through the Ordovician–Silurian boundary in the North Greenland Foldbelt, with Silurian turbidites at the top of the stratigraphic column. The folding there seems to be pre-Pennsylvanian, either late Caledonian or Variscan. However, isotopic determinations from folded Lower Paleozoic schistose rocks gave Tertiary ages attributed to a Laramidian regional reheating (Dawes, 1971).

Caledonian structures should continue in the Barents Shelf, where an important tectonic discontinuity may cross the shelf, separating the Norwegian and Svalbard orogenic belts (Harland, 1973*a*).

(5) The Devonian (or late Silurian–Devonian), predominantly fresh-water deposits (Old Red Sandstone) in Svalbard represent a late-orogenic intramontane molasse stage preceding the platform stage. The average rate of vertical motion during the Devonian amounts to 138 m/m.y. (Harland *et al.*, 1974).

(6) Renewed tectonic activity during late Devonian Svalbardian movements caused the formation of grabens with both vertical and horizontal displacements along the faults. The horizontal displacement along the Balliolbreen transcurrent fault is estimated to be between 200 to 1000 km (Friend and Moody-Stuart, 1972; Harland and Gayer, 1972; Harland *et al.*, 1974). There seems to be no evidence of any significant rotation caused by these movements (Storetvedt, 1972).

(7) The Svalbardian faulting (transcurrence) established a disjunctive pattern and defined the framework of the post-Caledonian platform in Svalbard and the Barents Shelf.

(8) The tectonic framework of the Devonian rift graben in Svalbard was subject to recurrent disturbances during Carboniferous through Paleogene times.

(8.1) In south Spitsbergen, two Variscan phases, a mid-Carboniferous Adriabukta Phase (= Erzgebirgian Phase ?) and a mid-Permian (Saalic) Phase produced unconformities. The first of these phases also expressed itself as thrusting of the eastern margin of the Devonian graben westward, and strong folding of the Devonian and Lower Carboniferous strata (Birkenmajer, 1964, 1975*b*). Several unconformities are recognized in the Mesozoic succession, particularly Eocimmerian, Mesocimmerian, and Neocimmerian, the latter mainly along the eastern edge of the Devonian basin (Balliolbreen Fault or Billefjorden Fault Zone—see Harland *et al.*, 1974) where basic volcanic activity accompanied the Neocimmerian movements.

(8.2) The western block bordering the Devonian graben assumed a generally passive attitude with respect to the platform as a source of clastics. The block itself underwent strong vertical displacement during the Lower Carboniferous, but from Middle Carboniferous until Lower Paleogene time it behaved as a leveled, low-profile landmass. Supposedly it was the northeastern tip of the Greenland block, seldom invaded by seas, neither rising nor subsiding appreciably, while the main subsidence took place further east in the central depression of Svalbard (Birkenmajer, 1972*b*). The central depression seems to continue for a considerable distance (at least 200 km) to the south as far as Spitsbergenbanken (Svalbardbanken) in the form of a gentle syncline. Study of gravel and clay fractions (Edwards, 1975; Björlykke and Elverhøi, 1975) suggests that the western and eastern limbs of the syncline consist of Triassic rocks, while in the flat core Jurassic and Lower Cretaceous sediments occur. Tertiary strata are missing.

(8.3) The rates of deposition and subsidence during Carboniferous through Paleogene time give low values for most stratigraphic units, which are characteristic of a relatively stable platform (Birkenmajer, 1972*a*, 1976; Harland, 1973*a,b;* Harland *et al.*, 1974; Livšic, 1974). The average rates of vertical motion were between 12 and 25 m/m.y. for Carboniferous through Lower Cretaceous time, and only during the Aptian–Albian (Carolinefjellet Fm.) do they increase to about 64 m/m.y. During the Tertiary (Van Mijenfjorden Gp.) they increase to 150 m./m.y. (Harland *et al.*, 1974).

(9) The breakup of Laurasia near the Cretaceous–Tertiary boundary proceeded north of Iceland along two major structural lines, the traces of which are now approximately the edges of the continental shelves of Greenland and Norway–Svalbard; the Harland Line to the south, and the De Geer Line to the north (Fig. 23). The De Geer Line fracture system is recognizable in Svalbard as a series of north–south-trending faults active since the late Devonian. The Harland Line fracture system is recognizable in the central East Greenland coastal zone as step faults parallel to the continental margin, repeatedly rejuvenated since the late Permian. The Great Glen Fault of Scotland and its northward prolongation in Shetland (Storetvedt, 1974) and still further north (Kvale, 1975) seem to fit well with the Harland Line.

In the predrift position of the continents (see Bullard *et al.*, 1965; Harland, 1967, 1969, 1973*a*; Birkenmajer, 1972*a*; Kvale, 1975) the junction of these systems formed an obtuse angle of 120°(Fig. 23), which is still recognizable in the sharp turns of continental margins at about 70°N on the eastern, and at 72–73°N on the western sides of the Greenland Sea (Fig. 22). At Andøya, North Norway, situated close to this junction, two systems of faults have been

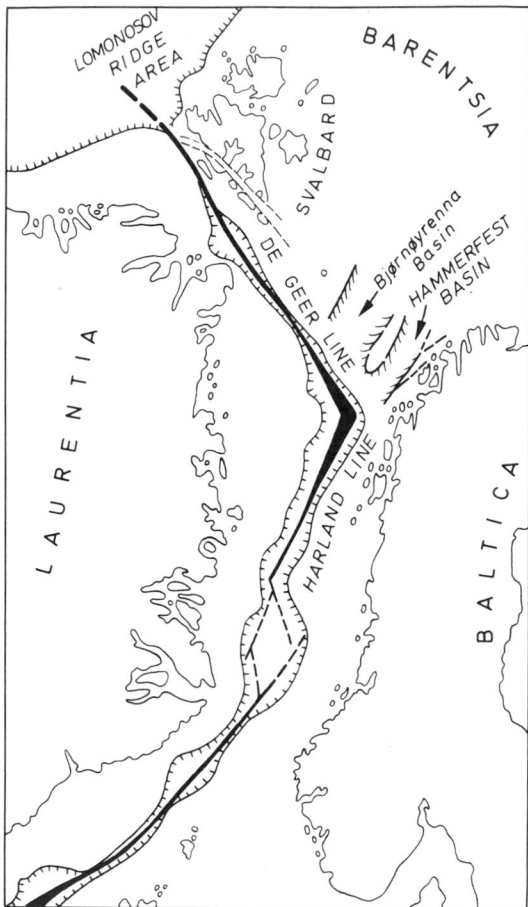

Fig. 23. Rupture of the Laurasia continent around 60 my. ago. The De Geer and Harland Lines represent active arms; the Hammerfest–Bjørnøy-renna Basins (probably Tertiary) apparently correspond to the failed arm of a triple junction on the Barents Shelf. (Modified from Birkenmajer, 1972a.)

distinguished (Dalland, 1975), of which the predominant, NNE–SSW-trending ones apparently parallel the De Geer Line, while the subordinate, WSW–ENE-running faults follow the Harland Line. The faulting in Andøya is mostly post-early Cretaceous in age, and is part of the Tertiary fault system of the Norwegian continental shelf, but is history goes back to Mid-Upper Jurassic and Lower Cretaceous (Dalland, 1975). Kvale (1975) argues that the Vøring Marginal Plateau was situated at the junction before its sinistral displacement some 350–450 km due south.

(9.1)   The separation of Norway and Greenland north of Iceland along the Harland Line began at about 60–70 m.y. ago, as indicated by the study of magnetic anomalies on the Reykjanes Ridge (Avery *et al.*, 1968; Vogt *et al.*, 1970). The then active sea-floor spreading axis, the Aegir Ridge, was located along the line of abyssal hills (seamounts) in the central Norwegian Sea (Johnson and Heezen, 1967; Johnson, 1975) and continued northward as a buried volcanic ridge on the Vøring Marginal Plateau (Talwani and Eldholm, 1972). The spreading rates in the Norwegian–Greenland Basin decreased from the initial 1.7 cm/yr 60 m.y. ago to 0.8–0.7 cm/yr 40 m.y. ago (Vogt *et al.*, 1970). About 60 m.y. ago the Lomonosov Ridge began to separate from the northern margin of the Eurasian Shelf due to the growth of the Gakkel Ridge (Pitman and Talwani, 1972; Vogt and Avery, 1974).

The phase of early Tertiary extension discussed earlier resulted in the formation of a huge pile of plateau basalts (tholeiites) in East Greenland, dated as 65 to 40 m.y. (Beckinsale *et al.*, 1970). In Svalbard, the main Tertiary sedimentary basin of Spitsbergen was formed subparallel to the De Geer Line. In the southwest part of the Barents Shelf, the Bjørnøyrenna and Hammerfest Basins with probable Tertiary sediment fill (Ronnevik *et al.*, 1975) apparently mark the failed arm of a triple junction, the other active arms being the De Geer and Harland Lines (Fig. 23).

(9.2) The succeeding stage of early Tertiary compression (or transpression) is well evidenced in Svalbard by strong folding and thrusting along the west coast of Spitsbergen due to dextral translation of the Greenland block relative to the Svalbard–Barents block along the De Geer Line (see Harland, 1967, 1969, 1973*a*; Birkenmajer, 1972*a,b;* Lowell, 1972; Harland and Horsfield, 1974). The foldbelt probably continues to the south as far as 71°N along the western margin of the Barents Shelf, as indicated by the highly disturbed deeper sediments of the Senja Ridge (Ronnevik *et al.*, 1975). The dating of the event is controversial: it is either late Paleocene-Eocene or younger (see discussion by Harland, 1975).

(9.3) The Forlandsundet Graben formed by extension (rift valley), filled with thick Tertiary molasse, and succeeded the main folding of Svalbard. The Tromsø Basin (possibly superimposed upon the older Hammerfest Basin) seems to be another rift graben of similar age and origin. Both structures are parallel to the De Geer Line. Their age is uncertain, either Paleocene-Eocene or younger.

(9.4) Wrench faulting subsequent to rifting, and transverse to the De Geer Line, has been related to minor west–east compression (Birkenmajer, 1972*a,b*). The age of the faulting is uncertain.

(9.5) Low rates of sea-floor spreading, 0.8 to 0.6 cm/yr north of Iceland, between 40 m.y. and 18–20 m.y. ago (Vogt *et al.*, 1970; Vogt and Avery, 1974), i.e., during Oligocene–Lower Miocene time, may reflect the stages of Alpine folding in Svalbard (9.2–9.4) or only the latest compressional stage (9.4) of the Spitsbergenian Phase.

(9.6) Between 20 and 10 m.y. ago, the spreading rates along the mid-ocean ridge between Greenland and Europe accelerated, and since 10 m.y. ago these rates measured in the direction of relative plate motion have ranged from 1.4 cm/yr in the Norwegian–Greenland Basin to possibly 0.55 cm/yr in the

Eurasia Basin along the Gakkel Ridge (see Pitman and Talwani, 1972; Vogt and Avery, 1974). Faulting of Oligocene–Miocene age parallel to the coast near Kap Brewster, East Greenland (Birkenmajer, 1972*d*) coincides with this extension stage. The Aegir Ridge probably became extinct before 30 m.y. ago and the spreading axis in the Norwegian–Greenland Basin shifted westward to an intermediate position (at about 18.5 m.y.), possibly splitting the Jan Mayen Ridge from the Greenland continental shelf, and jumping westward again (at about 9.5 m.y.) to form the Kolbeinsey Ridge (e.g., Vogt and Avery, 1974; Johnson, 1975; Laughton, 1975).

(9.7) Late Cenozoic activation of north–south-directed faults in Spits- bergen, some of which served as feeder veins for late Pleistocene–early Holocene basaltic volcanoes (see Harland, 1969; Birkenmajer, 1972*a*), may correlate with tension generated at the Knipovitch Ridge. Oblique spreading of the Greenland sea floor parallel to the Spitsbergen and Greenland Fracture Zones at a rate of probably less than 1 cm/yr is considered by Johnson (1975) as a possible explanation for the anomalous structure of the sea floor between Svalbard and North Greenland.

(9.8) No significant rotation of central Spitsbergen with respect to Scandinavia seems to have resulted from Cenozoic growth of the North Atlantic Ocean, but extensive displacements along the De Geer and Harland Lines are detectable (Storetvedt, 1972). Rotation of tectonic structures along the west coast of Spitsbergen is related to Tertiary folding (Birkenmajer, 1972*a,b*).

REFERENCES

Åm, K., 1975*a*, Magnetic profiling over Svalbard and surrounding shelf areas, *Nor. Polarinst. Arbok* 1973, p. 87–99.

Åm, K., 1975*b*, Aeromagnetic basement complex mapping north of latitude 64°N, Norway, *Nor. Geol. Unders. Publ.*, v. 29(316), p. 351–374.

Atkinson, D. J., 1962, Tectonic control of sedimentation and the interpretation of sediment alternation in the Tertiary of Prince Charles Foreland, Spitsbergen, *Bull. Geol. Soc. Am.*, v. 73, p. 343–364.

Atkinson, D. J., 1963, Tertiary rocks of Spitsbergen, *Bull. Am. Assoc. Petrol. Geol.*, v. 47, p. 302–323.

Atlasov, I. P., and Dibner, V. D., eds., 1964, *Tektonicheskaya karta Arktiki i Subarktiki* (1:5,000,000), Leningrad: Nauch. Issled. Inst. Geol. Arktiki.

Avery, O. E., Burton, G. D., and Heirtzler, J. D., 1968, An aeromagnetic survey of the Norwegian Sea, *J. Geophys. Res.*, v. 73, p. 4583–4600.

Backlund, H., 1907, Les diabases du Spitzberg oriental. Miss. Sci. pour le méridien au Spitzberg, 1899–1901 Mission Russe, *Uppsala Univ. Geol. Inst. Bull.*, v. 2(9B), p. 1–29.

Barbaroux, L., 1966, Contribution à l'étude tectonique de la Presqu'île de Brøgger (Spitsbergen), *Bull. Soc. Géol. Fr.*, v. 8, p. 560–566.

Barbaroux, L., 1967, *Étude géologique et sédimentologique de la Presqu'île de Brøgger, Baie du Roi, Vestspitsbergen*, Thèse, Fac. Sci. Univ. d'Aix, Marseille, Pts. I (185 pp.) & II (Annexe).

Barbaroux, L., 1968, Superposition des styles tectoniques et virgation forcée au Vestspitsbergen (79° lat. Nord), *C. R. Acad. Sci., Paris, v.* 266, p. 871–874.

Beckinsale, R. D., Brooks, C. K. and Rex, D. C., 1970, K–Ar ages for the Tertiary of East Greenland, *Medd. Dansk Geol. For.*, v. 20(1), p. 27–37.

Birkelund, T., Perch-Nielsen, K., Bridgewater, D., and Higgins, A. K., 1974, An outline of the geology of the Atlantic coast of Greenland, in: *The Ocean Basins and Margins*, v. 2, Nairn, A. E. M., and Stehli, F. G., eds., New York: Plenum Press, p. 125–159.

Birkenmajer, K., 1958, Preliminary report on the stratigraphy of the Hecla Hoek Formation in Wedel Jarlsberg Land, Vestspitsbergen, *Bull. Acad. Pol. Sci. Sér. Sci. Chim. Géol. Géogr.*, v. 6(2), p. 143–150.

Birkenmajer, K., 1959, Report on the geological investigations of the Hornsund area, Vestspitsbergen, in 1958. I. The Hecla Hoek Formation, *Bull. Acad. Pol. Sci. Sér. Sci. Chim. Géol. Géogr.*, v. 7(2), p. 129–136.

Birkenmajer, K., 1960a, Relation of the Cambrian to the Precambrian in Hornsund, Vestspitsbergen, 21st Int. Geol. Congr., Norden Copenhagen, 1960, Pt. 8, 64–74.

Birkenmajer, K., 1960b, Geological sketch of the Hornsund area (supplement to the guide for Excursion A 16, "Aspects of the Geology of Svalbard"), 21st Int. Geol. Congr., Norden, Copenhagen, 1960, 12 p.

Birkenmajer, K., 1960c, Raised marine features of the Hornsund area, Vestspitsbergen, *Stud. Geol. Pol.*, v. 5, 95 pp.

Birkenmajer, K., 1964, Devonian, Carboniferous, and Permian formations of Hornsund, Vestspitsbergen, *Stud. Geol. Pol.*, v. 11, p. 47–123.

Birkenmajer, K., 1972a, Tertiary history of Spitsbergen and continental drift, *Acta Geol. Pol.*, v. 22(2), p. 193–213.

Birkenmajer, K., 1972b, Alpine fold belt of Spitsbergen, 24th Int. Geol. Congr., Montreal, 1972, Sec. 3, p. 282–292.

Birkenmajer, K., 1972c, Cross-bedding and stromatolites in the Precambrian Höferpynten Dolomite Formation of Sørkapp Land, Spitsbergen, *Nor. Polarinst. Arbok* 1970, p. 128–145.

Birkenmajer, K., 1972d, Report on investigations of Tertiary sediments at Kap Brewster, Scoresby Sund, East Greenland, *Rapp. Groenl. Geol. Unders.*, no. 48, p. 85–91.

Birkenmajer, K., 1975a, Jurassic and Lower Cretaceous sedimentary formations of SW Torell Land, Spitsbergen, *Stud. Geol. Pol.*, v. 44, p. 7–44.

Birkenmajer, K., 1975b, Caledonides of Svalbard and plate tectonics, *Bull. Geol. Soc. Den.*, v. 24, p. 1–19.

Birkenmajer, K., 1977, Triassic sedimentary formations of the Hornsund area, Spitsbergen, *Stud. Geol. Pol.*, v. 51, p. 1–74.

Birkenmajer, K., and Morawski, T., 1960, Dolerite intrusions of Wedel Jarlsberg Land, Vestspitsbergen, *Stud. Geol. Pol.*, v. 4, p. 103–123.

Birkenmajer, K., and Narebski, W., 1960, Precambrian amphibolite complex and granitization phenomena in Wedel Jarlsberg Land, Vestspitsbergen, *Stud. Geol. Pol.*, v. 4, p. 37–82.

Birkenmajer, K., and Narebski, W., 1963, Dolerite drift blocks in marine Tertiary of Sørkapp Land, and some remarks on the geology of the eastern part of that area, *Nor. Polarinst. Arbok* 1962, p. 68–79.

Birkenmajer, K., and Olsson, I. U.,. 1971, Radiocarbon dating of raised marine terraces at Hornsund, Spitsbergen, and the problem of land uplift, *Nor. Polarinst. Arbok* 1969, p. 17–43.

Birkenmajer, K., and Pugaczewska, H., 1975, Jurassic and Lower Cretaceous marine fauna of SW Torell Land, Spitsbergen, *Stud. Geol. Pol.*, v. 44, p. 45–88.

Birkenmajer, K., and Wojciechowski, J., 1964, On the age of ore-bearing veins of the Hornsund area, Vestspitsbergen, *Stud. Geol., Pol.*, v. 11, p. 179–184.

Birkenmajer, K., Fedorowski, J., and Smulikowski, W., 1971, Igneous and fossiliferous sedimentary drift pebbles in marine Tertiary of Torell Land, Spitsbergen, *Nor. Polarinst. Arbok* 1970, p. 146–164.

Björlykke, K., and Elverhøi, A., 1975, Reworking of Mesozoic clayey material in the Northwestern part of the Barents Sea, *Mar. Geol.*, v. 18, p. M29–M34.

Blake, W., Jr., 1961, Radiocarbon dating of raised beaches in Nordaustlandet, Spitsbergen, in: *Geology of the Arctic*, Raasch, G. O., ed., Toronto: Toronto University Press, p. 133–145.

Buchan, S. H., Challinor, A., Harland, W. B., and Parker, J. R., 1965, The Triassic stratigraphy of Svalbard, *Nor. Polarinst. Skr.*, no. 135, 93 pp.

Bugge, T., Løfaldli, M., Maisey, G., Rokoengen, K., Skaar, F. E., and Thusu, B., 1975, Geological investigation of a Lower Tertiary/Quaternary core, offshore Trøndelag, Norway, *Nor. Geol. Unders. Publ.*, v. 29(316), p. 253–269.

Bullard, E., Everett, J. E., and Smith, A. G., 1965, The fit of the continents around the Atlantic, *Philos. Trans. R. Soc. London Ser. A*, v. 258, p. 41–51.

Challinor, A., 1967, The structure of Brøggerhalvøya, Vestspitsbergen, *Geol. Mag.*, v. 104(4), p. 322–336.

Corbel, J., 1965, Soulèvement isostatique et englacement ancien (Spitzberg et Mer de Barentz) in: *Vorträge des Fridtjof-Nansen-Gedächtnis-Symposions über Spitzbergen*, Büdel, J., and Wirthmann, A., eds., Wiesbaden: F. Steiner Verlag, p. 59–67.

Cutbill, J. L., and Challinor, A., 1965, Revision of the stratigraphical scheme for the Carboniferous and Permian of Spitsbergen and Bjørnøya, *Geol. Mag.*, v. 102(5), p. 418–439.

Dalland, A., 1975, The Mesozoic rocks of Andøya, northern Norway, *Nor. Geol. Unders. Publ.*, v. 29(316), p. 271–287.

Dawes, P. R., 1971, The North Greenland fold belt and environs, *Bull. Geol. Soc. Den.*, v. 20, p. 197–239.

Dietz, R. S., and Shumway, G., 1961, Arctic Basin geomorphology, *Bull. Geol. Soc. Am.*, v. 72, p. 1310–1330.

Dibner, V. D., 1962, Stratigrafija melovykh otlozhenij ostrovov Barencovo–Karskogo shelfa i gornogo Tajmyra, *Dokl. Akad. Nauk SSSR*, v. 144(5), p. 1113–1114.

Dibner, V. D., Razin, V. K., and Ronkina, Z. Z., 1962, Litologia i uslovija formirovanija mezozojskikh otlozhenij Zemli Franca-Josifa, *Tr. Inst. Geol. Arktiki*, v. 126, p. 44–74.

Dineley, D. L., 1958, A review of the Carboniferous and Permian rocks of the west coast of Vestspitsbergen, *Nor. Geol. Tidsskr.*, v. 38(2), p. 197–219.

Edwards, M. V., 1975, Gravel fraction on the Spitsbergen Bank, NW Barents Shelf, *Nor. Geol. Unders. Publ.*, v. 29(316), p. 205–217.

Efremova, E. S., 1972, Goterivskie ammonity ostrova Shpicbergen, in: *Mezozojskie otlozhenija Svalbarda*, Sokolov, V. N., and Vasilevskaja, N. D., eds., Leningrad: Inst. Geol. Arktiki, p. 90–99.

Eldholm, O., 1970, Seismic refraction measurement on the Norwegian continental shelf between 62° and 65°N, *Nor. Geol. Tidsskr.*, v. 50(3), p. 215–229.

Feyling-Hanssen, R. W., 1965a, Shoreline displacement in central Spitsbergen, in: *Vorträge des Fridtjof-Nansen-Gedächtnis-Symposions über Spitzbergen*, Büdel, J., and Wirthmann, A., eds., Wiesbaden: F. Steiner Verlag, p. 24–28.

Feyling-Hanssen, R. W., 1965b, A marine section from the Holocene of Talavera on Barentsøya in Spitsbergen with a record of the Foraminifera, in: *Vorträge des Fridtjof-Nansen-Gedächtnis-Symposions über Spitzbergen*, Büdel, J., and Wirthmann, A., eds., Wiesbaden: F. Steiner Verlag, p. 30–58.

Feyling-Hanssen, R. W., and Olsson, I. U., 1959–1960, Five radiocarbon dates of post-glacial shorelines in central Spitsbergen, *Nor. Geogr. Tidsskr.*, v. 17(1–4), p. 122–131.

Flood, B., Gee, D. G., Hjelle, A., Siggerud, T. and Winsnes, T. S., 1969, The geology of Nordaustlandet, north and central parts, *Nor. Polarinst. Skr.*, no. 146.

Flood, B., Nagy, J., and Winsnes, T. S., 1971a, *Geological map of Svalbard*, Sheet 1G, Spitsbergen, southern part, Oslo: Norsk Polarinstitutt.

Flood, B., Nagy, J., and Winsnes, T. S., 1971b, The Triassic succession of Barentsøya, Edgeøya, and Hopen (Svalbard), *Nor. Polarinst. Medd.*, no. 100, p. 1–20.

Forbes, C. L., Harland, W. B., and Hughes, N. F., 1958, Paleontological evidence for the age of the Carboniferous and Permian rocks of central Vestspitsbergen, *Geol. Mag.*, v. 95(6), p. 465–490.

Fortey, R. A., 1974, The Ordovician trilobites of Spitsbergen. I. Olenidae, *Nor. Polarinst. Skr.*, no. 160, 81 p.

Fortey, R. A., 1975, The Ordovician trilobites of Spitsbergen. II. Asaphidae, Nileidae, Raphiophoridae, and Telephinidae of the Valhallfonna Formation, *Nor. Polarinst. Skr.*, no. 162, 125 p.

Fortey, R. A., and Bruton, D. L., 1973, Cambro-Ordovician rocks adjacent to Hinlopenstretet, north Ny Friesland, Spitsbergen, *Bull. Geol. Soc. Am.*, v. 84, p. 2227–2242.

Frebold, H., 1929, Untersuchungen über die Fauna, die Stratigraphie und Paläogeographie der Trias Spitzbergens, *Skr. Svalb. Ishavet*, no. 26, 66 p.

Frebold, H., 1930, Verbreitung und Ausbildung des Mesozoikums in Spitzbergen, *Skr. Svalb. Ishavet*, no. 31, 126 pp.

Frebold, H., 1951, Geologie des Barentsschelfes, *Abh. Dtsch. Akad. Wiss. Berl. Kl. Math. Naturwiss.*, v. 5, p. 1–51.

Friend, P. F., 1961, The Devonian stratigraphy of north and central Spitsbergen, *Proc. Yorks. Geol. Soc.*, v. 33 (I), n. 5, p. 77–118.

Friend, P. F., 1967, Tectonic implications of sedimentation in Spitsbergen and midland Scotland, *International Symposium on the Devonian System, Calgary, 1967*, v.2, Oswald, D. H., ed., p. 1141–1147.

Friend, P. F., and Moody-Stuart, M., 1972, Sedimentation of the Wood Bay Formation (Devonian) of Spitsbergen: Regional analysis of a late orogenic basin, *Nor. Polarinst. Skr.*, no. 157.

Friend, P. F., Heintz, N., and Moody-Stuart, M., 1966, New unit terms for the Devonian of Spitsbergen and a new stratigraphical scheme for the Wood Bay Formation, *Nor. Polarinst. Arbok*, 1965, p. 59–64.

Gakkel, Ya. Ya., and Dibner, V. D., 1967, Bottom of the Arctic Ocean, in: *International Dictionary of Geophysics*, New York: Pergamon Press, p. 1–13.

Gayer, R. A., and Wallis, R. H., 1966, The petrology of the Harkerbreen Group of the Lower Hecla Hoek of Ny Friesland and Olav V Land, Spitsbergen, *Nor. Polarinst. Skr.*, no. 140.

Gayer, R. A., Gee, D. G., Harland, W. B., Miller, J. A., Spall, H. R., Wallis, R. H., and Winsnes, T. S., 1966, Radiometric age determinations on rocks from Spitsbergen, *Nor. Polarinst. Skr.*, no. 137, 39 p.

Gee, D. G., 1966, A note on the occurrence of ecologites in Spitsbergen, *Nor. Polarinst. Arbok*, 1964, p. 240–241.

Gee, D. G., 1972, Late Caledonian (Haakonian) movements in northern Spitsbergen, *Nor. Polarinst. Arbok*, 1970, p. 92–101.

Gee, D. G., and Hjelle, A., 1966, On the crystalline rocks of northwest Spitsbergen, *Nor. Polarinst. Arbok*, 1964, p. 31–45.

Gee, D. G., and Moody-Stuart, M., 1966, The base of the Old Red Standstone in central north Haakon VII Land, Spitsbergen, *Nor. Polarinst. Arbok*, 1964, p. 57–68.

Gee, E. R., Harland, W. B., and McWhae, J. R. H., 1952, Geology of central Vestspitsbergen. Pt. I. Review of geology of Spitsbergen, with special reference to Vestspitsbergen. Pt. II. Carboniferous to Lower Permian of Billefjorden, *Trans. R. Soc. Edinburgh*, v. 62(9), p. 299–356.

Gjelsvik, T., 1963, Remarks on the structure and composition of the Sverrefjellet volcano, Bockfjorden, Vestspitsbergen, *Nor. Polarinst. Arbok*, 1962, p. 50–54.

Gobbett, D. J., and Wilson, C. B., 1960, The Oslobreen Series, Upper Hecla Hoek of Ny Friesland, Spitsbergen, *Geol. Mag.*, v. 97, p. 441–457.

Grosswald, M. G., Devirts, A. L., Dobkina, E. L., and Semevskiy, D. V., 1967, Crustal movements and dating of glacial stages in the Spitsbergen region (translated from *Geokhimiya*, no. 1, p. 51–56, 1967), *Geochem. Int.*, v. 4(1), p. 30–35.

Grønlie, G., and Ramberg, I. B., 1971, Gravity indications of deep sedimentary basins below the Norwegian Continental Shelf and the Vøring Plateau, *Nor. Geol. Tidsskr.*, v. 50, p. 375–391.

Haller, J., 1971, *Geology of the East Greenland Caledonides*, New York: Interscience Publishers, 413 p.

Harland, W. B., 1959, The Caledonian sequence in Ny Friesland, Spitsbergen, *Q. J. Geol. Soc. London*, v. 114, p. 307–342.

Harland, W. B., 1960, The development of Hecla Hoek rocks in Spitsbergen, 21st Int. Geol. Congr. Norden, 1960, pt. 19, p. 7–16.

Harland, W. B., 1961, An outline structural history of Spitsbergen, in: *Geology of the Arctic*, v. I, Raasch, G. O., ed., Toronto: Toronto University Press, p. 68–132.

Harland, W. B., 1965, The tectonic evolution of the Arctic–North Atlantic region, *Philos. Trans. R. Soc. London Ser. A*, v. 258, p. 59–75.

Harland, W. B., 1966, A hypothesis of continental drift tested against the history of Greenland and Spitsbergen, *Cambridge Res.*, no. 2, p. 18–22.

Harland, W. B., 1967, Early history of the North Atlantic Ocean and its margins, *Nature (London)*, v. 216, p. 464–467.

Harland, W. B., 1969, Contribution of Spitsbergen to understanding of tectonic evolution of North Atlantic region, in: *North Atlantic—Geology and Continental Drift*, Kay, M., ed., *Mem. Am. Assoc. Petrol. Geol.*, v. 12. p. 817–851.

Harland, W. B., 1971, Tectonic transpression in Caledonian Spitsbergen, *Geol. Mag.*, v. 108(1), p. 27–42.

Harland, W. B., 1972, Early Paleozoic faults as margins of Arctic plates in Svalbard, 24th Int. Geol. Congr., Montreal, v. 3, p. 230–237.

Harland, W. B., 1973a, Tectonic evolution of the Barents Shelf and related plates, in: *Arctic Geology*, Pitcher, M. G., ed., *Am. Assoc. Petrol. Geol. Mem.*, v. 19, p. 599–608.

Harland, W. B., 1973b, Mesozoic geology of Svalbard, in: *Arctic Geology*, Pitcher, M. G., ed., *Am. Assoc. Petrol. Geol. Mem.*, v. 19, p. 135–148.

Harland, W. B., 1975, Paleogene correlation in and around Svalbard, *Geol. Mag.*, v. 112(4), p. 421–429.

Harland, W. B., and Gayer, R. A., 1972, The Arctic Caledonides and earlier oceans, *Geol. Mag.*, v. 109(4), p. 289–384.

Harland, W. B., and Horsfield, W. T., 1974, West Spitsbergen orogen, in: *Mesozoic–Cenozoic Orogenic Belts; Data for Orogenic Studies*, Spencer, A. M., ed., *Spec. Publ. Geol. Soc. Lond.*, no. 4, p. 747–755.

Harland, W. B., and Wilson, C. B., 1956, The Hecla Hoek succession in Ny Friesland, Spitsbergen, *Geol. Mag.*, v. 93, p. 265–286.

Harland, W. B., Cutbill, J. L. Friend, P. F., Gobbett, D. J., Holliday, D. W., Maton, P. I., Parker, J. R., and Wallis, R. H., 1974, The Billefjorden Fault Zone, Spitsbergen—the long history of a major tectonic lineament, *Nor. Polarinst. Skr.*, no. 161, 72 p.

Harland, W. B., Wallis, R. H., and Gayer, R. A., 1966, A revision of the Lower Hecla Hoek Succession in central north Spitsbergen and correlation elsewhere, *Geol. Mag.*, v. 103(1), p. 70–97.

Hjelle, A., 1969, Stratigraphical correlation of Hecla Hoek Succession north and south of Bellsund, *Nor. Polarinst. Arbok*, 1967, p. 46–51.

Hoel, A., and Holtedahl, O., 1911, Les nappes de lave, les volcans et les sources thermales dans les environs de la Baie Wood au Spitzbergen, *Vid. Selsk. Skr. (Kristiania) Mat. Naturvid.*, v. 1(8), p. 1–38.

Holtedahl, H., and Bjerkli, K., 1975, Pleistocene and recent sediments of the Norwegian continental shelf (62°N–71°N) and the Norwegian Channel area, *Nor. Geol. Unders. Publ.*, v. 29(316), p. 241–252.

Holtedahl, O., 1919, On the Paleozoic series of Bear Island, especially on the Heclahook system, *Nor. Geol. Tidsskr.*, v. 5, p. 121–148.

Holtedahl, O., 1937, On fault lines indicated by the submarine relief in the shelf area west of Spitsbergen, *Nor. Geol. Tidsskr.*, v. 6, p. 214–221.

Holtedahl, O., 1960, On supposed marginal faults and the oblique uplift of the land mass in Cenozoic time, in: *Geology of Norway*, Holtedahl, O., ed., *Nor. Geol. Unders. Publ.*, no. 208, p. 351–357.

Horn, G., and Orvin, A. K., 1928, Geology of Bear Island, with special reference to the coal deposits, and with an account of the history of the island, *Skr. Svalb. Ishavet*, no. 15, 152 p.

Horsfield, W. T., and Maton, P. I., 1970, Transform faulting along the De Geer Line, *Nature* (*London*),v. 226, p. 256–257.

Hyvärinen, H., 1969, Trullvatnet: A Flandrian stratigraphical site near Murchisonfjorden, Nordaustlandet, Spitsbergen, *Geogr. Ann.* (*Stockholm*) *A*, v. 51(1–2), p. 42–45.

Johnson, G. L., 1975, The morphology and structure of the Norwegian–Greenland Sea, Thesis, University of Copenhagen, 157 p., 95 figs.

Johnson, G. L., and Eckhoff, O. B., 1966, Bathymetry of the North Greenland Sea, *Deep Sea Res.*, v. 13, p. 1161–1173.

Johnson, G. L., and Heezen, B. C., 1967, The morphology and evolution of the Norwegian–Greenland Sea, *Deep Sea Res.*, v. 14, p. 755–771.

Johnson, G. L., Ballard, J. A., and Watson, J. A., 1968, Seismic studies of the Norwegian continental margin, *Nor. Polarinst. Arbok* 1966, p. 112–119.

Johnson, G. L., Freitag, J. S., and Pew, J. A., 1970, Structure of the Norwegian Basin: *Nor. Polarinst. Arbok* 1969, p. 7–16.

Korchinskaja, M. B., 1972, Biostratigrafija triasovykh otlozhenij Svalbarda, in: *Mezozojskie Otlozhenija Svalbarda*, Sokolov, V. N., and Vasilevskja, N. D., eds., Leningrad: Inst. Geol. Arktiki, p. 21–26.

Krasilshchikov, A. A., 1965, Nekotorye osobennosti geologicheskogo razvitija severnoj chasti archipelaga Spitsbergena, in: *Mater. Geol. Spitsb.*, Sokolov, V. N., ed., Leningrad: Inst. Geol. Arktiki, p. 29–44.

Krasilshchikov, A. A., 1973, Stratigrafija i paleotektonika dokembrija—rannogo paleozoja Shpitsbergena, *Tr. Inst. Geol. Arktiki*, v. 172.

Krasilshchikov, A. A., and Livshic, Ju. Ja., 1974, Tektonika ostrova Medvezhij, *Akad. Nauk SSSR Geotektonika*, no. 4, p. 39–51.

Krasilshchikov, A. A., Golovanov, N. P., and Milshtejn, V. E., 1965, K stratigrafii verchneproterozojskikh otlozhenij rajona Murchison-fjorda (Severo-vostochnaja Zemlja), in: *Mater. Geol. Spitsb.*, Sokolov, V. N., ed., Leningrad: Inst. Geol. Arktiki, p. 102–110.

Kulling, O., 1934, The "Hecla Hoek Formation" round Hinloopenstredet, *Geogr. Ann.* (*Stockholm*), v. 16, p. 161–254.

Kvale, A., 1975, Caledonides in Scandinavia compared with East Greenland, *Bull. Geol. Soc. Denmark*, v. 24, p. 129–160.

Laughton, A. S., 1975, Tectonic evolution of the northeast Atlantic Ocean; a review, *Nor. Geol. Unders. Publ.*, v. 29(316), p. 169–193.

Livšic (Livshic), Ju. Ja., 1967, Tretichnye otlozhenija zapadnoj chasti archipelaga Shpicbergen, in: *Mat. Strat. Shpicbergena*, Sokolov, V. N., ed., Leningrad: Inst. Geol. Arktiki, p. 185–204.

Livšic (Livshic), Ju. Ja., 1973, Paleogenovye otlozhenija i platformennaja struktura Shpicbergena, *Tr. Inst. Geol. Arktiki*, v. 174, 159 p.

Livšic (Livshic), Ju. Ja., 1974, Paleogene deposits and the platform structure of Svalbard, *Nor. Polarinst. Skr.*, no. 159, 51 p.

Lowell, J. D., 1972, Spitsbergen Tertiary orogenic belt and the Spitsbergen Fracture Zone, *Geol. Soc. Am. Bull.*, v. 83, p. 3091–3102.

Major, H., 1964, Geological Map of Svalbard C9G, Adventdalen, 1:100,000 (printed 1964, published in Major and Nagy, 1972), Oslo: Norsk Polarinstitutt.

Major, H., and Nagy, J., 1972, Geology of the Adventdalen map area, *Nor. Polarinst. Skr.*, no. 138, 58 p., 1 map.

Major, H., and Winsnes, T. S., 1955, Cambrian and Ordovician fossils from Sörkapp Land, Spitsbergen, *Nor. Polarinst. Skr.*, no. 106.

Manum, S., 1960, Some dinoflagellates and hystrichospherids from the Lower Tertiary of Spitsbergen, *Nor. Polarinst. Medd.*, no. 85 (reprinted from *Nytt Mag. Bot.*, v. 8, p. 17–26).

Manum, S., 1962, Studies in the Tertiary flora of Spitsbergen, with notes on Tertiary floras of Ellesmere Island, Greenland, and Iceland, *Nor. Polarinst. Skr.*, no. 125, 127 p.

Manum, S., 1966, Deposits of probable Upper Cretaceous age offshore from Andöya, Northern Norway, *Nor. Geol. Tidsskr.*, v. 46(2), p. 246–247.

Nagy, J., 1970, Ammonite faunas and stratigraphy of Lower Cretaceous (Albian) rocks in southern Spitsbergen, *Nor. Polarinst. Skr.*, no. 152.

Nathorst, A. G., 1910, Beiträge zur Geologie der Bären-Insel, Spitzbergens und des König-Karl-Landes, *Uppsala Univ. Geol. Inst. Bull.*, v. 10, p. 261–416.

Nicholson, R., 1974, The Scandinavian Caledonides, in: *The Ocean Basins and Margins*, v. 2, Nairn, A. E. M., and Stehli, F. G., eds., New York: Plenum Press, p. 161–203.

Oftedahl, C., 1972, A sideritic ironstone of Jurassic age in Beitstadfjorden, Trøndelag, *Nor. Geol. Tidsskr.*, v. 52, p. 123–134.

Ørvig, T., 1960, The Jurassic and Cretaceous of Andøya in Northern Norway, in: *Geology of Norway*, Holtedahl, O., ed., *Nor. Geol. Unders. Publ.*, no. 208, p. 344–350.

Orvin, A. K., 1934, Geology of the Kings Bay region, Spitsbergen *Skr. Svalb. Ishavet*, no. 57, 196 p.

Orvin, A. K., 1940, Outline of the geological history of Spitsbergen *Skr. Svalb. Ishavet*, no. 78.

Os Vigran, J., 1970, Fragments of Middle Jurassic flora from northern Trøndelag, Norway, *Nor. Geol. Tidsskr.*, v. 50(3), p. 193–214.

Parker, J. R., 1966, Folding, faulting, and dolerite intrusions in the Mesozoic rocks of the fault zone of central Spitsbergen, *Nor. Polarinst. Arbok*, 1964, p. 47–55.

Parker, J. R., 1967, The Jurassic and Cretaceous sequence in Spitsbergen, *Geol. Mag.*, v. 104(5), p. 487–505.

Pchelina, T. M., 1965, Stratigrafija i osobennosti veshchestvennogo sostava mezozojskikh otlozhenij centralnoj chasti zapadnogo Shpicbergena, in: *Mater. Geol. Shpicbergena*, Sokolov, V. N., ed., Leningrad: Inst. Geol. Arktiki, p. 127–148.

Pchelina, T. M., 1967, Stratigrafija i nekotorye osobennosti veshchestvennogo sostava mezozojskikh otlozhenij juzhnykh i vostochnykh rajonov Zapadnogo Shpicbergena, in: *Mater. Geol. Shpicbergena*, Sokolov, V. N., ed., Leningrad: Inst. Geol. Arktiki, p. 121–158.

Pitman, W. C., III, and Talwani, M., 1972, Sea-floor spreading in the North Atlantic, *Geol. Soc. Am. Bull.*, v. 83(3), p. 619–646.

Playford, G., 1962–1963, Lower Carboniferous of Spitsbergen, pts. 1 and 2, *Palaeontology*, v. 5(3–4), p. 550–678.

Ravn, J. P. J., 1922, On the Mollusca of the Tertiary of Spitsbergen, *Result. Nor. Spitsbergeneksped.*, v. 1(2), p. 1–28.

Ronnevik, H., Bergsager, E. I., Moe, A., Øvrebø, O., Narvestad, T., and Stangenes, J., 1975, The geology of the Norwegian continental shelf, in: *Petroleum and the Continental Shelf of Northwest Europe*, v. I: *Geology*, Woodland, A. W., ed., London: Applied Scientific Publications, p. 117–129.

Rosenkrantz, A., 1942, Das geologische Auftreten der Gattung *Thyasira*, *Dan. Geol. For. Medd.*, v. 10, p. 277–278.

Rozycki, S. Z., 1959, Geology of the north-western part of Torell Land, Vestspitsbergen, *Stud. Geol. Pol.*, v. 2, 96 p.

Sandal, S. T., and Halvorsen, E., 1973, Late Mesozoic paleomagnetism from Spitsbergen; implications for continental drift in the Artic, *Phys. Earth Planet. Inter.*, v. 7, p. 125–132.

Sandford, K. S., 1950, Observations on the geology of the northern part of North-East Land (Spitsbergen), *Q. J. Geol. Soc. London*, v. 105, p. 461–493.

Sandford, K. S., 1956, The stratigraphy and structure of the Hecla Hoek Formation and its relationship to a subjacent metamorphic complex in North-East Land (Spitsbergen), *Q. J. Geol. Soc. London*, v. 112, p. 339–362.

Sandford, K. S., 1963, Exposures of Hecla Hoek and younger rocks on the north side of Wahlenbergfjorden, Nordaustlandet (Svalbard), *Nor. Polarinst. Arbok*, 1962, p. 7–23.

Schytt, V., Hoppe, G., Blake, W., and Grosswald, M. G., 1967, The extent of the Würm Glaciation in the European Arctic. A preliminary report about the Stockholm University Svalbard Expedition 1966, *Medd. Naturgeogr. Inst. Stockholms Univ.* (Int. Assoc. Sci. Hydrol., Gen. Assoc. of Bern, 1967, Extr. Publ., 79), v. 20A, p. 207–216.

Sellevoll, M. A., 1975, Seismic refraction measurements and continuous seismic profiling on the continental margin off Norway between 62°N and 69°N, *Nor. Geol. Unders. Publ.*, v. 29(316), p. 219–235.

Siedlecka, A., 1975, Late Precambrian stratigraphy and structure of the north-eastern margin of the Fennoscandian Shield (East Finnmark–Timan region), *Nor. Geol. Unders. Publ.*, v. 29(316), p. 313–348.

Smith, D. G., Harland, W. B., and Hughes, N. F., 1975, Geology of Hopen, Svalbard, *Geol. Mag.*, v. 112(1), p. 1–23.

Smulikowski, W., 1965, Petrology and some structural data of lowest metamorphic formations of the Hecla Hoek Succession in Hornsund, Vestspitsbergen, *Stud. Geol. Pol.*, v. 18, 107 p.

Smulikowski, W., 1968, Some petrological and structural observations in the Hecla Hoek Succession between Werenskioldbreen and Torellbreen, Vestspitsbergen, *Stud. Geol. Pol.*, v. 21, p. 97–161.

Sokolov, D., and Bodylevsky, V., 1931, Jura- und Kreidefaunen von Spitzbergen, *Skr. Svalb. Ishavet*, no. 35, 151 p.

Sokolov, V. N., Krasiltshikov, A. A., and Livschitz, J. J., 1968, The main features of the tectonic structure of Spitsbergen, *Geol. Mag.*, v. 105(2), p. 95–115.

Storetvedt, K. M., 1972, Old Red Sandstone paleomagnetism of central Spitsbergen and the Upper Devonian (Svalbardian) phase of deformation, *Nor. Polarinst. Arbok 1970*, p. 59–69.

Storetvedt, K. M., 1974, A possible large-scale sinistral displacement along the Great Glen Fault in Scotland, *Geol. Mag.*, v. 111, p. 23–30.

Strand, T., and Kulling, O., 1972, *Scandinavian Caledonides, Pt. I, The Norwegian Caledonides*, New York: Wiley-Interscience, p. 1–145.

Sundvor, E., 1975, Thickness and distribution of sedimentary rocks in the southern Barents Sea, *Nor. Geol. Underst. Publ.*, v. 29(316), p. 237–240.

Talwani, M., and Eldholm, O., 1972, Continental margin off Norway: A geophysical study, *Geol. Soc. Am. Bull.*, v. 83, p. 3575–3606.

Tozer, E. T., and Parker, J. R., 1968, Notes on Triassic biostratigraphy of Svalbard, *Geol. Mag.*, v. 105(6), p. 526–542.

Tyrrell, G. W., and Sandford, K. S., 1933, Geology and petrology of the dolerites of Spitsbergen, *Proc. R. Soc. Edinburgh*, v. 53(21), p. 284–321.

Vallance, G., and Fortey, R., 1968, An Ordovician succession in North Spitsbergen, *Proc. Geol. Soc. London*, no. 1648, p. 91–97.

Vasilevskaja, N. D., 1972, Pozdnetriasovaja flora Svalbarda, in: *Mezozojskie Otlozhenija Svalbarda*, Sokolov, V. N., and Vasilevkaja, N. D., eds., Leningrad: Inst. Geol. Arktiki, p. 27–63.

Vogt, P. R., and Avery, O. E., 1974, Tectonic history of the Artic Basins: Partial solutions and unresolved mysteries, in: *Marine Geology and Oceanography of the Arctic Seas*, Herman, Y., ed., Berlin: Springer Verlag, p. 83–117.

Vogt, P. R., Ostenso, N. A., Johnson, G. L., 1970, Magnetic and bathymetric data bearing on seafloor spreading north of Iceland, *J. Geophys. Res.*, v. 75, p. 903–920.

Vogt, T., 1928, Den norske fjellkjedes revolusjons-historie, *Nor. Geol. Tidsskr.*, v. 10, p. 97–115.

Vonderbank, K., 1970, Geologie und Fauna der Tertiären Ablagerungen Zentral-Spitzbergens, *Nor. Polarinst. Skr.* no. 151, 119 p.

Werenskiold, W., 1922, Höyie strandlinjer paa Spitsbergen, *Norsk Geol. Tidsskr.*, v. 7, p. 7–12.

Wilson, C. B., 1958, The lower Middle Hecla Hoek rocks of Ny Friesland, Spitsbergen, *Geol. Mag.*, v. 95, p. 305–327.

Wilson, C. B., 1961, The upper Middle Hecla Hoek rocks of Ny Friesland, Spitsbergen, *Geol. Mag.*, v. 98, p. 89–116.

Wilson, C. B., and Harland, W. B., 1964, The Polarisbreen Series and other evidences of late Pre-Cambrian ice ages in Spitsbergen, *Geol. Mag.*, v. 101, p. 198–219.

Wilson, J. T., 1965, A new class of faults and their bearing on continental drift, *Nature (London)*, v. 207, p. 343–347.

Winsnes, T. S., 1965, The Precambrian of Spitsbergan and Bjørnøya, in: *The Geologic Systems*, v. 2, *The Precambrian*, Rankama, K., ed., New York: Wiley-Interscience Publishers, p. 1–24.

Worsley, D. L., 1973, The Wilhelmøya Formation—a new lithostratigraphic unit from the Mesozoic of eastern Svalbard, *Nor. Polarinst. Arbok 1971*, p. 7–16.

Chapter 7

GEOLOGY OF THE SOVIET ARCTIC: KOLA PENINSULA TO LENA RIVER

Michael Churkin, Jr., George Soleimani, Claire Carter, and Rhoda Robinson

U.S. Geological Survey
Menlo Park, California

I. INTRODUCTION

This chapter covers the geology of the western Soviet Arctic, extending from the Norwegian and Finnish borders on the west to the Lena River delta on the east. The original plan was to describe this vast region in three separately authored chapters. Members of the Institute of Arctic Geology (NIIGA or CEVMORGEO) in Leningrad were invited to write these three chapters. In the beginning of 1978, these promised papers were not forthcoming, so the editors decided to synthesize the geology of the region from the most recently published Soviet literature. Fortunately, since the Second Arctic Geology Symposium held in 1970, the U.S. Geological Survey in Menlo Park has routinely exchanged literature with the Soviet Institute of Arctic Geology. Using these data and our combined ability to assemble, translate, and assess Soviet literature, we attempt here to describe the geology of the western Soviet Arctic. Severe time limitations and spotty regional and topical data have made the subject coverage in this chapter somewhat uneven. Inaccuracies in translation and interpretation are also inevitable, and thus we have assembled a list of key references that are especially oriented to an English-speaking reader who may want to see the original data. The geology of the Kara, Barents, and Laptev Shelf Seas and the adjacent Eurasian Basin of the Arctic Ocean has been synthesized by Demenitskaya and Hunkins (1971). The geology of this area has also been reviewed by both Soviet and western geologists in the Proceedings of the Second Arctic Geology Symposium (Pitcher, 1973). Since then, very few new data have been published on the marine geology of the Arctic Ocean bordering the U.S.S.R.

George Soleimani was responsible for assembling the available literature, for compiling columnar sections and structure sections for each geologic province, and for translating some of the material. Michael Churkin and Claire Carter compiled the geology and wrote the text. Rhoda Robinson did most of the translating and much of the editing.

The major geologic provinces are described from west to east, focusing on the most recently available columnar and structure sections, accompanied by a minimum of descriptive text. The final section includes both a review of the main stages in the geologic development of the region as favored by many Soviet geologists and an alternative plate tectonic interpretation as favored by the authors.

II. KOLA PENINSULA–KARELIA—AREA I

The northwestern Soviet Arctic, including the Kola Peninsula, Karelia, and the adjoining parts of the Barents Sea (Fig. 1, area I), is underlain by a complex of crystalline rocks of the Precambrian Baltic Shield that is in turn overlain by the Proterozoic and Paleozoic platform cover (see Sec. III) of the Russian Platform. The region was extensively glaciated in the Pleistocene. The high-grade metamorphic rocks and associated plutons that form the crystalline basement of the Russian platform are of Archean or early Proterozoic age; this basement complex appears to continue west to form the metamorphic core of Scandinavia. The Russian platform deposits consist of essentially unmetamorphosed shelf strata that mantle the peneplained Baltic Shield and fill minor depressions cut in the shelf.

Gneiss, migmatite, amphibolite, schist, and, in places, marble of Archean and Proterozoic (3500–1900 m.y.) age make up the base of the Karelian section (Nalivkin, 1973, p. 45). These metamorphic rocks are intruded by syngenetic granodiorite, much of which is migmatite. Proterozoic rocks are widespread in Karelia and include schist, gneiss, quartzite, and altered lavas (Ladoga Suite, 1620–1820 m.y. old). Resting unconformably on these rocks are a basal metaconglomerate and sandstone with schist, quartzite, and *Collenia*-bearing algal dolomite (Segozerian and Onegian Suites). Unconformably overlying these strata are limestone and dolomite, which in turn are overlain by siliceous slate 300 m thick that contains prominent diabase bodies.

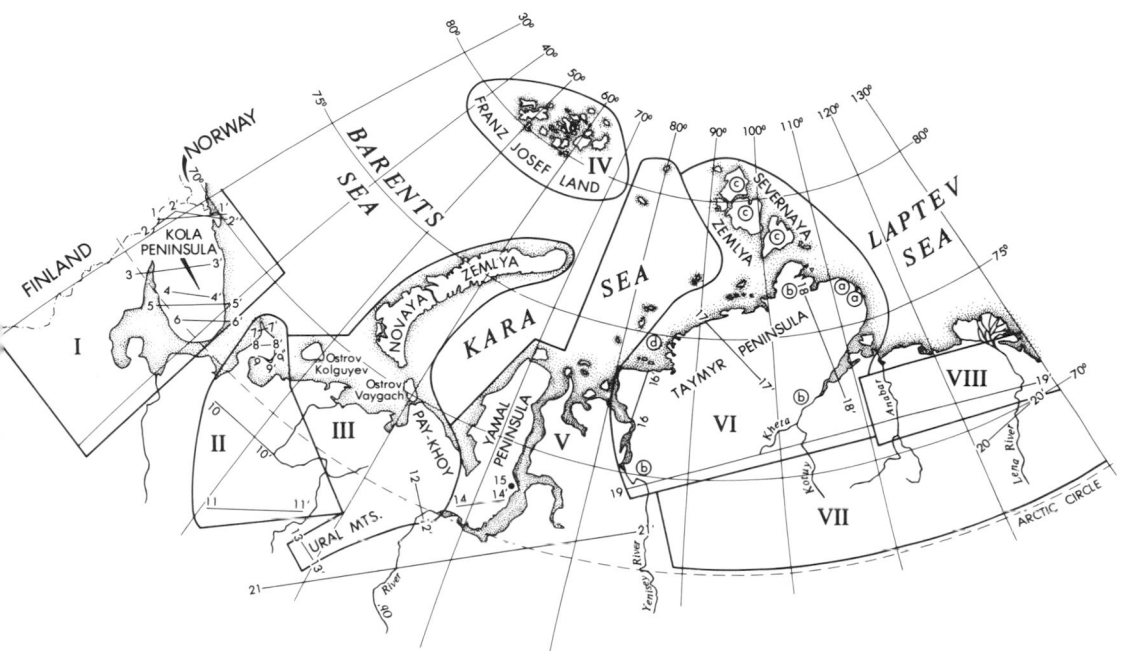

Fig. 1.   Index map showing major regions of the Western Soviet Arctic. (I) Kola Peninsula–Karelia; (II) Timan–Pechora; (III) Novaya Zemlya, Vaygach, Pay-Khoy, Polar Urals, and North Pechora Basin; (IV) Franz Josef Land; (V) Yamal Peninsula, West Siberian Lowlands, and Kara Sea Islands; (VI) Taymyr Peninsula and Severnaya Zemlya; (VII) North Siberian Lowlands; (VIII) Eastern North Siberian Lowlands. Small letters in circles refer to locations of detailed maps in subsequent figures.

Younger Proterozoic platform rocks (Jotnian Formation) can be traced west into Finland.

The Hyperborean Formation of Riphean age contains Upper Proterozoic red conglomeratic sandstone and carbonate rocks that crop out in the northwest part of the Kola Peninsula, in the Rybachiy Peninsula, and on an island offshore.

Magmatic activity is represented by several stages of granitic intrusion in Archean time that was followed by a development of mafic lava flows, sills, and dikes in Proterozoic time.

The economically important alkaline intrusive rocks of the Kola Peninsula have been dated as 300 m.y. old or late Paleozoic. Mesozoic and possibly younger basalts are widely distributed farther north in the Franz Josef Land Archipelago. The stratigraphic succession and structure of the Baltic Shield is quite varied owing to multiple periods of intrusion, intense metamorphism, and deformation. Figure 2 shows the main geologic units of the Kola Peninsula, which are similar to the units in Karelia. Figure 3 shows the major rock types

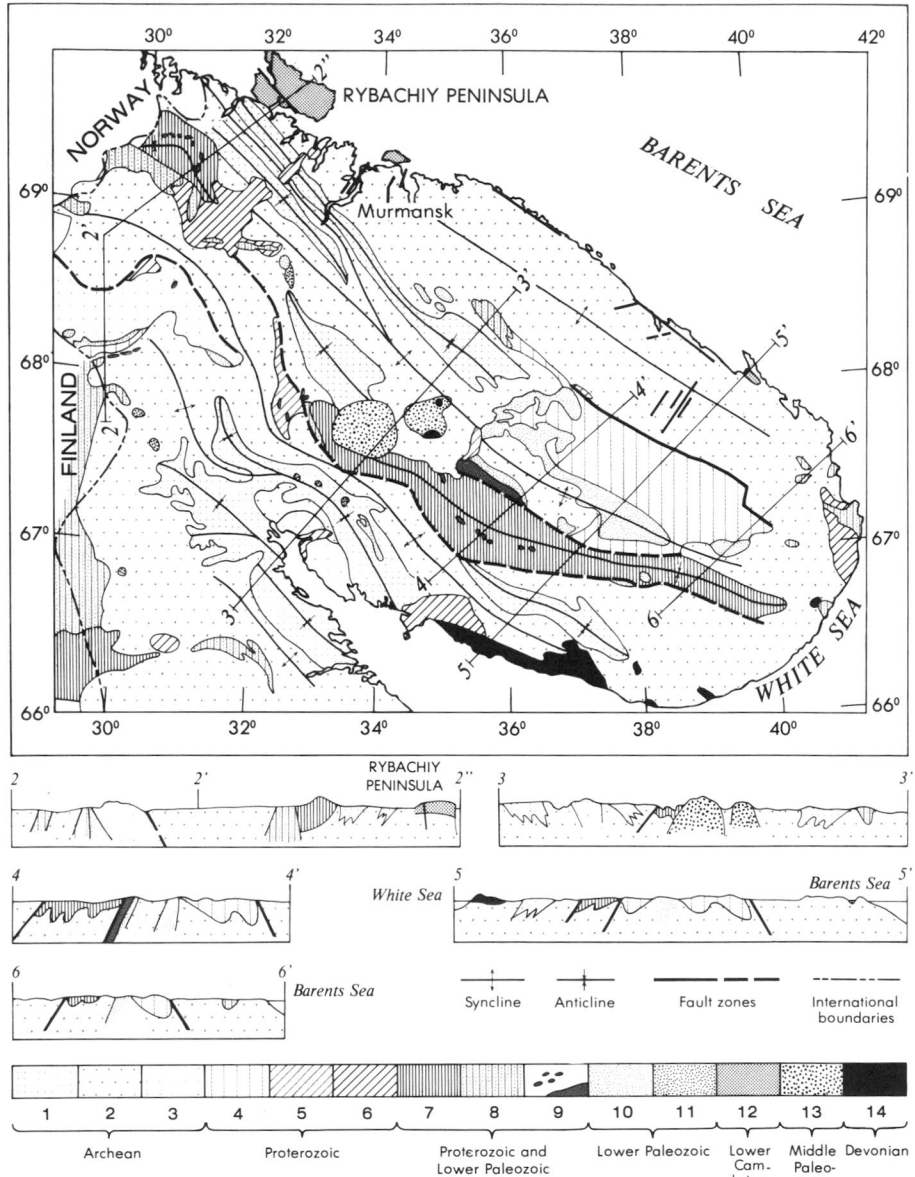

Fig. 2.   Generalized geologic map and cross sections of the Kola Peninsula. (1) Archean gneiss in synclinoria; (2) Archean gneiss, granite, migmatite, and minor basic intrusives forming anticlinoria and median massives; (3) Upper Archean basic rocks; (4) Proterozoic gneiss and schist in synclinoria; (5) Proterozoic basic rocks; (6) Proterozoic granite; (7) upper and (8) lower structural stage Proterozoic and Lower Paleozoic volcanic rocks forming synclinoria; (9) Proterozoic and Lower Paleozoic basic and ultrabasic rocks; (10) Lower Paleozoic alkali granite; (11) Lower Paleozoic basic and alkali rocks; (12) Lower Cambrian folded rocks; (13) Middle Paleozoic nepheline syenite; (14) Devonian rocks, weakly deformed.

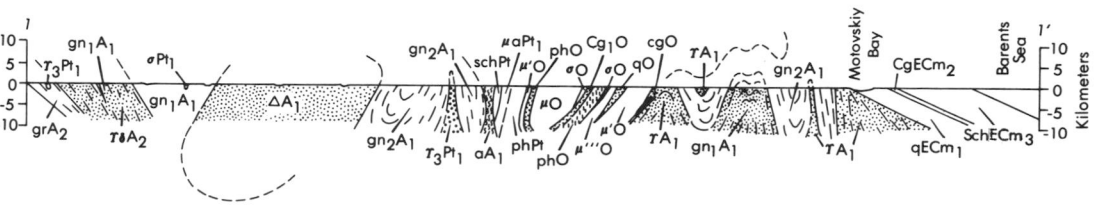

3. Detailed structure section across the Murmansk district, northwestern Kola Peninsula. See Fig. 1 for line of section. Units given from left to right (modified after Antropov and Khaeitonov, 1958). ($\gamma_3 Pt_1$) Microcline granite, partly gneissic; (grA_2) acid granulite: garnet–quartz–feldspar rocks; (gn$_1 A_1$) undifferentiated complex of garnet–biotite and biotite gneiss; ($\gamma\delta A_2$) basic granulites; gneiss–norite and hypersthene gneiss–diorite; (σPt_1) peridotite, pyroxenite, and gabbro; (ΔA_1) hypersthene diorite with subordinate basic and ultrabasic rocks, charnockite and pyroxene–magnetitic schist; (gn$_2 A_1$) undifferentiated complex of micaceous gneiss and migmatite; ($\gamma_3 Pt_1$) microcline granite porphyry; (aA_1) amphibolite and gabbro–amphibolite; (schPt) biotite–quartz schist with quartzite interbeds; ($\mu a Pt_1$) schistose amphibolite with subordinate biotite gneiss and schist; (phPt) calcereous chlorite–sericite graphitic schist, phyllite, and quartz–sericite schist; ($\mu' O$) diabasic and porphyritic greenstone; (phO) phyllite with subordinate carbonaceous phyllite, and schistose tuff; (μO) series of sheets of diabase, alternating with schist and schistose tuff; (σO) ultrabasic and basic rocks: peridotite, pyroxenite, and gabbro; ($c_3 O$) conglomerate; ($\mu''' O$) metabasalt and metaporphyry, agglomeratic tuff and schistose tuff; (qO) red arkosic sandstone, quartzite, conglomerate, limy quartzite, and dolomite; (CgO) conglomerate and arkose; (γA_1) oligoclase gneiss–granite with granodiorite and migmatite in gneissic host rocks; (gn$_1 A_1$) biotite gneiss with garnet; amphibolite; (gn$_2 A_1$) undifferentiated complex of micaceous gneiss and migmatite; (qECm_1) arkose, glauconite sandstone, quartz sandstone; quartzite with subordinate clay, sandy-clayey shale and dolomite; (CgECm_2) conglomerate; (SchECm_3) arkose with interbeds of conglomerate shale, argillaceous sandstone, and limy shale; (Pz) Paleozoic, undifferentiated; (Cm_1, Cm_2, Cm_3) Cambrian; (Pt, Pt$_1$) Proterozoic; (A_1, A_2) Archean.

and structures in the Murmansk region. A general review of the various stages of deformation and intrusion that occurred in the Murmansk region is provided by Antropov and Khaeitonov (1958), and a summary of isotopic dates is given by Polovinkina (1973). The most recent publications in English covering the tectonics of Karelia include Lazarev (1973) and Voytovich (1975); deep structure is described from drilling and geophysics by Akudinov *et al.* (1972).

Overlying the Karelian deposits with an angular unconformity are less deformed, weakly metamorphosed sedimentary rocks (Suisarian and Jotnian Complexes), which include distinctive pink and red sandstone (Shokshinian Sandstone), some volcanic rock, and thin horizons of phylitic schist. This overlying sequence is intruded by 1400- to 1600-m.y.-old Rapakivi granites. Weakly altered sandstone and shale range stratigraphically higher and, in places, underlie Lower Cambrian strata.

Late Archean gneiss, amphibolite, granite, and pegmatite form the basement of the Kola Peninsula. Lower Proterozoic rocks here are gneiss, amphibolite, and schist that are several kilometers thick (Tundra Suite in the eastern tip of the peninsula, and Keivy Suite in the east-central area).

The Upper Proterozoic, as in Karelia, consists of two complexes. The lower part is a sequence of marble, quartzite, arkose, slate, and phyllite and includes diabase, keratophyre, and quartz porphyry near the top of the sequence (Imandra–Varzuga Suite). The overlying complex contains gently dipping red arkose and a basal conglomerate (Ter' Suite), both of which were originally dated as Devonian but later found to contain Upper Proterozoic (Sinian) spores.

Along the northwestern edge of the Kola Peninsula, the Hyperborean Formation, up to 5–6 km of dolomite and argillaceous limestone, contains calcareous algae dated as 715–1020 m.y. old.

The northwestern edge of the Kola Peninsula is underlain by arkosic sandstone and associated unmetamorphosed gently dipping strata of Eocambrian age that dip north beneath the Barents Sea (Fig. 3). Weakly deformed strata occur along the White Sea in the southern part of the peninsula.

III. TIMAN–PECHORA REGION—AREA II*

The Timan–Pechora region (Fig. 1, area II), which includes the Kanin Peninsula,† is a platform area of post-Baikalian (late Precambrian) age that lies between the Russian Platform proper and the Urals. Probably, the main bedrock of the Barents Sea is a continuation of this post-Baikalian platform, as suggested by its bathymetry and geophysical characteristics (Demenitskaya and Hunkins, 1971) and by the presence of Paleozoic rock fragments in the Quaternary moraines covering the seafloor, which are identical to the Paleozoic strata of the Timan–Pechora region.

Timan is a low-lying mountain system with a maximum elevation of about 450 m. The mountains to the south pass into the Polyudovian Range of the Urals. All of the higher elevations of Timan and the Kanin Peninsula consist of metamorphosed Proterozoic rocks.

Stratigraphy. Geosynclinal Upper Proterozoic rocks and platform-type Paleozoic, Mesozoic, and Cenozoic rocks characterize the Timan–Pechora region. The Upper Proterozoic rocks here are very thick, strongly folded, and metamorphosed, in contrast to the thin, weakly folded, and unmetamorphosed Upper Proterozoic strata underlying the Russian Platform. Geology and subsurface features of the Timan–Pechora region are shown in Figs. 4 and 5.

In the Timan region, the Upper Proterozoic consists of 2000–2500 m of banded sericite-quartz-bearing phyllitic slate, up to 1000 m of transgressive red feldspathic quartzite and quartzite conglomerate near its base, 1000–2000 m of quartz-sericite slate and quartzite near the top, over 1000 m of dolomite and limestone marble with algal structures, and up to 1000 m of quartz–mica slate.

---

* Excerpted from Nalivkin (1973).

† Place names mentioned in text but not shown on the index map can be located on the map of the Arctic region (Pinther, 1975).

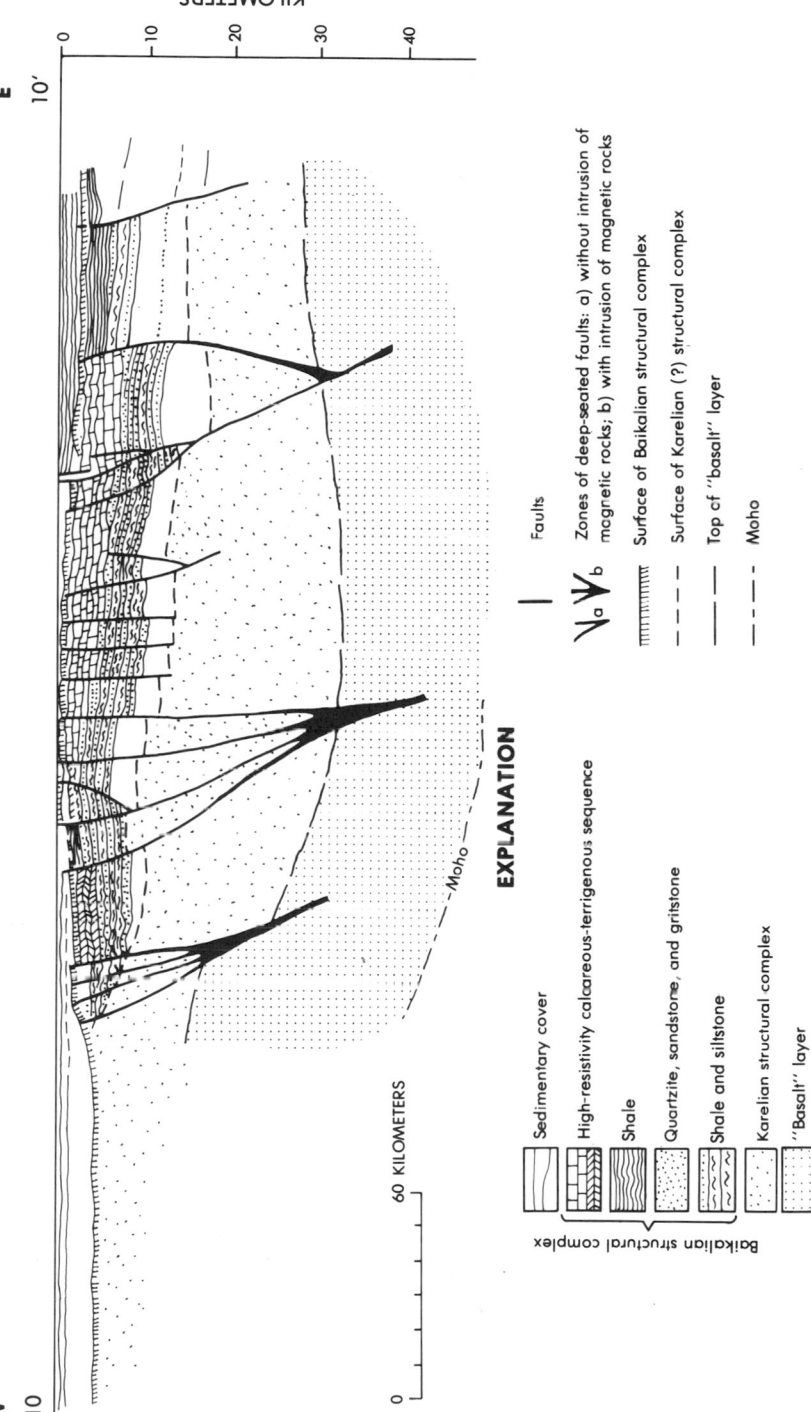

Fig. 4. Geologic cross section of part of the Timan region of the Russian Platform, based on borehole and geophysical data; location shown on Fig. 1 (modified after Shilov *et al.*, 1978, Fig. 1).

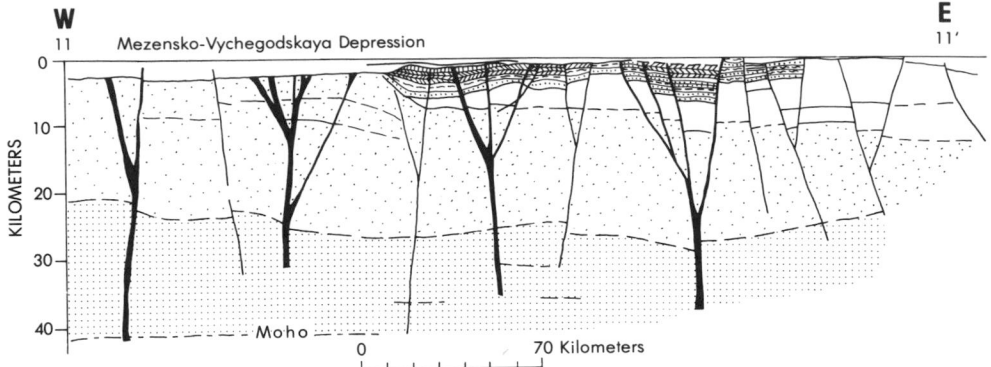

Fig. 5. Geologic cross section of the southern part of the Timan region of the Russian Platform, based on borehole and geophysical data; location shown on Fig. 1, explanation on Fig. 4 (modified after Shilov *et al.*, 1978, Fig. 3).

In the Kanin Peninsula (see Figs. 6 and 7), the Upper Proterozoic rocks occur in two belts. The northern belt (Kanin Kamen' Ridge) is composed of 2500 m of thinly bedded schist and phyllite, with interbeds of quartzite and marble at the base. Above this lies 1500 m of sandstone and quartzite with schist interbeds. The southern belt consists of 1000 m of dolomite marble with the calcareous algae *Collenia* and *Gymnosolen*.

Beds containing Cambrian or Ordovician fossils are unknown in Timan, the Kanin Peninsula, or the Pechora Depression.

Silurian rocks are exposed over a small area on the coast of northern Timan. They consist of a basal sandstone overlain by claystone and marl, which in turn is overlain by up to 25 m of argillaceous limestone containing large pentamerid brachiopods, *Favosites*, and *Leperditia*. Above these beds lie 25–30 m of brachiopod-bearing dolomite and limestone. The Silurian rocks here are weakly folded and nearly unmetamorphosed.

In the Pechora Depression, the Silurian is represented by 750 m of sandstone and shale, with dolomite and anhydrite at the top (Izhma-Omra Formation). The Silurian rhynchonellid brachiopod–coral fauna is similar to that of northern Timan. The Izhma-Omra Formation occurs throughout the Izhma–Pechora area, where Devonian rocks are present, but the Izhma-Omra Formation is missing in eastern and western Timan, where the Devonian rocks rest directly on Precambrian metamorphic rocks. Silurian sedimentary rocks, similar to the Silurian rocks in northern Timan, also occur on the eastern Kanin Peninsula.

The Devonian rocks of southern Timan and the Pechora Depression contain oil and gas and, consequently, have been extensively studied and drilled. These Devonian strata are widely distributed throughout the Timan–

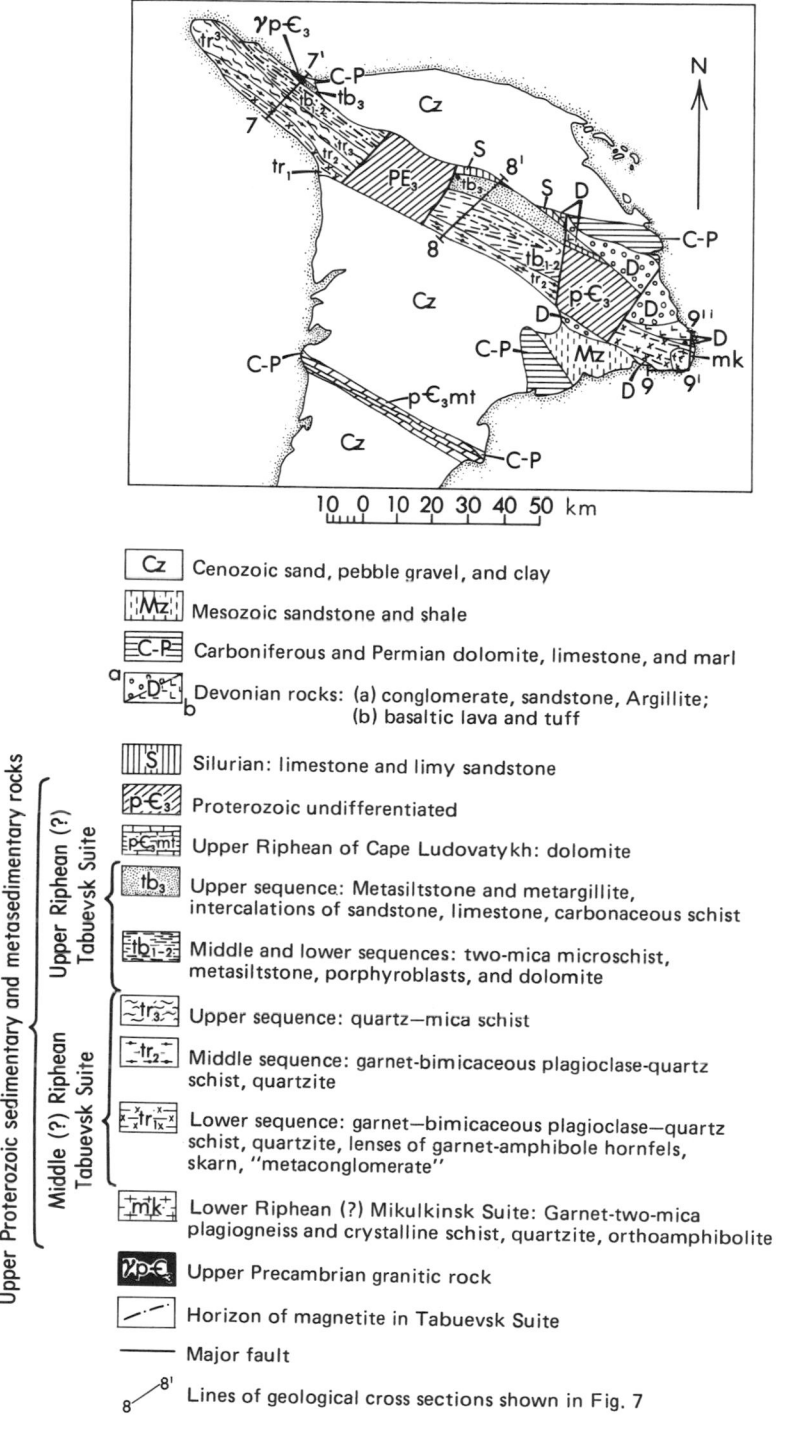

Fig. 6.   Geologic sketch map of the northern part of the Kanin Peninsula (modified after Pogrebitskiy and Kos'ko, 1977, Fig. 1).

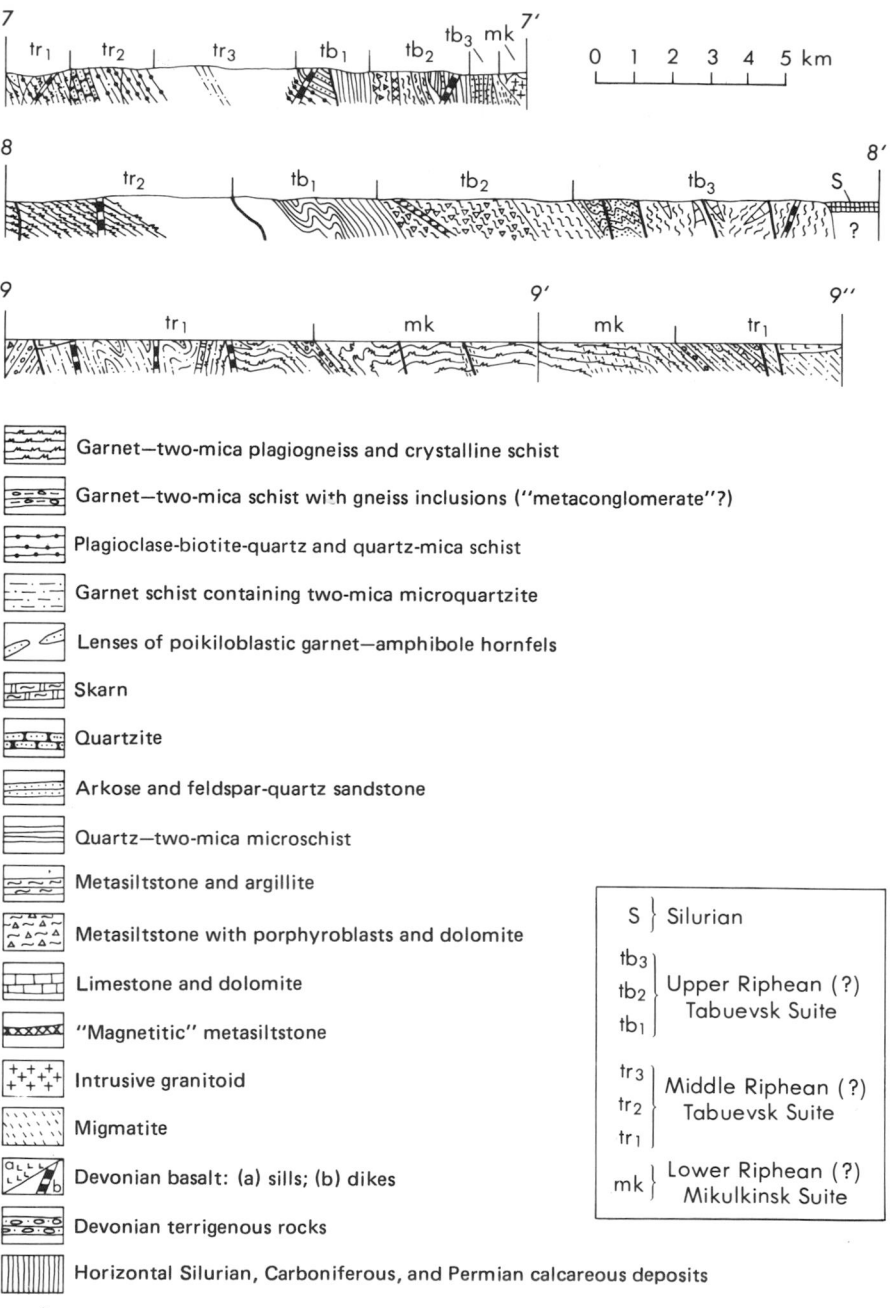

Fig. 7.   Geologic cross sections of Kanin Kamen' Ridge (modified after Pogrebitskiy and Kos'ko, 1977, Fig. 2).

Pechora region and the Kanin Peninsula, as well as along the western slope of the Urals. Throughout this huge platform area, the Devonian rocks are almost horizontal and nearly unaltered, are generally 1000–1200 m thick (up to 3000 m in the Pechora Depression), and are deformed into a number of large, gentle folds.

In western Timan, the Lower and Middle Devonian are absent, while the Upper Devonian consists almost entirely of limestone and dolomite with minor sandstone. In eastern Timan, the Lower Devonian is absent, the Middle Devonian includes shale and sandstone with some basalt and tuff, and the Upper Devonian has mainly argillaceous rocks in the lower part and mainly carbonate rocks in the upper. Central and northern Timan also lack the Lower Devonian, while the Middle and Upper Devonian consist of sandstone, shale, and limestone, with some basalt, diabase, and tuff in the upper Middle Devonian. In southern Pechora, the Lower and Middle Devonian consist of sandstone, limestone, and shale, while the Upper Devonian is predominantly limestone with lesser amounts of shale and sandstone. (See Nalivkin, 1973, Table 19, for a more detailed description of the Devonian successions of the Timan–Pechora province.)

Rapid and important facies changes are characteristic of the Devonian section. For example, from south to north in the successions of central and northern Timan, there is a decrease in number and thickness of marine sections and an increase in thickness of continental, plant-bearing formations. Another important feature of the Devonian deposits is the wide occurrence of basaltic lavas, which are interbedded with terrigenous rocks of Givetian and early Frasnian age.

The Upper Devonian of the Kanin Peninsula consists of continental red beds of Frasnian age, similar to those of northern Timan.

The Lower Carboniferous, though missing in western Timan, is well represented in eastern Timan and the Pechora Depression by 150–275 m of shale, sandstone, limestone, and dolomite. The Middle and Upper Carboniferous deposits represent an east-to-west transgression that inundated the Upper Proterozoic topographic highs and formed a direct connection between the seas covering the Russian Platform and the Pechora Depression. The resultant deposits constitute a single formation, 60–330 m thick, of carbonate rocks with rare claystone intercalations.

During the Permian, rapid facies changes again took place, and Permian deposits have a more limited distribution. Lower Permian marine limestone and dolomite, which resemble the underlying Upper Carboniferous, are succeeded by mixed facies of continental and lagoonal deposits—gypsum, dolomite, anhydrite, salt, sandstone, and mudstone. Toward the Pechora Basin, these rocks are coal bearing. In western Timan, the Upper Permian is a continental facies, but contains no red beds, as does the Upper Permian farther east. It also consists of continental near-shore or freshwater flagstone in the south, and marine shale, marlstone, and argillaceous limestone in the north. The uppermost Permian is represented by red shale, sandstone, and marlstone of continental origin, with rare interbeds of lacustrine limestone.

West of Timan, the Lower Triassic is widely distributed and consists of 100–150 m of red continental deposits, with fluviatile and lacustrine facies not uncommon. Triassic rocks are also found in the Pechora Coal Basin.

The Jurassic and Lower Cretaceous rocks have a limited distribution and are transgressive over older rocks. At the base of the Jurassic section are 10–15 m (150 m in southeastern Timan) of continental sandstone with shale layers, pebble beds, and wood fragments. The overlying Upper Jurassic marine shales with sandstone interbeds (often glauconitic) are best developed in the Izhma–Pechora area, where they measure 150 m in thickness. Ammonites and belemnites are common. The Lower Cretaceous occurs only to the west of Timan, where the section is represented by sandstone and black shale, overlain by micaceous shale, and followed by sandstone with predominantly continental clay interbeds. The total thickness of the Lower Cretaceous is up to 100–150 m.

Upper Cretaceous, Paleogene, and Neogene deposits have not been identified in the Timan–Pechora region. The Quaternary consists mainly of moraine and fluvioglacial, fluviatile, and lacustrine deposits.

The Timan–Pechora region lies between the Russian Platform proper and the Urals and is considered by Soviet geologists to be part of the post-Baikalian tectonic system that formed after the Baikalian Orogeny in the late Proterozoic. The Russian Platform has an older, folded basement (post Karelian) than Timan, but the Urals represent a younger foldbelt (post Hercynian). On the Russian Platform, Proterozoic strata are of the platform type, while in Timan, they are geosynclinal. The Middle and Upper Paleozoic strata in Timan, however, are of the platform type, as they are in the remainder of the Russian Platform; in the Urals, they are geosynclinal.

The metamorphic rocks of the Timan–Pechora area are intensely folded, forming numerous, complex, and often isoclinal folds, broken by thrusts to form nappes. The folds are cut by granite, alkaline rock, and other synorogenic intrusions. The Middle Paleozoic rocks are gently deformed into large-scale northwest-striking folds, in which dips usually are much less than 20–25°. The Upper Paleozoic, which is separated from the Middle Paleozoic by unconformities, is even more gently folded, while the Jurassic and Lower Cretaceous are not folded at all and lie transgressively and unconformably over the various Paleozoic units.

On the west side of the Urals, a basin filled with Permian deposits several kilometers thick borders the eastern margin of the Pechora Depression. This, the Pechora Coal Basin, is part of the Ural Foldbelt. The great section thickness (2.5–3.0 km) suggests the existence of an older basin in front of the Urals.

Large-scale faults are characteristic of the Timan–Pechora region. Most of them are normal faults of considerable length and throw. In Timan, the faults are deep and are interpreted by many to have served as conduits for basaltic magma throughout the entire length of the range. The faulting and magmatism occurred at the end of the Middle Devonian and the beginning of the Upper Devonian.

An older episode of magma generation, within the metamorphic rocks of northern Timan and the Kanin Peninsula, is marked by small intrusions of granite, gabbro, and alkaline rocks (nepheline syenites). The igneous rocks consist of basalt and diabase, which formed extensive lava flows 2–30 m thick. These flows occur within marine or, more rarely, continental Devonian deposits and form a belt 600 km long from southern Timan to the northern coast. In the south, their total thickness is 50–80 m, but reaches 240 m in northern Timan.

IV. NORTHERN PECHORA BASIN, POLAR URALS, PAY-KHOY, VAYGACH, AND NOVAYA ZEMLYA—AREA III

In the arctic region east of Timan, the lowland of the Pechora River Basin is located on the rocks of the Russian Platform and is separated from the West Siberian Lowland and tundra-covered Yamal Peninsula by the Polar Ural Mountains which, unlike the southern Urals, are a rugged mountain chain with elevations reaching 1 to 2 km. The geology of this region is dominated by a system that starts in the south with the Polar Urals and continues northward into the Pay-Khoy Range, Vaygach Island, and Novaya Zemlya (Table I). The foldbelt is thought to separate platform sequences beneath the Barents and Kara Seas.

The Precambrian and Paleozoic rocks of the Urals make up a geosyncline that grades westward into the Russian Platform beneath the Pechora Basin. On the east side of the Urals, a major suture separates the geosynclinal rocks from the Siberian Platform rocks that lie below the swampy West Siberian Lowland (Hamilton, 1970).

The Paleozoic rock types of the Polar Urals differ greatly on the western and eastern slopes. The rocks along the western slope (Fig. 8) are mainly carbonates that grade into clastics toward the core of the Urals (Puchkov, 1973). The same west-to-east facies change occurs in the Pay-Khoy. The eastern slope of the Urals, on the other hand, has widespread volcanic units with lenses of marble. A major fault along the axis of the Urals separates the miogeosynclinal rocks of the western part from eugeosynclinal rocks on the east.

TABLE I
Generalized Columnar Section for Polar Urals and Pay-Khoy, Vaygach, and Novaya Zemlya[a]

Age	Thickness	Description
Quaternary	90–140 m	Marine periglacial deposits: clay (1.5 m); moraine of maximal glaciation—sandy clayey loam with boulders (20–30 m); moraine interglacial deposit—sand, clay (30–40 m); continental formations—coarse sand and clayey loam (10 m); limnite (15 m); moraine of second glaciation—sandy clayey loam with boulders (5–10 m); postglacial deposit—fluvioglacial sand (3–15 m); limnite (thickness unknown). Recent: limnetic swamp. In Novaya Zemlya: marine, glacial, and alluvial deposits.
Tertiary	?	In all rivers discharging into Kara Sea: tuff breccia with clasts 2 m in size (thickness unknown). In Novaya Zemlya: thin conglomerate and breccia.
Cretaceous	>200 (?) m	Lower part: sand, clay with brown coal layers (160 m). Upper part: sand, marl (5–10 m); cherty siltstone (about 5 m). In Novaya Zemlya: Middle Jurassic and Early Cretaceous boulder conglomerate.
		Major Break
Permian	3000–5000 m	In Polar Ural Foothills: conglomerate, limestone (several hundred meters). Lower Pay-Khoy and Pechora Sequence: Vorkuta Series-Marl (12–31 m); argillite (50–80 m); sandstone (120–400 m); siltstone (500–1000 m); sandstone, siltstone (350–500 m). Pechora Series: many layers of coal, polymictic conglomerate, argillite (870 m); sandstone (710–720 m); Upper Pay-Khoy and Pechora Sequence: siltstone, coal layers (3000 m); basalt (27 m); conglomerate (600 m); sandstone (300 m); siltstone (100 m). Vaygach: sandstone, siltstone, and argillite (1500 m). In Novaya Zemlya: clastic strata (1.5–2 km) transgressively overlying Carboniferous, Late Permian, and Early Triassic; small granitic intrusions.
Carboniferous	500–1400 m	Lower: western slope of Polar Urals—Tournaisian limestone (150 m); Visean–Namurian limestone (1000 m). On southwest Pay-Khoy—limestone (100–150 m); limestone, dolomite (800 m); limestone (100 m); Vaygach—limestone and cherty limestone (450–1200 m). In western Novaya Zemlya: carbonate strata grade into clastics eastward (250–750 m).
		Middle: Polar Urals and Pay-Khoy—Middle and Upper Carboniferous mostly carbonate but thicknesses are poorly known. Vaygach—conglomerate, cherty limestone, limestone (40–120 m); Novaya Zemlya—conglomerate with siltstone and sandstone (100 m), carbonate strata (250–450 m).
		Upper: Vaygach—was not observed. In Novaya Zemlya: carbonate (100 m); conglomerate and sandstone, phyllite, and shale (350 m).

TABLE I (*Continued*)

Age	Thickness	Description
Devonian	300–900 m	Lower: western slope of Polar Urals—limestone and dolomite (350 m); eastern slope—limestone (600 m); southwest Pay-Khoy—limestone (30 m); Vaygach—limestone, marbled limestone; Novaya Zemlya—rhythmic alternating sequence of dolomite and dolomitic limestone (600–900 m).
	400–1500 m	Middle: Polar Urals and Pay-Khoy—limestone, sandstone, shale, limestone (250 m); shale (400 m); cherty shale, tuffaceous sandstone, tuff (150 m); shale (30 m); limestone (400 m); tuff pebble conglomerate, tuffaceous sandstone, cherty shale (200 m); Vaygach—reef limestone, limestone intercalated with siltstone and sandstone (>250 m); Novaya Zemlya—carbonates with terrigenous clastics increasing towards the north; limestone (600–900 m), terrigenous clastics (1000–1500 m).
	1200 2500 m	Upper: Polar Urals—volcanics, carbonates, and terrigenous clastics (>1500 m); Pay-Khoy—siliceous shales (100 m). Vaygach—shale and sandy limestone (50 m) limestone and shale (170 m), limestone and dolomite (400 m); Novaya Zemlya—clastic, carbonates, and volcanics (1200–2500 m).
Silurian	300–2000 m	Western slope of Polar Urals: limestone, marble (1000 m). Eastern slope of Polar Urals: volcanic sequence—effusive porphyry, tuff, tuffaceous conglomerate (500 m); limestone (200 m); tuff, conglomerate (200 m). Pay-Khoy: shale (175 m); coaly clayey shale (4.5 m), limestone with shale (65 m) (total thickness of Silurian in Pay-Khoy—less than 300 m). Vaygach: limestone (600–1300 m) overlain by fine clastics (200–700 m). Novaya Zemlya: 300 m of carbonate rock overlain by sandstone and shale 800 m thick; thin sheets of mafic rocks.
Ordovician	600–2200 m	Transgressive on Cambrian. basal conglomerate, quartzitic sandstone with shale, gritstone, conglomerate (>1000 m); sandstone, shale, polymictic sandstone, gritstone (600–700 m). In western Polar Urals the sedimentary deposits are replaced by a sequence of volcanic rocks (1000–1300 m). Eastward the sedimentary sequences are replaced with volcanic formations (100 m). Cherty schist with limestone (1000 m); dolomitized limestone, dolomite (50 m). Pay-Khoy: limestone (40–100 m); shale. Total thickness—900–1000 m. Vaygach: limestone and shale. Conglomerate at base rests unconformably on Cambrian (600–2500 m). Novaya Zemlya: up to 2200 m of coarse conglomerate and sandstone of Ordovician to Lower Silurian age.
Cambrian	400–2000 m	In Polar Urals: carbonaceous quartzite schist, phyllite, marble, and limestone (1000 m); quartzite, sandstone, quartzitic schist with effusive volcanic rocks, arkosic

(Continued)

TABLE I   (*Continued*)

Age	Thickness	Description
Cambrian (*continued*)		sandstone, gritstone, conglomerate (200 m); chloritic schist, carbonaceous schist, quartzite, siltstone, sandy and recrystallized limestone, tuff, tuff conglomerate (800–1300 m); Pay-Khoy: limestone (400 m); quartzose sandstone, polymictic sandstone, gritstone, conglomerate (700 m); Vaygach and Novaya Zemlya: mainly shale and sandstone with minor conglomerate and tuff (500–3000 m).
Precambrian	1000–2000 m	Well-developed gneiss–amphibolite complex forms axial part of the Urals. Thickness more than 1000 m. Proterozoic sequence in Pay-Khoy is 2000 m of conglomerate, sandstone, and chlorite schist overlying mafic lavas that are all in a greenschist facies. Vendian age limestone and dolomite 2000 m thick lie higher and are locally replaced by a rhyolite and basalt association that is 1500 to 2000 m thick. Proterozoic section is capped by variegated carbonate–clastic strata 1500–2000 m thick. At Vaygach, volcanic and terrigenous rocks are greater than 1500 m thick.

[a] In this and subsequent tables, descriptions within each system are from oldest to youngest.

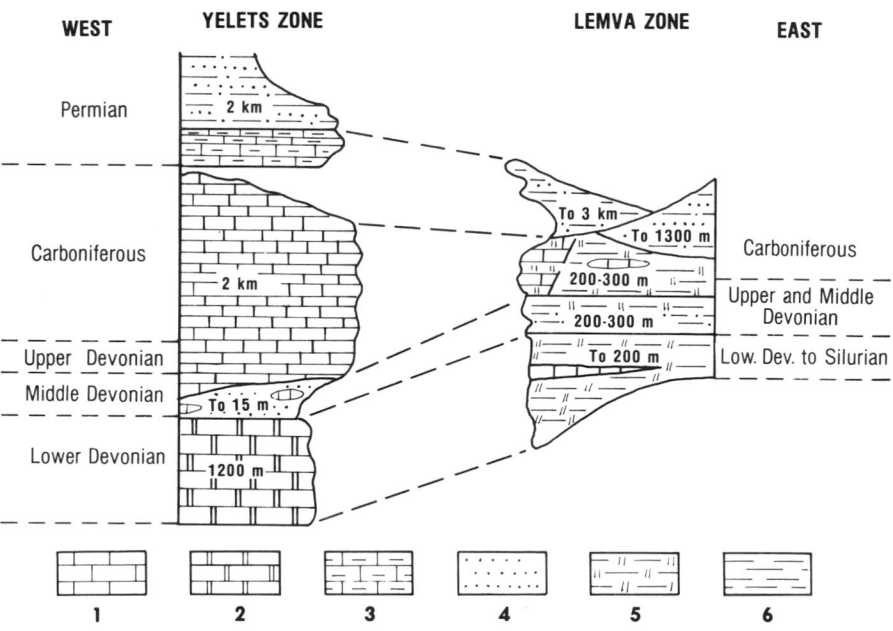

Fig. 8. Correlation of facies across the Polar Urals: (1) limestone, (2) dolomite, (3) silty limestone, (4) sandstone, (5) siltstone, (6) shale.

Proterozoic and Paleozoic igenous rocks of ultramafic to granite composition occur in the core and eastern limb of the Polar Urals, where they are associated with important thrust faults (Fig. 9).

Major basins filled with thick Permian and Triassic clastic strata lie on the west and east sides of the Pay-Khoy and Polar Ural Foldbelts. In the Korotaikha Basin, Permian marine deposits grade up into continental strata to form a section 6 to 8 km thick. On the west, Permian basalt flows, followed by Triassic continental deposits totaling 2000 m, are overlain by a Pliocene

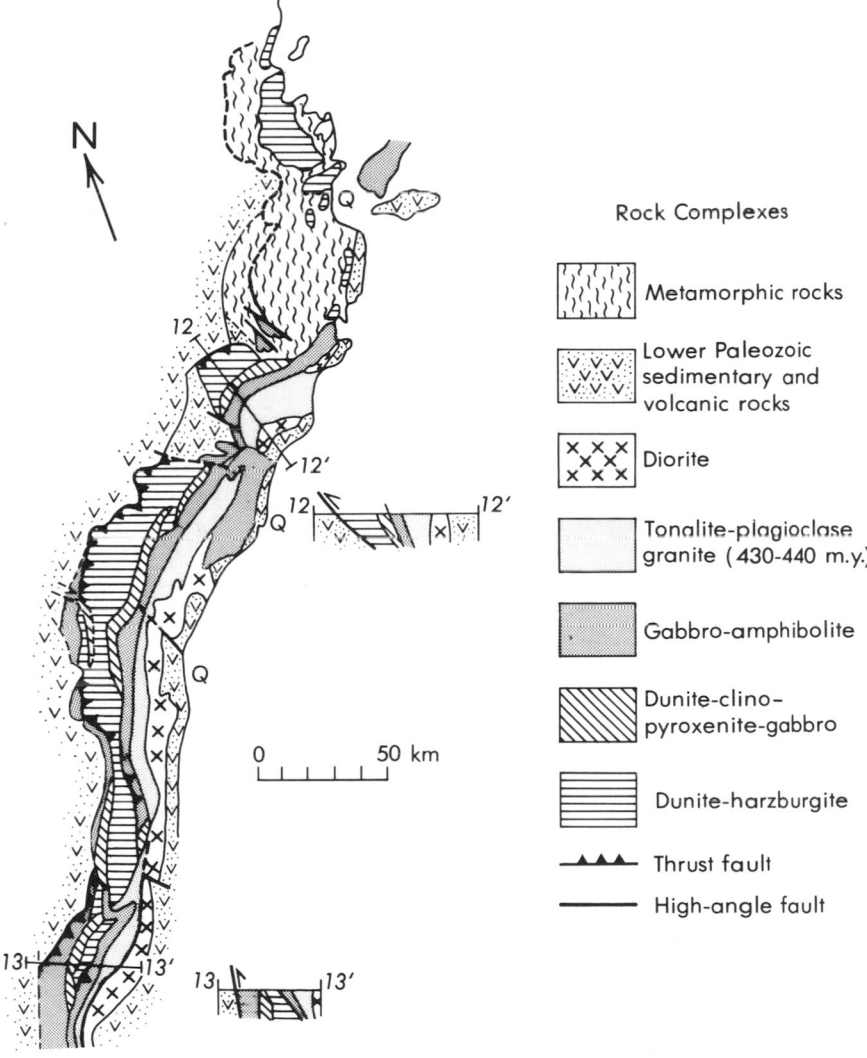

Rock Complexes

Metamorphic rocks

Lower Paleozoic sedimentary and volcanic rocks

Diorite

Tonalite-plagioclase granite (430-440 m.y.)

Gabbro-amphibolite

Dunite-clino-pyroxenite-gabbro

Dunite-harzburgite

Thrust fault

High-angle fault

0   50 km

Fig. 9.   Structure and tectonic position of gabbro–ultramafic massifs of the Polar Urals (after Dergunov and Moldavantsev, 1976, Fig. 1).

through Holocene marine section. Glacial marine sand and clay up to 200 thick cap the section. On the east side of the Polar Urals and Pay-Khoy Fold-belt trend, Permian strata are overlain by an extensive Upper Cretaceous sandstone section that is up to 250 m thick. A Pliocene through Holocene section similar to that on the west side of the foldbelt also occurs on the east side.

The Pechora Basin forms another large depression on the west side of the Polar Urals, south of the Korotaikha Basin described above. In the Pechora Basin, a platform cover of 4–6 km of Paleozoic and Mesozoic rocks lies on a crystalline basement that is probably Precambrian. Vendian (late Proterozoic) through Ordovician clastic rocks appear to overlie the basement, followed by a carbonate–clastic sequence of Ordovician through Triassic age, and by Triassic continental deposits, similar to but thinner than the sequence fronting on the northeast side of the Polar Urals. Middle Jurassic to Lower Cretaceous shale and sandstone in the Pechora Basin total less than 450 m in thickness and are overlain by glacial and marine sediments, as in the Korotaikha Depression.

An early Paleozoic ophiolite suite including ultramafic rock, gabbro, plagiogranite, and volcanic and sedimentary rock sequences is thrust along the eastern slope of the Polar Urals over westward-extending miogeosynclinal rocks (Dergunov and Moldavantsev, 1976). This suggests obduction of oceanic crust onto the eastern edge of the Russian plate (Efimov *et al.* 1978)., Also, a reconstruction of structurally juxtaposed facies in the Polar Urals indicates a collapsed continental margin composed of thick, carbonate-rich rocks that graded eastward into a condensed section of siliceous shale with pelagic fossils (Puchkov, 1973). Further east are volcanic arc rocks (Efimov *et al.*, 1978).

V. FRANZ JOSEF LAND—AREA IV*

Franz Josef Land is a large archipelago lying to the north of Novaya Zemlya on the edge of the Eurasian Basin part of the Arctic Ocean. The islands have been largely glaciated, even though the highest point in the archipelago is only 735 m. A composite stratigraphic column is given in Table II, and a geologic map of Cape Ganza on Zemlya Vil'cheka is shown in Fig. 10.

The oldest rocks in the archipelago are shale and coal beds of Carboniferous age. The next oldest strata are Upper Triassic (Karnian) limestone and shale on Zemlya Vil'cheka, containing a marine fauna of ammonites and lamellibranchs. They are overlain by 300 m of Upper Triassic and Lower Jurassic continental sandstone, siltstone, and argillite with fossil plants. The Middle Jurassic consists of about 30 m of shale and siltstone. Upper Jurassic

---

* Excerpted from Nalivkin (1973).

TABLE II
Composite Cross Section of Franz Josef Land and Viktoria Island

Age	Thickness	Description
Quaternary	3 m minimum (?)	Sea terrace, glacial, and eolian sediments.
Cenozoic	>25 m	Paleogene—reworked marine sediments with diatoms and algae (400 m); Neogene—siltstone, silty sandstone, and moraine (25 m).
Cretaceous	870 m	Terrigeneous sand and sandstone with brown coal layers (35 m); basalt sheets, sand with sandstone (175 m); effusive sequence—8 to 10 basalt flows (2–15 m thick) with tuff beds; above them 5 to 7 basalt sheets (to 60 m each) with intercalated tuff; siltstone, clay, carbonaceous shale with coal layers (200 m); basalt sheets with sandstone, sand, clay, and coal (60 m). Approximate thickness of basalt sheets and sedimentary deposits (Aptian–Albian)-260 m, and thickness of effusive sequence-600 m. Cenomanian: quartzitic sandstone (at least 15 m); unconsolidated silty clay (40 to 45 m).
Undifferentiated middle and late Jurassic	At least 400 m	Shale, siltstone, and basalt (about 400 m).
Jurassic	>530 m	Sand and sandstone with intercalated siltstone, limestone, pebble conglomerate, coal layers (220 m). On Northbrook and Rainer Islands: siltstone and argillite (30 m); shale (24 m); silty shale (about 35 m); shale, clayey sandy limestone, basalt sheet (25 m); sand with sandstone (170 m); sandy silty limestone (25 m).
Triassic	540 m	Zemlya Vil'cheka Island: slaty siltstone with intercalations of limestone (20 m) Karnian shaly siltstone with intercalations of silty, fossiliferous limestone (about 20 m thick). Kheisu and Vil'cheka Islands: continental sandstone and siltstone (240 m). Viner-Neishtoadt Island: sandstone (50–60 m); layer of brown coal (220 m).
Carboniferous	Unknown	Tsiglera, Gallya (Hall), and Zemlya Georga Islands: sandy coal-bearing deposits, basalt, sandstone and loose shale with coal layers, foliated pyroschist. Grab samples of sea floor near Zemlya Vil'cheka: limestone with coal. Grab samples of sea floor near Victoria Island: pebbles of carbonate rocks.

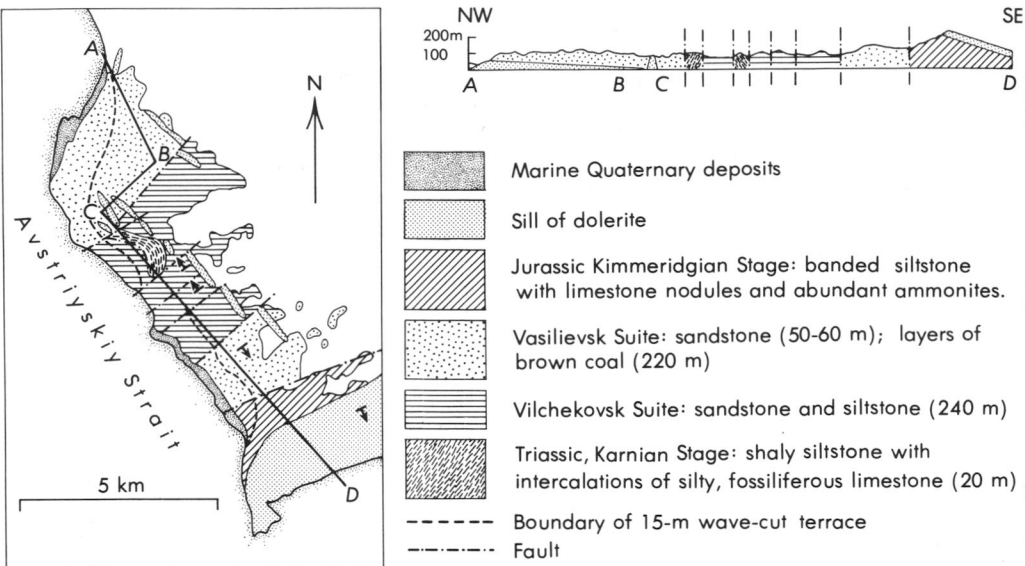

Fig. 10.   Geologic sketch map of Cape Ganza, Zemlya Vil'cheka, Franz Josef Land (after Tkachenko and Egia-zarov, 1970, Fig. 2).

shale and sandy limestone with concretions are at least 200 m thick and contain a rich and varied fauna of ammonites and belemnites. The Lower Cretaceous consists of dolerite sills and flows (2–60 m thick), separated by sandstone and carbonaceous shale with lenses of coal. Upper Cretaceous rocks are absent.

VI. WESTERN SIBERIAN LOWLAND AND ISLANDS OF THE KARA SEA—AREA V*

Between the Ural Mountains and the Taymyr Peninsula lies the vast tundra-covered plain of the Western Siberian Lowland. The region includes the low-lying areas of the Yamal and Gydan Peninsulas and the islands of the Kara Sea. These tundra-covered plains are underlain by nearly horizontal Mesozoic and Cenozoic strata.

The Western Siberian Lowland is a 3000-m-deep basin of Jurassic, Cretaceous, and Cenozoic marine strata, mainly sandstone, shale, and siltstone with some interbeds of continental, plant- or coal-bearing rocks. The basement consists of folded and metamorphosed Precambrian or Lower Paleozoic rocks cut by some isolated plutons. Figure 11 shows the subsurface geology of the region as determined from boreholes.

* Excerpted from Nalivkin (1973).

TABLE III
Composite Stratigraphic Column, Yamal Peninsula[a]

Age	Thickness	Description
Quaternary	Maximum thickness drilled—300 m	See Fig. 12.
Paleocene	>70 m	Novyi Port Village—in borehole: Marine terrigeneous sediments, occurring between coal-bearing deposits and opokalike clay of lower Eocene (7 m). Lower Eocene at a depth of 146–214 m(?): opokalike clay, opoka (minimum thickness 68 m).
Late Cretaceous	280 m	Novyi Port Region: Santonian and Campanian Stages: in boreholes at a depth of 387–504 m; claystone with intercallations of limey sandstone, (total thickness 117 m). Maestrichtian Stage: at a depth of 221–387 m; siltstone, claystone alternating with clayey siltstone, siltstone, sandstone (166 m).

[a] From German et al. (1963).

The Yamal Peninsula (Fig. 1) is covered by Quaternary terrace and marine deposits (see Fig. 12). A borehole (Table III) near the peninsula's southeastern shore has penetrated 283 m of Upper Cretaceous shale, sandstone, and siltstone below approximately 220 m of Paleocene and Quaternary sediments.

The oldest rocks in the islands of the Kara Sea are found on Pologiy and Troynoy Islands and consist of Upper Proterozoic(?) schist (Figs. 13 and 14). Cretaceous unconsolidated deposits are found on Ushakova, Vise, and Uedineniya Islands, the Arkticheskogo Instituta Islands, Sverdrup Island, and the Sergeya Kirova chain. The islands, just off the coast of the Yamal and Gydan Peninsulas, consist of Pleistocene and Holocene deposits, including glacial and fluvioglacial sand, gravel, and loam.

VII. TAYMYR AND SEVERNAYA ZEMLYA—AREA VI*

The Taymyr Peninsula (Fig. 1) consists of a mountain range reaching elevations of 1000–1200 m. The rocks are strongly folded and metamorphosed geosynclinal strata of Precambrian and Paleozoic age that are intruded by synorogenic plutons. The large archipelago of Severnaya Zemlya lies north of Taymyr and is heavily glaciated, despite its low maximum elevation of only 300 m above sea level. Figure 15 shows the main tectonic elements of Taymyr

* Excerpted from Nalivkin (1973).

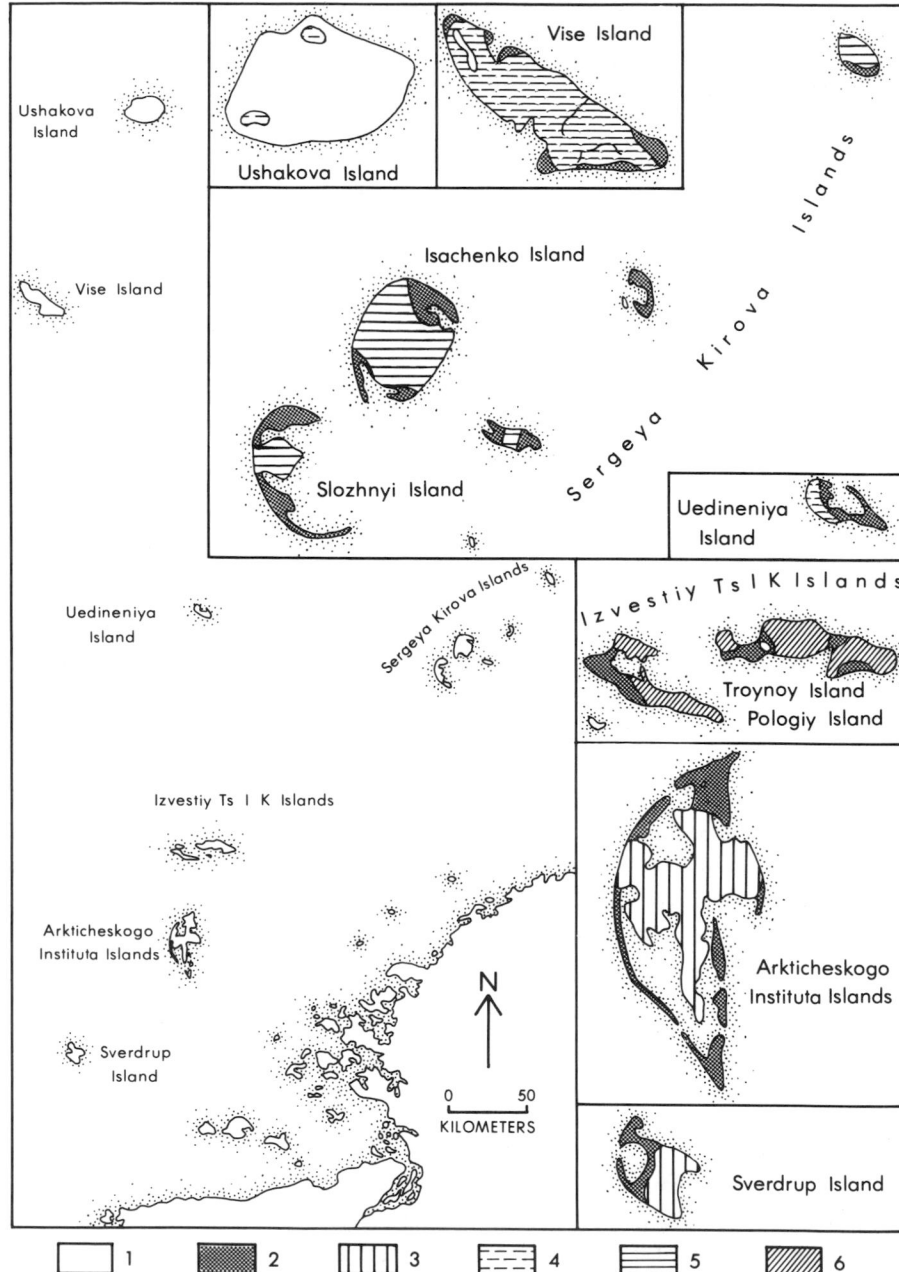

Fig. 13.   Generalized geologic map of the islands of the central Kara Sea: (1) glaciers; (2) modern marine deposits; (3) Cenomanian–Santonian; (4) Albian; (5) Aptian; (6) Riphean.

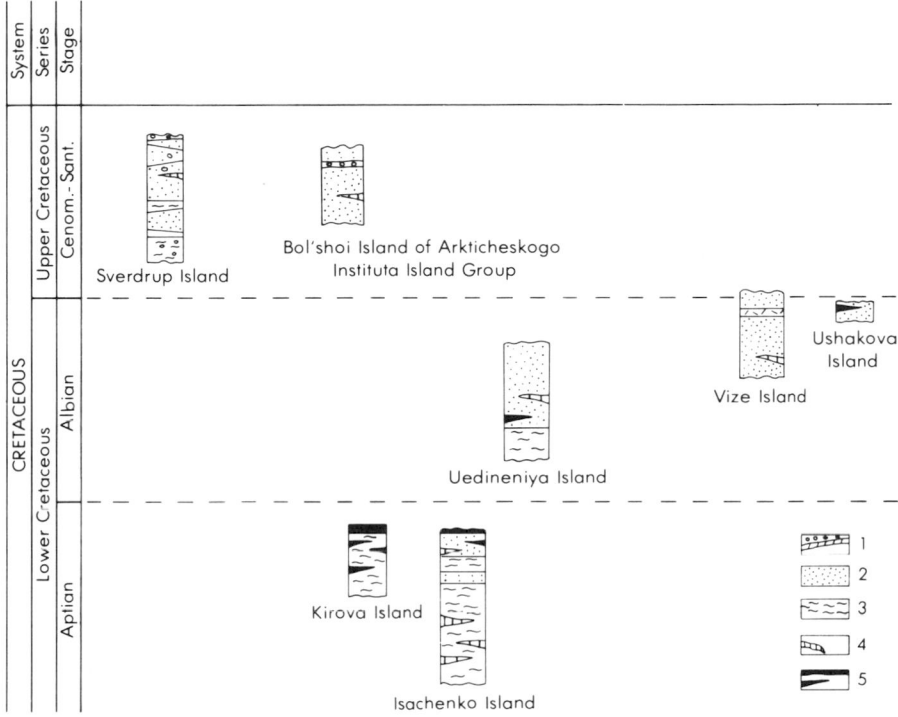

Fig. 14. Columnar sections of the Kara Sea islands: (1) pebble conglomerate; (2) sand and sandstone; (3) siltstone; (4) limestone; (5) coal and carbonaceous shale.

and the locations of the cross sections in Fig. 16. Composite stratigraphic sections of Taymyr are described in Tables IV and V. A diagrammatic cross section along southern Taymyr, from the Yenisey River to the Lena River, is shown in Fig. 17.

The oldest rocks in Taymyr consists of Lower Proterozoic crystalline and metamorphic rocks. Unconformably overlying these is an Upper Proterozoic complex of intrusive and metamorphosed rocks.

The Lower Cambrian in Taymyr is generally nonfossiliferous, so there is some question as to its extact age, but limestones in eastern Taymyr have yielded Middle Cambrian trilobites. Ordovician rocks are exposed in narrow belts in the central part of the peninsula. Thick limestone and graptolitic shale are important facies of the Ordovician here.

The Silurian rocks of Taymyr resemble those of Novaya Zemlya and the Urals in general lithology and fauna. Limestone predominates in western Taymyr, while shale and siltstone, some with Lower and Middle Silurian graptolites, are most often found in the north and northeast. The overlying Devonian rocks are rich in fossils and also are similar to those of Novaya

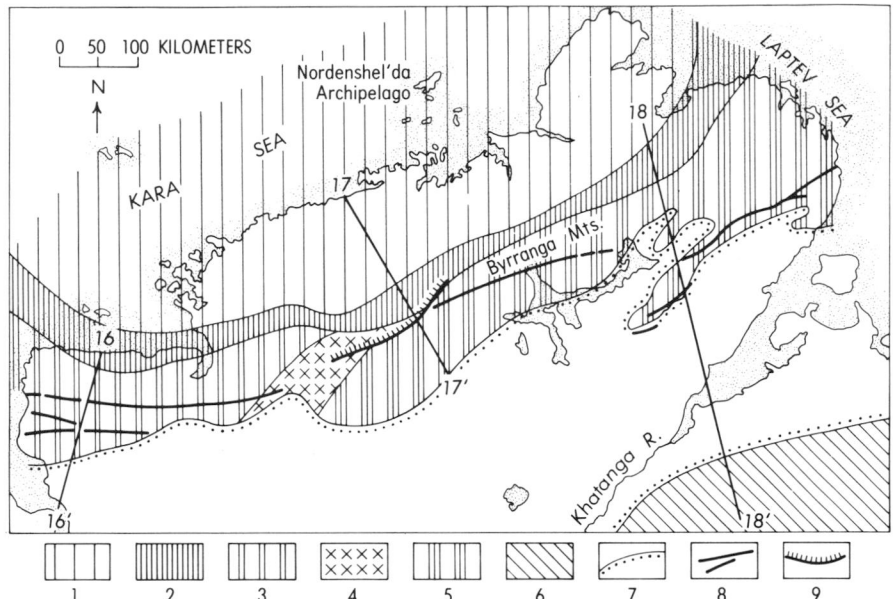

Fig. 15. Tectonic zones of the Taymyr Foldbelt (modified after Pogrebitskiy, 1971, Fig. 2): (1) the arched Karsk Uplift; (2) flexure margin; (3) the inverted Taymyr Trough; (4) the Tarei Arch; (5) the East Taymyr Basin; (6) north slope of the Anabar Anticlinorium; (7) boundary of the Mesozoic–Cenozoic Leno-Yenisey Trough; (8) major faults in the trough area; (9) zone of Upper Taymyr imbricate thrusts. Sections 16–16′, 17–17′, and 18–18′ are shown in Fig. 16.

Zemlya. Lower Devonian limestone, bituminous shale, and marlstone are succeeded by Middle Devonian lagoonal shale, marlstone, and sandstone with gypsum beds, and a terrestrial section with sandstone, shale, and conglomerate. Upper Devonian marine limestone, marlstone, and shale completes the succession. The Lower Carboniferous is closely associated with the Upper Devonian and consists of the same richly fossiliferous marine limestone and shale facies. The Middle and Upper(?) Carboniferous are found as isolated outcrops of limestone and shale.

Permian rocks are widespread in southern Taymyr and resemble the Permian sequence of the Urals. Lava flows and mafic sills are common in the Permian rocks of Taymyr, which can be divided into two sedimentary sequences, beginning with marine fine-grained sedimentary rocks and ending with continental coal-bearing deposits.

The Mesozoic section is not as thick or as widespread as the Paleozoic. Isolated exposures of Upper Jurassic and Lower Cretaceous strata occur in central and southern Taymyr. Sandstone and shale predominate, but some coal-bearing formations are present in the upper Lower Cretaceous. During the late Cretaceous, a major marine transgression advanced from the south

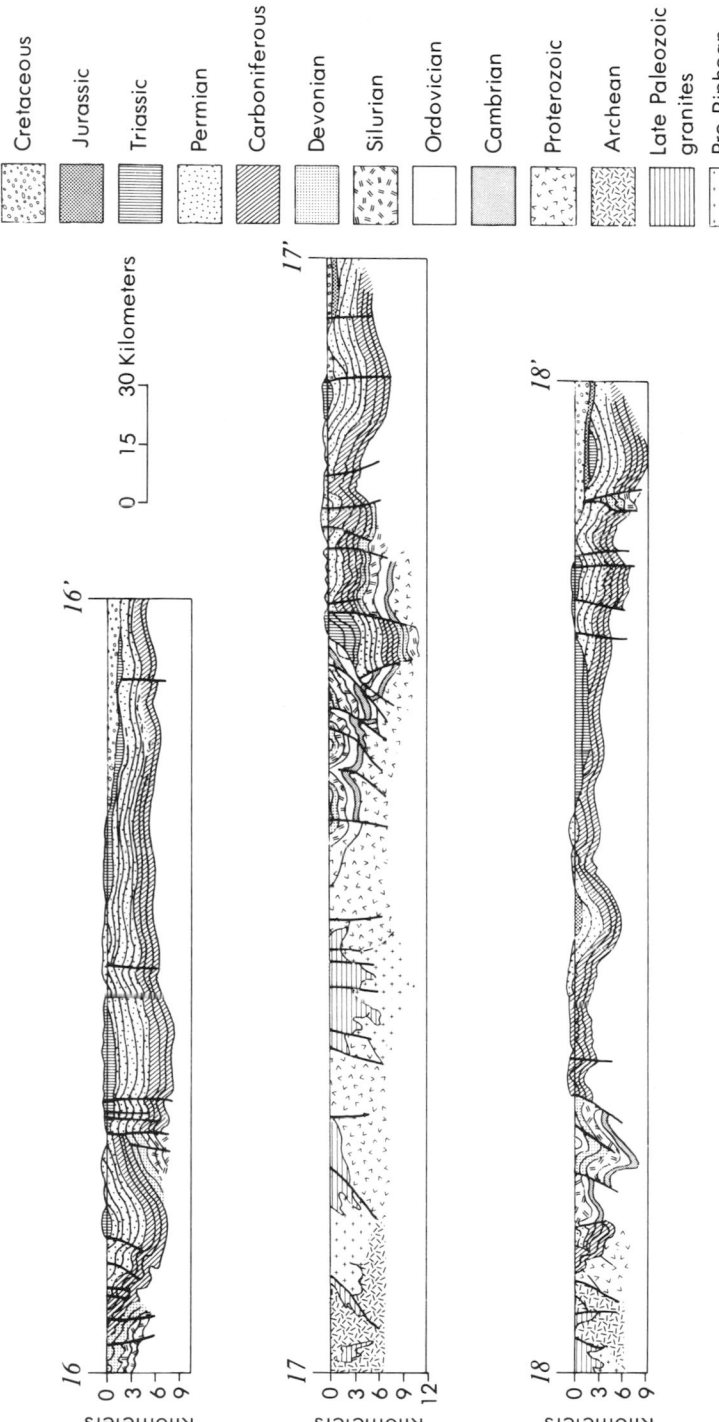

Fig. 16.   Geologic cross sections across the Taymyr Peninsula; locations shown on Figs. 1 and 15 (modified after Pogrebitskiy, 1971, Fig. 3).

TABLE IV
Composite Columnar Section for the Taymyr Peninsula[a]
(Area *a* in Fig. 1)

Age	Thickness	Description
Quaternary	45–75+ m	Reworked moraine—boulders with clayey loam (10–15 m). Marine and alluvial interglacial: clay, clayey loam, sandy loam, sand with pebbles, moraine and fluvioglacial deposits—clayey loam with sand and gravel, pebbles, and boulders (35–60 m). Karginsk transgression-formed terraces: sand, clay, clayey loam, pebbles. Holocene: sandy clayey deposits with pebbles.
Cretaceous	>600 m	In zone of transition into Taymyr Lowland: clayey, sandy siltstone, arkosic sandstone (400 m); sandy coal-bearing deposits (200 m); sandy limestone(?).
Jurassic	1070–1100 m	Mostly in eastern part of the peninsula: Laptev Sea shore—siltstone with conglomerate at its base (420 m); friable sandstone (26 m); clayey siltstone with sand and argillite layers (220–250 m); conglomerate (several meters), friable sandstone (280 m); unconsolidated sand and sandstone (about 120 m).
Triassic	1300–1400 m	Lower (in Eastern Taymyr): sandstone, shale (290 m), siltstone alternating with polymictic sandstone (160 m). Middle: siltstone with sandstone layers (390 m); polymictic sandstone, siltstone, alternating siltstone and shale with sandstone and coal (190 m). Upper: sandy clayey deposits (230 m); siltstone, sandstone, sandy argillaceous rocks (135–140 m).
Permian	3400–4800 m	Lower Permian often occurs transgressively on Carboniferous, Devonian, and Silurian: sandstone, argillite, siltstone, alternating argillite and siltstone (1100–1200 m); coal-bearing sandstone, siltstone, argillite (800–1100 m). (Total thickness of Lower Permian: 1900–2300 m.) Upper Permian: sandy clayey deposits, tuff conglomerate, basic lava and tuff (Upper Permian: 1500–2500 m).
Carboniferous	1000–1100 m	Southern Kara Sea shore: shale and limestone (550–600 m). Eastern Taymyr: Lower Carboniferous deposits (700–800 m). Middle Carboniferous: foliated argillite with intercalations of argillite and sandstone (150 m); organic limestone (50 m); shale (250 m). Upper Carboniferous: absent.

TABLE IV (*Continued*)

Age	Thickness	Description
Devonian	900–1000 m	Lower Devonian: Tareya River valley—bituminous shales, limestone, organic limestone with shale intercalations, shale, clayey limestone, limestone with marl and gypsum, sandy clayey deposits with conglomerate (600–700 m). Middle Devonian: Eastern Taymyr—limestone (250–300 m); in Zhdanova, total thickness of Middle Devonian 700 m. Upper Devonian: Premorskay Anticline—shale with limestone and limy sandstone (400–800 m).
Silurian	1500–1700 m	Pyasinsk region: limestone with shale (1500–1700 m). In eastern Taymyr: shale and limestone (800–1000 m). Toward the north and northeast, shale and siltstone predominate.
Ordovician	2400–3000 m	Pyasinsk Bay shore: dolomite, limestone intercalated with shale (500 m) [Ritakeia (1500 m); Chelyskin Peninsula (2800 m)]; sandy shale with conglomerate and limestone (from 80–100 to 600–800 m); limestone with intercalations of shale (400–500 m); limestone (450–500 m); shale (500–650 m); organic limestone.
Cambrian	3000–3700 m	Foliated sandstone, shale (1000 m in west, 1400–1500 m in east); dolomite, dolomitized limestone (1000–1100 m); siltstone, sandstone, shale, conglomerate, foliated clayey dolomite rocks (600 m); dolomitized limestone, dolomite with intercalations of shale (400–500 m).

Disconformity

Age	Thickness	Description
Late Proterozoic	4300–6800 m	In Eastern Taymyr region: volanic tuff and lava. In Western Taymyr region: tuff and terrigeneous arkosic sandstone with horizons of tuff, with sheets of porphyrite and minor basalt, greenschist with marbled dolomite and limestone, spilite (500 m); quartzite and phyllite alternating with schist (1000 m); schist with dolomite, phyllitic sandstone with sills of metamorphosed porphyry (1500–1700 m); conglomerate, metamorphosed sandstone, marble, sills of metamorphosed schistose felsite porphyry and tuff (about 2000 m). Angular disconformity.
Early Proterozoic	5000–8000 m	Bottom to top: Western Taymyr region: tuff and terrigeneous rocks, arkosic sandstone with tuff horizons and minor basalt (1200–2000 m). Okt'yabrsk Group: metamorphosed sandstone deposits (1500 m). Zhdanovsk Group: phyllite,

(Continued)

TABLE IV   (*Continued*)

Age	Thickness	Description
Early Proterozoic (*continued*)		schist, with marble and dolomitic schist (2600 m). Laptev Group: conglomerate alternating with schist and arkosic sandstone (400–1200 m). Eastern Taimyr region: Pronchishchevsk Group: sequence of greenschists—epidote-actinolite, actinolite-chlorite, and chlorite-sericite (1200–2000 m). Okt'yabrsk Group: Chelyuskin Peninsula—quartzite and phyllite alternating with schist (1000 m). Zhdanovsk Group: Chelysukin Peninsula—schist with marbled dolomite and limestone (1700 m). Laptev Group: metamorphosed and schistose sandstone with conglomerate beds, alternating with acidic volcanic rocks, felsite porphyry in places (1000 m).
		Disconformity
Late Archean	Unknown	Nizhne-Karskaya series: plagiogneiss (3000+ m). Verkhney-Karskaya series: plagiogneiss (2000 + m), gneisslike schist (2000 m). Fadeevskaya series: metamorphosed, schistose tuff porphyry (1800 m).

[a] After Markov *et al.* (1957).

into the southern parts of the western Arctic, leaving Upper Cretaceous deposits only on the southern margin of Taymyr. Cenozoic deposits are also poorly represented in Taymyr.

The Precambrian of Severnaya Zemlya is lithologically similar to that of Taymyr (Table VI and Fig. 18). It consists of two Upper Proterozoic formations, a predominantly argillaceous one below and a sandier one above. Above these lie barren formations provisionally considered to be Lower and Middle Cambrian. Higher in the section are Ordovician sandstone, limestone, dolomite, and marlstone with gypsiferous layers near the top. The Silurian rocks of Severnaya Zemlya have a rich fauna of tabulate corals and pentamerids in beds unconformably overlying the Ordovician. Limestone and sandstone predominate and gypsiferous beds occur near the top of the Silurian section.

The Devonian of Severnaya Zemlya differs from the Devonian of Taymyr and the Urals but is very similar to the Devonian of Svalbard and other parts of the northern Caledonides. A characteristic feature of the Devonian in Severnaya Zemlya is the predominance of the Old Red Sandstone facies, including red and variegated sandstone, siltstone, shale, marlstone, and

TABLE V

Generalized Columnar Section of the Arctic Lowland between Taymyr Peninsula on the North and the Mid-Siberian Upland on the South[a]

(Area *b* in Fig. 1)

Age	Thickness	Description
Quaternary	12–125 m	Anabar-Khatanga area: boulders and pebble gravel (12 m); on salt domes, sandy clayey deposits with boulders and pebbles (125 m); interglacial deposits (Yeniseisk Port, 75 m), alluvial, lake beds, etc. (? m).
Tertiary	12 m	Khatangsk depression: sand with thin coal layers and lenses of peat and pebbles (about 12 m).
Cretaceous	2250–3000 m	Total thickness in Ust'Yeniseisk Depression, 3 km. In Malokhetsk Anticline: shale, siltstone (204 m); alternating shale, siltstone, and sandstone (150 m). Total thickness in Khatangsk Depression, 2 km: sandstone, shale, siltstone, coal, conglomerate, gritstone (220 m); coal-bearing clayey silty rocks (270–490 m); coal-bearing argillaceous siltstone (320 m); argillaceous siltstone, coal-rich strata (170–205 m). Undifferentiated Albian and Senomanian: sandstone (310–575 m); sand, sandstone (200 m); clay (90 m); silt with clay, sand and sandstone (150–300 m); clay with silt (90 m); and sand (80 m).
Jurassic	3550–4030 m	In Ust'Yeniseisk Depression: encountered at a depth of 4–5 km in boreholes. In Malokhetsk Anticline region at a depth of 265–2500 m (2235 m thick): sand, sandstone and conglomerate (140 m); sand, sandstone (50–90 m); clay with sandstone and siltstone (70–100 m); sandy shale with conglomerate (40–100 m); argillite (50–105 m); sandstone (82–159 m); shale, argillite (63–110 m); sandstone (74–120 m); argillite, shale (48–82 m); sandstone with coal (70–108 m); shale, siltstone (95–165 m); sandstone (143–155 m); shale (30–44 m); sandy shaley rocks (45–57 m); shale (340 m).
Triassic	1425–1680 m	In boreholes in Malokhetsk Anticline: tuffaceous argillite, argillite with sandstone, siltstone, and tuff (300 m); flows alternating with tuff (160–227 m); argillite, sandstone (40–50 m) (in Khatangsk Depression, 300 m); siltstone, sandstone (80–130 m, in Nordvik region, 390–700 m); sandstone with siltstone and conglomerate (70–140 m); argillite, shale (50–230 m); sandstone alternating with siltstone and shale (25–100 m, in Tsvetkov Cape 135–400 m).
Permian	1710–1910 m	Nineteen cycles of rhythmically bedded rocks: alternating argillite, siltstone, and sandstone (892 m); argillite and siltstone alternating with sandstone (450 m) (In northern part of Khatangsk Depression: coal-bearing rhythmically alternating sandstone, siltstone, and argillite (2400–2500 m), and coal-bearing sandstone and siltstone (900–1000 m); alternating argillite, siltstone, and sandstone (270–460 m); conglomerate, argillite, siltstone (100–108 m).

(Continued)

TABLE V   (*Continued*)

Age	Thickness	Description
Carboniferous	950 m	In boreholes 1200 m below surface: limestone, limestone conglomerate (640 m); biogenic limestone (140 m); limestone with argillite and marl (70 m); limestone (at least 100 m).
Devonian	>620 m	In salt domes in Khatangsk depression: gypsum, limestone (several hundred m); gypsum with dolomite, anhydrite, diabase, shale (300 m); dolomite, limestone (30 m); argillaceous limestone, marl (about 190 m).
Silurian	1370 m	Yeniseiskiy-Tochinskiy Uplift, in boreholes at a depth of 154 m down to 1457 m (actual thickness 1303 m): metamorphosed limestone and marble and from a depth of 119 m, mostly dolomite, tuff, and lava (35 m). The Silurian is overlain by Middle Jurassic deposits at this location.

[a] Abstracted from Saks and Egorova (1957).

dolomite. Gypsum- and salt-bearing beds are common. The Upper Devonian is overlain by Quaternary strata and is gently folded. No Upper Paleozoic rocks have been found in Severnaya Zemlya. Mesozoic deposits are relatively thin and consist mainly of sandstone and shale. Cenozoic deposits also are thin, being mostly glacial and fluvioglacial.

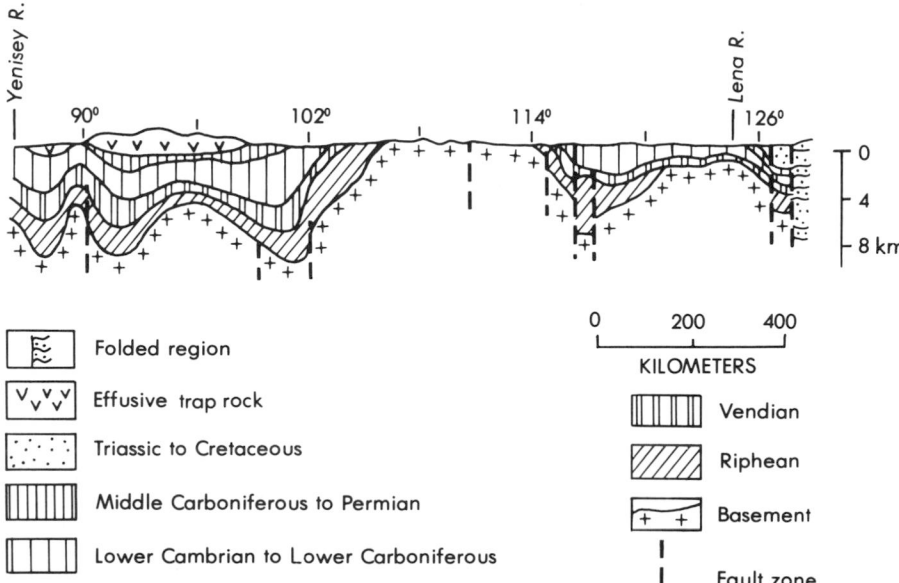

Fig. 17. Diagrammatic geologic cross section across the Siberian Platform from the Yenisey River to the Lena River (after Pyatnitskiy, 1974).

TABLE VI
Generalized Columnar Section for Severnaya Zemlya[a]
(Area c in Fig. 1)

Age	Thickness	Description
Holocene	Unknown	Marine, glacial, alluvial.
Quaternary	70–80 m	Bouldery marine deposits, moraine, marine interstadial deposits, glacial and fluvio-glacial deposits.
Paleogene	18–20 m	Sand, poorly cemented sandstone.
Cretaceous	30–40 m	Unconsolidated sandstone with shale intercalations (15 m); sand (15–20 m).
Late Jurassic	150–200 m	Unconsolidated sandstone, sandy shale, shale, sandy limestone (minimum thickness 150–200 m).
Early Jurassic Late Triassic	100–200 m	Sandstone, conglomerate (relicts).
Devonian	1400 m	Conglomerate (3 m); limestone with dolomite, marl, limy quartzose sandstone (140 m); limestone, dolomite (120 m); sandstone with conglomerate, limy sandstone, sandstone (300 m); dolomite with limy sandstone, marl (60 m); limestone (50 m); dolomite (40 m); sandstone, marl (20 m); dolomite with gypsum (130 m); limestone (60 m); marl, limestone (40 m); sandstone (100 m). Undifferentiated Middle and Upper Devonian: limestone, sandstone, marl (120 m); marl (35 m); dolomite (45–50 m); sandstone (195–220 m).
Silurian	1500–1700 m	Limestone (470–500 m); limestone, quartzose sandstone (100 m); limestone with gypsum-bearing limestone breccia (75–80 m); dolomitic limestone (600 m); limestone (30 m); limy quartzose sandstone (50 m).

Unconformity

Age	Thickness	Description
Ordovician	2250–2500 m	Basal conglomerate (7 m); sandstone (200–250 m); limestone (200 m); sandstone (500–550 m); dolomite, marl (100 m); gypsiferous limestone (100 m); dolomite, marl (150–200 m); limestone, dolomite, marl, sandstone (200–250 m); sandstone (200–250 m); gypsum, limestone (350–400 m); limestone (100 m); sandstone (100–150 m).
Cambrian	3000 m	Gritstone (12 m); shale (150–175 m); quartzose siltstone (800 m); shale (450 m); quartzose siltstone (350–400 m); shale with sandstone (1200 m).
Late Proterozoic	4000–4500 m	Chlorite-sericite-quartz schist with graywacke-like metasandstone and green phyllite; dark graphitic slate at top (2500 m). Quartzose and polymictic metasandstone with thin beds of slate, graywacke, and phyllite; meta-volcanic rocks at top (1500–2000 m).

[a] Abstracted from Tkachenko and Egiazarov (1970).

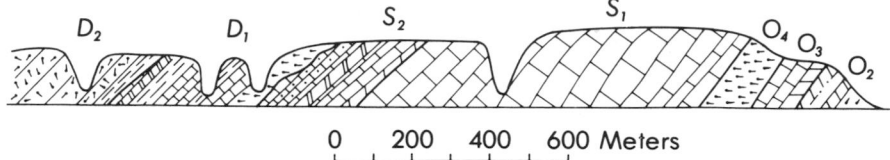

Fig. 18.   Structural cross section of Ordovician, Silurian, and Devonian strata of Severnaya Zemlya (after Nalivkin, 1973). Devonian: (D_2) red beds, with armored fish; (D_1) sandstone, grading upward into dolomite and limestone. Silurian: (S_2) limestone, sandy and flaggy toward top; (S_1) gray bedded limestone. Ordovician: (O_4) pale sandstone; (O_3) limestone and dolomite; (O_2) variegated strata, gypsiferous toward top.

The Laptev Sea Islands, off the east coast of Taymyr are mainly Lower Cretaceous sandstone, shale, and siltstone (Figs. 19 and 20). The Kara Sea Islands, off the west and north coasts of Taymyr (Fig. 21), exhibit a well-exposed section of Precambrian gneiss, schist, and metamorphosed sandstone and limestone. These rocks are overlain by Upper Silurian limestone, Permian sandstone and siltstone, Mesozoic coarse clastic rocks, and Quaternary unconsolidated deposits.

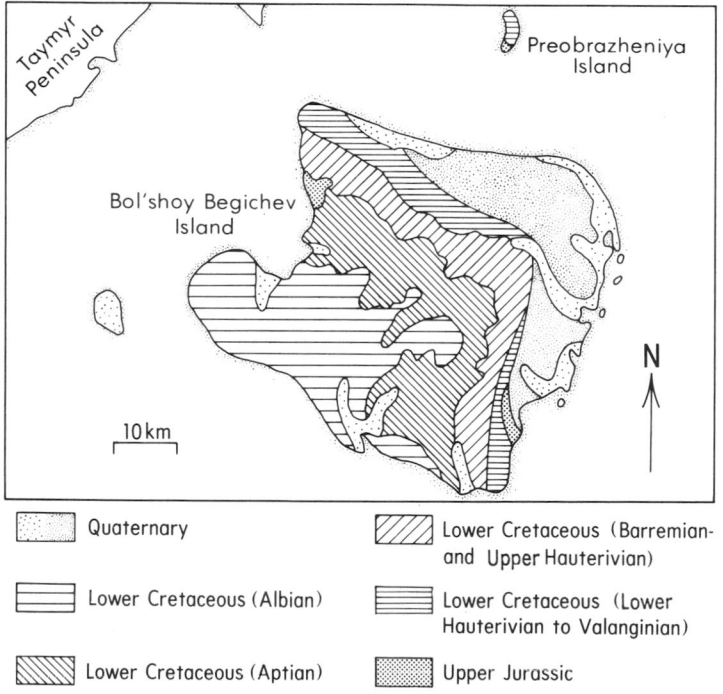

Fig. 19.   Geologic sketch map of Bol'shoy Begichev Island in the Laptev Sea (after Tkachenko and Egiazarov, 1970, Fig. 47).

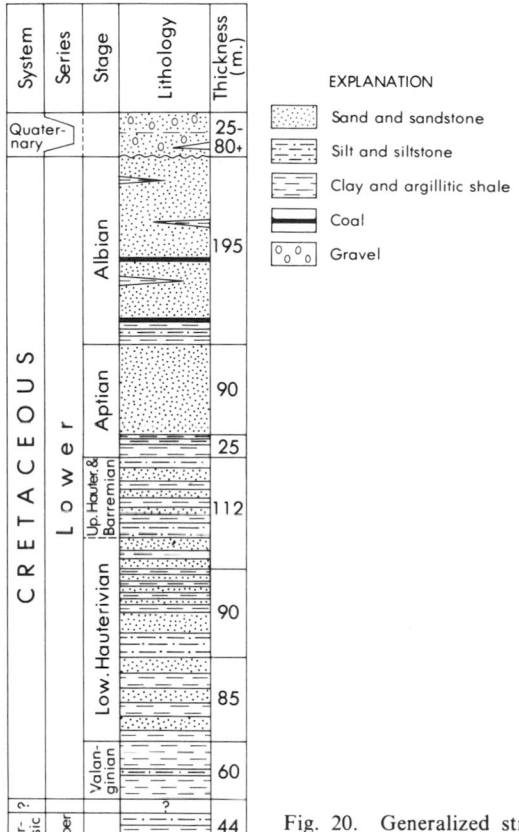

EXPLANATION

▒	Sand and sandstone
⊟	Silt and siltstone
▤	Clay and argillitic shale
▬	Coal
⬚	Gravel

Fig. 20. Generalized stratigraphic column for Bol'shoy Begichev Island (after Tkachenko and Egiazarov, 1970, Fig. 47).

Figure 22 is a tectonic diagram of the Taymyr–Severnaya Zemlya region. The Precambrian tectonic structures here form a large arc over 1000 km long, trending east–west in western Taymyr and gradually bending to trend nearly north–south in eastern Taymyr and Severnaya Zemlya. In Taymyr, these structures consist of small, isoclinal folds developed in Lower Proterozoic rocks and large, rather tight anticlines and synclines, tens of kilometers across, developed in Upper Proterozoic rocks that are further complicated by numerous isoclinal folds. In Severnaya Zemlya, all of the Proterozoic section is folded into a large anticlinorium with many smaller folds on its flanks.

The Lower Paleozoic section generally is unconformable on the Precambrian and has basal sandstones and conglomerate. The Cambrian and Ordovician structures border and parallel the trend of structures in the Proterozoic rocks. In Severnaya Zemlya, the Cambrian and Ordovician strata form linear structures of the same type as in the Proterozoic rocks.

Middle Paleozoic structures of the region are distinctly different from the

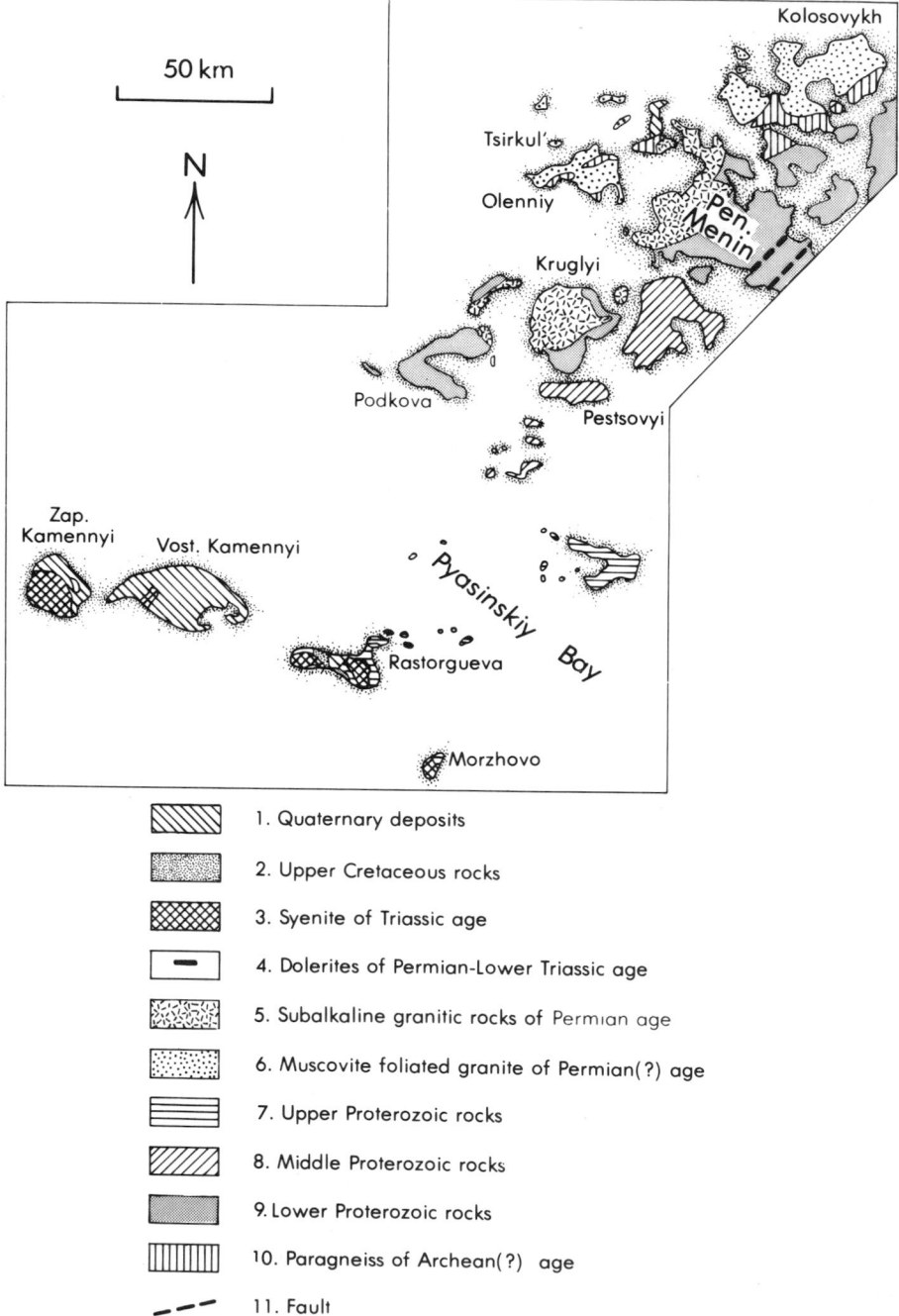

1. Quaternary deposits

2. Upper Cretaceous rocks

3. Syenite of Triassic age

4. Dolerites of Permian-Lower Triassic age

5. Subalkaline granitic rocks of Permian age

6. Muscovite foliated granite of Permian(?) age

7. Upper Proterozoic rocks

8. Middle Proterozoic rocks

9. Lower Proterozoic rocks

10. Paragneiss of Archean(?) age

11. Fault

Fig. 21.   Geologic map of the islands of Pyasinskiy Bay and of Menin Peninsula (area *d* on Fig. 1) (after Tkachenko and Egiazarov, 1970).

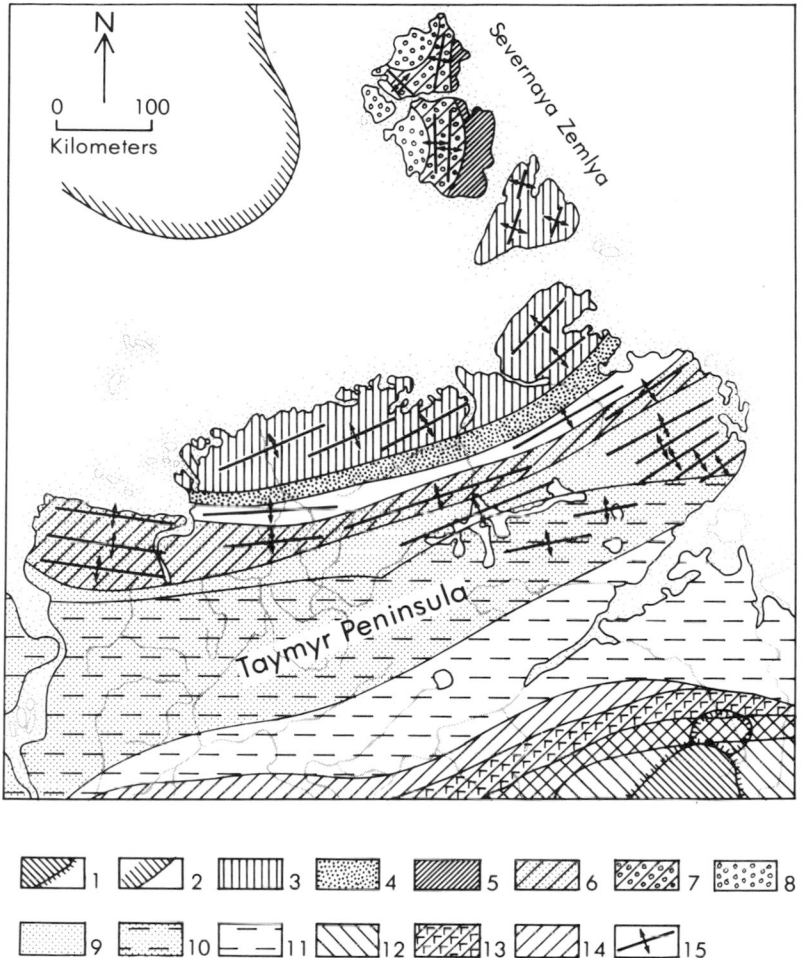

Fig. 22.   Tectonic map of Taymyr–Severnaya Zemlya area (generalized after numerous Russian authors). (1) Archean folded structure in the core of the Anabar Arch; (2) presumed Archean block of the Karsk Platform; (3) Proterozoic foldbelt; (4) Proterozoic structures reworked by Caledonian folding; (5) regions of Caledonian folding; (6) Caledonian structures reworked by Hercynian folding; (7) Caledonian structures reworked by late Devonian folding; (8) late Devonian structures; (9) region of Hercynian folding; (10) Hercynian structures buried under a cover of Mesozoic and Cenozoic deposits; (11) areas of Mesozoic–Cenozoic-internal structures; (12) the Anabar Arch; (13) the northern margin of the Siberian Platform; (14) major fault zones; (15) axes of major anticlines.

older structures. In Taymyr, they involve geosynclinal rocks broken by numerous thrusts parallel to the regional strike. In Severnaya Zemlya, the Devonian structures are of platform type, occurring as gentle domes and basins 4–16 km wide and affecting mainly continental beds of the Old Red Sandstone type. Therefore, Severnaya Zemlya, the southern Kara Sea, and the northern margin of Taymyr are included in the Caledonides, while the rest of Taymyr along with most of Novaya Zemlya, Pay-Khoy, and the Arctic Urals, according to Russian geologists, belong to the typical Hercynides.

The Upper Paleozoic is missing in Severnaya Zemlya, but in Taymyr, rocks of the Hercynian Foldbelt are deformed into linear, complex folds cut by thrusts and intruded by small Hercynian plutons.

Numerous batholiths of Precambrian and Caledoian age occur in Taymyr; associated volcanic rocks that are more or less metamorphosed into amphibolite and greenstone are widespread. Hercynian granites are found in some areas of Taymyr and so are the Permian and Triassic trap rocks, which are identical to the Siberian trap rocks. No younger volcanic rocks have been found. In Severnaya Zemlya the only plutonic rocks are Precambrian and Caledonian.

VIII. NORTH SIBERIAN LOWLAND—AREA VII

The region southeast of the Taymyr Peninsula is a tundra-covered lowland bordered on the north by the Laptev Sea and on the east by the Lena River flowing along the west side of the Verkhoyansk Mountains.

This region represents the northeast part of the Siberian Platform. The geology consists of an early Precambrian crystalline basement overlain by a thick Proterozoic, Paleozoic, and Mesozoic platform cover (Table VII).

The base of the section exposed in the Olenek Uplift is mainly gneiss, schistose quartzite, marble, and paragneiss. Radiometric dates on plutonic rocks cutting the metamorphic basement are in the range of 2000 m.y. (Vinogradov and Krasil'shchikov, 1963). The overlying Proterozoic section is a schistose sequence of siltstone, sandstone, and phyllite overlain by a thick nearly complete section of Paleozoic carbonate and clastic strata. Permian and Lower Triassic lava and tuff locally overlie Permian continental deposits and marine Paleozoic rocks. Higher in the section are thin deposits of Upper Cretaceous sandstone, followed by Quaternary alluvium and glacial deposits. An important summary of the Permian and Mesozoic geology of the arctic part of the Siberian Platform, including lithofacies and paleographic maps, is provided by Ronkina *et al.* (1977).

IX. EASTERN PART OF THE NORTH SIBERIAN LOWLAND AND LENA RIVER DEPRESSION—AREA VIII

In a northeast direction from the North Siberian Lowland, including the Olenek Uplift, the Precambrian crystalline basement of the Siberian Platform

TABLE VII
Generalized Columnar Section of the Northern Part of the Central Siberian Lowland[a]

Age	Thickness	Description
Quaternary	10–50 m	In Kareika River Valley: sand, sandy loam (50 m). Moiero–Olenek River valleys: bouldery gravel with sandy loam and sand (12 m); Interglacial alluvium forms a terrace 40 m above flood plain terrace. In Moiero–Olenek River Basin: conglomerate, sand (30 m); limnite (15–20 m); moraine (10 m); fluvioglacial (30–35 m).
Undifferentiated Upper Cretaceous and Paleogene rocks	25–30 m	Brown clay-sand and sandy loam (25–30 m) rest on eroded Paleozoic strata.
Lower Triassic volcanic rocks	100–1500 m	Lava series is confined to watershed areas of Enisei, Kotuy and Nizhnaya Tunguska Rivers. In some regions the lava series occurs not on tuff series but on a sequence of sedimentary rocks, continental Permian deposits, and marine Paleozoic deposits. The lavas form a thick sequence of basic flows (100–900 m). In Noril'sk region lavas are 720 m thick. In central part of Tungusk syncline lavas are 1500+ m.
Permian and Triassic volcanic rocks	35–800 m	Pure tuff to sedimentary strata with minor tuff; often the tuff has bands of sandstone and shale with coal layers. Thickness of the tuff series varies with distance from volcanic center: in Kotuy River 150 m, Khesa River 650–800 m, Noril'sk 35 m. (Thickness increases west to east and north to south.)
Undifferentiated Carboniferous Permian deposits	50–350 m	Noril'sk region: Tungusk Group. Regardless of very deep erosion at base of Tungusk Group no basal conglomerate was observed. Tungusk Group includes Aesekansk Formation: sandstone, argillite, siltstone, argillite with some coal layers (126 m); Rudninsk Formation: four cycles of sedimentation, in Rudnaya Mountains region—crossbedded sandstone, siltstone and argillite with coal beds (26 m), includes Kaerkunsk coal deposit (100 m); Shmidtin Formation: industrial coal layers, conglomerate, arkosic polymictic sandstone (100–135 m); Konierkansk Formation: conglomerate, arkosic sandstone (20 m); Ambarinsk Formation: argillite, tuffaceous sandstone. The Tungusk has major facies changes.
Permian	150–450 m	Lower part: overlies different horizons of Paleozoic marine rocks, from Carboniferous to Cambrian. In Kotuy River basin; polymictic sandstone with shale interbeds

(Continued)

TABLE VII (*Continued*)

Age	Thickness	Description
Permian (*continued*)		(60–65 m); limy sandstone (92 m). Total thickness of lower part: 120–125 m. Upper part: conglomerate, sandstone alternating with sandy shale with coal layers, tuff (60–250 m).
Undifferentiated Permian deposits	200–350 m	In Kotuy, Moiero, and Ulenek-Vilui River Basins, Permian deposits lie on eroded surfaces of Paleozoic rocks. In many sections: clay (3.5 m); sandstone, shale with interbeds of limy sandstone and sandy limestone (50–70 m). Total thickness on Kotuy River: 200–250 m; on Moiero River upstream, thickness: 300–350 m.

Disconformity

Age	Thickness	Description
Carboniferous–Lower Carboniferous	150–170 m	Two facies: marine—mostly calcareous and lagoonal; continental—coal-bearing. Noril'sk region strata (disconformably on marine Paleozoic rocks): Kolargon Group—limestone with marl, shale, and limy sandstone (30–50 m); limestone (90 m); dolomite (30 m).
Devonian	575 m	Lower part: Noril'sk region—marl and siltstone with interbeds of anhydrite and gypsum (Zubovsk Group, 250 m). Middle part: Razvedochnin Group—siltstone with interbeds of marl, sandstone, and limestone (225 m). Upper part: Fokinsk Group—marl, siltstone, dolomite, some secondary limestone with interbeds of sandstone, lenses of anhydrite (100 m).
Silurian	1260–1360 m	Lower part: Moiero River area—limy siltstone, limestone (30 m); limestone, clayey limestone, organic limestone (70 m); limestone with clay (38 m); organic clayey limestone with argillite and marl (42 m); limestone (35 m). Total thickness of Llandoverian: 215 m. Wenlockian Stage: limestone, marl, dolomite (350 m). Upper part: marl, dolomite with 6 m gypsum in lower part, dolomite alternating with gypsum (140–150 m). Thickness of Ludlovian Stage from west to east: 100 to 200 m; in Noril'sk region, upper part totals 450–500 m.
Ordovician	690 m	Ust'kutsk Stage: Moiero River area—limestone, conglomerate, less dolomitic limestone and dolomite with gypsum on top

TABLE VII (*Continued*)

Age	Thickness	Description
Ordovician (*continued*)		(140–145 m). Chun'sk Stage: Moiero River—dolomite, siltstone, argillite with widespread development of gypsum (93–95 m). Middle part: Krivolutsk Stage—limestone, argillite, marl, dolomite, conglomerate (41 m); Mangazeisk Stage—in Moiero River area; marl, argillite (45 m). In Moiero River area total thickness of Upper Ordovician: 365 m.
Cambrian	Up to 1930 m	Except for eastern and southeastern margins of Anabar shield, lies on erosion surfaces of Sinian strata. Lower Aldan Stage, two series: Kessusinsk and Erkeketsk. Kessusinsk Series: in Khorbuguonke River area: conglomerate grading into polymictic sandstone (25 m); polymictic sandstone (2 m); siltstone with argillite (60 m); sandy limestone, alternating with sandstone (25 m); glauconitic limestone (20 m); alternating siltstone, limestone (12 m). Total thickness: 144 m. Erkeketsk Series: argillaceous limestone (180–200 m). Undifferentiated: Lenskiy Stage—pyroschist with intercalated limestone on north and northeast. Schist grades into limestone (thickness of the schist 10–15 m, limestone 50 to 60 m on northeast and 700 m on northwest). Aneginsk Stage—pyroschist (up to 90 m). Middle Cambrian: Maisk Stage—argillaceous limestone, limy sandstone (300 to 600 m). Undifferentiated: limestone (1200 m); limestone marl (400 m). Upper Cambrian. in Olenek–Anabar watershed in north part of Siberian Platform and Siligir and Tunguska Rivers: sandy argillaceous limestone alternating with sandstone and argillite, conglomerate (140–150 m); sandy limestone with intercalation of bituminous limestone (70–80 m); thickness increases westward to 500–550 m of limestone, dolomite, argillaceous limestone.
Late Proterozoic	1620–1690 m	Siniiskiy Complex: underlying Lower Cambrian rocks (lower course of Olenek River and watershed with Lena River): Mukunsk Group with three subgroups; (1) Il'inskaya—sandstone, gritstone (320 m); (2) Burdinksk—quartz sandstone (200–230 m); (3) Labaztakh—quartz and arkosic

(*Continued*)

TABLE VII (*Continued*)

Age	Thickness	Description
Late Proterozoic (*continued*)		sandstone (180 m); siltstone, sandstone, shale, dolomite (60 m). Balliakhsk Series: in Anabar River basin—dolomite and limestone alternating with sandstone (in southern part 250 m; in north 50–100 m); Sooloolisk Group: conglomerate, sandstone (90 m), limestone alternating with siltstone layers with thickness 40–50 m. Turkntsk Group: three subgroups—(1) dolomite, sandstone; (2) sandy dolomite; (3) dolomite (330–380 m).
Proterozoic	1600–2000 m	Three sequences: (1) schistose siltstone with intercalations of schist (700–1000 m); (2) limy siltstone with intercalations of schist (150–250 m); (3) schistose silicified sandstone with intercalations of schist and phyllite (750 m).
Archean	20 km	Daldynsk Series: alternating plagiogneiss with gneiss and quartzitic schist-marble and paragneiss; Verkhne–Anabarsk Group: plagiogneiss, gneiss; Vekhnelomuisk Group: plagiogneiss, gneiss with paragneiss; Khapchansky Group: plagiogneiss, gneiss with marble (near Rassokhol River 3–4 km).

[a] After Tkachenko *et al.* (1957).

and its overlying Paleozoic platform cover are mostly covered by a thick Mesozoic flysch sequence that has Permian clastic rocks in its lower part. These rocks form the western edge of the much thicker clastic sequence in the Verkhoyansk Mountains along the eastern side of the Lena River.

In the region between the Lena and Anabar Rivers, the northeastern part of the Siberian Lowland has several thousand meters of Permian coal-bearing sandstone, siltstone, and shale, followed upward by 600 m of Triassic tuffaceous sandstone, sandy limestone, and shale. The Cretaceous section of coal-bearing alternating sandstone and shale, totaling 2000–3000 m, overlies the Triassic, followed by Quaternary alluvial cover (Table VIII).

In the Lena River delta region, toward the outer edge of the Siberian Platform, Permian strata in boreholes consist of similar sandstone and shale with some conglomerate. The Triassic appears to be absent here, and Jurassic sandstone and shale with conglomerate lenses are overlain by more than 4000 m of coal-bearing Cretaceous sandstone and siltstone (Table IX).

The diamond-bearing kimberlite pipes of the Lena River region are

located here along the contact of the northeastern edge of the Siberian platform and the Verkhoyansk Basin. According to Krivonos and Prokopchuk
(1971), the diamond pipes are concentrated along deep-seated fault zones that
developed at the juncture of the Siberian Platform and the Verkhoyansk
Geosyncline.

TABLE VIII
Generalized Columnar Section for the Northern Part of the Siberian Lowland between the Lena and Anabar Rivers[a]

Age	Thickness	Description
Quaternary	83–120 m	Lakebeds and alluvium: clayey sand with ice lenses, peat layers (1 m, in maritime lowland 80 m). Eluvial and deluvial: sandy-clay and gravel with boulders (3–5 m).
Cretaceous	1930–2900 m	In Checkanovsk Ridge: sandstone. Ulanchan–Yuriakh Anticline: sandstone, shale (400–500 m); clayey deposits (100–200 m); arkosic sandstone (30 m). Total thickness of Valanginian-200–220 m. Coal-bearing deposits: Ust'–Olenek region—sandstone with coaly siltstone (180–460 m); alternating sandstone and siltstone with coal layers (174 m decreasing to 20–30 m in the west); sandstone without coal (500–800 m); coal-bearing sands with coal layers and siltstone (Lena River, 100 m; Olenek River, 5–10 m); sandstone, siltstone, coal-bearing continental deposits, west to east 200 to 320 m. Olenek deposits: sandstone (250 m), arkosic sandstone (40–100 m), siltstone, sandy siltstone, argillite with coal layers, thickness 4–52 m (Buolkalakh River, 150–160 m; in Olenek, 250 to 300 m); arkosic sandstone (40 m to 200–300 m in Olenek); siltstone, sand with siltstone, argillite, sandstone with 3.5-m coal layers (20–100 m).
Triassic	640+ m	On arch of Ulachan–Yuriakh Anticline: tuff alternating with tuffaceous sandstone (80–190 m); sandy limestone (400 m), shale with limy sandstone (160 m). Checkanovsk Ridge: alternation of sandstone, siltstone, shale and conglomerate (80 m).
		Disconformity
Permian	2500 m	Chekanovsk Ridge and Severnyi Kharakhulakh: sandstone interbedded with siltstone and shale (268 m). Near mouth of Olenek River in axis of anticline: terrigenous deposits (1000 m); sandy-clayey siltstone, sand, sandstone; alternating sandstone, siltstone and coal (800–1000 m). From Popygai to Olenek Rivers and area between Lena and Olenek Rivers: sandstone, sand, gritstone, siltstone (from west to east—from 300–400 m to 300–350 m, on Olenek 200 m).

[a] After Siagaev (1957).

TABLE IX
Generalized Columnar Section of the Lena River Delta

Age	Thickness	Description
Quaternary	45–80 m	Sand, sandy loam, clayey loam, gravel, clay, and peat (45–80 m); glacial boulders on east side.
Cretaceous	3070–3670 m	Lenskaya Group: sandstone with intercalations of siltstone (130 m). In eastern part of area–Bulunskiy region: siltstone, sandstone (200 m); sandstone with conglomerate (20–25 m); sandstone with siltstone (60 m). Coal-bearing: in eastern part of Bulunskiy region—arkosic sandstone with siltstone and coal-layers (430 m); siltstone, sandstone (170 m); arkosic sandstone with sandy siltstone (in east 800 m, in west 520 m); siltstone, argillite (100 m); arkosic sandstone (500–550 m); siltstone, argillite (310–320 m). Lenskaya Group (2000–2400 m). Yngyrisk Group: coal-bearing (30 m); sandstone (30 m); coal-bearing (21 m); sandstone (51 m); coal-bearing (25 m). Olenek Series: coal-bearing sandstone (150 m); sandstone (46 m); coal-bearing (140–170 m); sandstone (50 m); 12 coal layers (total thickness of group 1300 m). Viluisk Group: northern part—sand, friable sandstone with coal layers (800–1000 m).
Jurassic	660 m	In northern part of the region: sandstone, siltstone with conglomerate lenses (30 m); argillite with limestone lenses (120 m); siltstone with argillite, limestone lenses (100 m). In southern part: argillite (80 m); siltstone, sandstone (70–100 m). Also in southern part: sandstone, siltstone, shale (100 m); sandstone (100 m). Undifferentiated Jurassic sandstone, siltstone, shale with coal (150 m); sandstone (50–60 m); marly sandstone (100 m).
Permian	200–240 m	Tumiati Village region: conglomerate, argillite, siltstone, sandstone, alternating sandstone with siltstone and argillite (200–240 m in borehole). Basement rock: Paleozoic and Triassic are reported in adjacent regions.

X. GEOLOGIC DEVELOPMENT OF THE WESTERN SOVIET ARCTIC

The main structural framework of the western Soviet Arctic is classified according to Tkachenko *et al.* (1973) as the Arcto-Atlantides Fold System, involving foldbelts and platforms that developed mainly on continental crust of Precambrian age. According to Tkachenko, whose interpretation represents a view shared by a large number of the Soviet geologists, the region consists from west to east of the Caledonian Foldbelt, the Russian Platform, the Ural–Pay Khoy–Novaya Zemlya Foldbelt, the Kara Platform, the Taymyr–

Severnaya Zemlya Foldbelt, and the Siberian Platform. The foldbelts are interpreted to be geosynclines that developed as narrow marine basins within a supercontinental block. These basins were later deformed *in situ* by Baikalian, Caledonian, and Hercynian orogenic episodes. A review of the stages in the development of the region, according to this theory, is abstracted below from Pogrebitskiy (1974).

From the Precambrian to the end of the Paleozoic, the whole vast expanse of northern Siberia from the Urals and Novaya Zemlya to the Lena River, including the Kara Sea, Severnaya Zemlya, Taymyr, and all of the Siberian Platform, including the West Siberia Lowland, represented a single structure. This superplatform, called the North Asian Platform, was destroyed by powerful tectonic activity at the end of the Permian and the start of the Triassic. Taymyr was inverted in to a folded block, and the Siberian Platform itself remained as a relic of the superplatform. At the same time, the territory of the West Siberian Lowland and the strip between Taymyr and the Siberian Platform was depressed and formed the Mesozoic to Cenozoic West Siberian Platform and the Yenisey–Khatanga Trough.

According to Pogrebitskiy, the disintegration and Hercynian rejuvenation of the ancient superplatform were predestined by its earlier history. At the end of the Archean, the territory represented a vast folded system; its primarily NNW-striking structures were reworked, uplifted, and deeply eroded to the amphibolite and granulite facies level. In Proterozoic time, this craton with submeridianal faults was broken into a series of plates separated by mobile belts that represent geosynclinal troughs and aulacogens. The processes of plate fragmentation and formation of the folded system took place in the Middle Proterozoic and Riphean. Thus by Cambrian time, the basement of the superplatform acquired a block structure from the combination of the heterogeneously folded Archean plates that became welded together through a process of consolidation of the narrow depressions and troughs into foldbelts.

Plate tectonic theory has provided a more mobilistic interpretation of the same region, suggesting that the Urals and their northern extensions into Pay Khoy and Novaya Zemlya consist of oceanic material scraped off against the edges of the Russian and Siberian Platforms as the two subcontinents approached and collided in the Permian and Triassic (Hamilton, 1970). There appears to be much stratigraphic and structural evidence, including paired miogeosynclinal–eugeosynclinal sequences that represent juxtaposed continental shelf and oceanic assemblages, Precambrian and Ordovician ophiolites and associated glaucophane schists, and convergent polar-wandering curves for the Russian and Siberian Platforms, to support Hamilton's model of convergence and collision of the two platforms.

Further analysis of paleomagnetic results from Eurasia (McElhinney, 1973) strengthens Hamilton's interpretation that the Urals mark a collision zone between the Russian and Siberian plates. McElhinney has reconstructed the polar-wandering curves for the Siberian and Kolyma plates, which suggest that the Verkhoyansk Mountains, along the east side of the Siberian plate, represent another collision zone of Cretaceous age (Hamilton, 1970; Churkin, 1972). More recently, Sears and Price (1978) have suggested that the Siberian plate formerly was part of the North American craton and was rifted off in the later Precambrian (Stewart, 1972). It is beyond our scope here to evaluate the fit of the geology of the Siberian plate with North America, but it is an idea worth pursuing. The review of the geology of the Siberian plate provided by Kosygin and Parfenov (1975) uses elements of both the vertical and plate tectonic theories to explain the geologic history of the region.

REFERENCES

Akudinov, S. A., Bolgurtsev, N. N., Litvinenko, I. V., and Poretova, G. A., 1972, Deep structure of the eastern part of the Karelian region, *Geotectonics (USSR)*, no. 5, p. 297, Fig. 2.

Antropov, P. Ya., and Khaeitonov, L. Ya., eds., 1958, Murmanskaya oblast', geologicheskoe opisanie, in: *Geologiya SSSR*, v. 27, Moscow: Ministerstvo Geologiy SSSR, 706 p.

Churkin, Michael, Jr., 1972, Western boundary of the North American continental plate in Asia, *Geol. Soc. Am. Bull.*, v. 83, p. 1027–1036.

Demenitskaya, R. M., and Hunkins, K. L., 1971, Shape and structure of the Arctic Ocean, in: *The Sea*, Maxwell, A., ed., v. 4, pt. 2, New York: J. Wiley and Sons, p. 223–249.

Dergunov, A. B., and Moldavantsev, Yu. E., 1976, Structure and tectonic position of gabbro and ultrabasic massifs of the Polar Urals, *Geotectonics (USSR)*, v. 10(3), p. 210–211.

Efimov, A. A., Lennykh, V. I., Puchkov, V. N., Savelyev, A. A., Savelyeva, G. N., and Yazeva, R. G., 1978, Ophiolites of Polar Urals, Fourth Field Ophiolite Conference, IGCP, Guidebook, Moscow.

German, E. V., Kisliakov, V. N., and Reinin, I. V., 1963, Geologia i geomorfologia poluostrova Yamal, in: Geologia i heftegazanosnost' severa zapadnoi Sibiri, *Tr. Vses. Neft. Nauchno-Issled. Geol.-Rasv. Inst.*, v. 225, p. 311–329.

Hamilton, Warren, 1970, The Uralides and the motion of the Russian and Siberian Platforms, *Geol. Soc. Am. Bull.*, v. 81, p. 2553–2576.

Kosygin, Yu. A., and Parfenov, L. M., 1975, Structural evolution of eastern Siberia and adjacent area, *Am. J. Sci.*, v. 275A, p. 187–208.

Krivonos, V. F., and Prokopchuk, B. I., 1971, The Usunku–Syungyude Zone of deep-seated faults, *Geotectonics (USSR)*, no. 1, p. 58–59.

Lazarev, Yu. I., 1973, Tectonic development of the early Karelides of Karelia, *Geotectonics (USSR)*, no. 5, p. 28.

Markov, F. G., Ravich, M. G., and Vakar, V. A., 1957, Geologicheskoe stroenie Taymyrskogo poluostrova, in: *Geologia Sovetskoi Arktiki: Gosgeoltekhizdat*, Markov, F. G., and Nalivkin, D. V., eds., *Tr. Nauchno-Issled. Inst. Geol. Arktiki*, p. 313–363.

McElhinney, M. W., 1973, Paleomagnetic results from Eurasia, in: *Implications of Continental Drift to the Earth Sciences*, v. 1, Tarling, D. H., and Runcorn, S. K., eds., New York: Academic Press.

Nalivkin, D. V., 1973, *Geology of the USSR*, translated by Rast, N., Edinburgh: Oliver and Boyd, 855 p.

Pinther, M., 1975, *Map of the Arctic Region*, New York: American Geographical Society, 1:5,000,000 scale.

Pitcher, M. G., ed., 1973, *Arctic Geology*, *Am. Assoc. Petrol. Geol. Mem.* 19, 747 p.

Progrebitskiy, Yu. E., 1971, Paleotektonicheskiy analiz Taymyrskoi skladchatoi sistemy, *Tr. Nauchno-Issled. Inst. Geol. Arktiki*, 156 p.

Progrebitskiy, Yu. E., 1974, Problemi geologiy poliarnikh oblastei zemli, *Tr. Nauchno-Issled. Inst. Geol. Arktiki*, p, 76–78.

Pogrebitskiy, Yu. E., and Kos'ko, M. K., 1977, Tektonika Arktiki, skladchatyi fundament shel'fovykh sedimentatsionnykh basseinov, Leningrad: NIIGA, 156 p.

Polovinkina, Yu. I., ed., 1973, Geochronology study outlining individual regions of the USSR, in: *The Eastern Part of the Baltic Shield*, Leningrad: "NEDRA", pt. 1, 348 p.

Puchkov, V. N., 1973, Paleotectonics of the Lemva zone, Polar Urals, *Geotectonics (USSR)*, no. 6, p. 342.

Pyatnitskiy, V. K., 1974, The relief of the basement surface and structure of the sedimentary mantle of the Siberian Platform, *Soviet Geol. Geophys.*, v. 15(9), p. 71–80.

Ronkina, Z. Z., Bro, E. G., Voitzekhovskaya, E. P., Kolokol'tzeva, E. P., Vishnevskaya, T. N., 1977, Epigenez Permsko-Mesozoiskukh terrigennykh tolsch severa Sibirskoi Platformy, in: *Geologiya i Neftegazonosnost' Mesozoiskikh Progibov Severa Sibirskoi Platformy, Sbornik Nauchnykh Trudov*, Sorokov, D. S., ed., Leningrad: NIIGA, p. 18–29, and appendix with maps.

Rostovtsev, N. N., ed., 1964, *Geologiya SSSR, Zapadno-Sibirskaya Nizmennost'*, Moscow: Ministerstvo Geologiy SSSR, v. 44, pt. 1.

Saks, V. N., and Egorova, I. S., 1957, Geologicheskoe stroenie severnosibirskoi nizmennosti (Taymyrskoi nizmennosti), in: *Geologia Sovetskoi Arktiki: Gosgeoltekhizdat*, Markov, F. G., and Nalivkin, D. V., eds., Moscow: NIIGA, p. 243–278.

Sears, J. W., and Price, R. A., 1978, The Siberian connection: A case for Precambrian separation of the North American and Siberian cratons, *Geology*, v. 6, p. 267–270.

Shilov, L. P., Kuznetsov, G. E., Kochetkov, O. C., and Podlovilin, E. S., 1978, Characteristics of the deep-seated structure of the Timan, *Izv. Vyssh. Uchebn. Zaved. Geol. Geofiz.*, no. 6, p. 32–40.

Siagaev, N. A., 1957, Geologicheskoe stroenie vostochnoi chasti severo-Sibirskoi nizmennosti (Leno-Anabarskow mezhdurechie), in: *Geologia Sovetskoi Arktiki: Gosgeoltekhizdat*, Markov, F. G., and Nalivkin, D. V., eds., Moscow: NIGGA, p. 290–306.

Stewart, J. H., 1972, Initial deposits in the cordilleran geosyncline: Evidence of a late Precambrian (850 m.y.) continental separation, *Geol. Soc. Am. Bull.*, v. 83, p. 1345–1360.

Tkachenko, B. V., and Egiazarov, B. Kh., eds., 1970, *Geologiya SSSR, v. 26, Ostrova Sovetskoy Arktiki*, Moscow: Ministerstvo Geologiy SSSR, 545 p.

Tkachenko, B. V., Rabkin, M. I., Denoridou, K. K., Vakar, V. A., Grozdilov, A. L., Butakova, E. L., and Strelkov., S. A., 1957, Geologicheskoe stroenie severnoi chasti sredne-sibirskogo ploskogoria, in: *Geologia Sovetskoi Arktiki: Gosgeoltekhizdat*, Markov, F. G., and Nalivkin, D. V., eds., Moscow: NIGGA, p. 133–203.

Tkachenko, B. V., Egiazarov, B. Kh., Atlasov, I. P., Lazurkin, V. M., Markov, F. G., Polkin, Y. I., Ravich, M. G., Romanovich, B. S., and Sokolov, V. N., 1973, Main geologic structure of the Arctic, in: *Arctic Geology*, Pitcher, M. G., ed., *Am. Assoc. Petrol. Geol. Mem.* 19, p. 336–347.

Vinogradov, V. A., and Krasil'shchikov, A. A., 1963, The age of the folded base basement of the Olenek Uplift in the Siberian Platform, *Dokl. Akad. Nauk SSSR*, v. 153(3), p. 687–689.

Voytovich, V. S., 1975, Generalized geologic map of the northeastern Karelia showing structure and thrusts, *Geotectonics (USSR)*, v. 9(2), p. 82.

Chapter 8

TECTONICS OF THE SOVIET FAR EAST

Y. A. Kosygin and L. M. Parfenov
Institute of Tectonics and Geophysics
Khabarovsk, USSR

I. INTRODUCTION

Mesozoic geosynclinal systems of different types, massifs, continental margin volcanic belts, and Cenozoic geosynclinal systems have been distinguished within the Soviet Far East between the Pacific Ocean and Siberian Platform, which is one of the oldest continental nuclei of Asia (Fig. 1). Different types of tectonic elements, discussed in the following sections, are revealed by their magnetic anomaly patterns. Thus, aeromagnetic data were used to identify these elements in "closed" regions, such as water covered areas and young depressions.

II. MESOZOIC GEOSYNCLINAL SYSTEMS

Mesozoic geosynclinal systems, occupying a large part of the territory, are quite varied. They differ from each other in the completeness of their stratigraphic columns, the composition of their sedimentary and magmatic assemblages, their size and form, and the date of termination of their geosynclinal development (early and late Mesozoic). Eugeosynclinal systems of types A and B and miogeosynclinal systems are the most significant among them.

Eugeosynclinal systems of type A (the Koryak, South Anyui, Mongolo-Okhotsk, Sikhote Alin, and East Sakhalin Systems) are separated by marginal

377

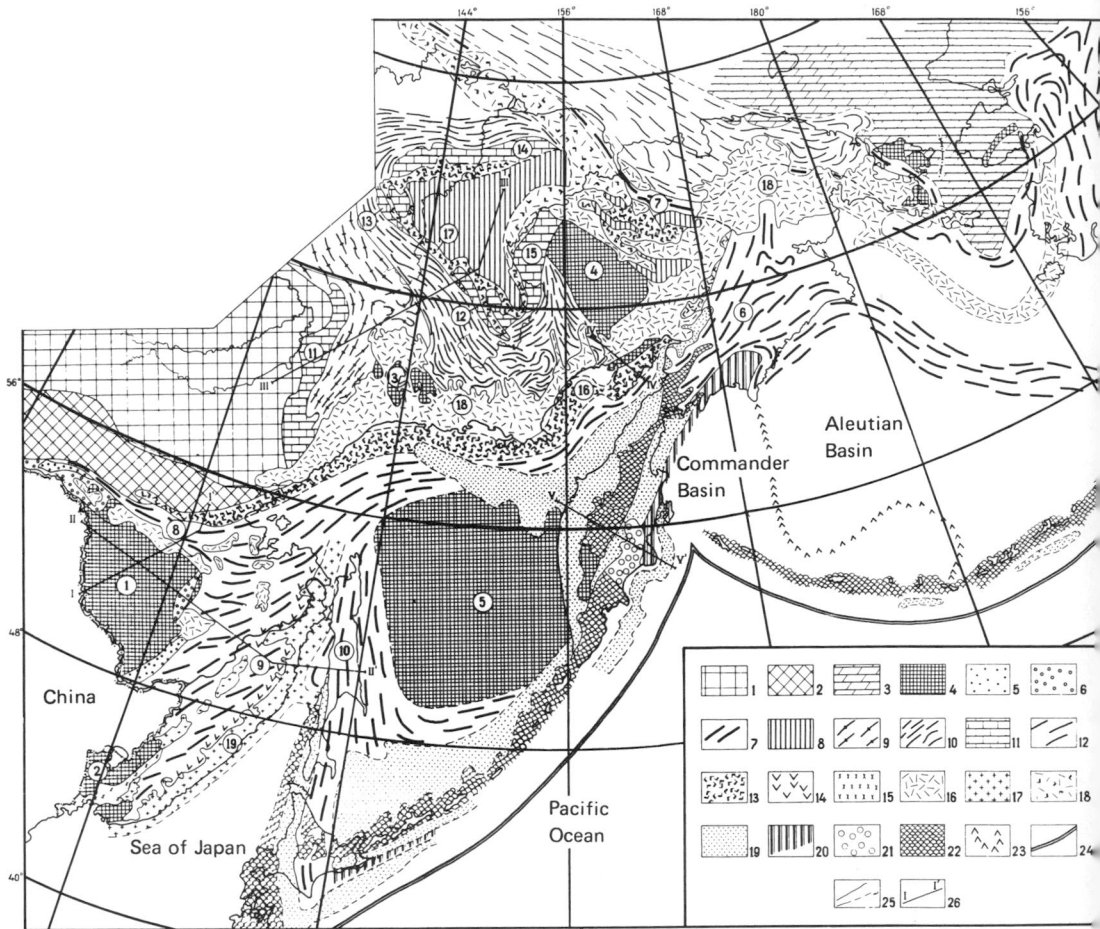

Fig. 1. Tectonic scheme of the Soviet Far East and the adjacent areas. (1) Siberian Platform; (2) Stanovoy area (o
than 2000 m.y.); (3) Middle Paleozoic folded rocks; (4) massifs of Precambrian blocks numbered on map (① Bu
Massif, ② Khanka Massif, ③ Okhotsk Massif, ④ Omolon Massif, ⑤ Sea of Okhotsk Massif); (5, 6) marg
troughs of the massifs of (5) Paleozoic age and (6) Mesozoic age; (7) Mesozoic eugeosynclinal systems of type A (
Koryak, ⑦ South Anyui, ⑧ Mongolo-Okhotsk, ⑨ Sikhote Alin, ⑩ Eastern Sakhalin); (8) Mesozoic Alazeia-
eugeosynclinal system of type B; (9, 10, 11) Yana-Kolyma Mesozoic miogeosynclinal system: (9) outer zone, (10) i
zone, (11) uplifts of Riphean and Lower-Middle Paleozoic rocks (⑪ Sette-Daban, ⑫ Omulyovaya, ⑬ T
hayakhtakh, ⑭ Ulakhans-Polousniy, ⑮ Prekolyma); (12) Chukchi Mesozoic miogeosynclinal system; (13,
Mesozoic volcanic geanticlines of island arc type of (13) Triassic-Jurassic and Jurassic age (⑯ Uda-Murgal,
Uyandina-Yasachnaya) and (14) late Cretaceous age; (15, 16, 17) continental-margin volcanic belt of (15) late Paleo
age, (16) late Mesozoic age (⑱ Okhotsk-Chukotsk Belt), and (17) late Senonian-Miocene age (⑲ Eastern Sik
Alin Belt); (18) volcanic piles or volcanic geanticlines and continental-margin belts of undifferentiated
Jurassic-Cretaceous age; (19-24) Cenozoic geosynclinal systems: (19) miogeosynclinal zones, interarc troughs, and tro
on slopes of deep-sea trenches, (20) eugeosynclinal zones, (21) volcanic ridges of Pliocene-Quaternary age; (22) volc
ridges of late Paleogene-Neogene age, (23) inferred volcanic ridges of Paleogene age, (24) modern deep-sea trenches;
boundaries of tectonic elements (solid—known, dashed—inferred); (26) position of the lithologic-stratigraphic cross
tions shown on Figs. 2, 3, 5, 7, and 8.

seas, fields of volcanic rocks, and young depressions. According to magnetic, structural, and paleogeographic data and the results of dredging, the systems converge with each other to form a single ocean-margin belt several hundred kilometers wide. The Mongolo-Okhotsk and South Anyui Systems, which jut far into the continent and form narrow strips tens of kilometers wide, are branches of this belt. These systems are mainly characterized by smooth, reduced, and nearly normal anisotropic magnetic anomaly patterns; the orientation of the anomalies coincides with the strike of the geosynclines.

The age of the main folding varies both in transverse and longitudinal directions. In the Koryak System, the age of the main folding ranges from Cenomanian (late Cretaceous) on the continent to Danian (early Tertiary) toward the ocean. In the same direction, the age of similar assemblages becomes younger and the amount of volcanic and siliceous rocks increases (Vinogradov *et al.*, 1974). The age of the main folding of the Sikhote Alin System changes from early Middle Triassic on the eastern margin of the Bureya and Khanka Massifs to Cenomanian on the coast of the Sea of Japan. The Koryak System extends to the southeastern edge of the Taigonos Peninsula and can be traced further in a southwestern direction to a connection with the Mongolo-Okhotsk System, which completed its geosynclinal development in late Jurassic time.

The systems consist mainly of siliceous, argillaceous, graywacke, and flysch deposits, including basic volcanic rocks, some of which formed in a deep-sea environment. The geosynclinal systems include Mesozoic, Paleozoic, and, in some places, late Precambrian sequences. Unconformities and superimposed folds indicate episodes of orogeny, but not a closure of the geosynclinal systems before the Mesozoic.

In the Mongolo-Okhotsk System, Upper Precambrian is assumed to exist. There are early Cambrian, Ordovician, Silurian, Devonian, Carboniferous, Permian, late Triassic, early and mid-Jurassic marine fossils. Similarity in the lithologic composition of the different age assemblages makes it difficult to distinguish and to map lithostratigraphic units, and the stratigraphic cross section (Fig. 2) is only approximate. Unconformities are known here at the base of the Devonian, Carboniferous, Permian, Upper Triassic, and Lower Jurassic.

The Sikhote Alin System is characterized by a wide distribution of Carboniferous and Permian argillaceous–siliceous rocks associated with basic volcanic rocks, including patches of graywacke near the western borders of the Bureya and Khanka Massifs (Fig. 3). These formations are underlain by graywackes and argillites with Devonian and Silurian fossils (Mishin, 1971). Lower to Middle Triassic rocks are known only in the southern part of the Sikhote Alin System, and it is assumed that the deformation and breaks in sedimentation correspond to this time interval. The sequence of Mesozoic strata starts with Upper Triassic argillites, cherts, and basalts. Jurassic and

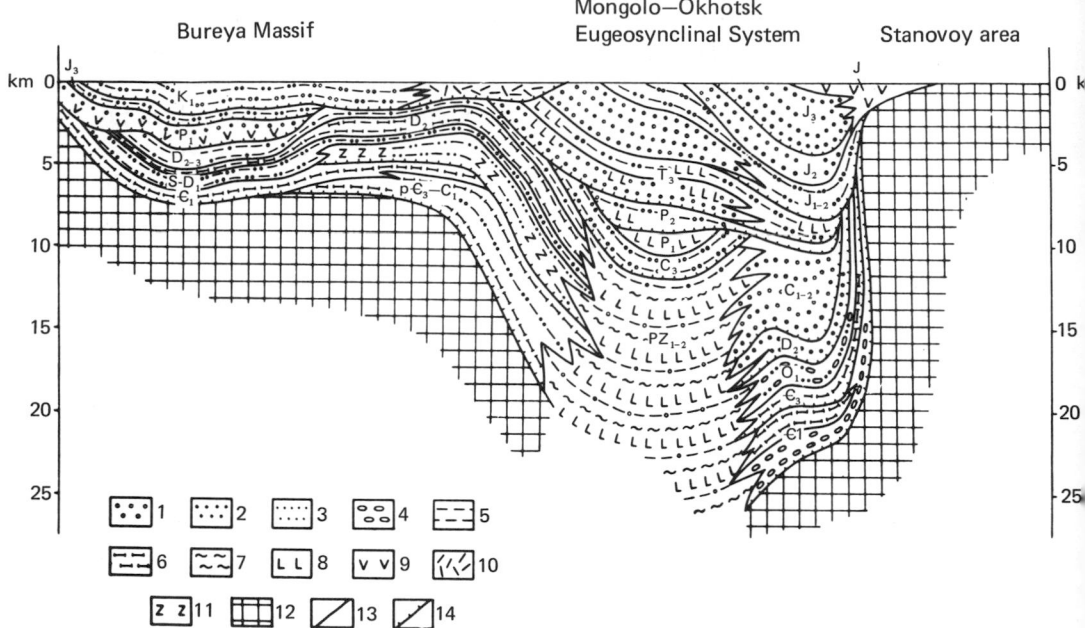

Fig. 2. Lithologic–stratigraphic cross section along line I–I′ (see Fig. 1). (1) Graywacke; (2) arkose; (3) orthoquartzite; (4) conglomerate; (5) argillite; (6) carbonate rock; (7) siliceous rock; (8) basalt; (9) andesite, andesite–basalt; (10) rhyolite, dacite; (11) trachyandesite; (12) Precambrian basement; (13) conformable geological boundaries; (14) unconformities. Associations of sedimentary rocks formed by a combination of rock types are shown by combining symbols.

Valanginian deposits are mainly of argillaceous–graywacke composition. Barremian and Lower Senonian deposits, mainly distributed in the eastern part of the system, form a thick complex of flysch. There are some conglomerates, breccias, and andesitic and andesitic–basaltic lavas and tuffs, the amount of which increases eastward to the eastern Sikhote Alin continental margin volcanic belt.

A significant discontinuity preceded the accumulation of the complex. The existence of Hauterivian sediments in the major part of the Sikhote Alin System is not proved. In the axial zone of the Sikhote Alin, the complex overlaps the Paleozoic rocks (Fig. 3).

On Sakhalin Island there are a few occurrences of Upper Permian foraminifers (Eliseyeva and Sosnina, 1964), which are found in limestone lenses within greenschists metamorphosed from basic volcanic, argillaceous, and siliceous rocks. Jurassic and Cretaceous deposits of the eastern regions of the island include jaspers, mudstones, graywackes, and basalts (Fig. 3).

In the Koryak Geosynclinal System there are volcanic-siliceous and argillaceous–graywacke sequences of Ordovician, Silurian, Devonian, Carboniferous, and Permian ages. Usually they form tectonic slices separated from the younger deposits by thrusts. Strongly folded volcanic–siliceous and graywacke complexes of late Jurassic–early Cretaceous age, which are complicated by thrusts, are typical of this region. Comparatively little-deformed Senonian and Paleogene deposits, with conglomerates in their basal part, usually lie unconformably on older rocks (Alexandrov *et al.*, 1975).

The South Anyui Geosynclinal System is composed of a thick series of Upper Jurassic volcanic–siliceous rocks and graywacke flyschlike assemblages of Lower Cretaceous age (Radzivil and Radzivil, 1975). Paleozoic rocks occur southeast of the system within the Alucha Uplift. They consist of basic volcanics, graywacke, and siliceous rocks of Carboniferous age (Dovgal', 1964).

Thick suites of basic volcanic rocks are widely distributed in the type A Mesozoic eugeosynclinal systems. They occur on several stratigraphic levels as follows: in the Mongolo-Okhotsk System, they are of late Precambrian–Cambrian, Silurian–Devonian, Carboniferous, Permian, and late Triassic ages; in the Sikhote Alin, they are of Silurian–Devonian, Carboniferous–Permian, and late Triassic ages; in eastern Sakhalin Island, they are of Permian-Triassic, Jurassic–early Cretaceous, and late Cretaceous ages. The oldest volcanic rocks are similar to tholeiitic basalts (Fig. 4). The content of titanium, iron, and calcium in the basalts decreases upward in the sequence and that of alkali (especially potassium), silica, and alumina increases. In younger assemblages, the amount of andesite increases and the complexes themselves become more differentiated. Within all of the above-mentioned stratigraphic levels of these complexes, one can trace similar trends of change in the volcanic composition. Intrusive suites, which accompanied the folding, are composed of ultramafic rocks, gabbro, and plagiogranites. The Alpine-type ultramafic rocks belonging to the dunite–harzburgite association usually occur together with siliceous–volcanic strata and, like these strata, they are found at different stratigraphic levels (Pinus *et al.*, 1973; Pinus and Sterligova, 1973; Zimin, 1973). However, they are distinctly dominant in the lower horizons. Belts of the Alpine-type ultramafic rocks are accompanied by glaucophane–schist zones, which become younger toward the ocean (Dobretsov, 1974). There are no large massifs of potash granites present, except in the Sikhote Alin System.

One of the characteristic features of all the systems under consideration is intensive folding, expressed in the extensive development of tight and recumbent folds, thrusts, and overthrusts. Serpentine melange zones have been found recently in many regions of the Koryak Mountains and on the east of Sakhalin Island (Ivanov and Baratov, 1974; Alexandrov, 1973; Rechkin, 1974). The superimposed folding of different ages complicates the initial structure considerably. The most intensive folding occurs within the Mongolo-Okhotsk and South Anyui Systems. The intensity of folding indicates that during sedimentation, the width of the geosynclinal systems was many times greater than

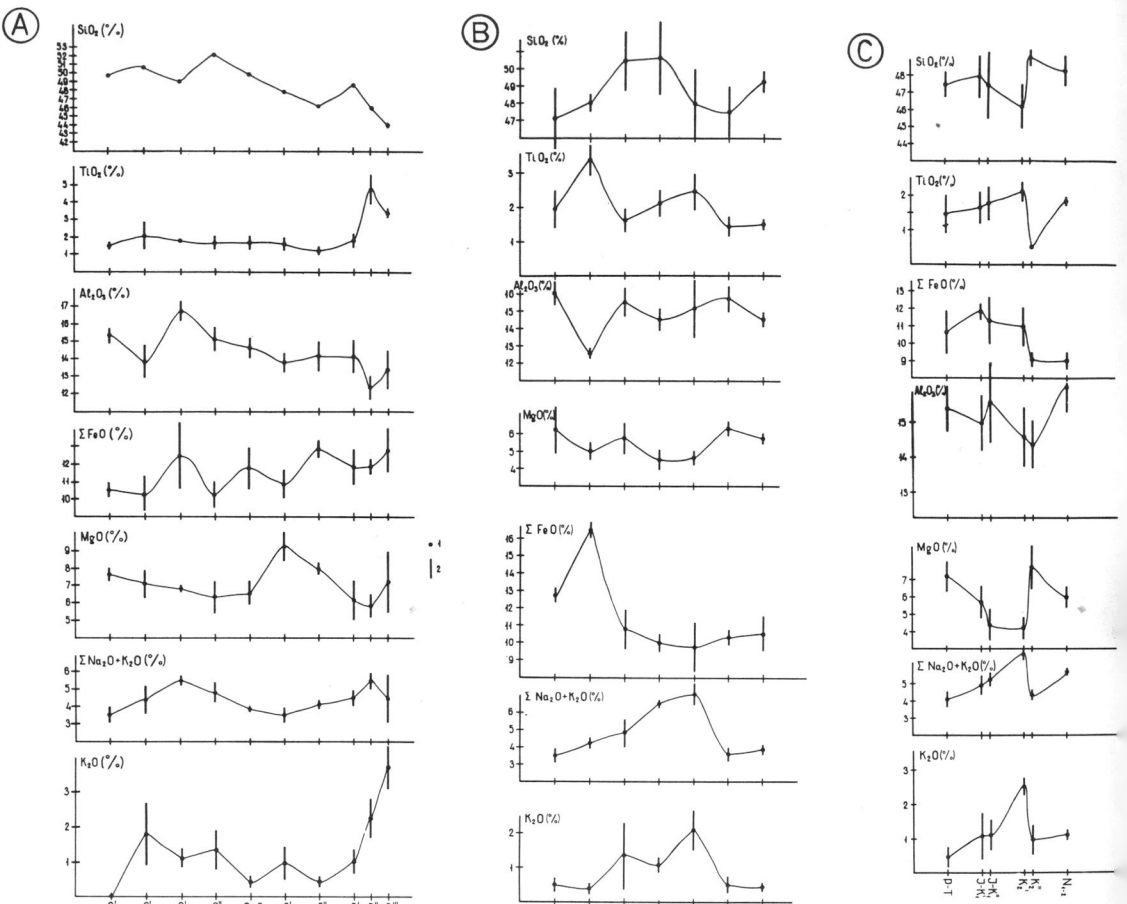

Fig. 4. Comparison of basalt chemical compositions from different stratigraphic levels: (A) the eastern part of the Mongolo-Okhotsk System; (B) the western part of the Mongolo-Okhotsk System; (C) Eastern Sakhalin (means are shown by points; dispersions are shown by vertical lines).

their present width. We assume that the paleogeography of these systems corresponds to deep marine basins.

A. The Eugeosynclinal Systems of Type B

The recently recognized Alazeia–Oloi System belonging to this type is located in the area formerly considered to represent the northern slopes of the Omolon and Kolyma Massifs (Rusakov and Vinogradov, 1969; Shilo and Merzlyakov, 1972).

A large part of this area is covered by a thin veneer of Neogene–Quaternary sediment that forms a vast swampy valley in the modern landscape. That is why the geology of this area is so controversial.

The most complete section of the system, including Paleozoic deposits, has been studied in recent years in the Alazeia Uplift. Table I shows the stratigraphic sequence in the area, as compiled by P. V. Gulyaev (1975).

Below the section is a thick (up to 1500–1700 m) nonfossiliferous suite of graywacke, basalt, andesite, and siliceous shale with limestone lenses. The stratigraphic relationships of this thick suite with the deposits of Middle and Upper Carboniferous age are unknown. It is tentatively assigned to the Carboniferous, but it could be much older. Glaucophane schists occur in the suite (Shilo *et al.*, 1973; Lichagin *et al.*, 1975).

The Ilin'–Tass Geosyncline occurs south of the Alazeia Uplift and is separated from it by the Ziryan Basin, which is filled with thick (up to 7000 m), coal-bearing, terrigenous sediments (Fig. 5). The geosyncline, reaching up to 200 km in width, extends northwestward for 700 km along the Paleozoic Omulyovaya Uplift, which is situated southward and is composed mainly of carbonate rocks of Ordovician, Silurian, and Devonian age. The geosyncline is composed of thick sequences (up to 5500–6000 m) of complexly folded Upper

TABLE I
Stratigraphic Column in the Alazeia Uplift[a]

Symbol	Description
K_1	Conglomerates and sandstones with plant fragments (200–270 m)
	Unconformity
	Rhyolite, rhyodacite, dacite, tuff, and ignimbrite (300 m)
	Andesite, andesite–basalt (280–350 m)
	Unconformity
$J_?$	Graywacke, andesite tuff (1300–1500 m)
J_2	Medium- and coarse-grained graywacke, argillite, and tuff
	Conglomerate (250–300 m)
	Gritstone, sandstone, and tuff (1200 m)
	Unconformity
J_1	Siltstone, tuff, sandstone, and gritstone (350–400 m)
T_3	Graywacke, gritstone, basalt, andesite, and tuff (250–260 m)
	Graywacke, tuff, and gritstone (450–500 m)
	Unconformity
P_2	Graywacke and conglomerate (450 m)
P_{1-2}	Trachydacite, rhyolite, tuff, volcanic breccia (600–700 m)
P_1	Siltstone and sandstone (770–800 m)
C_{2-3}	Graywacke, tuff, volcanic breccia of acid composition (750–800 m)

[a] After P. R. Gulyaev (1975).

Jurassic rocks (Krasniy, 1971). In the lower parts of the sequence, there are deep-sea siliceous argillites with basic volcanics that grade into argillites and sandy argillaceous sediments. Upward in the sequence, the sediments gradually change to shallower water deposits. Most of the plant fragments occur in the upper parts of the Jurassic sequence. These are shallow marine and coastal-marine deposits, which are followed upward by the Cretaceous strata of the Ziryan Basin.

Upper Jurassic deposits are also widely distributed between the Great Anyui and Oloi Rivers. They are composed of gently deformed marine and coastal-marine graywacke, gritstone, conglomerate, and basic and intermediate volcanic rocks (up to 2500 m) (Gulevich, 1975). The Triassic and Lower and Middle Jurassic deposits of this region consist of a thick series (6000 m thick) of marine graywackes and flyschlike argillite–graywackes, including andesite–basalts and andesites (Til'man et al., 1975). There is a disconformity at the base of the Middle Jurassic. Paleozoic rocks, represented by basalt, trachybasalt, andesite–basalt, andesite, dacite, rhyolite, siliceous shale, graywacke, and limestone with Devonian and Carboniferous fossils, crop out in some uplifts. In some places, a great amount of plant detritus occurs in the Carboniferous deposits, indicating shallow conditions of sedimentation. Thick sequences of Ordovician, Silurian, Devonian, and Carboniferous volcanic-argillaceous and argillaceous graywacke rocks are found in the inner zones of the Paleozoic Prekolyma, Omulyovaya, and Ulakhans–Polousniy Uplifts, which border the Alazeia–Oloi System. These thick eugeosynclinal-type complexes are replaced by carbonate rocks of the same age that form the central parts of the uplifts (Merzylakov and Lichagin, 1973; Sharkovskiy, 1975) (Fig. 6).

The intrusive formations of the Alazeia–Oloi System consist of small bodies of Paleozoic and Mesozoic gabbro, plagiogranites, and granosyenites.

Folding is not very intense and structural planes are gently dipping. Among the volcanic rocks of the Alazeia–Oloi System, in contrast with the volcanics of the eugeosynclinal system of type A, there are andesite–basalt, andesite, and acid volcanic rocks associated with the basalt. High alkalinity is a characteristic feature both for Paleozoic and Mesozoic volcanics (Seslavinskiy, 1973). Thick trachybasalt and trachyandesite series, with a K_2O content of up to 5%, occur in Ordovician deposits of the Omulyovaya Uplift (Merzlyakov and Lichagin, 1973). The Alazeia–Oloi Geosynclinal System is characterized by a combination of two types of anomalous magnetic fields: a mosaic field, similar to the magnetic field of the median massifs, and a field represented by linear anomalies oriented approximately parallel to the strike of the geosyncline.

The geosynclinal system's peculiarities can be explained by its formation

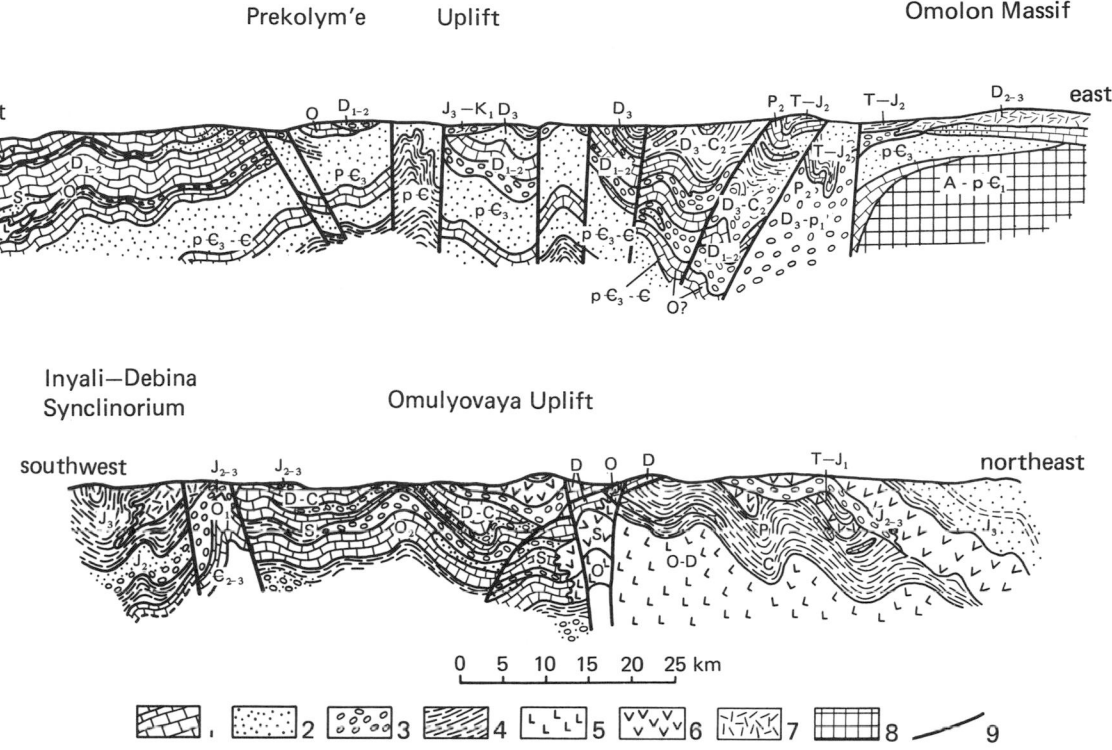

Fig. 6.   Geological cross sections of Prekolym'e and Omulyovaya Uplifts. (1) Carbonate rock; (2) sandstone and gritstone; (3) conglomerate; (4) argillite; (5) basalt; (6) andesite; (7) rhyolite; dacite; (8) Archaen–Lower Proterozoic crystalline schist and gneiss, (9) fault. (Simplified after Sharkovsky, 1975.)

through a process of breaking up of an old, probably Precambrian crystalline basement (Tilman *et al.*, 1975).

The combination of deep-sea siliceous–argillite and shallow coastal-marine sediments with conglomerates and plant fragments on the same stratigraphic horizons resembles the environment occurring in the modern Mediterranean Sea and Sea of Japan.

B. The Miogeosynclinal Systems

The Yana–Kolyma and Chukotsk Miogeosynclinal Systems are located on the border of the eugeosynclinal systems. The Yana–Kolyma System, bordering the Siberian Platform, is composed of thick, argillite–sandstone sequences of Carboniferous to Jurassic age, which are known as the Verkhoyanskiy Complex. The older Paleozoic and Riphean deposits crop out within the Sette–Daban Range on the border of the Siberian Platform and also in the Omulyovaya, Taskhayakhtakh, Prekolyma, and Ulakhans–Polousniy Uplifts.

Carboniferous and Permian deposits of the Verkhoyanskiy Complex crop out in the western part of the geosyncline near the Siberian Platform. Eastward they are replaced by Triassic and then Jurassic deposits, which can be seen clearly on all the geologic maps. These complexes have been considered to reflect the axial zones of late Paleozoic, Triassic, and Jurassic troughs that have been successively displaced northeastward. Detailed investigations of the Paleozoic and Triassic deposits (Arkhipov, 1974) have brought out that the Yana–Kolyma System consists of an Outer Zone bordering the platform and an Inner Zone, each with its own sedimentary sequence (Fig. 5).

In the Outer Zone, thick argillite–arkose and arkose sequences of marine, shallow-water, and coastal-marine deposits predominate. Westward they are replaced by lagoonal and alluvial–deltaic deposits that formed on the slope of the Siberian Platform. The platform is certainly the main source of the clastic materials. The compositon of these deposits is constant for hundreds of kilometers along the strike of the miogeosyncline and changes markedly across its strike.

Linear folds, usually symmetrical with the angles of inclination of the limbs up to 40°, are typical of the Outer Zone. In the southeastern part of the Outer Zone in an extension of the Okhotsk Median Massif, there are some areas of flat-lying and gently dipping beds. Orthoquartzite layers appear in these relatively undeformed areas within the Verkhoyanskiy Complex.

The Inner Zone of the miogeosynclinal system is characterized by the prevalence of Permian, Triassic, and Lower Jurassic argillaceous rocks. These are mainly very thick black argillites with intensive pyritization, wide distribution of subaqueous slump textures, and rare ammonoid, brachiopod, bivalve, and crinoid fossils (Arkhipov, 1974). Undoubtedly, these deposits are deeper water than those of the miogeosyncline's Outer Zone.

Within the In'yali–Debina Synclinorium, which occupies the northeastern part of the Inner Zone and extends along the Omulyovaya and Taskhayakhtakh Uplifts, the Triassic and Lower Jurassic argillaceous series are replaced upwards by Middle Jurassic flyschlike and Upper Jurassic shallow-water argillite–sandstone accumulations with plants (Chekhov, 1971).

The terrigenous rocks of the Inner Zone are represented predominantly by graywacke with fragments of basic volcanic rocks; horizons of basic and intermediate tuffs are also present.

Folding of the Inner Zone is very intense. Tight linear folds, overturned to the southwest and complicated by thrusts and reverse faults, are widespread. The intensity of the folding gradually decreases in the direction of the Outer Zone.

The accumulation of deposits within the Yana–Kolyma Miogeosynclinal

System probably took place under shelf conditions. The Outer Zone of the miogeosyncline corresponds to a shallow part of the shelf adjacent to the continent, and the Inner Zone corresponds to a deeper part of the shelf and continental slope.

Riphean and early to mid-Paleozoic deposits of the Yana–Kolyma System are characterized by a wide distribution of carbonate rocks in the section.

In the Sette–Daban Range, the Riphean is represented by an interbedded series of arkose, siltstone, mudstone, dolomite, and limestone. There are discontinuities and weathered zones in the section. Among the carbonate rocks, which include sandy, clastic, and oolitic varieties, stromatolites are abundant. The terrigenous rocks include both gray and thick red, cross-bedded suites, especially in the lower and upper parts of the section. They are mainly shallow marine and coastal-marine deposits. In the marginal parts of the platform, their thickness decreases greatly and orthoquartzites predominate.

The age of glauconites of the lower Riphean horizons has been determined by K-Ar dating as 1600 m.y., and the upper horizons of the Riphean as 700 m.y. (Komar *et al.*, 1970).

The Paleozoic complex (up to 3000 m thick) unconformably overlies the Riphean. This unconformity is clearly expressed on the platform by the cutting of different Riphean horizons and is less distinct within the area of the miogeosyncline. Carbonate rocks predominate in this complex. Limestones and dolomites are interbedded with argillaceous and siliceous slate, siltstone, and sandstone in the lower parts of the section (Vendian and Cambrian); they form a thick homogeneous Ordovician, Silurian, and Lower Devonian series and are interbedded with red, cross-bedded sandstone, grit, and basic lava in the upper parts of the section (Middle and Upper Devonian). The Middle and Upper Devonian deposits include some gypsum horizons, which are also found in thrust slices to the northwest of the Sette–Daban, within the Verkhoyanskiy Complex.

The Riphean and Lower and Middle Paleozoic deposits of the Sette–Daban Range form linear folds, complicated by reverse faults and thrusts. Ridge-forming anticlines, separated by wide flat synclines, are typical of the fold style.

In the Omulyovaya, Prekolyma, Taskhayakhtakh, and Ulakhans–Polousniy Uplifts, Lower and Middle Paleozoic and some isolated Riphean deposits are represented, as in the Sette–Daban, mainly by carbonate rocks (Komar *et al.*, 1970; Oradovskaya, 1966; Merazlyakov, 1971; Grebennikov *et al.*, 1974) (Fig. 6).

The Paleozoic deposits are characterized by the same fossil complexes as the Paleozoic series of the Sette–Daban, which suggests that they accumulated in the same marine basin.

Limestones greatly predominate over dolomites among the carbonate

rocks, especially in the Ordovician section. They consist of micritic, fossil-frag-
mental, oolitic, and, less frequently, of argillaceous and sandy varieties. Great
amounts of various fossils are present, including brachiopods, trilobites, ostracodes, crinoids, bryozoa, sponges, and tabulate and rugose corals. These prove
that the deposits were formed under open, shallow-marine conditions. Reef
formations have been noted. Southwestward (in the Omulyovaya Mountains)
the amount of micritic limestone in the Ordovician section increases while
organic remains decrease. This indicates a deepening of the basin to the
southwest. Lenses and beds of gypsum appear in the Middle and Upper
Devonian deposits as they do in the Sette–Daban.

The Chukotsk Miogeosynclinal System was formed mainly from a
Triassic marine argillite, graywacke, and arkose sequence about 3000 m thick.
In the eastern parts of the system, Triassic deposits are conformably underlain
by Permian argillites, and in the west they discordantly overlap Carboniferous
and Devonian carbonaceous–terrigenous deposits. Folding is not so intense and
the structure is nearly horizontal.

Gabbro–diabase sills occur throughout the Chukotsk System. They are
known in the Permian and Lower to Middle Triassic deposits but are not
found in the Upper Triassic. Andesite tuffs and lavas occur in some places
among Upper Devonian–Lower Triassic deposits (Ivanov and Milov, 1975) but
in insignificant amounts.

Along their outer boundaries, the eugeosynclinal systems are associated
with extensive volcanic geanticlines or highs that, according to rock composition and petrochemical properties, are similar to the volcanic ridges of the
modern island arcs, particularly to the inner parts of volcanic arcs.

The Uda–Murgal Volcanic Geanticline is postulated to exist along the
northern coast of the Sea of Okhotsk. It ranges in age from Triassic and, in
some places, Upper Permian to Hauterivian (Lower Cretaceous). The geanticline trends northeastward for 2500 km, separating the circumoceanic eugeosynclinal belt from the Yana–Kolyma Miogeosynclinal System and the
Alazeia–Oloi Eugeosynclinal System of type B (Fig. 7).

The triassic to Hauterivian successions of the geanticline are exposed on
the left bank of the Uda River (the northeastern margin of the Mongolo-
Okhotsk Geosynclinal System), on the Koni, P'yagina, and Taigonos Peninsulas, and in the basins of the Penzhina and Anadir Rivers. They consist
of thick (up to some 1000 m) deposits of andesite, andesite–basalt, and basalt.
They alternate with shallow-water marine and coastal-marine volcanic-bearing
sandstone, grit, and conglomerate (Voinova, 1975; Beliy and Milov, 1973a;
Beliy and Kotlyar, 1975; Nekrasov et al., 1971). In some regions, their accumulation undoubtedly took place in an island environment (Nekrasov, 1971).
Volcanic rocks of acid composition are found in lesser quantity. Tuffs and
lavas are widespread among the volcanic rocks. Greenschist facies metamor-

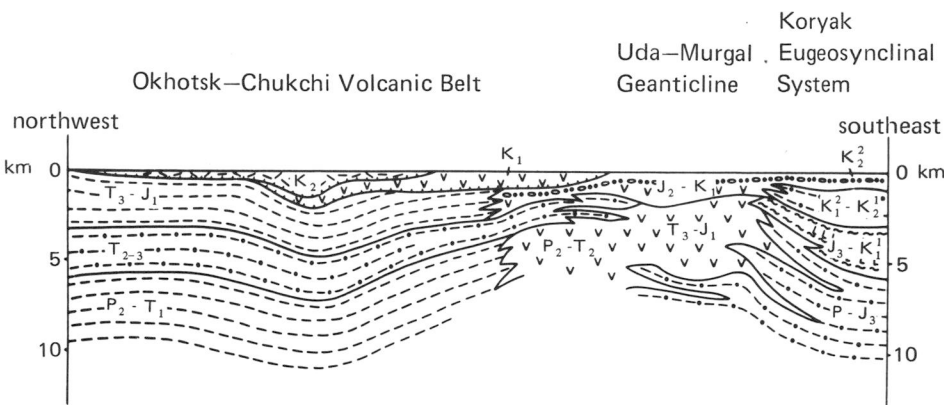

Fig. 7. Lithologic–stratigraphic cross section along the line IV–IV' (see Fig. 1). The legend is the same as in Fig. 2.

phism, expressed by the development of chlorite, epidote, and albite, occurs here. In this respect, they are similar to the green tuffs typical of the Miocene volcanic arc of the Japanese and Kurile Islands (Rotman, 1965).

The belt of late Jurassic volcanic rocks and associated volcanic-bearing clastic rocks of the Alazeia–Oloi Eugeosynclinal System of type B may possibly be a continuation of the Uda–Murgal Volcanic Geanticline. This belt, situated along the southwestern margin of the South Anyui Eugeosynclinal System of type A, stretches northwestward from the interfluve of the Oloi and Great Anyui Rivers to the area of the Svyatoi Nos Cape according to aeromagnetic data.

From the ocean side, the Uda–Murgal Volcanic Geanticline is associated with an ophiolitic belt which consists of Paleozoic and Mesozoic volcanic-siliceous rocks, Alpine-type ultramafics and gabbros, and is characterized by glaucophane–schist metamorphism. This belt trends almost unbrokenly from the Pekul'nei Range in the north to the southern edge of the Taigonos Peninsula (Alexandrov et al., 1975). From there it can be traced, using magnetic data, to a probable connection with the late Paleozoic ophiolites of the northeastern margin of the Mongolo-Okhotsk Geosynclinal System. This ophiolitic belt can be considered to represent the paleo-Benioff zone which, together with the Uda–Murgal Volcanic Geanticline, composes the Mesozoic island arc trending along the margin of the Mesozoic continent of Asia.

The Uyandina–Yasachnaya Volcanic Geanticline stretches for 1200 km along the Omulyovaya, Taskhayakhtakh, and Ulakhans–Polousniy Uplifts, separating the inner zone of the Yana–Kolyma Miogeosynclinal System from the Alazeia–Oloi Eugeosynclinal System (Fig. 5). The geanticline is composed of basalt, andesite–basalt, andesite, and rhyolite, alternating with argillite and clastic rocks bearing a late Jurassic marine fauna (Greenberg et al., 1974).

Data about the occurrence of volcanic rocks of Middle and probably Lower Jurassic age have been obtained recently (Konstantinovskiy, 1975).

III. MEDIAN MASSIFS

The Khanka, Bureya, Okhotsk, and Omolon Massifs stretch like a chain along the continental side of the eugeosynclinal system of type A.

These massifs are characterized by a wide distribution of potassium–sodium granitic rocks, including late Precambrian, early, middle, and late Paleozoic, and Mesozoic complexes. Within the Bureia massif, granitic rocks occupy almost 80% of the area; the other massifs have less.

Early Precambrian crystalline rocks form the framework of the massifs, among which are granulite complexes of the same type as within the Siberian Platform. Such complexes have been studied in the Khanka (Mishkin, 1969), Okhotsk (Greenberg, 1968), and Omolon (Gel'man and Terekhov, 1973) Massifs. In the Okhotsk massif, the age of pyroxene–amphibole gneisses has been determined by K-Ar dating as 2640 m.y. (Zagruzina et al., 1967), and the age of the basic crystalline schists is considered to be 3700 ± 500 m.y., using Pb isochron dating (Korol'kov et al., 1974). In addition to these old complexes, their basements also include younger gneisses and schists of Middle Precambrian age that are discordantly overlapped by Riphean strata.

In the Okhotsk and Omolon Massifs, late Precambrian and younger deposits have platform characteristics. In the Bureya and Khanka Massifs, Upper Precambrian–Lower Precambrian complexes are similar to geosynclinal formations. Nevertheless, nowhere do the deposits of this age possess eugeosynclinal features.

The Upper Precambrian assemblages of the Bureya Massif are not usually separated from those of the Cambrian. They include sandstone, grit, and, rarely, carbonaceous rocks with volcanic rocks of acid, intermediate, and sometimes basic composition. The thickness of the Upper Precambrian–Cambrian succession usually does not exceed 2000–3000 m (Figs. 2 and 3).

The younger, fossil-bearing deposits in the central areas of the massifs are made up of thick marine terrigenous and, rarely, carbonate strata of Silurian, Devonian, Carboniferous, and Permian ages. At the base of many of these complexes, there are a number of unconformities and conglomerates. Unconformities occur at the base of the Silurian, Middle Devonian, Carboniferous, and Upper Triassic. Granite pebbles are found in the basal conglomerates of all these systems where they directly overlie the granite massifs. Apparently, the discontinuities separating these complexes were of long duration, and sedimentation took place in a series of basins separated by uplifts. Shoreline facies have been recognized, but it is very difficult to outline the configuration of these old sedimentary basins because the massif is deeply eroded and poorly exposed. Folding of the Lower and Middle Paleozoic deposits is rather intensive in places with the development of compressed linear folds associated with cleavage. The Upper Paleozoic formations are more gently folded. The strikes of the fold zones are variable, being latitudinal, longitudinal, northeastern, and northwestern; they probably reflect the orientation of the primary sedimentary basins.

Along the margins of all of the massifs, there are in places late Precambrian, Paleozoic, and early Mesozoic troughs composed of terrigenous and carbonate or terrigenous strata. Their thickness exceeds several times the thickness of similar deposits in the inner areas of the massifs.

Volcanic rocks in varying amounts occur in all layers of the massifs. In some areas their quantity is quite large. Thus, in the Omolon Massif, they form almost the whole Kedonian Series (up to some kilometers thick) of Middle Devonian–Lower Carboniferous age (Lichagin, 1973). Acid rocks predominate among the volcanic rocks, but andesite and basalt are always present. All of them are characterized by high alkalinity and high potassium content. Pyroclastic formations usually predominate over lavas. Mosaic-patterned magnetic fields are found on these massifs. They are similar to the magnetic fields of the early Precambrian crystalline basement of the Siberian Platform on which linear anomalies of different orientations are superimposed by tectonic and magmatic reworking.

IV. CONTINENTAL MARGIN VOLCANIC BELTS

Thick accumulations (up to several kilometers) of intermediate, basic, acidic, and subalkaline continental volcanic rocks of Mesozoic and Cenozoic ages are associated with granitic rocks of the same age along the northeastern margin of the Asian continent. They form the giant Okhotsk–Chukotsk and Eastern Sikhote Alin Volcanic Belts, which trend for thousands of kilometers along the margin of the continent, and also small volcanic zones and fields in the inner continental areas.

The Okhotsk–Chukotsk Volcanic Belt for its entire enormous length (about 3000 km) is parallel to the outer side of the circumoceanic eugeosynclinal belt and the Uda–Murgal Volcanic Geanticline. The belt partially overlaps the Uda–Murgal Volcanic Geanticline on its inner (southeastern) edge (Fig. 7); on its outer edge, it discordantly overlies the following heterogeneous tectonic elements: the miogeosynclinal systems, the median massifs, and the Alazeia–Oloi Eugeosyncline and the South Anyui Eugeosyncline, which closed at the beginning of the early Cretaceous.

In the large northeastern part of the belt, the age of its lower strata is

Aptian–Albian. Hauterivian–Barremian formations apparently occur in the southwestern part of the belt. The development of the belt ended in late Cretaceous–early Paleogene.

From the paleogeographical point of view, the Okhotsk–Chukotsk Volcanic Belt and the Uda–Murgal Geanticline are thought to correspond to the Mesozoic continental margin with a deep marine basin adjacent to it on the southeast. This margin had already established itself in the early Paleozoic and remained stable during a long period of time.

The Eastern Sikhote Alin Belt, trending for 1500 km along the eastern side of the Sikhote Alin Eugeosynclinal System, is younger than the Okhotsk–Chukotsk Belt. Its formation took place from late Senonian up to Miocene time. The less extensive Western Sikhote Alin Belt of the same age is 120 km to the west. Continental volcanic rocks forming individual belts further westward are predominately of early and late Cretaceous age.

V. CENOZOIC GEOSYNCLINAL SYSTEMS

Included in the Cenozoic geosynclinal systems are geosynclines that developed during the Cenozoic, and most of them continue their development today with active volcanism and seismicity.

The Cenozoic geosynclinal systems (The Aleutian, Kurile–Kamchatka, Japan–Western Sakhalin) are distributed mainly in the transition zone between continent and ocean, in island arcs, in the Kamchatka Peninsula, and the southeastern part of the Koryak Highland.

A whole series of structures, which have much in common with structures within the older geosynclinal systems, can be distinguished in the Cenozoic Geosynclinal Systems (Fig. 8).

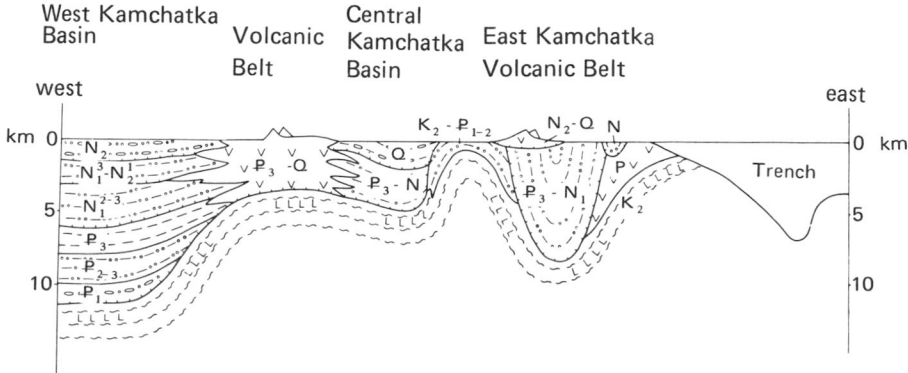

Fig. 8.   Lithologic–stratigraphic cross section along the line V–V′ (see Fig. 1). The legend is the same as in Fig. 2.

The volcanic geanticlines that form inner ridges or arcs are the axial elements of the Cenozoic systems. They originated at the end of Paleogene–early Miocene. They can be traced all along the strike of the geosynclinal systems in the Aleutian Islands, Greater Kurile Islands–Central Kamchatka, in the western part of Honshu, through Hokkaido to the shelf of western Sakhalin. As observed in Kamchatka, Honshu, and Hokkaido, they separate mio- and eugeosynclinal zones.

The miogeosynclinal zones are the back troughs of the island arcs. In western Kamchatka, the miogeosynclinal zone (the Western Kamchatka Trough) is formed of a thick (about 12,000 m) folded series of marine and coastal marine terrigenous rocks of Paleogene and Neogene age. The thick Tertiary series of western Sakhalin, trending to western Hakkaido (Pavlov and Parfenov, 1974), have an analogous tectonic position.

The eugeosynclinal zones (The Olutorskiy Trough, Eastern Kamchatka, and others) are composed of thick (up to several kilometers) basalt complexes with associated siliceous-argillaceous rocks and graywackes of Upper Cretaceous–Lower Miocene age. Usually they are strongly folded. Isoclinal linear folds and thrusts with the displacement towards the ocean are abundant. Intrusive rocks are represented by belts of Alpine-type ultramafics, small bodies of gabbro, and diorite. The zones under consideration do not differ from the Mesozoic eugeosynclinal systems of type A with respect to rock assemblages and fold style.

In places between the inner and outer arcs, throughgoing (about 1000 km) troughs composed of thick (up to several kilometers) argillite–graywacke series, including conglomerate, can be traced. Examples are the Central Kamchatka Trough of Miocene–Quaternary age and also the trough between the Greater and Lower Ridges of the Kurile Islands.

Similar troughs occur also between the outer arcs and the abyssal trenches, in particular, eastward from Kamchatka and the Lower Kurile Islands (Tuezov et al., 1975).

The Cenozoic geosynclinal systems are coupled along their edges with each other. The orientation of the systems is clearly discordant with the trends of the Mesozoic geosynclinal systems.

The volcanic ridges of the Cenozoic geosynclinal systems are uplifted in many areas, and the basement, upon which the systems rest, is exposed. In all these areas, the volcanic belts overlap geological complexes of different types and ages discordantly, such as the Paleozoic and Mesozoic folded complexes, and, in places, even Precambrian metamorphic formations and oceanic-type complexes (Minato et al., 1965; Jones, 1971; Marlow et al., 1973).

The majority of the tectonic elements of the Cenozoic geosynclinal systems are continuing their development into the present as indicated by modern topography. The volcanic belts for the most part are still active. They are represented by ridges of active volcanoes superimposed over older volcanic formations. The abyssal trenches, together with the adjacent parts of the ocean floor, are considered to be eugeosynclinal systems. It should be mentioned that, according to our scheme, the northern part of the Sea of Japan is regarded as a modern eugeosynclinal zone. The miogeosynclinal zones are represented by modern subsidence of the inner sides of island arcs (Tinro Basin, South Okhotsk Basin, and others).

Nevertheless, some discrepancy is noted between the modern tectonic schemes and the older structures. This discrepancy is clearly seen in the distribution of the modern volcanic zones. In some parts of the geosynclinal systems, volcanism is absent. There is no modern volcanism in the western part of the Aleutian System, in the northern part of the Kurile–Kamchatka System (north of the latitude of the Aleutian Islands), nor within the northern end of the Japan–Western Sakhalin System. In Kamchatka, the modern volcanic belt is displaced 200 km eastward and parallel to the Neogene belt.

VI. NATURE OF DEEP BASINS OF MARGINAL SEAS

The Aleutian, Commander, South Okhotsk, and Japan Deep Basins (deeper than 3000 m) are all characterized by oceanic-type crust. Different points of view exist concerning their origin. In this paper the origin is considered from the point of view of the tectonic scheme outlined here.

According to our scheme, these basins can have different origins. The Aleutian and Commander Basins are possibly parts of the Pacific Ocean floor separated from the ocean by volcanic arcs (Menard, 1964). If it is correct that the Bowers and Shirshov Ridges are volcanic arcs similar to the Aleutian Islands, but older, say of Paleogene age, then the Aleutian Basin should be older than the Commander Basin. The greater thickness of deposits in the Aleutian Basin in comparison to the Commander Basin (Scholl et al., 1974) supports this hypothesis.

The South Okhotsk Basin could be of similar origin. However, the great thickness of sediments within its area, about 5 km according to deep seismic-sounding data (Suvorov, 1975) and up to 4 km according to seismic-reflection data (Snegovskoi, 1974), and its position on the continuation of the Western Kamchatka Trough, makes it possible that this basin is a miogeosyncline or back trough (Vlasov, 1964). According to gravimetric and magnetometric data, folded eugeosynclinal deposits occur in the basement of the basin's western part; they can be traced in a longitudinal direction from eastern Sakhalin using geophysical anomalies (Kosygin and Mastulin, 1975).

There are no large sedimentary basins in the Sea of Japan. In this regard, the Sea of Japan differs greatly from the above-mentioned basins, which are characterized by consistently thicker deposits. The troughs, which have deposits (up to 2 km) within the Sea of Japan, are not large (several kilometers in diameter). These troughs are linear in form and can be considered to be grabens filled with sediments (Hilde and Wegeman, 1973).

VII. BASEMENT OF MESOZOIC AND CENOZOIC EUGEOSYNCLINES

Some data indicate the presence of old, probably Precambrian, sialic blocks of different sizes within the eugeosynclinal basement. Granulites of acid composition, up to 8% K_2O content, are found in eastern Kamchatka in the Ganalskiy Range (German, 1973). They are older than 487 m.y. based on Rb-Sr dating. The occurrence of sillimanite, cordierite, almandine, and zirconium resorbed crystals, which are typical of the metamorphic rocks in the pumice and ignimbrites of the Avachin and other volcanoes, proves the same metamorphic complex exists under the Cretaceous and Tertiary deposits. In this particular area, Cretaceous basalts have a high content of K_2O (almost 3%) (Markovskiy and Rotman, 1968). Xenoliths of granite gneisses and granulites with hypersthene and cordierite are found on the Paramushire Peninsula (Rodionova and Fedorchenko, 1971).

Large depressed blocks of metamorphic acid rocks seem to exist in the basement of the Sikhote Alin System. There is some evidence in favor of this assumption: for example, the occurrence of outcrops of migmatized metapelites with cordierite, staurolite, and almandine dated by the K-Ar method to be as old as 500–600 m.y.; wide distribution of high alumina granites which in places contain xenoliths of metamorphic rocks; the presence of high-potassic basalts (up to 5–7% of K_2O) in the lower parts of the Mesozoic geosynclinal complex. Together with graywackes, there are arkosic products in the composition of the Paleozoic and Mesozoic geosynclinal complexes. These complexes, unlike those of the Mongolo-Okhotsk System, exhibit a higher content of tin (2–3 times) in Clarke values (Mikhailov et al., 1974).

VIII. TECTONIC POSITION AND NATURE OF VOLCANIC BELTS

On the northeastern margin of the Asian continent, as has been mentioned above, there are two types of volcanic belts; continental margin belts (the Okhotsk–Chukotsk and Eastern Sikhote Alin) formed mainly under continental conditions, and island arc belts (the volcanic ridges of modern island arcs, the Uda–Murgal Volcanic Geanticline of the Triassic–Jurassic arc,

TABLE II
Average Composition of Magmatic Rocks of the
Okhotsk–Chukotsk Belt and Kurile Islands

Components	Okhotsk–Chukotsk Belt[a]	Kurile Islands[b]
SiO_2	65.52	59.26
TiO_2	0.69	0.75
Al_2O_3	15.74	17.43
Fe_2O_3	2.43	3.47
FeO	3.19	4.18
MnO	0.06	0.13
MgO	2.21	3.47
CaO	4.04	7.24
Na_2O	3.24	2.85
K_2O	2.88	1.22
Total	100.00	100.00

[a] After E. K. Ustiev (1965).
[b] After E. K. Markhinin (1967).

and others) formed under nonmarine island arc conditions. These belts possess both similar and dissimilar features. They consist of calc-alkali series rocks and are characterized by a similar sequence of eruptions. However, the differences in their chemical compositions are clearly noticeable (see Table II).

The genetic connection of the volcanic ridges of modern island arcs with the Benioff zones is now undoubted. It appears that the continental margin volcanic belts of northeastern Asia were connected with paleo-Benioff zones dipping under the continent and now outcropping along their outer (oceanic) boundaries. The following data support this supposition: (1) the connection of the Okhotsk–Chukotsk Belt with the Uda–Murgal Volcanic Geanticline of the island arc type; (2) the distinctive transverse zonation of the belt as expressed by the increased thickness of the volcanic rocks towards its near-oceanic margin, the decreased amount of acid volcanic rocks in the same direction, and the decreased content of alkali and potassium oxide and others (Shilo et al., 1974; Filatova and Dvoryankin, 1974).

In a transverse direction, the Okhotsk–Chukotsk Volcanic Belt is divided into a series of segments by the differing average composition of the constituent magmatic rocks: that of Central Chukotsk and Okhotsk, characterized by widespread acid and moderately acid rocks (mainly rhyolite and dacite); and that of Anadyr, formed on predominately andesite–basalt and andesite (Shilo et al., 1974; Kharkevich, 1971). The acid volcanics clearly gravitate to the inner (continental) margin part of the belt. The origin and the distinctive longitudinal and transverse zonation of the volcanic belt begins with the eruption of andesite–basalt, basalt, and andesite, followed by andesite–

basalt–rhyolite and rhyolite associations. Basic magma, indicative of mantle origin, is thought to have been the initial deposit of the belt. The appearance of acid rocks of both volcanic and intrusive origin is due to the melting of sialic crust from the heat of basalt magma.

IX. SOME CHARACTERISTICS OF GRANITIC MAGMATISM

The distribution of granitic rocks in the different tectonic zones of the Soviet Far East is very remarkable (Fig. 9). The largest granitic masses occur within the median massifs, sometimes occupying up to 80% of their territory. Granitic magmatism took place there during a long period of time lasting from

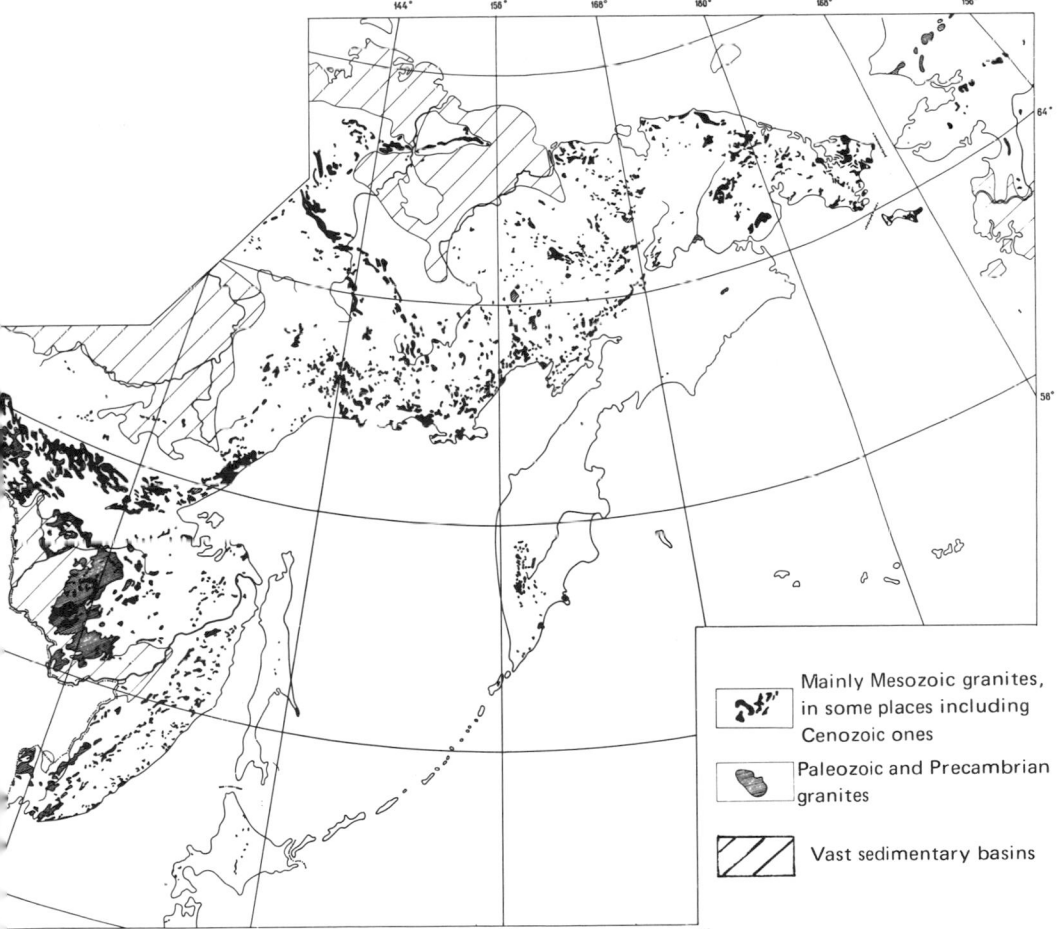

Fig. 9.   Granitic rocks of different ages in the Soviet Far East and adjacent areas.

the Precambrian up to the Mesooic. Large granitic masses of different ages are characteristic of the Stanovoi area, which has much in common with the massifs and differs from them only in having a linear form and a position on the border of the Siberian Platform.

Magmatic rocks of the Mesozoic miogeosynclinal systems are predominantly of granite composition; the average SiO_2 content is about 70% (Ustiev, 1959). Usually they form extensive belts and chains that suggest a connection of the granitic rocks with cracks in the pre-Riphean metamorphic basement that underlies the miogeosynclines.

Peculiarities in the distribution of acid rocks within the Okhotsk–Chukotsk Volcanic Belt have been mentioned above. The strong zonation, in conjunction with the overall transverse zonality of the belt, determines the distribution of granitic rocks of different types. Plagiogranites and tonalites are typical of the outer (near-oceanic) part of the belt, while potassium and sodic granodiorites of similar age occur in the inner zone (Milov, 1971; Beliy and Milov, 1973b). Toward the continent, they are followed by similar age granitics of the Yana–Kolyma Miogeosynclinal System, which are represented by granites and granodiorites with a high K_2O content (Greenberg, 1973). The composition of these granitic rocks is constant within the vast territory of the system and does not vary in spite of the compositional change of Paleozoic and Mesozoic country rocks.

Normal and potassium granites are usually absent in the eugeosynclines and especially in the eugeosynclines of type A. Small amounts of plagiogranites and diorites, associated with ultrabasites and gabbro, occur there. All sizable masses of normal and potassium granitic rocks appear only in zones where blocks of the old metamorphic rocks occur or, at least, are thought to occur (the Sikhote Alin, the central massif of Kamchatka).

X. MAIN FAULT SYSTEMS

A great number of faults have been mapped in the Soviet Far East. Some of them, which displace the geological contours on maps of a scale 1:2,000,000, are shown in Fig. 10.

A fault system is defined as an aggregate of normal faults, thrusts, reverse faults, and strike-slip faults of similar age formed under common kinematic conditions. Some types of fault systems are described and discussed in this paper (Fig. 11).

A. Island Arc Fault Systems

Modern island arcs are characterized by a regular system of faults. Frontal thrusts are typical of the outer nonvolcanic arcs adjoining the abyssal

Fig. 10.   Fault net of the Soviet Far East and adjacent areas.

trenches. They are found in Kamchatka (in the Oligocene–Miocene arc), in the Aleutian arc, and other regions (Vlasov, 1964; Shapiro and Seliverstov, 1975; Grow, 1973).

Frontal thrusts situated near the trench are associated with isoclinally folded flysch deposits, unlike those situated further from the trench, which are associated with different rocks, including ophiolite assemblages. These thrusts can be considered the continental thrusts of the Benioff zone.

The inner volcanic arc is characterized by the circular faults of volcanic depressions and also by normal faults and fractures that form a longitudinal linear tension zone. The latter have been studied in Kamchatka and are well known in some other arcs (Erlich *et al.*, 1974; Ermakov *et al.*, 1974). The tension zones near the frontal thrusts, where compression occurs, form because of the splitting action of intruding magmatic bodies.

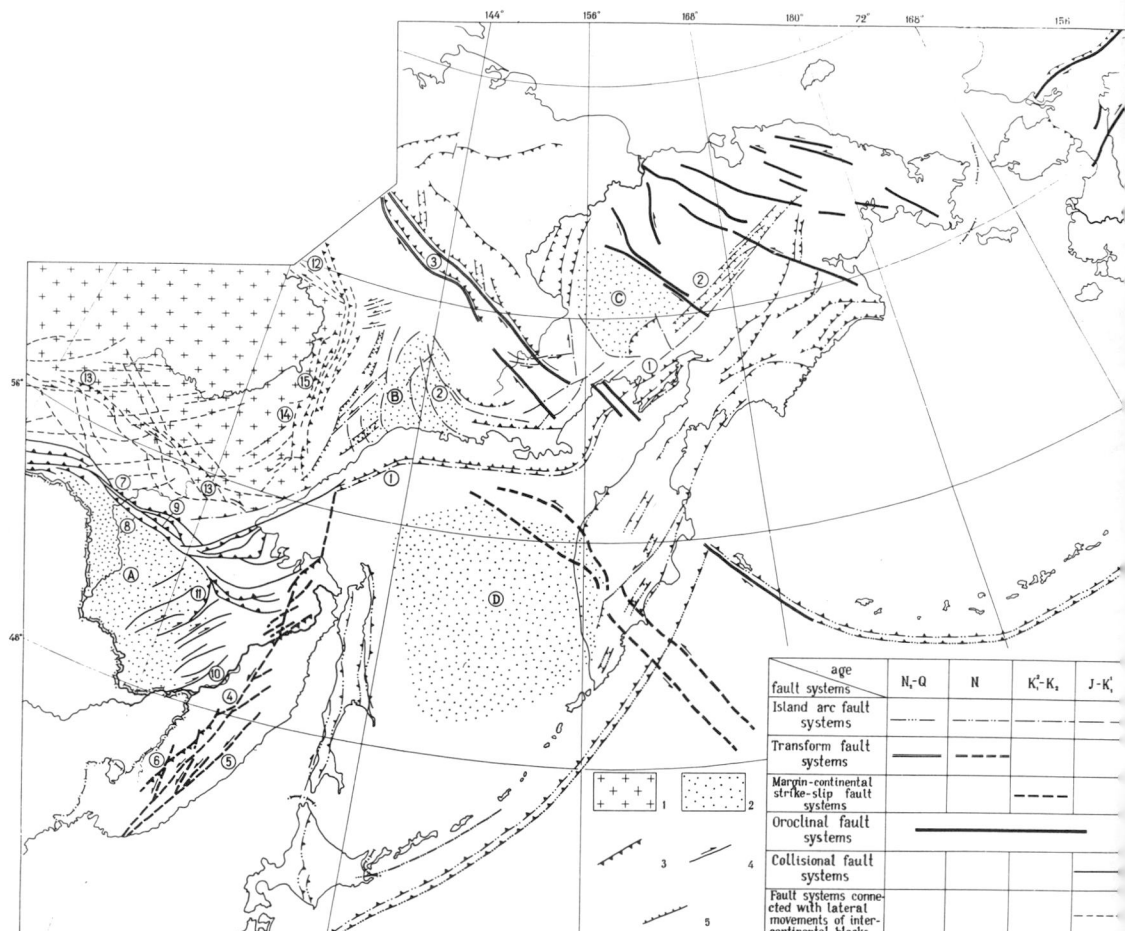

Fig. 11.   Main fault systems of the Soviet Far East and adjacent areas. (1) Aldan–Lena block of the Siberian Platform; massifs (Ⓐ Bureya, Ⓑ Okhotsk, Ⓒ Omolon, Ⓓ Okhotsk Sea); (3) thrusts and reverse faults, (4) strike-slip fau... (5) normal faults. Numbers in circles indicate: ① North Okhotsk Fault System; ② fault system of Okhotsk–Chukotsk Volcanic Belt; ③ fault system of Uyandina–Yasachnaya arc; ④–⑥ Sikhote Alin strike-slip fau... ④ Central, ⑤ Eastern, ⑥ Longitudinal; ⑦–⑨ Dzhagdi Collision Fault System: ⑦ Tukuringra, ⑧ South Tuk... ingra, ⑨ Champula faults; ⑩ Kukan Fault; ⑪ Tastakh Thrust; ⑫ Aldanian System of strike-slip faults; (... South Yakutian System of strike-slip faults; ⑭ Nel'kan Thrust; ⑮ Kyllah Thrust.

Longitudinal strike-slip faults occur in the rear areas of many island arcs. One of them can be traced for 600 km along the eastern edge of the Central Kamchatka Depression (Legler, 1976). Left-lateral displacement along the fault totals 10–20 km in the last 5 m.y. Characteristically, thrusts and normal faults coexist with the main strike-slip fault.

Longitudinal strike-slip faults of great extent are also thought to exist between the Greater and Lesser chains of the Kurile Islands (Strel'tsov, 1971), in Japan (Fitch, 1972), and in eastern Sakhalin (Rozhdestvenskiy, 1969). Modern arc elements are superimposed over the Oligocene–Miocene arc on the Kamchatka Peninsula. As a result, we have the situation shown in Fig. 11. In the back troughs, there are reverse faults and thrusts dipping towards the ocean. Such faults are known in western Kamchatka. In a zone of reverse thrusts, there is much less general dislocation of sediments than in a frontal thrust zone.

Fault systems of the same age, belonging to adjacent island arcs, are separated by transform faults or possess an end-to-end connection. The end-to-end connection occurs between the Kurile and Japan–Sakhalin arcs. The western part of the Aleutian Ridge is an example of a modern transform fault (Grow and Atwater, 1970). Here, a combination of the structural elements of both a strike-slip displacement and a compression developed with the general predominance of the first. The fault systems of the old island arcs are of similar construction.

The Northern Okhotsk Fault System, extending along the Uda–Murgal Geanticline, comprises frontal thrusts occurring in the southern part of the Taigonos Peninsula (Nekrasov, 1971), the Koni and P'yagina Peninsulas (Beliy and Kotlyar, 1975), and the Penzhina–Pekul'nei area (Alexandrov et al., 1975), and large-scale strike-slip faults. The Uligdan strike-slip fault, for example, can be traced for a hundred kilometers as a linear zone of schistose and crushed rocks a hundred meters wide; it goes in a northeasterly direction on the right bank of the Uda River (Roganov and Vizgalov, 1972) and probably stretches further to the northeast along the Dzhugdzhur coast of the Sea of Okhotsk. The inner volcanic arc, composed of Triassic–Jurassic volcanic rocks like modern island arcs, is characterized by comparatively mild deformations. The back faults of the Northern Okhotsk System are covered by the fields of volcanic rocks of the Okhotsk–Chukotsk Belt.

The late Mesozoic–early Cenozoic fault system of the Okhotsk–Chukotsk Belt is divided in a longitudinal direction into several en echelon sections. In places there are thrusts bounding the belt on both sides (the Chelomdzhn–Yama, Armanskiy, and others). Circular faults of different diameters are typical of the belt. They seem to have originated mainly in connection with magmatic processes. Circular faults about 40–60 km in diameter occur most often. They form chains stretching along the belt (Fig. 12).

The Uyandina–Yasachnaya arc of Jurassic age is bounded on both sides by thrusts and reverse faults that are inclined to its axis; frontal faults associated with the Alazeia Eugeosyncline (the Arga–Tass Fault), and back faults associated with the Inyali–Debina Synclinorium (the Darpir, Ulakhan, and other faults) (Chekhov, 1971; Mokshantsev et al., 1968).

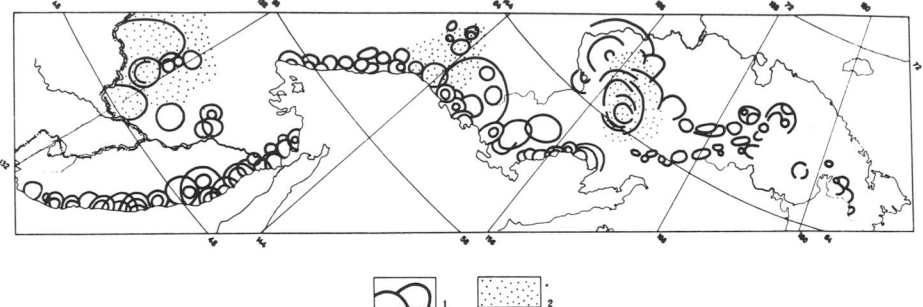

Fig. 12.   Systems of circular faults. (1) Faults, (2) massifs.

B. Transform Fault Systems

The western part of the Aleutian Ridge, as has already been mentioned, can be considered a modern system of arc–arc-type transform faults. A Mesozoic–Cenozoic fault system is located on the continuation of these faults in the Cherskiy Range, the middle course of the Yana River, and the Olenek Displacement Zone. The faults in these areas have a similar northwestern strike and lie along the strike of one another. They are considered to be late Mesozoic–Cenozoic, right-lateral strike-slip faults that developed along Jurassic thrust and reverse faults (the Darpir, Chai–Yur'inskiy, Adychanskiy, and other faults) (Chekhov, 1971; Izmailov and Chekhov, 1971; Mokshantsev *et al.*, 1968). In places, these strike-slip faults are associated with Cenozoic basins, the orientation of which suggests that they may be prefaulting tension zones.

The transform fault system is characterized by high seismicity along its entire length. Study of the stress field orientation of earthquake epicenters reveals that a horizontal compression takes place there, and its orientation does not contradict the assumption about the modern right-lateral strike-slip movement (Mokshantsev *et al.*, 1975). The transform fault system possibly extends northwestward to the rift of the Gakkel Ridge and southeastward to the Komandorskiy Islands.

The older transform fault system, trending northwestward, is found in southern Kamchatka. It is clearly expressed both on land and under water by linear magnetic anomalies of high intensity trending northwestward, and by the displacement of northeastern linear anomalies in the Pacific Ocean. According to magnetic data, the fault system stretches 500 km into the Pacific Ocean, crossing the continental slope and abyssal trench and extending to the Sea of Okhotsk (Shimaraev, 1971). The Miocene–Pliocene volcanic belt of the inner arc of Kamchatka is displaced 200 km along these faults in a left-lateral direction.

The age of the faults is identified as approximately late Pliocene, since the now-active volcanic belt of the Kurile–Kamchatka arc, which formed at the end of Pliocene–early Pleistocene time, and the abyssal trench cross this fault system without any visible displacement.

C. Continental Margin Strike-Slip Faults

A system of continental margin strike-slip faults is represented by the late Cretaceous left-lateral strike-slip faults of the Sikhote Alin Fold System (the Central, Eastern, Longitudinal, and others). The displacement along these faults ranges from several kilometers to 150 km (Ivanov, 1972). The strike-slip faults are accompanied by associated thrusts and in many ways influence the structures of the Sikhote Alin System.

The formation of these strike-slip faults is caused by the displacement of the oceanic plate approximately parallel to the continental edge toward the still active Okhotsk–Chukotsk Volcanic Belt (Fig. 13). The closing up of the geosynclines in the northern area of the Sea of Okhotsk led to the formation of the Kurile–Kamchatka Island Arc, where a new Benioff seismic zone evolved.

D. Oroclinal Fault System

In northeastern USSR and Alaska, extensive faults trending respectively northwest and northeast are widespread. These faults extend relatively symmetrically to the meridian that passes approximately through the Bering Strait. The Alaskan faults are right-lateral strike-slip faults (the Kaltag, Iditarod–Nixon Fork, Denali, and others) (Patton and Hoare, 1968; Lathram, 1972). The displacement along the faults from late Cretaceous to recent time is estimated in tens to a few hundred kilometers. The faults in Chukotka are left-lateral strike-slip faults (Sutygin and Tibilov, 1969).

The origin of these fault systems can be linked to the formation of the Bering Sea Orocline. This orocline is a gigantic (about 1200 km in diameter) bend of the folded systems of Chukotka, the Koryak Mountains, and Alaska, with its convex side facing southward. These folded systems join together within the Bering Sea Shelf (Fig. 14).

There are definite correlations between the orocline and the fault system as follows: (1) the axis of symmetry of the orocline coincides with that of the fault system, (2) the movement along the faults is mainly left-lateral on the west and right lateral on the east of the orocline, and (3) the faults are approximately parallel to the limbs of the orocline. The foregoing data raise the possibility that these fault systems are unusual areas of interlayer sliding like those occurring during the formation of flexure folds.

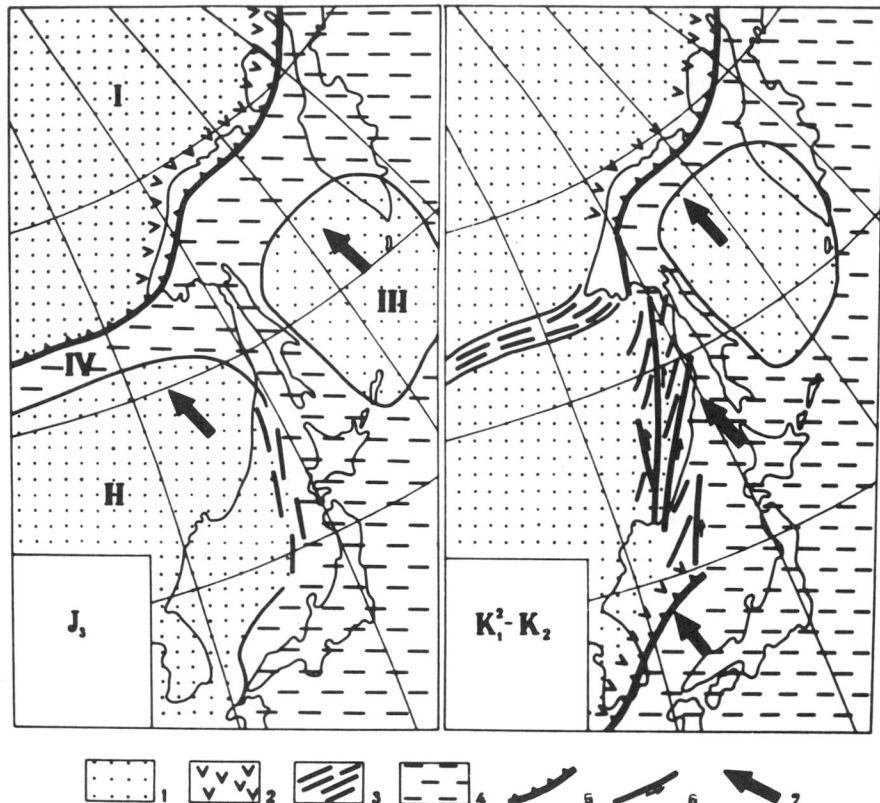

Fig. 13.   Origin of the Sikhote Alin strike-slip faults. (1) Continents [(I)—Northeastern Asia)] and microcontinents [(II) Bureya and (III) Okhotsk Sea Massifs)]; (2) volcanic arcs; (3) zones of foldings; (4) ocean and geosynclines [(IV) Mongolo-Okhotsk Eugeosynclinal System]; (5) inferred Benioff zones; (6) strike-slip faults; (7) orientation of movement.

E. Collision Fault Systems

Collision fault systems are formed from a collision of continental plates. The Dzhagdin System of late Jurassic–early Cretaceous thrusts, reverse faults, and strike-slip faults, occupying the eastern part of the Mongolo-Okhotsk Geosynclinal System, belongs to this type. These thrusts, reverse faults, and strike-slip faults were created when the Bureya Massif and Aldan–Stanovoy block converged and crushed the geosynclinal filling of the Mongolo-Okhotsk Geosynclinal System between them.

In the area opposite the Bureya Massif, the Dzhagdin System is represented by a series of subparallel thrusts and reverse faults that have been compressed together and that are mainly dipping northward (the Tukuringr, South Tukuringr, Champulin, and other faults). In places, a reverse direction of dip of the fault planes can be observed (Zabrodin and Turbin, 1970). East of the eastern margins of the Bureya Massif, these faults branch, making a wide fan with the convex side facing southward.

Paleozoic and early Mesozoic deposits of the Mongolo-Okhotsk Belt, in the area opposite the Bureya Massif, are characterized by very complicated, superimposed minor folds and boudinage representing several stages of deformation. Eastward from the Bureya Massif, the folding is not so intense.

Additional fault systems associated with the collision fault systems formed as a result of the movement of the continental plates. Thus, sub-parallel, strike-slip faults trending in a northeastern direction developed along the eastern margin of the Bureya Massif (the Kukan Fault and others). They sometimes join with longitudinal thrusts (the Tastakhskiy Fault and others). The origin of this fault system may be related to the northward displacement of the Bureya Massif.

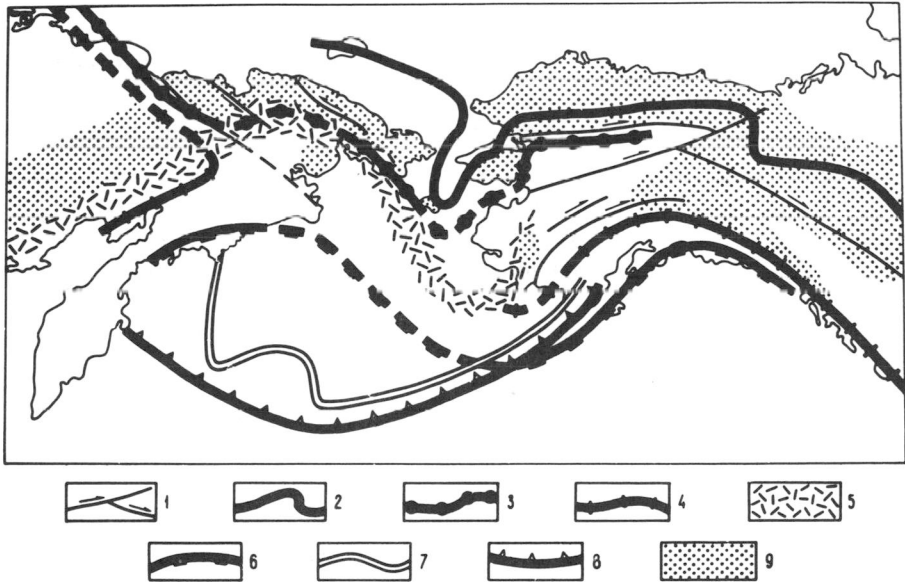

Fig. 14. Bering Sea orocline. (1) Strike-slip faults; (2) Paleozoic folded zone; (3) early Mesozoic eugeosynclinal systems; (4) Mesozoic volcanic arcs; (5) Okhotsk–Chukotsk Volcanic Belt and its inferred continuation in Alaska; (6) late Mesozoic eugeosynclinal systems; (7) Paleogene volcanic arc; (8) Neogene volcanic arc; (9) Mesozoic continents and microcontinents.

F. Fault Systems Connected with Horizontal Displacement within the Continental Blocks

Much geological data can be explained if a considerable displacement to the east is assumed (on the order of hundreds of kilometers) for the Aldan–Lena block of the Siberian Platform, relative to the northern part of the platform and the Stanovoy area, in late Jurassic–early Cretaceous time. The latitudinal boundaries of the Aldan–Lena block are the Aldan strike-slip faults on the north and the south Yakut strike-slip faults on the south.

The Aldan System of left-lateral strike-slip faults trends along the southern latitudinal angular bend of the Verkhoyansk Anticlinorium and extends further westward within the limits of the Siberian Platform (Mezhvilk, 1970).

The South Yakutsk Fault System stretches along the southern margin of the Aldan shield from East Zabaikal'e to the coast of the Sea of Okhotsk. The fracture pattern within late Jurassic intrusions in the Stanovoy area indicates right-lateral displacements along the southern border of the Aldan Shield (Fig. 15). These intrusions can be interpreted as filling the gigantic tension fractures associated with the strike-slip displacement.

In the frontal part of the Aldan–Lena block there is a system of steeply dipping, late Mesozoic thrusts and reverse faults, trending in a longtitudinal direction within the Sette–Daban Range (the Nel'kan, Killakh, and other faults). Gravimetric data indicate that these thrusts and eastward-dipping faults flatten out at depth and do not affect the Archean basement. The formation of these faults probably is related to the underthrusting of the Aldan–Lena block under the Paleozoic strata of the South Verkhoyansk region.

Spherical and concentric structures which accompany circular faults are widely distributed in the Soviet Far East. The wide distribution of circular faults within the Okhotsk–Chukotsk Volcanic Belt has been noted earlier. They are also readily recognizable in the Eastern Sikhote–Alin Belt (Fig. 12). Circular faults are typical of the median massifs. Their origin is still insufficiently clear. Magmatic processes seem to have been the main factor in their origins.

The main fault systems of the Soviet Far East have various origins. Faults associated with lateral displacements of oceanic and continental plates are the most widespread; these include the island arc, transform, continental margin strike-slip, and oroclinal fault systems. Faults of these types have been forming there during all of the Mesozoic and Cenozoic, and many of them are still active. The existence of similar fault systems in earlier stages of Pacific margin development, say in Paleozoic time, seems quite probable. However, recognition of these older faults is quite difficult and requires further study. Large lateral displacements also take place within the continental regions, both along

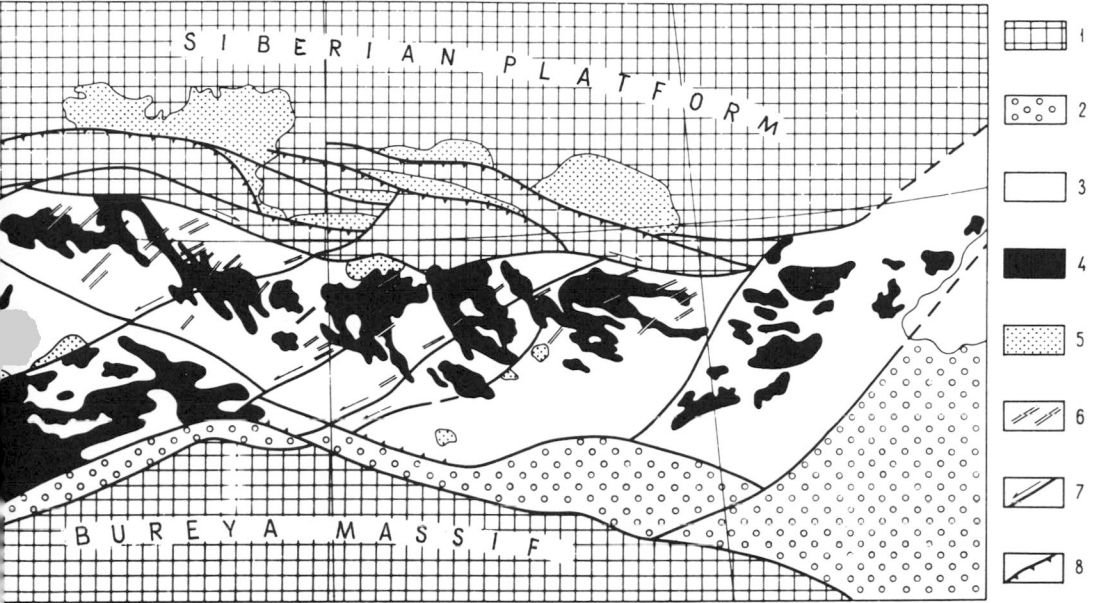

Fig. 15.   Position of late Jurassic granitic rocks of Stanovoy area. (1) Stable blocks; (2) Mongolo-Okhotsk Eugeosynclinal system; (3) Stanovoy area; (4) late Jurassic granitic rocks; (5) Jurassic basins filled by thick carbonate deposits; (6) Cretaceous dike swarms; (7) strike-slip faults; (8) thrusts and reverse faults.

boundaries with geosynclines and also within already consolidated ancient continental blocks.

ACKNOWLEDGMENTS

This paper draws information from the work of many geologists. Essential references have been given, but many others have been omitted for lack of space. Special thanks for helpful discussions are particularly due to S. M. Til'man, L. P. Karsakov, B. A. Natl'in, V. M. Merzlyakov, and D. F. Semenov. Miss I. S. Korotkova translated this paper into English, and Mrs. L. N. Nosacheva helped greatly with the references.

REFERENCES

Alexandrov, A. A., 1973, Serpentinitic melange in the upper course of the River Chirinai (Koryak Upland) (in Russian), *Geotektonika*, no. 4, p. 84–93.

Alexandrov, A. A., Bogdanov, N. A., Byalobzhevskiy, S. G., Markov, M. S., Til'man, S. M., Khain, V. E., and Chekhov, A. D., 1975, New data on the tectonics of the Koryak Upland (in Russian), *Geotektonika*, no. 5, p. 60–72.

Arkhipov, U. V., 1974, *Stratigraphy of Triassic deposits of Western Yakutsk* (in Russian), Yakutsk: Yakutsk, p. 271.

Beliy, V. F. and Kotlyar, I. N., 1975, New data on the geology of the western P'yagina Peninsula (the inner Zone of the Okhotsk–Chukotsk Volcanic Belt) (in Russian), in: *Proceedings on Geology and Mineral Resources of the Northeastern USSR*, v. 22, Magadan, p. 74–85.

Beliy, V. F., and Milov, A. P., 1973a, Structure and development of the inner zone of the Okhotsk–Chukotsk Belt in the basin of the Penzhina River (in Russian), *Sov. Geol.*, no. 1, p. 86–99.

Beliy, V. F., and Milov, A. P., 1973b, Petrological zonation of gabbro-granite series in the Okhotsk-Chukotsk Volcanic Belt (in Russian), *Dokl. Akad. Nauk SSSR Seria. Geol.*, no. 10, p. 49–57.

Chekhov, A.D., 1971, Comparative analysis of the In'yali–Debina and Olnoi synclinoriums (in Russian), in: *Mesozoic Tectogenesis*, Bogolepov, K. V., and Til'man, S. M., eds., Magadan: Northeast Interdisciplinary Science Research Institute, p. 87–96.

Dobretosov, N. L., 1974, Glaucophane schist and eclogite–glaucophane schist complexes of the USSR (in Russian), Novosibirsk: Nauka, 425 p.

Dovgal', Y. U. M., 1964, Ophiolite formations of the Alucha Uplift (in Russian), in: *Proceedings on Geology and Mineral Resources of the Northeastern USSR*, v. 17, Magadan: Magadan, p. 149–158.

Eliseyeva, V. K., and Sosnina, M. I., 1964, Occurrence of Upper Permian deposits on Sakhalin Island (in Russian), *Geol. Geofiz.*, v. 10, p. 159–161.

Erlich, E. N., Melekestsev, I. V., and Shanster, A. E., 1974, New tectonics (in Russian), in: *Kamchatka, Kurile, and Kommandorskiy Islands*, Luchitsky, I. V., ed., Moscow: Nauka, p. 345–368.

Ermakov, V. A., Milanovskii, E. E., and Tarakanov, A. A., 1974, The role of riftogenesis in the formation of the Kamchatka Quaternary volcanic zones (in Russian), Vestnik Moscow State University, *Geologia*, v. 3, p. 3–20.

Filatova, N. I., and Dvoryankin, A. I., 1974, Peculiarities of the structure and development of volcanic formations in the central part of the Okhotsk–Chukotsk Belt (in Russian), in: *Evolution of Volcanism in the History of the Earth*, Luchitsky, I. V., and Frend, G. M., eds. Moscow: USSR Academy of Science, p. 164–170.

Fitch, T. J., 1972, Plate convergence, transcurrent faults, and internal deformation adjacent to southeastern Asia and the Western Pacific, *J. Geophys. Res.* v. 77(23), p. 4432–4460.

Gel'man, M. L., and Terekhov, M. I., 1973, New data about the Precambrian crystalline complex of the Omolon Massif (in Russian), in: *Metamorphic Complexes of the Eastern Part of the USSR*, Smirnov, A. M., and Shuldiner, V. I., eds., Vladivostok: USSR Academy of Science, Far Eastern Science Center, p. 66–73.

German, L. L., 1973, Metamorphic rocks of granulitic facies on the Ganal' Ridge of Kamchatka (in Russian), *Dokl. Akad. Nauk USSR*, v. 209(3), p. 680–682.

Grebennikov, G. A., Rabotnov, V. T., and Spector, V. B., 1974, The Upper Precambrian stratigraphy of the Selennyakhskiy Ridge (in Russian), in: *Precambrian and Paleozoic of the Northeastern USSR*, Nikolaev, A. A., and Simakov, K. V., eds., Magadan: Magadan, 14 p.

Greenberg, G. A., 1968, *Precambrian of the Okhotsk Massif* (in Russian), Moscow: Nauka, p. 187.

Greenberg, G. A., 1973, Late Mesozoic granitoid sequences of Northeastern USSR (in Russian), in: *Magmatism of the Northeastern USSR*, Apeltsin, F. E., ed. Moscow: Nauka, p. 9–26.

Greenberg, G. A., Bakharev, A. G., Nedosekin, U. D., and Surnin, A. A., 1974, Volcanic complexes of the central part of the Uyandina–Yasachnaya Volcanic Belt (in Russian), in: *New Data on Magmatism of the Yakutsk*, Yakutsk: Yakutsk, p. 50–107.

Grow, J. A., 1973, Crust and upper mantle of the central Aleutian arc, *Geol. Soc. Am. Bull.*, v. 84(7), p. 2169–2192.

Grow, J. A. and Atwater, T., 1970, Mid-Tertiary transition in the Aleutian arc, *Geol. Soc. Am. Bull.*, v. 81, p. 3715–3722.

Gulevich, V. V., 1975, Late Jurassic volcanism of the upper reaches of the Great Anyui River (in Russian), in: *Magmatism of Northeast Asia*, Part 2, Shatalov, E. T., ed., Magadan: Magadan, p. 81–88.

Gulyaev, P. V., 1975, Tectonics of the Alazeia Uplift (in Russian), *Geotektonika*, no. 6, p. 30–43.

Hilde, T. W. C., and Wegeman, J. M., 1973, Structure and origin of the Sea of Japan, in: *The Western Pacific Island Arcs, Marginal Seas, Geochemistry*, University of Western Australia Press.

Ivanov, B. A., 1972, Central Sikhote Alin Fault (in Russian), Vladivostok: Far Eastern, 116 p.

Ivanov, O. N., and Baratov, Sh. Kh., 1974, Serpentine melange of the Khatirka River basin (Koryak Upland) (in Russian), *Dokl. Akad. Nauk SSSR*, v. 2, p. 404–406.

Ivanov, O. N., and Milov, A. P., 1975, Diabase succession of the Chukchi folded system and its connection with the basic magmatism of the northern sector of the Pacific mobile belt (in Russian), in: *Magmatism of Northeast Asia*, Part 2, Shatalov, E. T., ed., Magadan: Magadan, p. 155–159.

Izmailov, L. I., and Chekhov, A. D., 1971, The main systems of fracture dislocations of the southeastern part of the In'yali–Debina Synclinorium (in Russian), in: *Mesozoic Tectogenesis*, Bogolepov, K. V., and Til'man, S. M., eds., Magadan: Northeast Interdisciplinary Science Research Institute, p. 295–298.

Jones, J. G., 1974, Aleutian enigma: A clue to transformation in time, *Nature (London)*, v. 229.

Komar, V. L. A., Semikhatov, M. A., Serebryakov, S. N., and Voronov, B. G., 1970, New data on the stratigraphy and history of Riphean development in southeastern Siberia and Northeastern USSR (in Russian), *Sov. Geol.*, no. 3, p. 37–53.

Konstantinovski, A. A., 1975, Old blocks in the Jurassic deposits of the Cherskiy mountain range (northeastern USSR) (in Russian), *Geotektonika*, no. 6, p. 61–67.

Korol'kov, V. G., Rudnik, V. A., and Sobotovich, E. V., 1974, Late Archean age of the most ancient rocks of the Okhotsk Massif (in Russian), *Dokl. Akad. Nauk USSR*, no. 6, p. 1441–1444.

Kosygin, V. U., and Mastulin, L. A., 1975, Structure and development of the consolidated crust in the southern part of the Sea of Okhotsk in the light of new data (in Russian), *Dokl. Akad. Nauk USSR*, v. 224, v. 4, p. 898–901.

Krasniy, L. I., 1971, Tectonics of the Ilin–Tass Zone (the Yana–Kolyma System) (in Russian), in: *Mesozoic Tectogenesis*, Bogolepov, K. V., and Til'man, S. M., eds., Magadan: Northeast Interdisciplinary Science Research Institute, p. 298–301.

Lathram, E. H., 1972, Nimbus IV view of the major structural features of Alaska, *Science*, v. 175, p. 1423–1427.

Legler, V. A., 1976, Deformation of the subsiding lithospheric plate and longitudinal shifts of the Kurile–Kamchatka island arc (in Russian), in: *Plate Tectonics*, Moscow, p. 103–147.

Lichagin, P. P., 1973, Petrochemistry and petrology of Devonian volcanic rocks of the Omolon Massif (in Russian), *Geol. Geofiz.*, v. 8, p. 62–69.

Lichagin, P. P., Merzlyakov, V. M., Ponomaryova, L. G., Terekhov, M. I., and Khmel'nikova, O. S., 1975, Glaucophane schist metamorphism of the Alazeia Upland (in Russian), in: *Proceedings on Geology and Mineral Resources of Northeastern USSR*, v. 22, Tsopanov, O. Kh., ed., Magadan: Magadan, p. 112–119.

Kharkevich, D. S., ed., 1971, Map of magmatic formations of the USSR (in Russian), Scale 1:2,500,000, Leningrad: All-Union Geological Institute.

Markhinin, E. E., 1967, The role of volcanism in the formation of the earth's crust, as exemplified by the Kurile island arc (in Russian), Moscow: Nauka, 265 p.

Markovskiy, B. A., and Rotman, V. K., 1968, Types of geosynclinal basaltic magmas, as illustrated by those of Kamchatka (in Russian), *Dokl. Akad. Nauk USSR*, v. 182(3), p. 674–676.

Marlow, M. W., Scholl, D. W., Buffington, E. C., and Alpha, T. R., 1973, Tectonic history of the central Aleutian Arc, *Geol. Soc. Am. Bull.*, v. 84.

Menard, H. W., 1964, *Marine Geology of the Pacific*, New York: McGraw-Hill.

Merzlyakov, V. M., 1971, Stratigraphy and tectonics of the Omulyovaya Uplift (in Russian), Moscow: Nauka, 151 p.

Merzlyakov, V. M., and Lichagin, P. P., 1973, Ordovician volcanism of the northeastern USSR (in Russian), in: *Magmatism of the Northeastern USSR*, Apeltsin, F. E., ed., Moscow: Nauka, p. 207–212.

Mezhvilk, A. A., 1970, Lena suture zone (in Russian), in: *Tectonika Sibiri*, v. IV, Bogolepov, K. B., ed., Moscow: Nauka, p. 53–60.

Mikhailov, M. A., Moiseenko, V. G., and Sakhno, V. G., 1974, Some peculiarities of tine and gold distribution in the volcanic–sedimentary geosynclinal and continental complexes with examples from the Sikhote Alin and Okhotsk fold regions (in Russian), in: *Volcanic–Sedimentary Lithogenesis, Yuzhno–Sakhalinsk*, Markhinin, E. K., and Fremd, G. M., eds., Far Eastern Science Center, Institute of Geology, p. 221–222.

Milov, A. P., 1971, The influence of tectonic conditions on the peculiarities of Late Mesozoic granitoidal magmatism of Chukotka (in Russian), in: *Mesozoic Tectogenesis*, Bogolepov, K. V., and Til'man, S. M., eds., Magadan: Northeast Interdisciplinary Science Research Institute, p. 291–295.

Minato, M., Gorai, M., and Hunahashi, M., 1965, *The Geologic Development of the Japanese Islands*, Tokyo: Tsukiji Shokan Co., 442 p.

Mishin, V. P., 1971, Various ages of eugesynclinal development of the main anticlinorium of the Sikhote Alin (in Russian), in: *Mesozoic Tectogenesis*, Bogolepov, K. V., and Til'man, S. M., eds., Magadan: Northeast Interdisciplinary Science Research Institute, p. 114–118.

Mishkin, M. A., 1969, *Petrology of Precambrian Metamorphic Complexes of the Khanka Massif of Primorya* (in Russian), Moscow: Nauka, 182 p.

Mokshantsev, K. B., Gornshtein, D. K., Gudkov, A. A., Gusev, G. S., Den'gin, E. V., and Shtekh, G. I., 1968, *Deep-Sea Structure of the Eastern Part of the Siberian Platform and the Adjacent Folded Successions of the Verkhoyan'e–Chukotsk Area* (in Russian), Moscow: Nauka, 170 p.

Mokshantsev, K. B., Gusev, G. S., Koz'min, B. M., and Tret'yakov, F. F., 1975, Seismicity, modern and old areas of stress in northeastern Asia in connection with the Momsk Rift (in Russian), in: *Problems of Riftogenesis*, Florensev, N. A., and Logachev, N. A., eds., Irkutsk: USSR Academy of Science, Siberian Branch, p. 78–80.

Nekrasov, G. E., 1971, The place of ultrabasites, radiolarites, and basic effusives in the developmental history of the Taigonos Peninsula and Penzhina Ridge (in Russian), *Geotektonika*, v. 5, p. 37–44.

Nekrasov, G. E., Zaborovskaya, N. B., and Gel'man, M. L., 1971, Tectonics of the zone of mesozoide transition to the structures of the Koryak–Kamchatka fold region as exemplified by the Taigonos Peninsula (in Russian), in: *Mesozoic Tectogenesis*, Bogolepov, K. V., and Til'man, S. M., eds., Magadan: Northeast Interdisciplinary Science Research Institute, p. 80–87.

Oradovskaya, M. M., 1966, Two types of Ordovician sections in the Kolyma Massif (in Russian), in: *Proceedings on Geology and Mineral Resources of the Northeastern USSR*, v. 18, Magadan, p. 56–66.

Patton, W. W., and Hoare, J. M., 1968, The Kaltag Fault, west–central Alaska, *U.S. Geol. Surv. Prof. Pap.* 600–D, p. D147–D153.

Pavlov, U. A., and Parfenov, L. M., 1974, The geologic nature of the Hokkaido–Sakhalin gravity low(?) (in Russian), *Dokl. Akad. Nauk USSR*, v. 217(6), p. 1390–1393.

Pinus, G. V., and Sterligova, V. E., 1973, New Alpine-type ultrabasic rock belt of northeastern USSR with reference to some geological peculiarities in the formation of ultrabasic belts (in Russian), *Geol. Geofiz.*, v. 12, p. 109–111.

Pinus, G. V., Velinskiy, V. V., Lesnove, F. P., Bannikov, O. L. and Agafonov, L. V., 1973, Alpine-type ultrabasic rocks of the Anadyr–Koryak fold system (in Russian), Novosibirsk: Nauka, 318 p.

Radzivil, A. Ya., and Radzivil, V. Ya., 1975, Late Jurassic magmatic successions of the South Anui Trough (in Russian), in: *Magmatism of Southeastern Asia*, Part 2, Magadan, p. 71–80.

Rechkin, A. N., 1974, Ophiolites of the eastern ridge of the Shmidt Peninsula (Sakhalin Island) (in Russian), in: *Structural Analysis of Dislocations*, Parfenov, L. M., and Popeko, V. A., eds., Khabarovsk: Institute of Tectonics and Geophysics, p. 116–130.

Rodionova, R. I., and Fedorchenko, V. I., 1971, Xenolites in the lavas of Kurile Islands and some problems of deep-sea geology of this region (in Russian), in: *Volcanism and Depth of the Earth*, Vlodavets, V. I., and Fedotov, S. A., eds., Moscow: Nauka, p. 141–147.

Raganov, G. V., and Vizgalov, V. I., 1972, The developmental history of the Baladek Suture Zone (in Russian), in: *Geology of the Far East*, Parfenov, L. M., ed., Khabarovsk: Institute of Tectonics and Geophysics, p. 132–141.

Rotman, V. K., 1965, The formation of green tuffs and some associated problems (in Russian), *Geol. Geofiz.*, v. 12, p. 54–62.

Rozhdestvenskiy, V. S., 1969, Faults of the Eastern Sakhalin Mountains (in Russian), *Dokl. Akad. Nauk USSR*, v. 187(1), p. 156–390.

Rusakov, I. M., and Vinogradov, V. A., 1969, Eugeosynclinal and miogeosynclinal areas of the northeastern USSR (in Russian), *Scientific Papers of the Scientific Research Institute of Arctic Geology*, v. 15, pp. 5–27.

Scholl, D. W., Buffington, E. C., and Marlow, M. S., 1975, Plate tectonics and the structural evolution of the Aleutian-Bering Sea region, in: *Geology of the Bering Sea and Adjacent Regions*, Geol. Soc. Am. Spec. Pap. 151.

Seslavinskiy, K. B., 1973, Volcanism in the development of the northern part of the Omolon Massif (in Russian), in: *Evolution of Volcanism in the History of the Earth*, Moscow, p. 107–109.

Shapiro, M. N., and Seliverstov, V. A., 1975, Morphology and age of folded structures of Eastern Kamchatka at the latitude of the Kronotskiy Peninsula (in Russian), *Geotektonika*, v. 4, 85–94.

Sharkovskiy, M. B., 1975, The tectonics of the Kolyma–Indigirka (in Russian), *Geotektonika*, v. 6, p. 44–60.

Shilo, N. A., and Merzlyakov, V. M., 1972, Eugeosynclinal zones of the central part of the mesozoids in northeastern USSR (in Russian), *Dokl. Akad. Nauk USSR*, v. 204(5), p. 1202–1206.

Shilo, N. A., Gel'man, M. L., Merzlyakov, V. M., Terekhov, M. I., and Til'man, S. M., 1973, A new zone of glaucophane metamorphism in the Pacific belt (in Russian), *Dokl. Akad. Nauk USSR*, v. 213(6), p. 1385–1388.

Shilo, N. A., Beliy, V. G., and Sidorov, A. A., 1974, Volcanogenic belts of eastern Asia as applied to the problems of tectonics, magmatism, and metallogeny (in Russian), *Geol. Geofiz.*, v. 5, p 70–88.

Shimaraev, V. N., 1971, Tectonic structure of the eastern part of the Sea of Okhotsk (in Russian), in: *Geophysical Research Methods in the Arctic*, Leningrad, p. 55–57.

Snegovskoi, S. S., 1974, *Investigations by the Reflection Seismic Method and Tectonics of the Southern Part of the Sea of Okhotsk and the Adjacent Pacific Margin* (in Russian), Novosibirsk: Nauka, 88 p.

Strel'tsov, M. I., 1971, The horizontal component of crustal movements in the southern part of the Kurile island arc (in Russian), *Dokl. Akad. Nauk USSR*, v. 196(2), p. 425–428.

Sutygin, G. N., and Tibilov, I. V., 1969, Deep-sea displacements of Central Chukotsk and some problems of metallogeny (in Russian), in: *Mesozoic Tectogenesis*, Bogolepov, K. V., and Til'man, S. M., eds., Magadan: Northeast Interdisciplinary Science Research Institute, p. 100–101.

Suvorov, A. L., 1975, The deep structure of the earth's crust in the South Okhotsk sector, according to seismic date (in Russian), Novosibirsk: Nauka, 103 p.

Tectonic map of Japan, 1965, scale 1:2,000,000, distribution of elementary tectonic constituents. Geological Survey of Japan.

Til'man, S. M., Byalobzheskiy, S. G., Chekhov, A. D., and Krasniy, L. L., 1975, Specific features in the formation of the continental crust in the northeastern USSR (in Russian), *Geotektonika*, v. 6, p. 15–29.

Tuezov, I. K., Krasniy, M. L., Vasil'ev, B. I., Kulikov, A. A., and Mikhailov, V. I., 1975, Geological structure of the southern member of the Kurile island arc (in Russian), *Geol. Geofiz.*, v. 12, p. 63–72.

Ustiev, E. K., 1959, The Okhotsk tectonomagmatic belt and some related problems (in Russian), *Sov. Geol.* v. 3, p. 3–26.

Ustiev, E. K., 1965, Composition of primary magmas as exemplified by Cretaceous and paleogene formations of the Okhotsk Volcanic Belt (in Russian), *Dokl. Akad. Nauk USSR Ser. Geol.*, v. 3, p. 3–19.

Vinogradov, V. A., Gaponenko, G. I., Rusakov, I. M., and Shimaraev, V. N., 1974, *Tectonics of the East Arctic Shelf of the USSR* (in Russian), Leningrad: Nauka, 144 p.

Vlasov, G. M. ed., 1964, *Geology of the USSR*, v. XXXI: *Kamchatka, Kurile, and Kommandorskiy Islands*, part I (in Russian), Moscow: Nedra, 730 p.

Voinova, I. P., 1975, Magmatic formations of the Uda Volcanic Belt (in Russian), in: *Problems of Magmatism and Tectonics of the Far East*, Parfenov, L. M., and Popeko, V. A., eds., Vladivostok: Institute of Tectonics and Geophysics, p. 179–189.

Zabroidin, V. N., and Turbin, M. T., 1970, Large faults in the western part of the Dzhagdin Range (in Russian), *Geotektonika*, v. 3, p. 104–114.

Zagruzina, I. A., Gorbov, V. V., and Shnai, G. K., 1967, Geochronology of magmatic and ore-bearing formations of the Northeast (in Russian), in: *Proceedings of the Northeast Complex Research Institute*, Magadan: Siberian Branch, Academy of Sciences, USSR, v. 30, p. 69–80.

Zimin, S. S., 1973, Paragenesis of ophiolites and upper mantle (in Russian), Moscow: Nauka, 251 p.

Chapter 9

THE TECTONIC ZONES OF NORTHEASTERN USSR AND THE FORMATION OF ITS CONTINENTAL CRUST

N. A. Shilo and S. M. Til'man

Northeast USSR Scientific Research Institute
Siberian Division, Academy of Sciences of the USSR
Magadan, USSR

I. INTRODUCTION

Geosynclinal development has been envisioned as a complicated and many-stage process similar to that of the transformation of oceanic-type crust into continental crust (Peive, 1969; Peive *et al.*, 1971, 1972) and takes place both in primary and in secondary or regenerated eugeosynclines. Three main stages of eugeosynclinal development—oceanic, transitional, and continental—are recognized by their characteristic sets of magmatic and sedimentary formations. The oceanic stage is often, but not always, preceded by graben formation (Dewey and Bird, 1971). The grabens appear when geosynclines are formed in rift zones evolving within preexisting continental crust. Graben-stage rock complexes include mottled shallow-water conglomerates of arkosic composition and trachybasaltic, trachyandesitic, rhyolitic, and dacitic lavas and tuffs. They are sometimes associated with tholeiitic basalts and diabases. Together they clearly point to the processes of crushing, melting, and reworking of the sialic crust.

The oceanic-stage rock complexes are composed of basalts and spilites, turbidites, siliceous shale, chert, and inorganic limestones, characteristically

413

deep-water formations. Ordinarily, they occur stratigraphically above a melanocratic basement, the later being represented by ultramafic rocks, layered gabbro, plagiogranite, and metamorphic rocks. Together with oceanic basalts, these formations make up an ophiolite whose individual members, particularly plagiogranites and metamorphic rocks, have not been adequately identified and are discussed below.

The transitional stage is recognized by the first appearance and subsequent wide development of underwater or sometimes subaerial volcanic rocks of predominantly intermediate composition, thick slate and flysch layers, graywacke, chert, limestone, and dolomite formed in shallow water and (or) island arc conditions. The transitional stage is the, longest one and may embrace several geological periods. Its termination is related to the formation of sedimentary and volcanic analogues of lower molasse.

Indicative of the continental crust are upper molassic complexes and subsequent volcanic rocks and granitic intrusions with higher than normal K_2O content, the latter two being stratigraphically equivalent to upper molasse. All these formations occur together with the folded structure formed earlier, which has an independent tectonic framework.

However, the development of geosynclinal systems is not simply a change from one main stage to another since prolonged oceanic crust transformation in different parts of the region is attended by repeated formation of the granitic–metamorphic layer, whose characteristic rock complexes are formations developed during island arc uplift; lower molasse, and gabbro–plagiogranitic and granitic intrusions. For example, the regions with a late Cretaceous continential crust and the areas with a granitic–metamorphic layer are recognized as having been formed during the mid or late Paleozoic, at the end of the Jurassic, or at the beginning of the Cretaceous. In regions where continental crust formation is still in progress, there are areas with a granitic–metamorphic layer that was formed during late Senonian, Maestrichtian, Paleogene, and Neogene time.

The areas of earlier granitic–metamorphic-layer rocks appear as uplifts, serving as sites for the accumulation of geanticlinal formations. Sedimentary and volcanic analogues of lower molasse are formed in the more extensive areas of the later-developed granitic–metamorphic layer. The geanticlinal uplifts are characterized by predominantly gabbro–plagiogranitic magmatic development, while the later intrusions are typified by a wider range of com position including normal granite.

The eugeosynclinal changes, as well as the growth of the enclosed granitic–metamorphic layer, are attended by folding and faulting, including intrusions of mantle material by thrusting into upper structural levels. These events occurred repeatedly depending on specific developments in the individual zones. As a result, complexly foliated nappe structures, melange, and olistostromes develop in the eugeosynclines.

Tectonic processes proceed differently in miogeosynclines. These are shelf formations with thick carbonate and terrigenous deposits. Little or no volcanism accompanied the geosynclinal development. What little there was depended on magmatic processes occurring in adjacent eugeosynclines. It seems natural that no stages reflecting crustal evolution in eugeosynclines can be recognized in miogeosynclines. Within them, if one excludes secondary rifting, the crust remained essentially continental, even though it might undergo some thinning during sedimentary events or thickening during orogeny.

The Soviet Northeast territory contains various tectonic zones (Til'man *et al.*, 1974) within which continental crust formation took place at different times (Til'man *et al.*, 1975) (Fig. 1). This paper concentrates on where and how this process developed and what geological events accompanied it.

II. SIBERIAN PLATFORM AND MIOGEOSYNCLINAL ZONES

Pre-Riphean continental crust forms the basement of the Yana–Kolyma and Anyui–Chukotsk Miogeosynclinal Systems, the Arctic Belt, the median massifs, and the eastern edge of the Siberian Platform. This crust's development may be traced through an analysis of the basement structure underlying the Siberian Platform. The basement is exposed in the Anabar Massif and the Aldan Shield and is represented by strongly metamorphosed Archean and Proterozoic formations. In the Archean section of the Anabar Massif, according to B. G. Lutz (1964) and M. I. Rabkin (1968), the lowest stratigraphic position is occupied by the Daldy Series, which includes igneous rocks of mainly basaltic composition. Next follows the Anabar–Lomui Series, generated as a result of deep metamorphic processes that transformed andesite–dacite lavas and subordinate tuffaceous, terrigenous, and carbonate rocks. The Upper Khaichan Series comprises paragneisses and schists formed from sandy and argillaceous limestone. This thick complex (nearly 20 km thick) underwent metamorphic changes to granulite and amphibolite facies. It underwent folding and complex fracturing along a persistant WNW strike. The Khaichan Series includes gabbro sills and numerous intersecting granitic massifs.

Frequent facies change is characteristic of various parts of the Archean rocks in the Aldan Shield (Salop, 1968). It is thought that the Zverev Series, composed entirely of metaultramafic and mafic rocks, is the oldest member within the Archean Aldan Shield. The Iengr Series, which correlates with the Timpton and Dzheltulin Series, is located higher in the section. Quartzites

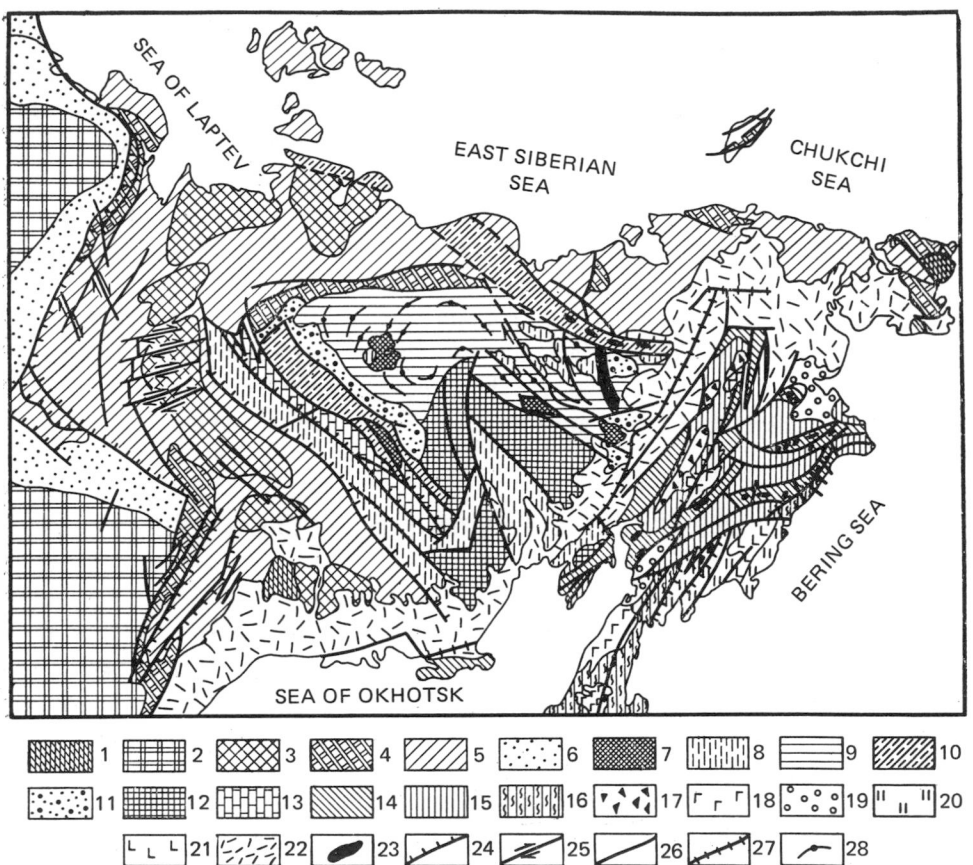

Fig. 1. Tectonic diagram of the northeastern USSR Continental crust of pre-Riphean age: (1) folded pre-Riphean basement. Complexes of subsequent developmental stages of pre-Riphean crust: (2) Siberian Platform cover; (3) gently deformed Verkhoyansk Complex; (4) miogeosynclinal folded Paleozoic deposits $(O-C_1)$; (5) folded Verkhoyansk Complex (C_2-J_2); (6) molasse of the pre-Verkhoyansk Trough (J_3-K_1). Continental crust of late Mesozoic age: (7) complexes of graben and oceanic stages (Pz); (8) rock complexes of the transitional stage $(T-J_2)$; (9) lower sedimentary and volcanic molasse (J_3-K_1); (10) late Mesozoic rifts (J_3-K_1); (11) upper molasse. Blocks of pre-Riphean continental crust within eugeosynclines: (12) blocks of gently deformed cover (R–J); (13) blocks of folded miogeosynclinal complexes (O–T). Regions with unfinished development of the continental crust; granitic and metamorphic layer of Oligocene–Miocene age (Koryak Upland): (14) complexes of oceanic and transitional stages $(Pz-K_1)$; (15) complexes of early transitional stage (K_1-K_2); (16) complexes of late transitional stage (K_2D-P_{1-2}); (17) molasse $(P-N_1)$; (18) plateau basalts (N–Q); (19) neotectonic depressions. Crust of unfinished consolidation: (20) oceanic and transitional complexes (K_2Sn-N_1); (21) plateau basalts (N–Q). Okhotsk–Chukotsk Volcanic Belt: (22) volcano-plutonic associations (K_1Al-K_2Cm) indicating the Mesozoic continental crust complexes. Other symbols: (23) melange zones, ultramafic inclusions, outcrops of mafic basement; (24) thrusts; (25) strike-slip faults; (26) undifferentiated fractures; (27) boundaries between inner and outer zones of the Okhotsk–Chukotsk Belt; (28) axes of aeromagnetic anomalies.

occur in the Iengr Series and are interbedded in places with gneisses and schists formed from basalt and basaltic tuff. In the Timpton and Dzheltulin Series, the rocks are predominantly of volcanic origin (mainly basalt).

The deformation of the Archean rocks within the Aldan Shield differs from that in the Anabar Massif. The rocks form a system of cupolalike uplifts and oval-shaped depressions, divided by narrow zones of tightly compressed folds. The Archean rocks exhibit a variable strike—from sublatitudinal to submeridional—thus reflecting the presence of circular structures (Sudovikov et al., 1962; Salop, 1968). The Aldanian Archean also contains ancient intrusive formations, including alaskite granites; however, there is less granitization of strata here than in the Anabar Massif.

Many age dates on gneisses and schists from the Anabar Massif and the Aldan Shield represent minimum ages because of metamorphic episodes (1750–2600 m.y.), and only a few determinations fall into the probable true age bracket of 2980–3900 ± 300 m.y. In the southern Aldan Shield, rock samples taken from strata equivalent to the Zverev Series yielded an age date of more than 4000 m.y. (Glukhovskiy and Pavlovskiy, 1973).

The above-mentioned data can be interpreted as follows. The lithostructural complex, corresponding to the Zverev Series of the Aldan Shield, resembles the magmatic basement of oceanic crust. Its counterpart is unknown in the Anabar Massif. If it is true that the abundant quartzites of the Iengr Series represent chemical sediments formed as a result of the disintegration of rocks of basic composition (Frolova, 1951), then the whole sequence of the Aldan Shield appears to be a complex derived from volcanogenic chert, graywacke, and other less common terrigenous formations. Considering the range of basaltic rocks in the section, down to and including the Dzheltulin Series, there is a firm foundation for assuming that the Aldan Shield has undergone only one Archean "proto-oceanic" stage of evolution, which probably differed from the Phanerozoic stage. A "protobasaltic layer" of primary crust marked this very long-lasting stage.

A different kind of picture may be reconstructed for the Anabar area. A type of geosynclinal process can be recognized here since the "proto-oceanic stage" was replaced by the "prototransitional" stage in which the primary sedimentary–volcanic layer formed.

Although different tectonic processes occurred in various portions of the Siberian Platform, they were never followed immediately by an orogenic stage during Archean time, as there is no evidence of any development of corresponding structural elements, formations, and magmatic complexes. It is inferred that toward the end of the Archean, the continental crust had not yet been formed in the Verkhoyansk–Chukotsk region or in the area occupied by the Siberian Platform. That event occurred later in the Proterozoic (2600–1750 m.y. an interval), to which post-Archean rejuvenation is confined.

The following direct and indirect data indicate the presence of continental crust as the basement of the Yana–Kolyma and Anyui–Chukotsk folded systems and the Arctic Belt.

The Archean rocks of the Okhotsk and Omolon Massifs, the Avekov block, and the Chukotsk Uplift are identical in composition to those of the Aldan Complex. The rocks generally include schist and orthogneisses formed from basic volcanic rocks and gabbro. They were subjected to metamorphism of granulite and amphibolite facies, magmatization, and subsequent less pronounced low-temperature silica–potassium metasomatism; they include granitic intrusions, the rubble of which is located in Riphean conglomerates.

In the Uchur–Maya area, the Gonam Suite of Lower Riphean age contains a gap and disconcordantly overlies the Ulkan Complex of mid-Proterozoic molasse, subsequent volcanic rocks of intermediate and acidic composition, and subvolcanic intrusions with an absolute age of 1600 m.y.

The Riphean rock sequences of the Uchur–Maya area, Okhotsk and Omolon Massifs, pre-Kolyma Uplift, and other Soviet northeastern areas generally include quartzitic, arkosic, and glauconitic sandstones formed from the disintegration of granitized crystalline basement in massifs—ancient interior uplifts that were subsequently submerged—and also from parts of the Siberian Shield. The facies in miogeosynclinal troughs were subject to change; the sandstone layers acquired the slight rounding of clastic material.

During pre-Riphean time, a folded zone, with features indicative of completed geosynclinal development, existed in the Arctic Foldbelt. Evidence of this zone may be seen in some structural features in the Lower and mid-Proterozoic sections of the Taimyr Peninsula, on Great Lyakhovsky and Wrangel Islands, and also in eastern Chukotka and northern Alaska. In the Arctic region, the folded Proterozoic belts "cemented" the Archean blocks. This led to the formation of a very stable structure here, characterized by typically continental, strongly granitized crust, which differed from the crust of the more southerly areas of the Pacific province. This is the reason why a reduction in the intensity of Mesozoic folding subsequently developed here (Til'man, 1973).

Further evidence that a pre-Riphean continental crust exists beneath the Verkhoyansk–Chukotsk region miogeosynclines is as follows: the lithologic content of Riphean–Paleozoic and Mesozoic troughs (carbonate and terrigenous deposits with insignificant volcanics), the germanotype tectonics, their structural position (located adjacent to ancient platforms), their crustal granitic magmatism, and their almost exclusively sialic metallogenesis.

III. ALAZEIA–OLOI SYSTEM

The Alazeia–Oloi System belongs to a group of tectonic elements with a late Mesozoic continental crust. It is situated in the central part of the Soviet

Northeast (Fig. 1) and consists of the Alazeia Zone in the west and the Oloi Zone in the east (Fig. 2).

The tectonic elements adjacent to the Alazeia and Oloi Zones (e.g., pre-Kolyma, Omulev and Ulakhans–Siss Uplifts, Omolon Massif) that overlie continental crust were subject to repeated transformations as revealed by their intensive disintegration, and by the type of Paleozoic and Mesozoic intrusive magmatism and subaerial and subaqueous lava eruptions.

It is known now that the breakup and reworking of the pre-Riphean continental crust affected the inner areas of the Alazeia–Oloi System to an even greater degree. Within the Oloi Zone, only small fragments of the ancient continental crust are preserved. These include, first of all, the Orlovkin block situated in the northern part of the zone on the right bank of the Great Anyui

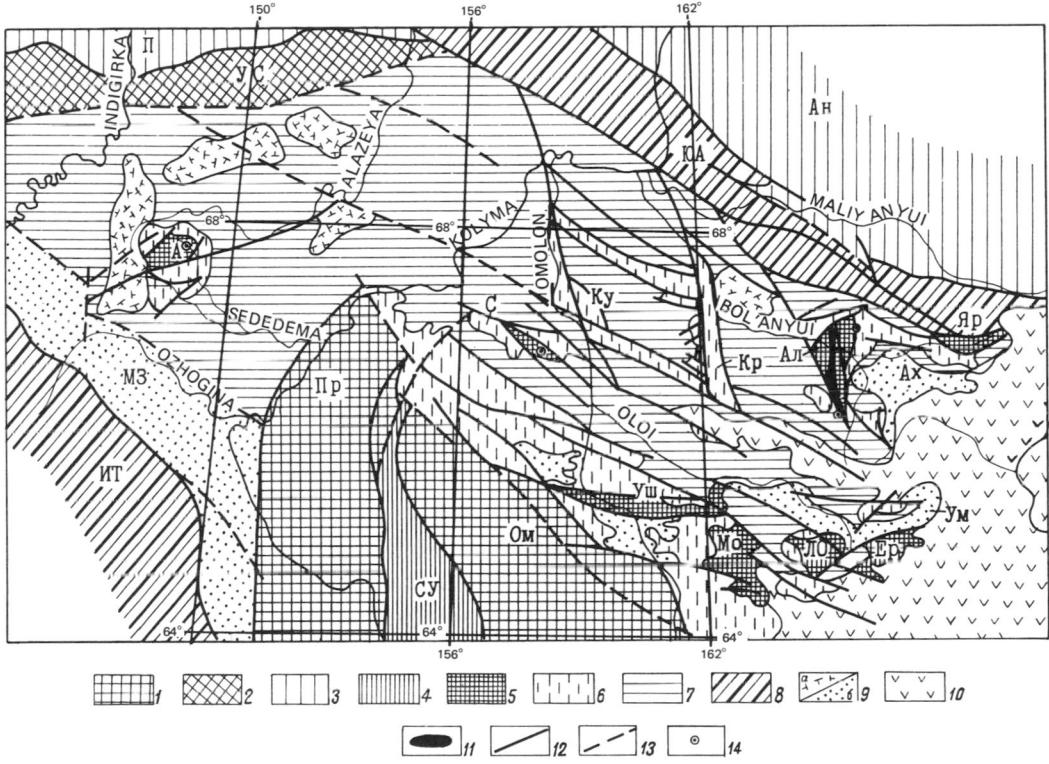

Fig. 2. Tectonic diagram of the Alazeia–Oloi system. Structural framework with original or slightly reworked pre-Riphean continental crust: (1) blocks with gently deformed cover: Omolon Massif (OM) and pre-Kolyma Uplift (IIp); (2) Ulakhans–Siss Uplift of Paleozoic beds (Cy); (3) miogeosynclinal folded zones: Polousniy (II) and Anyui (Ah); (4) Sugoi Trough (Cy) in thinned continental crust. Alazeia-Oloi System: (5) uplifts of Paleozoic rocks: Alazeia (A), Aluchin (Aц), Left Oloi (ЛО), Eropol (Ep); (6) distribution of early and middle Mesozoic rocks (T–J₂): Krichalskaya (Kp), Kur'ya (Ky), and others; (7) sedimentary and volcanic analogues of lower molasse (J₃–K₁Ap); (8) late Mesozoic continental rifts: South Anyui (ЮA) and Ilin'–Tass (ИТ); (9) upper molasse: (a) mainly volcanic and (b) mainly sedimentary; (10) volcanic nappes of Okhotsk–Chukotsk Belt; (11) Aluchin Ultramific Massif: (12) known fractures; (13) inferred fractures; (14) outcrops of metamorphic rocks, including glaucophane-schist facies. Scale: 1 : 5,000,000.

River. It is composed of metamorphic rocks; quartz–mica schist, mica–quartz–albite schist, garnet–biotite–quartz schist, epidote–quartz–actinolite schist, quartzite, and marble, apparently of Proterozoic age Radzivill, 1964; Dovgal' *et al.*, 1966). Another fragmentary remnant, known as the Siver Uplift, lies in the southern half of the Oloi Zone. Metamorphic rocks are exposed here together with gently dipping Middle Devonian volcanic and sedimentary formations, including rhyolite flows, tuffs, and seams and layers of marmorized limestone.

In the rest of the Oloi Zone, where wedges, blocks, and narrow bands of Middle Paleozoic rocks are exposed, the sections exhibit a great variety of structural composition. For example, the Devonian strata of the Ushurakchan Uplift include lavas and tuffs of basic, intermediate, and acidic composition, while trachybasalts and trachyandesites with bands of cherty rocks and brecciated limestone occur at the base of the section. In the Lower Carboniferous, carbonate–gypsum rocks developed as well as beds of flyschoid sandstone, siltstones, and slate (Simakov, 1974). Near the Omolon Massif at Molandzhin and some other uplifts, Devonian sections are predominantly rhyolite and rhyodacite, alternating with less common red carbonates and terrigenous and jasperlike rocks. A sequence of Devonian and Lower Carboniferous rocks, up to 2000 m thick, is exposed at the divide of the Oloi and Eropol Rivers (Left Oloi Uplift). In its lower part, the sequence includes tuffaceous gritstone and sandstone with bands of mottled rhyodacite flows and tuffs. Next come cherty shales, rhyodacite and felsite, and tuffs with interbeds of marmorized limestone (Til'man, 1962). The strata here dip monoclinally at an angle of 40–50° and are cut by Paleozoic granitic rocks.

In the southeastern part of the Oloi Zone (the Eropol River Basin), Devonian volcanic–sedimentary sequences change composition to andesite, spilite, and basalt. Cherty rocks begin to play a more important role here and the thickness of the section increases and is complicated by folds.

In the Yarakvaam Uplift of the Oloi Zone, a volcanic section is exposed consisting of deformed basalt, spilite, keratophyre, and tuffs, which include beds of graywacke and chert. These rocks are discordantly overlain by coastal-marine terrigenous and volcanic-terrigenous deposits with a Carboniferous–Permian fauna and flora.

A similar situation exists in the Alazeia Uplift where the basement of the Paleozoic section consists of the Kenkel'din Sequence, including graywacke, andesitic and basaltic tuff, cherty rocks with lenses of jasper, and argillaceous limestone. The age of the sequence is uncertain but because it is concordantly overlain by Lower Carboniferous deposits, it is probably of Devonian age. Glaucophane schist adjoins this sequence (Lichagin *et al.*, 1975). According to Lichagin *et al.*, the next higher sequence with a thickness of more than 1000 m consists of graywacke, cherty rocks, and andesitic–basaltic tuff with lenses of conglomerate. It is characterized by Lower Carboniferous brachiopods and bryozoans.

Thus, there are three different types of Middle Paleozoic sections within the Alazeia and Oloi Zones. The first type is characteristic of the blocks with a pre-Riphean continental crust without any significant reworking (Orlovkin block, Siver Uplift). The second type of section is characteristic of the so-called graben stage, during which the sialic crust was fractured by continental rifts (Ushurakchan, Molandzhin, Left Oloi Uplifts). The third type indicates a deep crustal exposure causing the formation of rock complexes resembling oceanic ones (Eropol, Yarakvaam, Alazeia Uplifts).

Upper Paleozoic volcanic–terrigenous sequences are widely distributed. Their facies are not sustained laterally, and their deposits represent varied sources. It appears, moreover, that some relationship can be observed between the Upper Paleozoic and the preceding Middle Paleozoic tectonic conditions. In places where the formation of oceanic-type rock complexes occurred during the Middle Paleozoic, coastal-marine and continental deposits accumulated during the Upper Paleozoic. On the other hand, in other places where graben facies formed during the Middle Paleozoic, further crustal exposure occurred in the Upper Paleozoic. In places, this exposure penetrated down to the mantle, and conditions were set up for the accumulation of either an oceanic complex or a nonvolcanic geosyncline.

For example, in the Yarakvaam Uplift, the Middle Paleozoic basalts, spilites, and keratophyres were replaced by the coastal-marine tuffaceous–terrigenous and molasselike deposits of the Upper Paleozoic, consisting of conglomerates, breccias, cross-bedded tuffaceous sandstone, coarse-grained tuff, and graywacke.

The Upper Paleozoic strata of the Eropol Uplift and adjacent areas are found in isolated blocks and include tuff, graywacke, sandy limestone, and conglomerate. The total thickness of the Upper Paleozoic deposits in these areas is estimated at only a few hundred meters.

The Upper Paleozoic complex in the Alazeia Uplift is composed mainly of gritstone, graywacke, and tuff of variable composition. In the Upper Permian section, acidic volcanic rocks play a significant part (Lichagin *et al.*, 1975). The Upper Paleozoic conglomerates are widely distributed throughout the section and contain pebbles of gabbro, plagiogranite, diorite, and volcanic rocks of various composition as well as sedimentary and pyroclastic rocks. These Upper Paleozoic deposits are cut and metamorphosed by dikes and minor gabbro–plagiogranitic intrusions.

It is possible to classify the rock complexes described as geanticlinal formations that fix the early development of the granitic–metamorphic layer in some parts of the Oloi and Alazeia Zones on one hand and, on the other hand, reflect the first appearance of uplift within the eugeosyncline. In the Yarakvaam, Eropol, and Alazeia Uplifts the most pronounced stratigraphic gaps and structural unconformities occurred during the Lower and part of the Middle Triassic.

A typical section of the Upper Paleozoic oceanic complex occurs on the right bank of the Great Anyui River, opposite the mouth of the Aluchin River. In an "ultramafic–gabbro layer" with unclear stratigraphic relationships, a 400-m sequence of basalt, spilite, diabase, keratophyre, tuff, and cherty and terrigenous rocks with limestone lenses and interlayers, a Middle to Upper Carboniferous fauna is found in the siltstone and limestone (Dovgal', 1964). This rock sequence is unconformably overlain by Triassic deposits. The whole range of oceanic rocks, including ultramafics, gabbro, and plagiogranite, is found in the clasts making up the conglomerates of Upper Triassic (Norian) age.

The oceanic complex is located in close spatial association with the ultramafic Aluchin Massif, in which hazburgites, peridotites, pyroxenites, and other ultramafic rock varieties are recognized. According to O. P. Timofeiev's data (unpublished), there is a narrow band of glaucophane schist formed from ultramafic rocks at the southern tip of the Aluchin Massif in the Teleneut River Basin.

The ultramafic Aluchin Massif together with the oceanic complex, described above, makes up an ophiolite association that is confined to the Oloi Zone. There are also data indicating that a mafic complex with glaucophane schists occurs in the Badyarikhin block of the Alazeia Zone (Lichagin *et al.*, 1975). However, its stratigraphic and structural position is unknown.

There is some question whether the Aluchin ophiolite association is a fragment of the primary oceanic crust or if its development was the result of a deep rift that breached the ancient continental crust and underlying mantle. The question may be answered by comparing the characteristics of the Oloi Zone and the Koryak Upland. The fact that the Aluchin ultramafic and oceanic complexes are in the vicinity of the Orloykin sialic block deserves attention because this association does not occur anywhere in the Koryak Upland. Within the Koryak Upland, outcrops of metamorphic rocks in the Penzhina Ridge and the Vayezh and Iomraut blocks are fragments of the malanocratic basement (Markov, 1975). The orientation of the ultramafic massif is discordant with the folds of the Oloi Zone, while in the Koryak Upland, all the melange zones and ultramafics are concordant with the imbri-cate-nappe structure (Alexandrov *et al.*, 1975). Ultramafics in the Aluchin Massif, particularly harzburgites, differ from the corresponding rocks of the Koryak Upland by a sharp decrease in K_2O content; thus they belong to the series of inner-continental Alpine-type ultramafics and not to the oceanic or oceanic margin ultramafics that occur in the Koryak Upland. These data indicate that the Aluchin Ophiolitic association formed under secondary, less deep, and lower-energy conditions than those of the Koryak ophiolite. It is possible to assume that the mantle underlying the Oloi Zone acquired its properties because it had already depleted itself in forming the pre-Riphean crust.

In the Alazeia and Oloi Zones, sections of Upper Paleozoic deposits appear that consist of marine geosynclinal formations: volcanics, cherts, terrigenous sediments, graywacke, and flysch (Shilo *et al.*, 1973). They are not like typical oceanic formations and apparently were formed on the continental slope or rise. Similar complexes occur in the upper reaches of the Oloi River and at the edge of the Alazeia Zone in the Ziryanka River Basin (Merzlyakov, 1971) where they form thick sequences that in places are intensly deformed and faulted. These complexes differ from geanticlinal formations in having distinctly and thinly layered sections without stratigraphic breaks. During late Paleozoic time in the Alazeia and Oloi Zones and sometimes beyond their borders, complex paleotectonic conditions existed that were characterized by simultaneous formations of land and sea uplifts, rift valleys, and narrow geosynclinal troughs.

Triassic marine transgressions covered wide expanses of northeastern USSR. In the Verkhoyansk–Chukotsk region at that time, the formation of the Verkhoyansk (Upper Yak) Complex was continuing, and, within the zones under discussion, a regime characteristic of the mature or transitional stage of eugeosynclinal evolution developed. The base of the Upper and part of the Middle Triassic deposits are represented by tuffaceous terrigenous formations, separated in places by layers of basaltic and andesitic lavas. On the geanticlinal uplifts in the Triassic sections, coarse-layered tuffaceous rocks, siltstones, slates, and volcanic sandstones predominate. They show comparatively gentle dislocation and form large-scale monoclines with complex flexure folds. They are distinguished by relatively small thicknesses.

For example, in the Yarakvaam Uplift at the base of the Triassic section, conglomerates and sandstone of the Upper Ladinian stage occur with a thickness of 100–120 m. They are concordantly overlain by a 170-m-thick band of volcanic sandstone, siltstone, and tuff of Karnian age. Norian deposits include similar rocks, as well as rare beds of basic lavas. They contain abundant brachiopods, echinoids, bryozoans, crinoids, and mollusks, including ammonoids and nautiloids. Norian deposits are as much as 500–550 m thick. Both the taxonomic composition and the ecologic specialization of this diverse fauna clearly testify to the existence of a direct relationship between the local marine basin and the world ocean system.

In the Aluchin Uplift, the Triassic rocks consist of a 100-m band of complexly intercalated sandstone, siltstone, gritstone, tuff, and basic lavas of Ladinian-Karnian age (Afitskiy, 1970). Isolated blocks of Norian siltstone, tuff, and intraformational conglomerates also occur.

In the Alazeia Uplift, the Karnian stage rocks occur in the Sededema River basin and its right-hand tributaries. They overlie the Upper Paleozoic deposits and are deeply eroded. They include tuffaceous sandstone, tuffaceous gritstone, and intermediate and basic lavas and tuffs. Their thickenss is 250–270 m. Norian beds, including tuffaceous and volcaniclastic sandstones and tuffaceous siltstones, with coquinoid limestone interlayers and lenses, concordantly overlie the Karnian deposits. Tuffs of intermediate composition, tuffaceous conglomerate, and tuffaceous gritstone occur in lesser amounts. The thickness of Norian deposits is 450–500 m (Gulyayev, 1975).

In basinal areas, the rock composition changes and the thickness of the Triassic deposits sharply increases; graywacke beds with flyschoid rhythmic bedding appear. Lava flows, basaltic and andesitic tuffs, and slates occur at the base of the sections. The intensity of faulting markedly increases. Such Triassic sections are characteristic of the Topolevka, Baimke, Krichalskaya, Innakh, and other areas of the central Oloi Zone (Gorodinskiy et al., 1974).

From exposures along the Bannaya, Svetlaya, and Topolevka Rivers, Afitskiy (1970) has compiled the following section (from the bottom upward):

1.   Pyroxene basalt, andesite basalt and tuff, intercalated with siltstone, sandstone, gritstone, and conglomerate. In the sedimentary rocks, *Halobia* and *Monotis* occur sporadically. Thickness is 500–600 m.

2.   Polymictic sandstone and gritstone, tuffaceous conglomerate, tuffaceous sandstone, and tuffaceous siltstone with *Monotis*. Thickness is 400 m.

3.   Graywacke and polymictic sandstone, gritstone, and conglomerate, tuffaceous siltstone, tuff, and breccia, and tuffaceous conglomerate. There are rare Norian pelecypods, nautiloids, and bryozoans in the sandstone. Thickness is 400–500 m.

4.   Amygdaloidal basalt, andesitic basalt, andesite and andesitic tuffs, rare beds of dacite and rhyolite, volcaniclastic flysch, tuffaceous gritstone with rare fossils of *Monotis*. Thickness is 150–200 m.

The total thickness of the Norian section here is at least 1700 m. The rocks are contorted into complex folds of various shapes from boxlike to isoclinal.

There are differences between the Lower Jurassic deposits developed in geanticlinal uplifts and those of downwarped areas. They are mainly composed of flysch in the downwarps. A. I. Afitskiy and V. I. Sizikh recognized this same kind of deposit in Hettangian rocks along the Krichalskaya River. Here rhythmic flysch is formed of volcaniclastic material, and in this respect, the sequence differs from the flyschoid sequences of the Anyui–Chukotsk Zone.

In eugeosynclinal uplifts, the Lower Jurassic flysch is poorly developed or totally absent; if absent, it is replaced by terrigenous and graywacke-like strata, or by tuff, as in the Yarakvaam and Alazeia areas. In the interfluve of the Egelyakh and Sededema Rivers, for example, siltstones and tuffaceous siltstones are exposed and show interbedded moderately brecciated tuff and conglomeratic sandstone with bands of small pebble conglomerate, tuff, and tuffaceous breccia. A Lower Jurassic fauna has been found in these rocks.

Middle Jurassic deposits occur in the Alazeia Zone and in the central and eastern portions of the Oloi Zone. Often at the base of the Middle Jurassic strata, there are thick beds (ranging from 80 to 300 m) of polymictic basal conglomerate and gritstone including ultramafic fragments in the Oloi Zone. These sections are characterized by an obvious increase in grain size; sandstone and siltstone predominate and flysch disappears almost completely. Tuffaceous material is dominant in the Middle Jurassic sections of geanticlinal uplifts, while terrigenous rocks formed in the downwarped areas.

In the eastern Oloi Zone, marine deposits bearing *Inoceramus* fossils make up a thick sequence (about 1500–1600 m) that transgressively overlaps the underlying strata. The age range of these deposits is Toarcian through early Callovian. Conglomerates occur at the base of the rock sequence (60–80 m) and are overlain by siltstones and sandstones that include tuff beds, gritstone, and graywacke. The Middle Jurassic in the Alazeia Uplift consists predominantly of volcaniclastic and tuffaceous sandstones of intermediate- to coarse-grain size, with interlayers of argillite, siltstone, tuff, conglomerate, and gritstone. The early and mid-Mesozoic deposits in the Alazeia and Oloi Zones present various complexes that range from molasse in the uplifts to flysch and basic volcanics in the troughs. The diverse composition, variable facies, changes in thickness, and different types of deformation all indicate that these deposits were formed under widely different tectonic conditions, characterized by island arcs separated by wide and narrow depressions.

In both zones, lower molasse comprises a thick volcanic-terrigenous complex of Oxfordian–Kimmeridgian–Neocomian age. The complex usually lies discordantly on the underlying beds and only in isolated areas is there a gradational contact. This sequence is composed of various rock types: basalt, andesite, rhyolite and tuff, conglomerate, gritstone, and sedimentary breccia, with interbeds of volcaniclastic sandstone mostly of marine origin but less commonly including some of continental origin. On the other hand, the upper half of the Lower Cretaceous is represented predominantly by continental coal-bearing deposits and volcanics, except in the Ainakhkurgen and Umkuveyem Depressions of the Oloi Zone, which have marine molasse deposits of Hauterivian, Barremian, and Aptian age (Til'man, 1962; Afitskiy and Palymskiy, 1971).

The rocks of the lower molasse complex occupy large areas in both zones; they have been tectonically deformed into troughs and downwarps, and they also include fields, belts, and isolated beds of lava. The rocks of the complex are far less deformed when compared with the underlying beds and they form large-scale brachyaxial folds and (or) buckle folds. The lower molasse is associated in time with the development of the gabbro–monzonite–syenite and diorite–granodiorite intrusions that make up the Oloi, Egdekgich, and other complexes.

The upper molasse also occurs within the zones under consideration and is of Albian–Cenomanian age. It is most completely developed in the Ziryanka Trough, where it occurs as continental coal-bearing deposits. In the vicinity of the Alazeia Uplift, there are separate sheets of acidic, intermediate, and basic lavas lying discordantly on the lower molasse and more ancient deposits. The Chimchememel' Suite from the Ainakhkurgen and Umkuveyem Depressions corresponds to upper molasse and includes conglomerate, coarse-grained sandstone, tuff, and lava of different composition together with coal. This suite overlies lower marine molasse and is replaced laterally by volcanic rocks that have the same composition and age as the extrusives from the outer zone of the Okhotsk–Chukotsk Belt. Massifs, stocks, and dikes of leucocratic potassium granites in the Topolevka Complex formed at the same time that the sedimentary and volcanic analogues of upper molasse were accumulating.

IV. LATE MESOZOIC RIFTS

Rock sequences stratigraphically equivalent to the lower molasse complex, but geosynclinal in nature, are involved in the structure of the South Anyui and Ilin'–Tass Continental Rifts. The South Anyui Rift lies on the northeastern border of the Oloi Zone, while the Ilin'–Tass Rift marks the southwestern border of the Alazeia Zone.

The South Anyui Rift stretches for about 1250 km from the upper reaches of the Great and Minor Anyui Rivers to the Novosibirsk Isles and its average width is 80–100 km. The rift can be traced by direct geological observations to the lower reaches of the Kolyma River and from there can be identified by positive magnetic anomalies of high intensity (1000–2000 gammas). The rift's basement outcrops at the crest of South Anyui Ridge and consists of ultramafic bodies (numbering more than twenty), which occur above deep-seated faults (Pinus and Sterligova, 1973). The ultramafic and gabbroic rocks are closely associated with a chert and volcanic sequence (1300–1400 m thick), that includes layers of spilite, albitized diabase, tuff, and cherty rocks with subordinate beds of siltstone and tuffaceous sandstone. In the upper part of the section, quartz keratophyre and tuff appear. The presence of a *Buchia* fauna in the rock sequence indicates an age corresponding to the Volgian stage (Radzivill, 1964). Stratigraphically higher there is a rhythmic interstratification of polymictic sandstone, siltstone, and shale nearly 1000 m thick. A Berriasian–Valanginian fauna has been found in the section. These deposits gradually change into a sequence (about 600 m) of coarse-grained, cross-bed-ded sandstone, alternating with gritstone, conglomerate, siltstone, and coaly clayey slate. This sequence is of Hauterivian age. Thus, the rocks of the South Anyui Rift make up four sequences that successively replace each other in time: ophiolites, cherty volcanics, flysch, and molasse.

Similar sequences are deposited in the Ilin'–Tass Rift except that the presence of ophiolites has not yet been proved. This rift is more than 700 km in length and up to 100 km wide. According to L. L. Krasniy (1969, 1971), a vol-canogenic shale and chert formation dominates the lower part of the late Mesozoic section here. Basalt and andesite are among the volcanic rocks in the central portion of the pre-Indigirka Rift, in the northeast acidic lavas and tuffs appear, and on the southwestern flank of the rift, basalts occur again in the Garmichan Graben. The lower Garmichan and Ilin'–Tass Suites change into the Bastakh slate and flysch formations. Here the thickness of the geosynclinal complex totals nearly 6000 m. It is overlain by Lower Neocomian and Upper Albian–Cenomanian molasse.

The Ilin'–Tass Rift differs from the South Anyui Rift in the shallower depth of crustal exposure, but the central part of the rift still may lie upon a mafic substratum.

The aforementioned rifts appear as highly compressed abnormal anticli-noria. The crests of these structures are complicated by a system of long, nar-row anticlines that either strike parallel to each other (South Anyui Rift) or occur in an en echelon pattern (Ilin'–Tass Rift). The spaces between the anti-clinal structures and their limbs are occupied by flyschoid deposits, and a molasse complex fills the synclinelike depressions.

In cross section, the rocks of the geosynclinal complex form a fanlike structure with the axial planes of the folds tilted from the crests towards opposite sides of the anticlinoria. For example, in the southern part of the South Anyui Rift, the folds are tilted to the side of the Yarakvaam and Aluchin Uplifts, but in the southwestern part of the Ilin'–Tass Rift, the tilting is toward the In'yali–Debina Synclinorium. Deformational features, including inclined and isoclinal beds, complicate the flanks of the rifts. Symmetrical folds are predominant along their crests. The inclination of the axial planes nearly conforms with the inclination of the fault planes, which are pre-dominantly thrusts. The average dip of fold limbs and of faults is 40–50°.

The rifts listed are closely associated with the late Mesozoic volcanic belts as follows: The Ilin'–Tass Rift with the Uyandina–Yasachnaya Belt and the South Anyui Rift with the Northern Oloi Belt. These paired features are ancient island arcs. The pairs were formed as a consequence of crustal spread-ing. Apparently, such paired tectonic elements reflect a structure intrinsic to marginal seas.

V. PROBLEM OF THE KOLYMA MASSIF

Tracing the structure sections of Paleozoic and Mesozoic deposits in this large territory, it is impossible to ignore how regularly their composition changes through time from graben and oceanic complexes, through volcanic, chert, terrigenous flysch, and graywacke formations, to lower and upper molasse. However, this sequence was disrupted with the creation of the late Paleozoic geanticlinal uplifts and the late Mesozoic continental rifts. The geanticlinal uplifts existed simultaneously with the geosynclinal troughs and the continental rifts with the superimposed depressions filled with lower molasse. Thereby, the Alazeia and Oloi Zones that display these phases of development are considered to be tectonic structures of the eugeosynclinal type.

At the same time, the Alazeia–Oloi System differs from formal eugeosynclinal systems in a number of specific features. First of all, its tectonic style is unusual; it does not resemble the folded structure of the Verkhoyansk region, and it cannot be correlated with the imbricated nappe structure of the Koryak Upland. Among the very extensive areas of Upper Jurassic–Lower Cretaceous deposits there are scattered narrow outcrop bands, fracture-related horsts, and individual blocks of Paleozoic and Lower Mesozoic rocks and their basements. The fractures include thrusts and normal faults, which control intrusive–extrusive belts, zones of hydrothermally altered rocks bearing gold sulfide, mercury, and polymetallic mineralization, and also relatively large bodies of ultramafics associated with platinum mineralization. On the whole, the Alazeia and Oloi Zones exhibit a mosaic pattern of structural blocks.

Besides these tectonic patterns, the composition of the fill of the early and late troughs also appears to be unique. The following features are distinctive: varied composition of the Paleozoic volcanic rocks, the graben facies at the base of sections, appreciable volcaniclastic rocks and graywacke, and rapid facies changes of sequences up section and laterally. If we also add the diverse intrusive magmatism—from mantle to crustal type—and the specific character of the ultramafics, then little doubt remains concerning the secondary nature of the Alazeia–Oloi System, which formed within the body of Verkhoyansk–Chukotsk continental block.

The Alazeia Zone, a continuation of the Oloi Zone, occupies a considerable part of the Kolyma Massif. Analysis of the data indicates that within the Kolyma–Indigirka interfluve, throughout a long geologic history, a fragment of pre-Riphean sialic crust existed, disintegrated, and, in places, was almost completely transformed into a secondary eugeosynclinal zone that developed beginning in the Devonian and in some areas in the Ordovician (the Rassoshir Zone of the Omulevka Uplift). If our foregoing correlation of the tectonics of the Alazeia and Oloi Zones is valid, then a likely corollary seems to be that an extensive bay of a marginal sea was located in this area through the Lower

Cretaceous: the bay opened eastward toward the Koryak–Kamchatka Basin. Cross sections and stratigraphic columns in Fig. 3 illustrate the structure of uplifts and troughs of the Alazeia and Oloi Zones.

VI. KONI–TAIGONOS ZONE

The continental crust was formed in the Koni–Taigonos Zone simultaneously, or almost simultaneously, with the Alazeia–Oloi System. The Koni–Taigonos Zones occupies the coast of the Okhotsk Sea and Gizhigin Bay and extends northeasterly along the Anadyr–Kolyma divide; it dies out in the Central Chukotsk area. The total length of the zone is about 1500 km. Melanocratic basement and the oceanic-type structural complexes have been identified at only two points. On the Koni Peninsula along thrusts inclined to the north, in some places a banded gabbro–ultramafic complex appears which may be assumed to be a mantle fragment, according to R. B. Umitbayev. Among Permian–Mesozoic rocks on the Taigonos Peninsula, there is a single block (possibly an allochthonous plate) composed of metabasalts and metadiabases with jasper and limestone interlayers and apparently of Lower Carboniferous age.

The transitional eugeosynclinal complex, ranging from Namurian to Neocomian, is very thick in the Koni–Taigonos Zone (up to 13 km). It includes terrigenous sediments, graywacke, cherty volcanics, and less commonly volcanic formations. It also contains gabbro–plagiogranite intrusions and ultramafic bodies in some areas. Geosynclinal deposits that are very intensely deformed and diversely folded occur, but they have a linear orientation and strike in a northeast direction as part of the general Pacific structural framework.

The available data indicate that the Koni–Taigonos Zone, during the epoch of geosynclinal formation, was an elongated island arc with predominantly andesitic volcanism. It was situated at the edge of the Verkhoyansk–Chukotsk continental block and also served as the southeastern border of the Oloi Zone, whose downwarps and uplifts were controlled by transform faults and interior rifts. Beyond the Koni–Taigonos Zone stretched the great Koryak–Chukotsk Oceanic Basin.

VII. OKHOTSK–CHUKOTSK VOLCANIC BELT

Over a considerable area, the structures of the Koni–Taigonos Zone are discordantly overlain by volcanic rocks of the Okhotsk–Chukotsk Belt. According to recent data of V. F. Beliy (1971) and other geologists, this belt

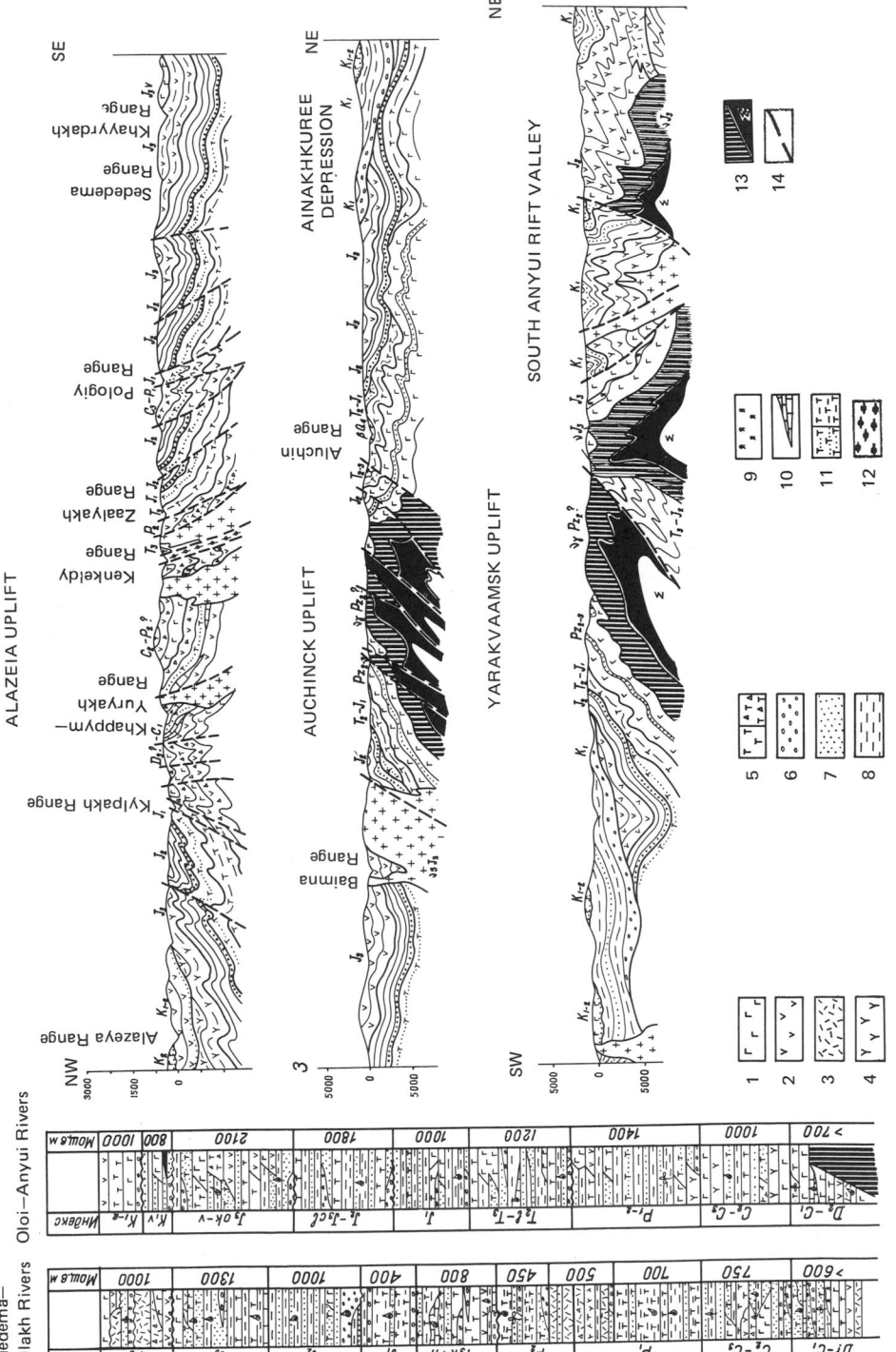

Fig. 3. Cross sections and stratigraphic columns of Alazeia and Oloi Zones. (1) Basalt; (2) mainly andesite; (3) extrusives of mixed composition; (4) extrusives of acidic composition; (5) tuff and tuffaceous breccia; (6) conglomerates; (7) sandstone; (8) siltstone; (9) jasper; (10) limestones; (11) tuffaceous sandstone and tuffaceous siltstone; (12) glaucophane schist; (13) mafic basement; (14) faults.

began forming during Albian time. More recent volcanic rocks in the belt have been dated as Cenomanian on the basis of plant remains. At the end of the Cretaceous and the beginning of the Paleogene, there were sporadic outbursts of volcanic activity; however, the lavas were superimposed on the already existing structural belt.

Contrasting assemblages make up the volcanic formations, including basalts, andesite–basalts and andesites, ignimbrites, rhyolites, and less commonly dacites. They compose a great many suites whose correlation urgently needs more work.

The belt is distinctly zoned. An outer zone can be clearly traced near the belt of Mesozoides, whereas next to the Koryak–Kamchatka region there is an inner zone. The belt's flanks are evident in the Okhotsk and Bering Sea areas where its typical features die out. The boundary between the belt's outer and inner zones, in the overwhelming majority of cases, is represented by a network of closely spaced fractures, sometimes in an en echelon pattern. The adjacent fractures of various kinds (faults, displacements, and thrusts), which make up this boundary, range in width from hundreds of meters to 2.5–3 km. All features indicate that the boundary corresponds to a continuous deep-seated fracture, clearly appearing in some places as topographic relief.

The structural differences between the inner and outer zones of the Okhotsk–Chukotsk Belt are as follows. In the outer zone, ignimbrite rock sequences are widely distributed, particularly in central Chukotka. The stratigraphic sections of the other areas (mid-Anady, Penzhina, and Okhotsk) reveal complex combinations of basic, intermediate, and acidic rocks. The thickness of these rocks may total up to 2500–3000 m. The outer zone is characterized by a mosaic structure. The equant and, less commonly, linear, volcanic depressions divided by uplifts (volcanic-cupola structures) are widely distributed here. The latter are remobilized uplifts, formed during the intrusion process. On the whole, the outer zone volcanics are faintly deformed or in places appear practically monoclinal. In the outer zone, the intensity, morphology, and anomaly orientation of the magnetic fields are identical to the fields of the Mesozoides.

The volcanic rocks of the inner zone are mainly andesite–basalts, and only in the uppermost sections of Cenomanian age are there some ignimbrites and tuffs of acidic and intermediate composition. A sharp increase in the thickness of the rock sequences is worth mentioning and is especially prominent in the grabenlike volcanic–tectonic structures, which are up to 6000 m thick. Deformation of the volcanic strata of the inner zone is far more complicated. Major anticlinal uplifts and adjacent grabenlike volcanic–tectonic depressions are present. At the crests of the uplifts, there are some exposures of the belt's basement rocks. The limbs of the uplifts and depressions are complex, with discontinuous folds of various sequences. The character of the magnetic field changes in the inner zone. Here there are very large-scale and sharp magnetic anomalies with northeasterly strike intersecting the field of the outer zone.

Granitic intrusions occur almost everywhere in the Okhotsk–Chukotsk Belt, except for the Central Chukotsk area, but there are also considerable differences between the granitic rocks of the outer and inner zones of the belt. Epizonal intrusions of granodioritic–granitic rocks are predominant in the outer zone, while the inner zone features mesozonal quartz diorites, tonalities, granodiorites, and granites. The inner zone rocks are characterized by much higher sodium content and the $K_2O:Na_2O$ ratio (percent by weight) varies from 1:4 to 1:2. In the outer zone, the alkaline ratio is about 1.

Sodium-rich granitic rocks are indigenous not only to the inner zone of the belt. They also appear in the orogenic intrusions of the Anadyr–Koryak System and in the eugeosynclinal zones of the Mesozoides of Eastern Asia. They obviously emphasize the differences in crustal type between areas adjacent to the Pacific Ocean and away from it.

Volcanic manifestations in the belt correspond to many features of the orogenic magmatism in the folded eugeosynclinal regions. Volcanic activity followed the termination of geosynclinal development in the Koni–Taigonos Zone and was localized right at the edge of the emerging late Mesozoic continent. Fracturing, instrumental in the infiltration of subaerial lavas, possibly was connected to an oceanic plate moving under a continental plate or the "collision" of two different crustal blocks. Such features as the lateral transition of the belt's volcanic rocks to the upper molasse of the Oloi Zone, the contrasting lava compositions, and the vigorous granitic magmatism, with accompanying hydrothermal metasomatism, all indicate that the continental stage began in Albian time in the Koni–Taigonos Zone and the neighboring Alazeia–Oloi System. By the beginning of late Cretaceous time, a mature continental crust had been formed in places (the outer zone of the belt and the Alazeia–Oloi System), while in other areas (the inner zone of the belt), the crust still was not completely "continentalized."

VIII. TECTONIC NATURE OF THE INYALI–DEBINA SYNCLINORIUM

In considering how the continental crust was formed in the northeastern USSR, the tectonic nature of the Inyali–Debina Synclinorium assumes major importance. We are inclined to think that the syclinorium formed between mid- and late Paleozoic time upon a thinned continental crust that had formed as a result of the stretching of a pre-Riphean continental block. The following facts support this idea.

The Mesozoic structural series of the synclinorium, unlike all the other Yana–Kolyma System structures, are clearly differentiated (Chekhov, 1971). Slate and flysch formations, characteristic of the transitional stages of eugeosynclinal development, were deposited here. In addition, Triassic and Jurassic rock sequences often contain isolated thin tuffs of intermediate and basic composition, as well as fragments of basalts, most probably of local origin.

It is possible that the Mesozoic strata of the Inyali–Debina Synclinorium are underlain by cherty–volcanic–terrigenous formations of Upper Paleozoic age since, at the southwestern tip of the Taskan Zone of the Omulevka Uplift, directly bordering the Inyali–Debina Synclinorium, there is an exposure of Middle and Upper Carboniferous and Permian rocks. They include slates and siltstones, cherty rocks, tuffs, and lavas of basic and intermediate composition (Merzlyakov, 1971).

The magnetic field of the synclinorium seems very unique and unlike the fields of the typical miogeosynclinal zones (Verkhoyansk, Taimyr, Anyui–Chukotsk, Millard Belt, and others). Linear reversed anomalies of high intensity stretch along the whole synclinorium in a northwest direction, reflecting the presence of complicating structures at depth, which may be the rocks of a metamorphic and melanocratic basement but also could be a body of basic and ultrabasic composition.* There are some grounds for this supposition. Within the granites of the Negyakh Massif, there are zenoliths of basic migmatized slates, while rounded, probably xenolithic inclusions of pyroxene-amphibole rocks have been found in the granites of the Mayak Massif. Both can be regarded as fragments of a metamorphic, melanocratic complex lying at the base of the Inyali–Debina Synclinorium. A minor fault, which controls the rocks of an ultramafic–gabbro association, parallels the folded structures along the western edge of the Mayak granitic pluton. According to V. A. Serebryakov (1966), banded gabbro, peridotite, pyroxenite, and serpentinous rocks are also found. This fact and others indicate the deep-seated nature of the synclinorium.

The final stage in continental crust formation in the Inyali–Debina Synclinorium corresponds with the epoch of Kolyma folding, which was accompanied by the development of molasse and by extensive granitic magmatism.

Rift zones of a similar type, formed on a thinned continental crust, are not rare in the northeastern USSR. The pre-Omolon and possibly the south Verkhoyansk Zones and the Kolyuchin–Mechigmen Zone (Chukotsk) can be numbered among them.

---

* There is another point of view concerning the nature of these anomalies. It has been suggested that they are related to extensive sulfide zones that are mirrored by these very anomalies everywhere in the Inyali–Debina Synclinorium.

IX. THE KORYAK UPLAND

In the east, the Mesozoide-type rocks in the Okhotsk–Chukotsk Volcanic Belt border the Koryak–Kamchtka geosynclinal folded region where continental crust is still forming. The principal structural features of the Koryak Upland have been published (Alexandrov *et al.*, 1975). According to new data, various compositional and structural complexes can be recognized in the area. These correspond to a mafic basement in its oceanic, transitional, and early orogenic stages. These complexes either overlap each other or replace each other laterally. Because of widely developed nappes many structural details still remain unknown, especially since the area is still in need of much more investigation.

The oceanic complexes have different ages in different regions. On the Penzhina Ridge, these formations, which overlie an "ultramafic–gabbro layer," were dated as Silurian; in the Ust-Belaya area, they were identified as Middle Paleozoic; and in the Khatyrka Zone, they are Upper Paleozoic. Where the mafic basement and the oceanic suite are exposed, the rocks are allochthonous. The oceanic formations are either basalts or rocks of spilite–diabase composition, which alternate with chert, jasper, and, less commonly, limestone. In these areas, volcanic–terrigenous rocks (graywacke, schist, sandstone, lava, and andesitic tuff) of a wide range of ages (early Mesozoic through mid-Albian) occur stratigraphically above the Paleozoic formations. The sequence varies in thickness, ranging from 6000 to 10,000 m. This complex is referred to the transitional type and often contains ultramafic bodies.

In other areas of the Koryak Upland, oceanic and transitional complexes include only late Mesozoic (Upper Jurassic-Senonian) rocks of various types including basalt and spilite–basalt, keratophyre, radiolarite, chert, graywacke, flysch, turbitite, slate, andesite, and terrigenous tuff. In the Koryak Upland, some analogous formations are the Kingiveyem, Inas'kvaam, Koiverelan, and other suites and sequences of Upper Jurassic–Valanginian age. They all are distributed in narrow, elongated zones containing ultramafic tabular bodies, among which are serpentinized rocks in tectonic contact with nonvolcanic (terrigenous) rocks of the same or younger age. It is not clear whether they contain a section of an older oceanic complex or begin a new tectonic cycle created by repeated rifting. The last assumption is more probable, as shown by observations in the Yagel River basin. Some suites and sequences of Senonian age developed in the Algan, Vanetat–Velikorechie, and other zones and possess exactly the same characteristics of composition and structure.

In the Penzhina–Anadyr and Koryak Zones, the continental stage is not as pronounced as it is in the eugeosynclinal systems of the Mesozoides. It is possible only to state the different periods of their stablization, marked by the

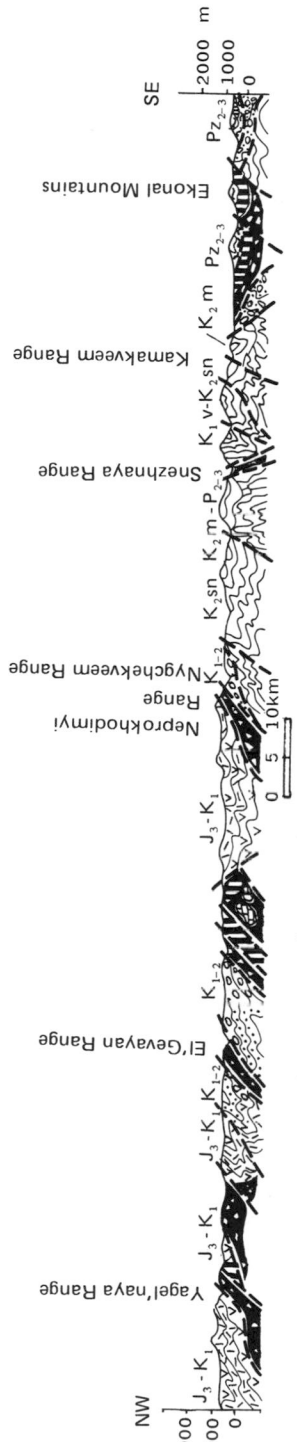

Fig. 4. Geological section across the strike of the northeastern Koryak Upland (Alexandrov et al., 1975). (1) Volcanic cherty formation (Pz_{2-3}); (2) volcanic-terrigenous deposits (J_3-K_1); (3) terrigenous deposits (J_3-P); (4) olistostrome formations (K_{1-2}); (5) marmorized limestone; (6) gabbro and plagiogranite; (7) serpentinite melange; (8) thrusts.

development of lower molasse: in the first case, the period is Danian–
Paleogene, and in the second, Oligocene–Miocene. These periods corresponded
to the formation of the Penzhina, Raritkin, and Opukha–Pekul'nei molassic
troughs, fissure eruptions of basic and acadic lava, and ultramafic recurrent
intrusions.

The established tectonic characteristics of the Penzhina–Anadyr and
Koryak Zones are: numerous bands of serpentinite melange of different types,
tectonic nappes, long active thrusts and dislocations that were accompanied by
the formation of olistostromes (Fig. 4).

The Olyutor Zone, south of the Koryak Zone, consists of the Vatin
Series, the Achaivayam, Vochvin, Goven, Il'pin, and several other suites and
sequences which form a thick (more than 16 km) geosynclinal complex ranging
from Senonian to Lower Miocene. The complex joins formations from both
the oceanic and transitional stages; spilite, basalt, and jasper dominate in the
section, except in the Il'pin suite.

Judging from the composition of the deposits filling the Olyutor Trough,
as well as from its structural position, it can be inferred that this zone
developed on primary oceanic crust; apparently it now is preserved in a stage
transitional to continental crust.

X. CONCLUSIONS

1. New data on the tectonics of the northeastern USSR quite convincingly support the theory that by Riphean time two large-scale blocks had
developed—the continental Verkhoyansk–Chukotsk block and the oceanic
Anadyr–Koryak block. The boundary between them had Atlantic-type fracture
zones and was located approximately along the present day Okhotsk–
Chukotsk Volcanic Belt. The Verkhoyansk–Chukotsk continental block disintegrated into eugeosynclinal and miogeosynclinal zones by the mid and, in
places, early Paleozoic; and from that time on, two trends developed. On the
one hand, a buildup of the granitic–metamorphic layer took place with a
thickening of the Precambrian continental crust, and, on the other hand, the
ancient continental crust was destroyed and new crust was created by late
Mesozoic time. The Anadyr–Koryak block, with the primary oceanic crust at
its base, has had a special development. Continental-type crust has still not
formed here. Instead, we are dealing with a special type of crust whose
geologic nature does not completely fit the three-layer geophysical model.

2. At least three types of tectonic formation can be recognized in
northeastern Asia and evolved as a result of the destruction of pre-Riphean
continental crust: (a) mosaic-block geosynclinal systems of long development
due to a varied depth of crustal exposure (the Alazeia–Oloi System, for

example); (b) rapidly developing geosynclinal zones due to deep crustal exposure down to the mantle (South Anyui and possibly the Ilin'–Tass Continental Rifts); (c) rapidly developing geosynclinal zones on thinned-out continental crust (the Inyali–Debina Synclinorium, for example).

3. Previous concepts about the existence of the Kolyma Median Massif in the central Verkhoyansk–Chukotsk region must be revised in the light of new data. In its place, there is a secondary eugeosynclinal zone with a complex and heterogeneous tectonic structure.

4. With new eugeosynclinal zones identified, the rules governing the distribution of mineral deposits are interpreted differently. Deposits of mafic and sialic profile will have an equivalent economic significance. The gold-bearing spectrum of plutonic–volcanic series is appreciably expanded. As yet poorly mapped melange zones, which may be directly linked to gold–platinum and mercury mineralization, are of growing interest.

REFERENCES

Afitskiy, A. I., 1970, *Biostratigraphy of the Triassic and Jurassic deposits from the Greater Anyui Basin*, Moscow: Nauka, 144 pp.

Afitskiy, A. I., and Palymskiy, B. F., 1971, Tectonic evolution of Late Mesozoic marginal depressions of Chukotka, in: *Mesozoic Tectogenesis. Proceedings of Session VII of the Scientific Council for Tectonic Studies of Siberia and the Far East*, Magadan: Northeast Interdisciplinary Science Research Institute, p. 68–73.

Alexandrov, A. A., Bogdanov, N. A., Byalobzheskiy, S. G., Markov, M. S., Til'man, S. M., Khain, V. E., and Chekhov, A. D., 1975, New data on the tectonics of the Koryak Upland, *Geotektonika*, v. 5, p. 60–72.

Beliy, V. F., 1971, Main types of volcanic belts in regions of Mesozoic tectogenesis in Eastern Asia, in: *Mesozoic Tectogenesis. Proceedings of Session VII of the Scientific Council for Tectonic Studies of Siberia and the Far East*, Magadan: Northeast Interdisciplinary Science Research Institute, p. 168–177.

Chekhov, A. D., 1971, Comparative characteristics of the In'yali–Debin and Ol'doi Synclinoria, in: *Mesozoic Tectogenesis. Proceedings of Session VII of the Scientific Council for Tectonic Studies of Siberia and the Far East*, Magadan: Northeast Interdisciplinary Science Research Institute, p. 87–96.

Dewey, J. F., and Bird, J. M., 1971, Origin and emplacement of the ophiolite suite: Appalachian ophiolites in Newfoundland, *J. Geophys. Res.*, v. 76(14), p. 3179–3206.

Dovgal', Yu. M., Radzivill, A. Ya., Titov, V. V., and Chasovitin, M. D., 1966, The tectonics of the Oloi–Anyui interfluve, in: *Proceedings on Geology and Mineral Deposits of Northeastern U.S.S.R.*, v. 18, Magadan: Magadan, p. 41–55.

Frolova, N. V., 1951, Sedimentation during the Archean era, *Trans. Irkutsk State Univ.*, v. 5, *Geol. Ser.* (2), *Gosgeoizdat*, p. 38–68.

Glukhovskiy, M. Z., and Pavlovskiy, E. V., 1973, The problem of the early stages of Earth development, *Geotektonika*, v. 2, p. 3–7.

Gorodinskiy, M. E., Gulevich, V. V., Neznanov, N. N., Palymskiy, B. F., and Radzivill, A. Ya., 1974, The geology and metallogeny of the Anyui and Oloi interfluve, in: *Proceedings on Geology and Mineral Deposits of Northeastern U.S.S.R.*, v. 21, Magadan: Magadan, p. 31–41.

Gulyayev, P. V., 1975, The tectonics of the Alazeya Uplift, *Geotektonika*, v. 6, p. 30–43.

Krasniy, L. L., 1969, The tectonics of the Moma–Zyryanka Trough, in: *Mesozoic Tectogenesis*, Theses, Magadan: Northeast Interdisciplinary Science Research Institute, p. 72–73.

Krasniy, L. L., 1971, Tectonics of the Ilin'–Tass Zone, in: *Mesozoic Tectogenesis. Proceedings of Session VII of the Scientific Council for Tectonic Studies of Siberia and the Far East*, Magadan: Northeast Interdisciplinary Science Research Institute, p. 291–301.

Lutz, B. G., 1964, *Petrology of the Anabar Massif's Granulitic Facies*, Moscow: Nauka, 124 pp.

Lychagin, P. P., Merzlyakov, V. M., Ponomareva, L. G., Terekhov, M. T., and Khmel'nikova, O. S., 1975, Glaucophane metamorphism of the Alazeya Upland, in: *Proceedings on Geology and Mineral Deposits of Northeastern U.S.S.R.*, v. 22, Magadan: Magadan, p. 112–119.

Markov, M. S., 1975, *Metamorphic Complexes and the "Basaltic" Crustal Layer of the Island Arcs*, Moscow: Nauka, 232 pp.

Merzlyakov, V. M., 1971, *Stratigraphy and Tectonics of the Omulevka Uplift*, Moscow: Nauka, 151 pp.

Peive, A. V., 1969, Oceanic crust of the geologic past, *Geotektonika*, v. 4, p. 5–23.

Peive, A. V., Shtreis, N. A., Knipper, A. L., Markov, M. S., Bogdanov, N. A., Perfil'yev, A. S., and Ruzhentsev, S. V., 1971, Oceans and the geosynclinal process, *Dokl. Akad. Nauk SSSR*, v. 196(3), p. 657–659.

Peive, A. V., Shtreis, N. A., Mossakovskiy, A. A., Perfil'yev, A. S., Ruzhentsev, S. V., Bogdanov, N. A., Burtman, V. S., Knipper, A. L., Makarychev, G. I., Markov, M. S., and Suvorov, A. I., 1972, Eurasian Paleozoides and some problems of the geosynclinal evolutionary process, *Sov. Geol.*, v. 12, p. 7–25.

Pinus, G. V., and Sterligova, V. E., 1973, New belt of Alpine-type ultramafics in the Soviet Northeast and certain geologic peculiarities in the formation of ultramafic belts, *Geol. Geofiz.*, v. 12, p. 109–112.

Rabkin, M. I., 1968, The Archean of the Anabar Massif, in: *Geologic Structure of the USSR*, v. 1, Shatalov, E. I., ed.-in-chief, Zhamoida, A. I., ed., Moscow: Nedra, p. 92–94.

Radzivill, A. Ya., 1964, New data on geology of the southeastern section of Southern Anyui Ridge, *Proceedings on Geology and Mineral Deposits of Northeastern U.S.S.R.*, v. 17, Magadan: Magadan, p. 57–62.

Salop, L. I., 1968, The Archean Aldan Shield, in: *Geological Structure of the USSR*, v. 1, Shatalov, E. T., ed.-in-chief, Zhamoida, A. I., ed., p. 94–99.

Serebryakov, V. A., 1966, Geological structure of Negyakh granitic massif, in: *Proceedings on Geology and Mineral Deposits of Northeastern U.S.S.R.*, v. 18, Magadan: Magadan, p. 147–164.

Shilo, N. A., Merzlyakov, V. M., Terekhov, M. I., and Til'man, S. M., 1973, The Alazeia–Oloi Eugeosynclinal System—A new element of northeastern USSR Mesozoides, *Dokl. Akad. Nauk SSSR*, v. 210(5), p. 1174–1176.

Simakov, K. V., 1974, Stratigraphy and the formation history of the Mid-Paleozoic deposits on the Ushurakchan mountain range, in: *The Precambrian and Paleozoic of USSR*, Theses, Reports at Interdepartmental Stratigraphic Conference, Magadan: Northeast Interdisciplinary Science Research Institute, p. 172–175.

Sudovikov, N. G., Drugova, T. M., Krylova, M. D., and Mikhailov, D. A., 1962, Characteristics of the tectonic structure of the Archean formations in the Aldan mining area, *Izv. Akad. Nauk SSSR Ser. Geol.*, v. 11, p. 95–100.

Til'man, S. M., 1962, Tectonics and evolution history of the northeastern fore-Kolyma area, *Trans. NEISRI Sib. Br. USSR Acad. Sci.*, v. 1, 190 pp.

Til'man, S. M., 1973, Comparative tectonics of the Mesozoides of the northern Pacific Ring, *Trans. NEISRI FESC USSR Acad. Sci.*, v. 40, 325 pp.

Til'man, S. M., Mokshantsev, K. B., Shilo, N. A., Merzlyakov, V. M., and Gusev, G. S., 1974, Mesozoic folded zones of the northeastern USSR, in: *General and Regional Tectonic Problems of the Pacific Belt: Proceedings of the Conference on Tectonics of the Pacific Belt*, Magadan: Northeast Interdisciplinary Science Research Institute, p. 61–71.

Til'man, S. M., Byalobzheskiy, S. G., Chekhov, A. D., and Krasniy, L. L., 1975, Characteristics of continental crust formations in the northeastern USSR, *Geotektonika*, v. 6, p. 15–29.

Chapter 10

GEOLOGY AND PHYSIOGRAPHY OF THE CONTINENTAL MARGIN NORTH OF ALASKA AND IMPLICATIONS FOR THE ORIGIN OF THE CANADA BASIN

Arthur Grantz, Stephen Eittreim, and Olive T. Whitney

U.S. Geological Survey
Menlo Park, California

I. INTRODUCTION

The continental margin north of Alaska is of Atlantic type. It began to form probably in early Jurassic time but possibly in middle early Cretaceous time, when the oceanic Canada Basin of the Arctic Ocean is thought to have opened by rifting about a pole of rotation near the Mackenzie Delta. Offsets of the rift along two fracture zones are thought to have divided the Alaskan margin into three sectors of contrasting structure and stratigraphy. In the Barter Island sector on the east and the Chukchi sector on the west the rift was closer to the present northern Alaska mainland than in the Barrow sector, which lies between them. In the Barter Island and Chukchi sectors the continental shelf is underlain by prisms of clastic sedimentary rocks that are inferred to include thick sections of Jurassic and Neocomian (lower Lower Cretaceous) strata of southern provenance. In the intervening Barrow sector the shelf is underlain by relatively thin sections of Jurassic and Neocomian strata derived from northern sources that now lie beneath the outer continental shelf.

The rifted continental margin is overlain by a prograded prism of Albian (upper Lower Cretaceous) to Tertiary clastic sedimentary rocks that comprises the continental terrace of the western Beaufort and northern Chukchi Seas. On the south the prism is bounded by Barrow Arch, which is a hinge line between the northward-tilted basement surface beneath the continental shelf of the western Beaufort Sea and the southward-tilted Arctic Platform of northern Alaska.

The Arctic Platform is overlain by clastic shelf and carbonate strata of Mississippian to Cretaceous age, and by Jurassic and Cretaceous clastic strata of the Colville Foredeep. Both the Arctic Platform and Colville Foredeep sequences extend from northern Alaska beneath the northern Chukchi Sea. At Herald Fault Zone in the central Chukchi Sea they are overthrust by more strongly deformed Cretaceous to Paleozoic sedimentary rocks of Herald Arch, which trends northwest from Cape Lisburne. Hope Basin, an extensional intracontinental sedimentary basin of Tertiary age, underlies the Chukchi Sea south of Herald Arch.

This paper presents an overview of the physiography and geology of the Arctic continental margin of Alaska (Fig. 1) and discusses some implications of the data for the origin of this margin and the adjacent Canada Basin of the Arctic Ocean. Many of the interpretations presented are preliminary, and the paper does not attempt a comprehensive analysis of the features discussed. Some details can be found in Grantz *et al.* (1970*a*, 1975) and Eittreim *et al.* (1979). Our data are mainly single-channel seismic reflection and sonobuoy refraction profiles obtained between 1969 and 1973 from U.S. Coast Guard icebreakers and a few of the multichannel CDP seismic reflection profiles obtained from the U.S. Geological Survey R/V *S.P. LEE* in 1977 (Grantz *et al.*, 1970*b*, 1971, 1972*a,b*, 1974, 1975).

II. PHYSIOGRAPHY

A. Major Features

The rim of the Arctic Basin in the western Beaufort and Chukchi Seas (see Fig. 2) consists of the narrow Western Beaufort Shelf, the broad Chukchi Shelf, and the Chukchi Continental Borderland. These face the Canada Basin (Fig. 3), which is that part of the deep Arctic Ocean Basin that lies between the Lomonsov Ridge, eastern Siberia, and North America. The continental shelf in the western Beaufort and northern Chukchi Seas is a subsiding, prograding terrace dissected by submarine canyons. In contrast, the continental shelf in the central and southern Chukchi Sea is a submerged marine-cut platform eroded into the North American and Eurasian continents. Many of the

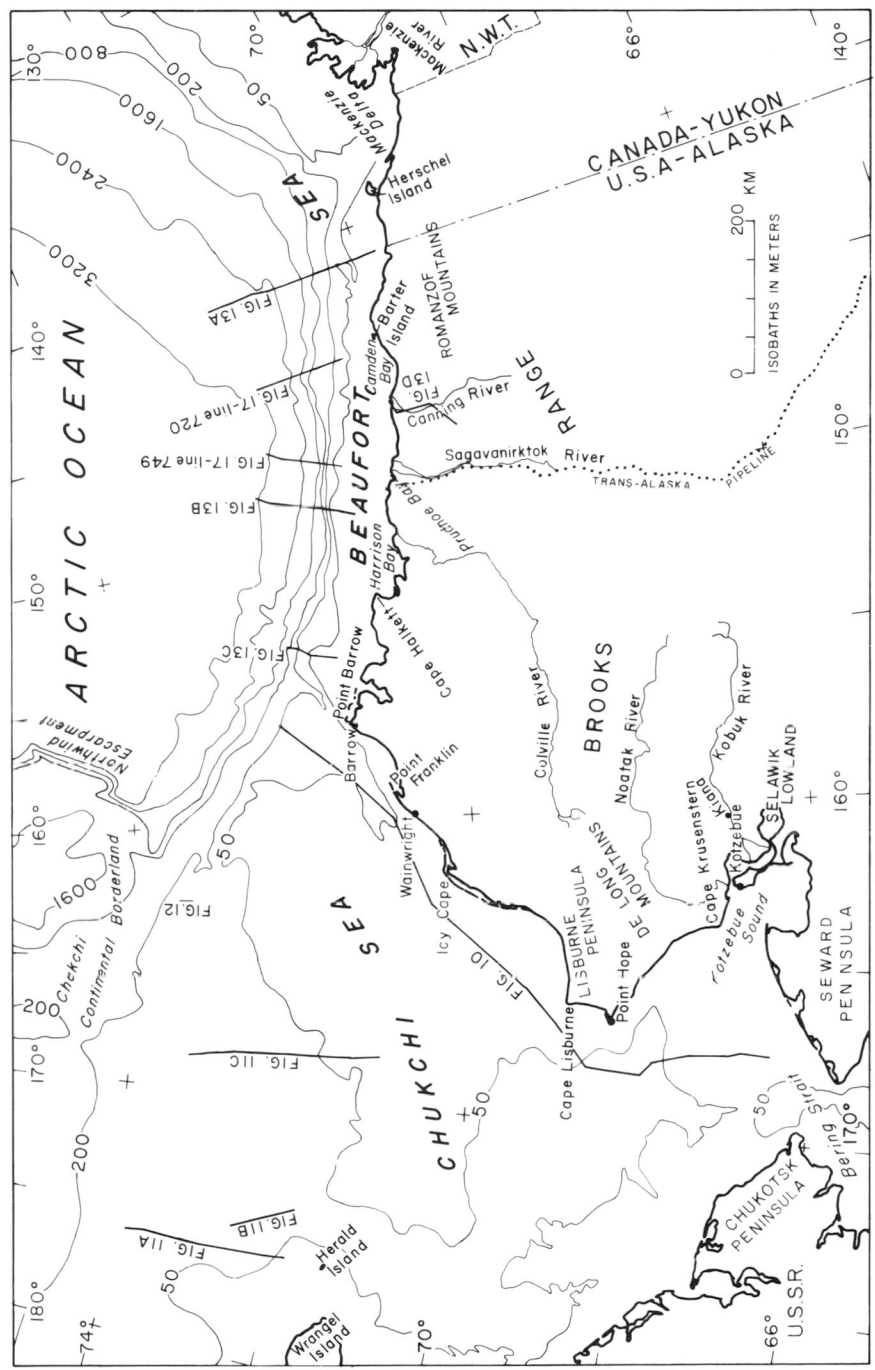

Fig. 1. Bathymetry and place names in the Beaufort and Chukchi Seas and vicinity and location of seismic profiles in Figs. 10, 11, 13, and 16.

Arthur Grantz *et al.*

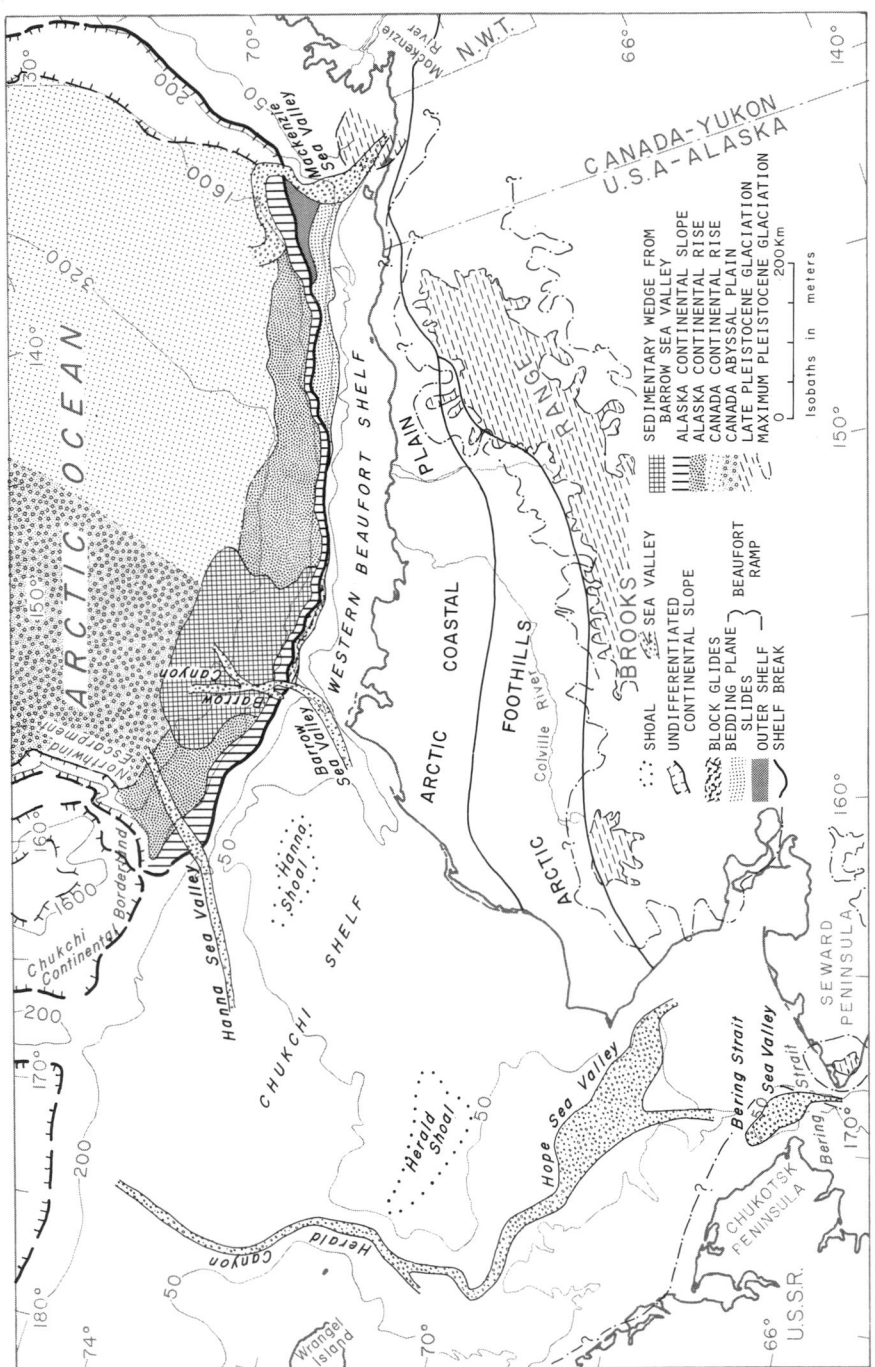

Fig. 2. Physiographic features of the Beaufort and Chukchi Seas. Extent of glaciation from Alaska Glacial Map Commitee (1965), Geological Survey of Canada (1968), Nelson and Hopkins (1972), and D. M. Hopkins (oral communication, 1978).

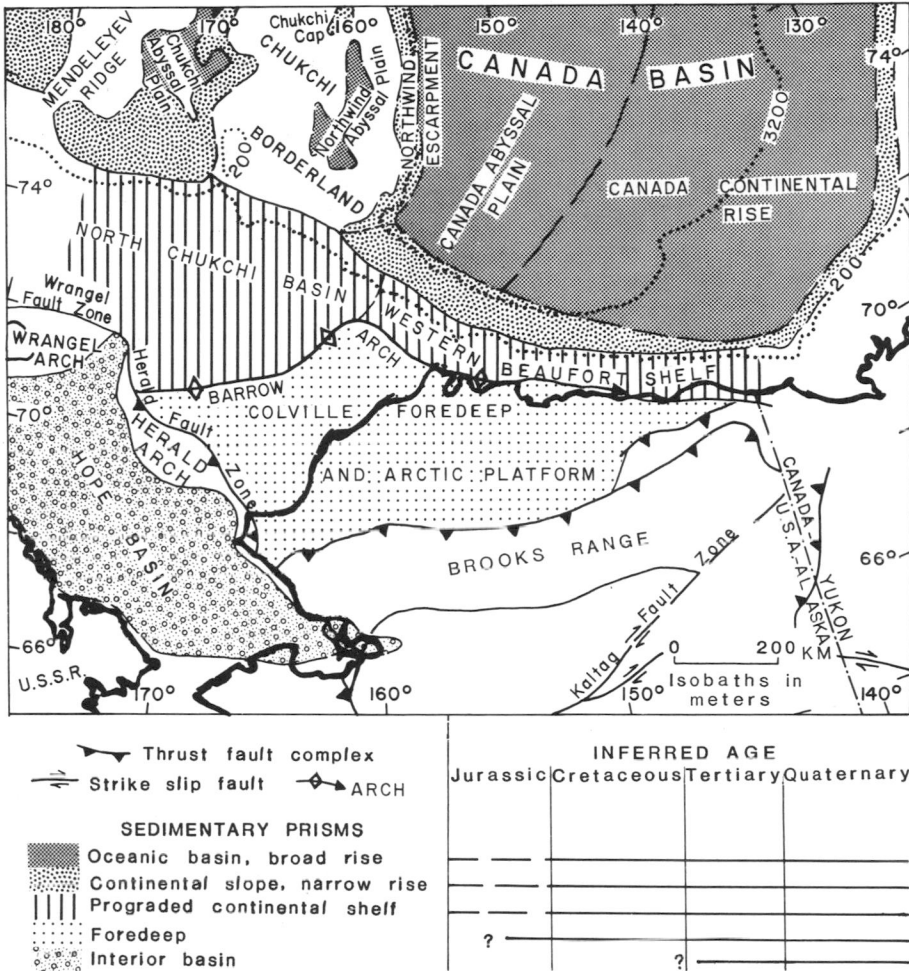

Fig. 3.   Post-Triassic tectonic provinces and sedimentary prisms of the Beaufort and Chukchi Seas and vicinity.

geologic formations and structures that underlie these shelves are coextensive with formations and structures in northern Alaska. The bathymetrically complex Chukchi Continental Borderland (Fig. 2), which juts into the Canada Basin, is inferred to contain fragments of continental crust. The southern Canada Basin, which lies adjacent to Alaska, is floored by an abyssal plain in the west and a continental rise built out from the Canadian Arctic Islands and Mackenzie River Delta in the east.

B. Western Beaufort Shelf

The relatively narrow continental shelf of the western Beaufort Sea extends between the Mackenzie and Barrow Sea Valleys, a distance of about 700 km (Figs. 1 and 2). The shelf is 70 to 120 km wide and extends seaward with a gentle gradient from barrier islands or low bluffs at the coast to a complex outer continental shelf break at depths of 200 to 800 m. East of 147° W longitude (Figs. 2 and 15) the shelf consists of an inner section with a gradient of about 1 m/km that extends from shore to a slump-controlled break in slope near the 60-m isobath, and an outer section with a gradient of about 16 m/km that extends from the 60-m isobath to the outer shelf break. The steeper outer section is here named the Beaufort Ramp.

The inner Beaufort Shelf west of the Colville River is basically a wave-cut surface, although predominantly fine-grained Quaternary clastic sediments, locally as much as 140 m thick, have been deposited upon it. Rocks as old as early Cretaceous are truncated at the base of the Quaternary cover. East of the Colville River, transportation and deposition of clastic sediment has been the dominant Neogene and Quaternary geologic process. Holocene sediment, dominantly lutite, has filled local depressions and smoothly prograded the continental shelf seaward. Sand and gravel occur in bars near shore, and in scattered accumulations across the entire shelf (Barnes and Reimnitz, 1974). The gravel must include dropstones from ice islands (tabular glacial icebergs from Ellesmere Island) that drift across the shelf.

Local relief on the Beaufort Shelf occurs at nearshore sand bars and ice gouges. Pleistocene glaciers (Fig. 2) apparently reached the shelf only east of Barter Island. The ice gouges are generally parallel to the coast and extend from nearshore to at least the 75-m isobath (Reimnitz and Barnes, 1974). Those related to the present stand of sea level are abundant to water depths of 30 m. The gouges are produced by grounding of deep-draft sea-ice pressure ridge and ice islands that are driven westward along the shelf by a wind-driven clockwise current, the Beaufort (Pacific) gyre of the Canada Basin. The deepest ice gouge measured north of Alaska is 5.5 m (Reimnitz and Barnes, 1974), but most are 1 to 3 m deep.

C. Chukchi Shelf

The Chukchi Shelf (Fig. 2) lies between Bering Strait on the south, Barrow Sea Valley on the east, and the longitude of Wrangel Island (180°) on the west. This broad shelf was eroded far into the continent and transects the principal mountain ranges of northern Alaska. Its north–south and east–west dimensions are both about 900 km. Although the self is characterized by low bathymetric gradients, underlying geologic control of its major physiographic features is evident. Hope Sea Valley–Herald Canyon, the largest sea valley system on the shelf, overlies Tertiary basins, and Herald and Hanna Shoals overlie structural highs that expose Paleozoic and Mesozoic rocks at the seabed. The shallow southern Chukchi Sea and Kotzebue Sound owe their shape and extent to Quaternary marine abrasion in the soft Tertiary strata of Hope Basin (Fig. 3).

Bottom sediment in the Chukchi Sea is dominantly silt and very fine sand, although sand and gravel occur near shore and on shoals. The sand and gravel on shoals may be lag deposits winnowed by ice gouging and currents (Toimil and Grantz, 1976). Gravel in the northern Chukchi Sea also includes glacial erratics dropped from ice islands. The Holocene sediment blanket on the Chukchi Shelf ranges in thickness from 0 to 12 m (Creager and McManus, 1967). On the southern Chukchi Shelf, much of this young sediment was swept northward from the Yukon River in the Alaskan Coastal Water, which flows north through Bering Strait (Nelson and Creager, 1977). Other sources are the Noatak and Kobuk Rivers and the smaller rivers that enter the Chukchi Sea north of Kotzebue Sound.

Small-scale bathymetric features on the Chukchi Shelf include nearshore sand bars, ice gouges, and large sediment dunes. Pleistocene glaciers reached the shelf only near Chukotka and Kotzebue Sound (Figs. 1 and 2). Well-defined ice gouges have been found where the seabed is as deep as 60 m, but they are most abundant in depths of 25 to 40 m (Toimil, 1978). Most are 1 to 3 m deep. The deepest ice keels that have been measured in the western Arctic do not extend more than 50 m, and most do not extend more than 25 m below sea level. Thus the gouges found at depths of 60 m in the north-central Chukchi Sea were probably made prior to 3000–4000 years ago, when sea level was eustatically lower than at present. Sand waves with wavelengths as large as 400 m are common bottom features near Bering Strait, and sediment dunes as much as 15 m high with 250 m between crests occur near Barrow Sea Valley. Several short cores and sediment grabs from the dunes yielded only lutite.

D. Sea Valleys and Submarine Canyons

Several large sea valleys and canyons cut the Chukchi and Beaufort Shelves and slopes. Mackenzie Sea Valley, the largest, is the major conduit of detritus to the Canada Basin. Ice lobes and melt water of the Pleistocene Laurentide ice sheets were major factors in carving this sea valley (Fig. 2).

1. Barrow Sea Valley and Barrow Canyon

The axis of Barrow Sea Valley (Fig. 2) is incised into the northern Chukchi Shelf about 20 km off the northeast-trending Alaska coast north of

Wainwright. It is a broad, flat-bottomed channel 200 km long and 2 to 8 km wide with an overall axial gradient of about 1 m/km. The east bank of the valley is higher than the west bank owing to preferential sedimentation on that side, but the gradient of the west bank is generally steeper. About 20 km south of the continental shelf break, the sea valley merges into Barrow Canyon. The canyon attains its maximum relief of 900 m where it crosses the projection of the shelf break. Its average axial gradient is 30 m/km but a steep upper segment slopes 48 m/km. A second canyon, comparable in relief to Barrow Canyon, lies 55 km to the east and a third, much smaller canyon lies 110 km to the east. The second, and possibly the third, appear to be abandoned extensions of the Barrow Sea Valley. These canyons may have been beheaded by the present Barrow Canyon, which affords a more direct route from Barrow Sea Valley to the abyssal plain.

Although it is the only sizable sea valley entering the Canada Basin west of the Mackenzie, Barrow Sea Valley has only a modest sedimentary wedge and fan complex on the continental slope and rise. The volume of sediment in the wedge and fan is estimated to be of the order of 10^4 km^3, compared to roughly 3×10^5 km^3 for the Mackenzie Cone, suggesting that Barrow Sea Valley has not been a major sediment conduit to the basin. Additional Barrow Sea Valley-derived sediment underlies the abyssal plain, and a little is stored on the continental shelf.

The Barrow Sea Valley heads on the northeast Chukchi Shelf, which receives sediment only from relatively small streams with a total drainage area of approximately 35,000 km^2. The only large sediment sources in the Chukchi Sea are the Kobuk and Noatak Rivers, which drain about 100,000 km^2 of northwest Alaska, and the Alaskan Coastal Water, which introduces silt-laden Yukon River water via Bering Strait. These sources are semi-isolated from the Barrow Sea Valley by Herald Shoal and its extension to Cape Lisburne. In contrast, the Mackenzie Sea Valley receives sediment from the 2,000,000-km^2 Mackenzie River drainage.

Barrow Sea Valley is thought to have been incised into the shelf and slope by subaerial streams during Pleistocene regressions in sea level and by the northeast-flowing Alaskan Coastal Water during interglacial times. Barrow Canyon is presumed to have been cut by turbidity currents. The principal influence of the Alaskan Coastal Water may have been as a source of clastic sediment for the turbidity currents and for the submarine fan complex at the mouth of the canyon. The Alaskan Coastal Water flows northeast along the coast from south of Cape Lisburne to north of Point Barrow and its axis follows the Barrow Sea Valley. Presumably the current is held against the coast by the Coriolis effect. Garrison and Becker (1965) argue, on the basis of physical oceanographic data, that this canyon also acts as a conduit for exchange of water masses between the Chukchi Shelf and the Canada Basin. They postulate that cold saline water generated on the shelf in winter drains north into the deep basin, and that under certain barometric conditions deep Canada Basin water wells up onto the shelf via the sea valley.

2. *Hope Sea Valley and Herald Canyon*

The Pleistocene and early Holocene Noatak and Kobuk Rivers crossed the Chukchi Shelf via Hope Sea Valley and Herald Canyon (Creager and McManus, 1965). According to these workers, this sea valley system (Fig. 2) consists of a series of channels interrupted by areas of deltaic deposition representing still stands during Pleistocene sea-level fluctuations. Bering Strait Sea Valley (Hopkins *et al.*, 1976), which flows north through Bering Strait, may be a tributary of Hope Sea Valley that was similarly modified by deltaic deposition. Because the deltaic areas lack clearly defined channels, the axis of the sea valley system is difficult to map. Herald Canyon (Fig. 3) apparently enters Canada Basin at the Chukchi Abyssal Plain between 170° and 180° W (Beal, 1969). This plain, a perched basin at about 2200 m depth, may have received much of its sedimentary fill from the Noatak and Kobuk Rivers and northern Chukotka via the Hope Sea Valley–Herald Canyon drainage system during Pleistocene glacial maxima.

3. *Hanna Sea Valley*

Hanna Sea Valley (Fig. 2), a broad, ENE-trending feature north of Hanna Shoal, may be an abandoned late Pleistocene–early Holocene course of the Hope–Herald Sea Valley System. The Noatak–Kobuk effluent and an ancestral northeast-flowing Alaskan Coastal Water, if it existed during the late Pleistocene–early Holocene, must have flowed west of then-emergent Herald Shoal to reach the Arctic Basin. The current may have been deflected northeastward and held against the north side of Herald and Hanna Shoals by the Coriolis effect. Hanna Sea Valley might have been related to this postulated current much as Barrow Sea Valley appears to be related to the present northeast-advecting Alaskan Coastal Water.

E. Chukchi Continental Borderland

The Chukchi Borderland (Figs. 2 and 3), 400 km wide and 600 km long, extends northward from the Chukchi Shelf between 160° and 170° W longitude (Beal, 1969; Hunkins *et al.*, 1962). The borderland consists of a cluster of shallow, flat-topped plateaus and ridges with an intervening deep basin, Northwind Abyssal Plain, and sea valleys. Chukchi Cap (Fig. 3), the largest high-standing plateau in the borderland, has flat shelves at 270 m and 400 to 500 m below sea level and rises locally to 246 m below sea level. The east margin of the borderland is the Northwind Escarpment, a steep continental slope that abuts the deep Canada Basin at depths near 3900 m. Gentler slopes, irregular in plan, characterize the other margins. Bathymetry and geophysical characteristics (Shaver and Hunkins, 1964) suggest that the high-standing parts of the borderland consist of continental crust.

F. Continental Shelf Break and Beaufort Ramp

The shelf break in the Chukchi and western Beaufort Seas (Fig. 2) has large regional variations in character and depth that reflect the interplay between tectonic processes and Quaternary sedimentation and slumping (Fig. 17). The main (outer) shelf break lies at the headwalls of submarine slumps on the continental slope. Its depth is 200 to 300 m near 150° W, 450 m near 160° W, and about 800 m at 139° W longitude.

The Beaufort Ramp, a transition zone between the low-gradient inner shelf and the continental slope, is the product of late Cenozoic subsidence and folding. Its inner edge, near the 60-m isobath, lies at or seaward of the crests of broad mid-shelf arches and anticlines expressed in Neogene beds. The slope of the ramp and the dip of the underlying strata are generally similar, typically 0.5 to 1.5°, which suggests that the strata and seabed were tilted together. The fold crests apparently acted as dams behind which shelf sediments ponded and smoothed the seabed of the inner shelf, but sediment also overtopped the fold crests and prograded a late Cenozoic sedimentary wedge seaward at the head of the ramp. Headwalls of bedding plane slides developed within this wedge determine the specific position of the inner shelf break.

On some of our seismic profiles (Fig. 15), Tertiary beds beneath deeper parts of the Beaufort Ramp are truncated at the base of the Quaternary deposits even though parts of the ramp are deeper than 800 m, well below the deepest Pleistocene regression. This truncation of bedding suggests that the ramp was once above wave base and subsequently subsided and tilted seaward. The outer shelf break along the Beaufort Ramp also increases in depth from 200 to 300 m near 150° W to 830 m near 139° W longitude (Fig. 16), a rate of about 1.2 m/km. Deepening of the shelf break corresponds closely with the easterly increase in thickness of the sedimentary prism of the Mackenzie Cone and Canada Continental Rise. The origin of the ramp may be explained by its spatial relation to the diapiric field that underlies the adjacent continental slope (Fig. 5) and the Mackenzie Cone. Tilting of the ramp may be a result of late Cenozoic flow of diapiric material from beneath the ramp to the diapiric terrane beneath the slope. Eastward deepening of the shelf break and northward tilting of the Beaufort Ramp may also be in part an isostatic subsidence due to crustal loading by sediment on the cone and rise. The later hypothesis is somewhat supported by the pocket of earthquakes reported at the head of the Mackenzie Cone and Canada Rise by Wetmiller and Forsyth (1978). These workers suggest (p. 15) that the earthquakes "are possibly related to isostatically uncompensated loads of recent sediments."

G. Alaska Continental Slope and Northwind Escarpment

The continental slope north of Alaska follows a gentle arc from the Mackenzie Sea Valley to the Chukchi Borderland (see Fig. 2). Its junction with the continental rise lies at depths of 1100 to 2000 m and its average slope ranges from 4° to 12°, with local steep pitches of 16°. The steep, nearly linear slope along the east face of the Chukchi Borderland, the Northwind Escarpment (Figs. 2 and 3), has slopes as steep as 23° (Fischer *et al.*, 1958). Numerous slumps, landslides, and incised channels characterize the Alaska Continental Slope. The largest slide masses head at or near the shelf break. Along many crossings the vertical drop at the headwall of these slides is 900 to 1200 m, and head-to-toe length of some large slides exceeds 35 km. Slumps and irregular prograded sedimentary wedges similar to these surficial features also occur within the Tertiary section beneath the outermost Western Beaufort Shelf (see Fig. 13B,C).

Seismic and bathymetric profiles show that many canyons incise the Alaska Continental Slope. Some appear to be slump scars; others contain bedded sediments, are bordered by levees, and appear to be turbidity current channels. In these channels the upper surface of the sedimentary fill is commonly tilted up toward the eastern bank owing to the Coriolis effect on the depositing current.

H. Canada Continental Rise and Abyssal Plain and Alaska Continental Rise

The Canada Abyssal Plain, which underlies the western two fifths of the Canada Basin, is about 3700 m to more than 4000 m deep (Fig. 2). Its smooth floor and sparsity of abyssal hills indicate that it overlies a thick sedimentary fill. The Canada Continental Rise underlies the eastern three fifths of the basin. It is about 3700 m deep at its gradational boundary with the abyssal plain on the west and slopes gently upward toward the Arctic Islands and Mackenzie Delta, where it abuts the continental slope between the 1400- and 1800-m isobath. Its morphology indicates that the Canada Rise is underlain by a thick sedimentary fill derived in large part from the Mackenzie Valley and Arctic Islands. Because these areas were covered by the Laurentide ice sheet, it is likely that glacial detritus is a major constituent of the sedimentary fill beneath the Canada Rise. The rise is as much as 500 km wide from the continental slope to the 3700-m isobath; the other rises fringing the Canada Basin, including the Alaska Rise, are much narrower. Sedimentation on the

Canada Rise has thus overwhelmed that on the Alaska Rise, and its sediments abut, or lap onto, the surface of the Alaska Rise (Fig. 2). As the Canada Abyssal Plain adjoins the Alaska Rise at depths as great as 4000 m, and the upper part of the Canada Rise laps onto it at depths of 1400 to 1800 m, sedimentation on the Canada Rise has reduced the vertical relief of the eastern part of the Alaska Rise and Slope by more than 2200 m. Concomitantly, the width of the Alaska Rise narrows from about 100 km near 150° W longitude to a wedge edge near Herschel Island. Westward, the Alaska Continental Rise wedges out again at the base of the steep Northwind Escarpment, where terrigenous sedimentation was slight. In places the escarpment appears, on reconnaissance charts, to plunge directly to the Canada Abyssal Plain. The disparity in size between the Canada Rise and the other continental rises of the Canada Basin demonstrates that the Arctic Islands and Mackenzie Valley were much more important sources of late Cenozoic detritus to the Canada Basin than northern Alaska or the Chukchi Shelf.

Smooth slopes characterize the Alaska Continental Rise and gradients range from 0.9° to more than 2.2°. Bedding in the underlying sediment is generally parallel to the seabed and presumably consists of turbidity current and disaggregated slide deposits. Channels and levees can be seen on bathymetric profiles and, vaguely, on bathymetric charts of the Alaska Rise. A few refraction measurements and estimates of sediment thickness beneath the nearby Canada Basin and Western Beaufort Shelf (Fig. 5) suggest that sediment beneath the Alaska Rise may be 4 to 8 km thick. Because it borders an Atlantic-type margin, the structural simplicity observed in its upper beds is inferred to characterize the entire Alaskan Rise sedimentary prism. However, many large diapiric folds deform the Alaska Rise and Slope east of 146° W longitude.

III. REGIONAL STRATIGRAPHY AND TECTONIC SETTING

The geologic formations of northern Alaska and vicinity are divided conveniently into four tectonic–stratigraphic sequences (Franklinian, Ellesmerian, Brookian, and Hope Basin) that correlate well with the major geologic events of the region. The character of these sequences is shown in Fig. 4, and their distribution in Figs. 5, 6, 8, and 9. The Franklinian, Ellesmerian, and Brookian Sequences were defined by Lerand (1973) from rock sequences in the Arctic Archipelago of Canada and the Brooks Range of Alaska. The Hope Basin Sequence is defined here from the Tertiary Hope Basin of the southern Chukchi Sea. The onshore stratigraphic and structural data presented in this paper include data and concepts from Alaska Geological Society (1971, 1972,

1977); Anonymous (1970); Armstrong and Bird (1976); Beikman and Lathram (1976); Bird (1978); Brosgé and Tailleur (1971); Campbell (1967); Chapman and Sable (1960); Detterman *et al.* (1975); Jones and Grantz (1964); Kameneva (1977); King (1969); Lerand (1973); Martin (1970); Norris (1977); Tailleur and Brosgé (1970); Tailleur *et al.* (1972); U.S. Geological Survey (1978); and Young *et al.* (1976).

A. Franklinian Sequence and Precambrian Rocks

Basement for standard seismic reflection and sonobuoy refraction surveys in northern Alaska and its continental shelves occurs within the Franklinian Sequence or, in places, Precambrian rocks. The Precambrian rocks are mainly slate, phyllite, schist, gneiss, marble, and metavolcanic rock. The Franklinian Sequence was named for the Franklinian Geosyncline of the Arctic Archipelago. In the archipelago the sequence is of Middle Cambrian to Devonian age and was derived mainly from northern sources. Its southern facies are mainly carbonate. In northern Alaska the Franklinian Sequence consists of generally strongly deformed Cambrian to Devonian miogeoclinal and eugeoclinal rocks (Fig. 4) strongly disrupted by thrust faults in the Brooks Range and in the Seward and Chukotsk Peninsulas. In places the sequence is mildly metamorphosed, and locally it is intruded by plutonic rocks. It is thought to extend seaward to the continental slope, and it may underlie the Chukchi Borderland.

B. Arctic Platform and Ellesmerian Sequence

The Franklinian rocks were strongly deformed and truncated by regional erosion during the late Devonian–early Mississippian Ellesmerian (Antler) Orogeny, creating a stable shelf, the Arctic Platform of northern Alaska. The platform maintained a remarkable stability for 200 million years, from early Mississippian to Cretaceous time. During this interval a diverse suite of mature marine and nonmarine clastic and carbonate rocks, the Ellesmerian Sequence, was deposited on the platform. The sequence is named for the Ellesmerian Orogeny. It is typically exposed in the Sverdrup Basin of the Arctic Archipelago, where it consists of Mississippian to Jurassic rocks of mainly easterly and southerly provenance. In northern Alaska, however, the Ellesmerian Sequence, which is also called the Arctic Alaska Basin (Tailleur and Brosgé, 1970), was derived mainly from northerly source terranes. Its constituent rock units (Fig. 4) in general thin, coarsen, and lap toward this source terrace, which is called Barrovia (Tailleur, 1973). On the Barrow Arch

GEOLOGIC AGE		PROVINCIAL SEQUENCES	EASTERN STRATIGRAPHIC UNITS	NORTH SLOPE LITHOLOGY	THICKNESS (METERS)
CENOZOIC	QUATERNARY	BROOKIAN	GUBIK FORMATION (South) / (North)	Marine sand, silt, gravel and clay.	10-200
	TERTIARY (NEOGENE / PALEOGENE)	BROOKIAN	SAGAVANIRKTOK FORMATION	Poorly consolidated nonmarine and shallow marine shale, sandstone, and conglomerate with some carbonaceous shale, lignite, and bentonite.	0-2500
MESOZOIC	CRETACEOUS UPPER	BROOKIAN	COLVILLE GROUP	Upper beds - Marine and nonmarine sandstone, siltstone, shale, conglomerate, coal, and tuff. Lower beds - Marine sandstone, siltstone, organic shale, and tuff.	900-3600
	CRETACEOUS LOWER	BROOKIAN	NANUSHUK GP. (east of Canning River)	Marine graywacke, siltstone, shale (Bathtub Graywacke, Tuktu Fm.).	>200-1200
			"PEBBLE SHALE" - KONGAKUT FM.	Marine organic siltstone and shale with rounded quartz grains and chert pebbles. Quartzose sandstone (Kemik member) near base locally.	40-700
	JURASSIC		KINGAK SHALE (includes Kuparuk River Sandstone at top locally)	Marine shale, siltstone, and organic paper shale.	0-1200
	TRIASSIC	ELLESMERIAN	SHUBLIK FORMATION	Fossiliferous marine argillaceous limestone and dolomite, shale, and sandstone. Upper beds locally quartzose sandstone (Sag River and Karen Creek sandstones).	0-140
	PERMIAN	ELLESMERIAN	SADLEROCHIT GROUP	Upper beds - Marine and nonmarine quartz and chert-rich sandstone, conglomerate, mudstone, and shale. Lower beds - Marine sandstone and siltstone with interbeds of shale; local conglomerate and chert.	0->700
PALEOZOIC	PENNSYLVANIAN	ELLESMERIAN	LISBURNE GROUP	Marine limestone and dolomite with chert, sandstone, and siltstone.	0-1200
	MISSISSIPPIAN	ELLESMERIAN	ENDICOTT GROUP	Marine and nonmarine sandstone, shale, conglomerate, limestone, and coal.	0->1000
	PRE-MISSISSIPPIAN	FRANKLINIAN	(PRE-MISSISSIPPIAN ROCKS)	Argillite, graywacke, limestone, dolomite, chert, quartzose sandstone, and shale and their metamorphic equivalents.	Thousands of meters

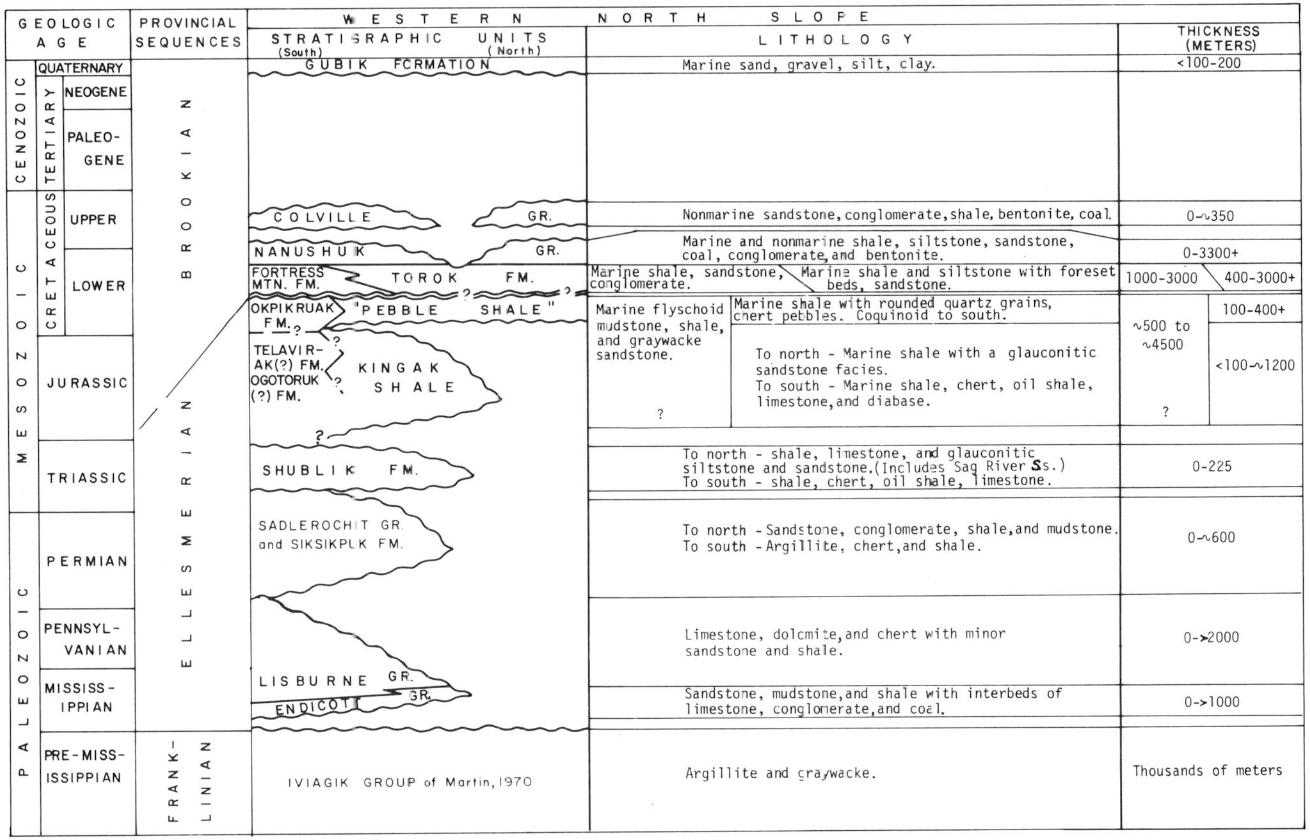

Fig. 4. Phanerozoic stratigraphy of the western and eastern North Slope of Alaska.

Mississippian to Jurassic beds of the sequence thin to a wedge edge owing in part to sedimentary thinning and in part to erosional truncation, and they are overstepped by the Neocomian "Pebble Shale." This shallow-water organic shale is the youngest unit in northern Alaska with a northern source. Outcrops of Jurassic and Neocomian strata in northern Yukon Territory and on the Arctic Coastal Plain of northeastern Alaska (Young *et al.*, 1976; and Reiser *et al.*, 1978) and the interpreted presence of 1800 m of Kingak Shale (Jurassic and lowest Cretaceous) beneath the Beaufort Shelf north of Yukon Territory (Norris, 1977) suggest that the Jurassic and earliest Cretaceous strata of these areas become shalier to the north and are of southern provenance. In northern Yukon Territory the Kingak Shale oversteps Mississippian to Triassic beds of northerly provenance lower in the Ellesmerian Sequence.

The boundary between the Ellesmerian and the overlying Brookian Sequence, according to Lerand (1973, p. 373), could be placed either at the oldest beds (Middle Jurassic) that received sediment from southern sources or at the youngest beds (Neocomian) that received sediment from northern sources. Lerand arbitrarily placed the boundary at the base of the Cretaceous. In this report the boundary is placed between the "Pebble Shale"–Kongakut Formation (Neocomian) of northern provenance and the Torok Formation–Nanushuk Group (Albian) of southern provenance beneath the Arctic Coastal Plain, northern Arctic Foothills, and Chukchi and Western Beaufort Shelves of northern Alaska (Figs. 2 and 4). In the southern Arctic Foothills and northern Brooks Range the boundary is placed between Upper Triassic beds of northern provenance and Middle Jurassic tuffaceous sandstone to Neocomian (Okpikruak Formation) turbidite of southern provenance. Since the boundary lies at the top of the "Pebble Shale" over most of northern Alaska and its continental shelves, the term Brookian in this report refers to post-Neocomian rocks unless otherwise specified. The character of the Jurassic and Neocomian beds of northern provenance suggests that the northern source terrane was then senescent.

C. Colville Foredeep and Brookian Sequence

The long period of tectonic quiescence recorded by the Ellesmerian Sequence ended during Jurassic and earliest Cretaceous time, when deformation began that created the Brooks Range orogen along the southern part of the Arctic Platform. Paleozoic and Mesozoic rocks were thrust northward onto the southern part of the platform and created a southward deepening, asymmetric foreland basin, the Colville Foredeep, north of the ancestral Brooks Range. The load of the thrust sheets and of the Jurassic and Cretaceous sedimentary rocks subsequently deposited in the foredeep caused it to deepen and broaden and tilted the Arctic Platform southward. Concomitantly, southern sources first supplemented then supplanted northern sources of sediment to the newly formed basin.

The oldest strata deposited in the Colville Foredeep were the Middle Jurassic tuffaceous beds and the Upper Jurassic to Neocomian turbidites of southern provenance now exposed in the northern Brooks Range and adjacent foothills. These are coeval with condensed shale and coquinoid shale deposited on an intrabasin high and with euxinic basin and shelf deposits of northern provenance (Jones and Grantz, 1964) in the upper part of the Ellesmerian Sequence to the north. Higher beds of the foredeep—namely Albian, Upper Cretaceous, and Tertiary strata of the Brookian Sequence, all of southern provenance—extend from the Brooks Range across the entire Arctic Platform to the Western Beaufort Shelf.

The Cretaceous sedimentary rocks of the Colville Foredeep exceed 6000 m in thickness on the south, where they consist mainly of turbidites and paralic–deltaic wedges. These strata thin northward toward the Barrow Arch by onlap and downlap (Fig. 10). Their thickness on the arch ranges from 500 m to 2000 m. The Cretaceous beds on the arch, and probably on the Chukchi Shelf, are of paralic and neritic to upper bathyal facies. In northwestern Alaska and on the Chukchi Shelf, they include deltaic deposits with thick coal beds. In the northern Chukchi Sea and the Canning–Sagava-nirktok Rivers area, wedges of northward-thickening Tertiary sedimentary strata overlie the Cretaceous rocks of the northern part of the Colville Foredeep and Barrow Arch. Nonmarine facies in the Tertiary and Cretaceous sequences appear to give way northward to marine facies, and the northward prograded sedimentary strata beneath the Western Beaufort Shelf are probably mainly marine. However, some nonmarine beds of both sequences reach the coast and appear on seismic records to extend offshore. Seaward, the prograded Brookian-equivalent beds thicken from about 500 to 2000 m on the Barrow Arch to more than 6000 m in places on the outer shelf.

Cessation of sedimentation from northern terranes that geographically occupied the present area of the southern Canada Basin argues for a rifted origin of the Canada Basin. Tailleur (1969, 1973) suggests, in addition, that the early phase of the Brooks Range, orogeny was produced along the leading edge of the southward-rotating northern Alaska plate in response to opening of the Canada Basin. However, late Jurassic to mid-Cretaceous Brooks Range tectonic events are generally similar and synchronous with those in the Cordillera south of Alaska. Accordingly, we believe that the ancestral Brooks Range and its structures are a foreland thrust belt related to a broader subduc-tion and collision zone between crustal plates of the Pacific region and the North American plate. Opening of the Canada Basin possibly increased the rate of subduction south of the Brooks Range, but any specific effects of this increased rate in the geologic record have not been documented.

D. Hope Basin Sequence

The young sedimentary rocks of Hope Basin in the southern Chukchi Shelf have not been dated, and their correlation with the bedded rocks of adjacent land areas is conjectural. Outcrops of late Neogene marine sedimentary rocks along the margins of the basin (Hopkins and MacNeil, 1960; Belevich, 1969) and scattered outcrops of Paleogene and Neogene nonmarine sedimentary rocks on the northern Seward Peninsula and in the Selawik Lowland (Hudson, 1977; Patton and Miller, 1968) belong to formations that probably extend into Hope Basin. Nonmarine rocks of late Cretaceous or early Tertiary age on the eastern Seward Peninsula (Sainsbury, 1976) possibly also extend into the basin. The Hope Basin Sequence was deposited in local intracontinental basins south of the Brooks Range, and is tectonically distinct from the Brookian Sequence of northern Alaska.

IV. GEOLOGY

A. Structural Provinces

The continental shelf and slope of the western Beaufort and northern Chukchi Seas is underlain by a thick prism of Cretaceous to Quaternary, and locally Jurassic, sedimentary strata that prograded seaward on a subsiding shelf. A thick correlative sedimentary sequence also underlies the adjacent continental rise and abyssal plain of the Canada Basin. In two places the prograded prism widens into sizable sedimentary basins. A southwestward bend of the Barrow Arch in the northern Chukchi Sea creates an embayment wherein lies the North Chukchi Basin (Fig. 3), which is filled with inferred Cretaceous, Tertiary, and perhaps older sedimentary rocks; and differential subsidence between Harrison Bay and Camden Bay has produced the Cretaceous and Tertiary Camden Basin (Fig. 5). This basin underlies both the continental shelf and a large area of coastal plain near the Colville and Sagavanirktok Rivers.

The central and southern Chukchi Shelf is dominated structurally by two positive (anticlinorial) and two negative (synclinorial) tectonic elements (Figs. 3 and 8–10). The positive elements, which bring Paleozoic rocks close to the surface, are the Barrow and Herald–Wrangel Arches. The negative elements are the Colville Foredeep, which contains thick accumulations of Jurassic, Cretaceous, and locally Tertiary sedimentary rocks, and Hope Basin, which is filled with Tertiary sedimentary strata. The basement underlying the Colville Foredeep is the Ellesmerian Sequence. In Hope Basin, the basement is composed of Precambrian to Lower Cretaceous rocks that trend beneath it from the western Brooks Range and Chukotsk Peninsula.

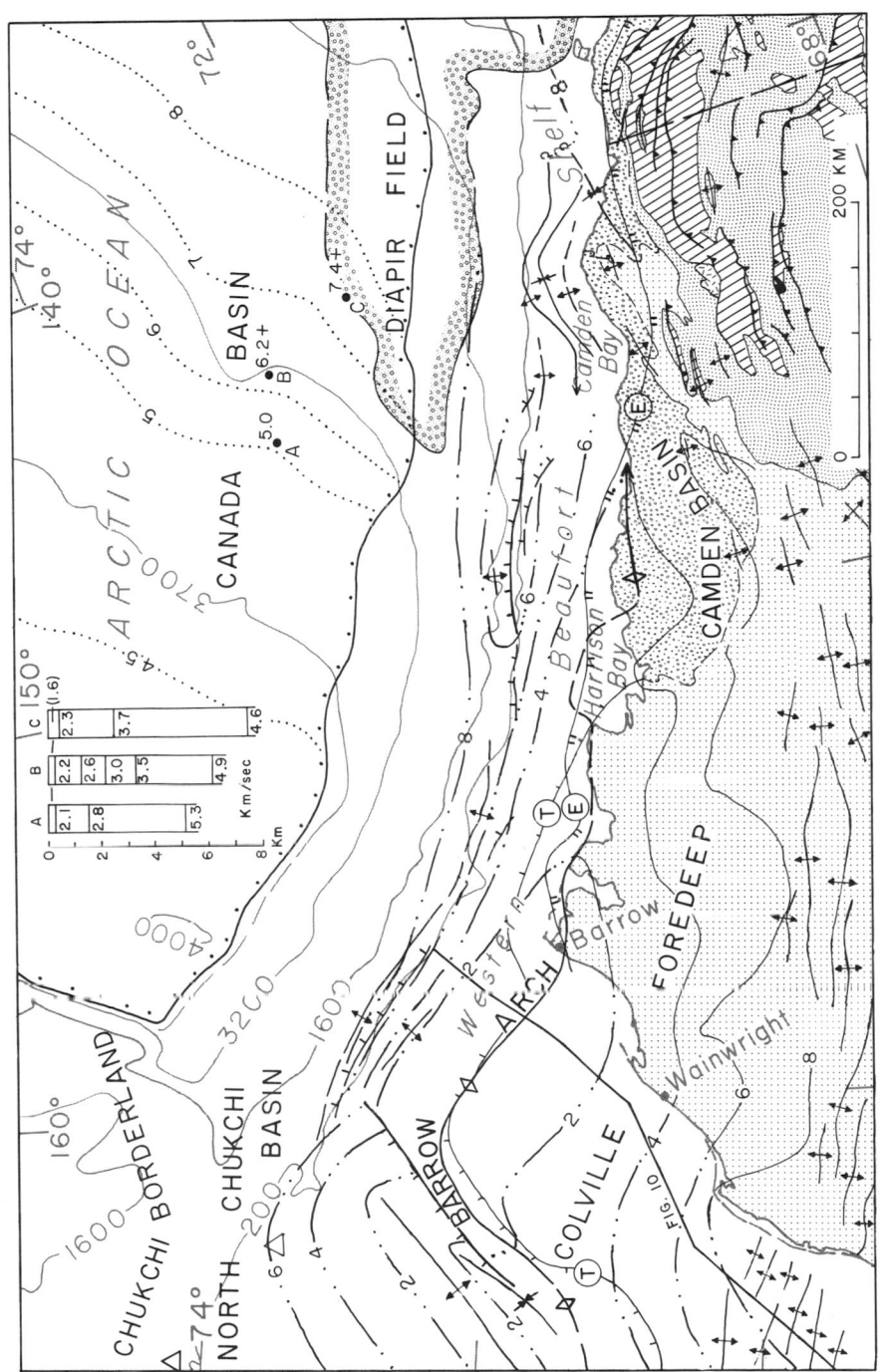

Fig. 5. Geologic structure and thickness of sedimentary rock beneath the Western Beaufort Shelf and southern Canada Basin. See Fig. 6 for explanation.

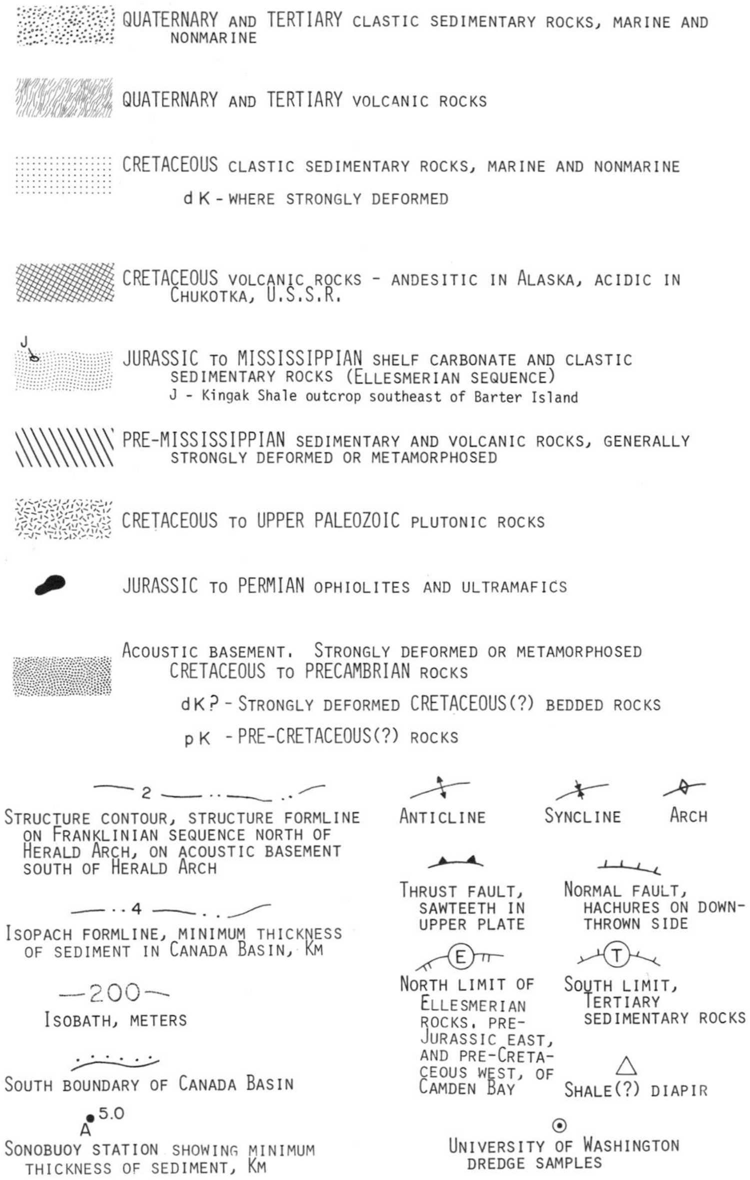

QUATERNARY AND TERTIARY CLASTIC SEDIMENTARY ROCKS, MARINE AND NONMARINE

QUATERNARY AND TERTIARY VOLCANIC ROCKS

CRETACEOUS CLASTIC SEDIMENTARY ROCKS, MARINE AND NONMARINE

d K - WHERE STRONGLY DEFORMED

CRETACEOUS VOLCANIC ROCKS - ANDESITIC IN ALASKA, ACIDIC IN CHUKOTKA, U.S.S.R.

JURASSIC TO MISSISSIPPIAN SHELF CARBONATE AND CLASTIC SEDIMENTARY ROCKS (ELLESMERIAN SEQUENCE)
J - Kingak Shale outcrop southeast of Barter Island

PRE-MISSISSIPPIAN SEDIMENTARY AND VOLCANIC ROCKS, GENERALLY STRONGLY DEFORMED OR METAMORPHOSED

CRETACEOUS TO UPPER PALEOZOIC PLUTONIC ROCKS

JURASSIC TO PERMIAN OPHIOLITES AND ULTRAMAFICS

ACOUSTIC BASEMENT. STRONGLY DEFORMED OR METAMORPHOSED CRETACEOUS TO PRECAMBRIAN ROCKS

d K ? - STRONGLY DEFORMED CRETACEOUS(?) BEDDED ROCKS

p K - PRE-CRETACEOUS(?) ROCKS

STRUCTURE CONTOUR, STRUCTURE FORMLINE ON FRANKLINIAN SEQUENCE NORTH OF HERALD ARCH, ON ACOUSTIC BASEMENT SOUTH OF HERALD ARCH

ISOPACH FORMLINE, MINIMUM THICKNESS OF SEDIMENT IN CANADA BASIN, KM

ISOBATH, METERS

SOUTH BOUNDARY OF CANADA BASIN

SONOBUOY STATION SHOWING MINIMUM THICKNESS OF SEDIMENT, KM

ANTICLINE   SYNCLINE   ARCH

THRUST FAULT, SAWTEETH IN UPPER PLATE

NORMAL FAULT, HACHURES ON DOWN-THROWN SIDE

NORTH LIMIT OF ELLESMERIAN ROCKS, PRE-JURASSIC EAST, AND PRE-CRETA-CEOUS WEST, OF CAMDEN BAY

SOUTH LIMIT, TERTIARY SEDIMENTARY ROCKS

SHALE(?) DIAPIR

UNIVERSITY OF WASHINGTON DREDGE SAMPLES

Fig. 6.   Explanation for geologic structure maps, Figs. 5, 8, and 9.

B. Barrow Arch

Barrow Arch is the dominant structure of the Western Beaufort and northern Chukchi Shelves. Emerging near the Canning River from beneath a thick Tertiary section in the Camden Bay area, its crest (Figs. 3, 5, 7, 8, 10, and 11C) follows the coast westward to Barrow, then cuts seaward to 70° N, 160° W in the northern Chukchi Sea. From here the arch turns southwest and persists as a more complex, but still broad structure to its termination in the west near the crosscutting Herald Fault Zone.

East of 160°W longitude the Barrow Arch trends westerly, is only slightly asymmetric (steeper flank north), and has low flank dips and generally simple structure (Fig. 10). West of 160°W longitude, the arch trends southwest and

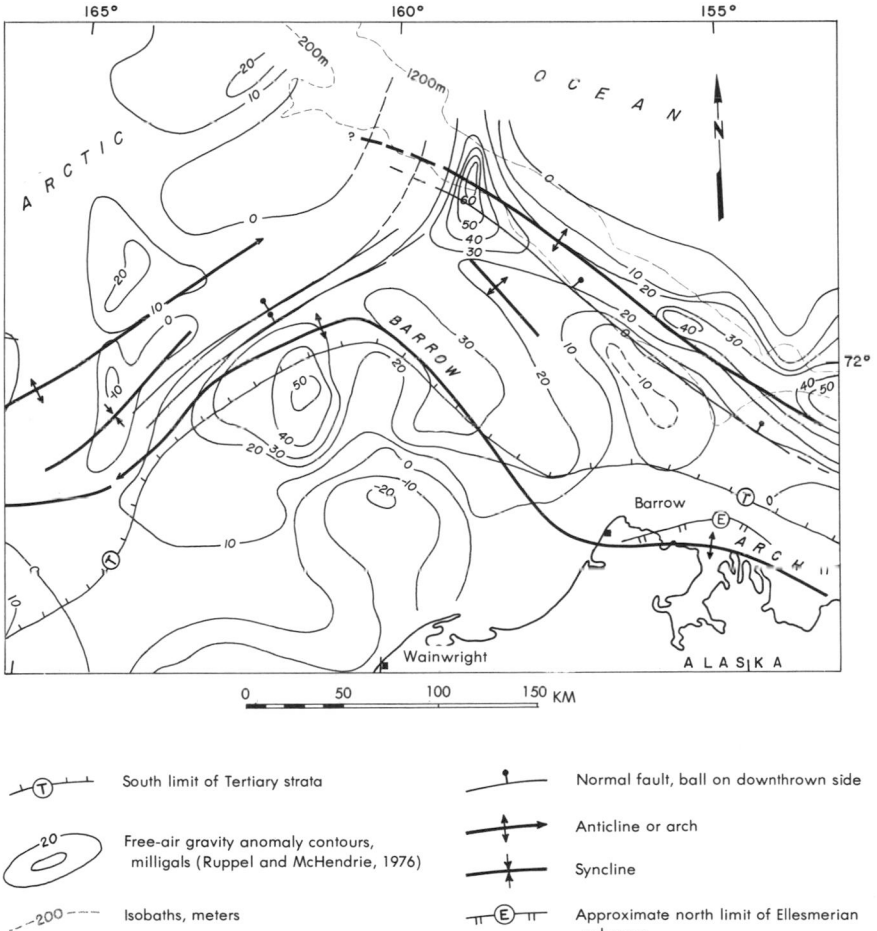

Fig. 7. Seismic reflection structure and free-air gravity anomalies near the change in trend of the Barrow Arch at 72° N latitude, 160° W longitude.

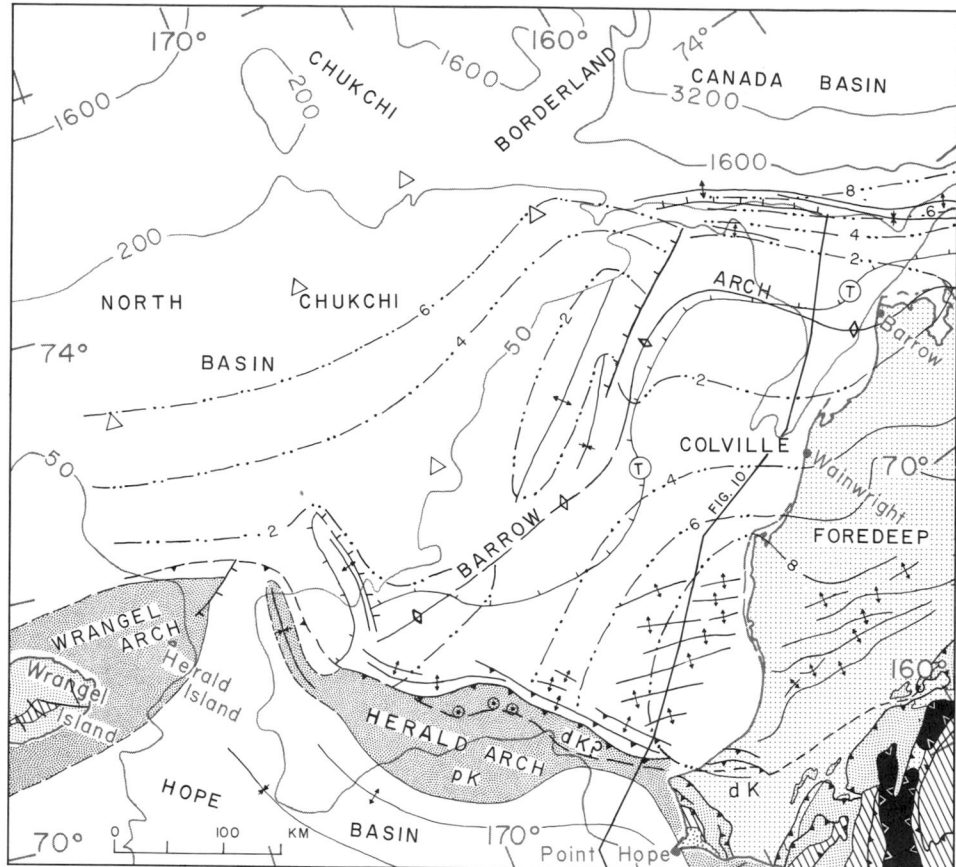

Fig. 8. Geologic structure and thickness of sedimentary rock beneath the northern Chukchi Sea. See Fig. 6 for explanation.

the northwest flank is considerably steeper than the southeast flank (Fig. 7). The northwest flank west of 160°W longitude contains large normal faults and is parallel to a strong gravity gradient that strikes northeast. Free-air anomaly values descend from 50 mgal over the arch to 0 to −10 mgals over the North Chukchi Basin. These features suggest that the northwest flank of the Barrow Arch may have resulted from truncation of the Arctic Platform west of 160°W longitude by a northeast-striking structure related to formation of the North Chukchi Basin.

The Barrow Arch is the broad structural culmination of the rifted and southward-tilted Arctic Platform of northern Alaska. Crustal heating associated with the initial Jurassic and early Cretaceous rifting of the Canada Basin is thought to have uplifted and tilted the northern part of the platform to the south along the arch (Rickwood, 1970). The south tilt is also partly a

Mississippian to Jurassic paleoslope and partly tilt produced by Cretaceous loading of the southern part of the Arctic Platform by nappes and voluminous detritus from the nascent Brooks Range. The north flank of the arch is the modified rift scarp reduced in relief by down-to-the-north normal faulting, Jurassic and early Cretaceous erosion, and regional post-rift subsidence. Such subsidence, commonly observed along rifted continental margins (see, for example, Watts and Ryan, 1976) may be a consequence of post-rift cooling of the crust beneath the rifted zone and subsequent sedimentary loading on and near the newly created continental margin. These processes are thought to have shifted the crest of the Barrow Arch south by 50 to 100 km to the vicinity of the present coastline from an original position beneath the continental slope (Fig. 5).

C. Arctic Platform and Colville Foredeep

Between Cape Lisburne and Barrow, the bedded rocks of northern Alaska trend northwest beneath the Chukchi Sea. Here the Ellesmerian and Brookian rocks are limited on the southwest by the Herald Arch and Fault Zone and thin northward toward the Barrow Arch. Our seismic data do not show to what extent Ellesmerian strata continue across the crestal zone of the arch to underlie the North Chukchi Basin. Brookian rocks do extend across the Barrow Arch and dominate the sedimentary prisms of the Western Beaufort Shelf and North Chukchi Basin.

A sedimentary sequence at least 3 km thick, speculatively designated "E-Fr" on the cross section in Fig. 10, has been recognized in the northeastern Chukchi Sea, although its extent has not been determined. The sequence is tentatively interpreted to be overlain by Albian or older early Cretaceous rocks in the north and by Mississippian beds low in the Ellesmerian Sequence in the south. It yields seismic interval velocities (preliminary) of 5.0 km/sec or more and rests on bedded rocks with preliminary interval velocities of 6.0 km/sec or more. If the overlying beds are correctly identified, the sequence may be either a local thickening of Lower to Middle Mississippian clastic sedimentary rocks (Endicott Group) at the base of the Ellesmerian Sequence or beds in the upper part of the Franklinian Sequence that are structurally more orderly than the Ordovician and Silurian argillite and graywacke found in boreholes in northern Alaska (Carter and Laufeld, 1975). The Endicott Group is represented in the subsurface of northern Alaska by more than 550 m of conglomerate and sandstone of the Itkilyariak Formation, Kekiktuk Conglomerate, and Kayak Shale. These beds rest with angular unconformity on the strongly deformed Ordovician and Silurian beds.

In northwestern Alaska, the Ellesmerian Sequence thickens from a wedge edge in the crestal region of the Barrow Arch to at least 4000 m, and perhaps more than 8000 m, in the southern part of the North Slope and the western Brooks Range (Fig. 4). Offshore, single-channel seismic reflections and sonobuoy refractions were received from the Ellesmerian rocks only north of Icy Cape (unit J-M in Fig. 10). In this area their seismic velocity is about 3.7 to 4.3 km/sec, and they consist of a gently south-dipping, southward-thickening packet of strong reflectors. An interpretation of the structure of these rocks in the eastern Chukchi Sea, which is speculative south of 70°50′ N latitude, is presented in Fig. 10.

The Brookian Sequence of northwest Alaska consists of delta front foreset beds and shallow marine and paralic topset clastic beds and coal, all of southerly source and late early Cretaceous (Albian) age. Tertiary beds of the Brookian Sequence overlie the crestal zone of the Barrow Arch and the northern part of the Arctic Platform everywhere except in the region between 153° W and 158° W longitude (Figs. 5 and 8). The Brookian Sequence, thinned by erosion, is 400 to 600 m thick and of Albian age in the vicinity of South Barrow gas field, on the crest of the Barrow Arch. Southward, as noted previously, it thickens to at least 6000 m near the Brooks Range. Northward, beneath the Western Beaufort Shelf, the Brookian Sequence contains Cretaceous and Tertiary beds and thickens to at least 6000 m.

The Cretaceous rocks of the central Chukchi Shelf have stratigraphic features, interpreted from seismic reflection profiles, that permit some of them to be correlated with subsurface strata of northern Alaska as described, for example, by Brosgé and Tailleur (1971), and Woolson (1962). On the Chukchi Shelf a thick seismic unit with irregular weak to moderately strong reflectors and many north or northeasterly dipping clinoform beds is well developed north of the foreland folds of the southern part of the Colville Foredeep (Fig. 8). This unit is correlated with the Torok Formation of Albian age, a thick lutaceous unit containing turbidite beds. An overlying unit typified by strong parallel reflectors underlies the region of the foreland folds. It is correlated with the paralic and deltaic Albian and Cenomanian(?) Nanushuk Group, which consists of sandstone, lutite, conglomerate, and coal. West of the foreland folds lies an extensive terrane characterized by weak reflectors, small irregular folds, and poor seismic penetration. This terrane also has a relatively high positive free-air gravity anomaly (see Ruppel and McHendrie, 1976). Taken together, these features suggest that the terrane of weak reflectors is structurally elevated lutite low in the Brookian Sequence (Torok Formation) or high in the Ellesmerian Sequence ("Pebble Shale" or Kingak Shale).

Structurally the rocks of the Colville Foredeep beneath the eastern Chukchi Shelf resemble those onshore. North of Icy Cape they are almost unbroken by faults or folds and dip gently and uniformly south. South of Icy Cape they are folded into long, east–west-striking anticlines and broad, shallow synclines (Fig. 8). The amplitude and tightness of folding increase to the south, and the southern folds show thrusting, incipient core diapirism, and northern vergence—features that indicate relative northward movement of the upper beds. Detachment in these foreland folds is inferred to occur within the early Albian Torok Shale and the youngest rocks affected are the Nanushuk Group of late Albian age (see Woolson, 1962; Chapman and Sable, 1960). Apparently related structures in the Ogotoruk Creek area, southeast of Point Hope, appear to trend westward beneath Paleogene(?) strata in Hope Basin. The detachment folds are therefore inferred to have formed during Laramide (late Cretaceous–early Tertiary) time. Along the Herald Arch and Fault Zone, which bound deposits of the Colville Foredeep on the southwest, the detachment folds are intersected obliquely by large thrust folds (elongate folds with thrust-faulted cores) of the Herald Fault Zone (Fig. 8). These folds and related faults extend farther west than the detachment folds and appear to be younger.

D. Herald and Wrangel Arches

Herald Arch, the dominant structural feature of the central Chukchi Sea, is a belt of strongly reflective rock 20 to 100 km wide that trends N 50°W from Cape Lisburne toward Herald Island (Figs. 3, 8, and 9). Near 70° 30' N latitude, 173° W longitude the arch merges with a zone of north–south structures that connect it with the broad zone of shallow acoustic basement called Wrangel Arch (Figs. 3 and 8). Because the Brooks Range and Wrangel Island are underlain by similar rocks of the Ellesmerian and Franklinian Sequences (Kameneva, 1977), Herald and Wrangel Arches are thought to form a structural connection between the two areas. Wrangel Arch resembles Barrow Arch because it is underlain by similar correlatives of the Franklinian and Ellesmerian Sequences and bounds the North Chukchi Basin on the south (Figs. 3 and 8). Unlike Barrow Arch, but similar to Herald Arch, it is bounded on the north by south-dipping reverse faults.

The strongly reflective rocks in Herald Arch are onlapped from the south by Tertiary rocks in Hope Basin. The south-dipping unconformity at the base of the Tertiary rocks constitutes the south flank of the arch. The north flank is formed by the northeast-verging, northwest-striking Herald Fault Zone, which brings the strongly reflective rocks of the arch against well-bedded, less strongly reflective rocks of the Colville Foredeep.

The character of "acoustic basement" in Herald and Wrangel Arches can be inferred from outcrops on Lisburne Peninsula and Wrangel Island, a reported occurrence of Jurassic(?) plutonic rocks on Herald Island (N. A. Bogdanov, oral communication, 1970), and three dredge hauls taken from the north side of Herald Arch near 70°N latitude, 168° to 170° W longitude by the University of Washington (Fig. 8). The dredge samples are predominantly well-indurated graywacke sandstone and siltstone (Platt, 1975) that most resemble Lower Cretaceous graywacke and shale in the western Brooks Range and Lisburne Peninsula, but a positive correlation could not be made. On the northeast side of the arch, "acoustic basement" produces scattered, dismembered patches of steeply dipping but coherent reflectors suggesting strongly deformed bedded rocks (dK? on Figs. 8, 9, and 10). The strongly deformed beds of Lower Cretaceous graywacke and shale in the western Brooks Range and Lisburne Peninsula (dK in Figs. 8 and 9), which strike northwest from the north side of Cape Lisburne into the area of "dismembered" reflectors, could produce this observed acoustic pattern. "Acoustic basement" southwest of the belt of dismembered reflectors has few internal reflectors and is inferred to comprise lower Ellesmerian clastic and carbonate rocks and strongly deformed Franklinian clastic rocks (Iviagik Group of Martin, 1970) as crop out on the Lisburne Peninsula. Limited acoustic data over the Wrangel Arch suggest that a similar boundary may underlie its northern part.

E. Herald and Wrangel Fault Zones

Herald and Wrangel Arches are bounded on the north by the Herald and Wrangel Fault Zones (Figs. 8 and 9). Although both fault zones exhibit reverse slip and border the north sides of similar large belts of acoustic basement, they appear to differ in character and age. Herald Fault Zone consists of several thrust folds and faults that dip southwest at low angles (about 15°?). The thrust folds are best developed at the southeast end of the fault zone where they involve well-bedded competent rocks interpreted to represent the Cretaceous Nanushuk Group. Northwestward the dip of the fault zone becomes difficult to map, but may be steeper, and the large thrust folds along the northeast side of the fault zone diminish in amplitude and apparently lack thrust-faulted cores. This difference in structural response may correspond to a change in lithology. Seismic reflections suggest that northwestward from Cape Lisburne the rocks of the Colville Foredeep probably change from well-bedded sandstone and lutite of the Nanushuk Group in the region of the foreland folds to harder, poorly bedded rocks belonging to the older Torok or Kingak Shales farther west. Alternatively, the fault style may change from dominantly easterly directed low-angle reverse slip near the Lisburne Peninsula to moderate or high-angle reverse slip with a large component of strike-slip west of 169° W longitude for reasons not directly related to lithology.

Offshore the Herald Fault Zone separates moderately deformed Cretaceous sandstone and lutite to the northeast from more strongly deformed Cretaceous clastic rocks to the southwest (Figs. 8 and 9) and strikes toward a similar contact on the north side of the Lisburne Peninsula. A more important break, however, may lie to the southwest, between the area of acoustic base-

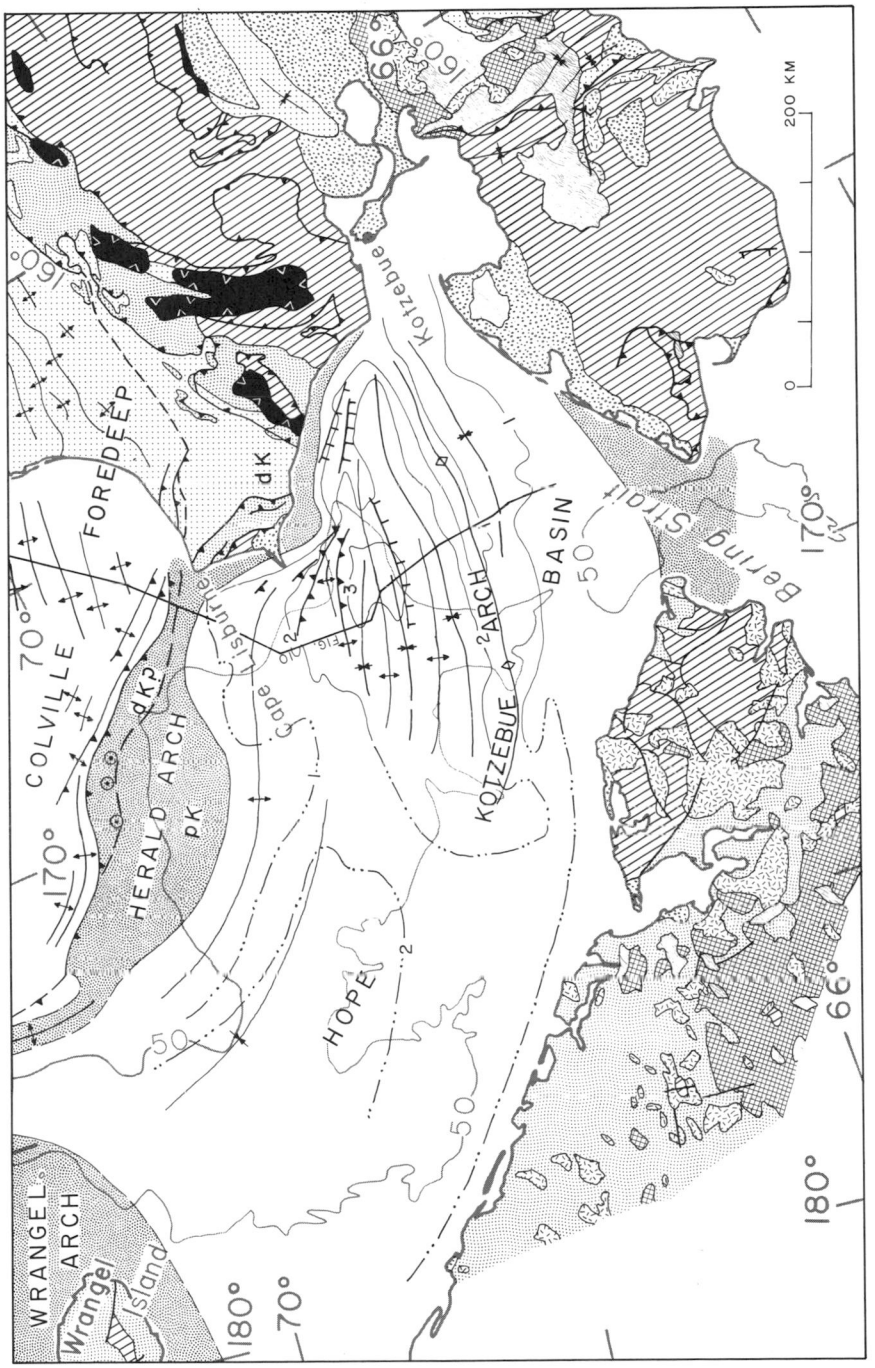

Fig. 9.   Geologic structure and thickness of sedimentary rock beneath the southern Chukchi Sea. See Fig. 6 for explanation.

ment with many bedding traces [Cretaceous(?) map unit dK?] and acoustic basement with few bedding traces [pre-Cretaceous(?) map unit pK]. This boundary strikes toward a southwest-dipping thrust-fault zone on the Lisburne Peninsula between Ellesmerian carbonates to the west and deformed younger rocks, mainly Cretaceous shale and graywacke to the east. Herald Fault Zone is younger than the Nanushuk Group strata (Albian), which it offsets, and older than the Paleogene(?) beds in Hope Basin, which overlie it. Wrangel Fault Zone, in contrast (Figs. 8, 11A,B), separates presumed Paleogene and (or) Cretaceous strata in the North Chukchi Basin from strongly reflective rocks in the Wrangel Arch. A younger Tertiary unit, perhaps Neogene, overlies the fault zone and is not deformed. The fault zone may therefore be of late Cretaceous or early Tertiary age, but definitive evidence for the age of the limiting rocks is lacking. Although it was seen on only a few profiles, the fault appears to be a steep south-dipping reverse fault with a number of splays on the north side (Figs. 11A,B). The splays bound subsidiary upthrown fault slivers that resemble horsts on seismic sections.

F. Chukchi Syntaxis

The Chukchi Syntaxis (Tailleur and Brosgé, 1970) is the right-angle junction of the westerly striking thrust faulted Paleozoic and Triassic terrane of the Brooks Range with the north and northwest-striking Lisburne Hills and Herald Arch (Fig. 3). Tailleur and Brosgé suggest that the Lisburne Hills and Herald Arch are a simple extension of the Brooks Range structural trend that was rotated into its present position, relative to the Brooks Range, by oroclinal bending at the Chukchi Syntaxis. In this model, Herald Fault Zone is a seaward extension of the frontal thrust faults of the Brooks Range beyond a relatively minor kink at the Lisburne Peninsula. Grantz *et al.* (1970a) consider the Herald Arch and Fault Zone to be the leading edge of an easterly directed, somewhat younger thrust fault system that crosscuts the Brooks Range and its thrusts. In this case, the Chukchi Syntaxis was created by the superposition of the Herald Fault Zone across the western Brooks Range and the detachment folds that lie in front of it.

The Herald Fault Zone trends across the detachment folds of the Colville Foredeep at an angle of 45°, yet there appears to be no faulting between the outer thrust folds of the fault zone and the intersected detachment folds (Figs. 8 or 9, and Grantz *et al.*, 1970a). These relations are most simply explained if the Herald Fault Zone and its inferred onshore extension, the thrust fault system that bounds the east side of the Lisburne Hills, postdate and are superimposed across the detachment folds (see also Chapman and Sable, 1960; and Martin, 1970). The suggested superposition could have created the Chukchi Syntaxis. Additional evidence bearing on the origin of the syntaxis comes from paleomagnetic poles in Mississippian (lower Lisburne Group) beds on the Lisburne Peninsula and in the western Brooks range (Newman *et al.*, 1979). Preliminary data suggest that these poles have a common orientation even though they occur on both flanks of the syntaxis. This result is difficult to reconcile with the orocline hypothesis but it is at least compatible with the superposition hypothesis for the origin of the syntaxis. The age of the syntaxis is late Cretaceous or early Tertiary since Albian strata are strongly deformed in its core and its apex is blanketed by Paleogene(?) beds in Hope Basin.

G. Hope Basin

Hope Basin (Figs. 9 and 10) lies between the presumed Paleozoic rocks of the Herald Arch to the northeast and the Paleozoic and Precambrian rocks of Seward and Chukotsk Peninsulas to the southwest. It overlies strongly deformed Lower Cretaceous (Albian) and older rocks (Campbell, 1967) of the Brooks Range Orogen. Low seismic velocities (V_p = 1.7 to 3.3 km/sec), the age of the underlying rocks, and the character and age of nearby onshore sedimentary basins suggest that Hope Basin is filled with Tertiary and possibly some Upper Cretaceous sediments. A series of basement ridges subdivides the basin into a number of east–west troughs in which sediment thicknesses locally exceed 3000 m. The largest of the basement ridges is Kotzebue Arch, a structural high that trends westerly across the southern part of the basin (Fig. 9). This arch, which appears to have existed since the early Tertiary, is overlain by several hundred meters of Tertiary sediments. The smaller ridges that lie between the Kotzebue and Herald Arches are more deeply buried. Kotzebue Arch is alined at Cape Krusenstern with the E- and ENE-trending Igichuk Hills, which may be its onshore extension. Indeed, an easterly striking positive linear gravity anomaly along the arch (Ostenso, 1968) trends onshore and follows the western Igichuk and Kiana Hills to the Kobuk River near Kiana (Barnes, 1976), 150 km inland.

During Neogene time, a deep, east–west elongate subbasin developed in eastern central Hope Basin north of Kotzebue Arch. The axis of the subbasin is defined by a thickening of the sedimentary section above a key regional seismic reflector that is believed to approximate the Neogene–Paleogene boundary. Subsidence of the subbasin was accommodated by numerous antithetic and normal faults (Fig. 10), and the arch itself was concurrently uplifted several hundred meters. Between Point Hope and Cape Krusenstern the Neogene subbasin is bounded by normal faults parallel to the coast that also form the western boundary of the De Long Mountains and the western Brooks Range. The northern boundary of the subbasin is a series of monoclines and normal faults that bring older rocks to the surface in Herald Arch. Westward,

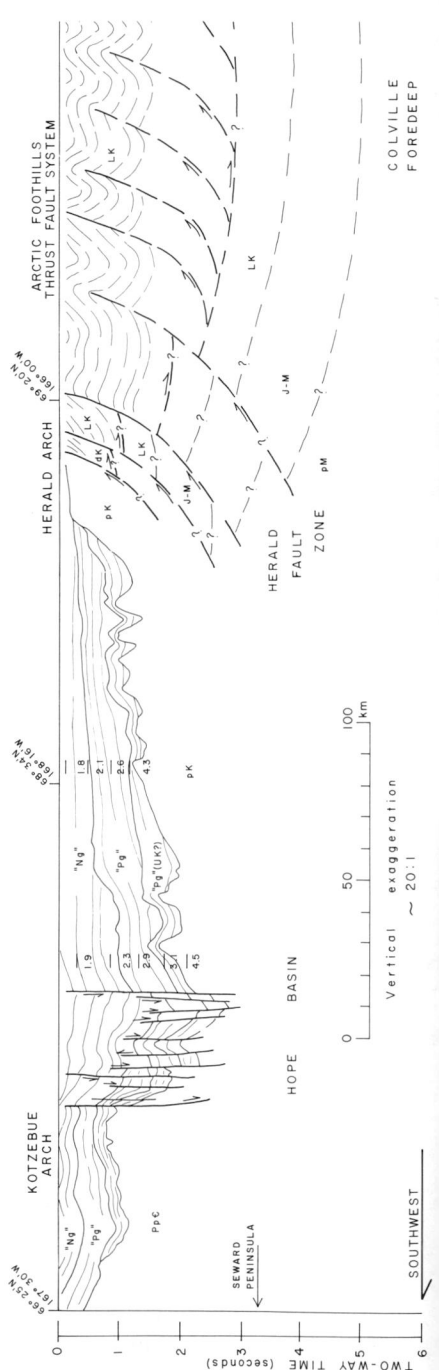

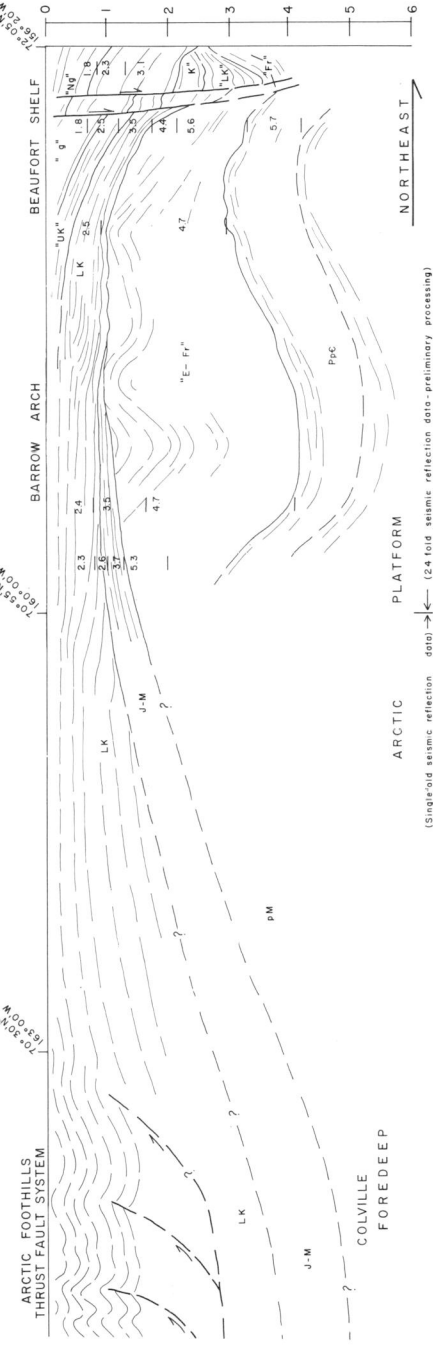

Fig. 10. Geologic cross section tentatively interpreted from seismic reflection data in the eastern Chukchi Sea from the continental shelf break to southern Hope Basin. See Figs. 1, 8, and 9 for location. Sedimentary strata shown are (Ng) Neogene(?) clastic rocks; (Pg) Paleogene(?) clastic rocks; (UK) Upper Cretaceous(?) marine clastic rocks; (LK) Lower Cretaceous marine clastic rocks; (J–M) Mississippian to Jurassic (Ellesmerian) marine and nonmarine clastic and carbonate rocks; (E–Fr) Lower Ellesmerian or Franklinian clastic rocks; and (dK) strongly deformed Cretaceous clastic rocks. Undifferentiated rock units are (pK) pre-Cretaceous; (pM) pre-Mississippian; (Pp€) Paleozoic and (or) Precambrian. Preliminary seismic CDP interval velocities n km/sec.

the Neogene subbasin gradually diminishes in depth, and near 171°W longitude its ridge and trough structures die out entirely.

An episode of volcanic and tectonic activity strongly affected the Seward Peninsula and lower Kobuk Valley beginning in late Miocene time. Plio-Pleistocene basalts flooded a large area south of Kotzebue Sound and tectonic warping and faulting offset Miocene gravels and Pleistocene glacial deposits. These displacements were accompanied by the formation of sizable nonmarine sedimentary basins and by block faulting in and adjacent to the Kigluaik and Bendeleben Mountains of the Seward Peninsula and the Waring Mountains of the lower Kobuk Valley (Hudson, 1977; Patton, 1973). This tectonism may also have been responsible for the Neogene subsidence, arching, and faulting in Hope Basin.

Hope Basin developed in two steps. A Neogene subbasin less than 100 km wide resulting from late Tertiary extensional tectonism formed within a broader mid-Tertiary or Paleogene basin more than 200 km wide that was produced by late Cretaceous–early Tertiary subsidence. Regional onshore stratigraphy suggests that both basins consist at least partly, and perhaps largely, of nonmarine rocks. However, a few marginal outcrops of Neogene marine strata and the periodic migration of Neogene marine faunas across Bering Strait (Hopkins, 1967) indicate that some marine beds must occur in the Neogene subbasin, which subsided rapidly. In both the older and younger basins marine rocks may replace nonmarine rocks away from shore.

H. North Chukchi Basin

The North Chukchi Basin (Grantz *et al.*, 1975, p. 687–690) underlies the northern Chukchi Shelf between the western Barrow and Wrangel Arches on the east and south and the Chukchi Continental Borderland on the north (Fig. 3). The westerly extent of the basin is not known.

The North Chukchi Basin contains a thick prograded sequence of Brookian clastic sedimentary rock. Our seismic reflection data and a long refraction line by Hunkins (1966) indicate that the total section is more than 6 km thick (Figs. 8 and 11). Seismic profiles show that some beds in the basin onlap the Barrow Arch on the south, but we lack data from their contact with the Chukchi Borderland to the north. Although both nonmarine and marine beds occur in the Brookian Sequence in northwest Alaska, the equivalent rocks in the North Chukchi Basin are probably mainly, and perhaps entirely, marine. The basin fill can be divided into two to four stratigraphic packets on the basis of unconformities seen on single-channel seismic reflection profiles. Sonobuoy refraction velocities and correlations with the stratigraphy of northern Alaska suggest that these units are Neogene, Paleogene, Cretaceous, and pre-Cretaceous.

Strata in the southern part of the North Chukchi Basin dip as much as 15° northerly. Basinward, to the north, dips decrease to between 0° and 1° northerly. The absence of beds with a southerly dip component, even on lines near 74° N latitude within 50 km of the Chukchi Borderland, suggests that the North Chukchi rocks were deposited on thinned continental or oceanic crust rather than in a subsiding intracontinental basin. If so, these rocks may onlap high-standing blocks of pre-Cretaceous [Franklinian(?)] rocks in the Chukchi Borderland.

Well-developed diapirs (Figs. 5 and 8) pierce Tertiary beds in the North Chukchi Basin to within a few tens of meters of the seabed (Grantz *et al.*, 1975, p. 689–693). Low sonobuoy velocities, apparent lack of strong gravity or magnetic anomalies (Fig. 12), and regional stratigraphy suggest that the diapirs are probably shale, rather than salt, gypsum, or igneous rock. Possibly the diapirs consist of soft Jurassic(?) or Lower Cretaceous prodelta shale, an early deposit in the basin subsequently deeply buried by rapidly deposited slope, shelf, or deltaic sediment. Such shale, being weak and of relatively low density, might have risen buoyantly toward the seabed under the load of thick overlying sediments, piercing and bending the adjacent beds upwards in the process. Five diapirs (two crossings, three near-misses) have been identified in the basin to date. The best studied is about 2 km in diameter and extends to 3 km or more beneath the seabed. Judging from the number found and the density of our line coverage, something like 30 or 40 diapirs may be present.

I. Western Beaufort Shelf Sedimentary Prism

1. *Sediments*

A relatively narrow northward-prograded sedimentary prism consisting primarily of Brookian rocks underlies the continental terrace of the western Beaufort and northeast Chukchi Seas (Fig. 3). Between the North Chukchi Basin and Camden Bay, the prism overlies Franklinian rocks on the north flank of the Barrow Arch. Beneath the southern part of the shelf, the sequence includes locally thick sections of upper Ellesmerian Jurassic and Neocomian beds of northern provenance (Fig. 13B) and, in places near shore, the northern wedge edges of pre-Jurassic Ellesmerian formations. Between Harrison and Camden Bays the Tertiary beds of the Brookian Sequence thicken and extend more than 100 km inland to form the Camden Basin.

West of Harrison Bay the sedimentary prism consists mostly of Lower Cretaceous marine rocks with a Tertiary wedge of unknown character beneath the outer shelf and slope. The thickness of this sequence is 1 to 3 km near the coast and 6 to 8 km beneath the outer shelf. Between Harrison and Camden Bays, however, the prism consists mainly of Upper Cretaceous and Tertiary

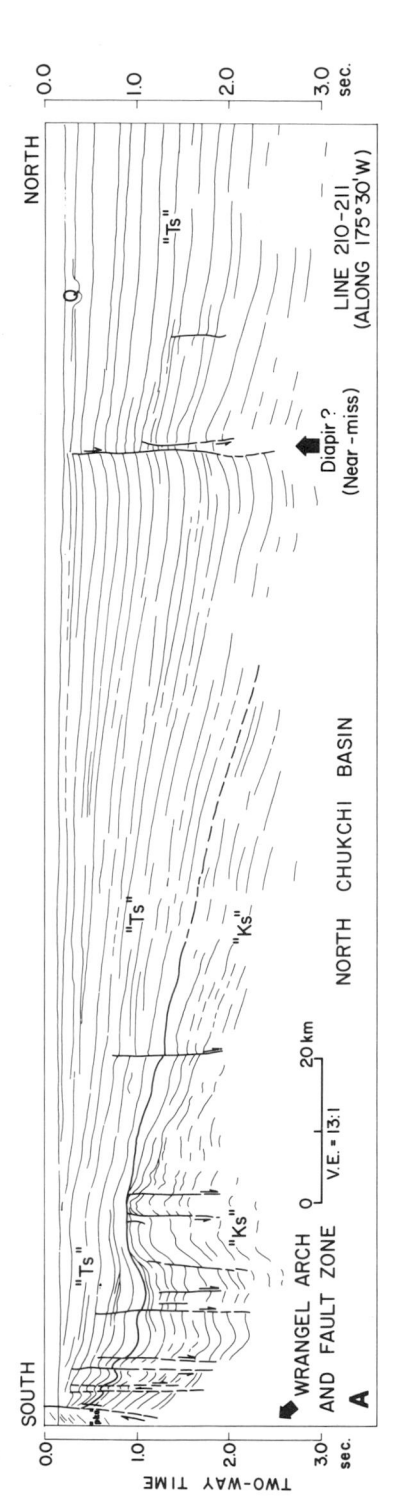

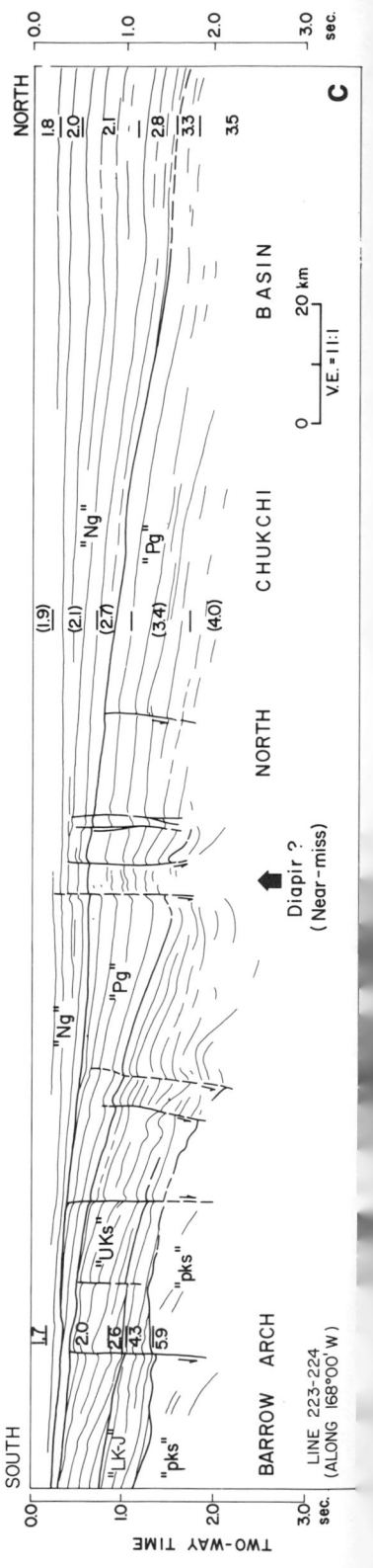

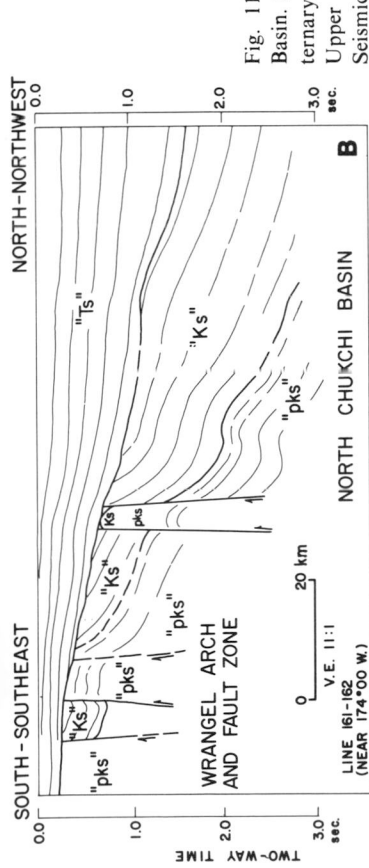

SOUTH-SOUTHEAST   NORTH-NORTHWEST

"Ts"

"Ks"

"pks"

"Ks"

"pks"

"Ks"

"pks"

"Ks"

"pks"

ks

pks

NORTH CHUKCHI BASIN

WRANGEL ARCH
AND FAULT ZONE

0   20 km

V.E. 11:1

LINE 161-162
(NEAR 174°00 W.)

TWO-WAY TIME

B

Fig. 11. Geologic interpretation of seismic sections across the North Chukchi Basin. See Fig. 1 for location. Inferred clastic sedimentary rock units are (Q) Quaternary; (Ts) Tertiary; (Ng) Neogene; (Pg) Paleogene; (Ks) Cretaceous; (UKs) Upper Cretaceous; (LK–J) Jurassic–early Cretaceous; (pKs) pre-Cretaceous(?). Seismic sonobuoy interval velocities (Fig. 13C) in km/sec.

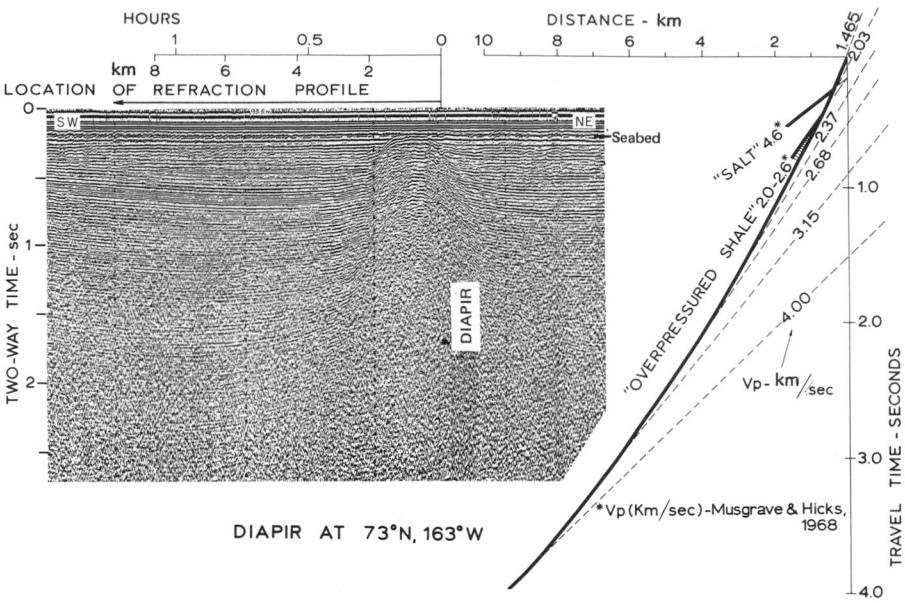

Fig. 12.   Seismic reflection and refraction profiles across a shallow shale(?) diapir at 73° N latitude, 163° W longitude, North Chukchi Basin. The lines labeled "Salt" and "Overpressured shale" show the expected positions of early refractions if the diapir were composed of salt or shale. The observed refractions lie within the V_p range of overpressured shale and below the V_p range of salt.

beds, with the Tertiary becoming dominant near Camden Bay. Here the prism is 3 to 6 km thick at the coast and 6 to 8 km thick on the outer shelf. Onshore correlatives of these rocks are both marine and nonmarine. The lower part of the Tertiary and probably the upper part of the Upper Cretaceous section in the Camden embayment contain coal beds (Alaska Geological Society, 1977). Both the Tertiary and Cretaceous beds are presumed to become more marine and finer grained seaward.

East from Camden Bay, Jurassic and Neocomian deposits as well as later Cretaceous and Tertiary deposits crop out on the Arctic Coastal Plain and are interpreted to project offshore beneath the Western Beaufort Shelf. Dark gray Kingak Shale with Middle Jurassic (Bajocian) marine fossils (Reiser *et al.*, 1978) crops out on the coastal plain 30 km southeast of Barter Island. The outcrops are surrounded by early Cretaceous (Neocomian) and late Cretaceous (Turonian) marine shale. Farther east, Norris (1977) interprets about 1800 m of Kingak Shale (Jurassic and lowest Cretaceous) to extend northward beneath the Beaufort Shelf from outcrops near the Yukon coast, where the formation overlaps pre-Ellesmerian rocks. The outcrops of Kingak Shale near Barter Island lie 30 km north of the easterly projection of the truncation edge (zero isopach) of the northward-thinning Kingak Shale. This isopach trends from near Barrow on the west to the eastern Sadlerochit Mountains, 50 km SSW of Barter Island. Thus, a thick and extensive section of the Jurassic Kingak Shale appears to be included at the base of the Western Beaufort Shelf sedimentary prism beneath the shelf to the east, but not to the west, of eastern Camden Bay.

2. Structure

The Western Beaufort Shelf contains three terranes of contrasting geologic structure. In the first structural terrane, which extends from Camden Bay to the Alaska–Yukon boundary, the inner half of the shelf is underlain by two large, compound structural arches or anticlines and several much smaller anticlines (Figs. 5 and 13A). The largest arch exceeds 200 km in length and 10 to 15 km in width; its maximum amplitude exceeds 4 km. The arches and anticlines are subparallel to the adjacent arcuate coastline and to long, large-amplitude onshore folds, such as Marsh Anticline, that lie in front of the northward salient of the Brooks Range in northeasternmost Alaska. The folds terminate along the projection of the structural front that bounds the west face of this salient (Fig. 5). This feature was informally named the Shaviovik Front by the late Sankey L. Blanton (oral communication, 1977). The outer half of the shelf in the first terrane is underlain by a monocline that dips seaward subparallel to the seabed out to the main shelf break, where the beds are dropped by slumping onto the continental slope.

The crests of the large arches are underlain at depths as shallow as 0.5 km by strata with seismic velocities of 3.5 km/sec or more, which are appropriate for the Cretaceous and Jurassic clastic sedimentary rocks of the region. The folding is as young as Neogene, and locally Quaternary, since beds of these ages are affected in Marsh Anticline, which underlies the Arctic Coastal Plain near Camden Bay.

The arches and associated folds are somewhat similar in geometry and tectonic position to the long, west-trending, décollement-related thrust folds of the Colville Foredeep north of the Brooks Range and west of the Shaviovik Front, although the folding occurred at very different times in these two areas. The eastern folds are thought to be related to late Cenozoic uplift and northernward translation of the northeast salient of the Brooks Range, whereas the folds of the Colville Foredeep are related to Laramide (Cretaceous and early Tertiary) uplift and relative northward translation of the Brooks

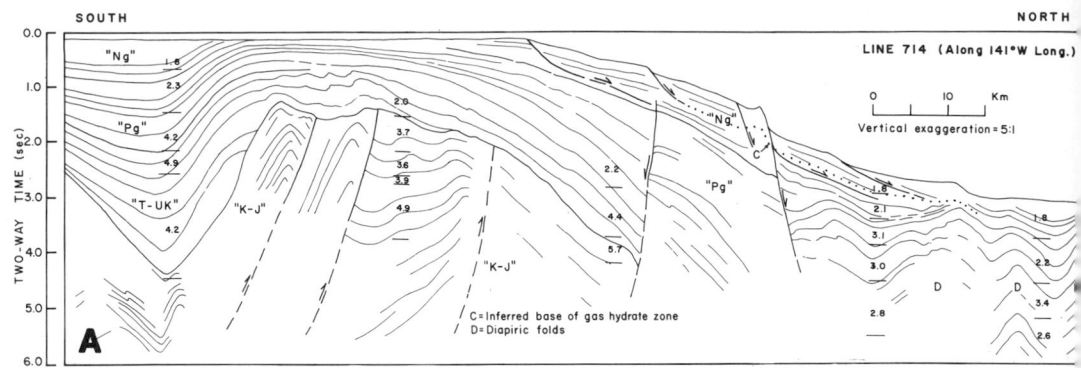

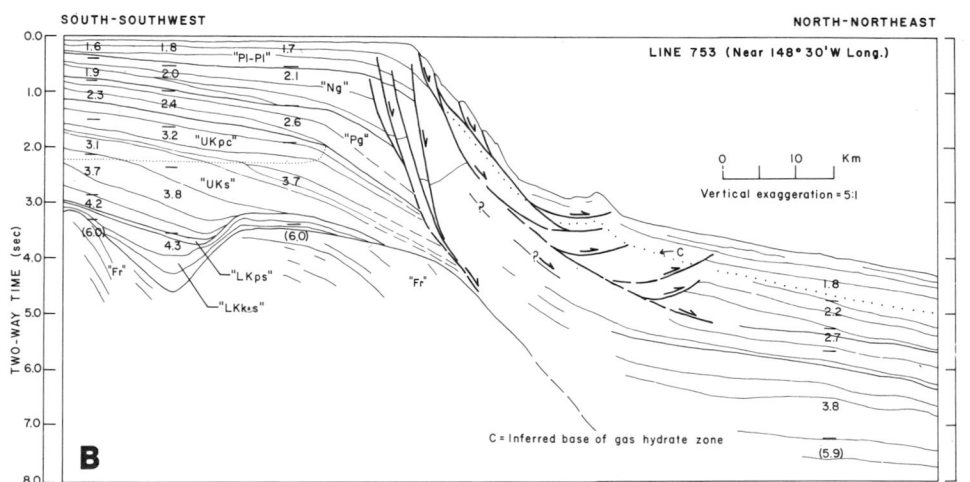

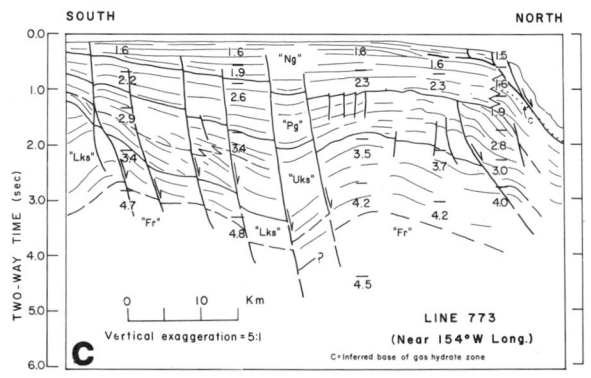

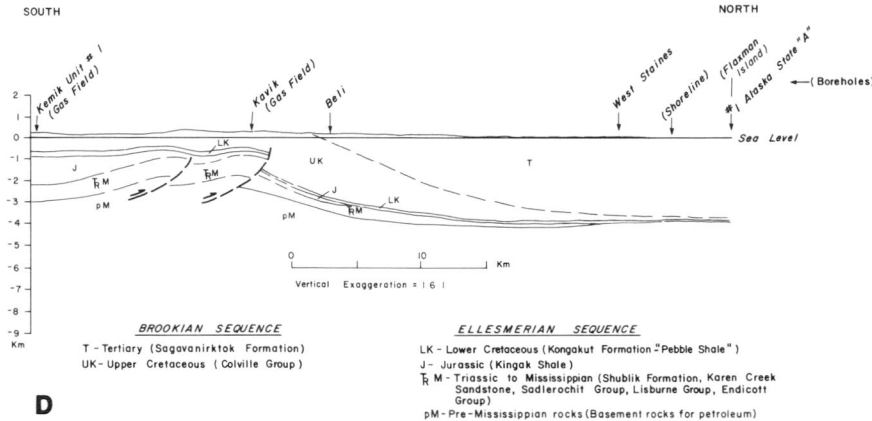

Fig. 13. Geologic interpretation of seismic sections across the Western Beaufort Shelf (A–C) and a cross section near the Canning River based on correlated test wells (D). See Fig. 1 for location. Seismic CDP interval velocities in km/sec. A) Inferred clastic sedimentary rock units are (Ng) Neogene; (Pg) Paleogene; (T–UK) Tertiary and (or) Upper Cretaceous; (K–J) Cretaceous and Jurassic, possibly only Lower Cretaceous and Jurassic. (B) Inferred clastic sedimentary rock units are (Pl–Pl) Plio-Pleistocene; (Ng) Neogene; (Pg) Paleogene; (UKpc) Upper Cretaceous Prince Creek Formation; (UKs) Upper Cretaceous Schrader Bluff Formation; (LKps) Lower Cretaceous "Pebble Shale"; (LKkrs) Lower Cretaceous Kuparuk River Sandstone; (Fr) Franklinian Sequence. (C) Inferred clastic sedimentary rock units are (Ng) Neogene; (Pg) Paleogene; (UKs) Upper Cretaceous; (LKs) Lower Cretaceous; (Fr) Franklinian Sequence. (D) Generalized structural cross section from Kemik gas field to Flaxman Island. Data interpreted from a compilation of geologic formations encountered in boreholes between the Canning and Colville Rivers by Tailleur *et al.* (1978). The control points for the cross section (boreholes) would not suffice to delineate any small-scale geologic structures that might be present.

Range west of the Shaviovik Front. A Laramide age is inferred because the thrust folds deform Albian and Upper Cretaceous beds north or the Brooks Range and are in turn deformed in the Chukchi Syntaxis, which is onlapped by Paleogene(?) beds in Hope Basin.

The northern, and apparently larger, of the offshore arches dies out near western Camden Bay. Off Canning River, 15 km west of our westernmost crossing of this structure, only down-to-the-basin normal faults with displacements of 100 or 200 m are seen on seismic profiles. Shallow water limited our profiles here to the outer two-thirds of the continental shelf, so we have no data as to whether the southerly arch or related structures continue westward on the inner shelf.

The second structural terrane of the Western Beaufort Shelf, which lies between the Canning River and Cape Halkett, is characterized by a monocline that dips typically 20 to 60 m/km seaward (Fig. 13B). This feature is broken on the inner and mid-shelf by small down-to-the-basin normal faults that are more or less parallel to the shelf edge. The faults become more numerous toward the shelf edge, where large down-to-the-basin gravitational failures dominate the structure.

The third structural terrane lies west of Cape Halkett. Here normal faults are prominent and a long structural high (Fig. 13C) underlies the outermost shelf. The sedimentary section beneath the outer shelf contains intraformational slumps and growth faults at least as far west as 160° W longitude (Figs. 10 and 13C). The outer shelf structural high is syn- and post-depositional, and it therefore did not serve as a dam that trapped the deposits of the Western Beaufort Shelf sedimentary prism. The high is 20 to 25 km wide and extends as a continuous or semicontinuous feature from western Harrison Bay to at least 160° W, a distance of 350 km (Figs. 5 and 7). Its amplitude is typically 0.3–0.5 km at depths of about 1 km, and in places as much as 1.5 km at depths of 1.5 to 2.0 km. Preliminary interpretation suggests that the feature was produced by rotation along normal or growth faults (Fig. 13C) during late Cretaceous and Tertiary time. On one CDP seismic section the growth fault dips about 50° N and displacement is down to the north about 1.5 km. An analogous but smaller and discontinuous feature, locally obscure, extends from Harrison Bay to Camden Bay.

3. *Block Glides*

The surface of the outermost continental shelf between 155° W longitude on the east and beyond 158° 30′ W longitude on the west is broken by multiple open cracks that become deeper toward the shelf break (Figs. 2 and 14). Water depth in the affected areas is 70 to at least 350 m. Similar features buried beneath Holocene sediment and low scarps border the zone on the south. The cracks and associated features are thought to be the result of Holocene block gliding. The affected sediments are flat lying, as much as 140 m thick, and readily penetrated by low-energy acoustic signals. They probably consist of poorly consolidated Quaternary deposits. Reflection characteristics and the ability to support deep open cracks suggest that the upper part of the section, which is well bedded, consists of lutite and fine sand. The lower part is more homogeneous and probably consists of lutite, except for intervals characterized by broken, irregular reflectors. These intervals are inferred to be slip zones on which the large blocks bounded by the observed cracks glide toward the shelf break. The gentle seaward dip of the slip zones, about 1°, and the inferred parallelism of the slip zones and bedding suggest that the slippage occurs within beds that are weak or susceptible to liquefaction. The block-glide terrane is adjacent to Barrow Sea Valley and the northeast-flowing Alaskan Coastal Water, which may be supplying sediment in the silt and very fine sand grades typical of liquefiable sediments.

The open cracks are commonly as deep as 8 to 17 m, and one exceptional crack is 37 m deep. Downward, they persist as filled cracks and closed frac-

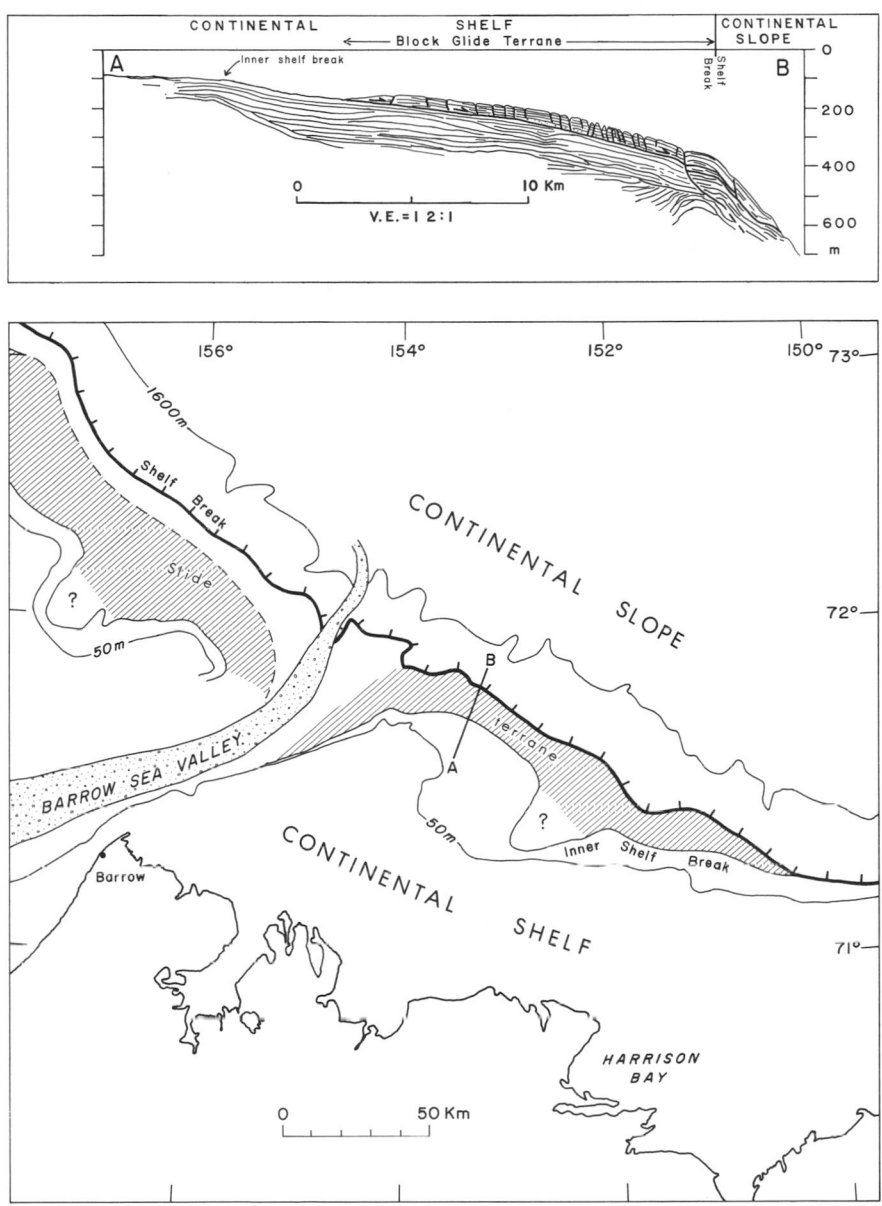

Fig. 14. Block glides on the Western Beaufort Shelf, mapped from high-resolution seismic profiles.

tures to depths of 40 or 50 m, and locally to 75 m below the seabed, ending at the inferred slip zones. The cracks, presumably aligned subparallel to the shelf break, are spaced 100 m to 500 m apart. Their spacing in general decreases seaward. The fact that many of the cracks are unfilled indicates that the block gliding is still an active, or potentially active, process. Three underlying slip zones and glide sheets have been recognized in places beneath the active, or quasi-active surficial glide sheet on high-resolution seismic records.

4. *Bedding Plane Slides*

A mosaic of bedding plane slides similar in origin and character to the block glides underlies the upper half of the Beaufort Ramp between 148° W longitude and the Mackenzie Sea Valley, a distance of about 300 km (Figs. 2 and 15). The slides are 10 to 38 km long in the direction of slip and 20 to 230 m thick. The dip of the basal slip planes is typically 0.5° to 1.5°. The slides

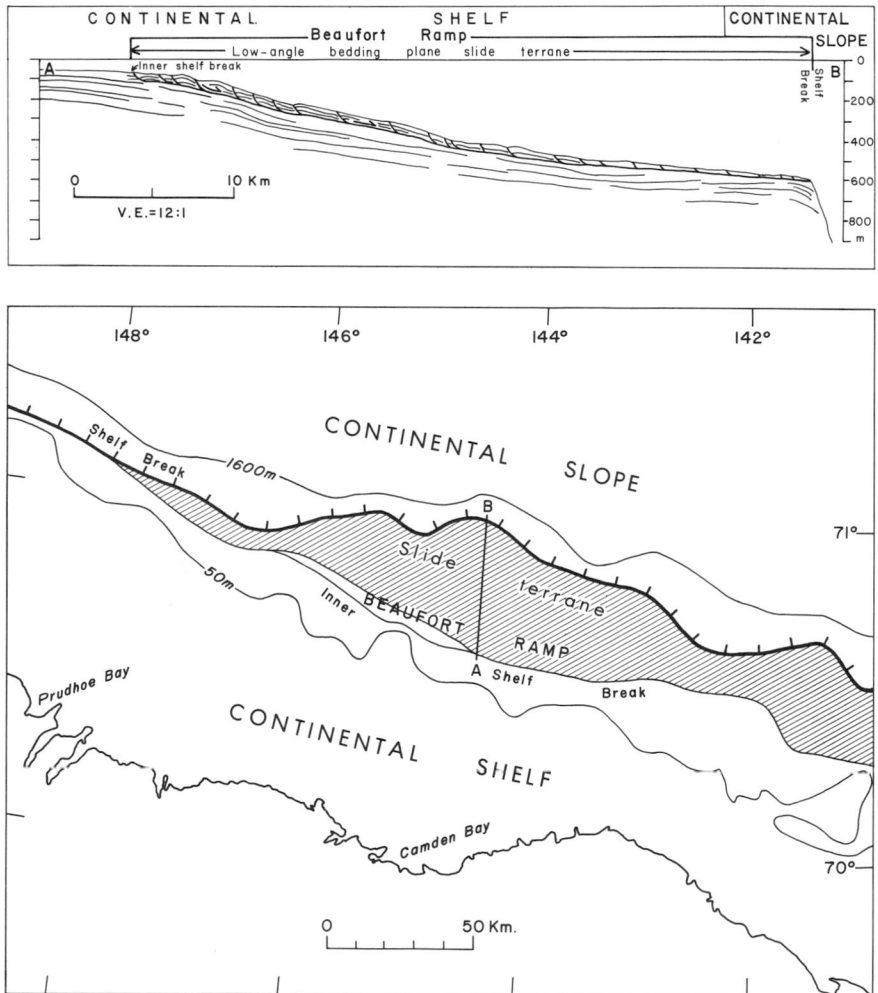

Fig. 15. Bedding plane slides on the Western Beaufort Shelf, mapped from high-resolution seismic profiles.

occur in a prograded wedge of Quaternary sediment beneath the Beaufort Ramp seaward from the crestal zones of midshelf anticlines. Over most of their length the upper parts of the slide masses are slabs of soft sediment broken by many extensional fractures. The lower beds appear on seismic profiles to be churned and dismembered. Variation in thickness of the slides of about 20 to 50 percent, caused by internal flowage in the lower beds, is superimposed upon a general downslope thinning of the wedge of sediment in which the slides developed.

At the landward margin of the Beaufort Ramp, near the 60-m isobath, the slides are bounded by well-defined slip-plane scarps and pull-apart grabens. The slide toes are usually low bulges that merge with the gently sloping seabed. Presumably, the slip planes dissipate their displacement downslope mainly by internal distortion within the lower part of the slide. The slip planes are subparallel to bedding, a geometry that suggests the shearing occurred in relatively weak or sensitive strata. Estimates of the horizontal slip at the head of well-developed slides range from 0.2 to 2.3 km, with a median value of about 1 km. Near 145° W longitude the lower 15 km of the slide is only 20 to 30 m thick and consists of surficial sediment with well-developed bedding that is parallel to the seabed and the slide plane. This thin part of the slide is characterized by numerous extension cracks 0.25 to 1.5 km apart, measured down the slope, and closely resembles the block glides west of 151° W longitude. The thin slide extends to the shelf edge, where it slumps down the upper continental slope.

J. Diapiric Folds beneath the Continental Rise

Diapiric folds disrupt bedding in the sedimentary rocks of the lower Alaskan Continental Slope and Rise and adjacent parts of the Canada Rise east of 146° W longitude (Figs. 5, 13A, and 16). The folds are 5 to 10 km apart along lines normal to the slope and as much as 0.5 km in amplitude. Some appear to be elongate parallel to the slope, but for most of them our lines are too widely spaced to distinguish between ridge or dome shapes. The geometry and position of the diapiric folds suggest that they rose as a consequence of sedimentary loading of the outer shelf and the movement of large slide masses on the continental slope and rise. Equally thick sediment and large slides are found, however, on the shelf and slope west of the diapirs. This distribution suggests that an easily deformable sedimentary unit susceptible to diapirism underlies the continental shelf, slope, and rise east, but not west, of 146° W longitude. Seismic interval velocities in the diapiric material vary from 2.6 to 3.0 km/sec, appropriate to shale, whereas seismic velocities in the neighborhood of 4.6 km/sec, appropriate to salt, were not observed. On the basis of their apparent stratigraphic position (Fig. 13A), we tentatively suggest that the diapiric beds may be lower Tertiary.

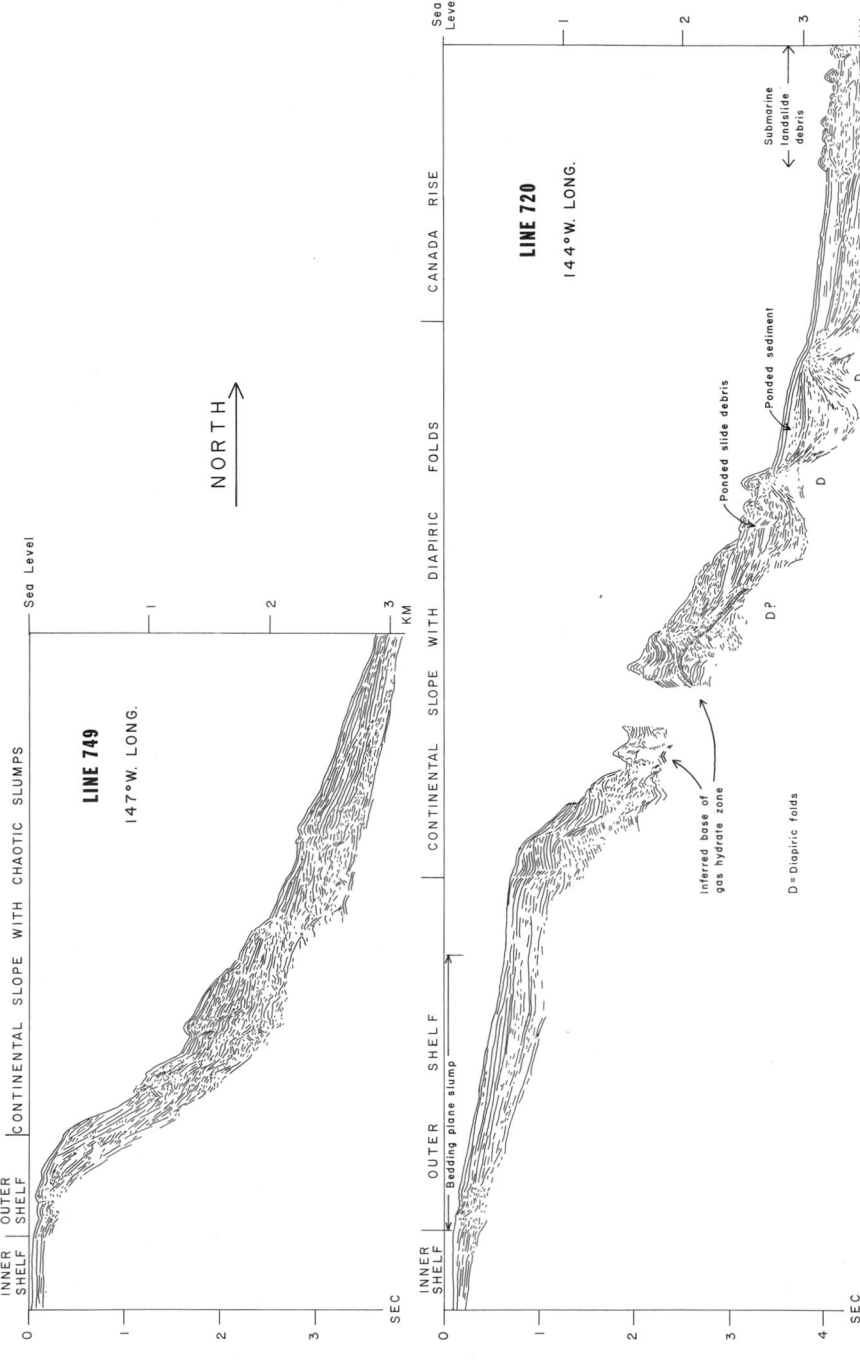

Fig. 16. Shallow structure interpreted from seismic profiles across the continental slope in the Western Beaufort Sea where diapiric folds are present (Line 720) and absent (Line 749).

Because the diapiric folds buckle the seabed and pond the youngest (Holocene) sediments and slumps on the continental slope and rise, it is likely that diapirism is active now, or at least was active during the Holocene. The physiographic distinction between continental slope and rise is obscured in the area of the diapiric folds because the folds have locally transformed the Alaska Rise from a smooth sedimentary apron to a series of bathymetric steps. In places the diapiric folds have so thoroughly disrupted the Alaska Rise sedimentary prism that bedding is obscure. Beneath adjacent parts of the Mackenzie Cone, however, some diapiric folds underlie undisturbed beds more than 1500 m thick. Low seismic velocities and estimated sedimentation rates indicate that the undisturbed beds are Neogene and Pleistocene. Thus, some of the diapirs appear to have been inactive since the Neogene.

K. Solid Gas Hydrate

Much of the continental slope and rise north of Alaska is underlain by an unusually strong seismic reflector (Grantz et al., 1976) that crosses bedding planes at many places. The reflector mimics, or simulates, the bathymetry of the sea floor, yet lies some 200 to 600 m beneath it (Figs. 13A,B and 16). This bottom-simulating reflector (BSR) occurs only where the sea is deeper than 400 to 600 m and was recognized beneath waters as deep as 3200 m. The BSR was identified beneath some 60 percent of our seismic lines on the Alaska Slope and Rise where the water depth exceeded 600 m. It is most strongly developed beneath bathymetric highs and weakest, indeed commonly absent, beneath bathymetric lows. In a few places where it is particularly well developed beneath highs, the underlying reflectors are bowed downward. These observations suggest that the BSR represents the interface of a zone of free gas in the sediments and an overlying impervious cap. Because the BSR lies at the approximate pressure–temperature boundary at which solid gas hydrates of methane and some multicomponent natural hydrocarbon gases break down to a gas + water phase, we postulate that it is produced by free gas trapped beneath an impermeable cap of sediment cemented with gas hydrate.

L. Canada Basin

Water depths exceeding 3500 m and dispersion patterns of Lg-phase seismic waves (Oliver et al., 1955) indicate that the Canada Basin is underlain by oceanic crust, and was therefore formed by sea-floor spreading. Some workers (e.g., Vogt and Ostenso, 1970; Hall, 1973) have postulated the spreading center to be the Alpha–Mendeleyev Ridge, which crosses the basin from Greenland to eastern Siberia. Carey (1958), Tailleur (1969, 1973), and Rickwood (1970) suggest that the Canada Basin opened by a relative rotation of northern Alaska away from the Canadian Arctic Islands about a pole near the Mackenzie Delta or in Alaska. Herron *et al.* (1974) postulate that the Canada Basin formed by rifting of the Kolyma Massif of eastern Siberia away from the Canadian Arctic Islands along a northerly trending spreading axis, and Yorath and Norris (1975) propose a northerly trending axis of assymetric spreading beneath the head of the continental rise in the eastern Beaufort Sea.

The inferred thickness of sediment in the southern part of the Canada Basin is shown by dotted isopach formlines in Fig. 5. The thicknesses are based on three sonobuoy refraction profiles and the assumption that differences in depth of the seabed reflect differences in the thickness of sediment that has accumulated on oceanic crust in the basin. As the age of the Canada Basin is inferred from stratigraphic relations to be Jurassic or early Cretaceous, differences in age from the center to the sides are probably not significant to thickness estimates. We therefore assumed that all points in the basin are in isostatic equilibrium, and generated isopach formlines from the bathymetric contours by correlating the isobaths with sediment thickness at the three sonobuoy refraction stations (Fig. 5). Sediment density for our model was based on the sonobuoy-derived velocities, which suggest that the average density of the sedimentary fill beneath the Canada Rise is about 2.2 g/cm^3. We assumed a mantle density of 3.4 g/cm^3 and that the 5.3 km/sec, 4.9 km/sec, and 4.6 km/sec refractors are from the top of oceanic layer 2 (basalt). On the basis of these assumptions, about 4.5 km of sediment underlies the Canada Abyssal Plain. Hall (1973), on the basis of seismic reflection data, reports that more than 2 km of sediment underlies the northern part of this plain.

An estimate of the age of the Canada Basin can be made from the thickness of sediment that it contains. If we assume that oceanic crust beneath the basin was isostatically depressed by the overlying sedimentary prism, we can calculate the depth of an "unloaded" crust and compare it with empirical age–depth curves for the Atlantic and Pacific sea floor. Such curves were compiled most recently by Parsons and Sclater (1977). Using mantle density 3.4 g/cm^3, sediment density 2.2 g/cm^3, and water density 1.03 g/cm^3, our data from the three sonobuoys in deep water give a "corrected" depth to basement of 6.4 km for the southern Canada Basin. This depth corresponds to a crustal age greater than 120 m.y.b.p. on the Parsons–Sclater curves. In this region, however, the curves have a low slope, and small uncertainties in the unloaded depth produce large uncertainties in age of sea floor. Thus the significance of the "unloaded" depth we obtained (6.4 km) is that it argues for a Mesozoic age for the crust of the Canada Basin.

M. Segmentation of the Alaskan Continental Margin

The continental margin north of Alaska consists of three sectors of contrasting geology separated by onshore extensions of possible oceanic fracture zones. The *Chukchi sector* is characterized by the North Chukchi Basin and the Chukchi Continental Borderland; the *Barrow sector* is characterized by the Barrow Arch, where Cenozoic sedimentary rocks are relatively thin or absent and where Mississippian to Neocomian sedimentary rocks coarsen, thin, and wedge out to the north; and the *Barter Island sector* is characterized by thick Cenozoic as well as Jurassic and Cretaceous clastic sedimentary rocks that appear to have been derived from the south and that become thicker and finer grained to the north. Large Neogene folds in Jurassic to Neogene strata and the northward salient of the northeastern Brooks Range also characterize this sector.

The Chukchi and Barrow sectors are postulated to be separated by a fracture zone along the northwest flank of the Barrow Arch west of 160° W longitude. The Chukchi Borderland may consist of fragments of continent that were transported relatively NNE, away from the Chukchi Shelf, along the fracture zone to create the North Chukchi Basin. The Barrow and Barter Island sectors may be separated by a fracture zone near eastern Camden Bay, but its location and trend have not been recognized. The northern coastal plain and shelf east of this postulated fracture zone is apparently underlain by a thick section of Jurassic sedimentary rocks that is thin or absent to the west (Fig. 13A,D).

The character and distribution of the Jurassic and Neocomian bedded rocks are critical for understanding the geometry and age of the rifted Alaskan margin. On the east, in the Barter Island sector, Jurassic and Neocomian sedimentary rocks in northern Yukon Territory (Kingak Shale) appear to have southern sources and to fine northward (Young *et al.*, 1976; Figs. 4, 5, and 7). These rocks are interpreted to be 1800 m thick under the adjacent Beaufort Shelf (Norris, 1977). As noted above, the Kingak Shale and the Neocomian Kongakut Formation also crop out on the Arctic Coastal Plain 30 km southeast of Barter Island. In the Barrow sector, in contrast, Jurassic and Neocomian strata have northern sources and fine southward. If projected E or ESE along the general trend of the depositional strike and the truncation edge of Jurassic strata on the Barrow arch, the source area would strike into the area of thick Jurassic and Neocomian marine strata inferred to underlie the coastal plain and shelf in the Barter Island sector. If projected WNW, the source area would strike into the deep North Chukchi Basin of the Chukchi sector. These relations, and the fact that the Upper Triassic rocks of both the Barrow and Barter Island sectors were derived from northern sources (Tailleur, 1969, 1973; Brosgé and Tailleur, 1971; Detterman *et al.*, 1975; Norris, 1977) indicate that rifting of the Alaskan margin began in early Jurassic time.

A northern source for the Kingak Shale and Neocomian strata of the Barrow sector is supported by subsurface stratigraphic data, but a southern source for these beds beneath the coastal plain and shelf in the Barter Island sector is less firmly based. It is inferred from a few outcrops in the Arctic Coastal

Plain, preliminary interpretation of seismic records on the continental shelf, and extrapolation of facies maps of Jurassic and Cretaceous sedimentary rocks in the Mackenzie Delta region. If this distribution of source terranes is correct, it implies that in the Barter Island and Chukchi sectors Barrovia, the northern source terrane for the Ellesmerian Sequence of the Arctic Platform, was rifted away from the platform in early Jurassic time. In the Barrow sector, however, the rift would have lain farther north, within Barrovia, and a large east–west-trending welt or island of the Barrovian Highlands would have remained attached to the Arctic Platform. The south shore of the island would have faced the epicontinental sea that covered the Arctic Platform in Jurassic and Neocomian time. We propose that the postulated island was the waning northern provenance area for the Jurassic and Neocomian sedimentary rocks (Kingak Shale, Kuparuk River Sandstone, and "Pebble Shale") of the Barrow sector of northern Alaska. If additional work shows that the Jurassic and Neocomian beds beneath the coastal plain and shelf of the Barter Island sector were derived from the north, rather than from the south, then the age of rifting would have been early Cretaceous. It would have accompanied or post-dated deposition of the Neocomian "Pebble Shale" of the Barrow sector, when sedimentation from northern sources was waning, and predated the arrival of south-sourced (Brookian) clastic sediments on the Western Beaufort Shelf in Albian time.

V. SUMMARY OF TECTONIC IMPLICATIONS AND CONSTRAINTS

1.   The continental margin north of Alaska is of Atlantic type and was formed by rifting. Our data are compatible with the proposals of Carey (1958), Tailleur (1969, 1973), Rickwood (1970), and Newman *et al.* (1979) that the Canada Basin was formed by rifting involving a relative rotation of northern Alaska away from the Canadian Arctic Islands. The pole of rotation was probably in the region of the Mackenzie Delta, as suggested by Rickwood.

2.   Northern sources for the Mississippian to Lower Cretaceous clastic sediments along the Barrow arch of northern Alaska diminished in importance in Jurassic and Neocomian time and were removed by the close of Neocomian time. Removal is attributed by Tailleur (1969, 1973) to opening of the Canada Basin by rifting in post-Triassic, probably early Jurassic time. Rickwood (1970) proposes opening in the late Jurassic and early Cretaceous. Our tentative interpretation of data on the character and distribution of Jurassic and Neocomian strata in the Barter Island sector in Young *et al.* (1976), Norris

(1977), Reiser *et al.* (1978), and the present report suggests that rifting began during the early Jurassic. A submarine canyon cut more than 1.4 km into seaward-prograded Albian marine clastic deposits between Barrow and Harrison Bay and filled with Turonian marine sediments (Collins and Robinson, 1967, p. 183) indicates that the Canada Basin was open by mid-Cretaceous time.

3. Continental extensions of oceanic fracture zones related to the geometry of rifting are postulated to have divided the northern Alaska continental margin into three sectors of contrasting structure and stratigraphy. The northeasterly trend of the postulated fracture zone between the Chukchi and Barrow sectors is compatible with proposals that rifting occurred about a pole of rotation near the Mackenzie Delta. Stratigraphic contrasts in the three sectors suggest that the postulated rift lay farther south in the Chukchi and Barter Island sectors than in the Barrow sector. These relations also imply that in early post-rift (Jurassic and Neocomian) time Barrovia was reduced to an easterly-trending island beneath the present outer continental shelf and slope in the Barrow sector. East and west of the island lay deep marine embayments of the Chukchi and Barter Island sectors.

4. The Western Beaufort Shelf is underlain by a sedimentary prism of Albian (late early Cretaceous) to Quaternary age that was prograded across a subsiding shelf underlain mainly by bedded rocks of the Franklinian Sequence.

5. The great thickness of seaward-prograded sediment in the North Chukchi Basin and its position between the Chukchi Borderland and the Barrow and Wrangel Arches suggest that the basin is floored by oceanic crust or thinned continental crust.

6. If continental outliers of the Chukchi Borderland originally occupied what is now the North Chukchi Basin, it would remove a major obstacle to a simple geometric fit of opposite margins of the Canada Basin. Such a fit would be required if the basin formed by rotation in the manner proposed by Carey (1958), Tailleur (1969, 1973), and Rickwood (1970).

7. The shale(?) diapirs that underlie the North Chukchi Basin and the continental slope and rise off the Barter Island sector occur in the deepest basins along the present northern Alaskan Continental Margin.

8. Pre-Mississippian basement (the Franklinian Sequence) underlies the Barrow sector of the Western Beaufort Shelf and extends seaward to the upper continental slope (Fig. 13B). This distribution suggests that the slope may approximate the position of the initial rift that is postulated to have formed the Canada Basin. A zone of subsided continental crust seaward of the boundary is not precluded.

9. Barrow Arch may be thought of as a hinge line created by Jurassic to Neocomian rifting of the Arctic Platform. The north flank of the arch was created by subsidence, normal faulting, and erosional truncation near the rifted margin. The south flank is a compound of paleoslope and of tilting of the

southern part of the Arctic Platform in response to sediment loading. The sediments consisted of Jurassic(?) and early Cretaceous nappes of the Brooks Range Foreland Thrust Belt and the thick Brookian prism of Jurassic, Cretaceous, and Tertiary clastic sediment deposited in the Colville Foredeep and on the adjacent Arctic Platform.

10.   The Beaufort Ramp may have formed in response to movement of diapiric shale(?) from the outer shelf to the adjacent slope and rise and to subsidence in response to loading of the adjacent Canada Basin by sediment of the Mackenzie Cone.

11.   Thickness of sediment and depth to basement imply a Mesozoic or older age for the Canada Basin, assuming it is underlain by oceanic crust and that the age versus depth relations of Parsons and Sclater (1977) for the North Atlantic and Pacific Oceans are applicable.

12.   The broad region of Hope Basin, Kotzebue Sound, lower Kobuk Valley, and Seward Peninsula appears to have constituted a Tertiary province of regional north–south crustal extension characterized by basin subsidence and block faulting. Late Cretaceous(?) and Paleogene (Laramide) extension created Hope Basin athwart the western Brooks Range, a compressional orogenic belt formed principally during Jurassic, Cretaceous, and early Tertiary(?) time. Probably in part because of this extension, the belt of Brooks Range rocks is much wider across Hope Basin (400 km) than across the central Brooks Range (200 km). The paleoslope and normal faults that form the east edge of Hope Basin terminated the physiographic Brooks Range at Kotzebue Sound by mid-Tertiary time. Neogene–Quaternary extension created normal faults, subbasins, and ridges within Hope Basin, and small sedimentary basins and block faults elsewhere in the province.

13.   Chukchi Syntaxis, the sharp swing in trend of typical Brooks Range rocks and structures at the Lisburne Peninsula, is thought to be the result of the superposition of thrust plates of the Herald Fault Zone across western Brooks Range structures during late Cretaceous or early Tertiary time.

ACKNOWLEDGMENTS

We are indebted to many U.S. Geological Survey colleagues for helping obtain the data presented; to the Coast Guard and the Naval Arctic Research Laboratory, Barrow, Alaska, for logistic assistance; and to David W. Scholl, Leslie Magoon, David A. Dinter, and Irvin L. Tailleur for reviewing the manuscript and offering helpful suggestions. This report has not been reviewed for conformance to stratigraphic nomenclature used by the U.S. Geological Survey.

REFERENCES

Alaska Geological Society, 1971, *West to East Stratigraphic Correlation Section, Point Barrow to Ignek Valley, Arctic North Slope, Alaska*, Anchorage: Alaska Geological Society.

Alaska Geological Society, 1972, *Northwest to Southwest Stratigraphic Correlation Section, Prudhoe Bay to Ignek Valley, Arctic North Slope, Alaska*, Anchorage: Alaska Geological Society.

Alaska Geological Society, 1977, *North to South Stratigraphic Correlation Section, Beaufort Sea–Prudhoe Bay–Nora No. 1, Arctic Slope, Alaska*, Anchorage: Alaska Geological Society.

Alaska Glacial Map Committee, 1965, Map showing extent of glaciations in Alaska, *U.S. Geol. Surv. Misc. Geol. Invest. Map* I-415, scale 1:2,500,000.

Anonymous, 1970, Islands of eastern Siberia and the Chukchi Sea, in: *Geologia SSSR*, Tkachenko, B., and Egiazarov, B. Kh., eds., v. 26, p. 377–404.

Armstrong, A. K., and Bird, K. J., 1976, Facies and environments of deposition of Carboniferous rocks, Arctic Alaska, Alaska Geological Society Symposium, 1975, Anchorage, Alaska, p. A1–A16.

Barnes, D. F., 1976, Bouguer gravity map of Alaska, *U.S. Geol. Surv. Geophys. Invest. Map* GP-913, scale 1:2,500,000.

Barnes, P. W., and Reimnitz, E., 1974, Sedimentary processes on arctic shelves of the northern coast of Alaska, in: *The Coast and Shelf of the Beaufort Sea*, Reed, J. C., and Sater, J. E., eds., Arlington, Virginia: Arctic Institute of North America, p. 439–476.

Beal, M. A., 1969, Bathymetry and structure of the Arctic Ocean, Oregon State University, Corvallis, Ph.D. Thesis, 204 p.

Beikman, H. M., and Lathram, E. H., 1976, Preliminary geologic map of northern Alaska, *U.S. Geol. Surv. Misc. Field Stud. Map* MF-789, 2 sheets, scale 1:1,000,000.

Belevich, A. M., 1969, Marine Neogene diatoms at Cape Enmakai, the north coast of the Chukchi Sea, *Nauch-Issled. Inst. Geol. Arktiki Uch. Zap. Paleontol. Biostratigr.*, v. 28, p. 74–75.

Bird, K. J., 1978, New information on Lisburne Group (Carboniferous and Permian) in Naval Petroleum Reserve, Alaska, *Am. Assoc. Petrol. Geol. Bull.*, v. 62, p. 880.

Brosgé, W. P., and Tailleur, I. L., 1971, Northern Alaska petroleum province, in: *Future Petroleum Provinces of the United States—Their Geology and Potential*, Cram, I. H., ed., *Am. Assoc. Petrol. Geol. Mem.* 15, p. 68–99.

Campbell, R. H., 1967, Areal geology in the vicinity of the Chariot site, Lisburne Peninsula, northwestern Alaska, *U.S. Geol. Surv. Prof. Pap.* 395, 71 p.

Carey, S. W., 1958, A tectonic approach to continental drift, in: *Continental Drift, A Symposium*, Carey, S. W., ed., Hobart: Tasmania University, p. 177–355.

Carter, C., and Laufeld, S., 1975, Ordovician and Silurian fossils in well cores from North Slope of Alaska, *Am. Assoc. Petrol. Geol. Bull.*, v. 59, p. 457–464.

Chapman, R. M., and Sable, E. G., 1960, Geology of the Utukok–Corwin region, northwestern Alaska, *U.S. Geol. Surv. Prof. Pap.* 303-C, 164 p.

Collins, F. R., and Robinson, F. M., 1967, Subsurface stratigraphic, structural and economic geology, northern Alaska, *U.S. Geol. Surv. Open-File Rep.*, 250 p.

Creager, J. S., and McManus, D. A., 1965, Pleistocene drainage patterns on the floor of the Chukchi Sea, *Mar. Geol.*, v. 3, p. 279–290.

Creager, J. S., and McManus, D. A., 1967, Geology of the floor of Bering and Chukchi Seas—American studies, in: *The Bering Sea Land Bridge*, Hopkins, D. M., ed., Stanford: Stanford University Press, p. 7–31.

Detterman, R. L., Reiser, H. N., Brosgé, W. P., and Dutro, J. T., Jr., 1975, Post-Carboniferous stratigraphy, northeastern Alaska, *U.S. Geol. Surv. Prof. Pap.* 886, 46 p.

Eittreim, S., Grantz, A., and Whitney, O. T., 1979, Cenozoic sedimentation and tectonics of the Hope Basin, southern Chukchi Sea, in: *Relationship of Plate Tectonics to Alaskan Geology and Resources*, Sisson, A., ed., Proceedings of the Alaska Geological Society Symposium, Anchorage, Alaska, April 1977, p. B-1–B-11.

Fisher, R. L., Carsola, A. J., and Shumway, G., 1958, Deep-sea bathymetry north of Point Barrow, *Deep-Sea Res.*, v. 5, p. 1–6.

Garrison, G. R., and Becker, P., 1976, The Barrow Canyon; a drain for the Chukchi Sea, *J. Geophys. Res.*, v. 81, p. 4445–4453.

Geological Survey of Canada, 1968, Glacial map of Canada, *Geol. Surv. Can. Map* 1253A, scale 1:5,000,000.

Grantz, A., Wolf, S. C., Breslau, L., Johnson, T. C., and Hanna, W. F., 1970*a*, Reconnaissance geology of the Chukchi Sea as determined by acoustic and magnetic profiles, in: *Proceedings of the Geological Seminar on the North Slope of Alaska, Palo Alto, California, 1970*, Adkison, W. L., and Brosgé, M. M., eds., Los Angeles, California: American Association of Petroleum Geologists, Pacific Section, p. F1–F28.

Grantz, A., Hanna, W. F., and Wolf, S. C., 1970*b*, Chukchi Sea seismic reflection and magnetic profiles, 1969, between northern Alaska and International date line, *U.S. Geol. Surv. Open-File Rep.* 70-139, 1 sheet magnetic profiles, 25 sheets seismic profiles.

Grantz, A., Hanna, W. F., and Wallace, S. L., 1971, Chukchi Sea seismic reflection profiles and magnetic data, 1970, between northern Alaska and Herald Island, *U.S. Geol. Surv. Open-File Rep.* 71-125, 32 sheets seismic profiles, 2 sheets magnetic data, 1 location map.

Grantz, A., Hanna, W. F., and Wallace, S. L., 1972*a*, Chukchi Sea seismic reflection and magnetic profiles, 1971, between northern Alaska and Herald Island, *U.S. Geol. Surv. Open-File Rep.* 72-137, 38 sheets seismic profiles, 2 sheets magnetic profiles, 2 location maps.

Grantz, A., Holmes, M. L., Riley, D. C., and Wallace, S. L., 1972*b*, Seismic reflection profiles. Part 1. Seismic, magnetic, and gravity profiles—Chukchi Sea and adjacent Arctic Ocean, 1972, *U.S. Geol. Surv. Open-File Rep.* 72-138, 19 sheets seismic reflection profiles, 2 maps.

Grantz, A., McHendrie, A. G., Nilsen, T. H., and Yorath, C. J., 1974, Seismic reflection profiles, 1973, on the continental shelf and slope between Bering Strait and Barrow, Alaska, and Mackenzie Bay, Canada, *U.S. Geol. Survey Open-File Rep.* 74-42, 49 sheets seismic reflection profiles, 2 index maps.

Grantz, A., Holmes, M. L., and Kososki, B. A., 1975, Geologic framework of the Alaskan continental terrace in the Chukchi and Beaufort Seas, *Can. Soc. Petrol. Geol. Mem.* 4, p. 669–700.

Grantz, A., Boucher, G. and Whitney, O. T., 1976, Possible solid gas hydrate and natural gas deposits beneath the continental slope of the Beaufort Sea, in: *The United States Geological Survey in Alaska: Accomplishments during 1975*, Cobb, E. H., ed., *U.S. Geol. Surv. Circ.* 733, p. 17.

Hall, J. K., 1973, Geophysical evidence for ancient sea-floor spreading from Alpha Cordillera and Mendeleyev Ridge, *Am. Assoc. Petrol. Geol. Mem.* 19, p. 542–561.

Herron, E. M., Dewey, J. F., and Pitman, W. C., 1974, Plate tectonics model for the evolution of the Arctic, *Geology*, v. 2, p. 377–380.

Hopkins, D. M., 1967, A Cenozoic history of Beringia—A synthesis, in: *The Bering Land Bridge*, Hopkins, D. M., ed., Stanford: Stanford University Press, p. 451–484.

Hopkins, D. M., and MacNeil, F. S., 1960, A marine fauna probably of Late Pliocene age near Kivalina, Alaska, in: *Geological Survey Research 1960*, *U.S. Geol. Surv. Prof. Pap.* 400-B, p. B339–B342.

Hopkins, D. M., Nelson, C. H., Perry, R. B., and Alpha, T. R., 1976, Physiographic subdivisions of the Chirikov Basin, northern Bering Sea, *U.S. Geol. Surv. Prof. Pap.* 759-B.

Hudson, T., 1977, Preliminary geologic map of Seward Peninsula, Alaska, *U.S. Geol. Surv. Open-File Map* 77-167A.

Hunkins, K., 1966, The Arctic continental shelf north of Alaska, *Geol. Surv. Can. Pap.* 66-15, p. 197–205.

Hunkins, K., Herron, T., Kutschale, H., and Peter, G., 1962, Geophysical studies of the Chukchi Cap, Arctic Ocean, *J. Geophys. Res.*, v. 67, p. 235–247.

Jones, D. J., and Grantz, A., 1964, Stratigraphic and structural significance of Cretaceous fossils from Tiglukpuk Formation, northern Alaska, *Am. Assoc. Petrol. Geol. Bull.*, v. 48, p. 1462–1474.

Kameneva, G. I., 1977, The tectonic setting of Wrangel Island and its Paleozoic structural bonds with Alaska (in Russian), in: *Tektonika Arctiki Skadchatii Fundament Shel'fovikh Sedimentatsionnikh Basseinov*, Pogrebitskii, Yu. E., and Kos'ko, M. K., eds., Leningrad: NIIGA, p. 122–129.

King, P. B., 1969, The tectonics of North America—A discussion to accompany the tectonic map of North America, scale 1:5,000,000, *U.S. Geol. Surv. Prof. Pap.* 628, 94 p.

Lerand, Monti, 1973, Beaufort Sea, in: *The Future Petroleum Provinces of Canada—Their Geology and Potential*, McCrossan, R. G., ed., *Can. Soc. Petrol. Geol. Mem.* 1, p. 315–386.

Martin, A. J., 1970, Structure and tectonic history of the western Brooks Range, De Long Mountains and Lisburne Hills, northern Alaska, *Geol. Soc. Am. Bull.*, v. 81, p. 3605–3622.

Musgrave, A. W., and Hicks, W. G., 1968, Outlining shale masses of geophysical methods, *Am. Assoc. Petrol. Geol. Mem.* 8, p. 122–136.

Nelson, H. C., and Creager, J. S., 1977, Displacement of Yukon-derived sediment from Bering Sea to Chukchi Sea during Holocene time, *Geology*, v. 5, p. 141–146.

Nelson, H. C., and Hopkins, D. M., 1972, Sedimentary processes and distribution of particulate gold in the northern Bering Sea, *U.S. Geol. Surv. Prof. Pap.* 689.

Newman, G. W., Mull, C. G., and Watkins, N. D., 1979, Northern Alaska paleomagnetism, plate rotation, and tectonics, in: *Relationship of Plate Tectonics to Alaskan Geology and Resources*, Sisson, A., ed., Proceedings of the Alaska Geological Society Symposium, Anchorage, Alaska, April 1977, p. C1–C7.

Norris, D. K., 1977, Geological map of parts of Yukon Territory, District of Mackenzie and District of Franklin, *Geol. Surv. Can. Open-File Rep.* 399, 4 p., map and cross sections, scale 1:1,000,000.

Oliver, J., Ewing, M., and Press, F., 1955, Crustal structure of the Arctic regions from the Lg phase, *Geol. Soc. Am. Bull.*, v. 66, p. 1063–1074.

Ostenso, N. A., 1968, A gravity survey of the Chukchi Sea region, and its bearing on westward extension of structure in northern Alaska, *Geol. Soc. Am. Bull.*, v. 79, p. 241–254.

Parsons, Barry, and Sclater, J. G., 1977, An analysis of the variation of ocean floor bathymetry and heat flow with age, *J. Geophys. Res.*, v. 82, p. 803–827.

Patton, W. W., Jr., 1973, Reconnaissance geology of the northern Yukon–Koyukuk province, Alaska, *U.S. Geol. Surv. Prof. Pap.* 774-A, p. A1–A17.

Patton, W. W., Jr., and Miller, T. P., 1968, Regional geologic map of the Selawik and southeastern Baird Mountains quadrangles, Alaska, *U.S. Geol. Surv. Misc. Geol. Invest. Map* I-530, scale 1:250,000.

Platt, J. B., 1975, Petrography of University of Washington dredge samples from the central Chukchi Sea, *U.S. Geol. Surv. Open-File Rep.* 75-269.

Reimnitz, E., and Barnes, P. W., 1974, Sea ice as a geologic agent on the Beaufort Sea Shelf of Alaska, in: *The Coast and Shelf of the Beaufort Sea*, Reed, J. C., and Sater, J. E., eds., Arlington, Virginia: Arctic Institute of North America, p. 301–354.

Reiser, H. N., Brosgé, W. P., Detterman, R. L., and Dutro, J. T., Jr., 1978, Geologic map of the Demarcation Point quadrangle, Alaska, *U.S. Geol. Surv. Open-File Rep.* 78-526, 1 sheet, scale 1:200,000.

Rickwood, F. K., 1970, The Prudhoe Bay field, in: *Proceedings of the Geological Seminar on the North Slope of Alaska Palo Alto, California, 1970*, Adkison W. L., and Brosgé, M. M., eds., Los Angles, California: American Association of Petroleum Geologists, Pacific Section, p. L1–L11.

Ruppel, B. D., and McHendrie, A. G., 1976, Free-air gravity anomaly map of the eastern Chukchi and southern Beaufort Seas, *U.S. Geol. Surv. Misc. Field Stud. Map* MF-785.

Sainsbury, C. L., 1976, Geology, ore deposits, and mineral potential of the Seward Peninsula, Alaska, *U.S. Bur. Mines Open-File Rep.*, 108 p., 3 figs.

Shaver, Ralph, and Hunkins, Kenneth, 1964, Arctic Ocean geophysical studies; Chukchi Cap and Chukchi Abyssal Plain, *Deep-Sea Res.*, v. 11, p. 905–916.

Tailleur, I. L., 1969, Rifting speculation on the geology of Alaska's North Slope, *Oil Gas J.*, v. 67, p. 128–130.

Tailleur, I. L., 1973, Probable rift origin of Canada Basin, *Am. Assoc. Petrol. Geol Mem.* 19, p. 526–535.

Tailleur, I. L., and Brosgé, W. P., 1970, Tectonic history of northern Alaska, in: *Proceedings of the Geological Seminar on the North Slope of Alaska, Palo Alto, California, 1970*, Adkison, W. L., and Brosgé, M. M., eds., Los Angeles, California: American Association of Petroleum Geologists, Pacific Section, p. E1–E19.

Tailleur, I. L., Mamet, B. L., and Dutro, J. T., Jr., 1972, Revised age and structural interpretations of the Nuka Formation at Nuka Ridge, northwestern Alaska, *U.S. Geol. Surv. Open-File Rep.* 72-547, 10 p.

Tailleur, I. L., Pessel, G. H., Levorsen, J. A., and Engwicht, S. E., 1978, Maps showing land status and well locations and tables of well data, eastern North Slope petroleum province, Alaska, *U.S. Geol. Surv. Misc. Field Stud. Map* MF-928A, 5 sheets.

Toimil, L. J., 1978, Ice-gouged morphology on the floor of the eastern Chukchi Sea, Alaska: A reconnaissance survey, *U.S. Geol. Surv. Open-File Rep.* 78-693, 113 p.

Toimil, L. J., and Grantz, Arthur, 1976, Origin of a bergfield in the northeastern Chukchi Sea and its influence on the sedimentary environment, in: *Arctic Ice Dynamics Joint Experiment (AIDJEX) Bulletin No. 34*, Johnson, A., ed., Seattle, Washington: Division of Marine Resources, University of Washington, 42 p.

U.S. Geological Survey, 1978, Folio, eastern North Slope petroleum province, Alaska, *U.S. Geol. Surv. Min. Invest. Field Stud.*, Maps MF-928-A to MF-928-V.

Vogt, P. R., and Ostenso, N. A., 1970, Magnetic and gravity profiles across the Alpha Cordillera and their relation to sea-floor spreading, *J. Geophys. Res.*, v. 75, p. 4925–4937.

Watts, A. B., and Ryan, W. B. F., 1976, Flexure of the lithosphere and continental margin basins, *Tectonophysics*, v. 36, p. 25–44.

Wetmiller, R. J., and Forsyth, D. A., 1978, Seismicity of the Arctic, 1908–1975, in: *Arctic Geophysical Review*, Sweeney, J. F., ed., Publications of the Earth Physics Branch, Canadian Department of Energy, Mines and Resources, v. 45(4), p. 15–24.

Woolson, J. R., 1962, Seismic and gravity surveys of Naval Petroleum Reserve No. 4 and adjoining areas, Alaska, *U.S. Geol. Surv. Prof. Pap.* 304-A.

Yorath, C. J., and Norris, D. K., 1975, The tectonic development of the southern Beaufort Sea and its relationship to the origin of the Arctic Ocean, *Can. Soc. Petrol. Geol. Mem.* 4, p. 589–611.

Young, F. G., Myhr, D. W., and Yorath, C. J., 1976, Geology of the Beaufort–Mackenzie Basin, *Geol. Surv. Can. Pap.* 76-11, 65 p.

Chapter 11

THE GREENLAND—NORWEGIAN SEA AND ICELAND ENVIRONMENT: GEOLOGY AND GEOPHYSICS

P. R. Vogt, R. K. Perry, R. H. Feden, H. S. Fleming, and N. Z. Cherkis
Naval Research Laboratory
Washington, D.C.

I. INTRODUCTION

The Greenland–Norwegian Sea (Fig. 1) connects the northeast Atlantic and Arctic Oceans. Neither the plate tectonic evolution nor the paleooceanography of the Greenland–Norwegian Sea can be discussed effectively independently of the Eurasia Basin to the north, the northeast Atlantic to the south, or the Labrador Sea and Baffin Bay to the west. In oceanographic or sedimentological terms the northeast Atlantic and Greenland–Norwegian Sea area, hereafter referred to as the GNSA, have been the battleground between polar and subtropical water masses over the last 3 million years (Kellogg, 1975; Ruddiman and McIntyre, 1976; Schrader *et al.*, 1976). Since polar water penetrated southward almost to the Azores during glacial extremes, a volume dealing with the Arctic must also stray that far south. North Atlantic Deep Water, formed in the Greenland Sea, flows over the Faeroe–Iceland–Greenland transverse ridge. The subsidence and breaching of this previous land bridge was a major event in the paleooceanography of the North Atlantic (Vogt, 1972a; Nilsen, 1978, Thiede 1979, 1980).

In plate tectonic terms the GNSA basins opened by accretion from a single plate boundary running from the Azores triple junction northeastward

PHYSIOGRAPHIC PROVINCES

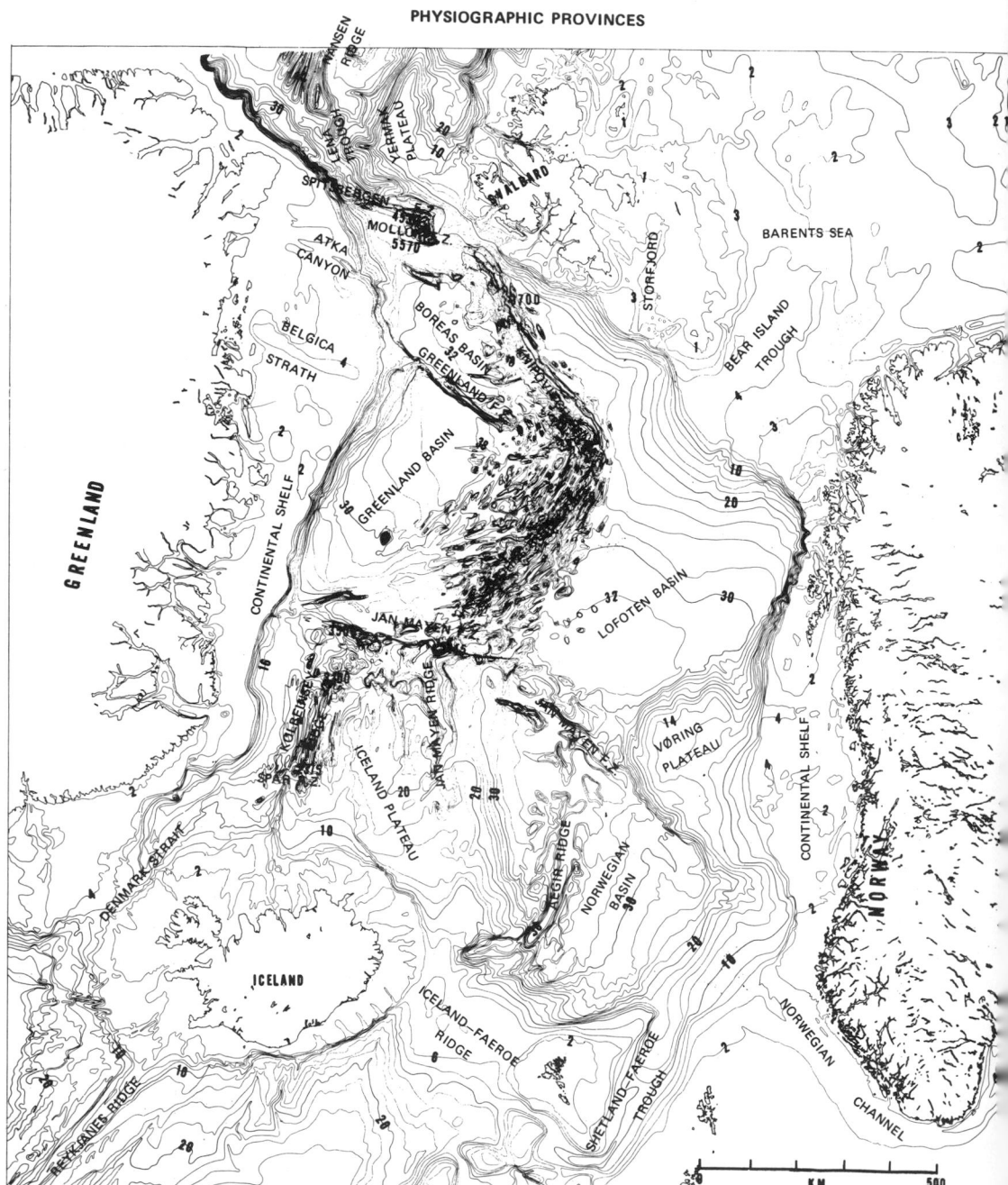

Fig. 1. Bathymetry of Greenland–Norwegian Sea area (GNSA) on polar stereographic projection (Perry *et al.*, 198
Contours labeled in units of 100 m (uncorrected).

through the Eurasia Basin into Siberia (Pitman and Talwani, 1972; Talwani and Eldholm, 1977; Phillips et al., 1980). Between about 60 and 40 m.y.b.p., active plate boundaries branched around Greenland. Our discussion of the GNSA is necessarily set in the context of this entire plate boundary system.

Iceland is discussed in some detail because it is the only midoceanic outcrop of a spreading axis. Many kinds of measurements (e.g., precision leveling) can be made only there. Other experiments (e.g., drilling and refraction profiles) have been conducted more extensively on Iceland then elsewhere in the oceans (Pálmason et al., 1979; Angenheister et al., 1979; Pálmason and Saemundsson, 1974).

The "mantle plume" concept (Morgan, 1972, 1973) and its possible corollaries (Vogt, 1971, 1972b, 1974, 1976, 1978, 1979a) have given new impetus to the study of Icelandic geology—particularly geochemistry (Schilling, 1973a,b, 1976; Sigvaldason, 1974). Iceland is merely the summit of a vast topographic anomaly on the Mid-Oceanic Ridge and its flanking basins (Figs. 1–5). Free-air gravity is positive over the whole of this region, implying anomalous densities and temperatures below the plates (Cochran and Talwani, 1978). Associated with these broader anomalies are peculiarities in seismicity, geochemistry, and morphology (Fig. 4). Some of the anomalies extend south of Iceland to the Charlie Gibbs Fracture Zone (F.Z.) at 52°N, and north to the Spitsbergen F.Z. and perhaps the central Eurasia Basin (Vogt, 1974). The GNSA cannot be discussed except in the context of the full range of these anomalies.

In recent years it has become apparent that both the initial volcanism and the line of separation between Greenland and Europe (and between Greenland and Canada) tended to follow a complex network of Mesozoic and earlier basins instead of breaking through older cratons (Talwani and Eldholm, 1977). Apparently the pre-Tertiary history of this region—and particularly the basins—played a vital "preconditioning" role in the large-scale plate displacements that were to follow.

Exploration for hydrocarbons has helped fuel a tremendous "knowledge explosion" about these now "marginal" basins bordering the Norwegian Sea (Rønnevik et al., 1979; Jorgensen and Navrestad, 1979; Gairaud et al., 1978) and connecting basins under the North Sea (Ziegler, 1977; Kent, 1975) and Barents Sea (Rønnevik and Motland, 1979).

In this text we review the discoveries of recent years, keeping the text relatively modest, relative to the subjects discussed, in order to offer a larger cross section of data and a large if not complete bibliography. For other reviews, data summaries, and additional details covering parts of the GNSA the reader is referred to Talwani and Eldholm (1977), Grønlie and Talwani (1978), Beloussov and Udintsev (1977), Pálmason et al. (1979), Heier (1977), and the Norwegian Sea Symposium (Norwegian Petroleum Society, Oslo, 1979).

II. PHYSIOGRAPHY AND MORPHOLOGY

The ocean basins and continental margins of the GNSA south and east of
the ice cover are among the most thoroughly sounded of the world's seas (Figs.
1 and 2). Detailed data are concentrated in the areas of the Reykjanes and
Kolbeinsey ridge crests (Talwani *et al.*, 1971; Vogt *et al.*, 1980; Meyer *et al.*,
1972), the Jan Mayen Ridge, The Vøring Plateau, the southeastern Barents
Sea, the North Sea (Caston, 1974), and the Iceland–Faeroe Ridge (Fleischer *et*
al., 1974). Mohns and Knipovich ridges and the Spitsbergen F.Z. area are
also much better known than a decade ago, although data are less extensive
than to the south. In the northwestern Greenland–Norwegian Sea and Arctic
Ocean north of the Spitsbergen F.Z., pack ice has generally excluded surface
ships, but a few data have been collected by drifting ice stations and submarines (Johnson and Heezen, 1967; Ostenso, 1968; Perry *et al.*, 1980; Feden
et al., 1979). Indirect bathymetric constraints are also provided by detailed
aeromagnetic surveys, especially near active plate boundaries not yet inundated by sediment (Johnson *et al.*, 1975a; Perry *et al.*, 1980; Kovacs and Vogt,
1979). On the inner continental shelves, detailed bathymetric charts have been
published by Norway, Iceland, and Denmark (Greenland) for purposes of
fisheries and navigation. A recent bathymetric chart (Fig. 1; Perry *et al.*, 1980)
incorporates most available data and previously unpublished results of cruises
by the Naval Research Laboratory in the Greenland Sea and represents the
best available bathymetry of the GNSA. This chart is, however, merely the
latest in a long series of bathymetric investigations (Ulrich, 1960; Litvin, 1964;
Eggvin, 1963; Johnson and Eckhoff, 1966; Johnson, 1974) and will no doubt
see future refinement. Excepting the complex axial topography of Mohns
Ridge and the Greenland continental margin north of 70°N, we consider the
chart by Perry *et al.* (1980) [like a similar one independently prepared by
Grønlie and Talwani (1978, 1979)] a good approximation to reality.
     The first-order topographic features of the greater GNSA (Figs. 1 and 2)
are the Mid-Oceanic Ridge (MOR), the flanking basins, and the continental
shelves. This first-order pattern is primarily the result of sea-floor spreading,
sedimentation, and glaciation of the continental shelves. Profiles transverse to
the MOR (Fig. 5) exhibit the characteristic increase of depth with crustal age
(Parsons and Sclater, 1977). To first order water depths over oceanic crust
depend on (1) initial depth at the accretion axis; (2) subsequent thermal subsidence; and (3) effects of sediment loading and its isostatic compensation.
Analysis of these factors makes posssible paleobathymetric (Grønlie, 1979;
Thiede, 1979, 1980) and depth anomaly (Fig. 3) charting.
     Superposed on the first-order regime are the Vøring marginal plateau, the
Jan Mayen Ridge, and the broad bulge of the oceanic basement centered on
Iceland (Figs. 2–4). The anomalous shoaling of the spreading axis (Fig. 4)

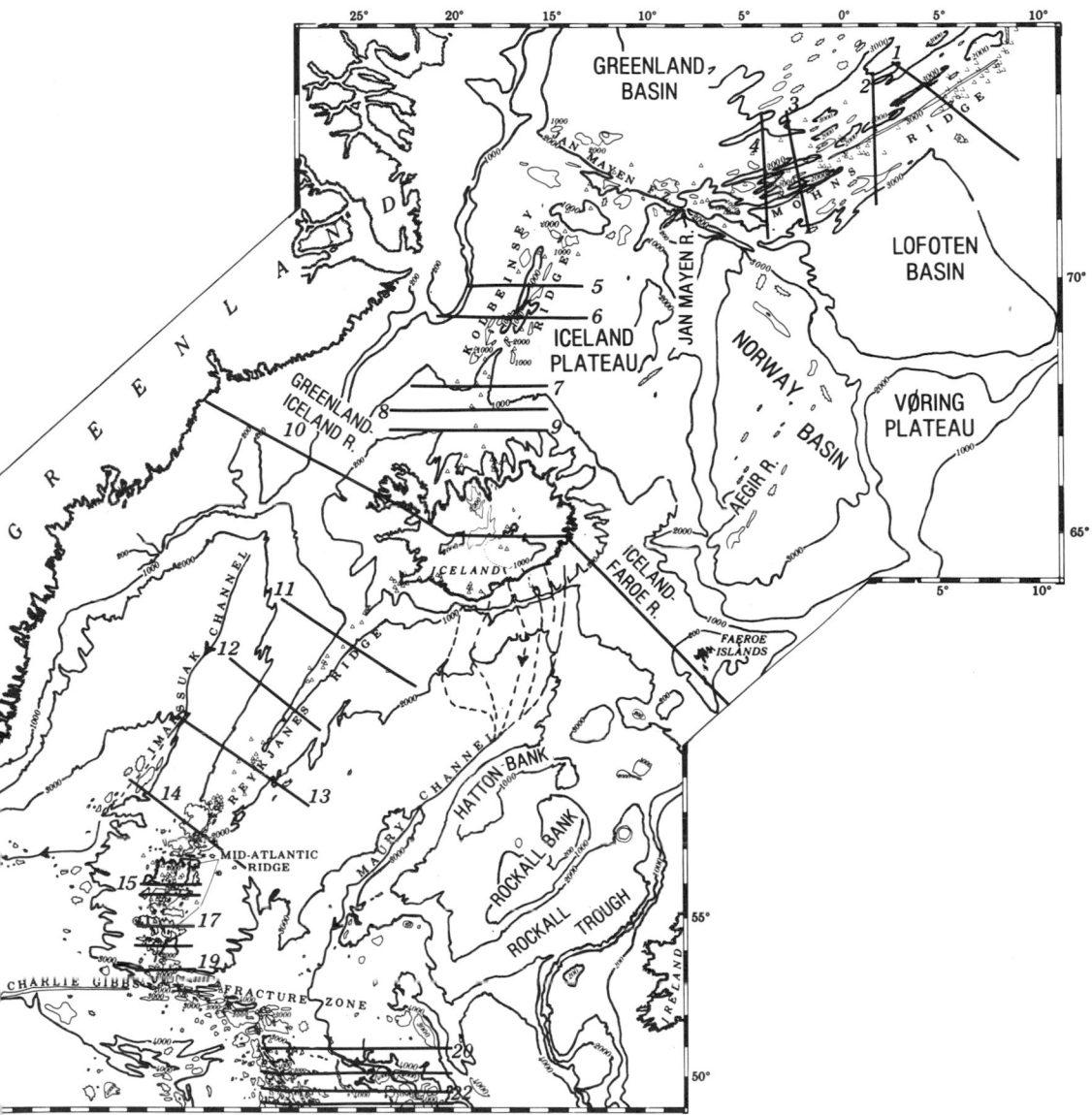

Fig. 2. Bathymetry of area surrounding Iceland, on Mercator projection (Johnson, 1974; Johnson and Campsie, 1974). Open triangles show earthquake epicenters (1961–1971). Chart also locates profiles 1–19 shown in Fig. 5.

begins at the Charlie Gibbs F.Z. (52.5°N) (Fleming *et al.*, 1970) and reaches a climax with volcanoes up to 2 km above sea level in central Iceland; northward the axis deepens progressively; depths exceeding 4 km are first reached at the ends of the short Molloy spreading center west of Svalbard (Fig. 1). The 4000-km-long "Iceland topographic anomaly" apparently characterizes all ocean

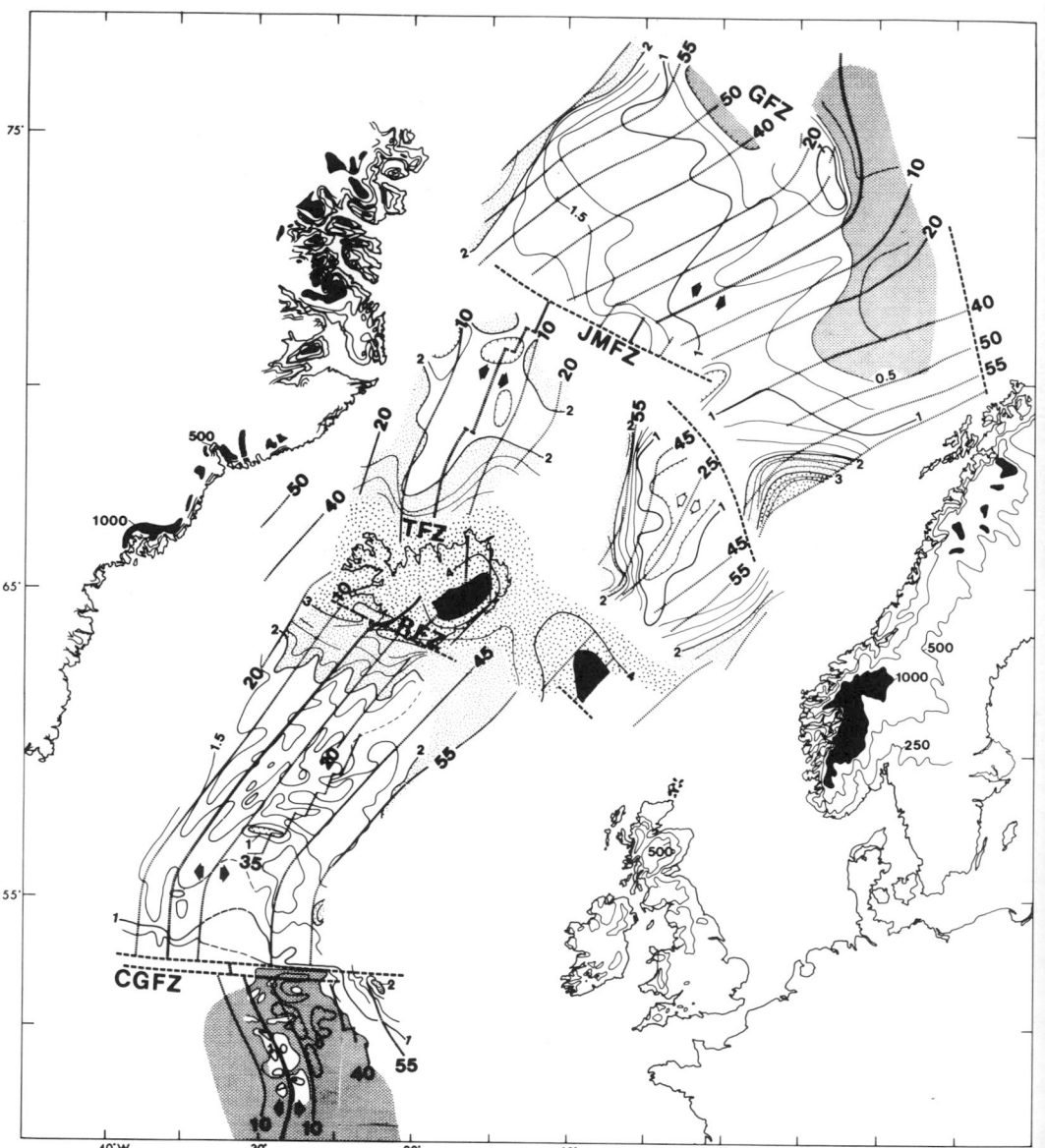

Fig. 3. Oceanic basement residual elevations (kilometers) with respect to North Pacific standard (Sclater *et al.*, 1
Parsons and Sclater, 1977); also shown are anomalously elevated coastal areas (present elevations) and oceanic cru
isochrons (m.y.b.p.). Oceanic anomalies > +4.5. km, black; +3–4.5 km, coarse stippling; +2–3 km, fine stippling; <
km, horizontal dashes. Continental elevations >1000 m, black. GFZ, JMFZ, TFZ, RFZ, and CGFZ refer to the Gr
land, Jan Mayen, Tjörnes, Reykjanes, and Charlie Gibbs Fracture Zones. From Vogt and Johnson (1975); revisions n
of Iceland–Faeroe Ridge based on data in Figs. 18 and 27.

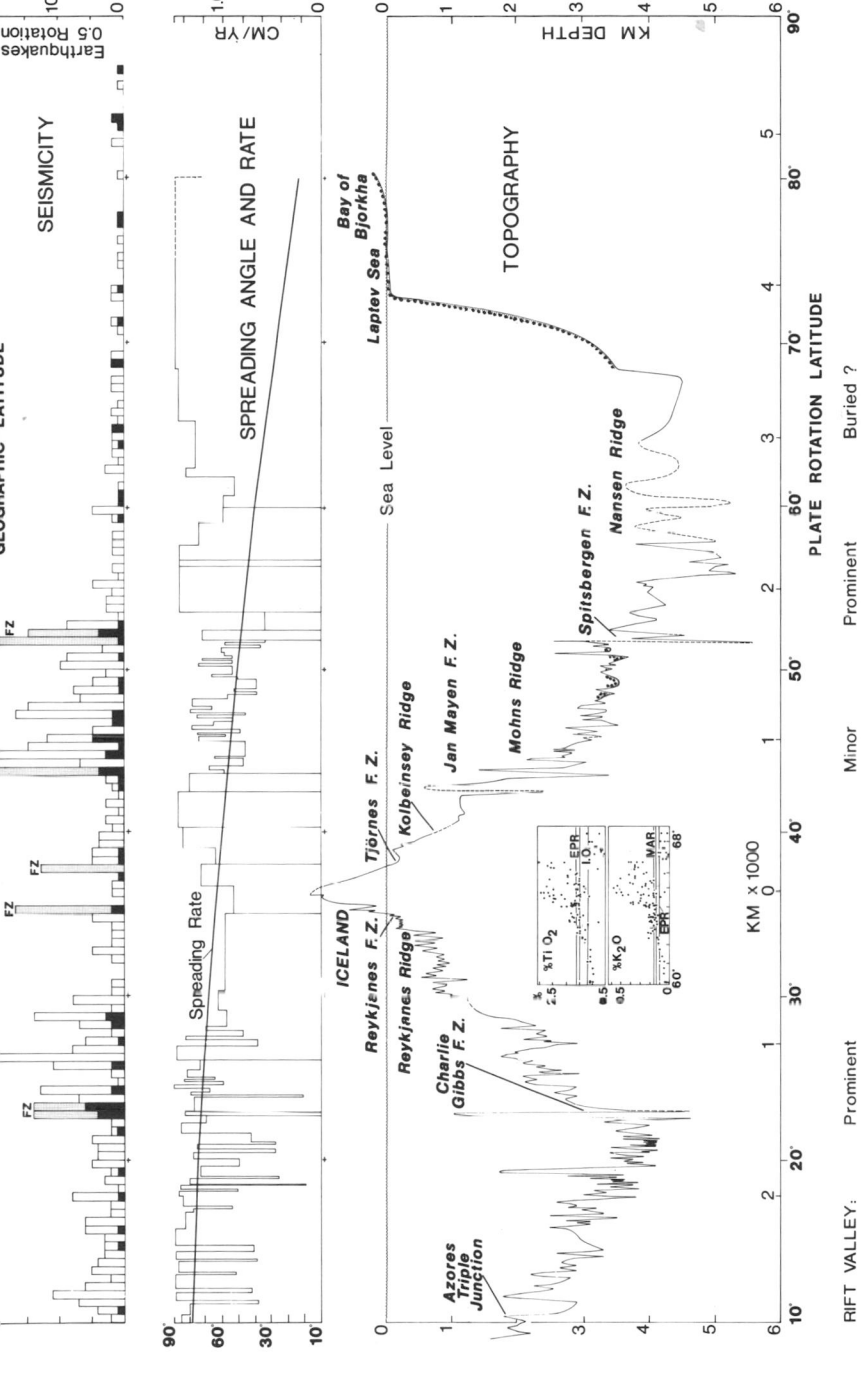

Fig. 4. Longitudinal profiles of seismicity, spreading angle (angle between spreading axis and direction of plate motion), spreading half-rate, water depth at rift axis, TiO_2 and K_2O concentrations, and rift morphology (schematic) along present plate boundary from the Azores triple junction to the Siberian coast. Black bar histogram (top) includes only magnitudes 5.0 and above. Cross-hatched bars represent seismicity attributed to transverse fractures. Dots above bathymetric profile indicate presence of thick sediments in rift zone. Concentration levels "IO," "MAR," "EPR" and unlabeled median line show average concentrations for Indian, Atlantic, and Pacific Oceans and for all oceanic samples (Melson et al., 1976). From Vogt et al. (1979a) and Feden et al. (1979).

crust back to 55–60 m.y.b.p., although the elevation anomaly may be lower and less extensive for mid-Tertiary crust and more complex for early Tertiary crust (Fig. 3; Vogt and Johnson, 1975; Cochran and Talwani, 1978). In a more general way, the highest areas of west Greenland (up to +3733 m), Norway (up to +2469 m) and the British Isles are those closest to the areas of greatest ocean crust elevation (Vogt and Johnson, 1975; Fig. 3). This is also true of the Baffin Bay–Labrador Sea (Vogt, 1974). The anomalously elevated ocean crust—and perhaps elevated continental margins as well—can be explained by anomalously high temperatures and concomitant thermal expansion in the upper 400 km of the mantle (Zielinski, 1977; Cochran and Talwani, 1978). However, adjacent to areas of continental uplift (Fig. 3) lie Mesozoic basins presumably floored by continental crust. These basins apparently did not share the Tertiary uplifts that formed the Iceland topographic anomaly.

Average longitudinal slopes along the spreading axis away from Iceland are in the range 5–25 parts per thousand (Vogt and Johnson, 1975; Fig. 4). Vogt (1976) suggested that the horizontal pressure gradients created by these slopes drive low-viscosity mantle materials, discharged by the Iceland plume, down the pipelike region of partial melting below the spreading axis.

Seen in cross section the MOR not only shoals steadily toward Iceland, but varies remarkably in roughness (Fig. 5). Rough flank topography, generally correlative with a rift valley, extends from south of the Charlie Gibbs F.Z. northward about to the crestal trend change at 57°N. The rift axis also exhibits bends and transverse fractures (Fleischer *et al.*, 1973). The floor of the rift valley is typically 1–1.5 km below adjacent peaks and flank relief is of the order 0.5–0.8 km, peak to trough. From 57°–58°N to Iceland a prominent rift valley is absent and peak-to-trough flank relief remains within 200–400 m, except for two pairs of escarpments prominent near the intersection of the Reykjanes Ridge and the Iceland Platform (Ulrich, 1960; Talwani *et al.*, 1971; Johnson, 1974). The older pair of escarpments has been substantially buried by sediment, particularly on the east flank. The escarpments and some lesser basement ridges (Vogt and Johnson, 1972) are "time-transgressive," occurring on progressively younger crust with increasing distance from Iceland. These features also gradually lose their relief and disappear south of about 60°N. Vogt (1971, 1974) attributed the features to asthenospheric "fronts" advancing southwestward below the Reykjanes Ridge away from Iceland at 10–20 cm/yr. The two escarpments may represent abrupt increases in plume discharge, which first affected the hot-spot center in Iceland about 7 and 17 m.y.b.p., respectively (Vogt, 1971, 1979*a*; Watkins and Walker, 1977).

The Reykjanes Ridge north of 57°N is markedly "oblique," making an angle of 35° with the N100°E spreading direction. A detailed study (Shih *et*

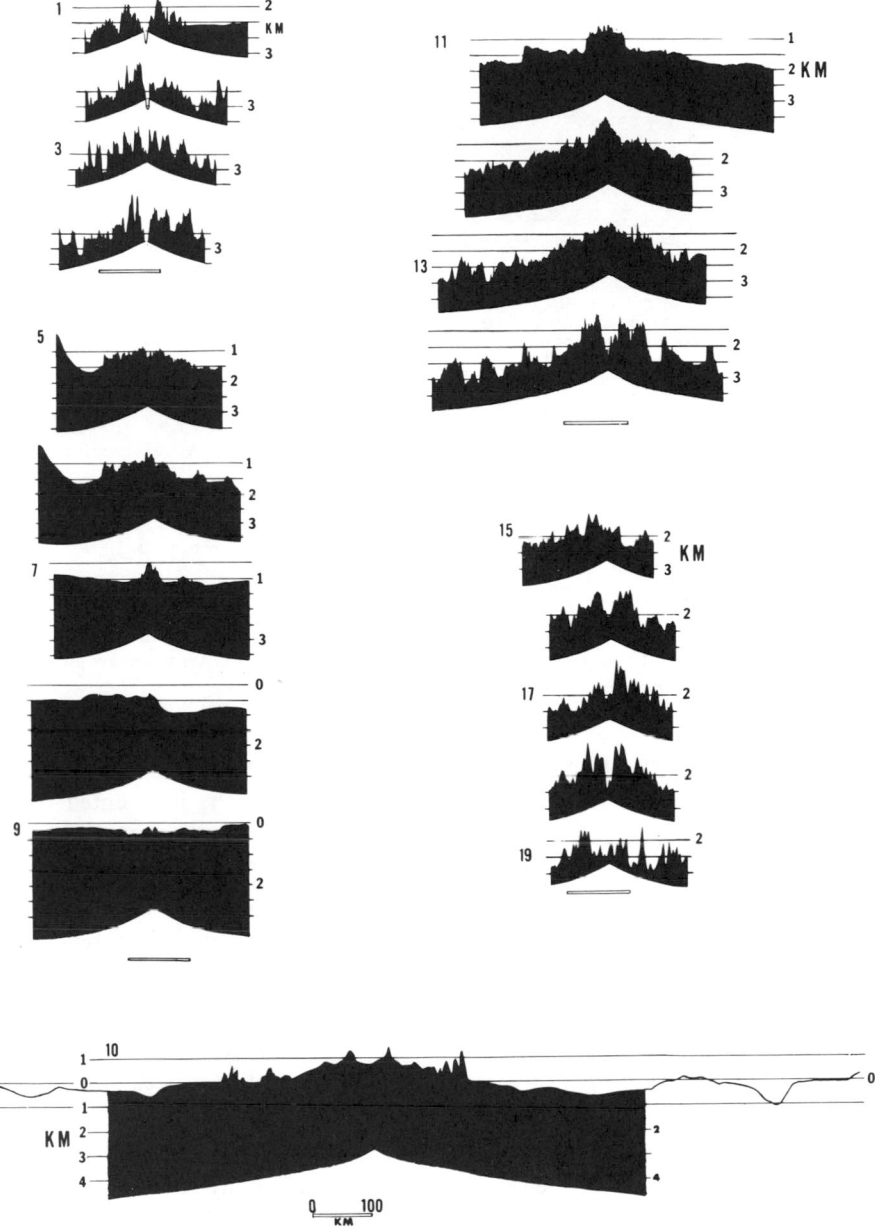

Fig. 5.   Bathymetric profiles across Mohns (1–4), Kolbeinsey (5–9), Iceland (10), and Reykjanes (11–19) portions of the Mid-Oceanic Ridge (indexed in Fig. 2). Adjusted for average sediment overburden and consequent isostatic depression, area shaded black shows degree of topographic excess above North Pacific Ocean crust (Sclater *et al.*, 1971). Adapted from Johnson (1974).

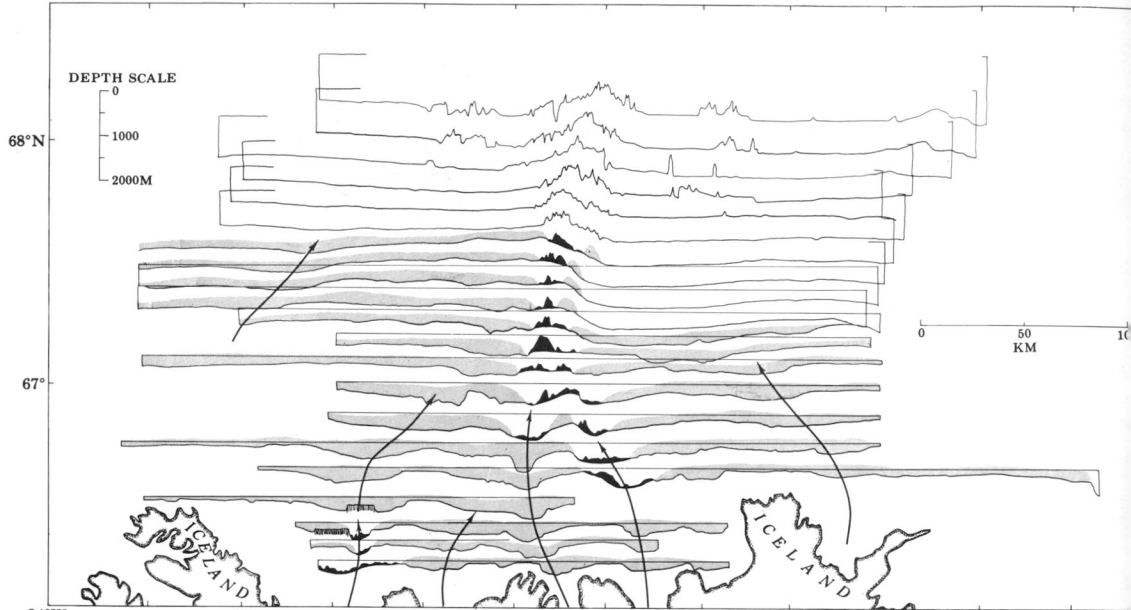

Fig. 6. Topographic profiles across Kolbeinsey Ridge just north of Iceland (Meyer *et al.*, 1972), showing areas poss covered by grounded glacier ice (gray) and postglacial submarine volcanism (black). Arrows show possible directions of flow and submarine sediment transport. From Vogt *et al.* (1980).

al., 1978) showed that the actual present center of spreading is locally formed of N15°E en echelon ridges, while fine flank topography is oriented N21°E. During the mid-Tertiary the northern Reykjanes Ridge—like the southern part today—was characterized by close-spaced transverse fractures and rough topography, rather than by oblique spreading (Vogt and Avery, 1974*a*; Voppel and Rudloff, 1980). The switch to oblique spreading began first near Iceland and then progressed southwestward (Vogt and Avery, 1974*a*).

Where the present spreading axis crosses the Iceland shelf it is a narrow band of rough topography composed of short en echelon spreading centers, perhaps most postglacial in age. Older crust has evidently been planed by ice or marine erosion and/or buried by sediment. Within the complex Tjörnes F.Z. (McMaster *et al.*, 1977), which offsets Iceland's northern Neo-Volcanic Zone from Kolbeinsey Ridge (Fig. 3), there is one (locally two) narrow band of axial topography (Fig. 6; see also Fig. 31 and later discussion). Gentle U-shaped valleys extending from Iceland northward across the shelf suggest that much of the latter was glaciated during climatic minima (Vogt *et al.*, 1980; Fig. 6).

Steep canyons incised in the shelf edge and continental rise off southern

Iceland (Fig. 2) probably were caused by slumping and suspension (turbidity) flows, although the canyons may also represent the seaward continuations of old glacial troughs. A high rate of present sediment input to the heads of the southeast Iceland canyons is probable, considering the proximity to the coast, the great topographic relief, the youthful volcanism, and the existence of an icecap (Vatnajökull). Occasional subglacial volcanic eruptions below Vatnajökull deliver floods of debris-laden meltwater (jökulhlaups) to the continental shelves. Sand-sized and finer volcanogenic material is then transported from the shelf-edge canyons into the Iceland Basin, down the Maury Channel, and perhaps ultimately into the Biscay Abyssal Plain (Ruddiman, 1972).

The topography of Iceland (Thorarinsson, 1974) should be mentioned, since it is exposed ocean crust, albeit anomalous and modified by glacial or fluvial erosion, or both. One might expect the youngest, postglacial topography in the Neo-Volcanic rift zones to resemble the submerged MOR most closely, since the youngest formations are least eroded. However, such subaerial lavas are more vesicular and fluid and, unlike submarine eruptions, flow over comparatively broad distances. Actually, older lavas that were extruded subglacially were surrounded by meltwater and thus more closely resemble submarine topography. There are two main kinds of subglacial basalt volcanoes: serrated ridges built along long fissures, and table mountains built up on short fissures, changing during th? early phase of eruptions into circular vents (Thorarinsson, 1974). Crater rows and shield volcanoes are the interglacial equivalents of these types. Although many faultlines run through the Icelandic rift zones, the linear valleys formed during glacial times are predominantly constructional volcanic features and not tectonic grabens (Fig. 7).

Although all the topographic types discussed above occur in the tholeiitic parts of the Neo-Volcanic Zone (Reykjanes–Langjökull line and the northern Neo-Volcanic Zone), table mountains and shield volcanoes are absent from the alkali-olivine zones (e.g., the southern East-Iceland Neo-Volcanic Zone). In the tholeiitic belts, table mountains and shield volcanoes seem to have formed preferentially during the closing millenia of glaciation and the early postglacial period, at a time of rapid isostatic uplift (Thorarinsson, 1974). This observation has been explained by rapid pressure release in the earth's mantle (Sigvaldason and Steinthorsson, 1974).

Post-10 m.y.b.p. sea-floor topography north of Iceland (Meyer et al., 1972; Vogt et al., 1980) exhibits some trends parallel to crustal isochrons, as expected, but also time-transgressive trends converging northward (Fig. 8). If these represent subaxial flow from the Iceland plume, the rate here ($\sim$ 1 cm/yr) is an order of magnitude weaker than the southwestward flow under

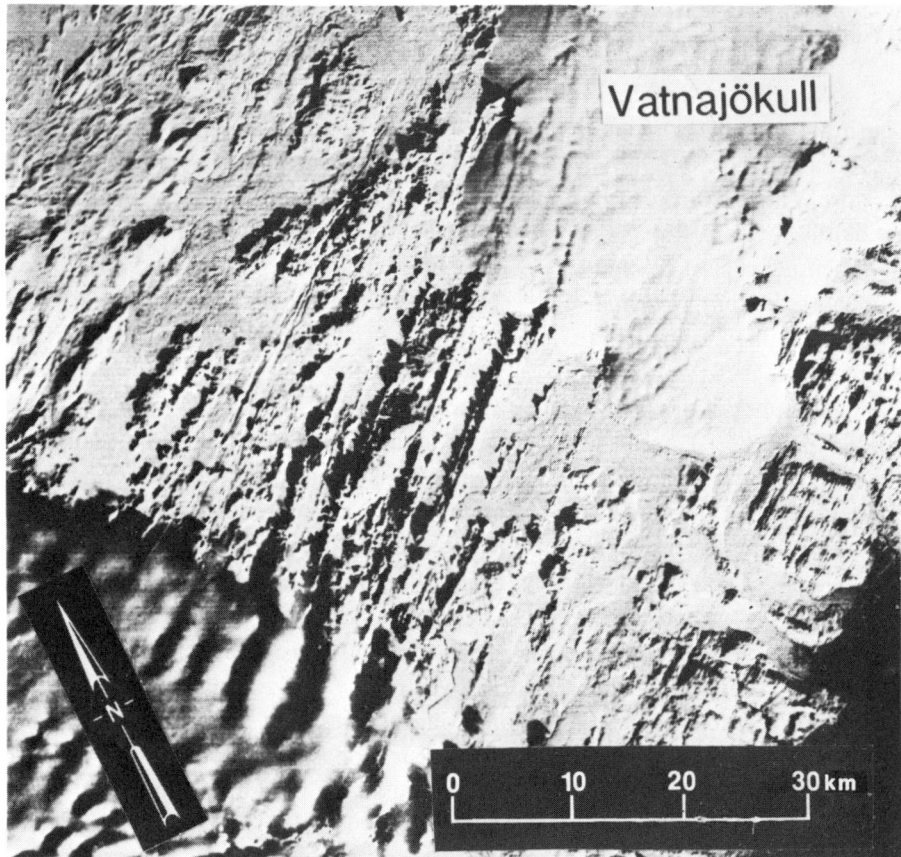

Fig. 7. ERTS-1 satellite photograph of Neo-Volcanic Zone west of Vatnajökull. Altitude 920 km; illumination angle 7°; ground snow-covered except bottom right. Clouds obscure ground in lower left. Lineated topography largely consists of serrated ridges and crater rows rather than horsts and grabens. From Thorarinsson (1974).

Reykjanes Ridge (Vogt, 1971, 1974). Kolbeinsey Ridge also lacks the younger of two pairs of escarpments found on the Reykjanes flanks (Fig. 5) and is depleted in LIL elements compared to the northern Reykjanes Ridge (Brooks and Jakobsson, 1974). These differences between the MOR north and south of Iceland might be caused by a damming effect of the Tjörnes F.Z. on northward flowing partially molten mantle below the northern Neo-Volcanic zone (Vogt and Johnson, 1975). The difference between Kolbeinsey and Reykjanes ridges may have been less, prior to 10 m.y.b.p., since there is evidence for the older pair of basement escarpments (labeled "E" by Vogt, 1971) on the flanks of Kolbeinsey Ridge (Vogt *et al.*, 1980).

Transform faults created during the last few million years offset the Kolbeinsey axis at 69°N (Spar F.Z.) and at 70.5°N (Johnson *et al.*, 1972; Meyer *et al.*, 1972; Talwani and Eldholm, 1977). Additional fractures of small offset, revealed by detailed aeromagnetic data (Fig. 26; Vogt *et al.*, 1980) have formed and disappeared repeatedly during the last few million years. Topography on Kolbeinsey Ridge resembles that of Reykjanes Ridge around 59°–61°N. However, there is a region of high fracture zone relief (up to 1 km) between 68.4° and 69°N on Kolbeinsey Ridge (Fig. 8). Starting at 69.5°N, a slight median rift valley (500 m relief) appears at the axis. The ridge topography from 70°N to its junction with the Jan Mayen F.Z. at 71.7°N is complex and not well known (Fig. 9). The existence of numerous seamounts and banks in this region, along with volcanically active Jan Mayen Island

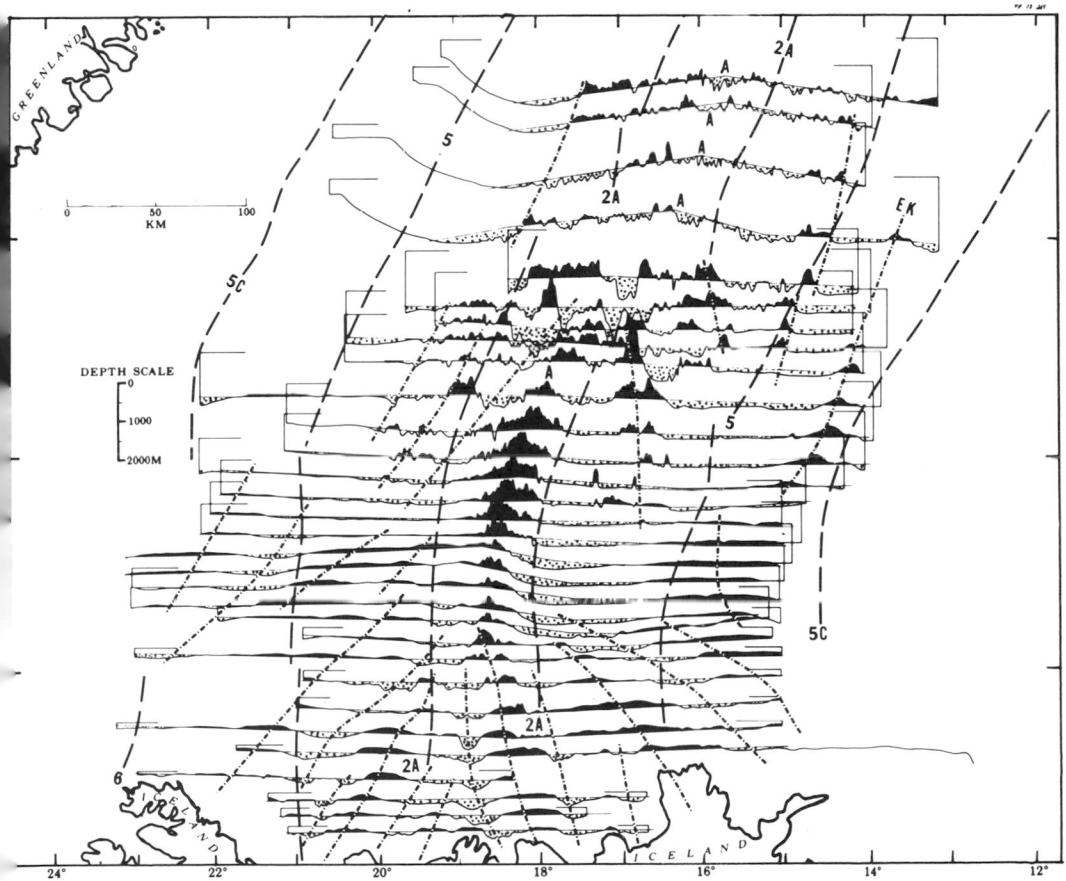

Fig. 8. Bathymetric profiles across Kolbeinsey Ridge (Meyer *et al.*, 1972). Local relief with respect to smoothed profile is black (positive) or stippled (negative). As shown by dash–dot lines, some bathymetric features parallel crustal isochrons (magnetic lineations, shown by thick dashed lines), while others are conspicuously diachronous, perhaps reflecting northward asthenosphere flow. After Vogt *et al.* (1980).

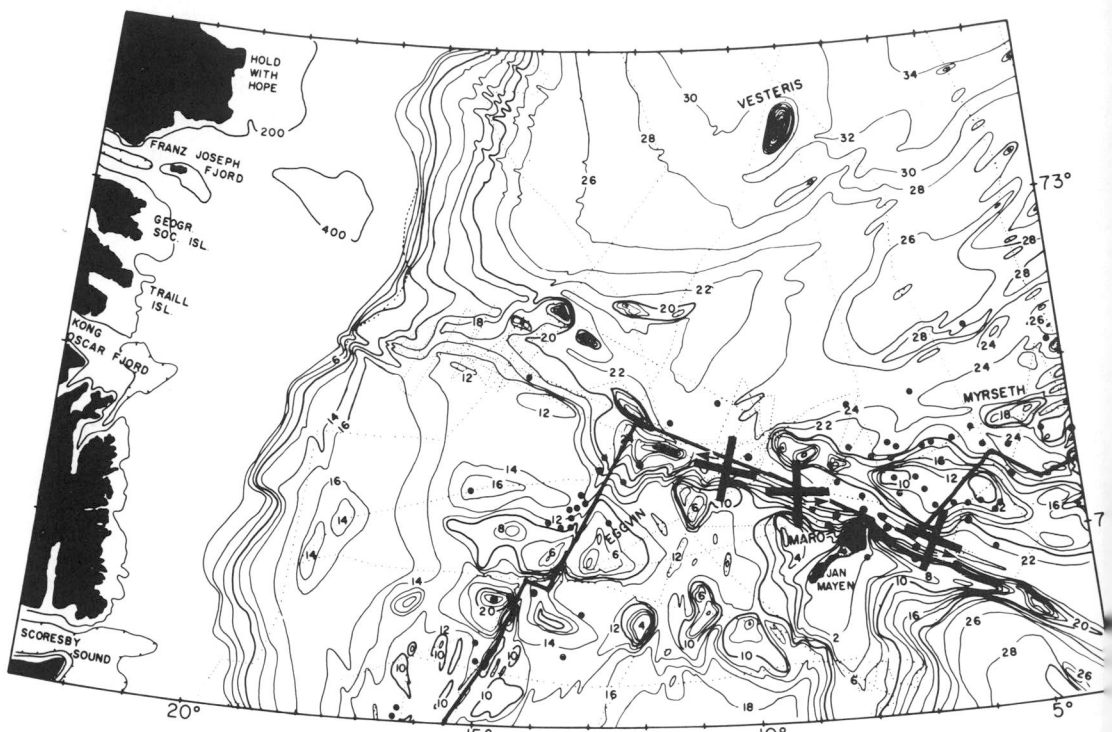

Fig. 9. Bathymetry in hundreds of meters, near western end of Jan Mayen F. Z. (Johnson, 1975). Postulated prese spreading axis shown by solid line. Solid circles show earthquake epicenters (1955–1975) compiled by Bungum a Husebye (1977); crosses show fault-plane solutions.

(Sylvester, 1975), suggests the existence of a separate Jan Mayen hot spot. Crossing Eggvin Bank near 71°N, the spreading axis shoals to < 1 km and then descends to below 3400 m at the Jan Mayen F.Z. Junction. This deep and a 3845-m hole at the southeastern end of the active F.Z. may reflect viscous head loss of the ascending magmas (Sleep and Biehler, 1970). The fracture zone consists of a NE-facing escarpment, most spectacular at Jan Mayen Island (highest point +2277 m), which merges into a ridge east of about 7°W. Additional areas of shoal topography exist north of the F.Z. (Fig. 9), e.g., Vesteris Seamount and a short spreading center offset 100 km from Kolbeinsey Ridge and 15 km from Mohns Ridge (Fig. 9). Bathymetric and magnetic data suggest that Mohns Ridge is essentially continuous. The topography may consist of short, interconnected spreading axes with varying degrees of obliqueness and only occasional true transform faults (Fig. 4), as in the area of detailed survey of the Mid-Atlantic Ridge at 47°–51°N (Johnson and Vogt, 1973). However, magnetic anomalies over southwestern Mohns Ridge have been interpreted as a "staircase" of short spreading axes and transform faults

(Fedhynskii *et al.*, 1975). In topographical profile Mohns Ridge (Figs. 1 and 5) resembles the southern Reykjanes Ridge, 51°–58°N. Northeastern Mohns Ridge appears asymmetric in cross section because sediments of the Barents Fan have encroached on its flanks and may even locally spill into the rift valley itself. In addition, crestal mountains and older basement topography, as well as accompanying gravity anomalies, are systematically higher on the west side of the axis. The Mohns Rift Valley deepens irregularly from 2500–3000 m near the Jan Mayen F.Z. to 2800–3500 m near its eastern end (Fig. 4).

The Knipovich Ridge section of the MOR is somewhat more oblique to the spreading direction (40° to 70°) than is the Mohns Ridge (50°–80°) (Fig. 4). As shown by Talwani and Eldholm (1977) and Perry *et al.* (1980), the rift valley (3200–3400 m) bends into the Mohns Ridge rift valley without offset (Fig. 1). Topography and gravity tend to be higher west of the axis. Sediments of the Svalbard continental rise have encroached westward on the Knipovich axis, burying most of its flank topography and pouring into the rift valley itself (Figs. 19 and 20; Vogt *et al.*, 1978). If the Knipovich valley fill were stripped off, the rift floor would be deeper than along the Mohns median valley.

The Knipovich valley axis is not exactly straight; short sections are more nearly normal to the spreading direction (Fig. 3). These sections are associated with belts of higher, rougher topography (Fig. 1). Detailed surveys of the Mid-Atlantic Ridge at 47°–57°N also show that oblique spreading segments are associated with lower, smoother topography (Johnson and Vogt, 1973). The explanation may be found in the greater viscous head loss of mantle mush rising in an oblique conduit.

The asymmetric location of northernmost Knipovich Ridge close to the Svalbard margin suggests asymmetric spreading (slower to the east) or an eastward shift in the axis (Vogt *et al.*, 1978). The prominent Greenland Ridge and perhaps the Hovgaard Ridge west of Knipovich Ridge are apparently associated with transform fractures that have become extinct during the evolution of the plate boundary.

At its northern end Knipovich Ridge terminates against the 120-km-long Molloy F.Z., a transform fault. To the north the 60-km-long Molloy Ridge spreading center (Perry *et al.*, 1980) is essentially perpendicular to the spreading direction. Soundings of 4532 m and 5572 m near the ends of Molloy Ridge represent the greatest known depths in the GNSA (Figs. 1 and 3). With a 150-km offset the Spitsbergen F.Z. connects Molloy Ridge with the Lena Trough, a little-known oblique (~45°) spreading axis extending approximately from 80.5°N, 2.5°E to 82.5°N, 2°E. Northward, yet another transform fault of uncertain offset shifts the spreading axis WNW to the southwestern end of Nansen Ridge near 82.8°N, 6.5°W. The bathymetry of these ice-covered features is very poorly known; Nansen Ridge, although generally deeper, is morphologically similar to the MAR (Johnson and Heezen, 1967, Feden *et al.*,

1979). The few sounding lines collected by submarines suggest that the floor of the rift valley generally lies between a depth of 3500 and 5500 m (Feden *et al.*, 1979; Anonymous, 1979; Vogt *et al.*, 1979*a,b*).

For some years it was thought that a deep trough connected the GNSA basins with the Eurasia Basin. Scattered soundings collected in what we now call the Molloy F.Z., Molloy Ridge, Spitsbergen F.Z., and Lena Trough probably gave this impression. However, the revised bathymetry (Fig. 1) suggests a still depth of only ~2600 m, which occurs in the region between the southern end of Molloy Ridge and northern Knipovich Ridge.

Extinct spreading axes have been postulated to exist on the Iceland–Faeroe Ridge (Talwani and Eldholm, 1977), along the southeast margin of the Norway Basin (Talwani and Eldholm, 1977), along the margins of the Greenland and Lofoten basins (Eldholm *et al.*, 1977), in the central Norway Basin (Vogt *et al.*, 1970), on the Iceland Plateau between the Kolbeinsey and Jan Mayen Ridges (Vogt *et al.*, 1970; Johnson *et al.*, 1972; Grønlie *et al.*, 1979) and possibly in the Boreas Basin west of northern Knipovich and Molloy Ridges (Vogt *et al.*, 1978). Extinct axes created by minor eastward shifts of northern Kolbeinsey Ridge should also exist (Fig. 26; Vogt *et al.*, 1980). Little or no bathymetric expression has been found for any of these extinct axes except the Aegir Ridge in the central Norway Basin (Fig. 1), which has a broad rift valley with basement depths of 4000–4500 m. Detailed magnetic data imply that the Iceland Plateau and Iceland–Faeroe Ridge extinct axes may not exist (Vogt *et al.*, 1980).

Most ocean crust older than 10–20 m.y., including extinct spreading centers, has been inundated by sediment to the point that little or no bathymetric expression survives. Relatively high pelagic sedimentation rates, the effect of Plio-Pleistocene glaciations, and nearness to uplifted land masses help explain this rapid burial. Early Tertiary basement topography still finds bathymetric expression in the case of the Aegir and Iceland–Faeroe Ridges and the Jan Mayen and Greenland F.Z.s, however. The present depths of the Norway, Lofoten, Greenland, and Boreas basins are the combined result of sedimentation and crustal subsidence (Grønlie, 1979). The relatively greater depth of the Greenland Basin, compared with its coeval counterpart southeast of Mohns Ridge, seems to be the result of slower input of terrigenous sediment. The floors of the basins are only locally flat enough to be called "abyssal plains" (gradients of 1 : 1000 or less). More mature abyssal plains will develop in the GNSA as the ocean widens and its deep basins extend farther seaward from sources of terrigenous sediment.

The present shapes of GNSA continental rises and slopes have also been strongly conditioned by nearby land masses. Holtedahl and Sellevoll (1971) noticed that where the Norwegian Shelf is narrowest, the continental slope is steepest. This seems to be generally true of the GNSA. Maximum slopes averaged over a 1000-m drop range from 130:1000 (7.5°) off the Lofoten Islands and 75:1000 (4.3°) on the Greenland margin near 77°N, to 40:1000 (2.3°) further south on the wide Greenland margin at 76°N and 15:1000 (0.9°) west of the Bear Island Trough in the southern Barents Sea. Evidently more material has been eroded from the wider shelves with their larger glaciofluvial drainage areas. This greater sediment input would account for the more gentle continental slopes. In addition, the steep slopes may reflect different modes of margin subsidence (Eldholm and Sundvor, 1980).

Continental slope contours bulge outward toward the basins at several places. Generally these bulges, which we interpret as prograded "deltas" and attached fans (Vogt and Perry, 1978), occur at the mouths of glacial or glaciofluvial valleys crossing the continental shelves (Fig. 1). Shelf prograding ranges from <1 to >50 km for the largest deltas in the GNSA area, e.g., east of Greenland's Scoresby Sund and Kangerlugssuaq fjords (Vogt *et al.*, 1980). Broader and longer valleys generally correlate with the most extensive bulges, the most spectacular example of which is the Bear Island Trough and its attached fan. The bathymetry and morphology of the shelves themselves has been described by Holtedahl and Bjerkli (1975) (Norwegian, Barents, and Svalbard shelves) and Sommerhoff (1973) (southeast Greenland shelf). All or nearly all the GNSA shelves were covered by grounded glacier ice during the Plio-Pleistocene climatic minima.

Sommerhoff's study shows the inner shelves to be characterized by rugged and hummocky topography. These ice-scoured plains are 240 m deep on average, but troughs cutting through them are dissected to a maximum depth of 1060 m (Gyldenlöves Trough). The troughs are generally prolongations of fjords. On the outer shelves three types of glacial accumulation forms can be recognized: (1) ice-margin deposits with clearly expressed terminal moraines, (2) glacial till plains, and (3) glacial marine outwash fans. The outer shelf banks, with an average depth of 180 m, surround glacial basins up to 670 m deep. Breaks of the terminal moraines seaward of the glacial basins suggest the action of floating shelf ice intermitently penetrating seaward. Delta-shaped projections in front of the shelf basins resemble the slope bulges discussed above. The Greenland slope topography is further complicated by marginal plateaus, terraces, ridges and hills, canyons, and slump features.

The Iceland platform is indented by numerous broad depressions (Fig. 1), probably carved by ice streams flowing radially outward from the insular ice dome. A 20- to 30-m-high, 100-km-long moraine at present depths of 200–250 m on the Iceland shelf may represent the maximum advance of the Weichselian (Wisconsin) ice sheet (Olavsdottir, 1975).

The Norwegian shelf exhibits similar glacial terrain (Holtedahl and Bjerkli, 1975). Minimum average shelf depths (200 m) occur west of Bergen, off the Lofoten Islands, and in the Barents Sea south and southeast of Svalbard. Off central Norway (64°–67°N) the shelf simultaneously broadens and deepens to ∼300 m, with glacially scoured basins bottoming at 400–500 m. The Bear Island and Norwegian troughs have been dissected to depths locally exceeding 500 m. Confined by the topography, ice streams gouged out even deeper depressions in some of the fjords. At 1308 m depth the Sognefjord represents the most extreme case of glacial overdeepening in the northern hemisphere.

III. SEDIMENTS AND PALEOCLIMATE

A. Surface Sediments

1. *Basins*

Kellogg (1975) analyzed GNSA surface sediments on 169 short cores that sampled no more than the last 2000–4000 years. Excluding volcanic ash, which is only locally important, GNSA sediments can be considered admixtures of (1) glacial-marine detritus, (2) clay-sized material derived from erosion on land, and (3) biogenic material, mainly calcareous planktonic foraminiferal shells. The climatic–oceanographic regime controls the relative proportion of these components (Fig. 10). No evidence of turbidity current deposition was found in any of the cores anlayzed. However, turbidity flows probably were more common during the lowered sea levels of glacial maxima, e.g., at 18,000 years b.p.

The distribution of coarse material (>62-μm particle size) reflects both glacial marine and biogenic components. High coarse-fraction values south of Iceland and between Iceland and Norway result from plentiful foraminiferal debris; between Greenland and Spitsbergen, ice-rafted glacial sand forms much of the coarse fraction.

Most of the Norwegian Sea floor is covered by sandy clays, marls, and oozes (Fig. 10). The sandy material is largely ice rafted. No ice presently penetrates the warm Norwegian Current; this explains the lack of terrigenous sand below the current. Curiously, little or no ice-rafted material was found west of Svalbard and in a band from Svalbard to Greenland, both areas of extensive present winter ice (see Vinje, 1976, for present-day sea ice conditions). Terrigenous sands found in sediments off the coast of Norway, Greenland, and on the Iceland–Faeroe Ridge are probably relict Pleistocene deposits. Current scour has winnowed out the finer fraction, depositing it in the deeper basins.

Analysis of the $CaCO_3$ fraction in GNSA deepwater sediments shows that planktonic foraminiferal debris is the primary component; benthonic foraminifers are two orders of magnitude less abundant. The productivity of plank-

Fig. 10. Surface sediments of the GNSA, with schematic surface currents added. Compounded from Kellogg (1975).

tonic foraminifers is highest in the relatively warm Norwegian Current; this accounts for the high $CaCO_3$ values (30–60%) in that area. Relatively warm-water foraminiferal species and right-coiling *Globoquadrina pachyderma* are more plentiful constituents below this current (Fig. 11). Coccoliths are less important north of Iceland, since these surface waters, even during the current interglacial, are too cold. X-ray diffraction shows only trace amounts of dolomite. Detrital carbonate is not important in GNSA sediments.

Dissolution of $CaCO_3$ may be comparatively unimportant in the GNSA. Except for a few cores in deeper parts of the Greenland Basin, planktonic forams show little or no dissolution and benthonic fractions are small. The

512   P. R. Vogt *et al.*

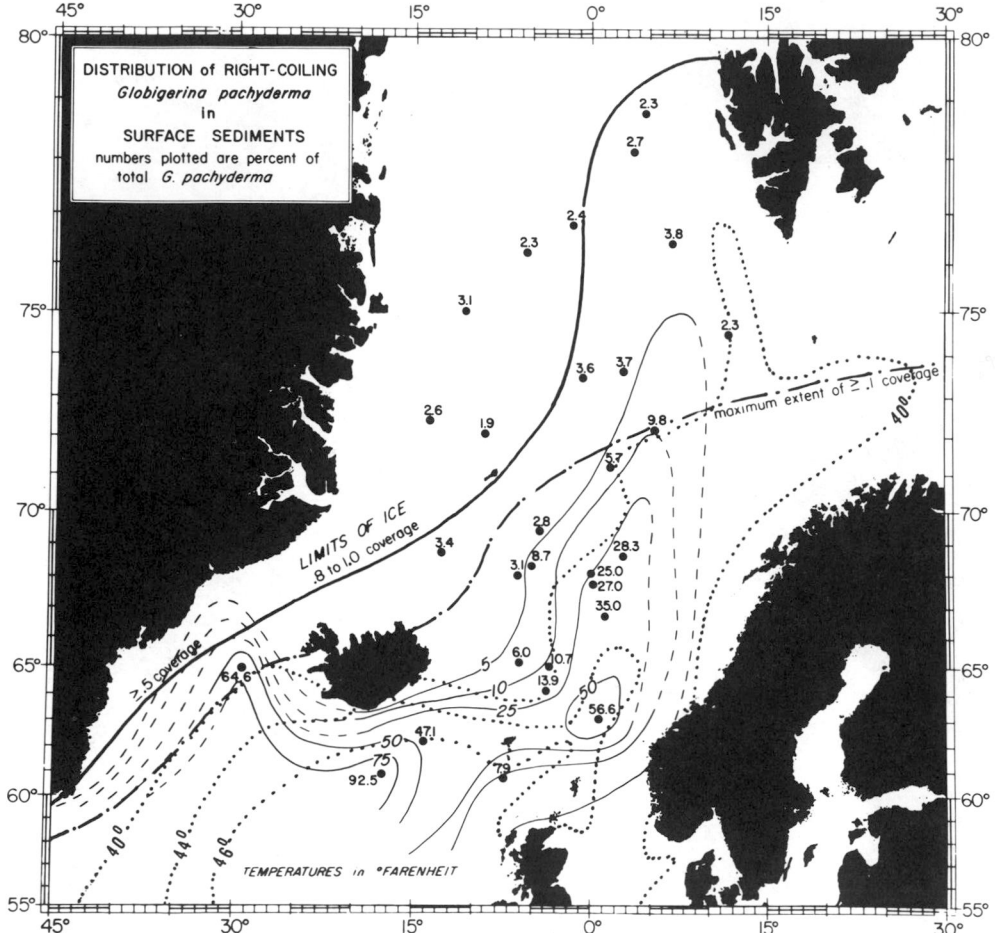

Fig. 11. Distribution of right-coiling (warm water) *Globoquadrina pachyderma*, February surface temperatures (°F), and ice limits. Compounded from Kellogg (1975).

high-CaCO₃ provinces cut across topographic provinces of varying depth, a feature that would not be expected if dissolution were important. In fact, the entire GNSA, with the exception of the Molloy Rift Valley west of Svalbard (Fig. 1) lies above the 4.5-km North Atlantic "carbonate compensation depth."

2. *Continental Margins*

Sediments along the continental margins are distinct from those in the basins. Within 5–30 km of the Norwegian coast, in fact, the sea floor consists of outcropping, glacially eroded crystalline and metamorphic rocks—the "strandflat" province of the inner shelf (Holtedahl and Bjerkli, 1975; Sundvor and Nysaether, 1975). Coarse sediments—sand, gravel, shells, stones, and blocks—next appear, probably the winnowed relicts of moraines, littoral deposits, and outwash. On the outer shelf the sediment tends to be finer, generally silt or silty clay below 400–500 m and in the Norwegian Channel. However, west of the steep, narrow shelves at 62°–63°N and 68°–69°N, the continental slope above 1000-m depth is covered by glacial-marine boulder clays; this sediment is associated with ridges paralleling the slope. The shallower ridges may be terminal moraines.

Where bottom currents are not strong, postglacial sediment would be expected to cover the sea floor. A core taken at about 64.5°N at 310 m on a submarine ridge revealed 1.5-m postglacial (Holocene) clay overlying glacial till (Holtedahl and Bjerkli, 1975). Off Troms and West Finnmmark, Holocene sediment ranges from generally less than 30 cm in areas shallower than 300 m, to as much as several meters in troughs (Vorren and Lind-Hansen, 1979).

The composition of glacial deposits on the Norwegian shelf is variable where the Quaternary cover is thin. Tertiary and Jurassic–Cretaceous sandstones, claystones, and limestones predominate. These materials were probably locally derived and redeposited by the ice sheet after relatively short transport. By contrast, thick glacial deposits are largely composed of igneous and metamorphic detritus, similar to that found in terrestrial deposits and derived from the mainland. Some rock fragments were demonstratably ice-rafted from the Oslo area.

Although much of the Greenland, Barents, and Iceland shelves has not been sampled in great detail, available data suggest the existence of relatively coarse, current-winnowed glacial sands and gravels, with substantial areas of glacially eroded, outcropping basement rock (e.g., Vogt, 1970) adjacent to the coasts. Gravel from the Spitsbergen Bank consists of reworked clasts of primarily Mesozoic sediment, transported over short distances by local glacier ice (Edwards, 1975; Nagy, 1973; Emelyanov et al., 1971). Basalt erratics on the southeast Faeroes shelf indicate transport by a thin Faeroes ice sheet as far as the shelf edge; erratics farther southeast include clasts from the British ice sheet (Waagstein and Rasmussen, 1975).

Clay, silt, and foraminiferal ooze, all mixed with coarser glacial detritus, is probably restricted to deeper valleys, glacially incised in the shelf, and to the outer shelves and continental slopes (Pálmason, 1974b; Johnson et al., 1975a,b; Sommerhoff, 1973). On the Southeast Greenland inner shelf, intermittent glacial deposits overlie outcropping crystalline rocks, while the outer shelf consists of terminal moraines paralleling the margin. The latter are locally breached by perpendicular troughs eroded by ice streams ground down to depths of 500–1000 m below present sea level. Coarse sediments of glacial origin are exposed on the moraines. The composition of these sands and gravels—gneiss, granite, and basalt—suggests derivation from the nearby coast; abundant basalt clasts may have been carried southwest by icebergs or, alternatively, were derived from hypothetical local outcrops on the inner shelf. In the shelf troughs fine-grained sediments occur mixed with gravels. Delta-shaped projections in front of the shelf basin suggest glacial marine sediment on the upper continental slope between 300- and 1300-m depths. Basalt fragments dredged from the southeastern Greenland margin were probably ice-rafted from the Scoresby Sund Tertiary igneous province (Rasmussen *et al.*, 1976).

B. Late Quaternary Climate

During the last half-decade much paleoclimatic/paleooceanographic information has been extracted from deep-sea sediment cores. Numerous cores in areas of microfossil deposition and preservation extend back 10^5-10^6 years in time and cover the latest glacial/interglacial cycles. The rather continuous marine record left by pelagic sedimentation in calm bottom environments stands in contrast to the confused, discontinuous sedimentary record left on adjacent continents. However, surviving continental ice sheets also yield information on paleoclimate: From ^{18}O analysis of ice recovered by a drillhole through the Greenland ice sheet, a paleotemperature record extending back to 125,000 years b.p. was inferred (Dansgaard *et al.*, 1969). This record suggests that the major Wisconsin/Weichsel ice age began about 70,000 years b.p., with occasional earlier cold periods, e.g., at 90,000 years b.p.

Applying transfer function analysis to down-core microfossil assemblage variations (Imbrie and Kipp, 1971), the CLIMAP team reconstructed world climate at 18,000 years b.p., during the latest period of extreme glaciation (CLIMAP Project Members, 1976).

As suggested by earlier studies of a small number of cores in the Norwegian Sea (Holtedahl, 1959; Ericson *et al.*, 1964), the North Atlantic from 40°N to 75°N, including the GNSA, has turned out to be a particularly sensitive "thermometer" of Late Quaternary climatic fluctuations. A north–south series of cores shows the alternating advance and retreat of polar water masses (Fig. 12; McIntyre *et al.*, 1972; Ruddiman and McIntyre, 1976, 1977). Judging by the last deglacial warming sequence, at least the northward retreat of the North Atlantic Polar Front was approximately synchronous with the retreat of the North American and Scandinavian ice sheets (Fig. 13; Ruddiman and McIntyre, 1973). [However, the Greeland ice sheet may have experienced its greatest expansion in *early* Wisconsin time (Weidick, 1976).] During glacial extremes, oceanographic conditions presently found only in the northwestern Greenland and Norwegian Seas extended southward almost to the Azores. Using as a time marker a volcanic ash horizon deposited about 9300 years b.p. over a wide area of the North Atlantic between Iceland and the Azores, Rud-

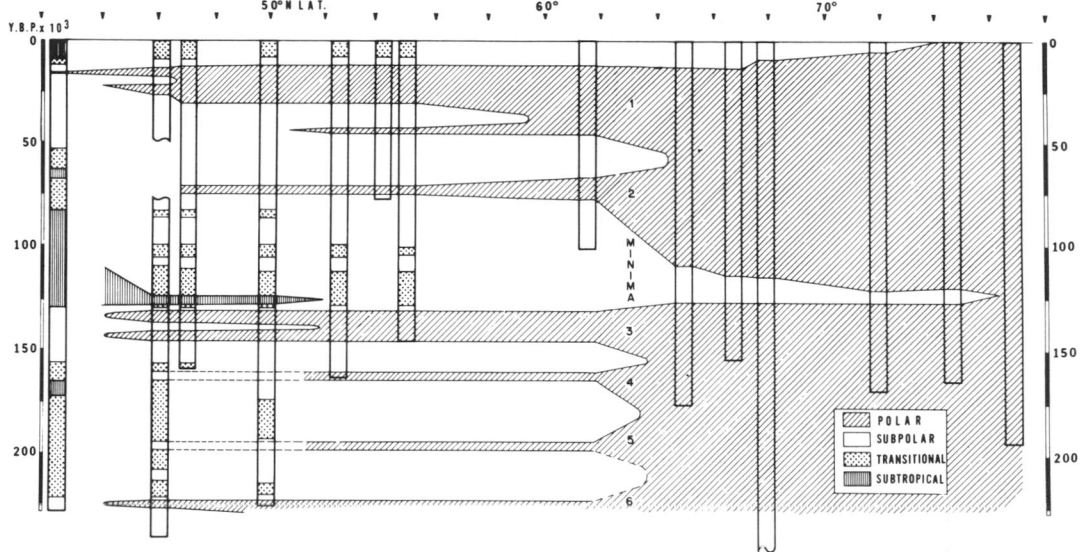

Fig. 12.   North–south series of pelagic sediment cores, showing intermittent southward advances of polar water recorded by faunal assemblages (Kellogg, 1975; Ruddiman and McIntyre, 1976).

diman and Glover (1975) were able to describe oceanographic conditions at a time when the retreat was in full swing.

   During warm interglacials, such as the last ~7000 years, subtropical, transitional, and subpolar planktonic assemblages spread northward as far as 42°–52°N, 55°–60°N, and 75°N, respectively. Superposed on this latitudinal pattern, especially during warmer interglacials, is east–west asymmetry: Cold, less saline Arctic water moves southwest along the East Greenland margin, while warmer, more saline waters flow northeastward along the eastern side of the basins (Figs. 10 and 11). This oceanographic regime is mirrored by the surface sediment distribution (Fig. 10). Another complexity is a southeast-protruding lobe of cool water, as defined by both the 9300 years b.p. and the present shape of the Subarctic Convergence, located around 50°–55°N (Ruddiman and Glover, 1975).

   During the 18,000-year-b.p. glacial extremum, the Polar Front and Subarctic Convergence appear to have been merged into a single, nearly east–west boundary (Fig. 13). At that time, no North Atlantic Drift water reached the GNSA. Without the saline Norwegian Current, the deep-water circulation of the North Atlantic must have been quite different from today. There was no Norwegian Sea overflow water, now formed by cooling, evaporation, and sinking of Norwegian current surface water east of Greenland. Paleontological data from the GNSA show that the Norwegian current was either much weaker or completely absent for at least 100,000 of the last

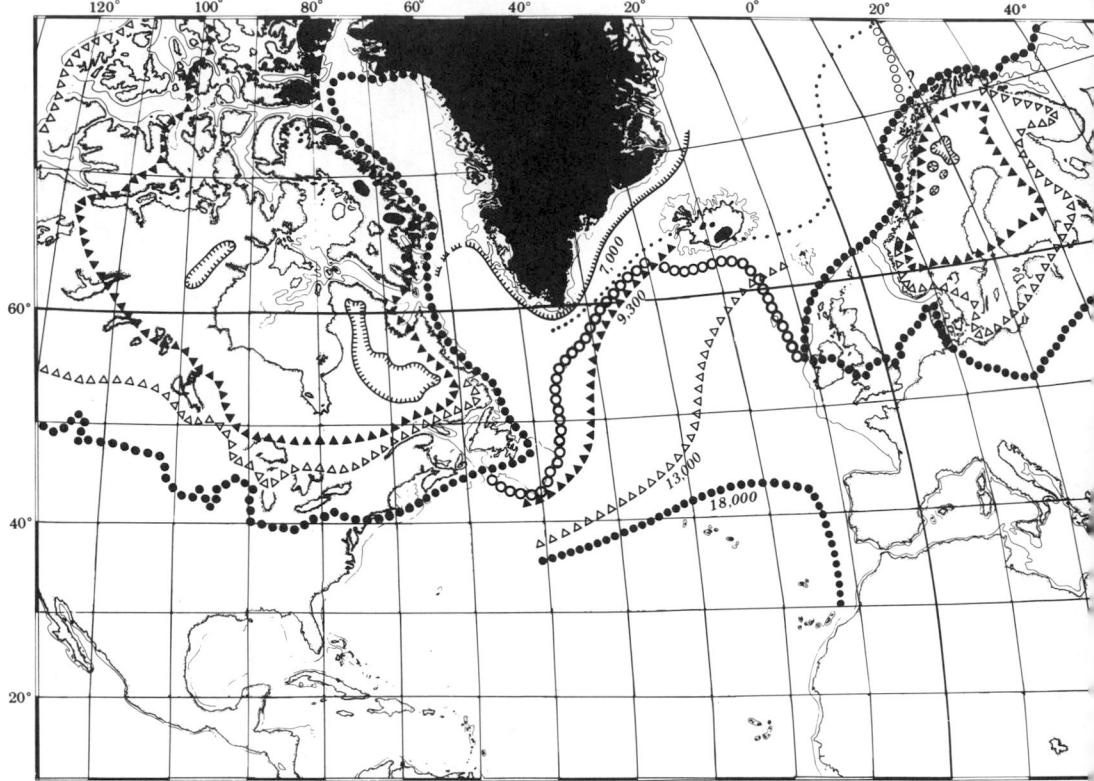

Fig. 13.   Deglacial retreat limits of continental ice and polar water, modified from Ruddiman and Glover (1975).
18,000-year b.p. polar front position in the Gibraltar area is after Thiede (1977) and Molina-Cruz and Thiede (1978).
areas are presently glaciated. Small dots show approximate maximum southeastward extent of pack ice since the early
century (Kellogg, 1975). Open circles show southern edge of the continuous glacial age American–Eurasian ice shelf
lated by Hughes *et al.* (1977).

127,000 years (Kellogg, 1975), and perhaps the entire period 450,000–127,000
years b.p. (Kellogg, 1977).

Although the sharply reduced faunal diversity characteristic of cold waters
makes quantitative results less reliable than elsewhere, winter temperatures of
$-2°–0°C$ and summer temperatures of $3°–6°C$ seem to have prevailed
throughout the Norwegian Sea for most of at least the last 450,000 years. Only
during the last 7000 years and about 120,000 years b.p., during a short part
(Stage 5e) of the longer "Eemian" interglacial, did climate ameliorate enough
to admit subpolar fauna to the Norwegian Sea (Fig. 12). Coevally, abyssal circulation in the North Atlantic was stimulated by sinking and southward overflow of cooled Norwegian Sea surface water (Ruddiman and Bowles, 1976).

Generally, carbonate content increases with calculated paleotemperature.
Holtedahl (1959) first identified the relatively barren clay below about 30 cm

in Norwegian Sea cores as glacial sediment. In Kellogg's northernmost cores, west of Svalbard, carbonate peaks occur at depths at which warm pulses would be expected, but where faunal compositon indicates only uniform cold. Evidently the productivity of left-coiling *Globoquadrina pachyderma*, the dominant species in this area, is more sensitive to climatic changes involving small changes in temperature near the freezing point of seawater than are the ecological responses of other species (Kellogg, 1975). This oceanographic regime—high carbonate productivity but low temperature estimates—prevailed in the entire GNSA during the "Eemian" interglacial, with the exception of only the short "warm" pulse at 120,000 years b.p. As observed at present, the regime is accompanied by partial ice cover. Low temperature *and* low carbonate characterized GNSA sediments deposited during both the Würm (Wisconsin) and preceding glacial periods.

Variations in oxygen isotope makeup of planktonic foraminifers (Kellogg *et al.*, 1978) have slightly revised the timing of interglacial pulses (124,000–115,000 and 13,000–0 years b.p.). The isotope record shows local temperature and salinity effects as well as global oceanic changes.

C. Late Quaternary Sedimentation Rates

Since the end of the last glaciation (ca. 11,000 years b.p.) sedimentation has proceeded at about 1.5–4.5 cm/1000 years within 400 km northeast of Iceland, 1.5–8 cm/1000 years under the Norwegian Current and, with considerably less certainty, 1–3 cm/1000 years in the Greenland Sea (from Kellogg, 1975). South of Iceland, a prominent ash horizon dated at 9300 ± 300 years b.p. (uncorrected ^{14}C age) forms a time-synchronous horizon detected in 35 cores between 43°N and 65°N (Ruddiman and Glover, 1975). Sedimentation rates are found to range from 0.7 to 11.8 cm/1000 years (mean, 3.7) since this ash was deposited. Both north and south of Iceland high local variability of sedimentation rates tends to mask regional patterns. Gradually, sedimentation rates decrease by a factor of about two, from the warmer waters of the North Atlantic, Irminger, and Norwegian Currents northward toward the Greenland and northern Norwegian Seas. This probably reflects a relationship between temperature and productivity: Ruddiman and McIntyre (1976) find higher absolute rates of biogenous input during interglacial climates. However, the increased delivery of terrigenous sediments during glacial conditions tends to compensate for the reduced biogenic production, thus maintaining the total sedimentation rate at a relatively constant value.

Kellogg (1975) identified the beginning (ca. 127,000 years b.p.) and end (ca. 75,000 years b.p.) of the last interglacial in six cores northeast of Iceland and 12 cores in the Norwegian Current. From his data we calculate average post-75,000-year-b.p. sedimentation rates of 1.4–3.7 cm/1000 years (mean,

2.3) for the Iceland cores and 2.6–5.7 (mean, 4.0) for the Norwegian Current cores.

Radiometric dating was used to derive an absolute sedimentation rate of 2.56 cm/1000 years in a core west of Iceland (Kellogg, 1975). This rate was maintained from about 150,000 to 30,000 years b.p., but then gradually rose to a value of 11.8 cm/1000 years for the period since the 9300-year-b.p. ash was deposited (Ruddiman and Glover, 1975).

The above discussion concerns sedimentation on a relatively calm ocean floor, where sampling has been concentrated for paleoclimatological purposes. However, bottom-touching currents have prevented or modified the accumulation of sediments in parts of the greater GNSA, especially on the continental shelves of Greenland, Iceland, and Norway, and in the Barents Sea, as well as in the deep Labrador Sea (Rabinowitz and Eittreim, 1974; Egloff and Johnson, 1975), and the northern Reykjanes Ridge (Talwani *et al.*, 1971). However, coring, bottom photography and 3.5-kHz echogram analysis show that large-scale contour-current activity has been unimportant in deeper waters north of the Iceland–Faeroe Ridge (Damuth, 1978). Local slumps and turbidity flows seem to have been more important in removing recent and older sediments from the steeper slopes of seamounts, continental margins, and crestal ridges of the MOR. The finer sediment may be redeposited in local drifts or swept into deeper, calmer environments.

Glacial erosion during the last 3 m.y. has removed most unconsolidated sediment from the inner continental shelves of the GNSA. Seismic reflection profiling reveals irregular deposits of "Quaternary" sediments unconformably overlying older consolidated sedimentary formations on the outer shelves of Norway (Floden, 1973; Floden and Sellevoll, 1972; Holtedahl and Sellevoll, 1971, 1972; Vorren and Lind-Hansen, 1979), Greenland (Johnson *et al.*, 1975a,b), Iceland (Pálmason, 1974b), and parts of the Barents Sea (Sundvor, 1974). A thickness of about 100 m is suggested. These deposits may well include late Pliocene sediment of glacial character; however, the unconformity below the "Quaternary" deposits has not been dated. Nor is it known what portion of the "Quaternary" deposits dates from the last glaciation. In the northern North Sea, much of the older "Quaternary" sediment was severely modified during the last glaciation, when ice reached the western margin of the Norwegian Channel (Egeberg, 1977). Deposits from earlier, greater advances may lie undisturbed outside the Weichsel limits in many parts of the GNSA.

D. Paleoclimate and Sedimentation during Earlier Plio-Pleistocene Glaciations

Deep drilling on Orphan Knoll (east of Newfoundland) and in the GNSA (Fig. 14) suggests that large-scale glaciations and southward penetration of polar water began sometime in the late Pliocene, about 3 to 3.5 m.y.b.p.

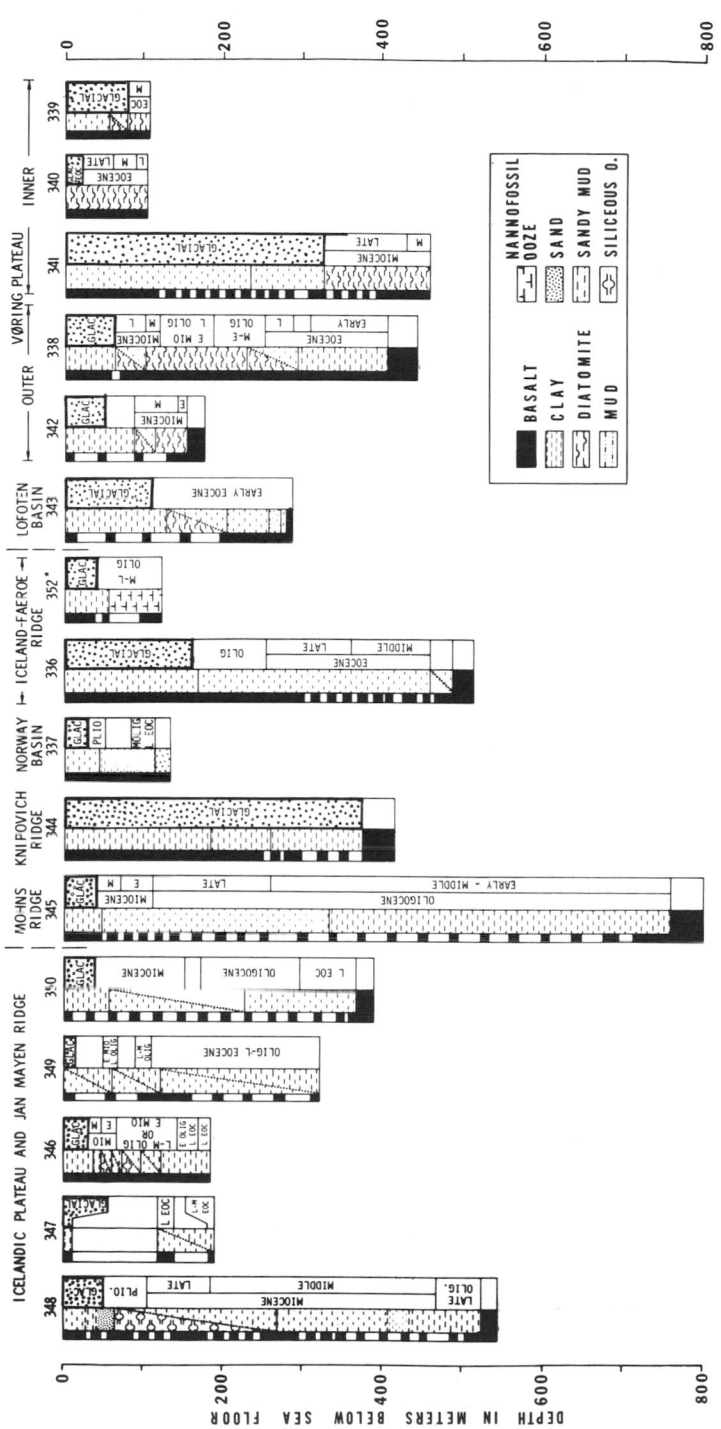

Fig. 14.   Summary of DSDP Leg 38 results (Scientific Staff, 1975; Talwani and Udintsev, 1976).   Cored intervals of drill holes (left column) are black.   Glacial section is coarsely stippled (right column).

(Schrader *et al.*, 1976; Berggren, 1972), and possibly as early as 5 m.y.b.p. (Warnke and Hanson, 1977). The cooling that eventually led to these glaciations seems to have begun in the earliest Pliocene or latest Miocene (Poore and Berggren, 1975). At that time, warm temperate flora (similar to what is presently found along the U.S. central Atlantic states) still existed on Iceland (Friedrich and Simonarson, 1974, 1976).

The late Pliocene arrival of ice-laden water over Orphan Knoll corresponds to the time extensive glaciations first affected Iceland and mountain ranges in western North America and Europe (Berggren, 1972), and marks the latest, most dramatic of a series of abrupt global declines in Tertiary bottom-water temperature (Savin *et al.*, 1975).

The cores described by Kellogg (1975) for the GNSA thus record less than 10% of the Plio-Pleistocene glaciations; one core in the Norway Basin extends to 450,000 years b.p. (Kellogg, 1977). Climatic fluctuations such as those well-documented back to 0.2 m.y.b.p. probably occurred as early as 0.7 m.y.b.p. (Ruddiman and McIntyre, 1976). Oxygen isotope measurements on benthonic foraminifera in a Pacific core extending back to 2.1 m.y.b.p. suggest that the first major northern climatic glaciation of middle Pleistocene character occurred about 0.8 m.y.b.p. (Shackleton and Opdyke, 1976). Prior to this, glacial events of higher frequency and possibly lower amplitude extend to 1.6 m.y.b.p., and with less regular frequency events down to the bottom of the core. Translated to the GNSA, these results suggest that ice cover was probably continuous or partial during most of the last 3 m.y., with perhaps fewer barren zones (indicating extremely cold conditions) prior to 0.8 m.y.b.p. If the Plio-Pleistocene transition occurred as late as 1.8 m.y.b.p., the influx of Pacific mollusks seen in the Tjörnes (Iceland) formation of this age (Einarsson *et al.*, 1967) conflicts with evidence for extensive ice cover in the Pacific and GNSA.

The glacial/preglacial boundary is generally abrupt in the GNSA deep drillholes (Schrader *et al.*, 1976) and permits estimates of average accumulation rates over the interval. Unfortunately the age of the boundary is poorly controlled; 3 m.y. is a reasonable estimate. The actual beginning of ice rafting may well have been 1 or 2 m.y. earlier, and time transgressive from north to south. The "glacial" (including interglacial) sediment formation ranges from 20 to 370 m in thickness in the 14 holes drilled (Fig. 14). The average thickness of the glacial formation in areas dominated by pelagic sedimentation is about 50 m (Warnke and Hansen, 1977). Allowing for compaction, we obtain an average accumulation rate of 2 cm/1000 years for the last 3 m.y. Considering the various uncertainties, this is not substantially different from values for the last 75,000 years, discussed earlier, and an average of 1.5 cm/yr for the last 450,000 years for a piston core site in the Norway Basin (Kellogg, 1977).

Three drill sites (336, 341, and 344) yielded anomalously thick (160–370 m) glacial sections. The sites were located on the eastern Iceland–Faeroe Ridge, the Vøring Plateau, and the continental rise just east of Knipovich Ridge. Proximity to the continental ice-sheet margins and to shallow-water areas, swept by bottom currents, may explain the high average sedimentation rate (more than 10 cm/1000 years at sites 341 and 344). The sediment diapirs east of the Vøring Escarpment seem to have been created by relatively lower density (1.32 g/cm³) impermeable preglacial ooze overlying higher density (1.85 g/cm³) glacial sediment (Caston, 1976).

Glacial sediments are locally thin or absent where glacial erosion predominated over deposition, for example in the "strandflat" province of outcropping crystalline rock on the inner shelves of Norway and Greenland.

The Greenland ice cap probably formed as early as 3 m.y.b.p., but complete disappearance during warm interglacials is possible (Weidick, 1976), while some nunataks escaped glaciation during the coldest glacials.

E. Preglacial, Post-Mesozoic Sediments

Since oceanic crustal accretion only began in the GNSA about anomaly 24–25 time (Talwani and Eldholm, 1977), we may conveniently divide the discussion of preglacial sediments at the Cretaceous/Tertiary boundary, which is only slightly older (anomaly 29; Tarling and Mitchell, 1976; La Brecque et al., 1977). [On DSDP Leg 38, the oldest post-opening sediment recovered was Lower Eocene, at Site 338 on the crest of the Vøring Outer Ridge (Schrader et al., 1976).]

According to one hypothesis (Gartner and Keany, 1978) the terminal Cretaceous extinction event was the initiation of rifting in the northern GNSA, admitting fresh or brackish water from a supposedly isolated Arctic Basin. Although this hypothesis might be accommodated by magnetic anomaly data from the Eurasia Basin (see Fig. 28 and later discussion), the anomalous iridium concentrations measured at the Cretaceous/Tertiary boundary and other data provide compelling evidence that the "boundary event" was actually the impact of an asteroid (Alvarez et al., 1980).

Paleoclimatic evidence from sediments within and bordering the GNSA suggests that temperate climates prevailed throughout the Tertiary, without significant glaciation, until the late Pliocene. Warm temperate climates are suggested by Lower Paleocene flora in west Greenland (Weidick, 1976) and by Middle Eocene vertebrate remains in northern Ellesmere Island (Dawson et al., 1976). Marine microfossils recovered on Leg 38 suggest cool–temperate seas, isolated from the Atlantic until Miocene time (Schrader et al., 1976).

Information about the post-opening sediment in the GNSA, as in other ocean basins, comes primarily from seismic reflection profiling, augmented by deep drilling and wide-angle reflection and refraction measurements (Fig. 15). In general, compressional wavespeed in sediment increases with depth (Fig. 16) and sediment age (Fig. 17). (In this chapter *velocity* is used as a synonym, for *speed*, by virtue of common usage.)

In most areas of the GNSA underlaid by postopening ocean crust, it is possible, using single-channel equipment, to obtain seismic reflections from the interface between igneous oceanic crust, created by axial accretion, and the overlying sediments. Near the continental margins the sediment overburdens exceed 2 km and multichannel techniques are required. Furthermore, over much of the Iceland–Faeroe Ridge and the Iceland Plateau between the Jan Mayen and Kolbeinsey Ridges the acoustic "basement" is an exceptionally smooth opaque horizon (Eldholm and Windisch, 1974; Talwani and Eldholm, 1977). This opaque reflector, perhaps similar to one known locally on the Reykjanes Ridge south of Iceland (Vogt and Johnson, 1973b), may be caused by coarse volcaniclastic sediment, thin basalt flows, erosional debris, or subsided sea-level surfaces. A similar reflector was formed in the early stages of spreading on the Voring Plateau and northeastward to 69.5°N (Eldholm *et al.*, 1979). In some areas (Fig. 18) coherent reflections from below this "basement" raise the possibility of strata below, even in areas of spreading-type magnetic anomalies.

Seaward of the boundary between continental and oceanic crust, the sediments are of post-opening age, late Paleocene and younger. The basins landward of the boundary also contain Mesozoic and perhaps older sediments, as discussed in the following section.

Post-opening thicknesses (Fig. 18) range from zero at the axis of the MOR and on some basement highs and steep escarpments, to more than 2 km along parts of the continental margin. The 0.2-km isopach lies near anomaly 5 (10 m.y.b.p.) on the flanks of Mohns Ridge; this yields a mean sedimentation rate of 2 cm/1000 years (Eldholm and Windisch, 1974), similar to average Quaternary values. If such a rate had obtained since 50–60 m.y.b.p., the present thickness on oldest GNSA oceanic crust would be 1.0–1.2 km. The measured thickness is about the same order (Fig. 18), except near the continental margin, where terrigeneous contributions have led to accumulations up to 6 km on possibly oceanic crust (Houtz and Windisch, 1977).

Factors governing the complexity of marginal sedimentation include primarily (1) the position of oceanic basement features such as fracture zones, marginal ridges, and the spreading axis relative to the flow of terrigenous material, and (2) the variability of terrigenous sediment input from different areas along the continental margin (Eldholm and Windisch, 1974). The primary role of basement features in the GNSA has been to dam or channel the seaward flow of terrigenous material. Thus, sediments from the East Green-

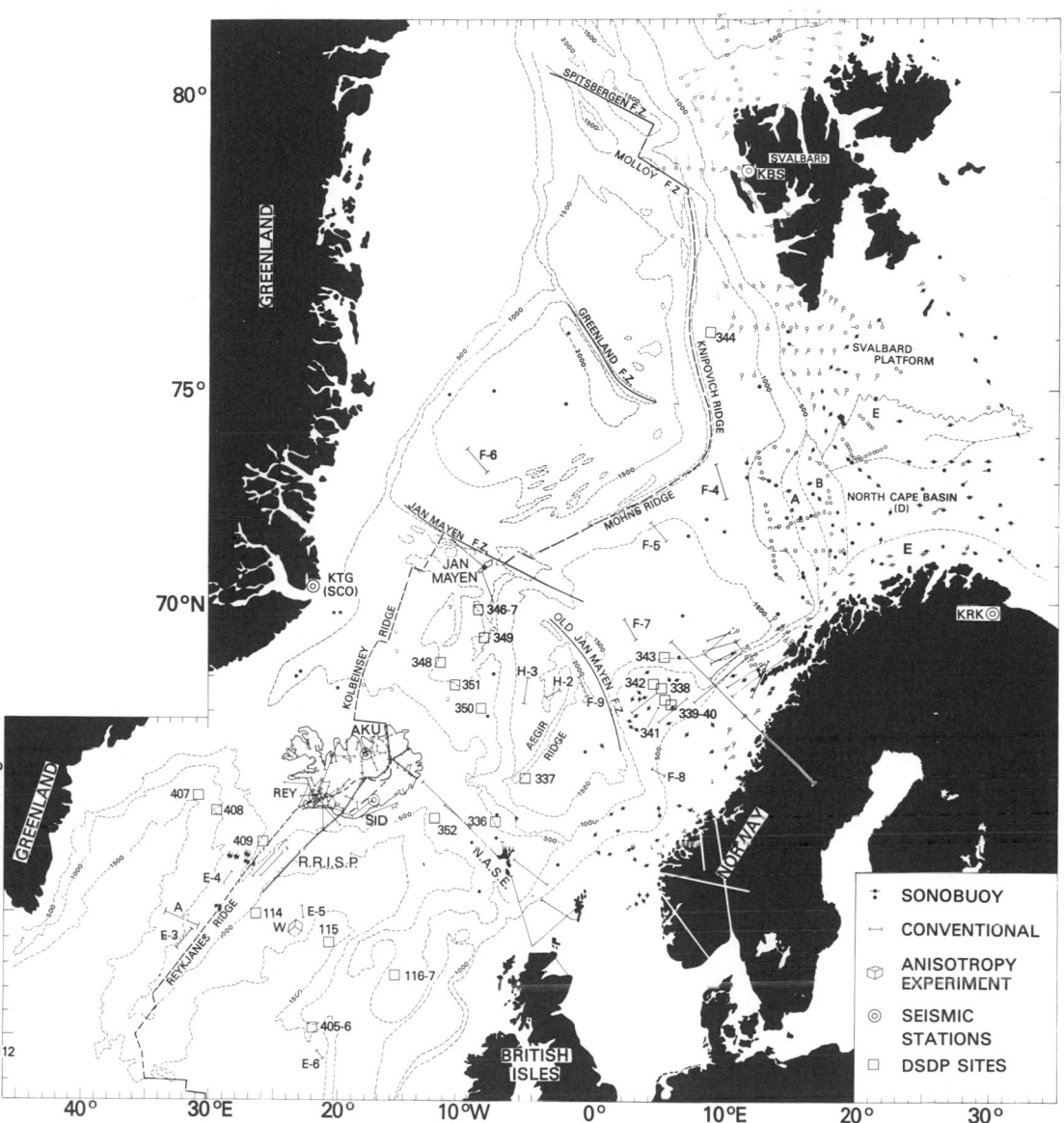

. 15.   Approximate locations of seismic refraction stations and DSDP sites in the GNSA, including Iceland and Reyk-
es Ridge, Compiled from Bott *et al.* (1975) (labeled N.A.S.E.), Pálmason (1971), Renard and Malod (1974), Sundvor
74), Hinz (1975), Sundvor and Nysaether (1975), Talwani *et al.* (1971), Whitmarsh (1971), Heier (1977), Sundvor *et al.*
75, 1977, 1979), Eldholm and Talwani (1977), Houtz and Windisch (1977), Angenheister *et al.* (1979) and Scientific
ff (1975–1977). Barred lines are conventional (two-ship) or long refraction lines; small filled circles, short expendable
obuoy profiles, showing direction where given by authors. Open circles show data published by Sundvor *et al.* (1975,
7, 1979) and Houtz and Windisch (1977). A, deep reflection line of Aric (1972). RRISP refers to Reykjanes Ridge–Ice-
d Seismic Project (Angenheister *et al.*, 1979). Polygon shows anisotropy experiment, utilizing bottom-mounted receivers
hitmarsh, 1971). Double circles show locations of teleseismic recording stations.

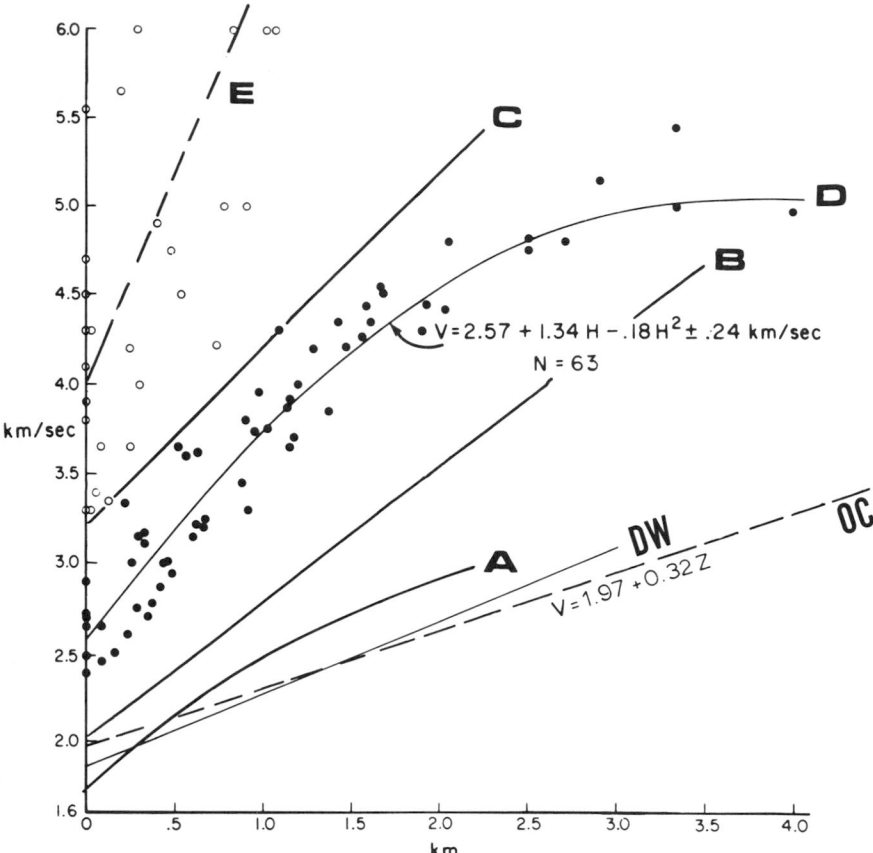

Fig. 16. Velocity–depth diagrams derived from sonobuoy stations for various Barents Sea provinces (A–D, from Houtz and Windisch, 1977; observed datum points shown only for provinces D and E); for all deep-water (>2 km) stations in the GNSA (line DW, from Eldholm and Windisch, 1974); and for all stations on oceanic crust (line OC, from Myhre, 1978). See Fig. 15 for data locations and boundaries of Barents Sea provinces.

land continental margin are ponded against the youthful Kolbeinsey Ridge, causing marginal thicknesses to exceed 2.4 km there (Fig. 18). Similarly, the Knipovich Ridge acts as a barrier to the westward transport of sediment originating on the Barents shelf and Svalbard. What material presently overflows the rift mountains east of the present spreading axis is trapped in the rift valley (Figs. 19 and 20). Thicknesses of well over 3 km characterize the margin east of Knipovich Ridge (Sundvor *et al.*, 1978, 1979; Hinz and Schlüter, 1978a; Schlüter and Hinz, 1978; Briseid and Mascle, 1975). To the south, Mohns Ridge axis acts as a two-day dam, isolating Greenland-derived turbidites of the Greenland Basin from Norwegian/Barents shelf material deposited in the

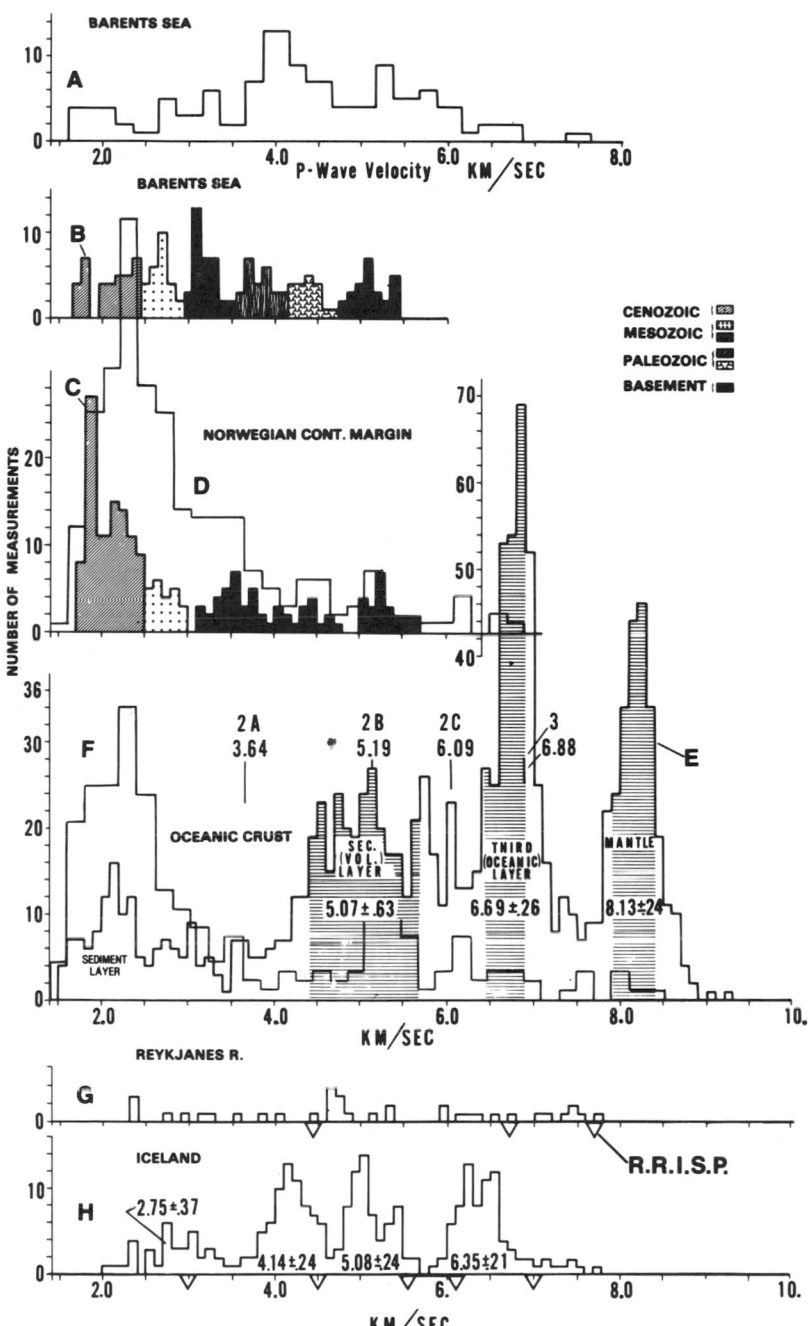

Fig. 17. Frequency distributions of P-wave velocities determined by seismic refraction stations in continental (upper) and oceanic (lower) parts of the GNSA, Iceland, and the Reykjanes Ridge. "Oceanic Crust" histogram shows data from all oceans for comparison. Compiled from the following authors (top to bottom): A, Renard and Malod (1974); B, Sundvor (1974); C, Sundvor and Nysaether (1975); E, Christensen (1974); G, Talwani *et al.* (1971); H, Pálmason (1971); D and F, Myhre, (1978); Layers 2A, 2B, 2C, and 3, Houtz and Ewing, (1976); RRISP values, Angenheister *et al.* (1979).

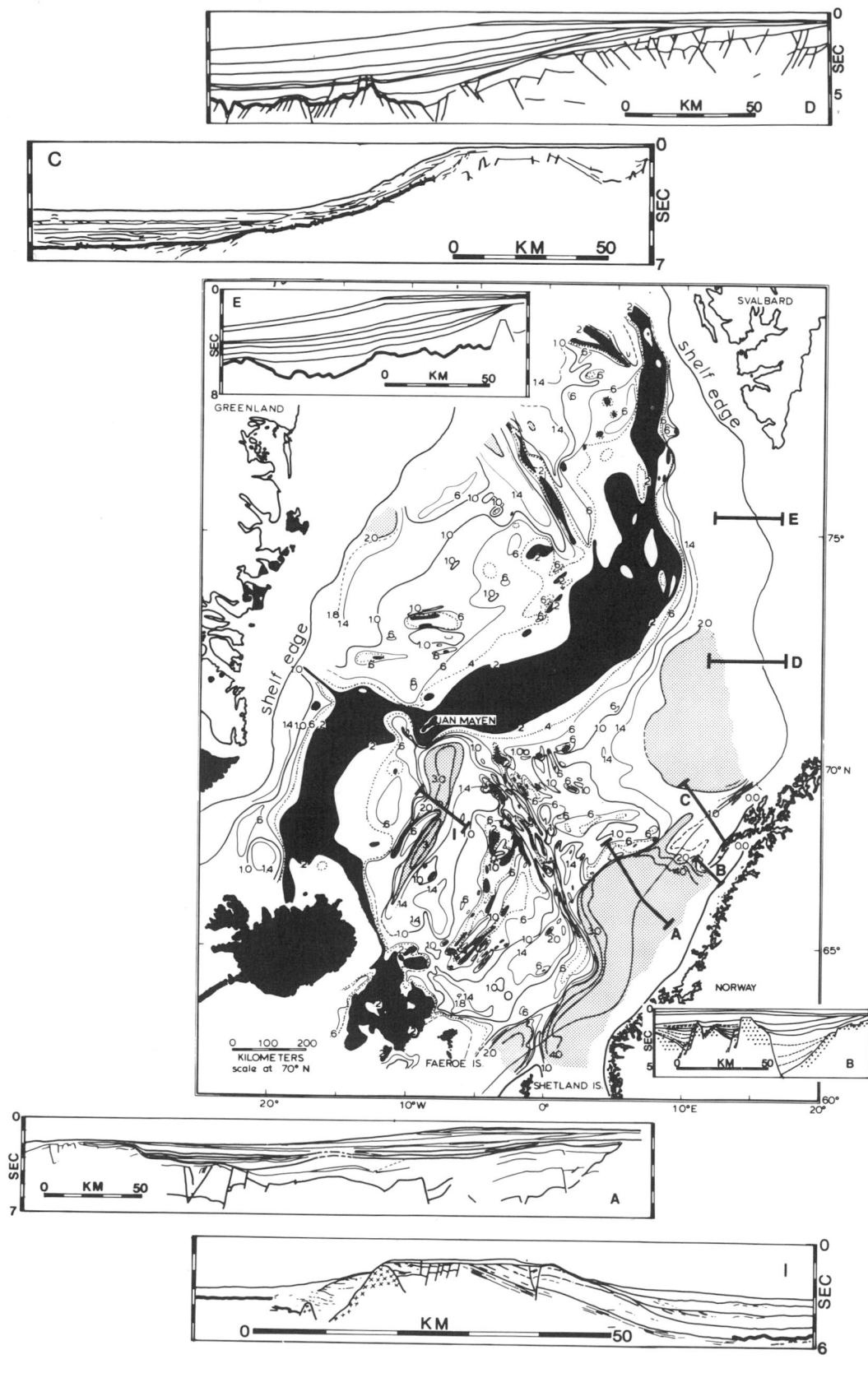

Lofoten Basin. However, at 73°– 73.5°N sediment from the Bear Island Fan has locally breached the Mohns Ridge crestal mountains and entered the rift valley.

Marginal basement ridges may also have hindered seaward transport of sediment during much of Tertiary time (Talwani and Eldholm, 1972; Eldholm and Windisch, 1974). Such ridges have been identified along the southeast margin of the Norway Basin (the Faeroe–Shetland Escarpment), the northwest margin of the Voring Plateau (the Voring Plateau Escarpment) (Fig. 1), and the Greenland margin just southwest of the Greenland F.Z. Basins landward of these marginal ridges continued their subsidence from earlier times, although mainly by downflexing rather than faulting after the early Tertiary. Up to 3.5 km of Cenozoic sediment has accumulated in the central North Sea (Ziegler, 1977) and 1 to 3 km of sediment overlies continental crust west of central Norway (Sundvor and Nysaether, 1975; Ronnevik and Navrestad, 1977).

Variability in sediment supply, both in time and space, has been the other important factor in controlling sedimentation along the margins. A major event was the uplift of western Norway, now thought to have occurred by early Eocene epeirogeny (Egeberg, 1977) rather than along faults in late Tertiary time (Holtedahl, 1960). The beginning of major glaciation and associated sea level falls was a second major event. Up to about 0.5 km interlayered "turbidite" and homogeneous strata overlie a lower homogeneous formation in parts of the Norway, Lofoten, and Greenland basins (Eldholm and Windisch, 1974). It seems reasonable to associate the upper formation with the Plio-Pleistocene glaciations.

Sedimentation west of the Barents Shelf seems to have been dominated by a drainage system, emergent through much of Tertiary time that crudely followed the axis of the present Bear Island Trough. Both bathymetry (Fig. 1) and seismic data (Fig. 21) suggest a massive sediment fan built out seaward from the mouth of the Bear Island Trough and delivering sediments as far as the present spreading axis. It appears likely that the entire Barents Shelf was glaciated at times during the Pleistocene; an active ice stream was grounded along the length of the Bear Island Trough—as in the Norwegian Channel to the south (Hughes et al., 1977). This ice stream would have severely modified any original fluvial topography and delivered additional sediment to the adjacent continental rise and slope. Refraction and reflection measurements

←——

Fig. 18.   Sediment thickness isopachs (Gronlie and Talwani, 1978) in seconds two-way travel time (approximately kilometers). Isopach interval is 0.2 sec; data were smoothed prior to contouring (see Fig. 25 for simplified isopachs in relation to magnetic anomalies). Black: Areas of Cenozoic igneous rocks covered by less than 200-m sediment. Multichannel profiles: A, Rønnevik et al. (1979); B, Jørgensen and Navrestad (1979); C, Eldholm et al. (1979); D and E, Rønnevik and Motland (1979); I, Gairaud et al. (1978). Heavy solid line in C, D, E, and I denotes interpreted oceanic (basaltic) basement.

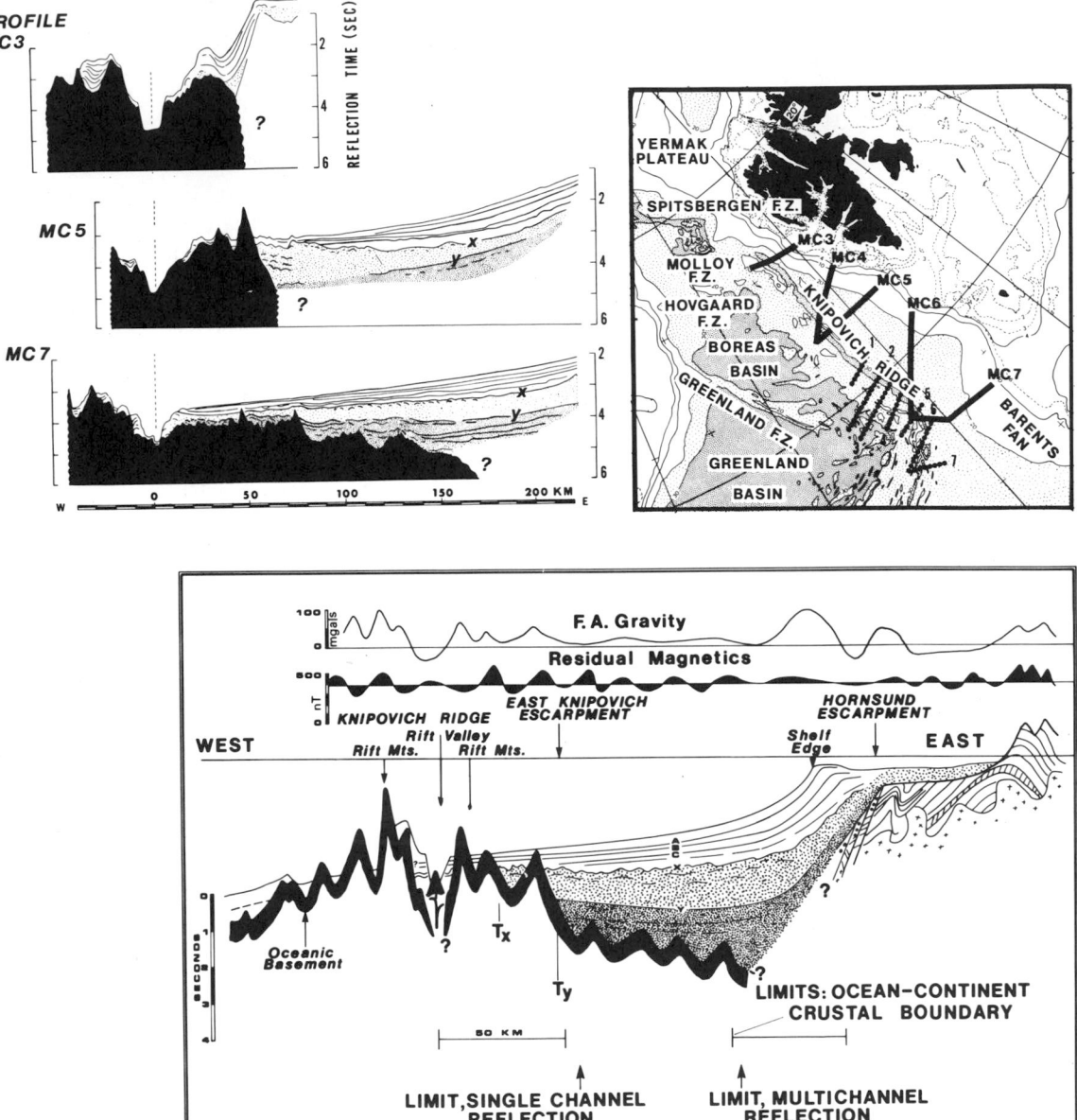

Fig. 19.   Upper right: Simplified bathymetry of northern Norwegian and Greenland Seas (from Perry *et al.*, 1980), show-ing locations of single-channel (1–6; Fig. 20) and multichannel profiles (upper left, from Briseid and Mascle, 1975). Bot-tom: Schematic profile showing representative geophysical characteristics of Knipovich Ridge and adjacent continental margin (Vogt *et al.*, 1978).

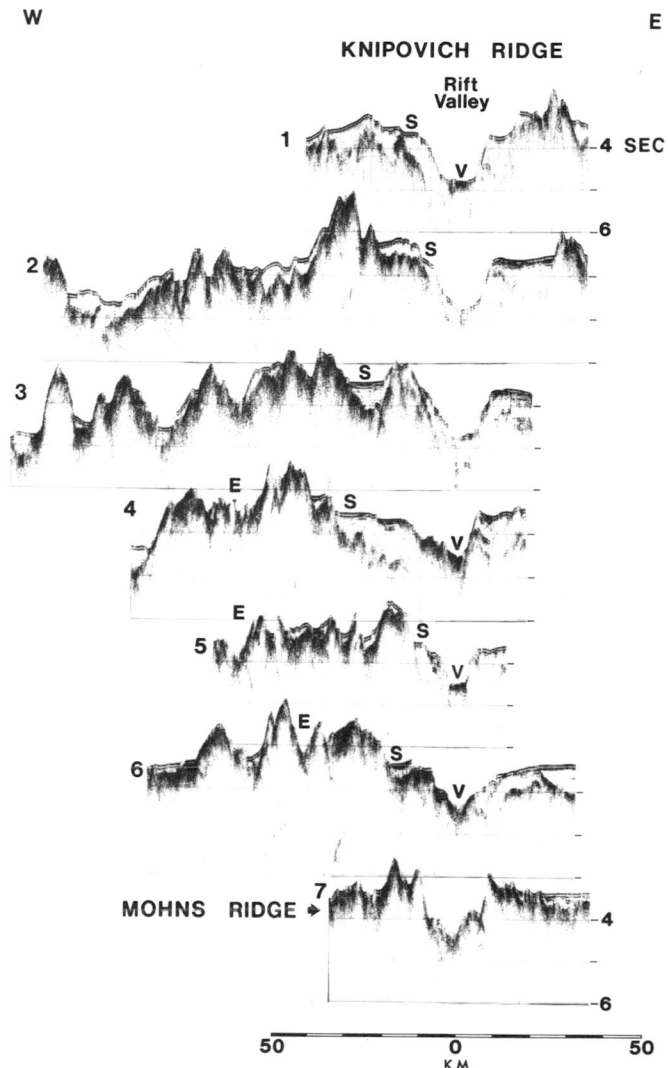

Fig. 20. Single-channel seismic reflection profiles 1–6 collected by Naval Research Laboratory across southern Knipovich Ridge (see Fig. 19 for locations). Horizontal scale is approximate, since profiles are plotted with respect to time. V, young volcanism on rift valley floor; S, thick sediment masses *west* of the present Knipovich valley axis; E, possible area of an extinct spreading axis.

(Eldholm and Ewing, 1971; Sundvor, 1974; Renard and Malod, 1974; Eldholm and Talwani, 1977; Houtz and Windisch, 1977; Hinz, 1978; Rønnevik and Motland, 1979) indicate that consolidated Mesozoic and Paleozoic sediments outcrop (or subcrop below thin Quaternary sediments) on most of the southwestern Barents Sea shelf; near the continental edge, however, a wedge of low-velocity (1.8–2.6 km/sec) Cenozoic sediment has been added by Cenozoic

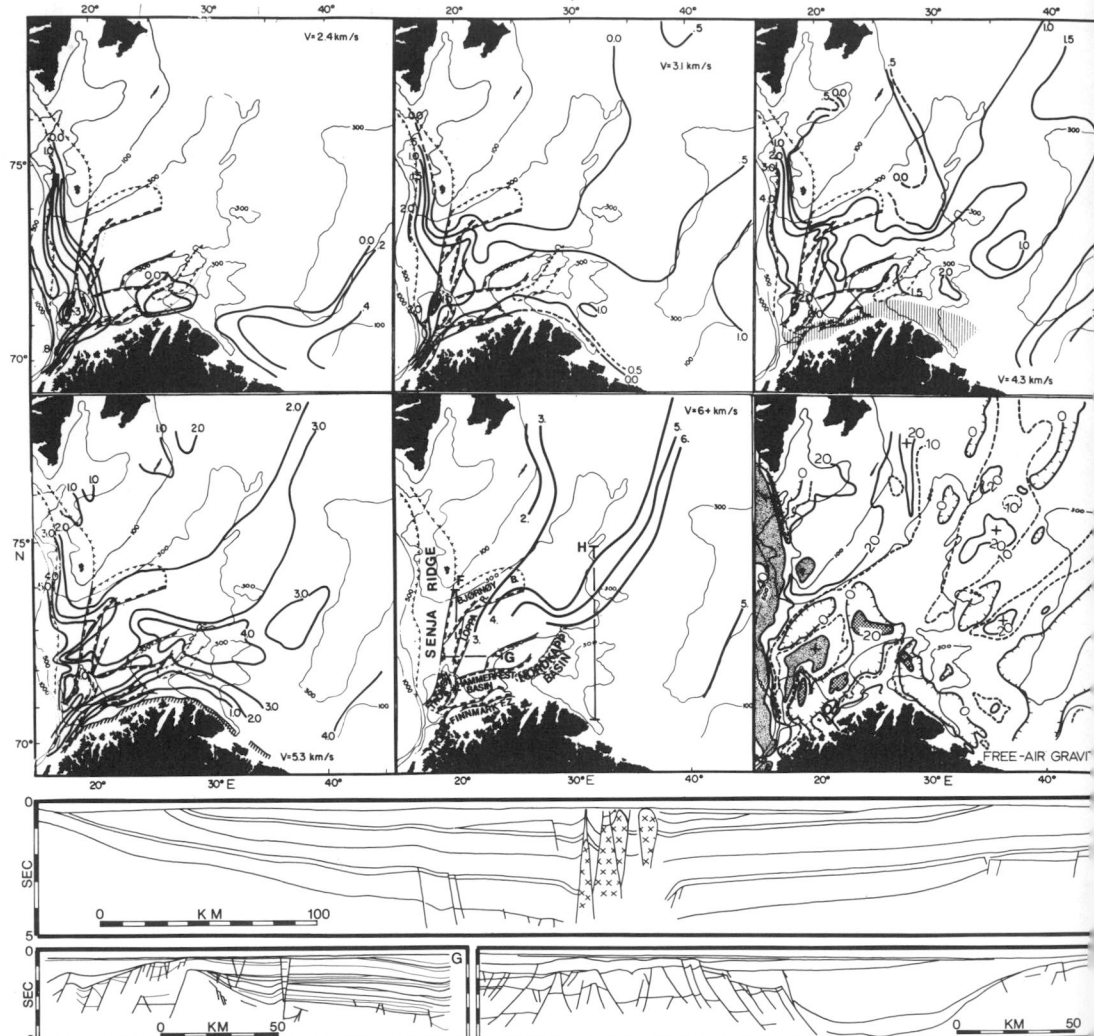

Fig. 21.   Contours: Isopachs above 2.4, 3.1, 4.3, 5.3, and 6+ km/sec refractor velocities in western Barents Sea, and air gravity anomalies (from Eldholm and Talwani, 1977). Seismic information was derived primarily from sonobuo tions (Fig. 15). Deep structures, inferred from multichannel profiling, from Øvrebø and Talleraas (1977); seismic profr G, and H from Rønnevik and Motland (1979).

fluvial and glacial drainage systems following the Bear Island and Storfjord Troughs. There is an imperfect correlation between the present Bear Island Trough and the axis of maximum late Paleozoic and Mesozoic deposition (Fig. 21). Cenozoic sediments are 0.5–1.0 km thick near the shelf break and thicken to perhaps 6 km under the adjacent continental rise and slope; the Paleogene strata appear generally undeformed except where they lie on continental crust (Rønnevik and Motland, 1979). Up to 0.5 km Cenozoic(?)

sediment may occur locally below the southern Barents shelf (Fig. 21). "Quaternary" deposits, up to 200 m thick near the shelf edge, truncate Tertiary foreset beds. Postrift shelf prograding locally exceeds 50 km (Vogt and Perry, 1978). The bathymetry (Fig. 1) suggests similar although less spectacular examples of prograding elsewhere along the continental shelf, slope, and rise seaward of major glacial channels incised into the shelves or adjacent landmasses, or both. Few of these other glaciofluvial deltas and fans have been seismically investigated, although Eldholm and Windisch (1974) measured at least 2.5 km of sediment at two sites east of Scoresby Sund, an area of apparent shelf outbuilding (Fig. 1). Multichannel profiles indicate sediments perhaps exceeding 3 km in thickness (Hinz and Schlüter, 1978). Although such sediment outlets into the GNSA basins may have existed for much of Cenozoic time, a substantial amount of the outbuilding probably occurred during the Plio-Pleistocene glaciations (Vogt and Perry, 1978; Vogt et al., 1980).

Sediment accumulations along the southwestern and northwestern margins of the Norway Basin (Fig. 18) may represent a special case of "fossil" continental margin sources. Thicknesses of 1–2 km are observed along the eastern margin of the Jan Mayen Ridge, and a pod up to 1.8 km thick lies on the northeast slope of the Iceland–Faeroe Ridge (Eldholm and Windisch, 1974). Since the Jan Mayen Ridge is believed to be a fragment of a former Greenland continental margin, detached about 28 m.y.b.p. (Talwani and Eldholm, 1977), the thick sediments east of this ridge may be largely a late Paleocene to Eocene continental margin deposit (Gairaud et al., 1978). A basement ridge of unknown composition lies below the western margin of the Jan Mayen Ridge; eastward dipping terrigenous sandstone (Talwani and Udintsev, 1976) of Paleogene and older age form a "fossil continental margin." Unconformably overlying these formations are 0.2–0.4 km flatlying marine sediments, mid-Oligocene and younger, postdating the separation of Jan Mayen Ridge from Greenland.

The crest of the Iceland–Faeroe Ridge was emergent—like present-day Iceland—until the middle or late Tertiary (Vogt, 1972b; Talwani and Udintsev, 1976; Nilsen, 1978). A fossil marine molluskan fauna on Iceland suggests that this land bridge may even have survived to the end of the Pliocene (Strauch, 1970). Thus, the thick sediments northwest of the ridge may be a type of marginal accumulation. However, the transparent, homogeneous character of the sediments suggest bottom current-controlled redeposition of pelagic sediment may be the dominant process (Eldholm and Windisch, 1974). Overflow of Norwegian Sea water probably explains the barren crest of the Iceland–Faeroe Ridge (Jones et al., 1970).

The composition of preglacial sediments in the GNSA was rather speculative until deep drilling began (Schrader et al., 1976) (Fig. 14). Older sediments generally do not crop out, and, where they do, have not been sampled. The lack of regional acoustic horizons (such as Horizon A in the central Atlantic and elsewhere) makes it difficult to identify such outcrops. The only "old" sediment previously recovered above the GNSA oceanic crust is the "Paleocene" volcanic lutite cored by Saito *et al.* (1967) at 66°21'N, 00°18'E. This lutite is now regarded as Eocene. One core from a diapiric structure landward of the Vøring Plateau Escarpment contained Upper Eocene clay (Bjørklund and Kellogg, 1972).

Drilling in the GNSA (Talwani and Udintsev, 1976) recovered primarily clays and muds, with substantial sections of diatomaceous ooze and diatomite. Numerous volcanic ash layers were cored, not a surprising result considering the proximity to present-day Iceland, Jan Mayen, and previous subaerial or shallow-extrusion centers. The CCD was exceptionally shallow throughout much of the Tertiary so that sediments were deposited near or below this inter-face (Schrader *et al.*, 1976). Calcareous ooze is not common in the GNSA; indeed, the only such sediment, a nannofossil ooze, was cored on the *southwest* flank of the Iceland–Faeroe Ridge (Site 352) and contains a rich microfauna and microflora characteristic of the North Atlantic. The ridge effectively barred these planktonic assemblages from the GNSA until the land bridge subsided in the later Tertiary (Vogt, 1972*b*), the first breach occurring about 25 m.y.b.p. (Talwani and Udintsev, 1976).

Sandy mudstones of terrigenous origin were cored below the uncon-formity on the Jan Mayen Ridge (Sites 346, 347, and 349). These primarily Oligocene sediments were apparently derived from the margin of Greenland prior to the development of the Kolbeinsey spreading axis. Terrigenous sedi-ments also occur in the Vøring Plateau area holes, but only in the Lower Eocene.

Although no drill holes were placed in "abyssal plains," seismic reflection profiling suggests that turbidity-flow-controlled provinces occur in the Lofoten, Greenland, and Boreas basins and between Kolbeinsey Ridge and the Green-land margin (Eldholm and Windisch, 1974). Turbidite deposition was probably most pronounced during glacial maxima.

Deep bottom currents, weak or absent in the Quaternary north of the Ice-land–Faeroe Ridge (Damuth, 1978), could have been more important in the past. An unconformity above the late Eocene at Sites 338, 346, and 349 may register the first access of Arctic deep water as a result of plate separation (Schrader *et al.*, 1976).

F. Mesozoic and Older Rocks

Although sea-floor spreading in the GNSA only got under way about 50–60 m.y.b.p., the line of opening generally followed a complex of preexisting sedimentary basins, fragments of which underlie the present North Sea (Ziegler, 1977), the continental margin off Norway (Rønnevik *et al.*, 1979; Jørgensen and Navrestad, 1979) and central East Greenland (Birkelund and Perch-Nielsen, 1976; Surlyk, 1978; Surlyk *et al.*, 1980), the Barents shelf (Hinz and Schlüter, 1978*a*; Rønnevik and Motland, 1979) and the Jan Mayen Ridge (Gairaud *et al.*, 1978; Johnson, 1975; Navrestad and Jørgensen, 1979) (Fig. 22). Sediments started to accumulate in this basin complex in the middle Paleozoic to early Mesozoic. Fluctuations of sea level, with corresponding transgressions and regressions (Vail *et al.*, 1977), persisted until the end of the Cretaceous, forming an epicontinental sea in which deposition generally kept pace with subsidence. The lack of known deep-water deposits suggests, but does not prove, that these seaways were never deep. The variable thickness of pre-Tertiary sediments indicates nonuniform subsidence, areas with high-density rocks having subsided the least (Talwani and Eldholm, 1972, 1974).

Mesozoic sediment accumulations locally exceed 4 km in the North Sea Central and Viking grabens (Ziegler, 1977). Drilling in the North Sea suggests that the late Paleozoic to Mesozoic basins developed episodically by normal faulting (Ziegler, 1977). The correlations between these rifting episodes and plate tectonic events in the North Atlantic remains unclear. Rifting movement in the North Sea diminished during the late Cretaceous and ceased after the Paleocene "Laramide" phase.

Since drilling north of 62°N did not begin until the 1980s, the tectonic/sedimentary history off central and northern Norway and in the Barents Sea must be considered provisional, particularly with regard to timing. Multichannel data do reveal regional seismic reflectors (Figs. 18 and 21), some of which can be traced southward to control wells in the North Sea (e.g., Rønnevik *et al.*, 1979). The hypothesis of rapid eustatic sea-level changes (Vail *et al.*, 1977) has also been used to estimate reflector age. In the Barents Sea, seafloor sampling and correlation with formations on Svalbard and Bjørnøya (Bear Island) has proved useful; finally, the observed increase of sound speed with sediment age (Fig. 17) provides some constraints.

A Mesozoic–Paleozoic sediment sequence, perhaps up to 10 km in thickness (Aam, 1975*a*) underlies the inner Vøring Plateau (Vøring Basin), landward of the Vøring Plateau Escarpment. Multichannel profiles (Rønnevik *et al.*, 1979) suggest the Cretaceous alone is 1–3 km thick; major tectonic movements occurred in the Lower or Middle Cretaceous, followed by lesser events in the lowest Tertiary and mid-Oligocene. The possibility of old continental crust below the Eocene basalts of the outer Vøring plateau cannot be entirely discounted. Also unlikely but unproven is the existence of old oceanic crust below the thick sediments of the inner Voring Plateau.

Seismic profiles across the Norwegian shelf west of the Vøring Basin are interpreted as follows: (1) an epeirogenically subsiding Post-Caledonian peneplain accumulated Carboniferous sediments and was then affected by a late

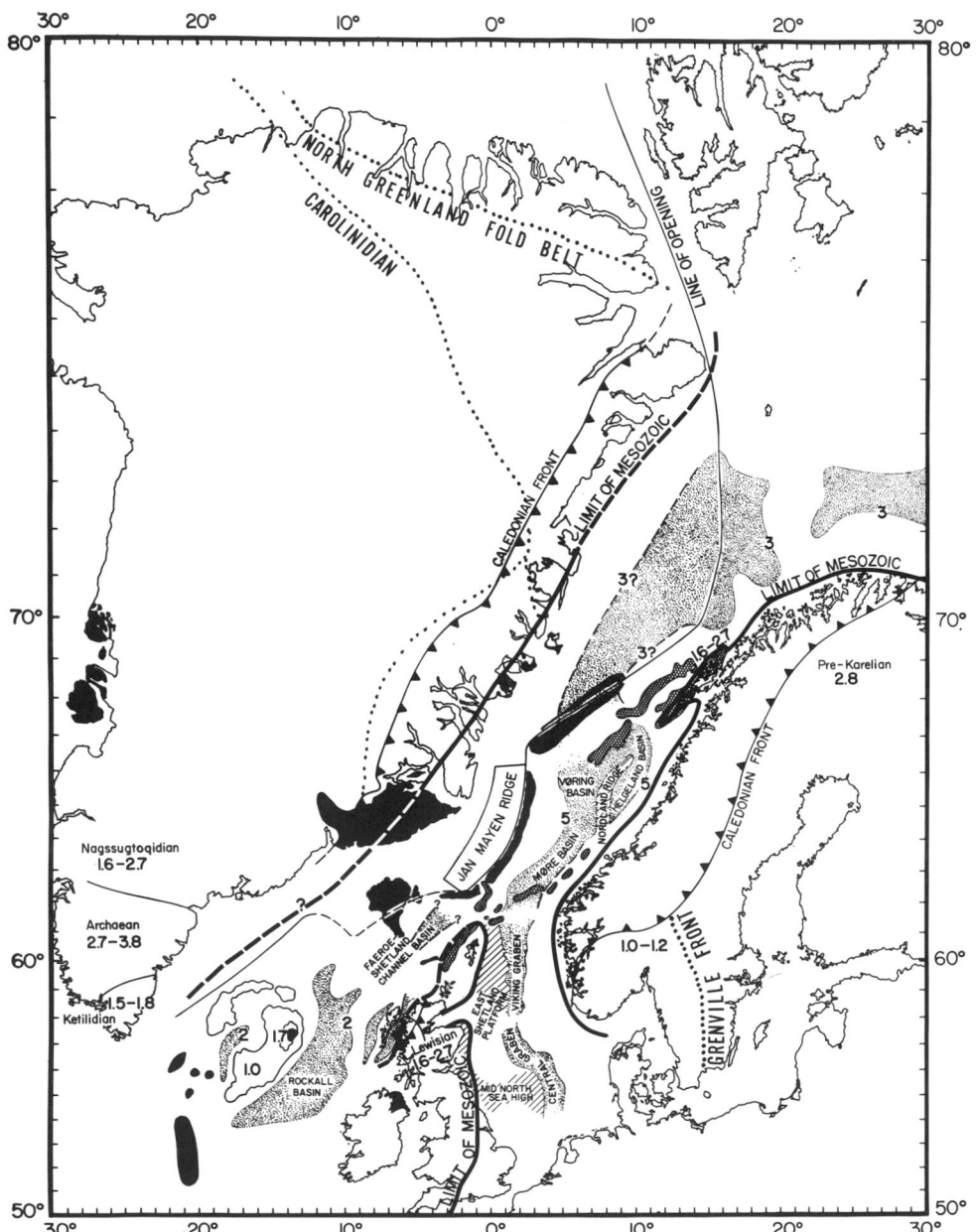

Fig. 22.   Reconstruction of GNSA at time of initial breakup (anomaly 24), with principal pre-Tertiary features including Mesozoic sediment basins (stippled; isopachs in kilometers). Black areas show known and probable early Tertiary volcanic and intrusive activity; diagonally ruled areas show structural highs in North Sea. Modified slightly from Talwani and Eldholm (1977).

Carboniferous tectonic phase creating rotated fault blocks. (2) In the late Paleozoic and early Mesozoic, sedimentation smoothed the relief, and the area behaved as one large basin. (3) A second tectonic phase began in the Middle Jurassic and culminated in the earliest Cretaceous. By the onset of sea-floor spreading the relief had been leveled by erosion and once more became a simple, epeirogenically subsiding basin. Total Mesozoic thicknesses are 0 to more than 2 km above the structural highs and 2–4 km or more in the basins.

Sediment accumulations of 2–4 km of Mesozoics, and similar thicknesses of Paleozoics characterize the southwestern Barents Sea (south of 75°N and west of 35°E). Mesozoic (and locally, Paleozoic) sediments outcrop in many parts of the area, or subcrop below a thin Quaternary unit. South of 75°N between 35°E and 25°E lies a broad basin whose generally epeirogenic subsidence, since Devonian time, has been complicated locally by salt tectonics. Between 20° and 25°E are the fault-bounded, Mesozoic Hammerfest and Bjør-nøya basins (Fig. 21). The Loppa Ridge, which separates these basins, is terminated on the northwest by a large salt dome. The complex structures between 25°E and the Tertiary oceanic crust date from the Lower Mesozoic, with tectonic modifications in mid-Cretaceous, base Tertiary, and mid-Oligocene.

East Greenland is unique in the GNSA for its exposure of an 800-km-long Mesozoic basin (70°–77°N), up to 140 km wide, containing up to 5 km sediment (Surlyk, 1978; Surlyk et al., 1980). During the Jurassic phase, a rifted basin propagated northward from the Scoresby Sund area; sediment infilling occurred in the opposite direction. By early Cretaceous the tectonic pattern here and in the then-contiguous Norwegian shelf basins had shifted to westward tilting blocks (half-grabens).

Although changing in details, most of the pre-Cenozoic basins seem to have maintained their overall configuration since the late Paleozoic. In the late Permian, shallow seas floored by evaporite deposits extended westward from Poland across Germany and Denmark (Ziegler, 1977). This "Zechstein Sea" is believed to have turned north through the present marginal areas of Norway and Greenland. At the present southwest edge of the Barents Sea, the Zechstein Sea branched "northwest" through the Greenland–Svalbard margin and "east" along the Bear Island Trough (Calloman et al., 1972). In East Greenland, the Foldvik Creek Formation contains typical Zechstein-type shallow water marine deposits (Calloman et al., 1972). Evaporite of unknown age was also dredged at 60°29′N, 41°40′W on the southeast Greenland continental margin (Johnson et al., 1975a). Diapiric structures—possibly caused by Permian salt—have been identified on seismic reflection profiles on the continental margin of southeast Greenland about 62°N, 40°W (Johnson et al., 1975a), and in the Tromsø, Hammerfest, and Nordkapp basins on the southwest Barents Shelf (Rønnevik et al., 1979). The Tromsø basin may have been created by salt diapirism in Jurassic–Cretaceous time. Some of the

Barents Sea salt may be Carboniferous–Devonian in age. Permian evaporites occur in Svalbard (Orvin, 1969). Refracting horizons with velocities in the range for evaporites (4.3–4.4 km/sec) exist locally below the Barents Sea (Sundvor, 1974; Renard and Malod, 1974), below the inner Vøring Plateau (Talwani and Eldholm, 1972, 1974), at 63.7°N on the Norwegian shelf (Sundvor and Nysaether, 1975) and below the North Sea (Collette *et al.*, 1970). Somewhat younger is the Permo-Triassic New Red Sandstone facies that extends from north Germany across to England, and occurs in the Oslo fjord area, in East Greenland, and along the southeast margin of the Barents Sea (Harland, 1969). These deposits are continental, however, and apparently were not interconnected.

If the late Paleozoic–Mesozoic basins seem to have controlled the line of initial sea-floor spreading in the Tertiary (Fig. 22; Talwani and Eldholm, 1977), so the basins themselves developed along the general line of Caledonian orogeny. This cycle of events began in the late Precambrian and early Paleozoic when a marine seaway—possibly a rifted ocean basin (Wilson, 1966)—developed from what is now Newfoundland through the British Isles, western Norway, Svalbard, and the Barents shelf, and northwest Greenland (Harland, 1969). Basin closure resulted in the Caledonian orogeny, which culminated in the Silurian; the relief developed by this orogeny was mostly eroded by Carboniferous time. Caledonian sediments survive, locally in great thicknesses, in the British Isles, Norway, Spitsbergen, and northeast Greenland (Henriksen and Higgens, 1976). A Caledonian basement complex connects British and Norwegian exposures at great depth below the northern and even central North Sea (Ziegler, 1977). Upper Devonian to Lower Carboniferous sedimentary rocks were dredged at three sites on the Greenland continental margin about 65.5°N, suggesting that Caledonian formations extend to the shelf edge there (Johnson *et al.*, 1975a).

Although the Caledonian orogeny affected Norway, Bear Island, and Svalbard, there is no clear geophysical evidence for buried structural connections across the southwestern Barents Sea. Instead, the principal tectonic grain is oriented NE–SW. The complex north-trending "Senja Ridge" is at least partly of early Tertiary origin (Øvrebø and Talleraas, 1977). The deepest "sedimentary" unit in this area (above a 6.2- to 6.4-km/sec basement) is characterized by a velocity of 5.3 km/sec (Eldholm and Talwani, 1977) and could consist of Lower Paleozoic metamorphic or well-lithified sedimentary rocks associated with the Caledonian orogeny, or Eocambrian sediment. The velocities are actually somewhat low compared with granitic or highly metamorphosed sedimentary rock of Caledonian or Precambrian age in Norway (6.0–6.3 km/sec), which may rather correlate with the Barents Sea basement. Renard and Malod (1974) show irregular basement relief of amplitude up to 3 km. Average basement depth in the central Barents Sea is 5–6 km (Eldholm and Talwani, 1977), shoaling northward to less than 2 km on the Spitsbergen Bank. It may outcrop in the form of the late Precambrian–early Paleozoic Hecla Hoek Formation exposed on Bear Island and Svalbard. Renard and Malod (1974) conclude that between the Mesozoic–Cenozoic layers (2.4–4.0 km/sec), and the basement there is a very well-defined continuous layer (4.2 km/sec). This layer is thought to be post-Caledonian, Carboniferous to Permian in age, possibly limestone. Sundvor (1974), in contrast, found no peak at 4.1 km/sec in the distribution of measured velocities (Fig. 17).

Because the east Greenland margin north of 69°N was connected to the Barents shelf and Svalbard until the Tertiary, Paleozoic rocks probably exist at depth below the Greenland shelf. At 69°N sonobuoy stations gave a total sediment thickness of 7 km overlying a 5.2-km/sec basement (Talwani and Eldholm, 1974). Magnetic basement may lie as deep as 6–10 km in the basins under the shelves (Birkelund, 1977; Kovacs and Vogt, 1979; Surlyk *et al.*, 1980).

Precambrian rocks around the GNSA are more igneous or metamorphic than sedimentary and their exposure along the sea coast north of the Greenland–Iceland–Faeroe Ridge (e.g., Krogh, 1977) represents survival as stable blocks through the Caledonian orogeny. The Lofoten Islands are a prominent example. Whereas Caledonian structures seem to follow earlier Precambrian ones farther south in the Atlantic, this may not be true of the GNSA area (Escher and Watt, 1976; Fig. 22). Radiometric dating and structural relationships suggest major orogenic activity at 1.0–1.2, 1.5–1.8, 1.6–2.7, and 2.7–3.8 b.y. Evidently igneous, metamorphic, and sedimentary rocks spanning virtually the entire post-4.0-b.y. history of the earth are preserved in the GNSA and the Precambrian shields surrounding it!

IV. GEOPHYSICAL SIGNATURES AND THE IGNEOUS CRUST

A. Magnetic Field Anomalies and Implications for the History of Plate Motion

The GNSA and adjacent marine areas of the northeast Atlantic and Arctic have been more densely covered by magnetic surveys (shipborne and airborne) than any other oceanic area of comparable size (Figs. 23–26). We discuss first the magnetic anomalies over oceanic crust and include here a review of plate kinematics (Figs. 27–29), primarily based on magnetic data.

1. *The Ocean Basins and Iceland: Implications for the History of Plate Motion*

The evolution of the greater GNSA in terms of sea-floor spreading (plate tectonics) has been developed by a number of authors (Johnson and Heezen,

1967; Avery *et al.*, 1968; Vogt *et al.*, 1970; Vogt and Avery, 1974*b*; Talwani and Eldholm, 1972, 1974, 1977; Kristoffersen and Talwani, 1977; LePichon *et al.*, 1977; Srivastava, 1978; Grønlie, 1979; Phillips *et al.*, 1980; Phillips and Tapscott, 1980). Several other plate-tectonic syntheses make predictions about the GNSA, directly or indirectly (Laughton, 1972, 1975; LePichon *et al.*, 1971; Pitman and Talwani, 1972). All these studies are based primarily on the identification of linear magnetic anomalies in terms of geomagnetic reversals and sea-floor spreading. For this reason we include the plate tectonic evolution of the GNSA in this section on magnetic anomalies. We note, however, that whereas the sea-floor spreading/continental drift concept is still rejected by a few modern authors (Beloussov and Melanovsky, 1977), the large horizontal movements between Greenland, North America, and Europe are not substantially different from those originally postulated by Alfred Wegener over 50 years ago.

Magnetic lineations caused by sea-floor spreading and geomagnetic reversals can now be traced from the northeast Atlantic through the GNSA and into the Eurasia Basin of the Arctic (Figs. 23 and 24). Spreading-type lineations have also been identified in the Labrador Sea (Vogt and Avery, 1974*a*; Kristoffersen and Talwani, 1977; Srivastava, 1978) (Fig. 23). Southwest of the Greenland–Iceland–Faeroe Ridge, between the Jan Mayen and Greenland–Senja F.Z.s, and in the Eurasia Basin the magnetic anomaly record is relatively easy to decipher, despite slow spreading rates. The ridge axis appears to have maintained its median position in these areas. Elsewhere the record is more confused, as discussed further below.

Northeast of a line connecting southern Greenland with southern Rockall/ Hatton Bank, the oldest clearly identifiable positive anomaly is 24, variously dated at 60 m.y.b.p. (Heirtzler *et al.*, 1968), 48 m.y.b.p. (Tarling and Mitchell, 1976), and 56 m.y.b.p. (La Brecque *et al.*, 1977). Anomaly 25 may locally be present between the Jan Mayen and Greenland F.Z.s (Phillips *et al.*, 1980). However, our best guess is that spreading began during the reversed period prior to anomaly 24, i.e., about 57–58 m.y.b.p. on the time scale of La Brecque *et al.* (1977). In the Eurasia Basin there is "room" for anomalies 25–27 (perhaps even 25–29: see Fig. 28), but a broad, shallow magnetic negative exists in their place (Karasik, 1974; Vogt *et al.*, 1979*a*,*b*). Either the anomalies were suppressed by sediment fill or some other process associated with initial rifting or the crust is

←───

Fig. 23.   Zebra-stripe depiction of residual magnetic lows (black) and highs on Mercator projection. Modified from Vogt and Avery (1974*b*) by additional data of Hall and Dagley (1970), Phillips *et al.* (1980), Johnson *et al.* (1975*a*), Laughton and Whitmarsh (1974), Williams (1975), Aam (1975*a*), Hagevang (1978), Fedhynskii *et al.* (1974, 1975), Voppel *et al.* (1979), Srivastava (1979), and unpublished aeromagnetic data west and southwest of Iceland kindly supplied by G. R. Lorentzen (Project MAGNET, U.S. Naval Oceanographic Office). No attempt was made to adjust data for match at survey boundaries. In case of overlap, only the more recent and/or more detailed data are shown.

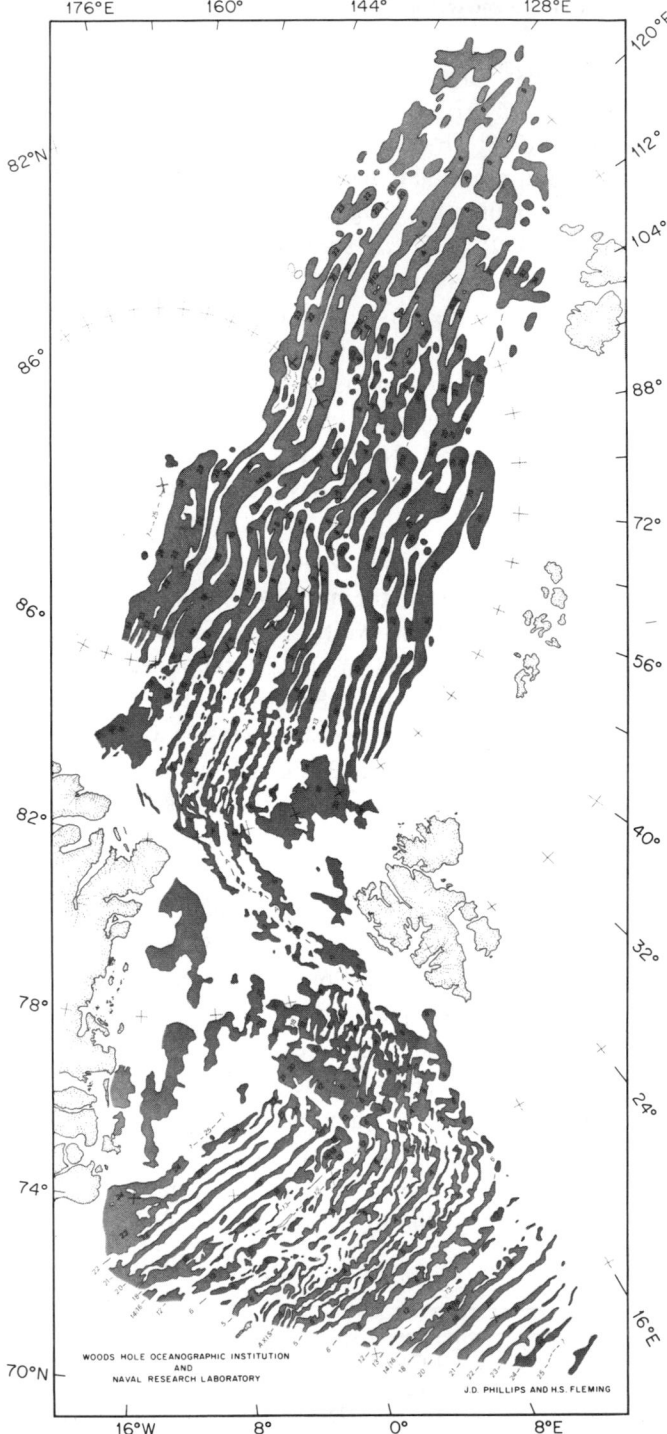

Fig. 24. Zebra-stripe depiction of residual magnetic anomalies in GNSA and Eurasia Basin (Perry *et al.*, 1980; Phillips *et al.*, 1980; Vogt *et al.*, 1978). Polar stereographic projection, with positive residuals black. Based on Fig. 24 and Soviet data (Karasik, 1974). Note overlap with Fig. 23.

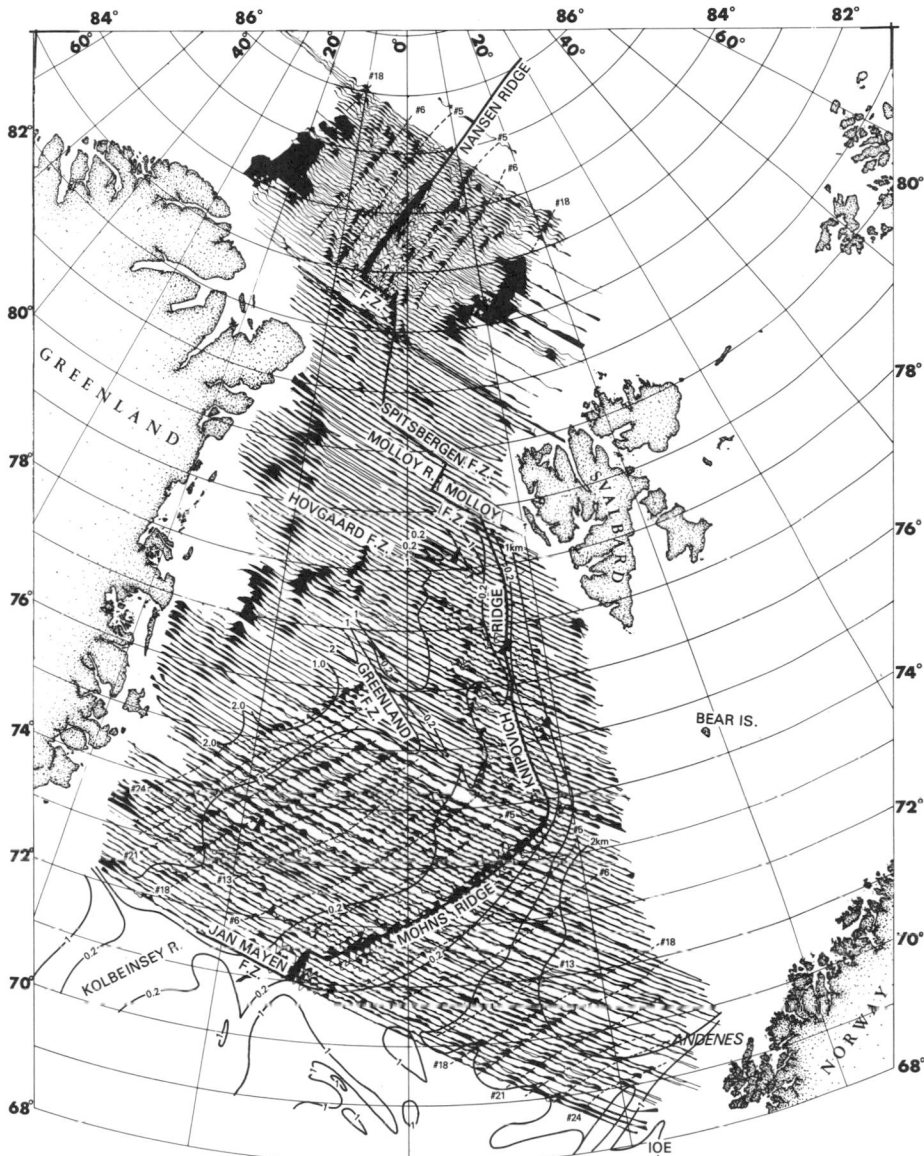

Fig. 25.   Magnetic anomalies (black: after Perry *et al.*, 1980), principal anomaly identifications and tectonic features, and simplified sediment isopachs (from Eldholm and Windisch, 1974).

subsided continental material. In the Lofoten and Greenland Basins, anomalies 20–24 form parallel strips but are markedly oblique (up to 20°–30°) to the nearby continental margins (Perry *et al.*, 1980; Talwani and Eldholm, 1977) (Fig. 27). There are two reasons for this: (1) the early rift sliced diagonally through the preexisting Voring Basin; and (2) the spreading axis became

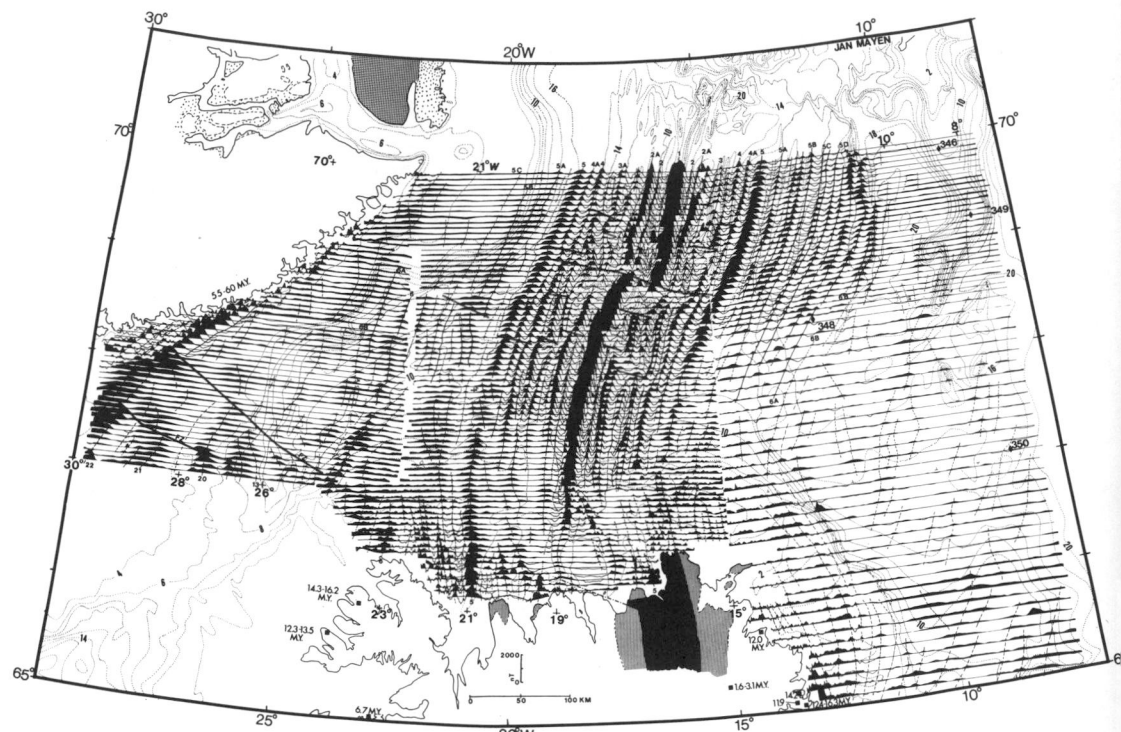

Fig. 26.   Aeromagnetic residual profiles and bathymetry (stippled contours) from Iceland to 70°N, adapted from Perry
al. (1980) and Vogt *et al.* (1980). Solid and dashed lines indicate magnetic lineations; some attributable to sea-floor sprea
ing and magnetic reversals are labeled. On Greenland, stippled denotes Precambrian and cross-hatched, Mesozoic. On I
land, black denotes volcanics <0.7 m.y. old; and vertically ruled, 0.7–3.0 m.y. old. Radiometric ages for Tertiary igne
rocks in Iceland and Greenland.

reoriented in the first 1–2 m.y. after rifting began (Eldholm *et al.*, 1979). Off
southwest Greenland there is, as in the Eurasia Basin, a gap between anomaly 24
and the physiographic continental margin (Johnson *et al.*, 1975a; Featherstone
et al., 1977). As the Iceland–Faeroe Ridge is approached from the south,
anomalies 23 and 24 disappear on the Greenland side and are believed to be
repeated west and southwest of the Faeroes, thus indicating a westward jump of
the spreading axis prior to anomaly 22 (Voppel *et al.*, 1979). Alternative
hypotheses have been advanced by Bott (1974). Fleischer *et al.* (1974), Talwani
and Eldholm (1977), and Larsen (1978). The latter author suggests that anomaly
24 is missing in the Iceland–Faeroe area because early plate opening was accommodated by dike injection along the east coast of Greenland.

The early opening problem has been disputed because the magnetic
anomaly pattern is highly complex on the entire Greenland–Iceland–Faeroe
aseismic ridge (Johnson *et al.*, 1975a; Kristjansson, 1976a; Serson *et al.*, 1968;

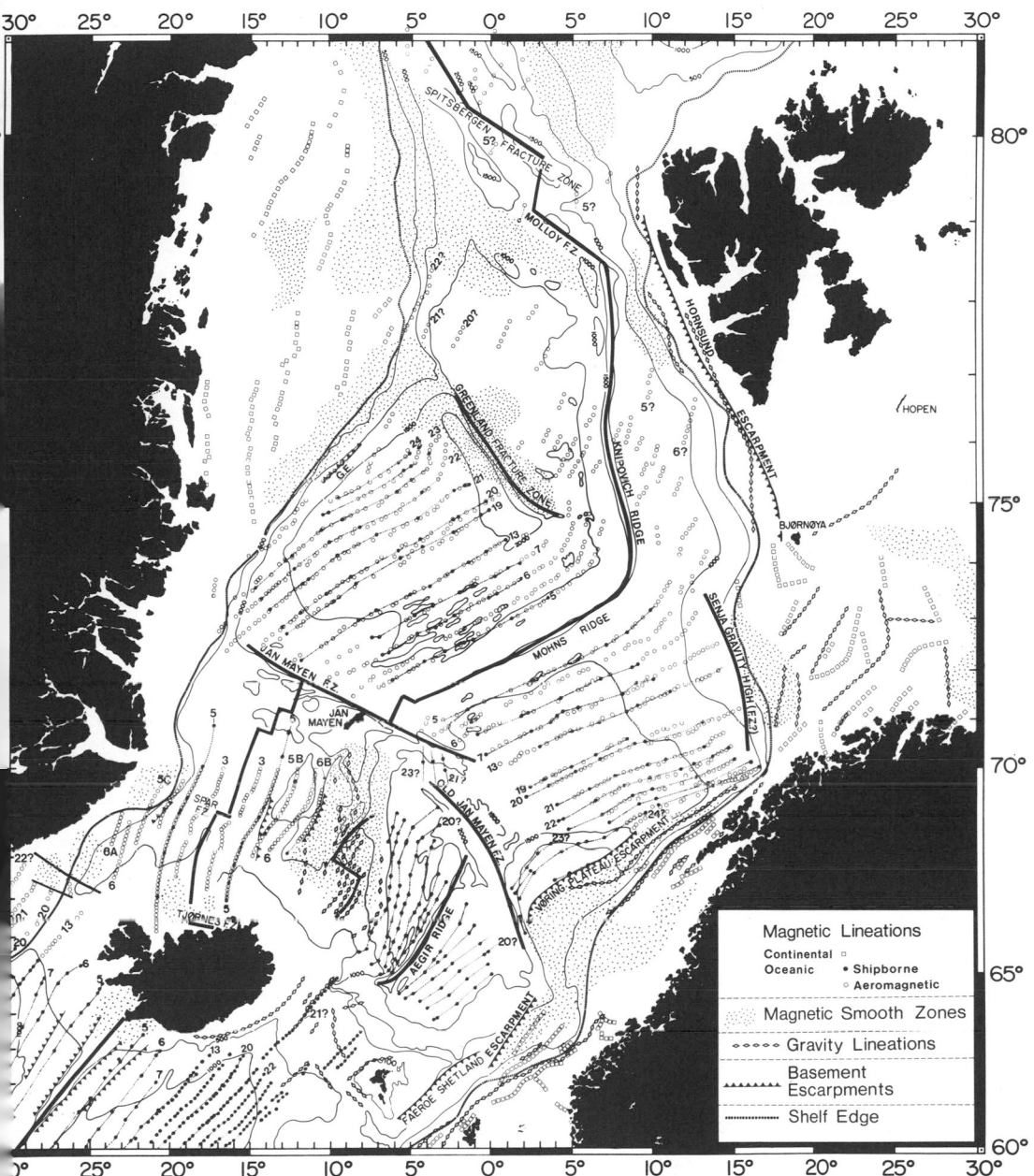

27. Summary of magnetic, gravity, and tectonic lineations in GNSA. Based on Talwani and Eldholm (1977),
lm and Talwani (1977), Grønlie and Talwani (1978), Aam (1975a), Gairaud *et al.* (1978), Voppel *et al.* (1979),
os *et al.* (1980), and Vogt *et al.* (1980). G.E., Greenland Escarpment.

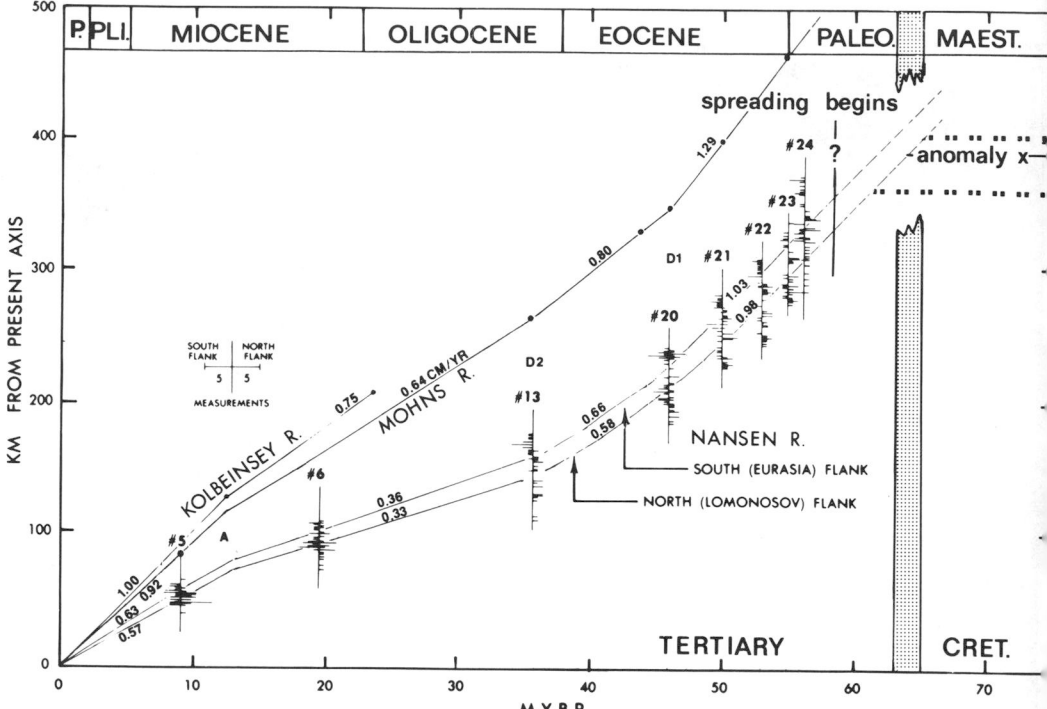

Fig. 28. Age-vs.-distance diagram, including average spreading half-rates across Kolbeinsey, Mohns, and Nan Ridges (Vogt *et al.*, 1980; Talwani and Eldholm, 1977; and Vogt *et al.*, 1979a,b, respectively). Based on rever time scale of La Brecque *et al.* (1977). Note spreading deceleration at points D1 and D2, and mid-Miocene accele tion at A. "Anomaly x" lies at the base of the Lomonosov Ridge.

Fleischer *et al.*, 1974) (Figs. 23, 26, and 27). Although there are broad, crudely NE–SW-oriented lineations, perhaps correlative with average magnetization polarity, identification of particular reversals is tenuous at best, probably because volcanic and intrusive activity has been distributed over a broad swath (as on Iceland) rather than confined to a narrow accretion axis as is typical of the MOR. Nevertheless, in eastern and southwestern Iceland, the lava stratigraphy can be correlated with aeromagnetic anomalies (Piper, 1973). On the closely surveyed Iceland–Faeroe Ridge (Figs. 23 and 27), a NE–SW magnetic anomaly pattern of possible "spreading" origin is interrupted by cross-lineations and local, equidimensional anomalies up to several thousand nT in amplitude (Fleischer *et al.*, 1974). The former may reflect transform fractures and the latter, extinct volcanic/intrusive centers analogous to the central volcanoes on Iceland (see also Kristjansson, 1976b, and Kristjansson *et al.*, 1977).

Although the wide extrusion zone and high lava production rates on Iceland have produced a confusing magnetic anomaly pattern, these factors make

Fig. 29. Sequential opening of the post-Mesozoic ocean basins of the GNSA by sea-floor spreading, and broad hot spot, according to Talwani and Eldholm (1977). Note, however, that longitudinal profiles along present plate boundary (Fig. 4) show Iceland hot-spot phenomenon to be sharply peaked in central Iceland.

the island ideal for absolute calibrations of mid-Miocene and younger magnetic reversals. Radiometric dating and polarity measurements on Icelandic lavas first dated the post-4-m.y. reversal sequence (McDougall and Wensink, 1966; Dagley *et al.*, 1967). The time scale was subsequently extended to 6.5 m.y.b.p. (McDougall *et al.*, 1977) and then to 13.0 m.y.b.p. (Harrison *et al.*, 1979). Such studies also shed light on geomagnetic/secular variations (Watkins *et al.*, 1977) and eruptive rates, e.g., 690 m/m.y. or 1 average lava flow every 13,000 years during anomaly 5 time (McDougall *et al.*, 1976). Kristjansson (1972) and Kristjansson and Watkins (1977) have summarized the magnetic properties of borehole samples and implications for the oceanic crust.

We have previously discussed the possible early westward jump of the spreading axis in the Iceland–Faeroes area (Voppel *et al.*, 1979). A more recent eastward displacement of the spreading axis on Iceland itself is indicated by the existence of three volcanic zones of Pliocene and younger age. The eastern zone was thought to have developed in northern Iceland about 6 m.y.b.p. (Aronson and Saemundsson, 1975). In south Iceland, crustal accretion has been mainly confined to the Reykjanes–Langjökull line; a jump to the eastern Neo-Volcanic Zone seems to be still in progress. An earlier spreading axis—highly oblique as is the present Reykjanes axis—existed in the Snaefellsnes area until about 7 m.y.b.p., when the Reykjanes–Langjökull line became activated. Residual volcanism has occurred on Snaefellsnes Peninsula until Recent times.

The history of spreading between Iceland and the Jan Mayen F.Z. has also been complicated by jumps of the accretion axis (Fig. 26). Fortunately, at least some of the events can be reconstructed from magnetic lineations. Further constraints derive from reconstructions of the entire plate boundary (Talwani and Eldholm, 1977). An extinct spreading axis, first postulated by Johnson and Heezen (1967) and Vogt *et al.* (1970), bisects the Norway Basin. A roughly symmetrical magnetic anomaly pattern was generated by this axis (Aegir Ridge) during its period of activity [anomaly 7 (?) to 23 times; Talwani and Eldholm, 1977]. The tentatively identified lineations (20–23) are fan shaped, diverging northward (Fig. 27). The fan-shaped pattern requires the existence of simultaneous spreading from one or more additional axes, e.g., the apparent rift valley trending obliquely across southern Jan Mayen Ridge (Fig. 27; Gairaud *et al.*, 1978), or else complex intraplate deformation.

The absence of anomaly 24, as well as the existence of a 150-km-wide strip of crust between anomaly 23 and the Faeroe–Shetland and Vøring Escarpments, which are presumed to mark the line of initial rifting, suggest an even earlier, short-lived axis associated with the "extra" crust (Talwani and Eldholm, 1977). Then, shortly after Aegir Ridge itself stopped spreading (anomaly 7 time, or 28 m.y.b.p.), a new axis had formed to the west, splitting off a segment of the Greenland continental margin. This segment became the physiographic Jan Mayen Ridge and crust immediately to the west and south (Figs. 27 and 29). The new spreading axis evolved into the present Kolbeinsey Ridge (Vogt et al., 1980); spreading half-rates increased from 0.75 to 1.0 cm/yr about 12–13 m.y.b.p. [Earlier authors had suggested an "Intermediate Iceland Plateau extinct axis" active approximately from anomaly 6B to 5D time (Talwani and Eldholm, 1977; Grønlie et al., 1979).] During the last 7 to 8 m.y. short transform faults have repeatedly formed and disappeared along Kolbeinsey Ridge; the Spar F.Z. at 69° achieved its present 30-km offset largely during the last 3 m.y. (Vogt et al., 1980; Meyer et al., 1972).

Mohns Ridge, located between the Jan Mayen and Greenland–Senja fracture zones, has maintained its median position since spreading started (Phillips et al., 1980; Talwani and Eldholm, 1977), with the possible exception of a short jump at 23A time (Eldholm et al., 1977). If there are fracture zones offsetting either the present axis (Fedhynski et al., 1975) or older isochrons, they are probably minor and transient. Like the late Tertiary Reykjanes Ridge, the Mohns spreading axis has been oblique, making angles of 50°–80° (rather than 90°) with the spreading direction (Fig. 4). However, between about 40 and 30–10 m.y.b.p., the Reykjanes Ridge was divided into a number of short, spreading centers and orthogonal transform faults (Vogt and Avery, 1974a). No similar "fracture episode" has so far been identified on Mohns Ridge.

Magnetic anomalies 2, 5, 6, 12, 13, 18, and 21–24 are clearly identifiable on both sides of the Mohns axis (Phillips et al., 1980; Figs. 23–25, 27). Using more widely spaced shipborne data, Talwani and Eldholm (1977) were previously able to identify anomalies 5, 6, 7(?), 13 and 19–24. The spreading history derived from these identifications is compared in Fig. 28 with that of Kolbeinsey (Vogt et al., 1980) and Nansen ridges (Vogt et al., 1979a,b). With the exception of Talwani and Eldholm's (1977) relatively fast spreading between anomaly 6 and 7 time—perhaps an artifact resulting from misidentification of anomaly 7—the spreading rate history parallels that observed on Reykjanes Ridge (Vogt and Avery, 1974a) and in the Eurasia Basin (Vogt et al., 1979a,b). Rates were high in the early and late Tertiary and lowest in middle Tertiary time, possibly in response to fluctuations in discharge from the Iceland plume (Vogt and Avery, 1974a).

Anomaly amplitudes are substantially lower on the northern part of the Norwegian side of Mohns Ridge; this might be caused by thermal demagnetization of basalts, in turn attributable to thick sediments of the Barents (Bear Island) Fan (Fig. 25). A similar amplitude asymmetry exists in parts of the Eurasia Basin (Vogt et al., 1979a).

Between the Greenland F.Z. and the southwestern termination of Nansen Ridge against the Spitsbergen F.Z., the evolution of the sea floor by spreading is poorly known despite detailed aeromagnetic data. The axial anomaly associated with Knipovich Ridge is much reduced in amplitude, possibly owing to the continental rise sediments being poured into the rift valley (Figs. 19 and 20), or to slow oblique spreading. Such processes might also explain the magnetically smooth regions surrounding Knipovich Ridge (Figs. 25 and 27).

For reasons unknown, the prominent Greenland F.Z. Ridge is also associated with a magnetic smooth zone (Fig. 27). A similar "fracture smooth zone" is found along parts of the Mendocino F.Z. in the northeast Pacific.

Phillips *et al.* (1980) have locally identified anomalies 5, 6, 12, 13, 18, and 20 west of Knipovich Ridge and 5 on the east. These identifications are tentative and others are possible (Fig. 27). The existence of anomalies older than 13 (36 m.y.b.p.) conflicts with Talwani and Eldholm (1977), who concluded that prior to 36 m.y.b.p. there was only shear motion between the margins of Svalbard and northeast Greenland. Probably only the *northern* half of this margin was "pure shear" (Eldholm and Sundvor, 1978), or even a combination of shear and compression (Phillips *et al.*, 1980). With the cessation of spreading in the Labrador Sea, the direction of spreading in the GNSA became more east–west (see also Vogt and Avery, 1974*a,b*). Only at anomaly 13 time did the northern part of the Greenland Sea begin to open significantly (Talwani and Eldholm, 1977; Fig. 29).

The plate motions derived by Phillips *et al.* (1980) differ from those of Talwani and Eldholm (1977) in other ways. Both authors agree that the period of early, more NNW–SSE motion lasted from anomaly 24 to 13 time and was succeeded by more east–west spreading. However, Phillips *et al.* (1980) infer a return to more NNW–SSE spreading at anomaly 6 time and lasting to the present.

Before anomaly 13 time, the Greenland F.Z. had a 100-km left-lateral offset. Subsequently, both it and the Hovgaard F.Z. to the north were replaced by more oblique spreading such as that associated with the present Knipovich axis. Since the latter is much closer to the Svalbard than to the Greenland margin, it is likely that the disappearance of the major fractures was accomplished by an eastward shift of the spreading axis, either by asymmetrical spreading, by jumps of the spreading axis, or both (Vogt *et al.*, 1978).

Magnetic lineations are again well developed in the Eurasia Basin (Feden *et al.*, 1979; Phillips *et al.*, 1980; Vogt *et al.*, 1979*a,b*; Karasik, 1974), particularly the section of Nansen Ridge from 82.9°N, 6°W to 84.5°N, 5°E (Fig. 25). A pair of massive basement ridges (the Yermak and Morris Jesup Plateaus), associated with complex high-amplitude (+400 to 1500 nT) magnetic anomalies, was created by this same section of spreading ridge about anomaly 12–18 time. The relatively high amplitude of the post-anomaly-5 spreading lineations (+500 to +1200 nT for the central anomaly) and the existence of large basement ridges suggests the existence of a previously unknown hot spot (Feden *et al.*, 1979). The high amplitudes might reflect highly fractionated FeTirich basalts, such as are found near some other hot spots (Vogt and Johnson, 1973*a*; Vogt, 1979*b*).

A connection between basalt chemistry and magnetic amplitudes might also explain amplitude variations near the Jan Mayen and Iceland hot spots. On a short spreading axis just northeast of Jan Mayen, the central anomaly reaches +1000 to +1300 nT, compared with +300 to +1000 nT elsewhere on this axis (Fig. 25). On Iceland, the greatest magnetic relief (at 3 km flight elevation; Serson *et al.*, 1968) occurs in the southeast, in the region of greatest basalt discharge and Fe concentrations (Jakobsson, 1972). The high amplitudes may reflect a combination of greater basalt thickness and greater Fe and magnetite content, and bulk magnetization (Vogt and Johnson, 1974).

As the spreading axis shoals toward Iceland from the north and south (Fig. 4), an increase in magnetic amplitudes would be expected because of (1) decreasing source depth, and (2) (only south of Iceland) increasing Ti and Fe (Schilling, 1973*a,b*). Actually, the axial magnetic relief peaks at 4000–5000 nT some 150 km short of the coast, then declines dramatically to less than 1500 nT before the spreading axis emerges from the sea. The decline might reflect thermal blanketing effects and demagnetization caused by reheating, or also glacial and other erosion during low sea-level stands (Vogt and Johnson, 1974). Another mechanism based on observed high vesicularity (Duffield, 1978) and low sulfur concentration (Moore and Schilling, 1973), at shallow depths involves degassing and therefore higher oxidation states of the titano-magnetites, and increased fragmentation. All these factors could contribute to the magnetic smooth zones presently being formed near Iceland (Fig. 27; Vogt *et al.*, 1980).

Magnetic anomalies and sea-floor evolution in the GNSA cannot be discussed without reference to the Labrador Sea and Baffin Bay. Prior to the separation of Greenland from Europe about anomaly 24–25 time, the Mid-Atlantic Ridge extended northwest into the Labrador Sea. Anomalies 25–32 have been identified in the southern part of that sea (Srivastava, 1978). In the northern Labrador Sea the oldest lineation is 28 (earliest Paleocene), whereas spreading did not commence in Baffin Bay until anomaly 24 time (Jackson *et al.*, 1979). The early geometry was succeeded at anomaly 24–25 time by a more N–S-oriented spreading direction. The Labrador–Baffin spreading axis (Ran Ridge) then became extinct about anomaly 13 time (Kristoffersen and Talwani, 1977), with the kinematic result that the GNSA began to open in a more E–W direction, Greenland now being fixed to the North America plate. To describe early Tertiary motion Phillips *et al.* (1980) introduce an additional "central Arctic plate," with a primarily compressional boundary along the Canadian Arctic Islands (Eurekan Orogeny) separating the Arctic and North American plates.

Plate kinematic modeling also suggests a certain amount of nonrigid behavior for the plates in the greater GNSA (Kristoffersen and Talwani, 1977; Srivastava, 1978; Phillips *et al.*, 1980).

2. *Continental Margins*

There is no magnetic slope or edge anomaly consistently present in the GNSA. However, several prominent long-wavelength lineations of 30- to 100-km wavelength and 300- to 1000-nT amplitude are known to follow the East Greenland shelf between 75°N and 81°N (Figs. 25 and 27; Perry *et al.*, 1980), and between 68° and 70°N (Johnson *et al.*, 1975*a*). The N–NNW strike of these anomalies is generally parallel to the Greenland coast and its Caledonian structural trends. Source depths [up to 10 km, according to Birkelund, 1976, and Kovacs and Vogt, 1979; 6–7 km according to more detailed analyses by Surlyk *et al.* (1980)] are consistent with the conclusion of Talwani and Eldholm (1977) that a thick Mesozoic sedimentary basin exists under the East Greenland shelf between 75° and 81°N. The sources of the anomalies are probably of Caledonian or Precambrian age. Magnetic basement dips abruptly in a seaward direction within 20 km of the Greenland coast, at least from 68°N to 70°N (Fig. 26; Vogt *et al.*, 1980).

Complex anomalies including intermediate wavelengths extend further out across the shelf adjacent to the Greenland–Iceland Ridge. The magnetic field over this part of the shelf probably includes large contributions by downflexed early Tertiary basalts and associated intrusives. The exact location of the boundary between continental and oceanic crust is unknown, but much of the Greenland shelf in the Denmark Straits–Scoresby Sund area may be underlain by oceanic crust (Fig. 26; Vogt *et al.*, 1980). A magnetic/bathymetric profile along the narrow Southeast Greenland shelf suggests basement outcrops, possibly basaltic dikes, 20–100 km offshore (Vogt, 1970). In 1977 the Greenland Geological Survey conducted a detailed aeromagnetic survey over the entire East Greenland shelf and upper slope (Larsen and Thorning, 1980).

The magnetic pattern over the continental margin west of Norway (Talwani and Eldholm, 1972; Vogt and Ostenso, 1973; Aam, 1975*a*; Hagevang, 1978) resembles that known from Greenland. Detailed aeromagnetic coverage extends over Norway and its continental shelf out to the 2-km isobath, from 62°N to 74°N (Aam, 1975*a*). Complex, high amplitude (several hundred to over 1000 nT) anomalies corresponding to the outcropping Caledonian or older basement (the "strandflat" province) occur only within 20–100 km of the coast. A nearly continuous belt of magnetic anomalies of several hundred nT amplitude lies just landward of the shelf edge and extends from northwest Scotland to the Lofoten–Vesterålen Islands (Fig. 27; Talwani and Eldholm, 1972). The magnetic maxima correlate well with gravity anomalies but are somewhat phase shifted. These and other data suggest (Talwani and Eldholm, 1972) that: (1) the magnetic and gravity anomalies are caused by Precambrian intrabasement structures; (2) the magnetization is largely remnant, with a low inclination consistent with a pre-Permian age; and (3) this belt of dense, magnetized rocks probably controlled the line of initial early Tertiary continental breakup. A difficulty with the latter point is that similar magnetic lineations on the West Greenland shelf are somewhat oblique to the early Tertiary spreading-type anomalies (Figs. 25 and 27).

Magnetic data have proved a useful complement to seismic methods in determining basement depth (and hence sediment thickness) along the Norwegian–Barents margin (Aam, 1975a) in the Barents Sea (Vogt and Ostenso, 1973), in the Svalbard area (Aam, 1975b), off East Greenland (Birkelund, 1977; Kovacs and Vogt, 1979; Larsen and Thorning, 1980), around the Jan Mayen Ridge (Navrestad and Jørgensen, 1979), and over oceanic crust northeast of the Jan Mayen F.Z. (Fig. 25; Kovacs and Vogt, 1979).

Landward of the marginal magnetic high, the inner Vøring Plateau is associated with a magnetic smooth zone (relief less than $\pm$ 100 nT) (Talwani and Eldholm, 1972). When Greenland and Europe are reconstructed in their predrift configuration, holding Europe arbitrarily fixed (Talwani and Eldholm, 1977), this smooth zone is due south of, and contiguous with, a magnetic smooth zone centered at 76°N, 10°W on the West Greenland margin (Figs. 26 and 27). This observation lends credence to the existence of a larger sedimentary basin, fragmented by sea-floor spreading (Talwani and Eldholm, 1977; Fig. 22). A magnetic smooth zone over and just west of Jan Mayen Ridge may also represent a portion of an old sediment-filled Mesozoic Basin (Fig. 27; Navrestad and Jørgensen, 1979).

The location of the continent–ocean boundary in the Vøring Plateau area is disputed: Talwani and Eldholm (1972) and Hagevang (1978) place the boundary (as we do) at the buried, east-facing Vøring Escarpment, whereas Roeser et al. (1975) explain the magnetic anomalies over the outer plateau in terms of basalts extruded over continental crust. Multichannel data have not settled the problem (Rønnevik et al., 1979).

A detailed aeromagnetic study of the southwestern Barents Sea (Fig. 23; Aam, 1975a) extends eastward from spreading-type lineations (west of the 1000-m isobath) across an 80-km-wide magnetic smooth zone and thence over an irregular pattern of anomalies probably due to the Precambrian or Caledonian basement. There is no magnetic evidence for structural continuity between the Norwegian Caledonides and Svalbard. This was earlier shown by reconnaissance shipborne data (Vogt and Ostenso, 1973). Shallow magnetic basement (<2 km) was found by the latter authors only within 30 km of Bear Island and near 74.5°N, 33°E. Basement is deep in the Kara Sea and highly variable in the eastern Barents Sea. Reconnaisance aeromagnetic data from the Svalbard area were interpreted by Aam (1975*b*); basement depths there range from 4 to less than 1 km.

B. Crustal and Mantle Structure: Seismic Evidence

In this section we review seismic refraction and reflection experiments (Figs. 15 and 17), with emphasis on the oceanic (i.e., post-anomaly-25) portion of the GNSA. The "first" or sediment layer lying above igneous basement has already been discussed. By *velocity* we shall henceforth mean P-wave speed unless otherwise noted.

Ewing and Ewing (1959) pioneered the investigation of the ocean basins by explosion seismology. Their GNSA profiles, all reversed, revealed a relatively thin (2.7–4.3 km) "second" layer (V_p = 5.0–5.8 km/sec) overlying an "anomalous" mantle with velocities of 6.94–8.04 km/sec (average, 7.5 km/sec). In crustal structure the GNSA basins and median ridges resembled the Mid-Atlantic Ridge, of which they are a continuation. In later years (e.g., Myhre, 1978) it has become possible to subdivide some of the layers into two or more sublayers (e.g., 2A–C; Fig. 17). Modern work, utilizing amplitudes as well as travel times, has also led to gradient models, which in some cases may lack velocity discontinuities altogether. Thus, the large body of solutions depicting the oceanic crust in terms of four discrete layers (Ludwig *et al.*, 1970) is only a rough approximation.

Crustal structure studies of the Reykjanes Ridge (Fleischer, 1974) were resumed a decade after Ewing and Ewing (Talwani *et al.*, 1971; Whitmarsh, 1971; Aric, 1972). The international Reykjanes Ridge–Iceland Seismic Project (RRISP) was carried out in 1977 (Figs. 15 and 17; Angenheister *et al.*, 1979).

The profiles of Talwani *et al.* (1971) showed surprisingly low basement velocities (2.4–3.7 km/sec) in a layer about 1 km thick. This unit was observed only on the axial block of Reykjanes Ridge; stations on older crust gave more "typical" basement velocities of 4.6–4.7 km/sec overlying a third layer (6.1–7.3 km/sec). Waves refracted from the top of this "oceanic" layer in the shear mode (3.5 km/sec) were observed as clear second arrivals.

Talwani *et al.* (1971) suggest that the 3-km/sec rock becomes transformed into the 4.7 km/sec material as a result of steady-state alteration of highly vesicular pillow basalt with age. However, the Reykjanes Ridge central block (Profile 11, Fig. 5) may also have an initially distinctive structure, a difficulty for steady-state models. The distinctive properties of the central block may be primary (Vogt, 1974) or a consequence of shallower extrusion depth, or both. Lower hydrostatic pressure and therefore greater vesicle content, and lower specific gravity (Moore and Schilling, 1973), would produce a significant reduction in seismic velocity. If the central block pillow layer is now being formed at 600-m depth, vs. 1000–2500 m for the flanking crust (Duffield, 1978; Luyendyk *et al.*, 1979) initial pillow densities of 2.8 g/cm^3 for the flanks vs.

less than 2.6 g/cm³ on the block, in turn, imply a 15–20% reduction in seismic velocity on the central block, of the same order as that observed (Talwani *et al.*, 1971). However, the observed worldwide increase in seismic velocity with plate age for the top of layer 2 (from 3.3 km/sec at the ridge crest to 5.2 km/sec for crust older than 30 m.y.; Houtz and Ewing, 1976) cannot be attributed to vesicularity differences.

Another feature of the central block crust (Talwani *et al.*, 1971) is anomalously low mantle velocity (7.4 km/sec) and the absence of an "oceanic" layer (V_p = 6.7 km/sec). Both of these features seem to be characteristic of the axial parts of the Mid-Atlantic Ridge; on the fast-spreading East Pacific Rise, anomalously low mantle velocities are found within 5 m.y. of the axis, and the third layer, although not absent, is only 3.5 km thick. It thickens gradually to about 5 km in the ocean basins. The oceanic layer is subdivided into a 6.7-km/sec upper layer of constant thickness (3 km) underlaid by a slowly thickening layer of average 7.3-km/sec velocity (LePichon *et al.*, 1973). This lower sublayer may form by low-temperature alteration of mantle. Although an oceanic layer may be absent near the spreading axis (Talwani *et al.*, 1971), absence of this layer throughout the GNSA (and elsewhere on the middle and lower flanks of the Mid-Atlantic Ridge) as reported by Ewing and Ewing (1959) may be partly an artifact of the method used. Talwani *et al.*, (1971) did find an oceanic layer near station E-4 (Fig. 15) as did Whitmarsh (1971) near station E-5 on the lower flanks of Reykjanes Ridge. Whitmarsh's subsedimentary crust consists of a 1.2-km second layer (4.74 km/sec) over a 4.9-km third layer (6.19 km/sec—rather low for this layer), overlying a low-velocity ($\geq$ 7.83 km/sec) mantle. It seems likely that earlier experiments (Ewing and Ewing, 1959) failed to detect enough first arrivals from layer 2 to determine a 4- to 5-km/sec line; using bottom receivers, Whitmarsh (1971) was able to observe first arrivals over a sufficiently greater range.

The hexagonal geometry of Whitmarsh's experiment (Fig. 15) was designed to detect seismic anisotropy. None was found, even in the upper mantle. Pacific and some Atlantic experiments have indicated mantle velocities 0.2–0.6 km/sec lower along the strike of crustal isochrons than perpendicular to them. The crustal layers appear isotropic in velocity.

Another information source on deep structure below the Reykjanes Ridge is a 120-km-long explosion-seismic reflection profile (Aric, 1972). The line lies between profiles E-3 and E-4 (Ewing and Ewing, 1959) and assumes the velocity structure deduced by those authors. Fairly continuous reflectors were interpreted in terms of interfaces between the 3.7- and 5.5-km/sec crustal layers, between the 5.5- and 7.3-km/sec (upper mantle?) layer, and at the top of a 8.1-km/sec mantle inferred to exist at a depth of 40 km. More irregular reflectors occur in intervening low-velocity mantle (7.3–7.7 km/sec) and tend to dip toward the Reykjanes Ridge axis.

The RRISP experiment (Angenheister *et al.*, 1979), conducted on 10-m.y.-old Reykjanes Ridge crust closer to Iceland, revealed a 2.4-km-thick second layer (V_p = 4.4 km/sec) and a 5.1-km-thick third layer (V_p = 6.7 km/sec), overlying anomalous mantle (V_p = 7.7 km/sec.).

Refraction data from oceanic crust in the Norwegian and Greenland Seas have been compiled and reinterpreted by Myhre (1978) (Fig. 17). Mean velocities are 3.79, 5.14, 6.07, 6.94, and 7.98 km/sec for layers 2A, 2B, 2C, 3 and 4. These values are similar to worldwide compilations (Houtz and Ewing, 1976). Layer 2A, only observed on crust less than 10 m.y. old (Myhre, 1978), seems to be "converted" to layer 2B-like material, probably through filling of voids and cracks by hydrothermal mineralization. The relatively large number of sediment layer velocity measurements in the GNSA (Fig. 17), compared with the ocean basins as a whole, reflects relatively great sediment thicknesses and low penetration of sonobuoy profiles.

Average upper mantle structure for the area north of Iceland was first deduced from Arctic earthquakes: V_p averages 7.4 km/sec below the crust and an increase to 8.2 km/sec occurs at 140-km depth (Tryggvason, 1961). Surface wave dispersion studies have also been carried out for paths crossing the Norwegian and Barents seas (Rygg, 1972), the Reykjanes Ridge (Girardin and Jacoby, 1979) and on Greenland, which was found to have a thick (40 km) continental crust (Gregersen, 1970).

The crustal structure of Iceland (Båth, 1960; Bath and Vogel, 1958; Tryggvason, 1962; Pálmason, 1971; Angenheister *et al.*, 1979), and the attached Iceland–Faeroe Ridge (Bott, 1974; Bott *et al.*, 1976) differ significantly from that of normal ocean crust, although the velocity of at least some of the layers is closely similar.

The RRISP recording stations (Fig. 15) generally follow the eastern Neo-Volcanic Zone, although a few were placed along the Reykjanes–Langjökull line and in the Tertiary basalt provinces of central and eastern Iceland (Angenheister *et al.*, 1979). The crustal structure model, derived by ray-tracing techniques along the eastern Neo-Volcanic Zone, comprises the following layers (Fig. 17): 3.0 km/sec, 0–2 km thick; 4.5 km/sec, 3–6 km; and 5.6–6.1 km/sec, 5–10 km. Below these crustal layers totaling 10–15 km in thickness, lies anomalous upper mantle (7.0 km/sec). A gradual downward velocity increase (± unknown reversals) takes place. Normal Moho velocities (8.0 km/sec) are not reached even at 60 km depth.

An average Icelandic crustal structure deduced from 80 refraction lines (Pálmason, 1971), some of which extend offshore up to 100 km across the south Iceland shelf (Figs. 15 and 17), was generalized in terms of five crustal layers. (The RRISP study may require changes in some of the following). Layer "0," which corresponds to Neo-Volcanic basalts, is a formation of high-porosity flows up to 1 km thick. Its mean velocity (2.75 km/sec) is even lower than the upper basement on the central block of Reykjanes Ridge (Talwani *et al.*, 1971). This is reasonable, since porosities of the subaerial basalts would be higher. Layer I (4.14 km/sec), about 1 km thick, corresponds to Tertiary plateau basalts and outcrops beyond the Neo-Volcanic Zone. The higher velocities may reflect filling of vesicles by calcite and zeolites, and healing of fractures; possibly volcanics with "Neo-Volcanic" velocities once existed but have been eroded off. Layer I velocities are about 0.5 km/sec less than the older flank basement on Reykjanes Ridge (Talwani *et al.*, 1971). Icelandic Layer II (5.08 km/sec; 2.15 km thick) may represent low-grade zeolite metamorphic facies basalts, and outcrops locally. Layer III seems to be the equivalent of the oceanic layer, except thicker and of slightly lower velocity (6.35 km/sec; given as 6.5 km/sec by Pálmason *et al.*, 1979). Like its oceanic counterpart it thickens with age, from 5 km in the Neo-Volcanic Zone to 13 km under southeast Iceland. Depth to the top of this layer is highly variable, ranging from 0.6 to 10 km over a few kilometers. Velocity may increase with depth in Layer III. The II/III interface is thought to correspond to the 350°–400°C isotherm, which governs the greenschist–amphibolite transition. Beyond some distance from the spreading axis the seismic interface would then represent a fossil isotherm. The II/III interface has not been penetrated, even in the 2820-m-deep Laugaland hole, more than 4 km below the original land surface. Layer IV, the anomalous upper mantle (7.2 km/sec), lies at 8–9 km below Reykjanes Peninsula and deepens to 14 km in southeast Iceland. A magnetotellurically determined increase in electric conductivity (Hermance and Grillot, 1970; Beblo and Björnsson, 1978), an anomalously high ratio of P- to S-wave velocities (2.0; Angenheister *et al.*, 1979), and extrapolations of borehole temperatures (Pálmason, 1971) all suggest that the III/IV interface may be near the melting point of basalt. Reflected waves from the III/IV boundary were observed by Pálmason on several long profiles crossing the south Iceland shelf.

Shear waves were recorded on many of Pálmason's (1971) profiles and Poisson's ratio (σ) could be calculated. The results for Layers I, II, and III are 2.34 km/sec (0.27), 2.78 km/sec (0.278), and 3.53 (0.269), respectively, where σ is in parentheses.

Pálmason's (1971) comprehensive study refined the more generalized results of earlier experiments: two refraction profiles (Båth, 1960), analyses of P-waves from Icelandic earthquakes (Tryggvason, 1959), and dispersion of surface waves (Båth and Vogel, 1958; Tryggvason, 1962). Tryggvason's structure has a crustal layer (V_p = 4.7 km/sec; V_s = 2.7 km/sec) about 3–5 km thick overlying, in the area within 500 km of Iceland, a pod of "oceanic" layer (V_p = 6.3 km/sec; V_s = 3.6 km/sec). This layer, up to 5 km thick under central Iceland, is absent from the ocean basin south of 59°N. An anomalous upper mantle (V_p = 7.4 km/sec; V_s = 4.3 km/sec) was found present in the entire area. Using travel times of body waves from earthquakes north and south of Iceland, Francis (1969) found upper mantle velocities to increase downward according to the relation $V_p = 7.0 \ (r_0/r)^{3.4}$ km/sec, where r_0 is 6370 km and V_p is the velocity at radius r. P-waves originating from teleseismic events are delayed 2–3 sec at Reykjavik, compared to Kiruna and Scoresby Sund (Tryggvason, 1964).

A delay of nearly 2 sec was also measured at Akureyri in north Iceland (Long and Mitchell, 1970); an *ScS* delay of +8.4 sec relative to average continents and +3 sec relative to average oceans was found at this station for a teleseismic event originating in the Okhotsk area (Sipkin and Jordan, 1975). The delays reflect exceptional intense partial melting associated with the nearby spreading axis and perhaps an Iceland mantle plume.

Anomalously high shear-wave attenuation under the MAR (Solomon, 1973) and, on Iceland, exceptionally rapid postglacial rebound [implying low subcrustal viscosity; Einarsson (1967)] are additional evidence for partial melting below the accretion axis.

Crustal and upper mantle structure below the Greenland–Iceland and Iceland–Faeroe Ridges and the Faeroes platform is important to assess the way in which Icelandic lithosphere "ages." Allowance must however be made for possible time variations in basalt discharge from the Iceland plume (Vogt, 1971, 1974) and shifts of the spreading axis (Talwani and Eldholm, 1977). The structure below the Faeroes plateau basalts ought somehow to reflect its continental (Bott *et al.*, 1974) or partially oceanic (Talwani and Eldholm, 1977) nature.

The Iceland–Faeroe Ridge was investigated by the North Atlantic Seismic Project in 1969 (Bott, 1974; Bott *et al.*, 1976). A four-layer model was derived: Layer 1 (3.2–4.6 km/sec) is 0–3 km thick and corresponds to basalt layers 0 and I of Pálmason (1971) and Layer 2A observed by Talwani *et al.* (1971) on Reykjanes Ridge. Layer 2, at least 5 km thick, exhibits V_p of 5.4–5.8 km/sec, comparable to, but slightly higher than, the typical oceanic "second" layer and Pálmason's Layer II on Iceland. Layer 3 (6.8 km/sec) was observed as a horizontal refractor at a depth of 7.5 km under the northwestern half of the Iceland–Faeroe Ridge. It is slightly "faster" than under Iceland, and its depth suggests no further thickening beyond the edge of the Iceland platform. A 7.8-km/sec mantle occurs at depths estimated as 22 km (Bott, 1974) and 27 km (Bott *et al.*, 1976). Apparently, the thickness of Icelandic Layer III increases further from southeast Iceland (13 km) to the Iceland–Faeroe Ridge (20 km).

Plateau basalts appear to overlie a continental crust under the Faeroes (Bott, 1974, 1975; but see Talwani and Eldholm, 1977). A thin upper basalt layer (3.9 km/sec) overlies a main basalt formation (4.9 km/sec) similar to Icelandic Layer II (Pálmason, 1965). These are thought to overlie 6.1-km/sec continental crust at 5-km depth. A 6.5-km/sec lower crustal layer extends to the 34-km depth of normal mantle (8.24 km/sec) (Bott et al., 1976).

Refraction profiles across Jan Mayen Island (Heier, 1977; Fig. 15) show <1 km sediment (2.5 km/sec) overlying 2–4 km of 3.14 km/sec [layer 2A(?)] overlying a 6.3-km/sec layer [2C(?)].

The nature of the oceanic–continental crustal transition remains a major unsolved problem, although multichannel profiling (Figs. 18, 19, and 21) has made it possible to trace the oceanic basement under the thick continental rise sediments (Briseid and Mascle, 1975; Hinz and Schlüter, 1978a,b; Surlyk et al., 1980; Eldholm et al., 1979; Rønnevik and Motland, 1979). Refraction data alone often lead to ambiguities, since the top of oceanic layer 2 exhibits velocities similar to those of compacted sediments.

As far as is known, the crustal transition seems to be generally abrupt, being associated with the landward-dipping Vøring Plateau, Faeroe–Shetland, and Greenland escarpments (Fig. 27; Talwani and Eldholm, 1977). The location and nature of the transition in the Vøring Plateau area has been disputed, however (Hinz, 1972, 1975; Hinz and Moe, 1971; Roeser et al., 1975). From Bear Island to northern Vestspitsbergen the crustal boundary may be associated with the seaward-dipping Hornsund Escarpment (Vogt et al., 1978); or it may lie farther seaward (Sundvor et al., 1978).

In some parts of the GNSA, e.g., the pre-anomaly-23 ocean crust west of the Lofoten Islands, the top of the oceanic crust is a smooth horizon locally underlain by deeper, seaward-dipping reflectors (Fig. 18; Eldholm et al., 1979). The nature of these subbasement reflectors remains speculative; volcaniclastic sediments, lava flows, and Paleocene terrestrial sediments buried by later volcanics are three possibilities.

Deep crustal refraction profiling has also been carried out over continental crust bordering the GNSA. For example, in southern Norway (Fig. 15; Heier, 1977) the main refractor velocities are 6.0–6.3, 6.6, 7.1, and 8.2 km/sec.

C. Gravity

1. Ocean Basins and Iceland

The anomalous free-air (F.A.) gravity field over the northeast Atlantic and GNSA can be described as the sum of two basic effects: (1) a long-wavelength positive, roughly centered (but not sharply peaked) on Iceland and reaching +30–40 mgals in amplitude (Fig. 30); and (2) anomalies with wavelengths 10–100 km and amplitudes up to ±50–100 mgals, generally associated with topographic or basement structures and their roots. The first effect can be subdivided into (1) an anomaly of +20 to +25 mgals over zero age crust and

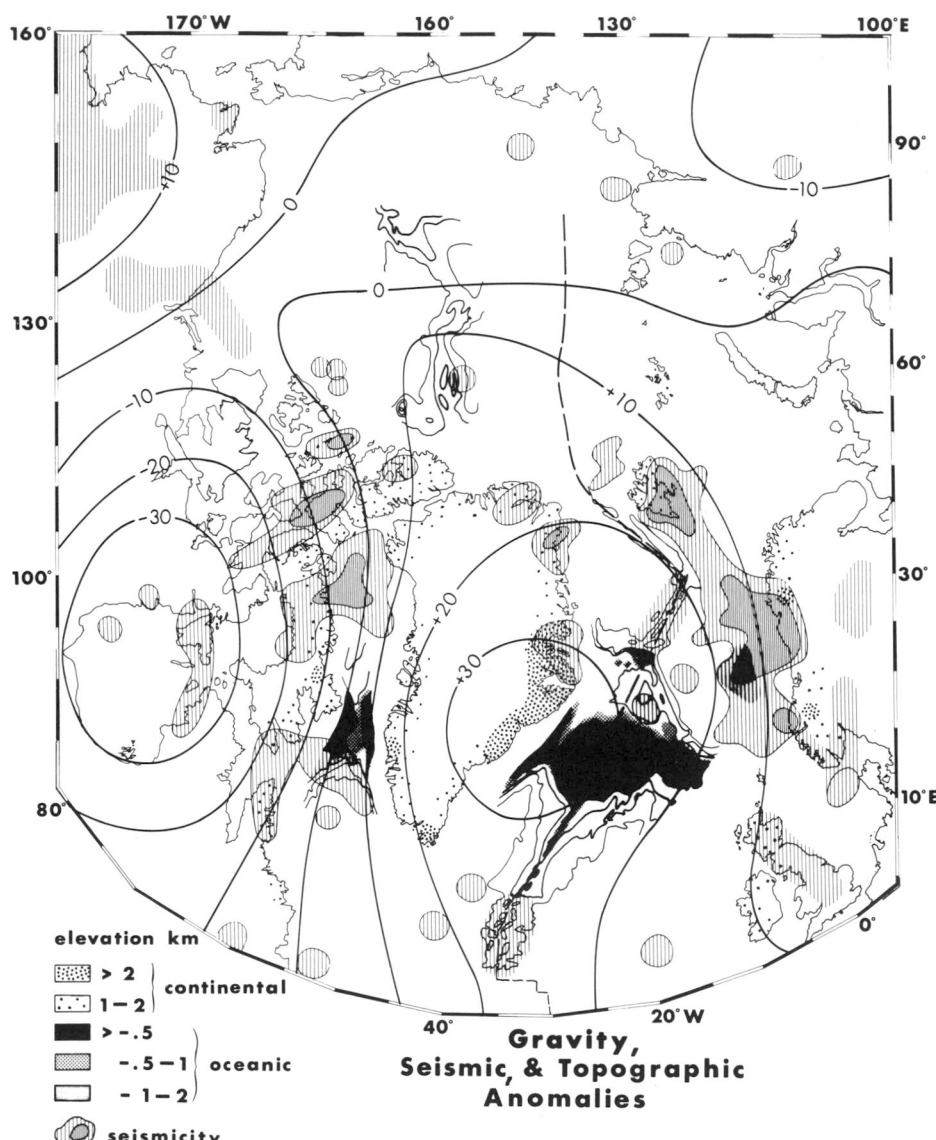

Fig. 30. Topography (with anomalous areas emphasized), regional free-air gravity anomalies, and areas of recent seismicity forming a loose halo around flanks of Icelandic gravity high (from Vogt, 1974, 1978).

gradually disappearing with increasing crustal age by about 40 m.y., and (2) a regional positive of +20 mgals extending from the northern Knipovich Ridge to 30°N (Cochran and Talwani, 1977, 1978).

The long-wavelength anomaly is best revealed by satellite-derived gravity measurements insensitive to the higher spatial frequencies (Gaposchkin and

Lambeck, 1971; Kaula, 1972; Fig. 30). It can also be obtained by smoothing shipboard measurements (Fleischer *et al.*, 1974). The anomaly is broad and the implied mass excess too great to be explained in terms of density variations in the lithosphere alone. Mass deficiencies extending to depths of several hundred kilometers, and probably reflecting a thermal anomaly of the order $+75°C$ in the asthenosphere, may account for the long-wavelength gravity and topographic anomalies (Cochran and Talwani, 1978; Zielinski, 1977). The upwelling (plume) needed to maintain this anomaly will also contribute to the gravity high (Anderson *et al.*, 1973; Sclater *et al.*, 1975). Since the topographic anomaly is peaked over Iceland (Figs. 3 and 4), whereas the F.A. positive is flat, asthenosphere densities must be lowest (and temperatures highest) below Iceland, as expected from a narrow plume model.

Since the F.A. anomaly is already positive in the GNSA, the simple Bouguer anomaly is still more so. Where the original F.A. anomaly is constant, the Bouguer anomaly is inversely related to regional water depth, an observation that reflects isostatic compensation of most broad topographic features of the earth's surface. Regional gradients of the Bouguer anomaly are sometimes related to the depth of compensation. F.A. anomalies of wavelength 10–50 km are greatly subdued in the Bouguer residual, thus reflecting a lack of compensation.

In their various papers on the GNSA (1972, 1974, 1977, and others), Talwani and Eldholm have also computed *isostatic* gravity anomalies. These anomalies, which are "two-dimensional" like the simple Bouguer, are computed on the assumption that the mass deficit caused by the water layer is compensated by topography on the crust/mantle interface. Water/bottom and crust/mantle density contrasts of 1.37 and 0.5 g/cm^3 were assumed by Talwani and Eldholm (1972); other authors may use somewhat different definitions (e.g., Grønlie and Ramberg, 1970).

Isostatic anomalies generally vary within ± 10–20 mgals and resemble F.A. anomalies in shape. However, over steep bathymetric escarpments between blocks of dissimilar elevation (e.g., between the Iceland platform and the Iceland–Faeroe Ridge) the F.A. "marginal high" is much reduced upon isostatic correction (Talwani and Eldholm, 1972). Thus, the marginal F.A. high is often largely an edge effect. It is caused by the relative closeness of the near-surface mass excess of the topographic step, compared to the deeper mass deficit in the compensating root. In areas of level sea floor, F.A. and isostatic anomalies do not differ except by a constant, hence the former may be used to infer lateral density variations. Local F.A. anomalies reflect sea-floor or basement relief. Where rough topography and a prominent rift valley characterize the spreading axis (e.g., southern Reykjanes, Mohns, and Knipovich ridges), F.A. anomalies range from 0 to over $+100$ mgals (Fleischer *et al.*, 1973; Talwani and Grønlie, 1976). The topographically smooth northern Reykjanes and southern Kolbeinsey Ridges exhibit lower gravity *relief* (50 to 80 mgals). Short wavelength F.A. highs also follow the time-transgressive basement lineations on the flanks of the Reykjanes Ridge (Talwani *et al.*, 1971) postulated to record southward asthenosphere flow from the Iceland plume (Vogt, 1971, 1974).

Over Iceland itself, regional gravity surveys indicate a bowl-shaped Bouguer minimum (Einarsson, 1954, and later unpublished work). The anomalously low density could extend to depths of 100–200 km (Bott, 1965). A relatively sharp F.A. maximum (+40 to +75 mgals), probably reflecting a step in the basaltic basement, has been charted just landward of the shelf edge off southeast (Pálmason, 1974*b*) and east (Fleischer *et al.*, 1974) Iceland. An ESE gravity trough associated with the complex Tjørnes F.Z. may indicate a basement depression filled with low-density, largely volcanogenic sediments (Pálmason, 1974*b*; Fig. 31).

Like Iceland, the Iceland–Faeroe Ridge (IFR) is associated with a +40-mgal F.A. high, about 20 mgals above values measured in the adjacent basins (Fleischer *et al.*, 1974). Local features, often correlating with magnetic anomalies, may reflect extinct volcanic/intrusive centers. A gravity trough intersecting the IFR may mark the southward continuation of the extinct Aegir rift valley, which also has a prominent F.A. expression (Fig. 32). Gravity (with seismic refraction data) has also played a role in the debate over whether the western Faeroe Islands are situated over oceanic (Talwani and Eldholm, 1977) or hidden continental (Bott *et al.*, 1974; Fleischer *et al.*, 1974) crust.

2. *Continental Margins*

The continental margins of the GNSA are characterized by locally prominent free-air (F.A.) gravity anomalies generally paralleling the margins (Figs. 27 and 32). In some instances where highs occur over the shelf edge, the anomaly is simply an edge effect. Other anomalies must represent crustal structures within the ancient continental crust, in the much younger ocean basins, or in the transition zone (Talwani and Eldholm, 1972, 1974, 1977; Roeser *et al.*, 1975).

A linear isostatic high just seaward of the base of the Greenland Continental Slope near 75.6°N, 5°–10°W may relate to a basement ridge and landward-facing escarpment just to the west (Fig. 27). Further south, an F.A. high near the shelf edge is primarily an edge effect (Talwani and Eldholm, 1974). Similar gravity lineations occur on the Svalbard margin. A prominent F.A. high runs along the base of the southwest Barents slope from 70° to 71°N and then cuts across the upper slope and shelf edge from 71° to 73.5°N (Fig. 32). This isostatic anomaly, up to 80 mgals in magnitude, is believed to reflect a deeply buried "Senja Fracture Zone," the eastern counterpart to the Greenland F.Z (Talwani and Eldholm, 1974). No seismic evidence for this pos-

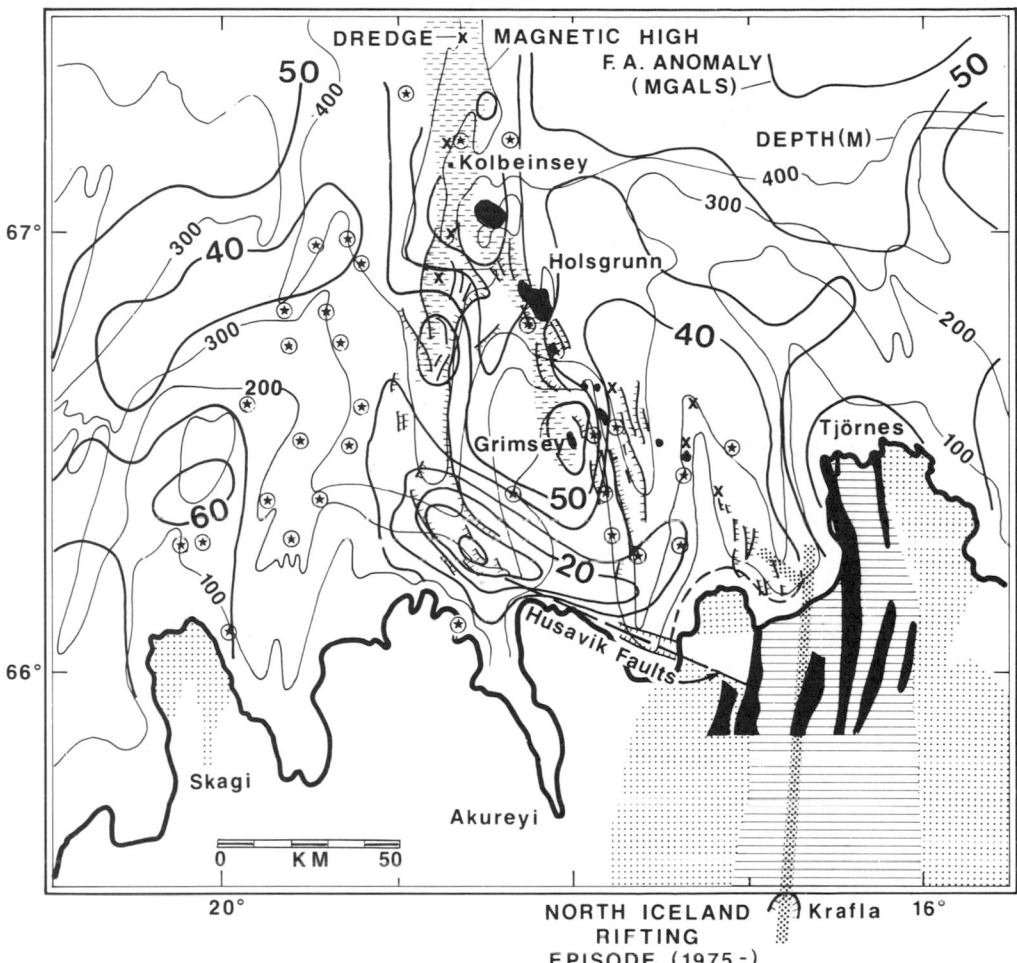

Fig. 31. Bathymetry (fine lines), free-air gravity contours, earthquake epicenters, North Iceland rifting episode (stippled band), and tectonic–volcanic structures at junction between Kolbeinsey Ridge and Iceland. Areas of young fissure eruptions and submarine volcanics black. Based on Johnson (1974), Pálmason (1974b), Björnsson et al. (1977, 1979), McMaster et al. (1977), and Einarsson (1979).

tulated basement structure has been adduced, however (Houtz and Windisch, 1977); the gravity high might also partly reflect uncompensated late Cenozoic sediments, as suggested for similar features in the Canadian Arctic (Sobczak, 1975).

Gravity and magnetic data together suggest an irregular basement relief in the western Barents Sea (Fig. 22; Eldholm and Talwani, 1977); eastwards the anomalies become smooth, averaging about +5 mgals east of 28°E. A pronounced high occurs at 73.2°N, 36.2°E (Eldholm and Ewing, 1971). The

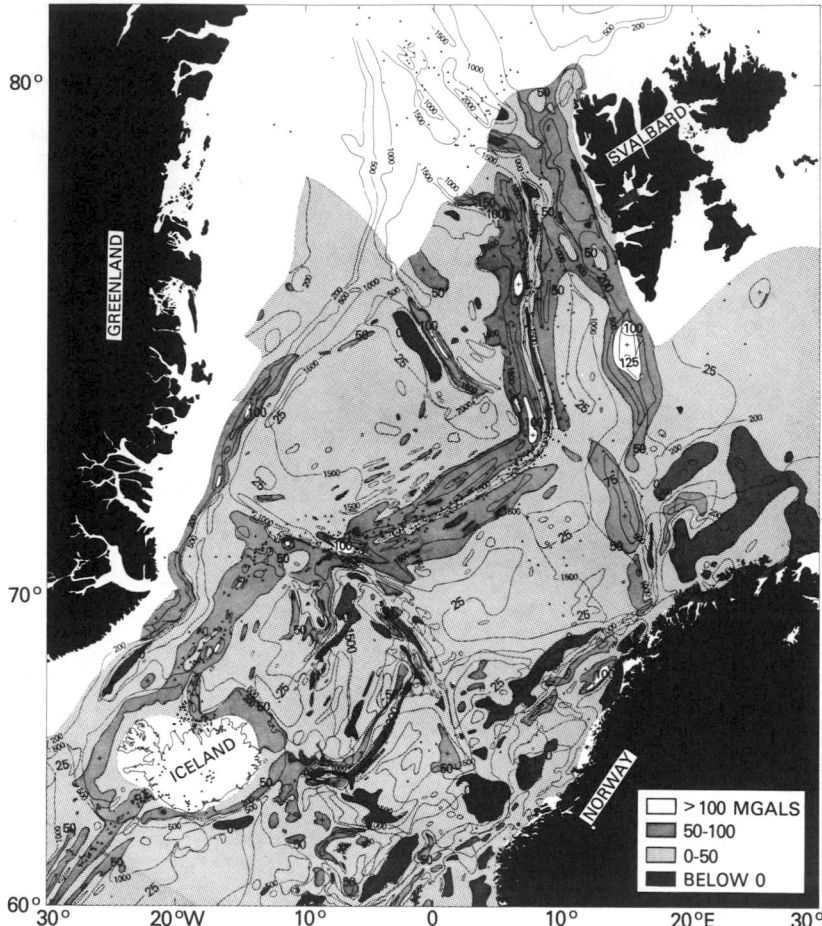

Fig. 32. Free-air gravity and bathymetric contours (redrawn from Talwani and Grønlie, 1976) with earthquake epicenters through May 1976.

Bouguer anomaly pattern shows good correlations with buried structural ridges and basins in the western Barents Sea (Rønnevik and Motland, 1979). Very few gravity data are available east of 35°E. Some point measurements south and east of Franz Josef Land (Vogt and Ostenso, 1973) suggest broad anomalies with dimensions of several tens to 100 km and amplitudes within 10–20 mgals of zero. Similar measurements in the Kara Sea indicate that a gravity low, about 10–20 mgals, follows the East Novaya Zemlya Trench.

Linear belts of gravity and magnetic anomalies lie in the vicinity of the Norwegian shelf edge (Fig. 27; Talwani and Eldholm, 1974). The gravity anomalies have been attributed alternatively to basement relief (Grønlie and Ramberg, 1970) and to intrabasement density contrasts (Talwani and Eldholm, 1972). One of the major gravity highs continues into the Lewisian (Pre-cambrian) basement of northwest Scotland (Watts, 1971); a similar high, over +130 mgals, is associated with Precambrian structures of the Lofoten block the associated Moho upwarp (Heier, 1977). Lithologically, many of the linear highs are connected with provinces of intermediate to ultrabasic intrusive rocks (Grønlie and Ramberg, 1970). Broad gravity minima between the ocean crust of the Norwegian Sea and the Norwegian continental shelf confirmed seismic evidence for the existence of deep, sediment-filled basins (Grønlie and Ramberg, 1970; Talwani and Eldholm, 1972, 1974, 1977). These basins have now been mapped by multichannel seismics (Rønnevik et al., 1979; Jørgensen and Navrestad, 1979).

D. Heat Flow

Most heat flow measurements in the GNSA north of 52°N are shown in Figs. 33 and 34. Not shown are three measurements in the Labrador Sea and seven in Baffin Bay (Pye and Hyndman, 1972; Hyndman, 1973). We shall summarize the results by crustal province, from the accretion axis toward the continental margins.

The crestal area, say within 100 km of the spreading axis, is characterized by extreme local variability and occasional very high values (Langseth and Zie-linski, 1974; Figs. 33 and 34). This pattern is now known from many other parts of the Mid-Oceanic Ridge. Between 50 and 100 km from the Reykjanes Ridge axis, but not on the flanks of Mohns, Knipovich, and Kolbeinsey Ridges—some heat-flow measurements are also low, locally even less than 0.1 HFU (μcal/cm^2·sec; 1 HFU = 41.84 mW/m^2).

Although high axial heat flow is predicted by simple models of cooling lithosphere (Sclater, 1972), early models did not consider that seawater circu-lating through crustal fractures can effectively remove heat otherwise exported by molecular conduction. Thus, the low measured heat flow on the Reykjanes Ridge is only part of the budget. A large but poorly known flux results from hydrothermal processes, which are highly localized in space and time. As continuous plate motion carries the crust out of the axial zone, a gradually thickening sediment blanket eventually seals the fractures, turning the heat export over to molecular conduction. Where sediments are thin, such as near the axis, the rough basement topography also causes spatial variability in heat flow. Yet another difficulty with observations in shallow areas, such as the Mid-Oceanic Ridge crest in the Iceland area, is the greater likelihood for time-variable bottom currents and small-scale cut-and-fill disturbances of the sedi-mentation process. Some of the ultralow heat flow values on the Reykjanes Ridge may reflect high local sedimentation rates caused by strong bottom cur-rents such as described by Talwani et al. (1971).

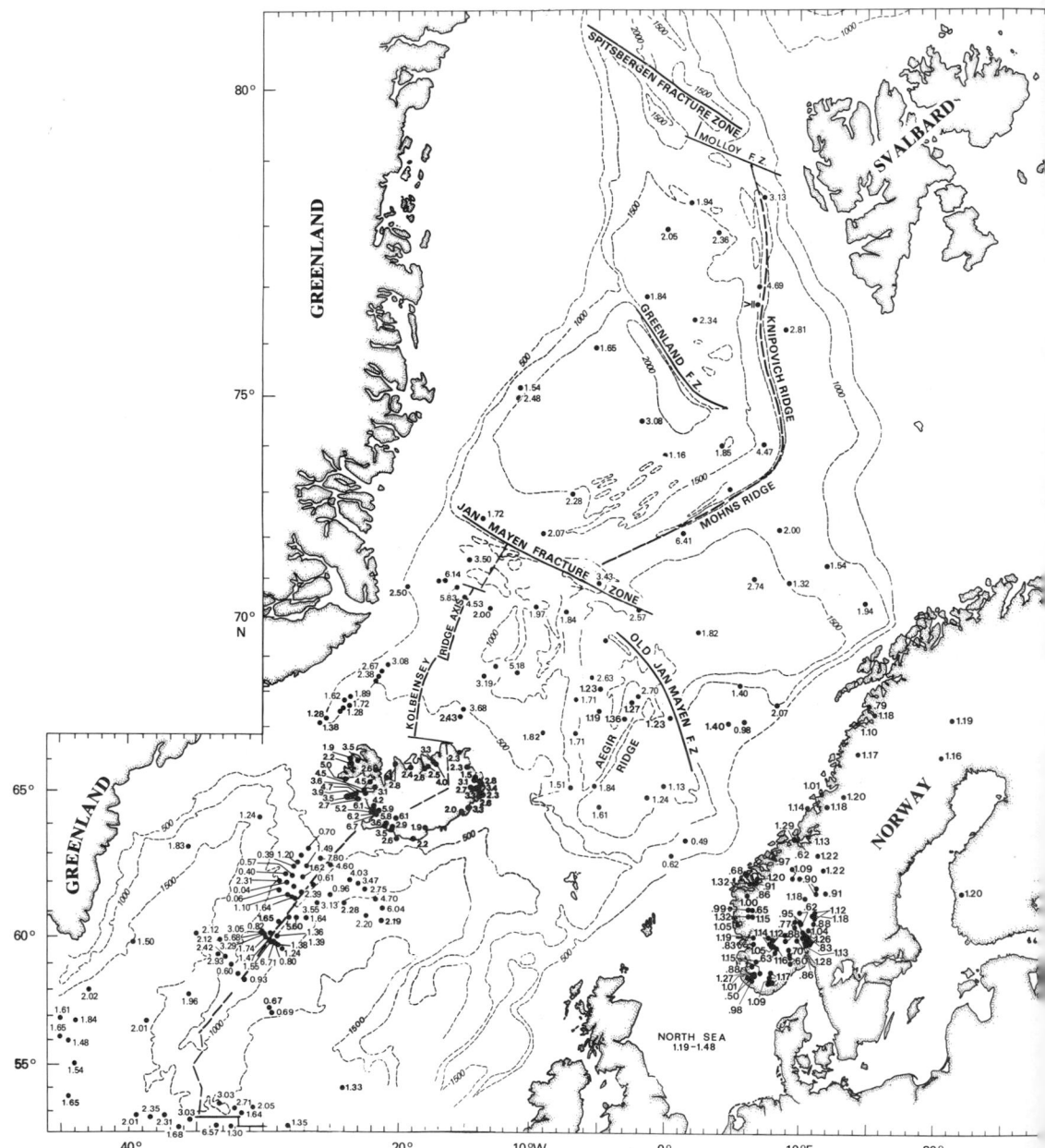

Fig. 33.   Locations and values (HFU) of heat flow measured in the northeast Atlantic, Iceland, and the GNSA. From Langseth and Zielinski (1974), with additional data from Lachenbruch and Marshall (1968), Horai *et al.* (1970), Talwani *et al.* (1971), Pálmason (1973), Zielinski (1977), Bellousov and Udintsev (1977), Pálmason *et al.* (1979), Harper (1971), Grønlie *et al.* (1977), and Hänel *et al.* (1974). Stations where no reliable value could be determined are also shown in oceanic areas.

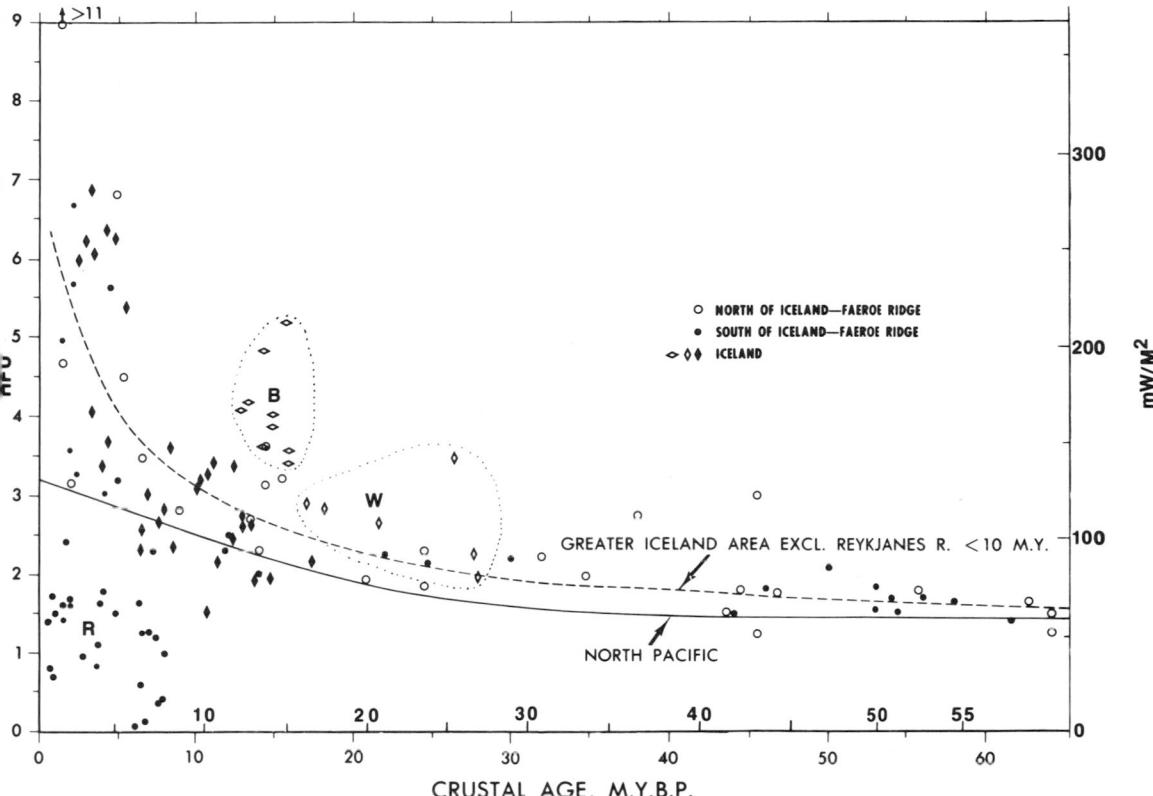

ig. 34.   Heat flow as a function of crustal age, in GNSA oceanic areas, including the Reykjanes Ridge and Iceland. From
angseth and Zielinski (1974), with additions from Lachenbruch and Marshall (1968), Talwani *et al.* (1971), Horai *et al.*
 970), and Pálmason *et al.* (1979). R, Reykjanes Ridge; B and W, Breidafjördur and western North Iceland heat flow highs
Pálmason *et al.*, 1979). "Crustal ages" for Iceland heat flow values estimated from spreading rate history on Reykjanes and
olbeinsey Ridges (Fig. 28; Vogt *et al.*, 1980). Crustal age scales after La Brecque *et al.* (1977) (top) and Heirtzler *et al.*
 968) (bottom).

The immediate spreading axis is generally devoid of sediments, precluding the measurement of heat flow. One exception is the southern part of
Knipovich Ridge (Figs. 19 and 20), whose rift valley is being filled by sediments in the manner of the Escanaba Trough at the southern end of the Gorda
Ridge. Three stations on the margins of Knipovich Rift Valley gave values of
3.13, 4.69, and >11 HFU (Langseth and Zielinski, 1974).

Heat flow has also been measured in boreholes from 500 to 2820 m deep
along the subaerial rift on Iceland (Pálmason *et al.*, 1979). The relative
importance of volcanism, hydrothermal activity, and heat conduction was estimated by Pálmason, (1973, 1974*a*): Total heat flow is about evenly divided
among these three processes. Total overall output for Iceland is 4000 MW
hydrothermal and 6000 MW extrusive (Pálmason *et al.*, 1979). Hydrothermal

activity, which extends to at least 3 km depth, is actually localized in about 17 high-temperature thermal areas, most of them associated with silicic volcanic centers (Arnorsson, 1974). This averages one thermal area per 30 km length of rift zone. Heat flow values for Iceland (Figs. 33 and 34) may be relatively too high compared to marine measurements, owing to neglect of basalt erosion and a possibly too-high conductivity used for the Iceland borehole data (Pálmason, 1973, 1974*a*). High heat flow in western North Iceland and around Breidafjördur Bay (Fig. 34) may reflect previously unsuspected extinct spreading axes (Pálmason *et al.*, 1979).

If there is a mantle hot spot under the Iceland area, one might expect abnormally high heat flow, all else being equal. Langseth and Zielinski (1974) showed that at least for ocean crust older than 10 m.y., GNSA heat flow averages about 0.2–0.5 HFU above that reported by Sclater and Francheteau (1970) for North Pacific crust of the same age (Fig. 34). Subsequently, Zielinski (1977) argued that the North Pacific curve is *below* normal because of hydrothermal heat loss, and that the GNSA heat flow is approximately normal. Although the regional topographic anomaly (Figs. 2 and 3) is ascribed to thermal expansion and abnormal warmth in the uppermost 200–400 km, the increased thermal gradient (about 1°C/km per 1 km topographic anomaly) will produce a regional heat-flow anomaly of 0.2 HFU at most, too low to be resolved from the present data base (Zielinski, 1977).

There are probably at least two extinct spreading axes between Iceland and the Jan Mayen F.Z. (Fig. 27). The Aegir Ridge in the Norway Basin exhibits values of about 2.0–2.5 HFU, about as expected for a ridge that ceased spreading around 30 m.y.b.p. Even higher values (3.2 and 3.7 HFU) were measured not far west of a supposed younger extinct axis located on the Iceland Plateau (Grønlie *et al.*, 1978); the existence of this ridge has been questioned, however (Vogt *et al.*, 1980). Yet another extinct axis (Ran Ridge), between Greenland and Canada, exhibits values in the range 1.16–1.54 HFU (Pye and Hyndman, 1972), rather low for a ridge that stopped spreading some time during the period 30–50 m.y.b.p. However, the high sedimentation rates in Baffin Bay and the Labrador Sea are not accurately known and the heat-flow measurements may therefore not have been accurately corrected for this effect. There is some indication that heat flow is about 0.2–0.3 HFU higher at the immediate extinct axis in Baffin Bay and the Labrador Sea, but this too may be an artifact of sedimentation (Pye and Hyndman, 1972).

Rapid and irregular sedimentation patterns and fluctuating bottom-water temperatures make the measurement of heat flow in the shallower marginal areas difficult. A transect across the Vøring Plateau Escarpment at 6°W (10 stations; Zielinski, 1977; Hänel, 1974) suggests an average heat flow of 1.4 HFU; any marginal heat-flow anomaly lies within measurement scatter (Zielinski, 1977). Two low values along the southeastern margin of the Norway

Basin (Fig. 33) are unreliable. Borehole measurements yielded values of 1.19–1.48 HFU for the North Sea (Harper, 1971). On land, heat flow measurements are most abundant in Norway, where 65 values give a mean of 1.0 ± 0.21 HFU. Weak heat flow highs occur along the eastern Caledonian Belt and in the Oslo Graben (Grønlie et al., 1977; Hänel et al., 1974.

E. Seismicity

1. General Results

The seismicity of the Greenland–Norwegian Sea area has been summarized by Husebye et al. (1975; see also Heier, 1977); compilations of historical events are given in Husebye et al. (1976) and Karnik (1971) (northwestern Europe), Keen et al. (1972) (northeastern Canada, Baffin Bay, and Labrador Sea) and Tryggvason (1973) (Iceland).

Local station networks have been utilized to investigate details of seismicity—mainly microearthquakes of magnitude 3 and less in Iceland (Ward and Björnsson, 1971; Klein et al., 1973, 1977, Björnsson and Einarsson, 1974), the Norwegian Channel (Båth, 1972), and Svalbard (Mitchell et al., 1979).

Other provinces, studied primarily with teleseismic data, include the present accretion axis north of the Azores triple junction (Sykes, 1967, 1970; Francis, 1973; Vogt and Johnson, 1973b, 1975; Vogt, 1974; Husebye et al., 1975; Einarsson, 1979) and intraplate seismicity (Sykes and Sbar, 1974; Sykes, 1978; Husebye et al., 1975; Vogt, 1978).

Some 529 relatively well-located (20 km or better) earthquakes occurred in the GNSA and its margins during the period 1955–1972. Of these, 15 events yielded fault plane solutions (Horsfield and Maton, 1970, Sykes and Sbar, 1974, Lazareva and Misharina, 1965, and others) and are tabulated in Husebye et al. (1975). Of those occurring along the MOR, all but one normal mechanism were strike-slip motions associated with transform faulting. Other fault-plane solutions have been published by Bungum (1977), Bungum and Husebye (1977), Einarsson (1979), and Mitchell et al. (1979).

The 529 events studied by Husebye et al. (1975) ranged between 3.5 and 6.0 in body-wave magnitude. The lower limit is determined by network sensitivity to teleseismic events. The worldwide seismic network is relatively sensitive to events in the northeast Atlantic, especially the Reykjanes Ridge (Francis, 1973). The rarity or absence of large-magnitude events in the GNSA is typical of relatively young oceanic lithosphere away from subduction zones. The most intense historical events, of estimated magnitude 7–8, occurred in or just offshore Iceland in 1755, 1784, 1896, and 1910. All 20 events of M ≥ 6.0 were restricted to the Tjörnes and Reykjanes "Fracture Zones" (Tryggvason, 1973).

Most or all of the GNSA earthquakes are shallow-focus, i.e., less than 70 km deep. If the axial lithosphere is only about 5 km thick (Vogt *et al.*, 1969), hypocenters along the spreading axis itself should be no deeper than this. Indeed, along the Reykjanes Peninsula spreading axis, most foci occur at 2–5 km depth and to date no earthquake has been located in seismic layer IV (V_p = 7.2 km/sec) (Björnsson and Einarsson, 1974). During the Heimaey eruption of 1973, earthquakes occurred at depths of 20–30 km, well within the mantle. This is not unreasonable, since the Vestmann Islands volcanoes appear to have developed very recently in preexisting 10-m.y.-old lithosphere produced at the Reykjanes accretion axis. The near absence of microearthquakes at depths less than 1.5–2 km (Klein *et al.*, 1973) may mean that numerous fissures, depriving the upper crust of any bulk strength, extend approximately to that depth before sealing under lithostatic pressure.

2. *Seismicity along the MOR, Including Iceland*

The increasing accuracy of epicenter location available by 1965 showed a relatively sharply defined, narrow band of seismicity that revealed the continuity of the MOR axis from the Azores northward through Iceland and along the axis of the Eurasia Basin (Sykes, 1965). The jogs of the seismic zone along fracture zones helped confirm Wilson's transform fault model (Wilson, 1965a), as did the fault-plane solutions of Sykes (1967), Horsfield and Maton (1970), and other workers. First motions of seismic waves intercepted by the teleseismic array were consistent with the model of normal faulting at spreading centers and strike-slip faulting along transform fractures.

A more complicated picture is indicated by earthquakes with nonorthogonal nodal planes on the southern Reykjanes Ridge, and two focal mechanisms with a component of thrusting, south of the Charlie Gibbs F.Z. (Einarsson, 1979). The latter could result from a regional compressive stress field caused by mantle plumes under Iceland and the Azores.

Despite its irregular trace through the GNSA, the MOR seismic zone seems to define a single, narrow plate boundary. Only in Iceland do two parallel accretion axes suggest the existence of one or perhaps two "microplates." The E–W-trending South Iceland Seismic Zone would then describe the motion between a South Iceland subplate—located between the two Neo-Volcanic Zones—and the Eurasia plate. Such a model might explain why the South Iceland Seismic Zone does not parallel the predicted Eurasia–America plate motion.

If the width of the seismic zone gives a measure of the width of the zone of crustal accretion, then this width must be variable through the greater GNSA area. The belt of microseismicity is only 2 km wide where the Reykjanes accretion axis enters Iceland. On morphologic grounds this width may apply to Kolbeinsey and northern Reykjanes Ridges as well. The seismic belt probably broadens where wide rift valleys exist, e.g., along the Mohns, southern Reykjanes, and especially Knipovich Ridges. A loose pattern of intra-plate earthquakes extends eastward from Mohns and Knipovich Ridges (Husebye et al., 1975; Heier, 1977), making the definition of what is "plate-boundary" seismicity difficult in that region.

Although the South Iceland Seismic Zone connecting the two volcanic belts is only about 25 km wide, perhaps consistent with the idea of a single transform fault, the postulated "Tjörnes Fracture Zone" north of Iceland has an epicenter belt at least three times wider (Björnsson and Einarsson, 1974; Fig. 31). According to Einarsson (1979), the transform motion is taken up by at least two parallel strike-slip faults.

On Iceland, studies of historical seismicity (Tryggvason, 1973) and microseismicity (Ward and Björnsson, 1971; Klein et al., 1973; Björnsson and Einarsson, 1974) corroborate and possibly extend studies of the submerged MOR. Intense earthquakes (magnitude 6.5 and greater) seem to be restricted to the two east–west seismic belts, probably transform fractures. Intense earthquakes are relatively more common (higher "b" values) on oceanic transform faults than along spreading centers (Sykes, 1970; Francis, 1973).

In both "ridge" and "fracture" provinces, a number of earthquakes may occur in a localized area within a relatively short time, on the order of 1–8 days. However, along fracture zones there is typically one major event followed by lesser aftershocks. In the east-trending South Iceland Seismic Zone, possibly a fracture zone, earthquakes sometimes begin at one end of the zone, with activity migrating along strike until in a period of days to many weeks the entire zone is activated. A similar migration of earthquake activity has been noted in conjunction with some volcanic eruptions of the volcano Hekla, located at the eastern end of the South Iceland Seismic Zone.

Generally, earthquake sequences along spreading centers are of the "swarm" type, i.e., without one outstanding main shock. The Tjörnes "Fracture Zone" seems to be an exception, since swarms occur there also (Björnsson and Einarsson, 1974). Sykes's (1970) worldwide treatment of teleseismic swarm activity along the MOR includes four earthquake sequences along the Charlie Gibbs F.Z. and Reykjanes Ridge, and four on Kolbeinsey and Mohns ridges. The sequences contained 5–31 events, mostly between magnitudes 3.5 and 5.5, and lasted on the order of 1–6 days. One sequence on Spar F.Z. (Figs. 1, 26, and 27) and one on the Charlie Gibbs F.Z. were probably of the foreshock–aftershock variety characteristic of fracture zones. The remainder were swarms. On Iceland, swarms occur on Reykjanes Peninsula and several other locales. Much of the short-wavelength "spikiness" observed on longitudinal profiles of seismicity along the MOR (Francis, 1973; Vogt and Johnson, 1973b; Vogt, 1974; Fig. 4) reflects the occurrence of swarms. It seems likely that most earthquake swarms are caused by magmatic, volcanic, or hydrothermal activity (Sykes, 1970). On Iceland at least, swarms may precede volcanism (Björnsson and Einarsson, 1974).

Reconnaissance surveys of microearthquake activity (magnitudes of 3 to −1 and less) on Iceland indicate a geographic correlation with zones in which swarms of larger shocks have occurred. There is also a good correlation with geothermal areas (Ward and Björnsson, 1971; Klein *et al.*, 1973). Microseismic activity was observed at 9 of about 16 high-temperature areas in Iceland. Nonseismic geothermal areas seem to be characterized by an absence of fissures and are structurally related to acidic intrusions.

Microearthquake activity is rather constant in most geothermal fields but sporadic elsewhere. In the constant areas the daily level of activity is on the order of 3–30 events of magnitude greater than −1 (Björnsson and Einarsson, 1974). However, in their Reykjanes Peninsula study area, Klein *et al.* (1973) found that average levels of seismicity can vary by two orders of magnitude in a few days.

Although Reykjanes Peninsula seems transitional between Iceland and the submerged MOR, it is effectively a very oblique spreading axis, consisting of en echelon N45°E-trending fissures making up the N75°E-trending peninsula and its associated seismic zone. The latter trend makes an angle of only 25° with the direction of relative plate motion. Both normal and strike-slip earthquake solutions occurred, often in close proximity.

The North Iceland rifting episode that began in 1975 is characterized by short pulses of rifting at intervals of several months. The pulses are accompanied by earthquake swarms and occasionally volcanism (Björnsson *et al.*, 1979).

Earthquake focal mechanism solutions in the axial rift zone of Iceland indicate a *minimum* horizontal compression in a direction perpendicular to the main rift trend. By contrast, on the flanks of the rift zone stress measurements by overcoring and hydrofracturing methods show a maximum compressive stress in that direction, from the surface at least to 375 m depth (Pálmason *et al.*, 1979).

Systematic regional variations in earthquake frequency seem to characterize the GNSA spreading axis (Fig. 4). Earthquakes are less common within about 900 km of Iceland, i.e., approximately the length of ridge axis devoid of prominent rift valleys and rough flank topography (Vogt and Johnson, 1973*b*; Francis, 1973). The normal faulting needed to maintain this tectonic "graben" structure qualitatively explains why high seismicity correlates with a rift valley. It may be the relatively hot crystal slush, flowing in a bidirectional pipe below the spreading axis away from the Iceland plume, that keeps the axial lithosphere thin and hot and thus unable to support a deep, wide rift valley (Vogt *et al.*, 1969; Vogt, 1974).

3. "Midplate" Seismicity

An unusual and poorly understood feature of the GNSA is the relatively high level of "midplate" seismicity, i.e., earthquakes not associated with the spreading axis and active transform faults. Of the 529 GNSA epicenters (1955–1972), about 94 events (18%) were located more than 200 km from the active plate boundary (Husebye *et al.*, 1975). Midplate seismicity also occurs in the Labrador Sea, Baffin Bay, and adjacent parts of Arctic Canada (Keen *et al.*, 1972; Sykes, 1978). When these two areas are considered together (Vogt, 1978) the midplate seismicity forms a loose "halo" surrounding Iceland and approximately correlative with the flanks of the regional gravity high (Fig. 30). The relatively high seismicity on Mohns and southern Reykjanes Ridges, noted earlier, occurs where this halo intersects the MOR. The seismic "halo," if real, may reflect stresses exerted by radial outflow of asthenosphere from the Iceland plume. However, midplate mechanism solutions are generally "strike-slip" or "thrust" (Sykes and Sbar, 1974; Lazareva and Misharina, 1965; Mitchell *et al.*, 1979); no regional pattern is evident, except that the scarcity of normal solutions suggests only crust younger than about 20 m.y. is under tension (Sykes and Sbar, 1974). Midplate events do not correlate convincingly with known crustal structures and probably do not have a single cause (Husebye *et al.*, 1975; 1977). However, the two areas of enhanced seismicity along the Norwegian margin lie seaward of the greatest elevations (Figs. 4 and 30; Vogt and Johnson, 1975; Kvale, 1960) perhaps indicating ongoing relative uplift caused by glacial rebound or differential cooling. Seismicity also appears concentrated along the shelf edge from south-central Norway to the Barents Sea (Husebye *et al.*, 1977), perhaps in response to differential sediment loading. Inland over Fennoscandia, seismic activity is mostly confined to the Telemark–Vänern, Bothnian, and Lappland zones. The stress field is probably complex, the cumulative effect of ancient orogenic/tectonic events, such as the Caledonian cycle and the Oslo Rift, plus more recent events such as Tertiary sea-floor spreading and Plio-Pleistocene glaciation. Added to this would be stresses caused by the present driving forces of plate motion.

Intraplate seismicity extends eastward from the Knipovich Ridge plate boundary (Austegard, 1976). The small Heerland seismic zone along the Storfjorden coast (southeast Vestspitsbergen) trends east–west, transverse to mapped structures; sinistral strike-slip motion along a vertical fault plane is indicated (Mitchell *et al.*, 1979).

V. IGNEOUS GEOCHEMISTRY AND THE ICELAND MANTLE PLUME

The chemical and mineralogical makeup of the GNSA ocean crust is poorly known except on Iceland (e.g., Sigvaldason, 1974; Bailey and Noe-

Nygaard, 1976; Jakobsson, 1972; Schilling *et al.*, 1978; Sigurdsson *et al.*, 1978), along the axis of the MOR—particularly Reykjanes (Schilling, 1973*a,b*) and Kolbeinsey Ridges—and in parts of the early Tertiary igneous provinces of east Greenland, the Faeroes, and the British Isles (Brooks and Jakobsson, 1974). Deep drilling has added a few samples from older crust (Schilling, 1976) whose sediment overburden, ice-rafted erratics, and weathered condition generally limits the value of sea-floor dredging operations. Except on Iceland, only the top of the crust (Layer IIA) has been sampled. Several drillholes on Iceland have penetrated more than 1000 m (up to 2820 m), into the 4–5 km/sec basalt layer. None of the holes has reached the third layer (Pálmason *et al.*, 1979). The composition of the deep crust can only be inferred from seismic data (Christensen, 1974). On Tertiary Iceland, erosion has been sufficient to expose basalts once buried 1–2 km (Walker, 1974); thus, merely sampling the surface is equivalent to deep drilling.

Chemical analyses of the sampled rocks suggest subdivision into three basic basalt types: (1) Mid-Oceanic Ridge basalts, (2) LIL-enriched basalts, and (3) alkali basalts.

1.   *Mid-Oceanic Ridge basalts (MORB)* are very depleted in "large ionic lithophile" (LIL) elements such as K, Ti, P, Rb, Cs, Sr, Ba, and Zr and the light rare earths (e.g., La). Such rocks are also low in radiogenic Sr and Pb. This basalt type is believed to originate by extensive ($\sim$30%) partial melting of an ultrabasic mantle already depleted in LIL elements by previous partial melting episodes. MORBs are commonly sampled along the Mid-Atlantic and Reykjanes Ridges as far north as 60°N. MORB-like basalts are rare—but not absent—on Iceland (Fig. 4), and MORB becomes the predominant rock type once again on Kolbeinsey Ridge (including Kolbeinsey Island) immediately north of the Tjörnes F.Z. (Brooks and Jakobsson, 1974), on the Iceland Plateau, and in the Norway Basin (Schilling, 1976). The MORBs north and south of Iceland appear even more depleted than do normal MOR basalts (Fig. 4), perhaps because they are generated by outward-flowing ultradepleted mantle stripped of lithophile elements under Iceland.

2.   *Icelandic basalts*, rather like those generated by some other hot spots, tend to be enriched in Fe, Ti, P, Sr, K, and the light rare earths, and depleted in Al and Mg relative to MORBs (Fig. 4). The Iceland "zero-age" LILE anomaly appears to peak (with large scatter!) around Kverkfjöll in the central part of the eastern Neo-Volcanic Zone (Sigvaldason and Steinthorson, 1974). Enriched basalts also occur along the plate boundary on both sides of the Jan Mayen F.Z. (Dittmer *et al.*, 1975), perhaps a reflection of a Jan Mayen hot spot, and for unknown reasons at DSDP Site 344 on Knipovich Ridge (Schilling, 1976). The observation that "hot spot" basalts are enriched to varying degrees in radiogenic Sr and Pb (e.g., Hart *et al.*, 1973), and in the light rare earths, led Schilling (1973*a,b*) to postulate a chemically distinctive primary hot

mantle plume (PHMP) source from which the Icelandic basalts are derived by partial melting. Subsequent studies of Ge-Si, Ga/Al (de Argollo and Schilling, 1978), Cl and Br (Unni and Schilling, 1978) tend to confirm this model. The two volatiles also reveal degassing effects at water depths of less than 500 m.

In the case of the Iceland hot spot, the postulated "PHMP" source must already have existed in the early Tertiary to explain the composition of the lower Faeroes and Vøring Plateau basalts (Schilling and Noe-Nygaard, 1974; Bollingberg et al., 1975; Schilling, 1976). The LIL-enriched basalt chemistry near Iceland grades smoothly toward MORB values along the northern Reykjanes Ridge. In the Schilling "binary mixing" model (see also Vogt, 1971, 1974; Brooks and Jakobsson, 1974), the enriched parent mantle flows southwestward under the spreading axis, gradually mixes with normal depleted mantle, and is used up in the accretion process. By contrast, O'Hara (1973, 1975) proposed derivation of all the magmas from a single parent by different degrees of fractional crystallization. O'Nions and Pankhurst (1973) postulated and later rejected (O'Nions et al., 1976) disequilibrium partial melting from a single mantle parent. They now favor a chemically heterogeneous mantle but rule out simple binary mixing. Sigvaldason's (1974) plume-type model involves partial melting over a substantial depth range under Iceland. Asthenosphere flowing outward along the MOR would melt at restricted but shallow depths to yield depleted MORB.

3. *Alkali basalt* is the third, volumetrically least significant, and most variable rock type. Such basalts are even more LIL-enriched than the Icelandic type but have more Al and less Si than do either of the other major categories. Alkali basalts are believed to originate as very low-degree partial melts at great depths. This hypothesis is confirmed by their occurrence at some distance from the spreading axis, i.e., at sites where the substantial plate thickness perhaps tends to keep partial melting at greater depths. Alkali basalts in the greater Iceland area occur in the Vestmannayjar and Snaefellsnes zones of Iceland, on Jan Mayen (Weigand, 1972), in Kong Oscars Fjord and Kangerdlugssuaq in East Greenland, and in the Hebridean province (Brooks and Jakobsson, 1974).

VI. EARLY TERTIARY VOLCANISM AND POSSIBLE ANTECEDENTS

Iceland's oldest outcrops (12–16 m.y.b.p.; Moorbath et al., 1968; Pálmason and Saemundsson, 1974) occur near the eastern and northwestern coasts. The Iceland–Faeroe Ridge (Talwani and Udintsev, 1976) and the Greenland–Iceland Ridge are formed of still older basalts, linking Iceland to the early Tertiary igneous rocks widely exposed on the adjacent continents.

Sea-floor spreading in the Norwegian–Greenland Sea area began to the accompaniment of widespread and relatively short-lived Paleocene to Eocene volcanism in an area extending from Baffin Island and west Greenland to east Greenland, the Faeroes, Rockall, and northern Ireland (Neo-Nygaard, 1974; Fig. 22). Much of the area of original subaerial vulcanism lies subsided below the continental margins or was removed entirely by erosion. This Thulean volcanic province may be divided into more or less isolated subprovinces; e.g., Baffin Island/West Greenland, two areas of East Greenland—possibly once contiguous with the Faeroes and Vøring Plateau, and the Hebridian province of Scotland. Volcanism apparently did not extend in a continuous belt across Greenland from east to west; for example, the East Greenland basalts thin from 9 km on the coast (stratigraphic thickness) to zero ~250 km inland (Soper *et al.*, 1975).

Extensive areas of Tertiary volcanic rocks occur in northwest Britain, in the Antrim Plateau of northern Ireland, and on the islands of Lundy, St. Kilda, and Rockall (Fitch *et al.*, 1974). Volcanism in the British Tertiary Province is thought to have begun at or shortly before 66 m.y.b.p., in earliest Paleocene time (Evans *et al.*, 1973). Tuffs of probable lower Thanetian (57 m.y.) to mid-Lower Eocene age occur in the North Sea area, with a peak in the Upper Paleocene (Jacque and Thouvenin, 1975). Ash layers 52–53 m.y. old in northern Denmark suggest sources in the Skagerrak (Pedersen *et al.*, 1975). The apparent ages of the major plutonic centers lie between 60 and 58 m.y., but the plutonic activity continued at various localities until at least 50 m.y., in the Eocene. Igenous activity on the Isle of Mull lasted from about 63 to 53 m.y.b.p. (Beckinsale, 1974).

Four main stages of igneous activity have been recognized by many—although not all—investigators of the British Tertiary province (Noe-Nygaard, 1974): (1) voluminous outpourings of plateau lavas, (2) establishment of central vents and agglomerate formation, (3) intrusion of plutonic and hypabyssal rocks at a number of centers, and (4) intrusion of regional dike swarms.

The Faeroe Islands are composed of plateau lavas extruded in the same period (Paleocene–Eocene), with apparent radiometric ages ranging from 29 to 62 m.y.b.p. (Noe-Nyggard, 1974; Tarling and Gale, 1968). In East Greenland (Noe-Nygaard, 1976; Deer, 1976; Surlyk *et al.*, 1980) the largest area of plateau basalts was extruded from 55 to 66 m.y.b.p. in Scoresby Sund (Beckinsale *et al.*, 1970). Ages of 52–74 m.y.b.p. (Soper *et al.*, 1975; Hailwood *et al.*, 1973, Tarling and Mitchell, 1976) were questioned by Brooks and Gleadow (1977), who report 55 m.y. for the Skaergaard intrusion and 55–60 m.y. for the surrounding basalts. The earliest lavas flowed from an easterly source into a shallow submarine environment (Surlyk *et al.*, 1980). A volcanic landmass then formed and the extrusion surface remained subaerial until near the end of the plateau basalt phase. Among the intrusive centers associated with the plateau basalts, the gabbro intrusions are older, and the syenitic and alkali granite complexes younger than the East Greenland coastal flexure, which is of post-Middle Eocene age. Except for a few dolerite dikes, the silicic plutons (latest dates, 50 m.y.b.p.) mark the end of igneous activity there. Local occurrences of fossiliferous sediments bracketing the East Greenland basalts (Soper *et al.*, 1975) and within the Faeroes basalts (Noe-Nygaard, 1974) generally confirm the radiometric dating as primarily Upper Paleocene to Lower Eocene.

As the spreading axis migrated away from the coast, post-igneous faulting and seaward fault-block rotation occurred between 66.5° and 70°N (Surlyk *et al.*, 1980). To the north, still more recent (Oligocene–Lower Miocene) normal faulting may be related to the separation of the Jan Mayen Ridge block from Greenland. Uplift of the coastal area (1–2 km; as much as 6 km around Kangerdlugssuaq) occurred at the same time, accompanied by 3- to 6-km offshore subsidence (see also Brooks, 1973; Hinz and Schlüter, 1978*b*).

In North Greenland, ages of 72.9 ± 9.0 and 66.0 ± 6.6 m.y. (late Cretaceous?) were reported for dolerite dikes at 82°N in southern Peary Land (Dawes, 1976).

In the West Greenland (Svartenhuk-Disko Island)–Baffin Island volcanic province, basalt dates range from 50 to 70 m.y.b.p. (mostly 56–63 m.y.b.p.; Clarke and Pedersen, 1976). Beckinsale *et al.* (1974) concluded the plateau basalts were extruded prior to 65 m.y.b.p. The province includes a minor basalt swarm in the Godhabsfjord area at 64°N (LePichon *et al.*, 1971; Fitch *et al.*, 1974). On Disko Island, a flow dated by Deutsch and Kristjansson (1974) has an age of 70 ± 4 m.y. Centrally located in the west Greenland basalt province is one locality (Nugssuaq) where the beginning of volcanism can be dated by marine fossils: Tuffs were first deposited in the Kangilia Formation of late Danian age (NP 3, or 60–63 m.y.b.p.) (Henderson *et al.*, 1976; Clarke and Pedersen, 1976). This is the earliest stratigraphically controlled evidence for "Thulean" volcanism. The tuffs were followed by initially submarine basalts whose total stratigraphic thickness may reach 10 km (Rosenkrantz and Pulvertaft, 1969). Perhaps as a prelude to the volcanism, the area of crustal subsidence had expanded in late Cretaceous/early Paleocene time. The principal West Greenland igneous province is thought to extend offshore from 73.3° to 68°N (Keen and Clarke, 1974; Clarke, 1975). On Baffin Island, the Cape Dyer basalts reach only 500 m in thickness; a single K–Ar age of 58 ± 2 m.y. has been reported (Clarke and Upton, 1971). Basalts west of Greenland are more picritic than east of Greenland, whereas the latter are associated with more syenitic plutons and coast-parallel dike swarms.

South of the British Tertiary province, extensive volcanic activity occurred about 60 m.y.b.p. on the Rockall–Hatton Banks, just south of the Charlie Gibbs F.Z., and on the west coast of Portugal (Fig. 22; Vogt and Avery, 1974*b*).

The submerged Faeroe–Shetland and Vøring Plateau Escarpments, and possibly similar features under the Greenland margin (Talwani and Eldholm, 1977), may be constructional volcanic lineaments, the submarine equivalents of the Scoresby Sund or Faeroes basalts, but extruded along the line of initial rifting (Fig. 22). Drill Site 343 on the Vøring Outer Ridge gave a basement age of early Eocene.

The almost entirely reversed magnetic polarity of early Tertiary Thulean basalts supports the concept of coeval, widespread effusion during a limited period of time (Tarling and Mitchell, 1976; Hailwood *et al.*, 1973; Deutsch and Kristjansson, 1974; Soper *et al.*, 1975; Athavale and Sharma, 1975). Perhaps this was the 2-m.y.-long reversed period just prior to anomaly 24, i.e., between 56–60 and 58–67 m.y.b.p. on the time scale of La Brecque *et al.* (1977). Thus maximum magmatism coincided or immediately preceded spreading, which also began during this reversed period, in the late Paleocene. However, the beginning of Thulean volcanism, at least in West Greenland, was about 5 m.y. earlier (Upper Danian; Clarke and Pedersen, 1976). Therefore, there is some suggestion that activity occurred earlier in West Greenland (prior to 65 m.y.b.p.) than in the Faeroes–East Greenland area (Beckinsale *et al.*, 1970, 1974). Volcanism in the British Tertiary province may have occurred in two pulses (Soper *et al.*, 1975), thereby complicating the correlation with initial spreading.

The nature of the mantle disturbance that initiated magmatism (and spreading) about 55–60 m.y.b.p. is conjectural. Perhaps there was a family of isolated diapirs (blobs) rising simultaneously from a broad hot spot (Fig. 29), such as postulated by Talwani and Eldholm (1977) and Cochran and Talwani (1978). However, contrary to those authors, the present "Iceland phenomenon" is sharply peaked (Fig. 4). North of the Iceland–Faeroe Ridge, lines of preexisting Mesozoic basins may have constituted zones of weakness that permitted magmatism to penetrate the lithosphere. However, the southeast Greenland–Rockall/Hatton Bank area had been a stable shield prior to the early Tertiary (Fig. 22). Perhaps the initial rift was controlled by fracture patterns such as those mapped in Greenland and Norway from satellite imagery (Frost and Dikken, 1977). Some of the igneous subprovinces may also relate to triple junctions among small plates (Burke and Dewey, 1973).

Did the Thulean igneous provinces first develop during the earliest Tertiary, or did they already exist in subdued form during the Mesozoic? As yet there are only scattered suggestions of such earlier activity, and no assurances that these were ancestral to the Thulean episode. Roberts *et al.* (1974) recovered microgabbros of Upper Cretaceous age (83 ± 3 m.y.)

immediately east of the early Tertiary granite (52 ± 9 m.y.) of Rockall Island; similarly, Upper Maestrichtian chalk, overlying volcanic rocks, was dredged from the Anton Dohrn Kuppe, a guyot situated in the central part of Rockall Trough (Jones *et al.*, 1974).

Basalts from Skåne in southern Sweden were once believed to be Tertiary. However, K/Ar ages range from 79 m.y.b.p. to around 150 m.y.b.p. The most probable age of this volcanism is 100–150 m.y.b.p. (Printzlau and Larsen, 1972).

Scattered early Cretaceous and Jurassic volcanic occurrences include a 134-m.y. phonolite and Bathonian tuff (160 m.y.b.p.) in southern England (Roberts, 1974); tholeitic basalts of Upper Jurassic and Lower Cretaceous age are also found in Svalbard, northeast Canada, and southwest Greenland (Noe-Nygaard, 1974). A thick Middle Jurassic volcanic sequence, consisting of as much as 740 m alkali-olivine basalt, was encountered in the Forties and Piper areas of the North Sea (Gibb and Kanaris-Sotiriou, 1976).

Profound volcanism and graben formation occurred in the North Sea area in the Permo-Carboniferous. Permian faulting, tholeiitic igneous activity, and nonmarine sediment deposition followed the extrusion of Carboniferous alkalic basalts in the Midland Valley of Scotland. Similar Permian faulting and alkaline igneous activity affected the Oslo graben area (Ramberg, 1976). Both occurrences have been linked to mantle hot spots (Burke and Dewey, 1973).

VII. DIRECT MEASUREMENT OF CRUSTAL MOTION ON ICELAND

Direct verification of continental motion was already contemplated by Alfred Wegener in 1928. Although we now know the kinematic history of drift rather well, magnetic anomalies and deep-sea drilling only yield long-term averages of spreading rate; short term fluctuations in plate motion, over time periods of 10^5 years and less, still cannot be resolved and direct measurement is desirable. Since Iceland is the only midoceanic outcrop of a spreading axis, there is considerable interest in direct measurement of horizontal distances across the Neo-Volcanic Zones, which must accommodate most or all of the "current" extension. In south Iceland, plate tectonics dictates that if there is extension across both Neo-Volcanic Zones, their sum should equal 2 cm/yr, provided the million-year average rate also applies for a few years.

Horizontal distance measurements were begun in 1938 by O. Niemczyk and colleagues in the Myvatn area (Gjastikki) of the Neo-Volcanic Zone in northeast Iceland (Gerke, 1974). The network west of Lake Myvatn revealed a total expansion of 6.5 cm between 1965 and 1971, or about 1 cm/yr. The Gjastikki network showed an apparent shortening of 44.5 cm in 27 years (1938–1965) followed by a minimally significant expansion of 5 cm over the following 6 years. The Thingvellir network showed no significant changes, 1965–1971 (Gerke, 1974). British work by R. Mason and his team (Brander *et al.*, 1976) consisted of networks at Myvatn and Thingvellir, near the volcano Hekla in the eastern Neo-Volcanic Zone, and at Reykjanesviti at the southwestern tip of Reykjanes Peninsula. The measurements, carried out during the period 1968–1972, showed no significant changes at the Myvatn and Hekla sites. Total expansion rates of 0.25 mm/yr across the Thingvellir Graben and 0.9 mm/yr on Reykjanes Peninsula were inferred. Other distance measurements (1967–1970) were carried out at Thingvellir Graben and northeast of Hekla (Decker *et al.*, 1971). The lack of significant changes in the first area limits motion to ±0.5 cm/yr at most. Extension at Hekla (6.5 ± 3.2 cm) occurred over a 3-yr period including the 1970 eruption. Resurveying in 1973 showed no significant subsequent changes (Brander *et al.*, 1976).

A rifting episode started in 1975 on the accreting plate boundary in North Iceland after 100 years of quiescence (Björnsson *et al.*, 1977, 1979). Horizontal extension of some 3 m has been observed in the Krafla caldera (Fig. 31) and the associated 80-km-long fissure swarm. The rifting occurs periodically in short active pulses at intervals of a few months. Between these active pulses, continuous inflation of 7–10 mm/day of the caldera is caused by 5 m^3/sec inflow of magma into a magma chamber at a 3-km depth. The active pulses are caused by a sudden east–west expansion of the fissure swarm and a contraction of zones outside the fissure swarm. Rapid flow of magma out of the magma chamber and into the fissures toward north or south is indicated. These pulses are accompanied by earthquake swarms, vertical ground movements of up to 2 m, and sometimes volcanic eruptions and formation of new fumaroles as well. The magma chamber below the Krafla caldera thus acts as a trigger for plate-margin activity in North Iceland. Of course, it is premature to conclude that accreting plate boundaries typically experience "Krafla episodes" every century or so at any particular site.

Measurement of *vertical* displacements of the Neo-Volcanic Zone with respect to adjacent "stable" regions is also of interest. Since the once-horizontal Tertiary flood basalts are now tilted 5°–15° (occasionally 20° to 40°), and faults with offsets up to 1000 m occur on Iceland, considerable vertical motion must occur (Tryggvason, 1974), perhaps sporadically as in the most recent episode. Boreholes drilled at the tip of Reykjanes Peninsula on "zero-age" crust and in Reykjavik on late Tertiary (or early Quaternary) crust reached 1730 and 2200 m, respectively, without penetrating below either subaerial or glacial/interglacial basalt flows. Precision leveling shows that the flanks of the rift zone are tilting toward the axis at 10^{-5}–10^{-6} rad/yr, indicating axial subsidence (2 cm/yr) of the same order as the spreading rate (Tryggvason, 1974; Gerke, 1974). That sinking rates of such magnitude have prevailed over the last few thousand years as well is suggested by the tilt of young lava flows the initial flow direction of which is known.

Thus, a given basalt parcel may sink more than 10 km in the rift province, then rebound 2 km in the following 10 m.y. (Schäfer, 1975) before thermal subsidence (Parker and Oldenburg, 1973; Sclater and Parsons, 1977) takes over as the dominant process. However, sinking in the rift axis during Neo-Volcanic time may have been rather higher than during Tertiary times. McDougall *et al.* (1976) estimated a lava accumlation rate of 690 m/m.y. for the period 9–10 m.y.b.p. If the subsidence rate had been any faster than 0.07 cm/yr, the land surface would soon have disappeared below the sea because lava extrusion could not have kept up with the subsidence!

The plate-tectonic meaning and "representativeness" of horizontal motions measured thus far (largely in the last two decades) is uncertain. Data collected prior to 1975 suggested different types of local motions, e.g., toward active geothermal wells (Brander *et al.*, 1976), near-active volcanoes (Decker *et al.*, 1971), or for other still unexplained reasons (Brander *et al.*, 1976; Gerke, 1974). Measured extensional motions across the Neo-Volcanic Zones were on the order of 0–1 cm/yr at various times during the last decade prior to 1975. However, it is not known how the total motion was distributed between the eastern and western Neo-Volcanic Zones, or whether there was any extension beyond these zones as well. Recent events (Björnsson *et al.*, 1977, 1979) suggest that the pre-1975 measurements in North Iceland may have been made during a time of more regional strain accumulation now being relieved in that area by local rifting.

VIII. CONCLUDING REMARKS

One or two decades of intensive exploration have revealed the northeast Atlantic, Iceland, and Greenland–Norwegian Basins as an unexcelled laboratory for the study of diverse geological processes.

Nowhere have the Pleistocene glaciations etched a finer climatic record (Kellogg, 1975; Ruddiman and Glover, 1975). The Reykjanes Ridge, for all its anomalous characteristics, was the first to exhibit an unambiguously symmetrical magnetic anomaly pattern (Vine, 1966) and perhaps the first to offer clues of mantle flow below the plates (Vogt, 1971, 1974). The same ridge was also the first to indicate the presence of a geochemical transition zone, with all its puzzling implications, between the LIL-enriched Icelandic basalts and the depleted "MORB" basalts of the Mid-Atlantic Ridge (Schilling, 1973a,b; Hart *et al.*, 1973; Sigvaldason, 1974; Brooks and Jakobsson, 1974). Iceland itself was probably the first segment of the MOR where crustal drift could be demonstrated (Bodvarsson and Walker, 1964).

The plate-boundary shoals from -3.5 km at 50°N toward a $+1$ km summit on Iceland and then descends to over 5-km depths in the Eurasia Basin (Fig. 4), thereby spanning the full range of MOR axial depths observed in the present world ocean. As a result, water-depth-(pressure-)dependent phenomena, such as basalt vesicularity and degassing, may be investigated as nowhere else (Moore and Schilling, 1973; Unni and Schilling, 1978). Anomalous though they are, the Icelandic spreading axes are the world's only mid-oceanic exposure, with unique opportunities for detailed refraction work (Pálmason, 1971; Angenheister *et al.*, 1979), geological mapping (Walker, 1974), calibration of the reversal time scale (Harrison *et al.*, 1979), precise distance measurement (Gerke, 1974), deep drilling (Pálmason *et al.*, 1979), or observation of crustal rifting "*in vivo*" (Björnsson *et al.*, 1977, 1979), to name just some examples. On the adjacent continents, the relationship among early Tertiary igneous activity, magnetic reversals, and initiation of spreading is nowhere so amenable to study (Larsen, 1978; Scrutton, 1973). The spreading process can be investigated under a wide range of water depths, obliquenesses, and morphological types. Spreading variously occurs below the Vatnajökull ice cap in central Iceland, subaerially, underwater, or under possibly as much as several kilometers of sediments (Figs. 4, 19, and 20). The continental margins include sheared, rifted, and intermediate types (Eldholm and Sundvor, 1979). The first break occurred within and/or along the margins of older sedimentary basins (Talwani and Eldholm, 1977) or—in the case of southeast Greenland and Rockall–Hatton Bank—within a stable Precambrian craton. In view of the regional extent of the anomalous "Iceland phenomenon" (Figs. 3 and 4; Vogt, 1974), care should be taken before extrapolating from the results of GNSA studies to other oceanic areas.

The knowledge explosion in the GNSA shows no sign of abating. Nevertheless, in areas such as bathymetry, gravity, magnetics, refraction studies of crustal structure, and ocean basin evolution by spreading, the future may well bring only refinements. In contrast, igneous geochemistry, "true" crustal structure, pre-late Pleistocene climatic history, precise distance measurement, and deeper mantle structure below the Iceland hot spot represent fields still in their infancy. Of these, the first three will advance only in relation to future deep drilling. Major unsolved questions include the following: What is the "primary" or "secondary" nature of the LIL enrichment on Iceland (e.g., Schilling, 1973*a,b* vs. Sigvaldason, 1974, or O'Hara, 1975)? Is there a mantle plume under Iceland, and from what mantle depths do the Iceland plume materials rise? Is the plume narrow (Morgan, 1972, 1973) or a broad zone of upwelling (Vogt, 1974; Vogt and Avery, 1974*a,b*; Talwani and Eldholm, 1977)? To what degree has the Iceland hot spot remained fixed in a mantle coordinate system (e.g., Minster *et al.*, 1974; Minster and Jordan 1978)? What caused the various axis shifts on Iceland and to the north, when otherwise the Mid-Atlantic Ridge has generally maintained a median position? What is the nature of the continent–ocean crustal boundary (e.g., Surlyk *et al.*, 1980; Eldholm *et al.*, 1979)? What makes oblique spreading on Reykjanes, Mohns, and Knipovich Ridges the preferred mode (Fig. 4) when it is so uncommon elsewhere? Why did the Iceland "plume" happen to develop below a string of preexisting Mesozoic basins? Finally, is the apparently fluctuating activity of the Iceland hot spot (Vogt, 1971, 1972*b*, 1974, 1979*a*) part of a global pattern, connected in some way to the causes of plate tectonics?

ACKNOWLEDGMENTS

We thank C. Fruik, E. L. Gladmon, B. Jolliffe, K. Kramer, J. Krause, D. Love, D. O'Neill, J. Peery, B. Schmitt, X. Toverud, and W. Worsley for help in preparing this chapter. W. Ruddiman, J. R. Phillips, M. Talwani, O. Eldholm, G. L. Johnson, and T. Kellogg kindly provided preprints or copies of illustrations. The authors are grateful to B. Hurdle for continuous support. The senior author benefited from review and comments by O. Eldholm, G. Grønlie, E. Sundvor, and J. Thiede.

REFERENCES

Aam, K., 1975*a*, Aeromagnetic basement complex mapping north of latitude 62°N, Norway, *Nor. Geol. Unders. Publ.*, v. 316, p. 351–353.

Aam, K., 1975*b*, Magnetic profiling over Svalbard and surrounding shelf areas, *Nor. Polarinst. Arbok*, v. 1973, p. 87–99.

Alvarez, W., Alvarez, L. W., Asaro, F., and Michel, H. V., 1980, Extraterrestrial cause for the Cretaceous–Tertiary extinction, *Science*, v. 208 (4448), p. 1095–1108.

Anderson, R. N., McKenzie, D., and Sclater, J. G., 1973, Gravity, bathymetry, and convection in the earth, *Earth Planet. Sci. Lett.*, v. 18, p. 391–407.

Angenheister, G., Gebrande, H., Miller, H., Weigel, W., Goldflam, P., Jacoby, W., Pálmason, G. G., Björnsson, S., Einarsson, P., Zverev, S., Loncarevic, B., and Solomon, S., 1979, First results from the Reykjanes Ridge Iceland Seismic Project 1977, *Nature (London)*, v. 279, pp. 56–60.

Anonymous, 1979, *General Bathymetric Chart of Oceans: Arctic, Sheet 5.17*, G. L. Johnson, D. Monahan, G. Grønlie, and L. Sobczak, eds., Ottawa, Canada: Canadian Hydrographic Office.

Aric, K., 1972, Der Krustenaufbau und die Tiefenstruktur des Reykjanes-Rückens nach Reflexions seismischen Messungen, *Dtsch. Hydrogr. Z. A*, v. 11, 42 p.

Arnorsson, S., 1974, The composition of thermal fluids in Iceland and geological features related to the thermal activity, in: *Geodynamics of Iceland and the North Atlantic Area*, Kristjansson, L., ed., Dordrecht, Holland: D. Reidel, p. 397–323.

Aronson, J. L., and Saemundsson, K., 1975, Relatively old basalts from structurally high areas in central Iceland, *Earth Planet. Sci. Lett.*, v. 28, p. 83–97.

Athavale, R. N., and Sharma, P. V., 1975, Paleomagnetic results on early Tertiary lava flows from West Greenland and their bearing on the evolution and history of the Baffin Bay–Labrador Sea region, *Can. J. Earth Sci.*, v. 12, p. 1–18.

Austegard, A., 1976, Earthquakes in the Svalbard area, *Norsk Polarinst. Årb.*, v. 1974, p. 84–99.

Avery, O. E., Burton, G. D., and Heirtzler, J. R., 1968, An aeromagnetic survey of the Norwegian Sea, *J. Geophys. Res.*, v. 73, p. 4583–4600.

Bailey, J. C., and Noe-Nygaard, A., 1976, Chemistry of Miocene plume tholeiites from northwest Iceland, *Lithos*, v. 9, p. 185–201.

Båth, M., 1960, Crustal structure of Iceland, *J. Geophys. Res.*, v. 65, p. 1793–1807.

Båth, M., 1972, Zum Stadium der Seismizität von Fennoskandia, *Gerlands Beitr. Geophys.*, v. 81, p. 213–226.

Båth, M., and Vogel, A., 1958, Surface waves from earthquakes in northern Atlantic–Arctic Ocean, *Geofis. Pura Appl.*, v. 39, p. 35–54.

Beblo, M., and Björnsson, A. J., 1978, Magnetotelluric investigation of the lower crust and upper mantle beneath Iceland, *J. Geophys.*, v. 45, p. 1–16.

Beckinsale, R. D., 1974, Rb-Sr and K-Ar age determinations, and oxygen isotopes for the Glen Cannel granophyre, Isle of Mull, Argyllshire, Scotland, *Earth Planet. Sci. Lett.*, v. 22, p. 267–274.

Beckinsale, R. D., Brooks, C. K., and Rex, D. C., 1970, K-Ar ages for the Tertiary of East Greenland, *Bull. Geol. Soc. Den.*, v. 20, p. 27–37.

Beckinsale, R. D., Thompson, R. N., and Durham, J. J., 1974, Petrogenetic significance of initial[87] Sr/[86] Sr ratios in the North Atlantic Tertiary igneous province in the light of Rb-Sr, K-Ar and ^{18}O-abundance studies of the Sarqata qaqa intrusive complex Ubekendt Ejland, West Greenland, *J. Petrol.*, v. 15, p. 528–538.

Beloussov, V. V., and Milanovsky, Ye. Ye., 1977, On tectonics and tectonic position of Iceland, *Tectonophysics*, v. 37, p. 25–40.

Beloussov, V. V., and Udintsev, G. B., eds., 1977, *Islandiya i sredinno-okeanicheskii behrebet: Stroenie dna okeana (Iceland and Mid-Oceanic Ridge: Structure of the Ocean Floor)* (in Russian), Moscow: Nauka, 204 p.

Berggren, W. A., 1972, Late Pliocene–Pleistocene glaciation, in: *Initial Reports of the Deep Sea Drilling Project*, v. 12, Laughton, A. S., and Berggren, W. A., eds., Washington, D.C.: U.S. Government Printing Office, p. 953–963.

Berggren, W. A., and Hollister, C. D., 1974, Paleogeography, paleobiography and the history of circulation in the Atlantic Ocean, in: *Studies in Paleooceanography*, W. W. Hay, ed., Spec. Publ. 20, Tulsa, Oklahoma: Society of Economic Paleontologists and Mineralogists, p. 126–186.

Birkelund, T., 1977, Geology of East Greenland: A review, in: *Offshore North Sea Conference*, Stavanger, Norway: Norwegian Petroleum Society, p. 3–23.

Birkelund, T., and Perch-Nielsen, K., 1976, Late Paleozoic–Mesozoic evolution of central East Greenland, in: *Geology of Greenland*, A. Escher and W. S. Watt, eds., Copenhagen, Denmark: The Geological Survey of Greenland, p. 305–339.

Bjørklund, K., and Kellogg, D., 1972, Five new Eocene radiolarian species from the Norwegian Sea, *Micropaleontology*, v. 18, p. 386–396.

Björnsson, A., Saemundsson, K., Einarsson, P., Tryggvason, E., and Gronvold, K., 1977, Current rifting episode in north Iceland, *Nature (London)*, v. 266, p. 318–323.

Björnsson, A., Johnsen, G., Sigurdsson, S., Thorbergsson, G., and Tryggvason, E., 1979, Rifting of the plate boundary in North Iceland 1975–78, *J. Geophys. Res.*, v. 84, p. 3029–3038.

Björnsson, S., and Einarsson, P., 1974, Seismicity of Iceland, in: *Geodynamics of Iceland and the North Atlantic Area*, L. Kristjansson, ed., Dordrecht, Holland: D. Reidel, p. 225–240.

Bodvarsson, G., and Walker, G. P. L., 1964, Crustal drift in Iceland, *Geophys. J. R. Astron. Soc.*, v. 8, p. 285–300.

Bollingberg, H., Brooks, C. K., and Noe-Nygaard, A., 1975, Trace element variations in Faeroes basalts and their possible relationships to ocean floor spreading history, *Bull. Geol. Soc. Den.*, v. 24, p. 44–60.

Bott, M. H. P., 1965, The upper mantle beneath Iceland, *Geophys. J. R. Astron. Soc.*, v. 9, p. 275–277.

Bott, M. H. P., 1974, Deep structure, evolution, and origin of the Icelandic transverse ridge, in: *Geodynamics of Iceland and the North Atlantic Area*, Kristjansson, L., ed., Dordrecht, Holland: D. Reidel, p. 33–48.

Bott, M. H. P., 1975, Structure and evolution of the Atlantic floor between northern Scotland and Iceland, *Nor. Geol. Unders. Publ.* v. 316, 195–199.

Bott, M. H. P., Sunderland, J., Smith, P. J., Casten, U., and Saxov, S., 1974, Evidence for continental crust beneath the Faeroe Islands, *Nature (London)*, v. 248, p. 202–204.

Bott, M. H. P., Nielsen, P. H., and Sunderland, J., 1976, Converted P waves originating at the continental margin between the Iceland–Faeroe Ridge and the Faeroe block, *Geophys. J. R. Astron. Soc.*, v. 44, p. 229–238.

Brander, J. L., Mason, R. G., and Calvert, R. W., 1976, Precise distance measurements in Iceland, *Tectonophysics*, v. 31, p. 193–206.

Briseid, E., and Mascle, J., 1975, Structure de la marge continentale nervegienne audebauche de la mer de Barentsz, *Mar. Geophys. Res.*, v. 2, p. 231–241.

Brooks, C. K., 1973, Rifting and doming in southern East Greenland, *Nature (London) Phys. Sci.*, v. 244, p. 23–25.

Brooks, C. K., and Gleadow, A. J. W., 1977, A fission-track age for the Skaergaard intrusion and the age of the East Greenland basalts, *Geology*, v. 5, p. 539–540.

Brooks, C. K., and Jakobsson, S. P., 1974, Petrochemistry of the volcanic rocks of the North Atlantic Ridge System, in: *Geodynamics of Iceland and the North Atlantic Area*, Kristjansson, L., ed., Dordrecht, Holland: D. Reidel, p. 139–154.

Brooks, C. K., Jakobsson, S. P., and Campsie, J., 1974, Dredged basaltic rocks from the seaward extensions of the Reykjanes and Snaefellsnes volcanic zones, Iceland, *Earth Planet. Sci. Lett.*, v. 22, p. 320–327.

Bungum, H., 1977, Two focal-mechanism solutions for earthquakes from Iceland and Svalbard, *Tectonophysics*, v. 41, p. T15–T18.

Bungum, H., and Husebye, E. S., 1977, Seismicity of the Norwegian Sea, The Jan Mayen Fracture Zone, *Tectonophysics*, v. 40, p. 351–360.

Burke, K., and Dewey, J. F., 1973, Plum-generated triple junctions: Key indicators in applying plate tectonics to old rocks, *J. Geol.*, v. 81, p. 406–433.

Burr, N. C., and Solomon, S. C., 1978, The relationship of source parameters of oceanic transform earthquakes to plate velocity and transform length, *J. Geophys. Res.*, v. 83, p. 1193–1205.

Calloman, J. G., Donovan, D. T., and Trumpy, R., 1972, An annotated map of the Permian and Mesozoic formations of East Greenland, *Medd. Grønl.*, v. 168(3), 135 p.

Caston, V. N. D., 1974, Bathymetry of the northern North Sea: Knowledge is vital for offshore oil, *Offshore*, Feb., p. 76–84.

Caston, V. N. D., 1976, Tertiary sediments of the Vøring Plateau, Norwegian Sea, recovered by Leg 38 of the Deep Sea Drilling Project, in: *Initial Reports of the Deep Sea Drilling Project*, v. 38, Talwani, M., and Udintsev, G., eds., Washington, D.C.: U.S. Government Printing Office, p. 1197–1212.

Christensen, N. K., 1974, The petrological nature of the lower oceanic crust and upper mantle, in: *Geodynamics of Iceland and the North Atlantic Area*, Kristjansson, L., ed., Dordrecht, Holland: D. Reidel, p. 165–176.

Clarke, D. B., 1975, Tertiary basalts dredged from Baffin Bay, *Can. J. Earth Sci.*, v. 12, p. 1396–1405.

Clarke, D. B., and Pedersen, A. K., 1976: Tertiary volcanic province of West Greenland, in: *Geology of Greenland*, Escher A., and Watt, W. S., eds., Copenhagen: Geological Survey of Greenland, p. 365–385.

Clarke, D. B., and Upton, B. G. J., 1971, Tertiary basalts of Baffin Island: Field relations and tectonic setting, *Can. J. Earth Sci.*, v. 8, p. 248–258.

CLIMAP Project Members, 1976, The surface of the ice-age Earth, *Science*, v. 191, p. 1131–1137.

Cochran, J. R., and Talwani, M., 1977, Free-air gravity anomalies in the world's oceans and their relationship to residual elevation, *Geophys. J. R. Astron. Soc.*, v. 50, p. 495–552.

Cochran, J. R., and Talwani, M., 1978, Gravity anomalies, regional elevation, and the deep structure of the North Atlantic, *J. Geophys. Res.*, v. 88, p. 4907–4924.

Collette, B. J., Laagay, R. A., Ritsema, A. R., and Schooten, J. A. 1970, Seismic investigations in the North Sea, Parts 1 and 2, *Geophys. J. R. Astron. Soc.*, v. 12, p. 363–373.

Dagley, P., Wilson, R. L., Ade-Hall, J. M., Walker, G. P. L., Haggerty, S. E., Sigurgeirsson, T., Watkins, N. D., Smith, P. J., Edwards, J., and Grasty, E. L., 1967, Geomagnetic polarity zones for Icelandic lavas, *Nature (London)*, v. 216, p. 25–29.

Damuth, J. E., 1978, Echo-character of the Norwegian–Greenland Sea: Relationship to Quaternary sedimentation, *Mar. Geol.*, v. 28, p. 1–36.

Dansgaard, W., Johnson, S. J., Moller, J., and Langway, G. C., 1969, Thousand centuries of climatic record from Camp Century on the Greenland icesheet, *Science*, v. 166, p. 377–381.

Dawes, P. R., 1976, Precambrian to Tertiary of northern Greenland, in: *Geology of Greenland*, Escher, A., and Watt, W. S., eds., Copenhagen: Geological Survey of Greenland, p. 248–303.

Dawson, M. R., West, R. M., Langston, W., Jr., and Hutchison, J. H., 1976, paleogene terrestrial vertebrates: Northernmost occurrence, Ellesmere Island, Canada, *Science*, v. 192, p. 781–782.

de Argollo, R. M., 1978, Ge/Si and Ga/Al variations along the Reykjanes Ridge and Iceland, *Nature (London)*, v. 276, p. 24–28.

Decker, R. W., Einarsson, P., and Mohr, P. A., 1971, Rifting in Iceland: New geodetic data, *Science*, v. 173, p. 530–532.

Deer, W. A., 1976, Tertiary igneous rocks between Scoresby Sund and Kap Gustav Holm, East Greenland, in: *Geology of Greenland*, Escher, A., and Watt, W. S., eds., Copenhagen: Geological Survey of Greenland, p. 403–429.

Deutsch, E. R., and Kristjansson, L. G., 1974, Paleomagnetism of Late Cretaceous–Tertiary volcanics from Disko Island, West Greenland, *Geophys. J. R. Astron. Soc.*, v. 39, p. 343–360.

Dittmer, F., Fine, S., Rasmussen, M., Bailey, J. C., and Campsie, J., 1975, Dredged basalts from the Mid-Oceanic ridge north of Iceland, *Nature (London)*, v. 254, p. 298–301.

Duffield, W. A., 1978, Vesicularity of basalt erupted at Reykjanes Ridge crest, *Nature (London)*, v. 274, p. 217–220.

Dypvik, H., and Nagy, J., 1978, Early Tertiary bentonites, a preliminary report, *Polarforschung*, v. 48, p. 139–150.

Edwards, M. B., 1975, Gravel fraction on the Spitsbergen Bank, NW Barents Shelf, *Norges Geol. Unders. Publ.*, v. 316, p. 205–217.

Egeberg, T., 1977, En undersokelse av norsk kontinental sokkel mellom 58°–62°N ved hjelp av geofysiske metoder, med saerlig vekt pa Tertiaer landhevning og dannelsen av Norskerennen (Thesis, in Norwegian), University of Oslo, 207 p.

Eggvin, J., 1963, Bathymetric chart of the Norwegian Sea and adjacent areas (chart), *Fiskeridir. Havforsk. Inst.*, Bergen, Norway.

Egloff, J., and Johnson, G. L., 1975, Morphology and structure of the southern Labrador Sea, *Can. J. Earth Sci.*, v. 12, p. 2111–2133.

Einarsson, T., 1954, A survey of gravity in Iceland, *Soc. Sci. Island*, v. 30, p. 1–22.

Einarsson, T., 1967, Subcrustal viscosity in Iceland, in: *Iceland and the Mid-Ocean Ridges*, Björnsson, S., ed., Reykjavik: Society Scientifica Islandica, p. 109–110.

Einarsson, P., 1979, Seismicity and earthquake focal mechanisms along the Mid-Atlantic plate boundary between Iceland and the Azores, *Tectonophysics*, v. 55, p. 127–153.

Einarsson, T., Hopkins, D. M., and Doell, R. R., 1967, The stratigraphy of Tjörnes, northern Iceland, and the history of the Bering Land Bridge, in: *The Barents Land Bridge*, Hopkins, D. M., ed., Stanford: Stanford University Press, p. 312–325.

Eldholm, O., 1980, The Norwegian–Greenland Sea–Plate tectonic development, *Norwegian Sea Symposium*, Oslo: Norwegian Petroleum Directorate, NSS 2–13.

Eldholm, O., and Ewing, J., 1971, Marine geophysical survey in the southwestern Barents Sea, *J. Geophys. Res*, v. 76, p. 3832–3841.

Eldholm, O., and Myhre, A. M., 1977, Hovgaard Fracture Zone, *Norsk Polarinst. Årb.*, v. 1976, p. 195–208.

Eldholm, O., and Sundvor, E., 1980, The continental margins of the Norwegian–Greenland Sea: Recent results and outstanding problems, *Phil. Trans. R. Soc. London*, v. 294, p. 77–86.

Eldholm, O., And Talwani, M., 1977, Sediment distribution and structural framework of the Barents Sea, *Geol. Soc. Am. Bull.*, v. 88, p. 1015–1029.

Eldholm, O., and Windisch, C. C., 1974, The sediment distribution in the Norwegian–Greenland Sea, *Geol. Soc. Am. Bull.*, v. 86, p. 1661–1676.

Eldholm, O., Sundvor, E., and Myhre, A., 1979, Continental margin off Lofoten–Vesterålen, Northern Norway, *Mar. Geophys. Res.*, v. 4, p. 3–35.

Emelyanov, E. M., Litvin, V. M., Levshenko, V. A., and Martynova, G. P., 1971, The geology of the Barents Sea, in: *The Geology of the East Atlantic Continental Margin*, Delaney, F. M. ed., Report No. 70/14, London: Institute of Geological Science, p. 5–15.

Ericson, D. B., Ewing, M., and Wollin, G., 1964, Sediment cores from the Arctic and subarctic seas, *Science*, v. 144, p. 1183–1192.

Escher, A., and Watt, W. S., eds., 1976, *Geology of Greenland*, Copenhagen: Geological Survey of Greenland, 603 p.

Evans, A. W., Fitch, F. J., and Miller, J. A., 1973, Potassium–argon age determinations on some British Tertiary volcanic rocks, *J. Geol. Soc. London*, v. 129, p. 419–443.

Ewing, J., and Ewing, M., 1959, Seismic refraction measurements in the Atlantic Ocean Basins, in the Mediterranean Sea, on the Mid-Atlantic ridge and in the Norwegian Sea, *Geol. Soc. Am. Bull.*, v. 70, p. 291–318.

Featherstone, P. S., Bott, M. H. P., and Peacock, J. H., 1977, Structure of the continental margin of southeastern Greenland, *Geophys. J. R. Astron. Soc.*, v. 48, p. 15–27.

Feden, R. H., Vogt, P. R., and Fleming, H. S., 1979, Evidence for the Yermak hot spot, Arctic Ocean, *Earth Planet. Sci. Lett.*, v. 4, p. 18–38.

Fedhynskii, V. V., Rassokho, A. U., Demenitskaya, R. M., Karasik, A. M., and Rozhdestbenskii, S. S., 1974, Magnitnye anomalii grebnya sredinno-atlanticheskogo khrebta (in Russian), *Dok Akad. Nauk SSSR*, v. 217, p. 1416–1419.

Fedhynskii, V. V., Rassokho, A. I., Demenitsikaya, R. M., Karasik, A. M., and Rozhdestvenskii, S. S., 1975, O strukture anomalhnogo magnitnogo polya yugo-zapadnoi chasti khrebta Mona (in Russian), *Dokl. Akad. Nauk SSSR*, v. 223, p. 726–729.

Fitch, F. J., 1965, The structural uniting of the reconstructed North Atlantic continent: A symposium on continental drift, *Trans. R. Soc. Phil.*, v. 258, p. 191–193.

Fitch, F. J., Miller, J. A., Warrell, D. M., and Williams, S. C., 1974, Tectonic and radiometric age comparisons, in: *The Ocean Basins and Margins*, v. 2, *The North Atlantic*, Nairn, A. E. M., and Stehli, F. G., eds., New York: Plenum Press, p. 485–538.

Fleischer, U., 1974, The Reykjanes Ridge—a summary of geophysical data, in: *Geodynamics of Iceland and the North Atlantic Area*, Kristjansson, L., ed., Dordrecht, Holland: D. Reidel, p. 17–32.

Fleischer, U., Korschunow, A., Schultz, G., and Vogt, P. R., 1973, Eine gravimetrische und erdmagnetische Vermessung des südlichen Reykjanes-Rückens mit F. S. "Meteor," 22.4–9.6. 1966, Endgültige Auswertung der Forschungsfahrt Nr. 4., *Meteor Forschungsergeb. Reihe C*, v. 13, p. 64–84.

Fleischer, U., Holzkamm, F., Vollbrecht, K., and Voppel, D., 1974, Die Struktur des Islands–Faröer–Rückens aus geophysikalischen Messungen, *Sonderdruck Dtsch. Hydrogr. Z.*, v. 27(3), p. 97–113.

Fleming, H. S., Cherkis, N. Z., and Heirtzler, J. R., 1970, The Gibbs fracture zone: A double fracture zone at 52°30′N in the Atlantic Ocean, *Mar. Geophys. Res.*, v. 1, p. 37–45.

Floden, T., 1973, Notes on the bedrock of the eastern Skagerrak with remarks on the Pleistocene deposits, *Stockholm Contr. Geol.*, v. 24, p. 79–102.

Floden, T., and Sellevoll, M. A., 1972, Two seismic profiles across the Norwegian channel west of Bergen, *Stockholm Contr. Geol.*, v. 24, p. 25–32.

Francis, T. J. G., 1969, Upper mantle structure along the axis of the Mid-Atlantic ridge near Iceland, *Geophys. J. R. Astr. Soc.*, v. 17, p. 507–520.

Francis, T. J. G., 1973, The seismicity of the Reykjanes Ridge, *Earth Planet. Sci. Lett.*, v. 18, p. 119–123.

Friedrich, W. L., and Simonarson, L. A., 1974, Bemerkungen zur Neogen-Flora Island, *Cour. Forsch. Inst. Senckenberg (Frankfurt am Main)*, v. 10, p. 5–6.

Friedrich, W. L., and Simonarson, L. A., 1976, Acer askelssoni N. sp., Grosse Neogene Teilfrüchte aus Island, *Paleontographica*, v. 155, p. 140–148.

Frost, R. T. C., and Dikken, A. J., eds., 1977, Fault tectonics in N. W. Europe, *Geologie en Mijnbouw*, v. 56, p. 273–372.

Gairaud, H., Jacquart, G., Aubertin, F., and Beuzart, P., 1978, The Jan Mayen Ridge: Synthesis of geological knowledge and new data, *Oceanolog. Acta*, v. 1, p. 335–358.

Gaposchkin, E. M., and Lambeck, K., 1971, Earth gravity field to the sixteenth degree and station coordinates from satellite and terrestrial data, *J. Geophys. Res.*, v. 76, p. 4855–4883.

Gartner, S., and Keany, J., 1978, The terminal Cretaceous event: A geologic problem with an oceanographic solution, *Geology*, v. 6, p. 708–712.

Gerke, K., 1974, Crustal movements in the Myvatn and the Thingvallavatn area, in: *Geodynamics of Iceland and the North Atlantic Area*, Kristjansson, L., ed., Dordrecht, Holland: D. Reidel, p. 263–278.

Gibb, F. G. F., and Kanaris-Sotiriou, R., 1976, Jurassic igneous rocks of the Forties field, *Nature (London)*, v. 260, p. 23–25.

Girardin, N., and Jacoby, W. R., 1979, Rayleigh wave dispersion along the Reykjanes Ridge, *Tectonophysics III*, v. 55, p. 155–171.

Gregersen, S., 1970, Surface wave dispersion and crust structure in Greenland, *Geophys. J. R. Astron. Soc.*, v. 22, p. 29–39.

Grønlie, G., 1979, Tertiary paleogeography of the Norwegian–Greenland Sea, *Nor. Pol. Inst. Skr.*, v. 170, p. 49–61.

Grønlie, G., and Ramberg, I. B., 1970, Gravity indications of deep sedimentary basins below the Norwegian continental shelf and the Vøring Plateau, *Nor. Geol. Tidsskr.*, v. 50, p. 357–391.

Grønlie, G., and Talwani, M., 1978, *Geophysical Atlas: Norwegian–Greenland Sea*, Vema Research Series v. 4, Palisades, New York: Lamont–Doherty Geological Observatory, 26 p.

Grønlie, G., and Talwani, M., 1979, Bathymetry of the Norwegian–Greenland Sea, *Nor. Polarinst. Skr.*, v. 170, p. 3–24.

Grønlie, G., Heier, K. S., and Swanberg, C. A., 1977, Terrestrial heat-flow determinations from Norway, *Nor. Geol. Tids.*, v. 57, p. 153–162.

Grønlie, G., Chapman, M., and Talwani, M., 1979, Jan Mayen Ridge and Iceland Plateau: Origin and evolution, *Nor. Pol. Inst. Skr.*, v. 170, p. 25–47.

Hagevang, T., 1978, En marinegeofysisk undersøkelse av kontinentalmarginen utenfor Helgeland (Thesis, in Norwegian), Oslo: University of Oslo, 151 p.

Hailwood, E. A., Tarling, D. H., Mitchell, J. G., and Lovlie, R., 1973, Preliminary observations on the paleomagnetism and radiometric ages of the Tertiary basalt sequence of Scoresby Sund, East Greenland, *Groenl. Geol. Unders. Rapp.*, v. 58, 43–47.

Haines, G. V., Hannaford, W., and Serson, P. H., 1970, Magnetic anomaly maps of the Nordic Countries and the Greenland and Norwegian Seas, *Publ. Dom. Obs.*, v. 39, p. 121–149.

Hall, D. H., and Dagley, P., 1970, Regional magnetic anomalies: An analysis of the smoothed aeromagnetic map of Great Britain and Northern Ireland, *Inst. Geol. Sci. Ann. Rept.*, v. 70(10), p. 1–8.

Hänel, R., 1974, Heat flow measurements in the Norwegian Sea, *Meteor Forschungsergeb. Reihe C*, v. 17, p. 74–78.

Hänel, R., Grønlie, G., and Heier, K. S., 1974, Terrestrial heat flow determinations from lakes in southern Norway, *Nor. Geol. Tidsskr.*, v. 54, p. 423–428.

Harland, W. B., 1969, Contribution of Spitsbergen to understanding of tectonic evolution of North Atlantic, in: *North Atlantic—Geology and Continental Drift*, Kay, M., ed., *Am. Assoc. Petrol. Geol. Mem.* 12, p. 817–851.

Harper, M. L., 1971, Approximate geothermal gradients in the North Sea Basin, *Nature (London)*, v. 230, p. 235–236.

Harrison, C. G. A.. McDougall, I., and Watkins, N. D., 1979, A geomagnetic field reversal time scale back to 13.0 million years before present, *Earth Planet. Sci. Lett.*, v. 42, p. 143–152.

Hart, S. R., Schilling, J. G., and Powell, J. L., 1973, Basalts from Iceland and along the Reykjanes ridge: Sr-isotope geochemistry, *Nature Phys. Sci.*, v. 246, p. 104–107.

Heier, K. S., ed., 1977, *The Norwegian Geotraverse Project*, Oslo: Norges Geologische Undersøkelse, 298 p.

Heirtzler, J. R., LePichon, X., and Baron, J. G., 1966, Magnetic anomalies over the Reykjanes Ridge, *Deep Sea Res.*, v. 13, p. 427–443.

Heirtzler, J. R., Dickson, G. O., Herron, E. M., Pitman, W. C., and LePichon, X., 1968, Marine magnetic anomalies, geomagnetic field reversals, and motions of ocean floor and continents, *J. Geophys. Res.*, v. 73, p. 2119–2136.

Henriksen, N., and Higgins, A. K., 1976, East Greenland Caledonian fold belt, in: *Geology of Greenland*, Escher, A., and Watt, W. S., eds., Copenhagen: Geological Survey of Greenland, p. 340–362.

Hermance, J. F., and Grillot, L. R., 1970, Correlation of magnetotelluric, siesmic, and temperature data from southwest Iceland, *J. Geophys. Res*, v. 75, p. 6582–6591.

Hinz, K., 1972, The seismic crustal structure of the Norwegian continental margin in the Vøring Plateau, in the Norwegian deep sea and on the eastern flank of the Jan Mayen Ridge between 66 and 58°N, *International Geological Congress*, Montreal, Sec. 8, p. 28–37.

Hinz, K., 1975, Results of geophysical surveys in the area of the Aegir Ridge, the Iceland Plateau, and the Kolbeinsey Ridge, *Nor. Geol. Unders. Publ.*, v. 316, p. 201–203.

Hinz, K., and Moe, A., 1971, Crustal structure in the Norwegian Sea, *Nature Phys. Sci.*, v. 232, p. 187–190.

Hinz, K., and Schlüter, H.-U., 1978a, The geological structure of the western Barents Sea, *Mar. Geol.*, v. 26, p. 199–230.

Hinz, K., and Schlüter, H.-U., 1978b, Der Nordatlantik-Ergebnisse geophysikalischer Untersuchungen der Bundesanstalt für Geowissenschaften und Rohstoffe an nordatlantischen Kontinentalrändern, *Erdoel-Erdgas Zeit.*, v. 94, p. 271–280.

Holtedahl, H., 1959, Geology and paleontology of Norwegian Sea bottom cores, *J. Sediment Petrol*, v. 29, p. 16–29.

Holtedahl, H., and Bjerkli, K., 1975, Pleistocene and recent sediments of the Norwegian continental shelf (62°N–71°N), and the Norwegian Channel area, *Nor. Geol. Unders. Publ.*, v. 316, p. 241–252.

Holtedahl, H., and Sellevoll, M. A., 1971, Geology of the continental margin of the eastern Norwegian Sea and of the eastern Norwegian Sea and of the Skagerrak, in: *The Geology of the East Atlantic Continental Margin*, Delaney, F. M., ed., *Inst. Geol. Sci. London Rep*. No. 70/14, p. 33–52.

Holtedahl, H., and Sellevoll, M. A., 1972, Notes on the influence of glaciation on the Norwegian continental shelf bordering on the Norwegian Sea, *Ambio Special Report* No. 2, p. 31–38.

Holtedahl, O., 1960, On supposed marginal faults and the oblique uplift of the land mass in Cenozoic time, in: *Geology of Norway*, *Nor. Geol. Unders. Publ.*, v. 208, p. 351–357.

Hood, P. J., and Bower, M. E., 1973, Low-level aeromagnetic surveys of the continental shelves bordering Baffin Bay and the Labrador Sea, *Can. Geol. Surv. Pap.* 71–23, p. 573–598.

Hood, P. J., and Bower, M. E., 1975, Aeromagnetic reconnaissance of Davis Strait and adjacent areas, in: *Canada's Continental Margins*, Yorath, C. J., Parker, E. R., and Glass, D. J., eds., Calgary: Canadian Society of Petroleum Geologists, p. 433–451.

Horai, K., Chessman, M., and Simmons, G., 1970, Heat flow measurements on the Reykjanes Ridge, *Nature (London)*, v. 225, p. 264–265.

Horsfield, W. T., and Maton, P. I., 1970, Transform faulting along the DeGeer line, *Nature (London)*, v. 226, p. 256–257.

Houtz, R., and Ewing, J., 1976, Upper crustal structure as a function of plate age, *J. Geophys. Res.*, v. 81, p. 2490–2498.

Houtz, R., and Windisch, C., 1977, Barents Sea continental margin sonobuoy data, *Geol. Soc. Am. Bull.*, v. 88, p. 1030–1036.

Hughes, T., Denton, G. H., and Grosswald, M. G., 1977, Was there a Late Würm ice sheet?, *Nature (London)*, v. 266, p. 596–602.

Husebye, E. S., Bungum, H., Fyen, J., and Gjøystdal, H., 1976, Earthquake activity in Fennoscandia between 1497 and 1975 and intraplate tectonics, *Nor. Geol. Tidsskr.*, v. 58, p. 51–68.

Husebye, E. S., Gjøystdal, H., Bungum, H., and Eldholm, O., 1975, The seismicity of the Norwegian–Greenland Sea, *Tectonophysics*, v. 26, p. 55–70.

Hyndman, R. D., 1973, Evolution of the Labrador Sea, *Can. J. Earth Sci.*, v. 10, p. 637–644.

Imbrie, J., and Kipp, N. G., 1971, A new micropaleontologic method for quantitative paleoclimatology: Application to a late Pleistocene Caribbean core, in: *Late Cenozoic Glacial Ages*, Turekian, K. K., ed. New Haven, Connecticut: Yale University Press, p. 71–191.

Jackson, H. R., Keen, C. E., Falconer, R. K. H., and Appleton, K. P., 1979, New geophysical evidence for sea-floor spreading in central Baffin Bay, *Can. J. Earth Sci.*, v. 16, p. 2122–2135.

Jacque, M., and Thouvenin, M., 1975, Lower Tertiary tuffs and volcanic activity in the North Sea, in: *Petroleum Geology of the Shelf of Northwest Europe, I*, Woodland, A. W., ed., London: Geology, Applied Science Publishers, p. 455–465.

Jakobbsson, S. P., 1972, Chemistry and distribution pattern of recent basaltic rocks in Iceland, *Lithos*, v. 5, p. 365–386.

Johnson, G. L., 1974, Morphology of the mid-ocean ridge between Iceland and the Arctic, in: *Geodynamics of Iceland and the North Atlantic Area*, Kristjansson, L., ed., p. 49–62.

Johnson, G. L., 1975, The Jan Mayen Ridge, in: *Canada's Continental Margins and Offshore Petroleum Exploration*, Yorath, C. J., Parker, E. R., and Glass, D. J., ed., *Can. Soc. Petrol. Geol. Mem. 4*, p. 205–224.

Johnson, G. L., and Campsie, J., 1974, Morphology and structure of the western Jan Mayen fracture zone, *Nor. Polar Inst. Årbok.*, v. 1972, p. 69–81.

Johnson, G. L., and Eckhoff, O. B., 1966, Bathymetry of the north Greenland Sea, *Deep Sea Res.*, v. 13, p. 1161–1173.

Johnson, G. L., and Heezen, B. C., 1967, Morphology and evolution of the Norwegian–Greenland Sea, *Deep Sea Res.*, v. 14, p. 755–771.

Johnson, G. L., and Vogt, P. R., 1973, Mid-Atlantic Ridge from 47° to 51° North, *Geol. Soc. Am. Bull.*, v. 84, p. 3443–3462.

Johnson, G. L., Southall, J. R., Young, P. W., and Vogt, P. R., 1972, The origin and structure of the Iceland Plateau and Kolbeinsey Ridge, *J. Geophys. Res.*, v. 77, p. 5688–5696.

Johnson, G. L., McMillan, N. J., and Egloff, J., 1975a, The continental margin of East Greenland, in: *Canada's Continental Margins and Offshore Petroleum Exploration*, Yorath, C. J., Parker, E. R., and Glass, D. J., eds., *Can. Soc. Petrol. Geol. Mem. 4*, p. 391–410.

Johnson, G. L., Sommerhoff, G., and Egloff, J., 1975b, Structure and morphology of the West Reykjanes basin, *Mar. Geol.*, v. 18, p. 175–196.

Jones, E. J. W., Ewing, M., Ewing, J. I., and Eittreim, S. L., 1970, Influence of Norwegian Sea water on sedimentation in the northern North Atlantic and Labrador Seas, *J. Geophys. Res.*, v. 75, p. 1655–1680.

Jones, E. J. W., Ramsay, A. T. S., Preston, N. J., and Smith, A. C. S., 1974, A Cretaceous guyot in the Rockall Trough, *Nature (London)*, v. 251, p. 129–131.

Jørgensen, F., and Navrestad, T., 1979, Main structural elements and sedimentary succession on the shelf outside Nordland (Norway), Norwegian Sea Symposium *Nor. Petrol. Soc.*, NSS/11, 20 p.

Karasik, A. M., 1974, Yevraziskii bassein cevernogo ledovitogo okeana s pozitsii tektoniki plit, in: *Problemi Geologii Polyarnikh Oblastei Zemli*, Leningrad: NIDRA, p. 24–31.

Karasik, A. M., and Rozhdestvenskii, S. S., 1977, Struktura osi razrastaniya okeanicheskovo dna i zakonomernosti yeyo formirovaniya (na primerÿ riftovio zonÿ severo-atlanti cheskovo metabasseina), in: *Osnovnie Problemi Riftogeneza*, Logachev, N. A., ed., Novosibirsk: Nauka, p. 167–175.

Karnik, V., 1971, *Seismicity of the European Area*, II, Dordrecht, Holland: D. Reidel.

Kaula, W. M., 1972, Global gravity and mantle convection, *Tectonophysics*, v. 13, p. 341–359.

Keen, C. E., and Keen, M. J., 1974, Continental margins of eastern Canada and Baffin Bay, in: *The Geology of Continental Margins*, Burk, C. A., and Drake, C. L., eds., New York: Springer-Verlag, p. 381–390.

Keen, C. E., Barrett, D. L., Manchester, K. S., and Ross, D. I., 1972, Geophysical studies in Baffin Bay and some tectonic implications, *Can. J. Earth Sci.*, v. 9, p. 239–256.

Keen, C. E., Keen, M. J., Ross, D. I., and Lack, M., 1974, Baffin Bay: Small ocean basin formed by sea-floor spreading, *Am. Assoc. Petrol. Geol. Bull.*, v. 58, p. 1089–1108.

Keen, M. J., and Clarke, D. B., 1974, Tertiary basalts of Baffin Bay: Geochemical evidence for a fossil hot spot, in: *Geodynamics of Iceland and the North Atlantic Area*, Kristjansson, L., ed., Dordrecht, Holland: D. Reidel, p. 127–138.

Kellogg, T. B., 1975, Late Quaternary climatic changes in the Norwegian and Greenland Seas, in: *Climate of the Arctic*, Weller, G., and Bowling, S., eds., Fairbanks, Alaska: Geophysical Institute, University of Alaska, p. 3–36.

Kellogg, T. B., 1977, Paleoclimatology and paleooceanography of the Norwegian and Greenland Sea: The last 450,000 years, *Marine Micropaleo*, v. 2, p. 235–249.

Kellogg, T. B., Duplessy, J. C., and Shackleton, N. J., 1978, Planktonic foraminiferal and oxygen isotopic stratigraphy and paleoclimatology of Norwegian Sea deep-sea cores, *Boreas*, v. 7, p. 61–73.

Klein, F. W., Einarsson, P., and Wyss, M., 1973, Microearthquakes on the Mid-Atlantic plate boundary, *J. Geophys. Res.*, v., 78, p. 5084–5099.

Klein, F. W., Einarsson, P., and Wyss, M., 1977, The Reykjanes Peninsula, Iceland, Earthquake swarm of September 1972 and its tectonic significance, *J. Geophys. Res.*, v. 82, p. 865–888.

Kent, P. E., 1975, Review of North Sea Basin development, *J. Geol. Soc. London*, v. 131, p. 435–468.

Kovacs, L. C., and Vogt, P. R., 1979, Depth-to-magnetic source analysis of the Greenland/Norwegian Sea and Arctic Ocean, *EOS-Trans. Am. Geophys. Union*, v. 60, p. 372.

Kristjansson, L., 1972, On the thickness of the magnetic crustal layer in southwestern Iceland, *Earth Planet. Sci. Lett.*, v. 16, p. 237–244.

Kristjansson, L., 1976a, A marine magnetic survey off southern Iceland, *Mar. Geophys. Res.*, v. 2, p. 315–326.

Kristjansson, L., 1976b, Central volcanoes on the western Icelandic shelf, *Mar. Geophys. Res.*, v. 2, p. 285–289.

Kristjansson, L, and Watkins, N. C., 1977, Magnetic studies of basalt fragments recovered by deep drilling in Iceland, and the "magnetic layer" concept, *Earth Planet. Sci. Lett.*, v. 34, p. 365–374.

Kristjansson, L., Thors., K., and Karlsson, H. R., 1977, Confirmation of central volcanoes off the Icelandic coast, *Nature (London)*, v. 268, p. 325–326.

Kristoffersen, Y., and Talwani, M., 1977, Extinct triple junction south of Greenland and the Tertiary motion of Greenland relative to North America, *Geol. Soc. Am. Bull.*, v. 88, p. 1037–1049.

Krogh, E., 1977, Evidence of Pre-Cambrian continent collision in western Norway, *Nature (London)*, v. 267, p. 17–19.

Kvale, A., 1960, Norwegian earthquakes in relation to tectonics, *Mat. Naturv. Ser. Bergen Univ.* 10, 17 p.

La Brecque, J. L., Kent, D. V., and Cande, S. C., 1977, Revised magnetic polarity time scale for Late Cretaceous and Cenozoic time, *Geology*, v. 5, p. 330–335.

Lachenbruch, A. H., and Marshall, B. V., 1968, Heat flow and water temperature fluctuations in the Denmark Strait, *J. Geophys. Res.*, v. 73, p. 5829–5842.

Langseth, M. G., and Zielinski, G. W., 1974, Marine heat flow measurements in the Norwegian–Greenland Sea and in the vicinity of Iceland, in: *Geodynamics of Iceland and the North Atlantic Area*, Kristjansson, L., ed., Dordrecht, Holland: D. Reidel, p. 277–296.

Larsen, H. C., 1978, Offshore continuation of East Greenland dyke swarm and North Atlantic Ocean formation, *Nature (London)*, v. 274, 220–223.

Larsen, H. C., and Thorning, L., 1980, Project EASTMAR: Acquisition of high sensitivity aeromagnetic data off East Greenland, *Rapp. Groenl. Geol. Unders.*, v. 100, p. 91–94.

Larsen, O., Dawes, P. R., and Soper, N. J., 1978, Rb/Sr age of the Kap Washington Group, Peary Land, North Greenland, and its geotectonic implication, *Rapp. Groenl. Geol. Unders.*, v. 90, p. 115–119.

Laughton, A. S., 1972, The Southern Labrador Sea, a key to the Mesozoic and early Tertiary evolution of the North Atlantic, in: *Initial Reports of the Deep Sea Drilling Project*, v. 12, Laughton, A. S., and Berggren, W. A., eds., Washington, D.C.: U.S. Government Printing Office, p. 1155–1179.

Laughton, A. S., 1975, Tectonic evolution of the northeast Atlantic Ocean: A review, *Nor. Geol. Unders. Publ.*, v. 316, p. 169–193.

Laughton, A. S., and Whitmarsh, R. B., 1974, The Azores–Gibraltar plate boundary, in: *Geodynamics of Iceland and the North Atlantic Area*, Kristjansson, L., ed., Dordrecht, Holland: D. Reidel, p. 63–82.

Lazareva, A. P., and Misharina, L. A., 1965, Stresses in earthquake foci in the Arctic seismic belt, *Izv. Earth Phys. Ser.*, v. 2, p. 5–10.

LePichon, X., Hyndman, R. D., and Pautot, G., 1971, Geophysical study of the opening of the Labrador Sea, *J. Geophys. Res.*, v. 76, p. 4724–4743.

LePichon, X., Francheteau, J., and Bonnin, J., 1973, *Plate Tectonics*, Amsterdam: Elsevier, 300 p.

LePichon, X., Sibuet, J. C., and Francheteau, J., 1977, The fit of the continents around the North Atlantic Ocean, *Tectonophysics*, v. 38, p. 169–209.

Litvin, V. M., 1964, Bottom relief of the Norwegian Sea (in Russian), *Tr. Polyarn. Nauchno-Issled. Proektn. Inst. Morsk. Rhybn. Khoz. Okeanogr. (PINRO)*, v. 16, p. 89–109.

Long, R. E., and Mitchell, M. G., 1970, Teleseismic P-wave delay time in Iceland, *Geophys. J. R. Astron. Soc.*, v. 20, p. 41–48.

Ludwig, W. J., Nafe, J. E., and Drake, C. L., 1970, Seismic refraction, in: *The Sea*, v. 4, Part 1, Bullard, E. C., and Maxwell, A. E., eds., New York: Wiley-Interscience, p. 53–84.

Luyendyk, B., P., Shor, A., and Cann, J. R., 1979, General implications of the Leg 49 drilling program for North Atlantic Ocean Geology, in: *Initial Reports of the Deep Sea Drilling Project*, Luyendyk, B. P., and Cann, J. R., eds., v. 49, Washington, D.C.,: U.S. Government Printing Office.

McDougall, I., and Wensink, H., 1966, Paleomagnetism and geochronology of the Pliocene–Pleistocene Lavas in Iceland, *Earth Planet. Sci. Lett.*, v. 1, p. 232–236.

McDougall, I., Watkins, N. D., Walker, G. P., and Kristjansson, L., 1976a, Potassium–argon and paleomagnetic analysis of Icelandic lava flows: Limits on the age of anomaly 5, *J. Geophys. Res.*, v. 81, p. 1505–1512.

McDougall, I., Watkins, N. D., and Kristjansson, L., 1976b, Geochronology and paleomagnetism of a Miocene–Pliocene lava sequence at Bessastadaa, eastern Iceland, *Am. J. Sci.*, v. 276, p. 1078–1095.

McDougall, I., Saemundsson, K., Johannesson, H., Watkins, N. D., and Kristjansson, L., 1977, Extension of the geomagnetic polarity time scale to 6.5 m.y.: K–Ar dating, geological and paleomagnetic study of a 3,500-m lava succession in western Iceland, *Geol. Soc. Am. Bull.*, v. 88, p. 1–15.

McIntyre, A., Ruddiman, W. F., and Jantzen, R., 1972, Southward penetrations of the North Atlantic polar front: Faunal and floral evidence of large-scale surface water mass movement over the last 225,000 years, *Deep-Sea Res.*, v. 19, p. 61–77.

McMaster, R. L., Schilling, J.-G., and Pinet, P. R., 1977, Plate boundary within Tjornes Fracture Zone on northern Iceland's insular margin, *Nature (London)*, v. 269, p. 663–668.

Melson, W. G., Vallier, T. L., Wright, T. L., Byerly, G., and Nelen, G., 1976, Chemical diversity of abyssal volcanic glass erupted along Pacific, Atlantic, and Indian Ocean spreading centers, in: *The Geophysics of the Pacific Ocean Basin and Its Margin*, Sutton, G. H., Manghani, M. H., and Moberly, R., eds., *Geophys. Monogr. Am. Geophys. Union*, v. 19, p. 351–367.

Meyer, O., Voppel, D. Fleischer, U., Closs, H., and Gerke, K., 1972, Results of bathymetric, magnetic and gravimetric measurements between Iceland and 70°N, *Dtsch. Hydrogr. Z.*, v. 25(5), p. 193–201.

Minster, J. B., and Jordan, T. H., 1978, Present-day plate motions, *J. Geophys. Res.*, v. 83, p. 5331–5354.

Minster, J. B., Jordan, T. H., Molnar, P. and Haines, E., 1974, Numerical modelling of instantaneous plate tectonics, *Geophys. J. R. Astron. Soc.*, v. 36, p. 541–576.

Mirlin, Ye., G., and Melikhov, V. R., 1976, New data on the nature of the magnetic anomalies in the North Atlantic, *Oceanology*, v. 16, p. 52–55.

Mitchell, B. J., Zollweg, J. E., Kohsmann, J. J., Cheng, C.-C., and Haug, E. J., 1979, Intraplate earthquakes in the Svalbard archipelago, *J. Geophys. Res.*, v. 84, p. 5620–5626.

Molina-Cruz, A., and Thiede, J., 1978, The glacial eastern boundary current along the Atlantic Eurafrican continental margin, *Deep-Sea Res.*, v. 25, p. 337–356.

Moorbath, S., Sigurdsson, H., and Goodwin, R., 1968, K–Ar ages of the oldest exposed rocks in Iceland, *Earth Planet. Sci. Lett.*, v. 4, p. 197–205.

Moore, J. G., and Schilling, J. G., 1973, Vesicules, water and sulfur in Reykjanes Ridge basalts, *Contrib. Mineral. Petrol.*, v. 41, p. 105–118.

Morgan, W. J., 1972, Deep mantle convection plumes and plate motions, *Am. Assoc. Petrol. Geol. Bull.*, v. 56(2), p. 203–213.

Morgan, W. J., 1973, Plate motions and deep mantle convection, in: *Studies in Earth and Space Sciences*, Shagam, R., ed., *Geol. Soc. Am. Mem.* 132, p. 7–22.

Myhre, A. M., 1978, Analyse av seismiske refraksjonsdata fra Norskehavet og omliggende kentinental marginer (Thesis, in Norwegian), University of Oslo, 135 p.

Nagy, J., 1973, Fossilførende blokker av mesozoisk alder fra Svalbardbanken, in: Publikasjon 42, *NTNF Kontinentalsokkel Prosjekt*, Oslo: Institutt for Geologi, 26 p.

Navrestad, T., and Jørgensen, F., 1979, Aeromagnetic investigations on the Jan Mayen Ridge, in: *Norwegian Sea Symposium*, Oslo: Norwegian Petroleum Society, NSS/9, p. 1–12.

Naylor, D., Pegrum, D., Rees, G., and Whiteman, A., 1974, The North Sea trough system, *Norwegian J. Oil Gas Matt.*, v. 2, p. 1–5.

Nicholson, R., 1974, The Scandinavian Caledonides, in: *The Ocean Basins and Margins*, v. 2, *The North Atlantic*, Nairn, A. E. M., and Stehli, F. G., eds., New York: Plenum Press, p. 161–203.

Nilsen, T. H., 1978, Lower Tertiary laterite on the Iceland–Faeroe Ridge and the Thulean land bridge, *Nature (London)*, v. 274, p. 786–788.

Noe-Nygaard, A., 1974, Cenozoic to Recent volcanism in and around the North Atlantic basin, in: *the Ocean Basins and Margins*, v. 2, *The North Atlantic*, Nairn, A. E. M., and Stehli, F. G., eds., New York: Plenum Press, p. 391–443.

Noe-Nygaard, A., 1976, Tertiary igneous rocks between Shannon and Scoresby Sund, East Greenland, in: *Geology of Greenland*, Escher, A., and Watt, W. S., eds., Copenhagen: Geological Survey of Greenland, p. 387–402.

O'Hara, M. J., 1973, Non-primary magmas and dubious mantle plume beneath Iceland, *Nature (London)*, v. 243, p. 507–508.

O'Hara, M. J, 1975, Is there an Icelandic mantle plume?, *Nature (London)*, v. 253, p. 527–529.

Olafsdottir, T., 1975, A moraine ridge on the Iceland shelf, west of Breidafjördur, *Natturufraedingurinn*, v. 45, p. 31–36.

O'Nions, R. K., and Pankhurst, R. J., 1973, Secular variation in the Sr-isotope composition of Icelandic volcanic rock, *Earth Planet. Sci. Lett.*, v. 21, p. 13–21.

O'Nions, R. K., Pankhurst, R. J., and Gronvold, K., 1976, Nature and development of basalt magma sources beneath Iceland and the Reykjanes Ridge, *J. Petrol.*, v. 17, p. 315–338.

Orvin, A. K., 1969, Outline of the geological history of Spitsbergen, *Nor. Svalbard Ishavs. Unders.*, v., 78, 57 p.

Ostenso, N. A., 1968, Geophysical studies in the Greenland Sea, *Geol. Soc. Am. Bull.*, v. 79, p. 107–132.

Øvrebø, O., and Talleraas, E., 1977, The structural geology of the Troms area (Barents Sea), *Geol. J.*, v. 1, p. 47–54.

Pálmason, G., 1965, Seismic refraction measurements of the basalt lavas of the Faeroe Islands, *Tectonophysics*, v. 2(6), p. 475–482.

Pálmason, G., 1971, *Structure of Iceland from Explosion Seismology*, Reykjavik: Science Institute, University of Iceland, 239 p.

Pálmason, G., 1973, Kinematics and heat flow in a volcanic rift zone, with application to Iceland, *Geophys. J. R. Astron. Soc.*, v. 33, p. 451–481.

Pálmason, G., 1974a, Heat flow and hydrothermal activity in Iceland, in: *Geodynamics of Iceland and the North Atlantic Area*, Kristjansson, L., ed., Dordrecht, Holland: D. Reidel, p. 297–306.

Pálmason, G., 1974b, The insular margin of Iceland, in: *The Geology of Continental Margins*, Burk, C. A., and Drake, C. L., eds., New York: Springer-Verlag, p. 375–398.

Pálmason, G., and Saemundsson, K., 1974, Iceland in relation to the Mid-Atlantic Ridge, *Annu. Rev. Earth Planet. Sci.*, v. 2, p. 25–50.

Pálmason, G., Arnorsson, S., Friedleifsson, I. B., Kristmannsdottir, H., Saemundsson, K., Stefansson, V., Steingrimsson, B., Tomasson, J., and Kristjansson, L., 1979, The Iceland crust: Evidence from drillhole data on structure and process, in: *Deep Drilling Results in the Atlantic Ocean*: Ocean Crust, Talwani, M., Harrison, C. G. A., and Hayes, D. E., eds., Maurice Ewing Series, Washington, D.C.: American Geophysical Union, v. 2, p. 43–65.

Parker, R. L., and Oldenburg, D. W., 1973, Thermal model of ocean ridges, *Nature Phys. Sci.*, v. 242, p. 137–139.

Parsons, B. P., and Sclater, J. G., 1977, An analysis of the variation of ocean floor bathymetry and heat flow with age, *J. Geophys. Res.*, v. 82, p. 803–827.

Pedersen, A. K., Engell, J., and Rønsbo, J. G., 1975, Early Tertiany volcanism in the Skagerrak: New chemical evidence from ash-layers in the mo-clay of northern Denmark, *Lithos*, v. 8, p. 255–268.

Perry, R. K., Fleming, H. S., Cherkis, N. Z., Feden, R. H., and Vogt., P. R., 1980, *Bathymetry of the Norwegian–Greenland and Western Barents Seas*, Washington, D.C.: U.S. Naval Research Laboratory.

Phillips, J. D., and Tapscott, C., 1980, A plate kinematic synthesis for the Greenland/Norwegian Sea and Arctic Ocean (in preparation).

Phillips, J. D., Feden, R., Fleming, H. S., and Tapscott, C., 1980, Aeromagnetic studies of the Greenland/Norwegian Sea and Arctic Ocean (in preparation).

Piper, J. D. A., 1971, Ground magnetic studies of crustal growth in Iceland, *Earth Planet. Sci. Lett.*, v. 12, p. 199–207.

Piper, J. D. A., 1973, Interpretation of some magnetic anomalies over Iceland, *Tectonophysics*, v. 16, p. 163–187.

Pitman, W. C., III, and Talwani, M., 1972, Sea-floor spreading in the North Atlantic, *Geol. Soc. Am. Bull.*, v. 83, p. 619–646.

Poore, R. Z., and Berggren, W. A., 1975, Late Cenozoic planktonic foraminiferal biostratigraphy and paleoclimatology of Hatton-Rockall basin: DSDP Site 116, *J. Foraminiferal Res.*, v. 5, p. 270–293.

Post, R. L., Jr., and Griggs, D. T., 1973, The earth's mantle: Evidence of non-Newtonian flow, *Science*, v. 181(4106), p. 1242–1244.

Printzlau, I., and Larsen, O., 1972, K/Ar determinations on alkaline olivine basalts from Skåne, southern Sweden, *Geol. For. Stockh. Forhandl.*, v. 94, p. 259–269.

Pye, G. D., Hyndman, R. D., 1972, Heat flow measurements in Baffin Bay and the Labrador Sea, *J. Geophys. Res.*, v. 77, p. 938–944.

Rabinowitz, P. D., and Eittreim, S. L., 1974, Bottom current measurements in the Labrador Sea, *J. Geophys. Res.*, v. 79, p. 4085–4090.

Ramberg, I. B., 1976, Gravity interpretation of the Oslo Graben and associated igneous rocks, *Nor. Geol. Unders. Publ.*, v. 325, 194 p.

Rasmussen, M. H., Campise, J., Dittmer, F. B., and Bailey, J. C., 1976, Basalts from the southeastern Greenland continental margin, *Bull. Geol. Soc. Denm.*, v. 25, p. 73–78.

Renard, V., and Malod, J., 1974, Structure of the Barents Sea from seismic refraction, *Earth Planet. Sci Lett.*, v. 24, p. 33–47.

Roberts, D. G., 1974, Structural development of the British Isles, the continental margin, and the Rockall Plateau, in: *The Geology of Continental Margins*, Burk, C. A., and Drake, C. L., eds., New York: Springer-Verlag, p. 343–360.

Roberts, D. G., Flemming, N. C., Harrison, R. K., and Binns, P., 1974, Helen's Reef: A Cretaceous microgabbroic intrusion in the Rockall intrusive center, *Marine Geol.*, v. 16, p. M21–M30.

Roeser, H. A., Plaumann, S., and Forstner-Ballheim, E., 1975, The crustal structure of the Norwegian continental margin and the Norwegian Basin according to magnetic and gravimetric measurements, *"Meteor" Forschungsergeb. Reihe C*, v. 21, p. 1–14.

Rønnevik, H. C., and Motland, K., 1979, Geology of the Barents Sea, in: *Norwegian Sea Symposium*, Oslo: Norwegian Petroleum Society, NSS/15, 34 p.

Rønnevik, H., and Navrestad, T., 1977, Geology of the Norwegian shelf between 62°N and 69°N, *Geol. J.*, v. 1, p. 33–46.

Rønnevik, H., Jorgensen, F., and Motland, K., 1979, The geology of the northern part of the Vøring plateau, in: *Norwegian Sea Symposium*, Oslo: Norwegian Petroleum Society, NSS/12, 21 p.

Rosenkrantz, A., and Pulvertaft, T. C. R., 1969, Cretaceous–Tertiary stratigraphy and tectonics in Northern Greenland, in: *North Atlantic—Geology and Continental Drift*, Kay, M., ed., *Am. Assoc. Petrol. Geol. Mem.* 12, p. 883–898.

Ruddiman, W. F., 1972, Sediment redistribution on the Reykjanes ridge: Seismic evidence, *Geol. Soc. Am. Bull.*, v. 83, p. 2039–2062.

Ruddiman, W. F., and Bowles, F. A., 1976, Early interglacial bottom-current sedimentation on the eastern Reykjanes Ridge, *Mar. Geol.*, v. 21, p. 191–210.

Ruddiman, W. F., and Glover, L. K., 1975, Subpolar North Atlantic circulation at 9300 yr BP: Faunal evidence, *Quat. Res.*, v. 5, p. 361–389.

Ruddiman, W. F., and McIntyre, A., 1973, Time-transgressive de-glacial retreat of polar waters from the North Atlantic, *Quat. Res.*, v. 3, p. 117–130.

Ruddiman, W. F., and McIntyre, A., 1976, Northeast Atlantic paleoclimatic changes over the last 600,000 years, *Geol. Soc. Am. Mem.* 145, p. 111–146.

Ruddiman, W. F., and McIntyre, A., 1977, Late Quaternary surface ocean kinematics and climatic change in the high-latitude North Atlantic, *J. Geophys. Res.*, v. 82, p. 3877–3887.

Russell, M. J., 1976, A possible Lower Permian age for the onset of ocean floor spreading in the northern North Atlantic, *Scott. J. Geol.*, v. 12, p. 315–323.

Rygg, E., 1972, Rayleigh wave dispersion and crustal structure: The Norwegian Sea and adjacent areas, *Nor. Polarinst. Arbok.*, v. 1970, p. 169–177.

Saemundsson, K., 1974, Evolution of the axial rifting zone in northern Iceland and the Tjörnes fracture zone, *Geol. Soc. Am. Bull.*, v. 85, p. 495–504.

Saito, T., Burckle, L. H., and Horn, D. R., 1967, Paleocene core from the Norwegian basin, *Nature (London)*, v. 216, p. 357–359.

Savin, S. M., Douglas, R. G., and Stehli, F. G., 1975, Tertiary marine paleotemperatures, *Geol. Soc. Am. Bull.*, v. 86, p. 1499–1510.

Schäfer, K., 1972, Transform faults in Iceland, *Geol. Rundsch.*, v. 61, p. 942–960.

Schäfer, K., 1975, Horizontal and vertical crustal movements in Iceland, *Tectonophysics*, v. 29, p. 223–231.

Schilling, J. G., 1973a, Iceland mantle plume: Geochemical study of Reykjanes ridge, *Nature (London)*, v. 242(5400), p. 565–571.

Schilling, J. G., 1973b, Iceland mantle plume, *Nature (London)*, v. 246, p. 141–143.

Schilling, J. G., 1976, Rare-earth, Sc, Cr, Fe, Co, and Na abundances in DSDP Leg 38 basement basalts: Some additional evidence on the evolution of the Thulean volcanic province, in:

Initial Reports of the Deep Sea Drilling Project, v. 38, Talwani, M., *et al.*, eds., Washington, D.C.: U.S. Government Printing Office, p. 741–750.

Schilling, J. G., and Noe-Nygaard, A., 1974, Faeroe–Iceland plume: Rare-earth evidence, *Earth Planet. Sci. Lett.*, v. 24, p. 1.

Schilling, J.-G., Sigurdsson, H., and Kingsley, R. H., 1978, Skagi and western Neovolcanic zones in Iceland: II, Geochemical variations, *J. Geophys. Res.*, v. 83, p. 3983–4002.

Schlüter, H. U., and Hinz, K., 1978, The continental margin of West Spitsbergen, *Polarforschung*, v. 48, p. 151–169.

Schrader, H. J., Bjørklund, K., Manum, S., Martini, E., and Van Hinte, J., 1976, Cenozoic biostratigraphy, physical stratigraphy, and paleooceanography in the Norwegian-Greenland Sea DSDP Leg 38 Paleontological synthesis, in: *Initial Reports of the Deep Sea Drilling Project*, v. 38, Talwani, M., *et al.*, eds., Washington, D.C.: U.S. Government Printing Office, p. 1197–1211.

Scientific Staff, 1975, Leg 38—Deep Sea Drilling Project, *Geotimes*, v. 20(3), p. 24–26.

Scientific Staff, 1976, Glomar Challenger sails on Leg 48, *Geotimes*, v. 21(12), p. 19–23.

Scientific Staff, 1977, "Young and hot" drilling, *Geotimes*, v. 22, p. 25–28.

Sclater, J. G., and Francheteau, J., 1970, The implications of terrestrial heat-flow observations on current tectonic and geochemical models of the crust and upper mantle of the earth, *Geophys. J. R. Astron. Soc.*, v. 20, p. 509–542.

Sclater, J. G., Anderson, R. N., and Bell, M. L., 1971, Elevation of ridges and evolution of the central eastern Pacific, *J. Geophys. Res.*, v. 76, p. 7888–7915.

Sclater, J. G., Lawver, L. A., and Parsons, B., 1975, Comparison of long-wavelength residual elevation and free-air gravity anomalies in the North Atlantic and possible implication for the thickness of the lithospheric plate, *J. Geophys. Res.*, v. 80, p. 1031–1052.

Scrutton, R. A., 1973, The age relationship of igneous activity and continental break-up, *Geol. Mag.*, v. 110, p. 227–234.

Sellevoll, M. A., 1975, Seismic refraction measurements and continuous seismic profiling on the continental margin off Norway between 60°N and 69°N, *Nor. Geol. Unders. Publ.*, v., 316, p. 219–235.

Sellevoll, M. A., and Sundvor, E., 1974, The origin of the Norwegian Channel—A discussion based on seismic measurements, *Can. J. Earth Sci.*, v. 11, p. 224–231.

Serson, P. H., Hannaford, W., and Haines, G. V., 1968, Magnetic anomalies over Iceland, *Science*, v. 162, p. 355–357.

Shackleton, N. J., and Opdyke, N. D., 1976, Oxygen-isotope and paleomagnetic stratigraphy of Pacific core V 28–239: Late Pliocene to latest Pleistocene, *Geol. Soc. Am. Mem.* 145, p. 449–464.

Shih, J. S. F., Atwater, T., and McNutt, M., 1978, A near-bottom traverse of the Reykjanes Ridge, *Earth Planet. Sci. Lett.*, v. 39, p. 75–83.

Sigurdsson, H., Schilling, J.-G., and Meyer, P. S. 1978, Skagi and Langjökull volcanic zones in Iceland: I—Petrology and structure, *J. Geophys. Res.*, v. 83, p. 3971–3982.

Sigvaldason, G., 1974, Basalts from the centre of the assumed Icelandic mantle plume, *J. Petrol.*, v. 15, p. 497–524.

Sigvaldason, G., and Steinthorsson, S., 1974, Chemistry of tholeiitic basalts from Iceland, and their relation to the Kverkfjoll hot spot, in: *Geodynamics of Iceland and the North Atlantic Area*, Kristjansson, K., ed., Dordrecht, Holland: D. Reidel, p. 155–164.

Sipkin, S. A., and Jordan, T. H., 1975, Lateral heterogeneity of the upper mantle determined from the travel times of ScS, *J. Geophys. Res.*, v. 80, p. 1474–1484.

Sleep, N. H., 1971, Thermal effects of the formation of Atlantic continental margin by continental break up, *Geophys. J.*, v. 24, p. 325–350.

Sleep, N. H., and Biehler, S., 1970, Topography and tectonics at the intersections of fracture zones with central rifts, *J. Geophys. Res.*, v. 75, p. 2782–2752.

Sobczak, L. W., 1975, Gravity anomalies and passive continental margins, Canada and Norway, in: *Canada's Continental Margins and Offshore Petroleum Exploration*, Yorath, C. J., Parker, E. R., and Glass, D. J., eds., *Can. Soc. Petrol. Geol. Mem.* 4, p. 743–761.

Solomon, S. C., 1973. Shear wave attenuation and melting beneath the Mid-Atlantic Ridge, *J. Geophys. Res.*, v. 78, p. 6044–6059.

Sommerhoff, G., 1973, Formenschatz und morphologische Gliederung des sudost grönlandischen Schelfgebietes und Kontinental-abhanges, "*Meteor*" *Forschungsergeb. Reihe C.*, v. 15, p. 1–54.

Soper, N. J., Higgins, A. C., Downie, C., Matthews, D. W., and Brown, P. E., 1975, Late Cretaceous–early Tertiary stratigraphy of the Kangerlugssuak area, East Greenland, and the age of opening of the northeast Atlantic, *Geol. Soc. London J.*, v. 132, p. 85–104.

Srivastava, S. P., 1978, Evolution of the Labrador Sea and its bearing on the early evolution of the North Atlantic, *Geophys. J. R. Astron. Soc.*, v. 52, p. 313–357.

Srivastava, S. P., 1979, Marine gravity and magnetic anomaly maps of the Labrador Sea, *Geol. Surv. Can. Open File Rep.* 627.

Strauch, F., 1970, Die Thule—Landbrücke als Wanderweg und Faunenscheide zwischen Atlantik und Skandik im Tertiar, *Geol. Rundschau*, v. 60, p. 381–417.

Sundvor, E., 1974, Seismic refraction and reflection measurements in the southern Barents Sea, *Mar. Geol.*, v. 16, p. 255–273.

Sundvor, E., 1975, Thickness and distribution of sedimentary rocks in the southern Barents Sea, *Nor. Geol. Unders.*, v. 316, p. 237–240.

Sundover, E., and Nysaether, F., 1975, Geological outline of the Norwegian continental margin between 60° and 68°N, in: *Canada's Continental Margins*, Yorath, C. J., Parker, E. R., and Glass, D. J., eds., Calgary: Canadian Society of Petroleum Geologists, p. 267–281.

Sundvor, E., Sellevoll, M., Gidskehaug, A., and Eldholm, O., 1978, Seismic investigations on the western and northern margin off Svalbard, *Polarforschung*, v. 48, p. 41–43.

Sundvor, E., Myhre, A., and Eldholm, O., 1979, The Svalbard continental margin, *Norwegian Sea Symposium*, Oslo: Norwegian Petroleum Society, NNS-6, p. 1–25.

Surlyk, F., 1978, Jurassic basin evolution of East Greenland, *Nature (London)*, v. 274, p. 130–133.

Surlyk, F., Clemmensen, L., and Larsen, H. C., 1980, Post-Palaeozoic evolution of the East Greenland continental margin, *Can. Assoc. Petrol. Geol.*, v. 7 (in press).

Sykes, L. R., 1965, The seismicity of the Arctic, *Bull Seism. Soc. Am.*, v. 55, p. 501–518.

Sykes, L. R., 1967, Mechanism of earthquakes and nature of faulting on the mid-Oceanic ridges, *J. Geophys. Res.*, v. 72, p. 2131–2153.

Sykes, L. R., 1970, Earthquake swarms and sea-floor spreading, *J. Geophys. Res.*, v. 75, p. 6598–6611.

Sykes, L. R., 1978, Intraplate seismicity, reactivation of preexisting zones of weakness, alkaliane magmatism and other tectonism postdating continental fragmentation, *Rev. Geophys. Space Phys.*, v. 16(4), p. 621–688.

Sykes, L. R., and Sbar, M. L., 1974, Focal mechanism solutions of intraplate earthquakes and stresses in the lithosphere, in: *Geodynamics of Iceland and the North Atlantic Area*, Kristjansson, L., ed., Dordrecht, Holland: D. Reidel, p. 207–224.

Sylvester, A. G., 1975, History and surveillance of volcanic activity on Jan Mayen Island, *Bull. Volc.*, v. 39, p. 1–23.

Talwani, M., and Eldholm, O., 1972, The continental margin off Norway: A geophysical study, *Geol. Soc. Am. Bull.*, v. 83, p. 3575–3608.

Talwani, M., and Eldholm, O., 1974, Margins of the Norwegian–Greenland Sea, in: *The Geology of Continental Margins*, Burk, C. A., and Drake, C. L., eds., New York: Springer-Verlag, p. 361–374.

Talwani, M., and Eldholm, O., 1977, Evolution of the Norwegian–Greenland Sea, *Geol. Soc. Am. Bull.*, v. 88, p. 969–999.

Talwani, M., and Gronlie, G., 1976, Free-air gravity field of the Norwegian–Greenland Seas, Chart MC-15, Boulder, Colorado: Geological Society of America.

Talwani, M., and Udintsev, G., 1976, Tectonic synthesis, in: *Initial Reports of the Deep Sea Drill-*

ing Project, v. 38, Talwani, M., *et al.,* eds. Washington, D.C.: U.S. Government Printing Office, p. 1101–1168.

Talwani, M., Windisch, C. C., and Langseth, M. G., 1971, Reykjanes ridge crest: A detailed geophysical study, *J. Geophys. Res.,* v. 76, p. 473–517.

Tarling, D. H., and Gale, N. H., 1968, Isotope dating and paleomagnetic polarity in the Faeroe Islands, *Nature (London),* 218, p. 1043–1044.

Tarling, D. H., and Mitchell, J. G., 1976, Revised Cenozoic polarity time scale, *Geology,* v. 4, p. 133–136.

Thiede, J., 1977, Aspects of the variability of the Glacial and Interglacial North Atlantic eastern boundary current (last 150,000 years), *"Meteor" Forschungsergebn. Reihe C.,* v. 28, p. 1–36.

Thiede, J., 1979, Paleogeography and paleobathymetry of the Mesozoic and Cenozoic North Atlantic Ocean, *Geol. J.,* v. 3, p. 263–272.

Thiede, J., 1980, Paleo-oceanography, margin stratigraphy, and paleo-physiography of the Tertiary North Atlantic and Norwegian–Greenland Seas, *Phil. Trans. R. Soc. London A,* v. 294, p. 177–185.

Thorarinsson, S., 1974, On the topography of the volcanic zones in Iceland, in: *Geodynamics of Iceland and the North Atlantic Area,* Kristjansson, L., ed., Dordrecht, Holland: D. Reidel, p. 203–206.

Tryggvason, E., 1959, Longitudinal wave velocity in the earth's crust in Iceland, *Natturufraedingurinn (Reykjavik),* v. 29, p. 80–84.

Tryggvason, E., 1961, Wave velocity in the upper mantle below the Arctic–Atlantic Ocean and northwest Europe, *Ann. Geofis.,* v. 14, p. 379–392.

Tryggvason, E., 1962, Crustal structure of the Iceland region from dispersion of surface waves, *Bull. Seism. Soc. Am.,* v. 52, p. 359–388.

Tryggvason, E., 1964, Arrival of P-waves and upper mantle structure, *Bull Seism. Soc. Am.,* v. 54, p. 727–736.

Tryggvason, E., 1973, Seismicity, earthquake swarms, and plate boundaries in the Iceland region, *Bull. Seism. Soc. Am.,* v., 63, p. 1327–1348.

Tryggvason, E., 1974, Vertical crustal movement in Iceland, in: *Geodynamics of Iceland and the North Atlantic Area,* Kristjansson, L., ed., Dordrecht, Holland: D. Reidel, p. 241–262.

Ulrich, J., 1960, Zur Topographie des Reykjanes-Rückens, *Kieler Meeresforsch.,* v. 16, p. 155–163.

Unni, C. K., and Schilling, J.-G., 1978, Cl and Br degassing by volcanism along the Reykjanes Ridge and Iceland, *Nature (London),* v. 272, p. 19–23.

Vail, P. R., Mitchum, R. M., Todd, R. G., Widmier, J. M., Thompson, S., Sangree, J. B., Bubb, J. N., and Hatlelid, W. G., 1977, Seismic stratigraphy and global changes of sea level, in: *Seismic Stratigraphy—Applications to Hydrocarbon Exploration,* Payton, C. E., ed. *Am. Assoc. Petr. Geol. Mem.* 26, p. 49–212.

Vann, I. R., 1974, A modified predrift fit of Greenland and western Europe, *Nature (London),* v. 251, p. 209–211.

Vine, F. J., 1966, Spreading of the ocean floor: New Evidence, *Science,* v. 154, p. 1405–1415.

Vinje, T. E., 1976, Sea ice conditions in the European sector of the marginal seas of the Arctic, 1966–1975, *Nor. Polarinstitutt Årbok,* v. 1975, p. 163–174.

Vogt, P. R., 1970, Magnetized basement outcrops on the southeast Greenland continental shelf, *Nature (London),* v. 226, p. 743–744.

Vogt, P. R., 1971, Asthenosphere motion recorded by the ocean floor south of Iceland, *Earth Planet. Sci. Lett.,* v. 13, p. 153–160.

Vogt, P. R., 1972a, The Greenland–Iceland–Faeroe aseismic ridge and the Western Boundary Undercurrent, *Nature (London),* v. 239, p. 79–81.

Vogt, P. R., 1972b, Evidence for global synchronism in mantle plume convection, and possible significance for geology, *Nature (London),* v. 240, p. 338–342.

Vogt, P. R., 1974, The Iceland phenomenon: Imprints of a hot spot on the ocean crust and implications for flow below the plates, in: *Geodynamics of Iceland and the North Atlantic Area,* Kristjansson, L., ed., Dordrecht, Holland: D. Reidel, p. 105–126.

Vogt, P. R., 1976, Plumes, sub-axial pipe flow, and topography along the Mid-Oceanic Ridge, *Earth Planet. Sci. Lett.*, v. 29, p. 309–325.

Vogt, P. R., 1978, Long-wavelength gravity anomalies and intraplate seismicity, *Earth Planet. Sci. Lett.*, v. 37, p. 465–475.

Vogt, P. R., 1979a, Global magmatic episodes: New evidence and implications for the steady-state mid-oceanic ridge, *Geology*, v. 7, p. 93–98.

Vogt, P. R., 1979b, Amplitudes of oceanic magnetic anomalies and the chemistry of oceanic crust: Synthesis and review of "magnetic telechemistry," *Can. J. Earth Sci.*, v. 16, p. 2236–2262.

Vogt, P. R., and Avery, O. E., 1974a, Detailed magnetic surveys in the northeast Atlantic and Labrador Sea, *J. Geophys. Res.*, v. 79, p. 363–389.

Vogt, P. R., and Avery, O. E., 1974b, Tectonic history of the Arctic basins: Partial solutions and unsolved mysteries, in: *Marine Geology and Oceanography of the Arctic Seas*, Herman, Y., ed., New York: Springer-Verlag, p. 83–117.

Vogt, P. R., and Johnson, G. L., 1972, Seismic reflection survey of an oblique aseismic basement trend on the Reykjanes Ridge, *Earth Planet. Sci. Lett.*, v. 15, p. 248–254.

Vogt, P. R., and Johnson, G. L., 1973, A longitudinal seismic reflection profile of the Reykjanes Ridge: Part II—Implications for the mantle hot spot hypothesis, *Earth Planet. Sci. Lett.*, v. 18, p. 49–58.

Vogt, P. R., and Johnson, G. L., 1974, Magnetic telechemistry is elegant but nature is complex—A reply, *Nature (London)*, v. 251, p. 498–499.

Vogt, P. R., and Johnson, G. L., 1975, Transform faults and longitudinal convection below the Mid-Oceanic Ridge, *J. Geophys. Res.*, v. 80, p. 1399–1428.

Vogt, P. R., and Ostenso, N. A., 1973, Reconnaissance geophysical studies in the Barents and Kara seas—Summary, in: *Arctic Geology*, Pitcher, M. G., ed., Tulsa, Oklahoma: American Association of Petroleum Geologists, p. 588–598.

Vogt, P. R., and Perry, R., 1978, Post-rifting accretion of continental margins in the Norwegian–Greenland and Labrador Seas: Morphologic evidence, *EOS Trans. Am. Geophys. Union*, v. 59, p. 1204.

Vogt, P. R., Schneider, E. D., and Johnson, G. L., 1969, The crust and upper mantle beneath the sea, in: *The Earth's Crust and Upper Mantle*, Hart, P. J., ed., Washington, D.C.: American Geophysical Union, p. 556–617.

Vogt, P. R., Ostenso, N. A., and Johnson, G. L., 1970, Magnetic and bathymetric data bearing on sea-floor spreading north of Iceland, *J. Geophys. Res.*, v. 75, p. 903–920.

Vogt, P. R., Feden, R. H., Eldholm, O., and Sundvor, E., 1978, The ocean crust west and north of the Svalbard Archipelago: Synthesis and review of new results, *Polarforschung*, v. 48, p. 1–19.

Vogt, P. R., Taylor, P. T., Kovacs, L., and Johnson, G. L., 1979a, Detailed aeromagnetic investigation of the Arctic Basin, *J. Geophys. Res.*, v. 84, p. 1071–1089.

Vogt, P. R., Kovacs, L. C., Johnson, G. L., and Feden, R. H., 1979b, The Eurasia Basin, in: *Norwegian Sea Symposium*, Oslo: Norwegian Petroleum Society, NSS-3, p. 1–29.

Vogt, P. R., Johnson, G. L., and Kristjansson, L., 1980, Morphology and magnetic anomalies north of Iceland, *J. Geophys.*, v, 47, p. 67–80.

Voppel, D., and Rudloff, R., 1980, On the evolution of the Reykjanes Ridge south of 60°N between 40 and 12 million years before present, *J. Geophys.*, v. 47, p. 61–66.

Voppel, D., Srivastava, S. P., and Fleischer, U., 1979, Detailed magnetic measurements south of the Iceland-Faeroe Ridge, *Deut. Hydrogr. Z.*, v. 32, p. 154–172.

Vorren, T. O., and Lind-Hansen, O. W., 1979, Quaternary sediments on the continental shelf off Troms and West Finnmark, in: *Norwegian Sea Symposium*, Oslo: Norwegian Petroleum Society, NSS/21, 13 p.

Wasgstein, R., and Rasmussen, J., 1975, Glacial erratics from the sea-floor southeast of the Faeroe Islands and the limit of glaciation, *Ann. Soc. Faeroensis*, v. 23, p. 101–119.

Walker, G. P. L., 1974, The structure of eastern Iceland, in: *Geodynamics of Iceland and the North Atlantic Area*, Kristjansson, L., ed., Dordrecht, Holland: D. Reidel, p. 177–188.

Ward, P. L., and Björnsson, S., 1971, Microearthquakes, swarms, and the geothermal areas of Iceland, *J. Geophys. Res.*, v. 76, p. 3953.

Warnke, D. A., and Hanson, M. E., 1977, Sediments of glacial origin in the area of operations of DSDP Leg 38 (Norwegian–Greenland Seas): Preliminary results from Sites 336 and 344, *Ber. Naturforsch. Ges. Freiburg im Breisgau*, v. 67, p. 371–392.

Watkins, N. D., and Walker, G. P. L., 1977, Magnetostratigraphy of eastern Iceland, *Am. J. Sci.*, v. 277, p. 513–584.

Watkins, N. D., McDougall, I., and Kristjansson, L., 1977, Upper Miocene and Pliocene geomagnetic secular variation in the Borgarfjordur area of western Iceland, *Geophys. J. R. Astron. Soc.*, v. 49, p. 609–632.

Watt, W. S., 1969, The coast-parallel dike swarm of southwest Greenland in relation to the opening of the Labrador Sea, *Can. J. Earth Sci.*, v. 6, p. 1320–1321.

Wegener, A., 1928, *The Origin of Continents and Oceans*, Biran, J., trans., New York: Dover Publications, 1966.

Weidick, A., 1976, Glaciation and the Quaternary of Greenland, in: *Geology of Greenland*, Escher, A., and Watt, W. S., eds., Copenhagen: Geological Survey of Greenland, p. 430–458.

Weigand, P. W., 1972, Bulk-rock and mineral chemistry of Recent Jan Mayen basalts, *Nor. Polarinst. Årbok*, v. 1970, p. 42–52.

West, R. M., and Dawson, M. R., 1978, Vertebrate paleontology and the Cenozoic history of the North Atlantic region, *Polarforschung*, v. 48, p. 103–119.

Whiteman, A. J., Rees, G., and Pegrum, R. M., 1975, North Sea troughs and plate tectonics, *Nor. Geol. Unders. Publ.*, v. 316, p. 137–161.

Whitmarsh, R. B., 1971, Seismic anisotropy of the uppermost mantle absent beneath the east flank of the Reykjanes ridge, *Bull. Seism. Soc. Am.*, v. 61, p. 1351–1369.

Williams, C. A., 1975, Sea-floor spreading in the Bay of Biscay and its relationship to the North Atlantic, *Earth Planet. Sci. Lett.*, v. 24, p. 440–456.

Wilson, J. T., 1965, A new class of faults and their bearing on continental drift, *Nature (London)*, 207, p. 343.

Wilson, J. T., 1966, Did the Atlantic close and then reopen?, *Nature (London)*, v. 211, p. 676–680.

Woodland, A. W., ed., 1975, *Petroleum and the Continental Shelf of North-West Europe*, v. 1, *Geology*, London: Applied Science Publishers, 501 p.

Ziegler, P. A., 1977, Geology and hydrocarbon provinces in the North Sea, *Geo. J.*, v. 1, p. 7–32.

Zielinski, G. W., 1977, Thermal history of the Norwegian Greenland Sea and its rifted continental margin, Ph.D. Thesis, New York: Columbia University, 160 p.

Chapter 12

GEOLOGY AND GEOPHYSICS OF THE AMERASIAN BASIN

David L. Clark

Department of Geology and Geophysics
University of Wisconsin
Madison, Wisconsin

I. INTRODUCTION

The deeper Arctic Ocean is conveniently divided into two unequal portions. The Amerasian Basin is the larger. It is separated from the smaller Eurasian Basin by the Lomonosov Ridge, a feature that forms a prominent backbone for the deeper Arctic Ocean (Fig. 1).

It was not until almost the beginning of the twentieth century that the true deep ocean nature of the Arctic was known. Prior to this it was believed that discontinuous landmasses extended from Greenland to Siberia. The permanent ice cover obscured the oceanic nature of the area. Fridtjof Nansen discovered the Arctic Basin's true nature in a unique manner. In 1893 he sailed the *Fram* into the ice pack off the New Siberian Islands and allowed the ship to become icebound. In this frozen state he completed a drift to Spitsbergen three years later in 1896. Eight wire soundings taken during this voyage proved the deep ocean nature of the Arctic Basin (Nansen, 1904). Other attempts at Arctic Ocean bathymetric studies include an incomplete sounding by Perry at the North Pole in 1908 and USSR icebreaker activity from 1935–1939 (Dietz and Shumway, 1961). Wilkins (1928) flew into the deeper Arctic in 1928 for a sounding and this method was expanded by the USSR just prior to World War II. Papanen (1947) landed on the ice and established a research station near

Fig. 1. Index map of Amerasian Basin and its relationship to other major Arctic Ocean physiographic features (modified from Ostenso and Wold, 1973, with permission of the authors and the American Association of Petroleum Geologists).

the North Pole. This was the beginning of a new mode of ocean study. In the 1950s both the USSR and the US occupied bases on pack ice as well as ice islands. The latter are icebergs from Greenland and Ellesmere Island glaciers that are incorporated into the ice pack. This work initiated a mode of investigation that continues to the present. The motion of the permanent ice pack is such that rather wide traverses of the Arctic Basin have been achieved. Most important for the Amerasian Basin has been the work from Fletcher's Ice Island T-3 (Fig. 2). This scientific station was occupied more or less continuously from the early 1950s until 1974. Geophysical and sedimentological investigations were made over the area covered during drift and more than 500 short sediment cores were recovered. Most of our direct information concerning Amerasian Basin geology has been obtained in this

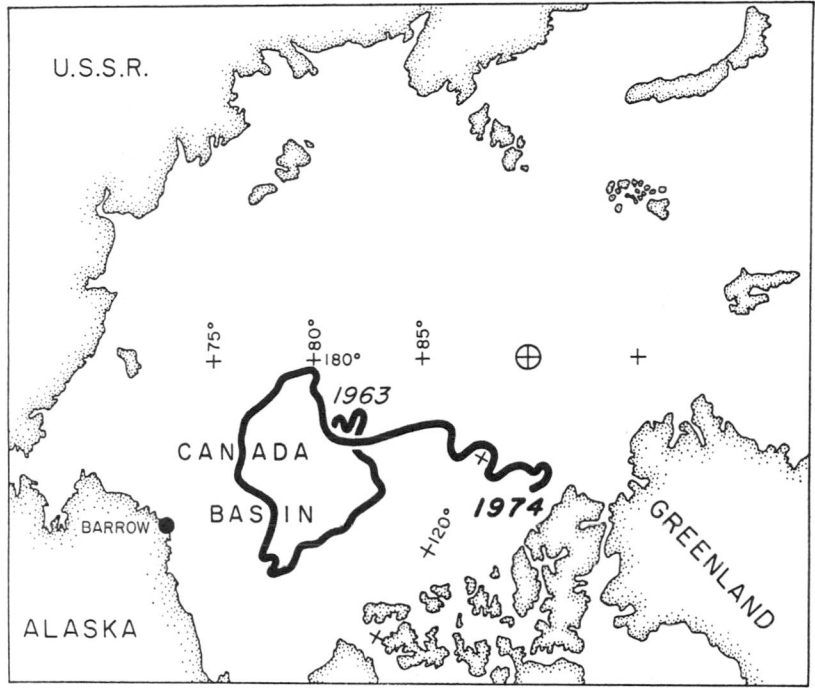

Fig. 2.   Trace of the drift of Ice Island T-3 from 1963 to 1974. Much of data for this chapter was collected along this traverse.

manner. In addition, aerial surveys have provided a variety of geophysical data during the past 15 years. More recently, nuclear submarines passing beneath the ice have gathered some bathymetric data. The differing techniques of exploration have all contributed to our present understanding of the Arctic. Almost every major structure known has been discovered during the past 25 years (Fig. 3).

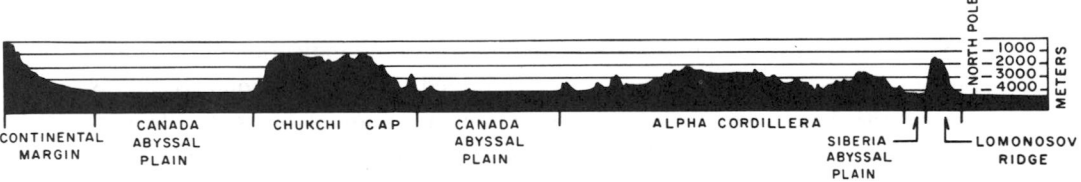

Fig. 3.   Profile across the Amerasian Basin from Alaska to the Lomonosov Ridge. Compare with Figs. 1 and 4 for major structural features. Profile based on SSN Nautilus soundings and taken from Demenitskaya and Hunkins (1970), by permission of the authors and John Wiley and Sons. Vertical exaggeration ×26.

II. AMERASIAN BASIN

The Amerasian Basin is an irregular triangular basin of approximately 3 million km², bounded on the north by the Lomonosov Ridge. The sides of the basin are formed by prominent escarpments bordering the shallower Siberian, Chukchi, and Beaufort Seas. The Lomonosov Ridge rises 3000 m from its base. Apparently, this ridge was derived from the Barents Shelf during the formation of the Eurasian Basin. The Amerasian Basin consists of several significant physiographic provinces: the Canada Basin, an abyssal plain averaging 3000 m in depth; the Alpha Ridge, an irregular high cordillera that extends the width of the basin; the Chukchi Plateau; and the Wrangel and Fletcher Abyssal Plains (Fig. 4). Differing nomenclature has been used to describe features of this area, although an attempt to stabilize terminology was published by Beal *et al.* (1966).

A. Canada Abyssal Plain

This plain is outlined by the 3000-m depth contour and comprises the largest portion of the Amerasian Basin. More than one hundred 3- to 4-m sediment cores have been taken here. Turbidity deposits comprise more than 27% of the cores in the northeastern and southeastern parts of the abyssal plain, but less than 3% in the north-central part of the plain. Turbidity-dominated sediment obscures older sedimentary and structural relationships.

The oldest sediment yet recovered in the Canadian Basin is 700,000 years old (Middle Pleistocene). An average sedimentation rate of 83 mm/1000 years is supported by a ^{14}C date of 700 y.b.p. for Foraminifera in a brown lutite layer recognized in 34 cores. This is the highest rate of sedimentation known in the Arctic Ocean (with the possible exception of that in the Wrangel Abyssal Plain) and compares to 2 mm/1000 years calculated for most other parts of the Amerasian Basin (Clark, 1970).

Little is known concerning the structure beneath this surficial sediment cover. Churkin (1969, 1973) has inferred that the Canada Plain may have an early Paleozoic age because of the relationships of Paleozoic geosynclines along its southern margin. Hall (1973) interpreted seismic profiles across the Canadian Plain as indicating 2 km of sediment with no indication of basement (Fig. 5). Gramberg and Kulakov (1975) reported that the Canadian Basin, as part of the Amerasian Basin, may represent a collapsed block of "Hyperborean Platform," a part of the North American–Greenland lithospheric plate. Hall (1973) suggested that the oldest sediment in the Canada Basin may be located near the Mackenzie Delta based on inferred directions of spreading along the Alpha Cordillera.

Wold *et al.* (1970) reported that the transition from continental to oceanic

ig. 4. Bathymetry of Amerasian Basin and major structural features, as of October 1971. Negative values represent epth in feet below sea level; positive values represent height in feet above sea level (Heezen and Tharp, 1971; © 1971 Jational Geographic Society).

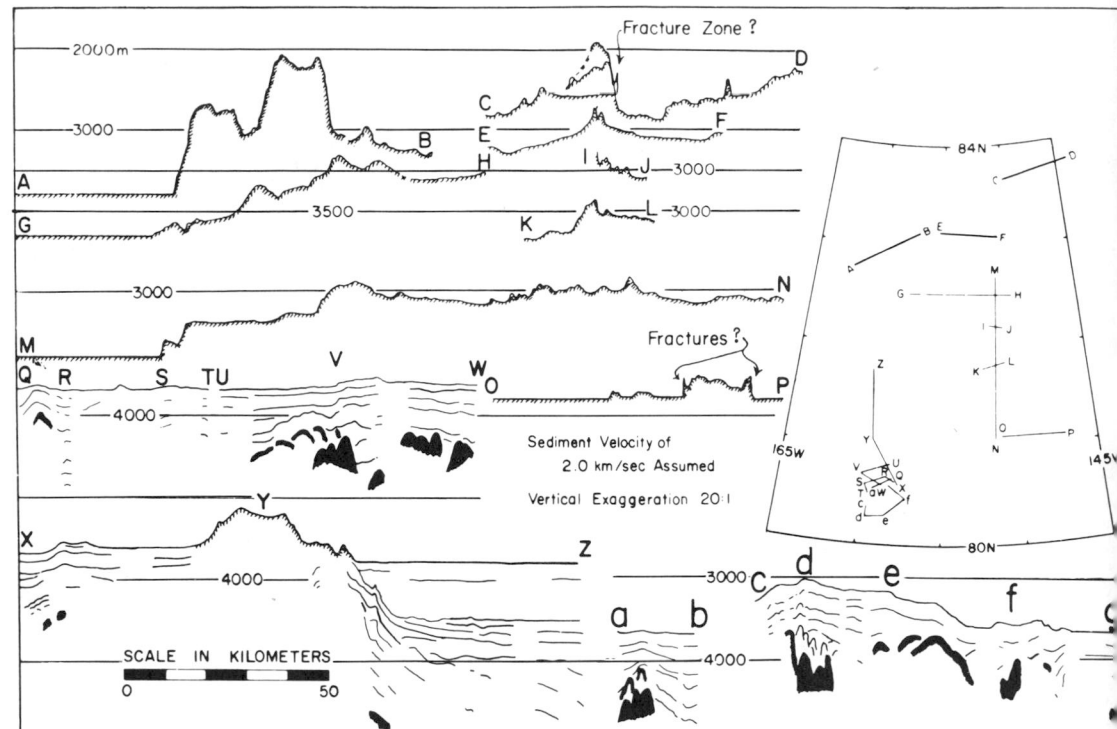

Fig. 5. Seismic profiles of Alpha Ridge, Canada Abyssal Plain, and Chukchi Rise (from Hall, 1973, with permission author and the American Association of Petroleum Geologists).

crust on the southern (Alaskan–Yukon) edge of the Canada Basin occurs in a narrower zone than is seen in such transitions elsewhere. Specifically they reported that the zone centers on the 200-m isobath, rather than the normal 2000-m isobath. From this observation they concluded that a very thick wedge of sediments may be present.

The Mackenzie and Colville Rivers probably are responsible for most of the sediment that is transported into the Canada Basin by turbidity currents. In excess of 30 million tons of sediment per year probably is delivered to this area (Wold *et al.*, 1970; Ritchie and Walker, 1974). Beal (1968) has reported numerous apparent slump structures on submarine bathymetric profiles in this area. Without any definite knowledge of older sediment or structure, our interest is primarily with the 700,000-year-old sediment cover for which there is good data.

1. *Turbidity Deposits of the Canada Plain*

Campbell (1973), Clark (1975), and Campbell and Clark (1977) have reported details of turbidites studied in more than 100 cores. These studies show the presence of turbidites in most parts of the Canada Basin (Fig. 6). All of the turbidites recognized are in cores from water deeper than of 3000 m. Campbell's detailed study confirmed the hypothesis of Hunkins and Kutschale (1967) and Hunkins (1968) that the Canada Abyssal Plain is covered with turbidity current deposits that were transported from the Canadian Archipelago and that very little turbidite was derived from the Chukchi area or elsewhere.

Evidently, the most common sedimentary structures of the turbidites are the five Bouma sequences, A to E. The plane-parallel laminated Bouma B or D sequences predominate. In addition, ripple cross-laminated Bouma C and massive Bouma A sequences are common.

Most of the Canada Basin turbidites are composed of fine silt-sized particles. The sand-sized portions of the turbidites are composed of well-sorted angular to subrounded fine sands (Fig. 7). Sand-size material is composed dominantly of quartz, chert, detrital calcite, K-feldspar, and polycrystalline quartz. Clays are composed of illite and kaolinite and lesser quantities of chlorite and smectite, including mixed layered, collapsed, and expandable varieties.

The fine-grained nature of the turbidites supports the idea that these sedi-

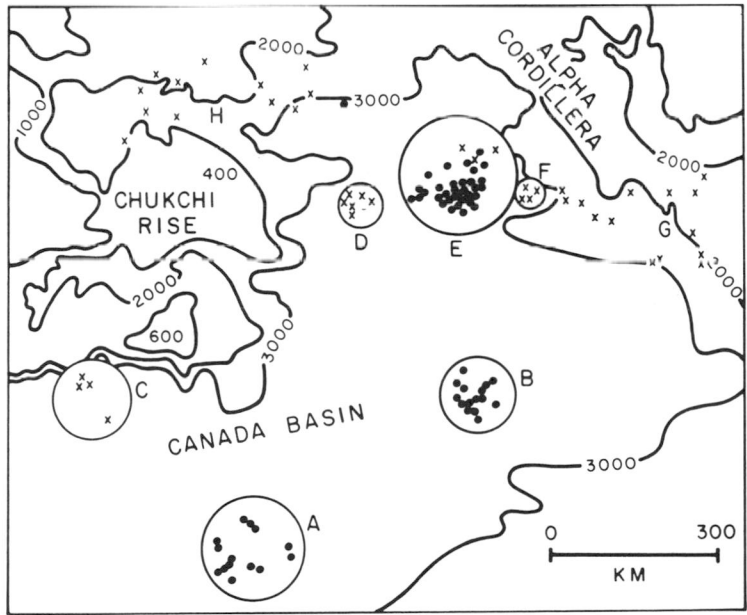

Fig. 6.   Distribution of turbidites in Canada Abyssal Plain. (A–F) areas of similar sediment types. Dots indicate presence of turbidites in T-3 cores; X indicates turbidites absent. Notice that all turbidites are within the 3000-m depth contour (modified from Campbell, 1973).

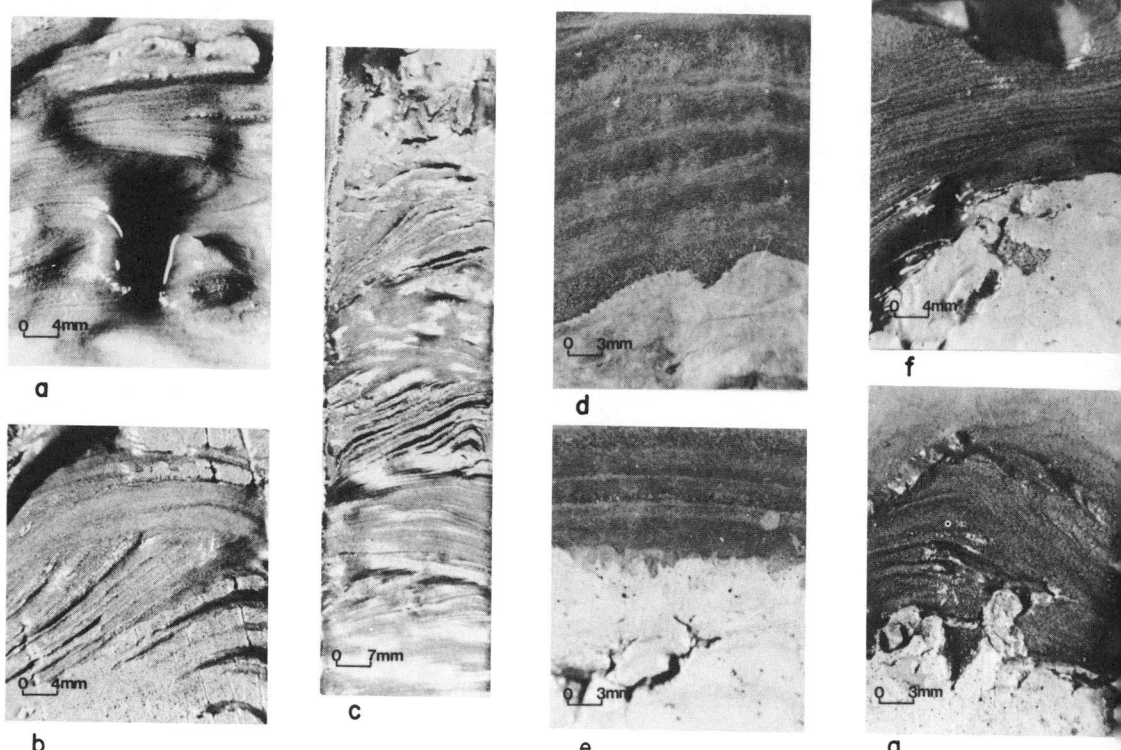

Fig. 7.   Turbidites in seven cores from Canada Abyssal Plain. (a) Ripple cross-laminae of Bouma C sequence of core F1
32 (area B of Fig. 6); (b) ripple cross-laminae of Bouma C sequence truncated above by Bouma D laminae in core F1-3
(area B of Fig. 6); (c) three sets of ripple cross-laminae of Bouma C truncated above by plane-parallel laminae of Bouma
of core F1-63 (area A of Fig. 6); (d) Bouma B or D sequence unconformably overlying lutite of core F1-30 (area B of Fig.
6); (e and f) load deformation of fine-grained turbidite of Bouma B or D on lutite (core 40, area B of Fig. 6); (g) loa
deformation along contact between turbidite (above) and glacial sediment (below) (core 36, area B of Fig. 6).

ments are distal turbidite deposits. Coarser deposits should be found closer to
the shelf but no data are available to confirm this.

Campbell (1973) determined that organic carbon ranged from less than
1% to more than 6%, while CO_3 ranged from 1 to 24 weight percent. These
results were applied to correlation among different turbidites in the cores but
the results were inconclusive. Correlation was most successful when marker
beds were used.

In area A (Fig. 6), the turbidites shown in Fig. 7 comprise from 3 to 68%
of 14 cores for an average content of 27%; 15 to 42% (average of 27% of 15
cores) in area B; and 0 to 6% (average of 2.7% of 9 cores) in area E. No turbidites were found in areas C, D, and F.

For 34 cores, the sedimentation rate ranged from 4.2 to 462.8 mm/1000
years, and averaged 83 mm/1000 years. This is several hundred times greater

than sedimentation rates in other parts of the Arctic and explains the thick sediment cover.

Probably the turbidity deposits were generated on the Canadian Continental Slope and Rise as a result of either seismic shocks or oversteepening of slopes because of glacial deposition and the influx of river sediment (30 million tons/year). It probably is safe to assume that this type of activity has been more or less continuous as long as the Arctic has been ice covered and drainage patterns have been similar to those of the present. If 30 million tons/year represents an average deposit from rivers alone, this could amount to 30 million tons/year times 20–30 m.y. or 6–9 trillion tons of sediment that have been available for the Canada Basin since the Middle Cenozoic.

2. *Origin and Age*

Figures 3 and 4 portray the physiography of the Canada Plain but the origin of this important part of the Amerasian Basin remains controversial. Age and tectonic relationships for the abyssal plain are more easily discussed along with other Amerasian structures in a later section.

B. The Alpha Cordillera

This prominent structural complex (Fig. 4) was discovered in August, 1957, from ice station Alpha. Its presence was reported later that year in several "anonymously" authored notes (*Science*, v. 126, p. 499, Sept. 13, 1957; *I.G.Y. Bull.* 6, Dec. 1957). Subsequently, additional details based on subsequent tracks were reported by Hunkins (1958, 1959). Hope (1959) published a general description and included a proposal to name the structure the Alpha Range. Further details were published in 1961 by Hunkins.

Work from the Russian NP-4 ice station a year or so earlier had detected a general high in the area adjacent to the East Siberian Continental Shelf but evidently the significance of the ridge was not realized at that time. Following the American reports, the Russians referred to this eastern portion of the ridge as the Mendeleyev Ridge. Thus, the broadest, most complex Amerasian Basin structure was discovered less than 25 years ago. This early work led Hope (1959) and Hunkins (1961) to suggest that the cordillera was of a fault-block origin. Today, however, there seems to be little evidence of a horst or graben nature. More recent work, discussed later in this chapter, has suggested alternate modes of origin.

The Alpha Cordillera is 1500 km long and has a maximum width of 600 km. Submarine profiles of the Alpha Cordillera (Beal, 1968) as well as aeromagnetic and gravity work by Vogt and Ostenso (1970) established the fact that the feature, called a "range" by Hope (1959) and a "rise" by Hunkins (1961), is topographically and structurally very complex. Hall (1970, 1973) compiled additional data, principally from T-3 work, and has provided the most comprehensive description of the complex, now appropriately referred to as a "cordillera" or ridge (Figs. 8 and 9).

1. *Description*

In general, the Alpha Cordillera is an irregular ridge that rises 2 km above the Canadian Basin. Hall's description (1973) included the idea expressed by Vogt and Ostenso (1970) that the structure was a fossil mid-ocean ridge, an early Cenozoic spreading site. He compared the rough basement physiography indicated by seismic profiles over the ridge with the physiography of the Mid-Atlantic Ridge (Fig. 10). According to this interpretation the spreading ridge origin of the structure was the controlling factor for Alpha physiography. Hall's data are limited to areas of T-3 drift. This information supplemented data by Beal (1968) from nine submarine profiles and was used in the Arctic Ocean floor map compiled by Heezen and Tharp (1971) (Fig. 4).

Hall (1973) also mapped six seamounts on the Alpha Cordillera. Four of these have depressed summits and were presumed to be volcanoes, an idea in

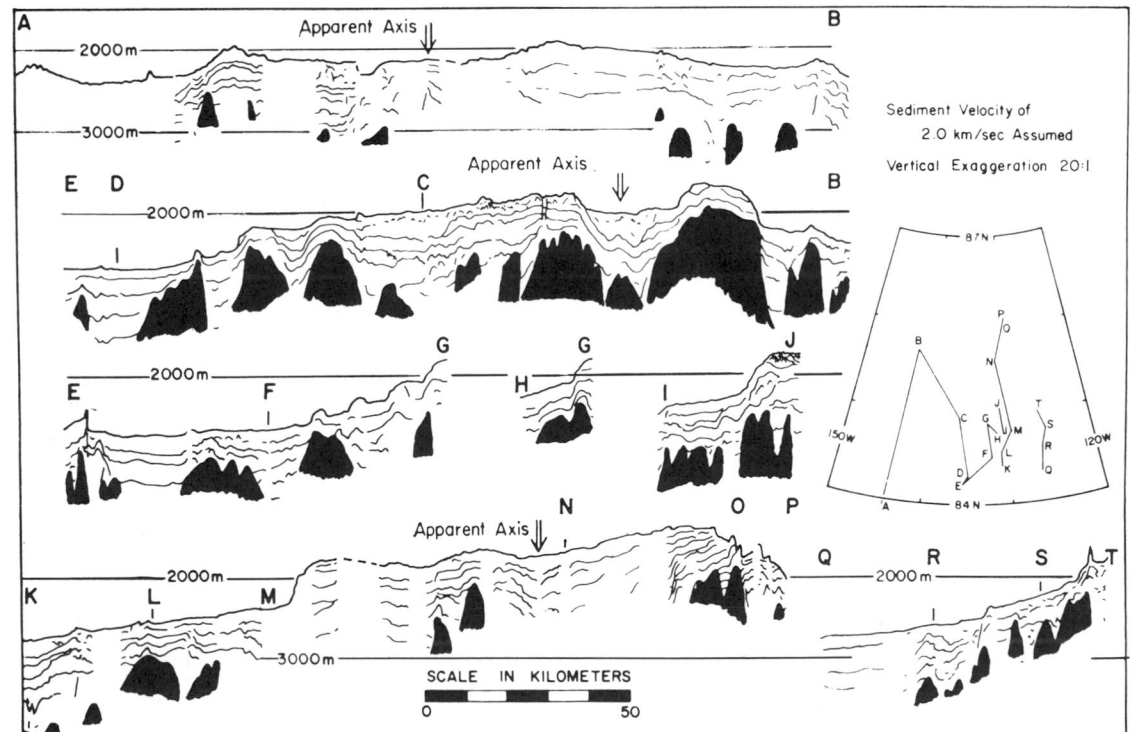

Fig. 8.   Seismic profiles of the western section of Alpha Ridge (from Hall, 1973, with permission of the author and the American Association of Petroleum Geologists).

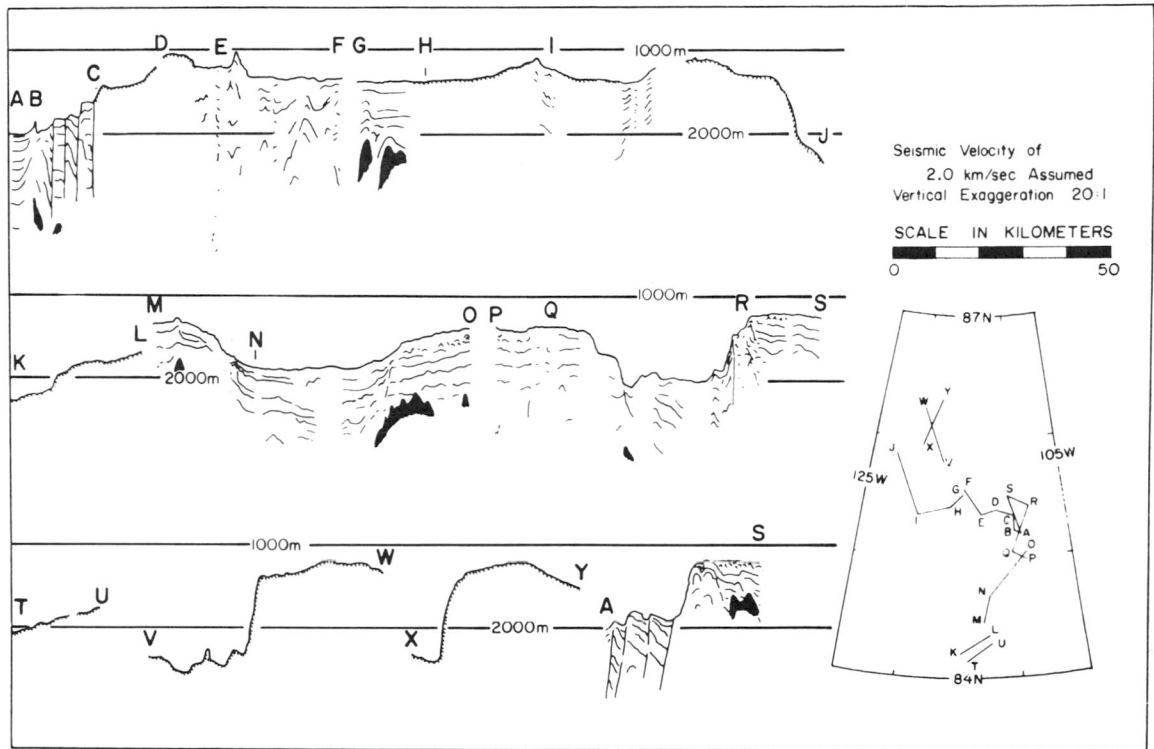

Fig. 9. Seismic profiles of the eastern section of Alpha Ridge (from Hall, 1973, with permission of the author and the American Association of Petroleum Geologists).

accord with earlier interpretations of Beal (1968). The fact that the four volcanic structures are in an area of low heat flow (Lachenbruch and Marshall, 1966) suggests that they are extinct, and is in harmony with the Vogt and Ostenso (1970) idea that the ridge is a fossil spreading site.

Hall (1973) observed "numerous lineations extending from the ridge crest in what appears to be a more or less systematic pattern" (p. 548). This fracture zone appears to have strong similarities to zones recognized along active spreading ridges (Fig. 1). These observations were interpreted as indicating a spreading ridge structure for the Alpha Cordillera.

2. Sediments

Hall reported that much of the rugged relief on the ridge was masked by a sediment cover of between 100 and 1200 m. The southeastern part of the Alpha Cordillera (the Mendeleyev Ridge area) is covered by 300 m of sediment. In contrast, the main portion of the crest of the Alpha Cordillera has sedimentary structures that Hall (1973) interpreted to be sediment hyperbolae

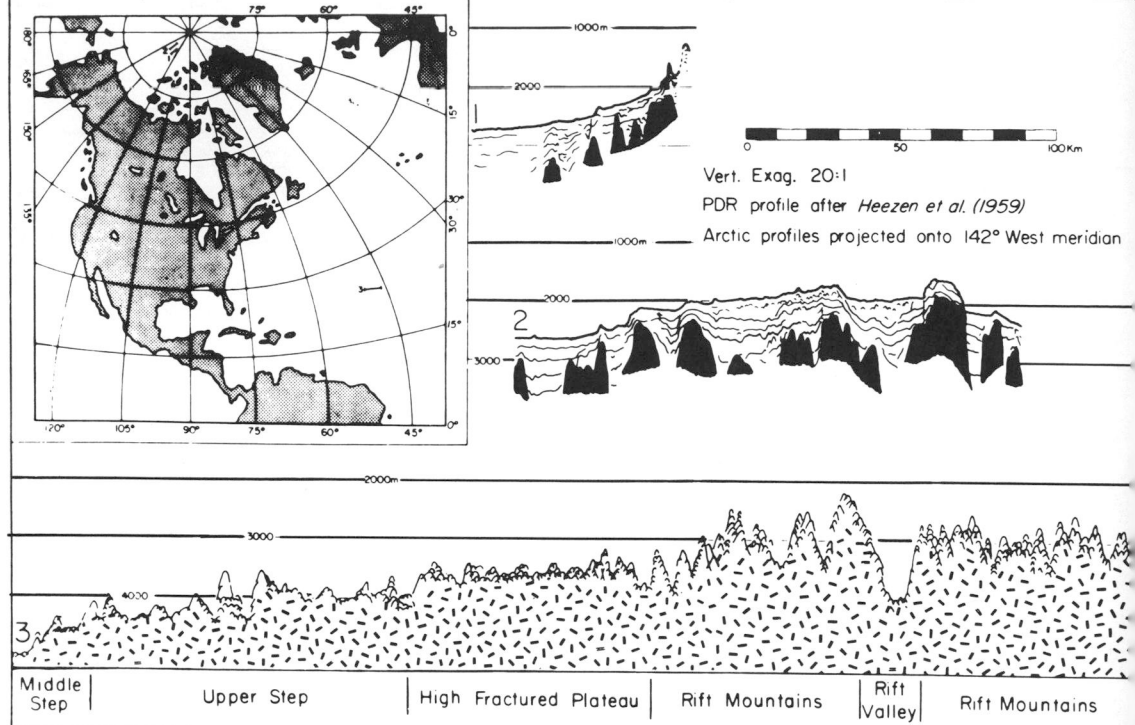

Vert. Exag. 20:1
PDR profile after *Heezen et al. (1959)*
Arctic profiles projected onto 142° West meridian

Middle Step | Upper Step | High Fractured Plateau | Rift Mountains | Rift Valley | Rift Mountains

Fig. 10.   Profiles of basement rock from the Alpha Ridge and Mid-Atlantic Ridge (from Hall, 1970).

or "sedimentary structures formed by bottom-current-related transport of fine sediment across the crest of the Alpha Cordillera from northwest to southeast" (p. 553). Hall speculated that the processes that formed the sediment hyperbolae are inactive. Structures interpreted as buried channels by Hall (1973) occur on the Mendeleyev portion of the ridge. According to Hall, these channels have a maximun depth of 700 m.

Since 1963, Ice Island T-3 has drifted across much of the Alpha Cordillera (Fig. 2). As a result, more than 400 sediment cores have been taken from this enigmatic submarine structure. The short cores average 3 m, but provide a good Recent to Pliocene and possibly older sedimentary record. In addition, two cores have recovered Cretaceous and early Cenozoic sediment (Clark, 1973).

Some 13 lithostratigraphic units can be differentiated in the late Cenozoic sediment. These sedimentary units can be recognized over a minimum of 120,000 km² of the Alpha Cordillera. A generalized section illustrating these units is shown in Fig. 11. The sediment of the units is glaciomarine and contains glacial erratics or drop stones as well as other characteristically glacially derived debris (Mullen *et al.*, 1972) (Fig. 12). The units can be distinguished on the basis of percentage of silt, sand, and characteristic "pink" layers. Most units are extensively burrowed and there is good correlation between coarseness and reduced bioturbation. How much of the finer clay-sized material has been derived from ice deposition is difficult to determine (Darby, 1975). It is possible that up to 10% of the clay material was derived from the atmosphere (Mullen *et al.*, 1972), although a figure of one percent has been determined at some localities (Darby *et al.*, 1975). On the average, the sediment contains approximately 50% illite, with chlorite and kaolinite consisting of approximately 20% each. Montmorillonite and other clay minerals comprise another 10%.

Benthonic and planktonic foraminifera range throughout these units but are geologically long-ranging Miocene to Recent taxa. No radiometric dates are available and the best magnetic stratigraphic correlations suggest that the base of unit A (Fig. 11) was deposited during Miocene Normal Polarity Epoch 5. This is a tenuous correlation but suggests that glacial activity was

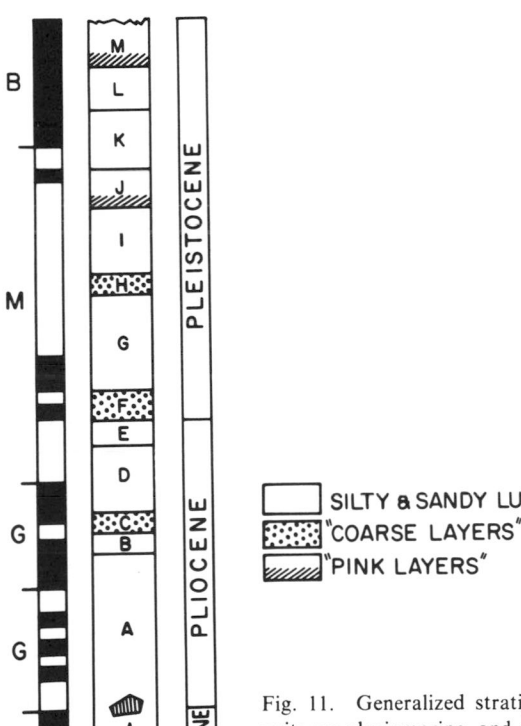

SILTY & SANDY LUTITES
"COARSE LAYERS"
"PINK LAYERS"

Fig. 11.   Generalized stratigraphy of the Alpha Cordillera. The 13 units are glaciomarine, and erratics (here shown in unit A) are present throughout. Magnetic correlations are tenuous but suggest that the base of unit A is >5 m.y. old.

Fig. 12.   Glacial erratics recovered from T-3 cores. Carbonate and quartzitic fragments probably were derived from Greenland or Ellesmere Island and transported to the Amerasian Basin. Upper left original is 3.5 cm in greatest dimension.

underway in the Amerasian Basin at least 5 m.y. ago. The implications of ice rafting occurring continuously in the central Arctic Ocean for this duration are profound, but lessened by the tenous nature of the Miocene magnetic determination.

Shells of microorganisms form a significant part of the sediment. The foraminiferan *Globorotalia pachyderma* predominates (Clark, 1971). Benthonic foraminifera are less abundant, but 73 species have been identified (Lagoe, 1975). Up to 15% by weight and 80% by volume of Alpha Cordillera sediment consists of foraminiferal tests. Sparse ostracodes (Joy and Clark, 1974), fish teeth, and holothurian sclerites are present in the cores. Organic carbon content of the sediment is less than 1% (Darby, 1971).

Older sediment in slump blocks has been found in two cores on the Alpha Cordillera (Clark, 1974). One core consists of Cretaceous and the other, early Cenozoic siliceous rich sediment. No Middle Cenozoic sediment has been recovered. The sequence of sedimentation indicates a change from siliceous sediment in the late Mesozoic–early Cenozoic to pelagic in the Plio-Pleistocene. The Middle Cenozoic must have been a critical period, for it was during this time that sedimentation changed.

The average sedimentation rate for the Alpha Cordillera has been 1.2

mm/1000 years for at least the last 3.5 m.y. (Clark, 1970). This is only 1% of the sedimentation rate normal in other oceans and reflects the low productivity and permanent ice cover that has existed since the early Pliocene.

3. *Origin*

When the Alpha Cordillera was discovered it was theorized that its origin was related to block faulting (Hope, 1959; Hunkins, 1961). This idea was advanced over theories of folding (Hope, 1959) or earlier ideas that the entire Amerasian Basin was a sunken "Hyperborean Shield" (e.g., Eardley, 1961). As details of the ridge structure began to accumulate during the 1960s, it became apparent that none of the theories were adequate. Vogt and Ostenso (1970) suggested that the Alpha Cordillera was an early Cenozoic spreading site and neatly wove its origin into the fabric of lithospheric plates and mid-ocean ridge tectonics (Fig. 13). Vogt and Ostenso (1970) and later Ostenso (1973) and Ostenso and Wold (1971, 1973) based their idea of an early Cenozoic spreading site on magnetics, the apparent correlation of anomaly patterns here with the better understood patterns of the Mid-Atlantic Ridge. These ideas, while based principally on aeromagnetic data, answered many questions not resolved by earlier theories. According to this interpretation, the Alpha Ridge was active from 60 to 40 m.y. ago and then became dormant as spreading shifted to the Nansen Ridge (in the Eurasian Basin) for the period from 40 m.y. ago to the present.

If the Alpha Cordillera is a fossil spreading site, a typical spreading ridge profile with central valley and transverse faults, modified by 40 million years of inaction and submarine deposition and erosion, could be expected. The Alpha Cordillera is much broader than many other ridges and no prominent central valley has been identified. Possible transverse faults have been suggested by Hall (1973).

A different interpretation has been proposed by Herron *et al.* (1974) who theorize that the Alpha Cordillera is a subduction zone (Fig. 14). This argument is supported by evidence that normal spreading ridges exhibit similar age vs. depth profiles (Sclater *et al.*, 1971), a configuration not present on the Alpha Cordillera. Herron *et al.* (1974) believe the subduction zone necessary in order to explain compression resulting from an opening Atlantic (Pitman and Talwani, 1972). This theory does not adequately deal with the magnetic anomalies reported by Vogt and Ostenso (1970). The idea that the Alpha Cordillera may be a subduction zone is a convenient way to explain the crustal shortening that had to occur as the North Atlantic opened but to date there is little evidence to support this hypothesis. These ideas are developed in some detail in the section devoted to the origin of the entire Amerasian Basin.

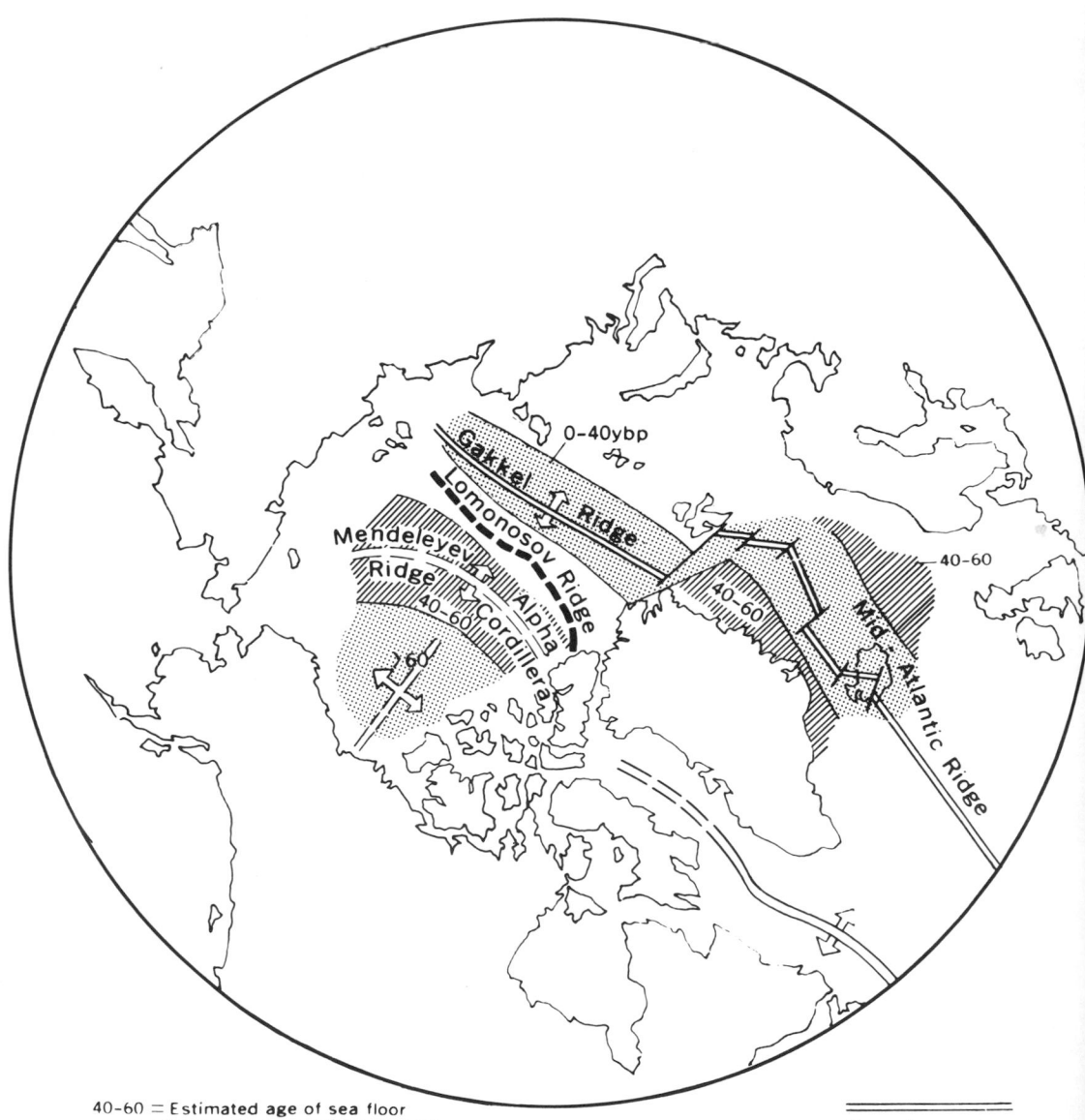

40-60 = Estimated age of sea floor
in millions of years before present

Spreading centers

Fig. 13.   Arctic Ocean sea-floor spreading. The primitive Amerasian Basin may have developed by spreading along an axis at right angles to the present shoreline ~60 m.y. ago, according to Tailleur. Spreading site shifted to present Alpha Ridge from 60–40 m.y. ago, in conjunction with the opening of Davis Strait, and then, finally, spreading occurred along the present Nansen (Gakkel) Ridge from 40 m.y. ago to the present, as an extension of North Atlantic rifting (from Churkin, 1973, with permission of the author and the American Association of Petroleum Geologists).

Fig. 14.   Subduction along Alpha Ridge. According to Herron *et al.* (1974), the pole of rotation for the development of the North Atlantic requires compression in the Arctic. This compression was manifested as subduction along the Alpha Ridge. (Black) plate positions 81 m.y. ago. (Stippled) plate positions at 63 m.y. ago (modified from Herron *et al.*, 1974, and used with permission of the author and the Geological Society of America).

III. CHUKCHI PLATEAU AND ADJACENT AREAS

A. Introduction

The Chukchi Plateau and adjacent escarpments (Fig. 4) occupy an area of 250,000 km² adjacent to the Chukchi Continental Shelf 1200 km north of the Bering Strait.

Bathymetric profiles obtained since 1950 indicate that the Chukchi Rise or Plateau consists of a complex of flat-topped plateaus and steep escarpments intersected by a series of troughs that open to abyssal plains on three sides (Figs. 4 and 5). The most prominent part of the complex is the Chukchi Cap or

Plateau. It was discovered by Soviet scientists from Ice Station NP-2 in 1950–1951 and reported in Somov (1955). The initial reports were based on wire soundings.

More extensive precision bathymetry was taken from the SSN *Nautilus* in 1958 (Dietz and Shumway, 1961). Additional bathymetry as well as seismic reflection and magnetic and gravity determinations were made during 1959 and 1962 from aerial surveys, from US Station Charlie, and Ice Island T-3. Details of this geophysical work were reported between 1962 and 1964 (Hunkins *et al.*, 1962; Ostenso, 1962; Shaver and Hunkins, 1964); thus, our knowledge of the Chukchi Complex has come from nuclear submarines, ice stations, and aerial surveys.

B. Physiography

The general pattern of the Chukchi Complex has been figured diagrammatically by Heezen and Tharp (1971) (Fig. 4). The main portion of the plateau rises to within 300 m of sea level and one ice station traverse recorded a minimum depth of 246 m (Hunkins *et al.*, 1962). Atop the plateau, relief ranges between 5 and 30 m. The sides of the Chukchi Plateau rise very steeply from the surrounding abyssal plains (3000 m). There is a shallow depression that serves as a small separation between the plateau and the Chukchi Continental Shelf on the south side (Shaver and Hunkins, 1964). One of the prominent troughs that intersect the southwest side of the plateau is at least 75 km long and 200 m deep (Shaver and Hunkins, 1964). The adjoining escarpments and seamounts are similar in overall physiography but are smaller in size.

C. Sediment

During the 1966, T-3 crossing of the Chukchi Complex, short sediment cores were recovered from the east and west flanks of the complex. The upper 2–3 m of sediment in these areas consists of grayish-brown lutite. A single unique pinkish sediment zone occurs in core 91 from the western flank, and 4 pink zones occur in core 87 from the eastern flank. These pink zones are similar to those found in Alpha Cordillera cores. Hoffman (1972) concluded that they are associated with decreases in free iron and manganese oxide and an increase in coarser sediment as compared to other intervals in the cores. The microfauna in the sediment is essentially the same as that discussed for other similar Arctic sediment (Clark *et al.*, 1975).

A sedimentation rate of 1.9 mm/1000 years has been calculated for the cores from the margins but no information is available for the shallower parts of the complex. Hunkins *et al.* (1962) reported sediment much like that of the T-3 core from the Chukchi Plateau taken during the 1959–1960 drift of Ice Station Charlie.

Seismic profiles indicate a considerable thickness of sediment (12 km?) on the Chukchi Plateau (Shaver and Hunkins, 1964).

D. Structure and Origin

Detailed information concerning the structure of the Chukchi Complex has been published, including seismic reflection and magnetic data (Hunkins *et al.*, 1962), aeromagnetic measurements (Ostenso, 1963), and magnetic and gravity measurements (Shaver and Hunkins, 1964). The various magnetic anomalies were interpreted to suggest that the Chukchi Plateau and high areas probably are composed of thick sedimentary rock. A prominent magnetic anomaly along the western flank of the plateau is thought to indicate a large ridge in basement rock of high susceptibility.

The crustal thickness is 18.5 km beneath the plateau and 21 km beneath the Inter-Chukchi Abyssal Plain. These data support the idea that originally the Chukchi Complex was a continental shelf, now somehow superimposed on the edge of a deep ocean basin. Shaver and Hunkins (1964) proposed a model for its origin that involves horizontal wrenching from the Alaskan Continental Shelf east of Point Barrow.

Later, Hunkins (1966) suggested that the Chukchi block may have been a piece of the continental shelf located "along the Alaskan Coast east of Barrow and then rotated a quarter-turn counterclockwise into its present position." This hypothesis could account for the absence of magnetic anomalies on the continental shelf east of Barrow.

Demenitskaya *et al.* (1973) regarded the complex as an original part of the Chukchi Continental Shelf that has been submerged recently. This interpretation calls for no horizontal movement. Ostenso and Wold (1973) interpret the Chukchi Plateau as having formed as a result of motion on the Alaskan orocline that caused tension on the northern edge of the Chukchi Shelf and resulted in the Chukchi Complex being "left behind as a remnant of continued rotation." Thus interpreted, the Chukchi Complex is a result of horizontal movement associated with Chukchi Continental Shelf modifications during plate adjustment. Its structure and history do not appear to be directly tied to the origin of the Canadian Basin or the Alpha Cordillera.

IV. WRANGEL AND FLETCHER ABYSSAL PLAINS

A. Description

Between the Alpha and Lomonosov Ridges lie two abyssal plains (e.g., Kutschale, 1966), the Wrangel Abyssal Plain at the base of the Siberian Continental Shelf and the Fletcher Abyssal Plain, parallel to and separating the Lomonosv and Alpha Ridges (Fig. 4). The Fletcher Abyssal Plain attains depths of more than 4000 m compared to 3000 m in the Wrangel Plain. These two plains occupy approximatcly 250,000 km². Fletcher's Ice Island (T-3) drifted across the northern part of the Fletcher Abyssal Plain in 1952. In 1958, the SSN *Skate* recorded precision depth soundings. Our ideas concerning these abyssal plains have developed principally from this work (Dietz and Shumway, 1961; Heezen and Laughton, 1963). In addition, Gramberg *et al.* (1974) have described Russian seismic work in this area.

B. Sediments

Kutschale (1966) interpreted magnetic and gravity readings to indicate the presence of a buried "basement" ridge between the Wrangel and Fletcher Plains. This ridge separates two distinctive sedimentary basins, the Wrangel with 3.5 km of horizontal strata and the Fletcher with 1.5 km. The great thickness of sediment in the Wrangel Basin probably was derived from the Siberian Continental Shelf during Pleistocene glaciation when lower sea level exposed the 600-km-wide shelf to weathering (the North Slope and the much narrower Canadian Continental Shelf provided the thick sediment for the Canada Abyssal Plain).

Kutschale (1966) speculated that the sediment of the deeper Fletcher Plain probably was derived from the "overflow" of turbidity current sediments from the shallower Wrangel Plain across the buried ridge and described four well-developed channels across it. These channels probably were cut by turbidity currents flowing south to north.

Kutschale (1966) published an isopachous map for part of the area (Fig. 15) and described sediment from six cores taken during the drift of Arlis II over the Wrangel Plain. The cores consist of lutite, at least some of which was deposited from turbidity currents. Foraminifera and ice-rafted glacial debris constitute the bulk of the sediment in the short cores. The seismic record of the sediment has been interpreted to indicate turbidites (Kutschale, 1966).

C. Structure

Gravity (Ostenso, 1963), seismic reflection, and magnetic data (Kutschale, 1966) suggest that the 3–5 km of sediment in the Wrangel Plain overlie a 6–8 km thick crust. This suggestion has been confirmed by Gramberg *et al.* (1974). The basement ridge that separates the Wrangel and Fletcher Plains has a crust that is 22 km thick.

The Wrangel and Fletcher Abyssal Plains lie between the Alpha Ridge, which has been interpreted alternatively as a fossil spreading center and a subduction zone, and the Lomonosov Ridge, widely thought to be a former piece of the Barents Continental Shelf separated from the present shelf by

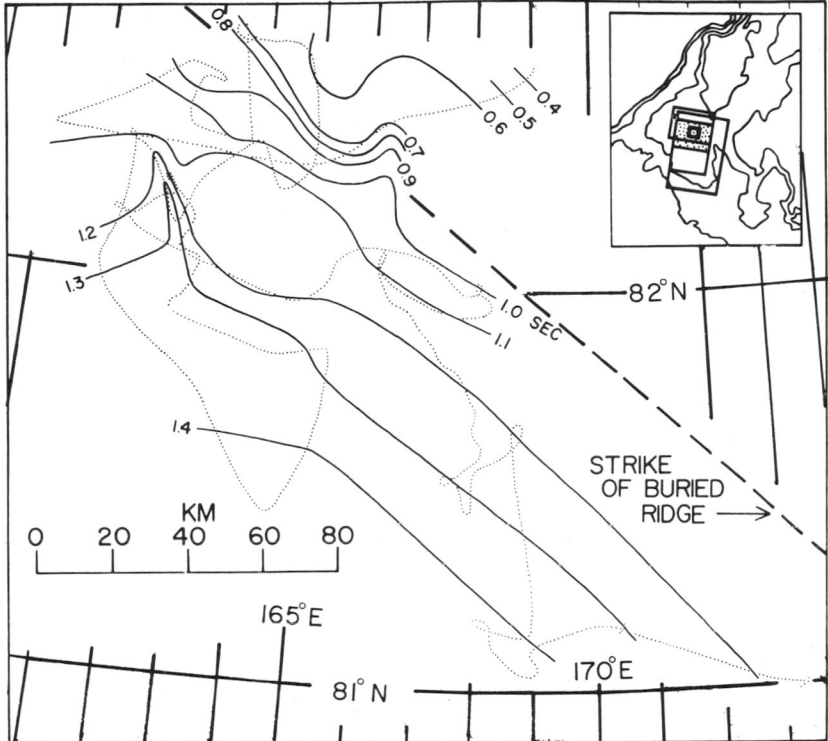

Fig. 15. Sediment isopach map in Wrangel Basin. Sediments measured above prominent reflector horizon. Sediment velocity of 2 km/sec assumed (from Kutschale, 1966, with permission of the author and *Geophysics*).

spreading along the Nansen Ridge. The presumed origin of the adjacent ridges suggests that early in their history most parts of the plains were located adjacent to continental shelves and the Asian continent from which the present sediment blanket probably was derived (Fig. 3, labeled "Siberia Abyssal Plain").

Gramberg and Kulakov (1975) have interpreted seismic velocities in the Wrangel Plain as indicating a layer of Lower–Middle Paleozoic carbonates. They suggest that the Wrangel Plain was a carbonate basin in the Paleozoic but was transformed into a clastic basin as the Arctic Ocean structure evolved. This is an interesting theory but unfortunately supporting data have not been published.

V. ORIGIN OF THE AMERASIAN BASIN

The two major parts of the Arctic Ocean (Eurasian and Amerasian Basins) probably had separate and distinct histories. It is thought that the

Eurasian Basin may have had the simpler history but this may be only because its interpretation is less controversial. If the Nansen Ridge is correctly interpreted as a spreading site (supported by seismic, heat flow, and bathymetric data) then its history probably is related to that of the North Atlantic. Spreading was initiated between 63 and 40 m.y. ago and has modified the former area of the Barents Shelf while carrying a piece of this shelf (Lomonosov Ridge) to the north. Thus, the present basin is bounded on the north by a piece of the shelf carried in this direction during spreading and by the Barents Continental Shelf margin to the south. The Nansen Ridge occupies a central position with abyssal plains on either side.

A more complex geologic history is suggested for the Amerasian Basin. Earlier work was summarized by Eardley (1961) who pointed out that the early theories of origin supposed that a "Hyperborean Shield" had sunk and that the Arctic was simply a subsided continental block (Fig. 16). However, seismic studies show that the Arctic Basin has an oceanic crust (Oliver et al., 1955). A simpler idea has been considered by Meyerhoff (1973) who outlined the view that the entire Atlantic–Arctic had been more or less stable (continually oceanic) since the late Proterozoic at least. A more tenable idea on tectonic history is the "Alaskan Orocline" theory championed by Carey (1955). This theory involves relative motion of the Canadian Arctic and Alaska with the latter rotating counterclockwise from the present Arctic Archipelago (Fig. 17).

Churkin (1973) proposed a Proto-Amerasian Basin in the early Paleozoic or even Precambrian with geosynclinal development by typical continental margin tectonics. He pointed out that the geology of the surrounding Siberian, Alaskan, and Canadian margins supports a Paleozoic Amerasian Basin. Ostenso and Wold (1973) believed that the modern Amerasian Basin had a Mesozoic origin and that it involved growth from a spreading ridge, the present Alpha Cordillera. Tailleur (1973) proposed that the "primitive" Canadian Basin formed by sea-floor spreading from a site perpendicular to the present North Slope during the Jurassic and Cretaceous (Fig. 13). According to Tailleur, the center of growth shifted to the Alpha Cordillera during the period 60–40 m.y. ago and ultimately to the Nansen Ridge in the Eurasian Basin approximately 40 m.y. ago. Thus, traditional theories of Amerasian Basin origin range from a stable Proterozoic origin with Paleozoic geosynclines, to Mesozoic sea-floor spreading at angles to, and later along the Alpha Cordillera.

A somewhat different interpretation has been proposed by Herron et al. (1974) who visualize the Paleozoic Amerasian Basin area as having been occupied by the Siberian Kolyma block. Movement of this block against the North American plate formed the Parry Island Foldbelt of the present Canadian Arctic Archipelago in the Middle Paleozoic. This theory further

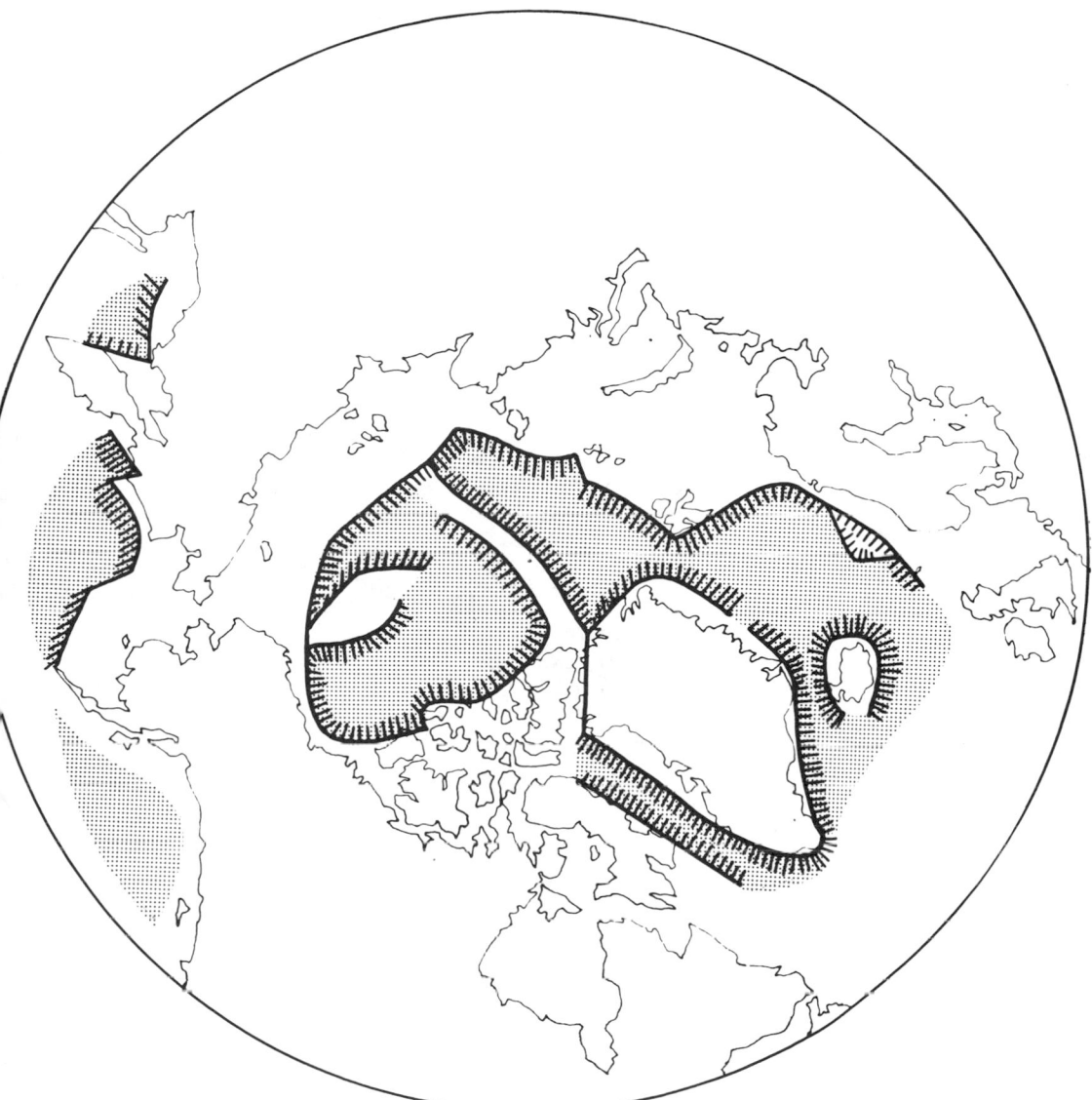

16. Fault block theory of Arctic Ocean formation. The simplest version of this theory assumes continental crust lapsing to form modern Arctic Ocean (from Churkin, 1973, with permission of the author and the American Association Petroleum Geologists).

considers that the Kolyma block remained in contact with the North American plate unti mid-Mesozoic time when it reversed its direction of movement and collided with the Siberian Platform, forming the Verkhoyansk Foldbelt (Fig. 18). This movement, incidentally, opened the present Amerasian Basin. Compression from an opening North Atlantic Basin at this time was expressed

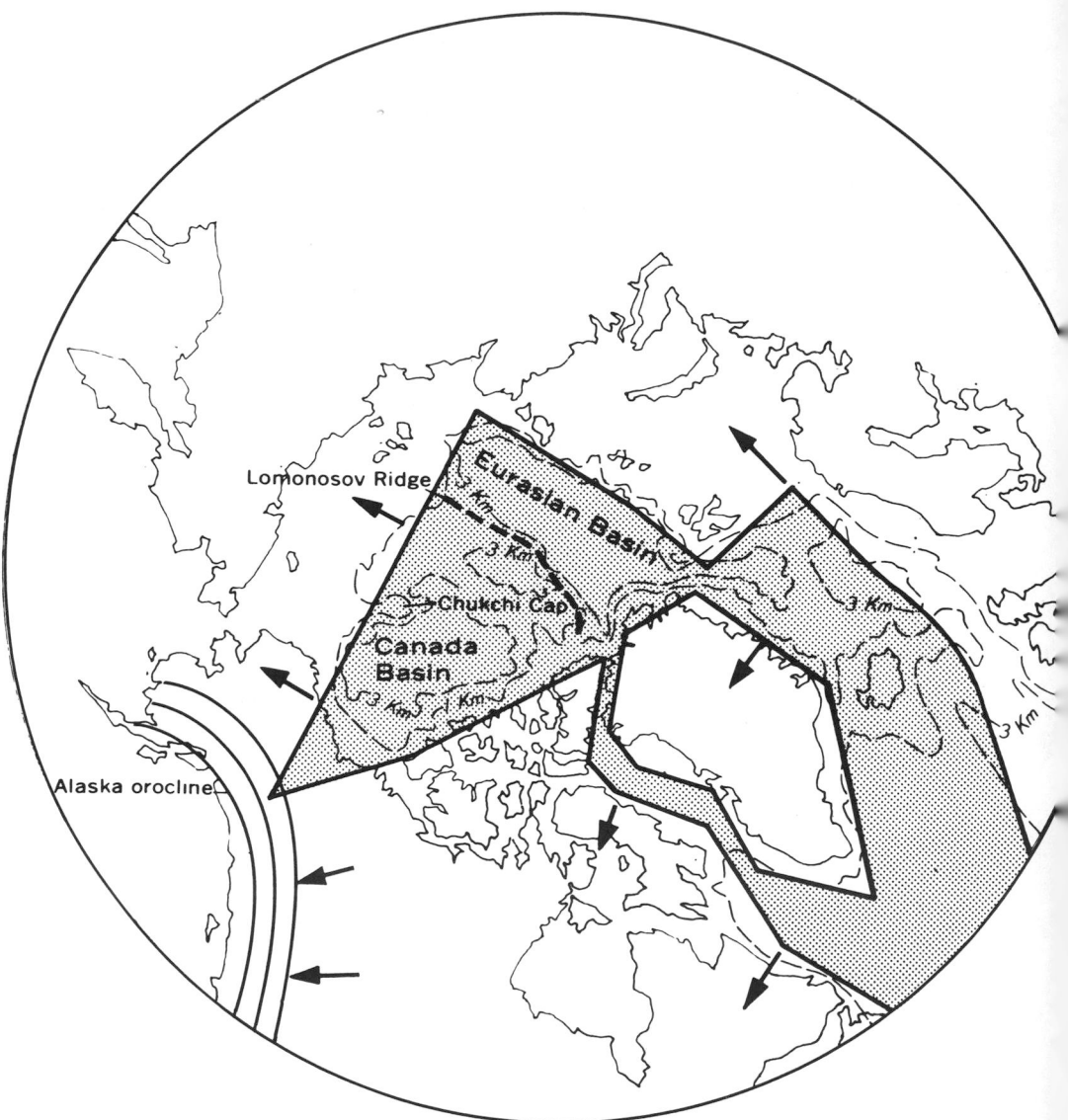

Fig. 17.   Orocline theory of Arctic Ocean formation. Oroclinal bend of Alaska produced rifting in the central Arc⸱
Churkin, 1973, with permission of the author and the American Association of Petroleum Geologists).

in the North American Laramide Orogeny and the Alpha Cordillera was the
subduction zone for much of this movement.

Evidence cited in support of this theory includes the fact that the Alpha
Ridge has Cretaceous and early Cenozoic siliceous sediment (Clark, 1974) and

the observation that the age vs. depth profile (Sclater *et al.*, 1971) normal for a spreading ridge is not present on the Alpha Ridge.

The mobile Kolyma block, alternately "filling" the Amerasian Basin and sliding back to the Siberian Platform, and an Alpha Ridge subduction zone are major differences of the Herron theory. Thus, beginning with a Proterozoic oceanic setting, the modern Amerasian Basin formed because of a Mesozoic–early Cenozoic spreading site (Alpha Ridge) or with the Kolyma block alternately deforming the present Canadian Arctic Archipelago and then the Siberian Platform, with the Alpha Ridge constituting a Mesozoic subduction zone. All of the theories apparently agree that the Amerasian Basin has escaped major tectonic modification since the mid-Cenozoic. Although the shifting Kolyma block theory is attractive, evidence for back and forth movement other than the deformed Canadian island chain and Verkhoyansk Foldbelt is lacking. The data presently at hand support a Paleozoic Amerasian Basin, described by Churkin (1973), modified during the Mesozoic by some tectonic activity along the Alpha Ridge. While there is some evidence that spreading (e.g., Vogt and Ostenso, 1970) along the ridge was possible, Mesozoic deformation and subduction related to the North Atlantic development may explain the structure (e.g., Pitman and Talwani, 1972; Herron *et al.*, 1975). The Canadian and Wrangel–Fletcher Abyssal Plains represent the undeformed parts of the Basin. The Chukchi Complex must be continental shelf material and its present position adjacent to continental shelf suggests little alteration.

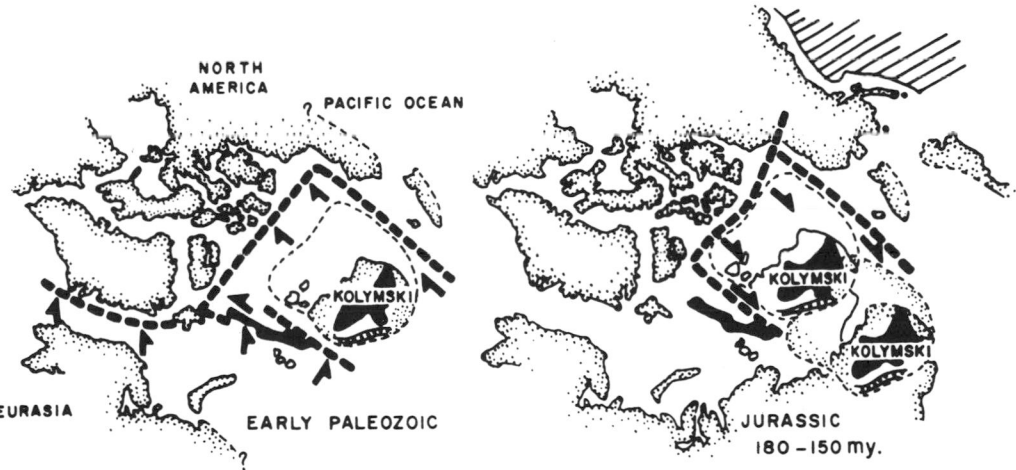

Fig. 18. Kolymski block movement across Arctic during Paleozoic and Jurassic (modified from Herron *et al.*, 1974, with permission of the author and the Geological Society of America).

VI. PALEONTOLOGY OF THE AMERASIAN BASIN

Sediments from the Amerasian Basin contain abundant Foraminifera. Although the abundance of these protozoans is not as great in the Arctic as in other oceans, they are clearly the most abundant and important fossils. Other fossils studied to date also are microscopic (Fig. 19). This probably is a result of sampling bias introduced by the use of 1.5-in. coring tubes. Larger samples such as box cores probably would yield at least some larger invertebrates as well as vertebrates.

A. Phylum Protozoa

Foraminifera. There are at least two important planktonic foraminiferal species in the Amerasian Basin and a considerably greater number of benthic species. As in other oceans, planktonic species are far more common than the benthic species. The most important planktonic form is *Globigerina pachyderma*. It is associated with less abundant *Globigerina quinqueloba*. The generic assignment of *G. pachyderma* apparently still is under consideration by taxonomists who may agree ultimately on its assignment to *Globorotalia* or *Turborotalia*.

Herman (1974) described several species not previously recognized from central Arctic Ocean cores. An earlier report of these additional planktonic species was challenged by Hunkins *et al.* (1971a,b), who had studied the same cores used by Herman. They suggested that Herman's additional species might not be valid. The species illustrated (Herman, 1974) include variants that many students have assigned to *G. pachyderma* and *G. quinqueloba* (Steuerwald and Clark, 1972; Clark *et al.*, 1975).

Abundance and Distribution of Planktonic Foraminifera. The two planktonic species were relatively abundant during the past 3 million years. All cores in which a complete Brunhes (700,000 years) record is present show a minimum of two abundance peaks. Other cores show a fluctuating pattern of abundance throughout the time interval (Clark, 1971). Comparison of the core data indicates that planktonic Foraminifera have never been more abundant than they are today. It also has been demonstrated that the highest percentage of dextral (warm water) shells occurs in the youngest sediment (Steuerwald and Clark, 1972). These two facts together support the idea that the Arctic Ocean currently is as warm as it has been at least in the present geographic setting.

Abundance and Distribution of Benthic Foraminifera. Some 73 species of benthic Foraminifera have been described in the upper 3 cm of some 100 central Amerasian Basin cores. There may be as many as 50 additional but rarer species. The samples were taken from depths ranging from 1069 to 3812

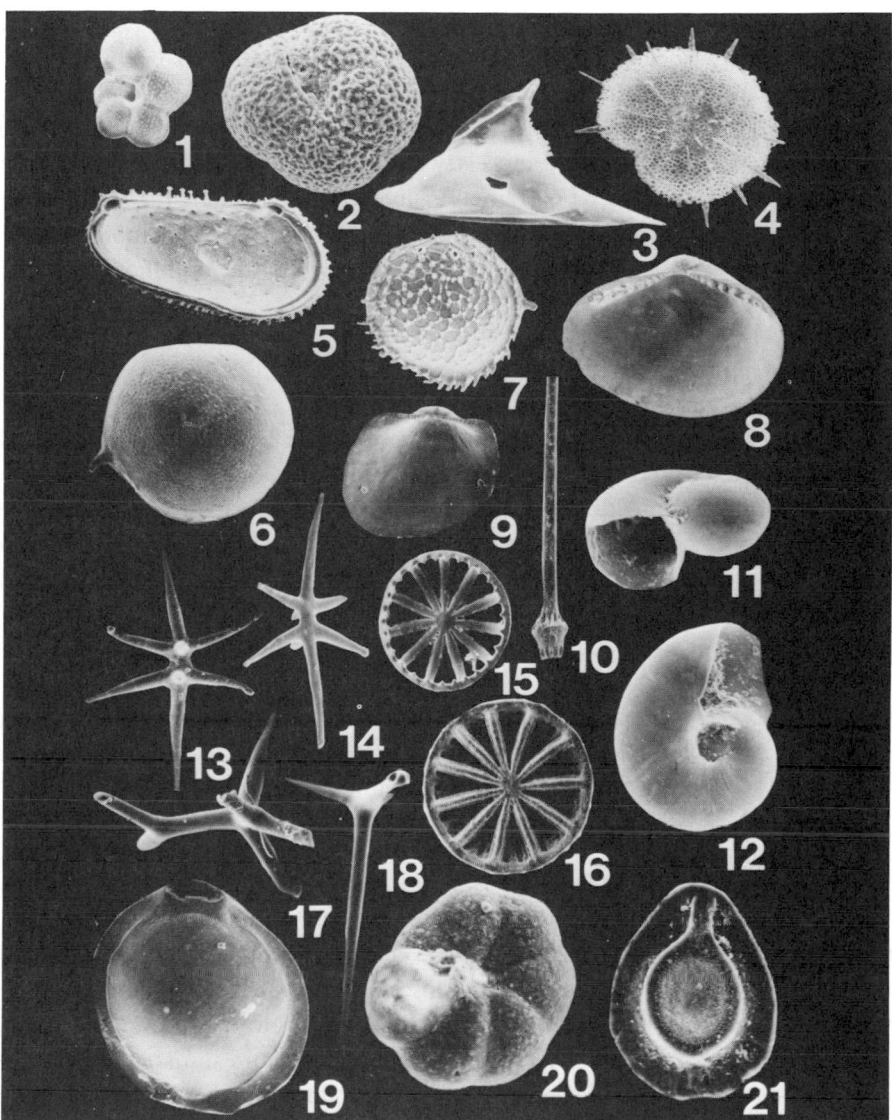

Fig. 19.   Representative microfossils from the Amerasian Basin. (1,2) Planktonic Foraminifera: (1) *Globigerina quinqueloba*, spiny form, ×200; (2) *Globorotalia pachyderma*, the dominant Foraminifera of the Arctic Pliocene to Recent, ×125. (3,5-7) Ostracoda: (3) *Cytheropteron alatum*, dorsal view of right valve, ×75; (5) *Echinocythereis dasyderma*, interior of left valve, ×75; (6) *Polycope punctata*, exterior of left valve, ×75; (7) *Polycope* sp., exterior of right valve, ×75. (4) Radiolarian, ×250. (8,9) Bivalves: (8) ×15; (9) ×50. (10) Echinoid spine, ×15. (11,12) Ptero- pods: *Limacina helicina*, dominant Arctic form; (11) sinistral coiled view; (12) umbilical view, ×125. (13-16) Holothurian sclerites: (13-14) *Elpidia* sp.; (13) with spines upright; (14) spines down, ×50; (15,16) holothurian wheels, ×100. (17,18) Sponge spicules, ×100. (19-21) Benthonic Foraminifera: (19) *Pyrgo* sp.; (20) *Eponides tumidulus*; (21) *Fissurina* sp.; all ×250. Figure reproduced at 75%.

m. Lagoe (1975) recognized three broadly overlapping depth associations. An *Epistominella* association was dominant from 1069 to 2250 m, a *Quinqueloculina–Planulina* association from 1700 to 3000 m, and a *Stretsonia* association from approximately 2500 to 3709 m. In the Arctic ecosystem, the diversity of benthic species decreases with depth, in contrast to diversity trends observed in the Atlantic, Gulf of Mexico, and elsewhere. Most of the ecologic parameters below 1000 m appear to be extremely stable and the reasons for the distribution of the Arctic fauna are not well understood (Lagoe, 1976). Approximately 95% of the benthic species are rotaliids, less than 5% are miliolids, and 0.1% are textularids.

B. Other Protistians

Tibbs (1967) reported on various protistians that he collected from Arlis I, including a single species of the silicoflagellate *Distephanus*, and several radiolarians, tintinnids, and dinoflagellates. None of these organisms is especially important in the bottom sediment at present. In two cores, one Cretaceous and the other early Cenozoic, taken from parts of the Alpha Cordillera, silicoflagellates have been found in abundance (Ling *et al.*, 1973; Clark, 1974). Silicoflagellates and diatoms occur almost to the exclusion of any other organism in these cores. The generic associations suggest that the Cretaceous–early Cenozoic Arctic Ocean was much warmer than at present, certainly it was ice-free. Today diatoms are present in the water column and in the bottom sediments. Fossil coccoliths have not been identified.

C. Ostracodes

Joy and Clark (1977) reported 19 species of ostracodes in the bottom sediment of the Amerasian Basin. Ostracodes are rarest in the Canadian Basin but were found at many coring sites at depths of 1351 to 3812 m. Only two species were found below 3000 m and the ostracode species of the Arctic constitute a single bathyal fauna. The Amerasian ostracodes are more similar to the faunas of Scandinavia than to those of the northern Pacific. This probably occurs because the broad shallow Chukchi Continental Shelf blocked the entrance of deep-water species from the Pacific while the North Atlantic fauna had a deeper-water migration route.

D. Miscellaneous Invertebrates

The pteropod *Limacina helincina* is abundant in the upper sediment of many parts of the Amerasian Basin. The aragonitic pteropods show evidence of carbonate solution and are virtually absent below depths of 40 cm in the cores. Other invertebrates known from core tops consist of six species of bivalves, one echinoid, five types of sponge spicules, and three holothurian species. The holothurians (*Kolga, Elpidia*, and *Myriotrochus*) are the most widely distributed of these groups (Gamber, 1976). The holothurians are deep-water echinoderms, some of which were described first from the Kara Sea. *Elpidia* is present in the deep Norweigan Sea, Baffin Bay, and the deep South Pacific at depths below 6000 m (Hansen, 1956). The sponge spicules are more enigmatic. Details of the invertebrates have recently been published (Gamber and Clark, 1978).

E. Fish Remains

Simple conical fish teeth and bone fragments are not uncommon in the Amerasian sediment but are not abundant enough to be a significant factor in Arctic paleontology. The Upper Cretaceous and Lower Cenozoic cores from the Amerasian Basin contain abundant fish bones and paleoniscoidlike teeth.

F. Trace Fossils

An abundance of trace fossils has been photographed on the floor of the Amerasian Basin by Dr. Kenneth Hunkins of the Lamont Geologic Observatory from T-3. Several dozen varieties of tracks, trails, burrows and meander patterns constitute a unique high latitude trace fossil assemblage representing a variety of echinoderms and worms, arthropods and fishes. A variety of studies based on the photographs is available (Kitchell, 1979; Kitchell and Clark, 1979).

VIII. PALEOECOLOGY OF THE AMERASIAN BASIN

The nature of the Paleozoic Amerasian Ocean is unknown. The oldest known sediment from the Arctic Ocean is late Cretaceous (Maestrichtian) and direct knowledge of the paleoecology of the Arctic Ocean begins at this time.

The single late Mesozoic Arctic Ocean core consists of 172 cm of orange–yellow tuffaceous sediment. It overlies and is overlain by Plio-Pleistocene sediment and quite clearly is part of a small slumped mass of Cretaceous sediment (Clark, 1974). The sediment has a high water content (77–85%) and consists of various clay minerals and a great concentration of organics, such as silicofglagellates (Ling *et al.*, 1973), diatoms, and fish fragments. Paleoecologic inferences are made on the basis of the silicoflagellate genus *Dictyocha*, a genus rare in water of less than 4°C. Abundant *Dictyocha* usually indicates relatively warm water. The concentration of this phytoplanktonic organism in the Cretaceous core (5 to 3 million individuals per cc of sediment) suggests productivity similar to that found in modern upwelling areas of the ocean (Dietrich, 1963). Although it might not be possible to substantiate that the Cretaceous material is a fossil "red tide," the concentration of organisms is similar to that found to latitudes of 55° or less in modern seas (Brongersma-Sanders, 1957). Certainly, the Cretaceous phytoplankton lived in warmer water than exists here today and paleomagnetic data indicate that the location of the sediment was 20° or more away from the Cretaceous north pole. Silicoflagellate abundance, present latitude distribution of taxa, and paleomagnetism all support the idea that the water of the Cretaceous Amerasian Basin was not ice covered but similar to that of the present North Atlantic in the Iceland and southern Greenland area (Fig. 20).

The early Cenozoic record also is based on a single core from the Alpha Ridge (Clark, 1974). This core consists of 160 cm of light orange–yellow tuffaceous sediment and 183 cm of dark brown to gray brown tuffaceous lutite. Montmorillonite, illite, kaolinite, and chlorite are the dominant clay minerals. Much of the tuff is amorphous. This core contains Eocene silicoflagellates and, like the Cretaceous core, is a slump mass resting in Plio-Pleistocene lutite.

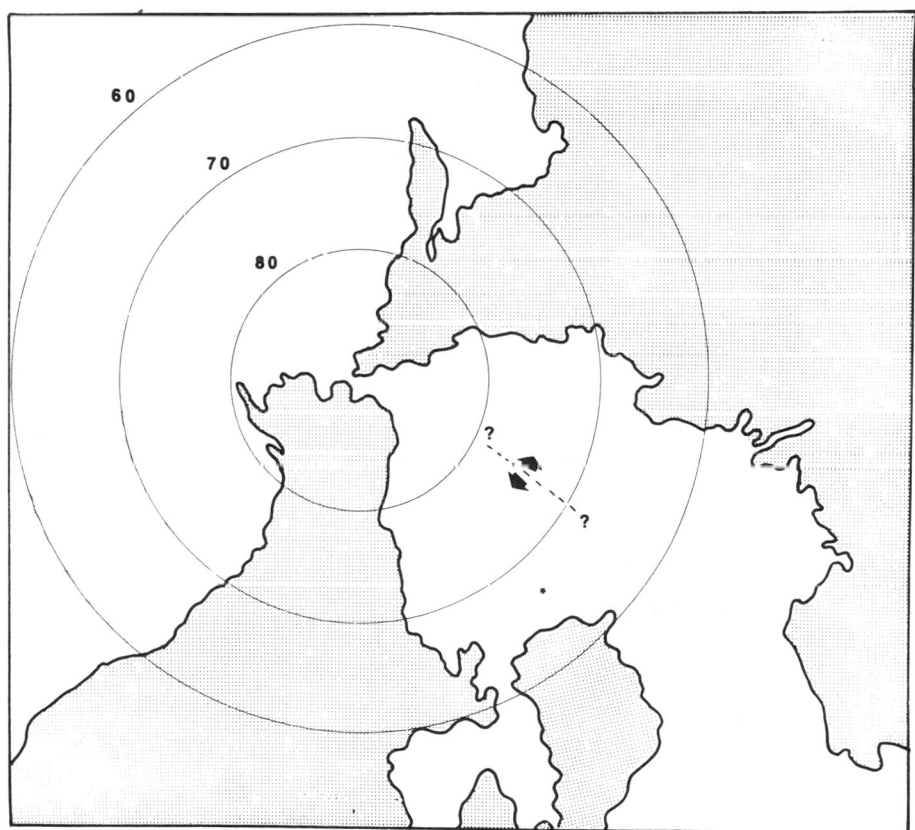

Fig. 20.   Possible late Cretaceous paleolatitudes and the Alpha Ridge as a spreading site during this time.

Dawson *et al.* (1976) have found a variable Eocene vertebrate fauna on Ellesmere Island, only a few hundred kilometers from the Alpha Ridge core. The fauna includes crocodiles, an occurrence supporting a warm climate interpretation. The same conclusion as to water temperature made from the Cretaceous sediment are valid for the Eocene, and the late Mesozoic to early Cenozoic sequence of events probably was similar. Phytoplankton in considerable abundance accumulated along with probable volcanic ash in the relatively warm water of the late Cretaceous and early Cenozoic ocean. No evidence of glacial activity (e.g., erratics exclusively of cold-water origin) is present in either core.

To date, no Oligocene or Miocene (mid-Cenozoic) sediment has been recovered from the Arctic Ocean. This interval is bracketed below by older Eocene evidence for warm, open water and above by Pliocene evidence that indicates an ice-covered Arctic Ocean (Clark, 1971). It seems safe to conclude that it was during this mid-Cenozoic interval of 30 million years that an initial freezing of the Arctic Ocean occurred and geologic events elsewhere support this idea. During this time, the northern hemisphere crustal plates attained approximately their present positions. Glaciers had begun to form on other parts of the Earth (Rutford *et al.*, 1972) and there is evidence of a Middle Cenozoic worldwide temperature drop. For example, Dury (1971) showed that paleobotanical data indicate a general Cenozoic cooling (following a warm Cretaceous and Eocene interval) interrupted only by a brief early Miocene warm peak. This trend has been substantiated from DSDP core data by Savin *et al.* (1975). Thus, if the Arctic did not freeze during the Oligocene, it probably did in the late Miocene, as the cooling trend continued.

Plio-Pleistocene paleoecology (Clark, 1969, 1970, 1971; Steuerwald and Clark, 1972), based on the abundance of Foraminifera and glacial erratics in cores representing the past 3–5 m.y., shows that the Arctic was ice covered by at least the Middle Pliocene. Since that time, the ice thickness may have varied but ice cover was probably continuously present. This argument is based on the fact that planktonic Foraminifera have not been more abundant during the last 3–5 m.y. than they are now (Fig. 21). If the present 3-m ice thickness is the minimum condition necessary to explain abundance peaks of Arctic Ocean plankton, then times of less abundant Foraminifera may indicate colder water, thicker ice, and restricted photosynthesis and productivity. In support of the biologic interpretation is sedimentary data showing that glacial erratics, delivered more or less continuously at present, have been more or less uniformly delivered into the Amerasian Basin during the past 3–5 m.y. This interpretation clearly challenges the Ewing–Donn glacial theory that demands an alternately ice-free and frozen Arctic Ocean during the past 2 m.y. If the Amerasian core data have been interpreted correctly, the warm water, late Mesozoic and early Cenozoic Arctic became cooler, even ice covered during

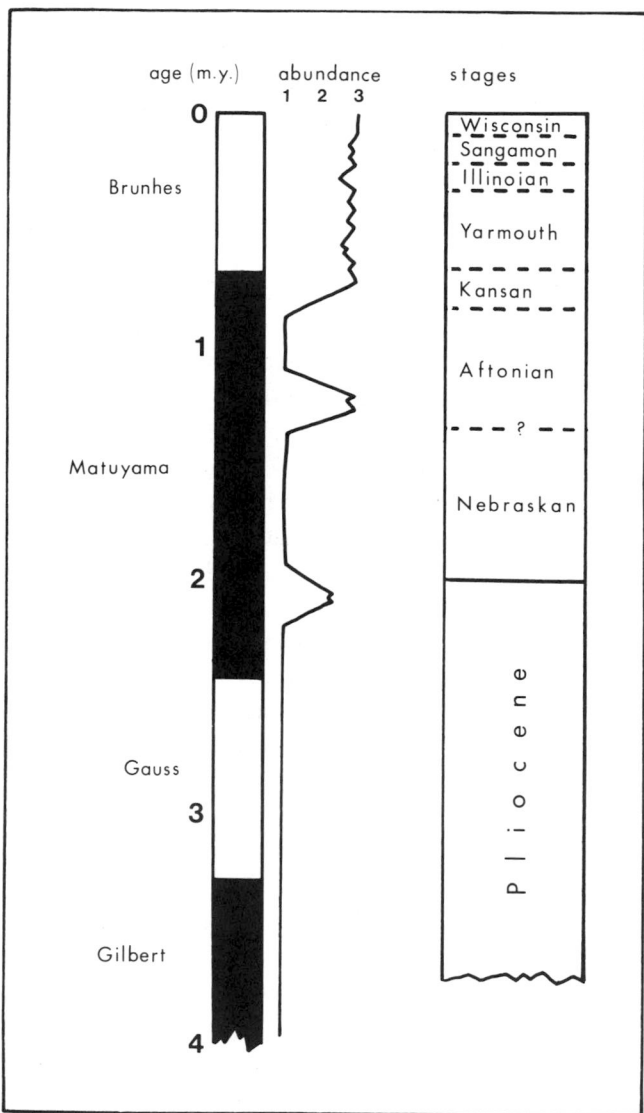

Fig. 21. Generalized planktonic Foraminifera abundance in Arctic cores compared to Pleistocene glacial stages; 3 = greatest abundance of Foraminifera (>10% of sediment by weight). Lack of correlation between the peak abundance and Pleistocene glaciation is apparent. Foraminifera abundances of Recent time suggest that the present ice-covered conditions represent as warm a condition as has existed since Early Pliocene (after Clark, 1971).

Fig. 22. The Arctic Ocean probably became an ice-covered ocean during the Oligocene or Miocene. It has retained its ice cover since that time.

the Middle Cenozoic, and this ice cover has been maintained to the present. The most profound changes, since at least the Middle Pliocene, have been in ice thickness (Fig. 22). The paleoecologic ramifications suggest that the potential for northern hemisphere climate modifications such as that now produced by the Arctic Ocean has been in existence at least since the Middle Pliocene and may have existed earlier.

REFERENCES

Beal, M. A., 1968, Bathymetry and structure of the Arctic Ocean, Ph.D. Thesis, Oregon State University, Corvallis, 204 p.

Beal, M. A., Edvalsen, F., Hunkins, K., Molloy, A., and Ostenso, N. A., 1966, The floor of the Arctic Ocean: Geographic names, *Arctic*, v. 19, p. 215–219.

Brongersma-Sanders, M., 1957, Mass mortality in the sea, *Geol. Soc. Am. Mem.* 67(1), p. 941–1010.

Campbell, J. S., 1973, Pleistocene turbidites of the Canada Abyssal Plain, Arctic Ocean, *Univ. Wisconsin Arctic Sed. Stud. Tech. Rep.* 13, 106 p.

Campbell, J. S., and Clark, D. L., 1977, Pleistocene turbidites of the Canada Abyssal Plain of the Arctic Ocean, *J. Sed. Petrol.*, v. 47, p. 657–670.

Carey, S. W., 1955, Orocline concept in geotectonics, pt. 1, *Pap. Proc. R. Soc. Tasmania*, v. 89, p. 255–288.

Churkin, M., Jr., 1969, Paleozoic tectonic history of the Arctic Basin north of Alaska, *Science*, v. 165, p. 549–555.

Churkin, M., Jr., 1973, Geologic concepts of Arctic Ocean Basin, in: *Arctic Geology*, Pitcher, M. G., ed., *Am. Assoc. Petrol. Geol. Mem.* 19, p. 485–499.

Clark, D. L., 1969, Paleoecology and sedimentation in part of the Arctic Basin, *Arctic*, v. 22, p. 233–245.

Clark, D. L., 1970, Magnetic reversals and sedimentation rates in the Arctic Ocean, *Geol. Soc. Am. Bull.*, v. 81, p. 3129–3134.

Clark, D. L., 1971, Arctic Ocean ice cover and its Late Cenozoic history, *Geol. Soc. Am. Bull.*, v. 82, p. 3313–3324.

Clark, D. L., 1973, Arctic Ocean studies progress, *Oil Gas J.*, v. 71, p. 104–106.

Clark, D. L., 1974, Late Mesozoic and Early Cenozoic sediment cores from the Arctic Ocean, *Geology*, v. 2, p. 41–44.

Clark, D. L., 1975, Geological history of the Arctic Ocean Basin, in: *Canada's Continental Margins and Offshore Petroleum Explorations*, Yorath, C. J., Parker, E. R., and Glass, D. J., eds., *Can. Soc. Petrol. Geol. Mem.* 4, p. 501–524.

Clark, D. L., Larson, J. A., Root, R. E., and Fagerlin, S. C., 1975, Foraminiferal patterns of the Arctic Ocean Pleistocene, *Univ. Wisconsin Arctic Sed. Stud. Tech. Rep.* 18, 94 p.

Darby, D. A., 1971, Carbonate cycles and clay mineralogy of Arctic Ocean sediment cores, *Univ. Wisconsin Arctic Sed. Stud. Tech. Rep.* 8, 43 p.

Darby, D. A., 1975, Kaolinite and other clay minerals in Arctic Ocean sediments, *J. Sed. Petrol.*, v. 45, p. 272–279.

Darby, D. A., Burckle, L., H., and Clark, D. L., 1975, Airborne dust on the Arctic pack ice, its composition and fallout rate, *Earth Planet. Sci. Lett.*, v. 24, p. 166–172.

Dawson, M. R., West, R. M., Langston, W., Hutchinson, S. H., 1976, Paleogene terrestrial vertebrates: Northernmost occurrence, Ellesmere Island, Canada, *Science*, v. 192, p. 781–782.

Demenitskaya, R. M., and Hunkins, K. L., 1970, Shape and structure of the Arctic Ocean, in: *The Sea*, v. 4, pt. 2, Maxwell, A. E., ed., New York: Wiley-Interscience, p. 223–249.

Demenitskaya, R. M., Gaponenko, G. I., Kiselev, Yu. G., and Ivanov, S. S., 1973, Features of sedimentary layers beneath Arctic Ocean, in: *Arctic Geology*, Pitcher, M. G., ed., *Am. Assoc. Petrol. Geol. Mem.* 19, p. 332–335.

Dietrich, Gunter, 1963, *General Oceanography—An Introduction*, New York: John Wiley, 558 p.

Dietz, R. S., and Shumway, G., 1961, Arctic Basin geomorphology, *Geol. Soc. Am. Bull.*, v. 72, p. 1319–1330.

Dury, G. H., 1971, Relict deep weathering and duricrusting in relation to the paleoenvironments of middle latitudes, *Geogr. J.*, v. 137, p. 511–522.

Eardley, A. J., 1961, History of geologic thought on the origin of the Arctic Basin, in: *Geology of the Arctic*, v. 1, Raasch, G. O., ed., Toronto: Toronto University Press, 607–621.

Gamber, J. A., 1976, Recent sponges, molluscs, and echinoderms in sediment cores from the central Arctic Ocean, *Univ. Wisconsin Arctic Sed. Stud. Tech. Rep.* 23, 79 p.

Gamber, J. A., and Clark, D. L., 1978, Distribution of microscopic molluscs, echinoderms and sponges in the Central Atlantic Ocean, *Micropaleontology*, v. 24, p. 422–431.

Gramberg, I. S., and Kulakov, Yu. N., 1975, General geological features and possible oil–gas provinces of the Arctic Basin, in: *Canada's Continental Margins and Offshore Petroleum Exploration*, Yorath, C. J., Parker, E. R., and Glass, D. J., eds., *Can. Soc. Petrol. Geol. Mem.* 4, p. 525–529.

Gramberg, I. S., Demenitskaya, R. M., and Kiseley, Yu. G., 1974, Sedimentary deposits in the deep water regions of the Arctic Ocean, in: *Problems of Geology in the Polar Regions of the Earth*, Leningrad: Arctic Geology Institute Press, p. 43–46.

Hall, J. K., 1970, Arctic Ocean geophysical studies: the Alpha Cordillera and the Mendeleyev Ridge, Ph.D. Thesis, Columbia University, New York, 125 p.

Hall, J. K., 1973, Geophysical evidence for ancient sea-floor spreading from Alpha Cordillera and Mendeleyev Ridge, in: *Arctic Geology*, Pitcher, M. G., ed., *Am. Assoc. Petrol. Geol. Mem.* 19, p. 542–561.

Hansen, B., 1956, Holothuroidea from depths exceeding 6000 meters, *Galathea Rep.*, v. 2, p. 33–54.

Heezen, B. C., and Laughton, A. S., 1963, Abyssal plains, in: *The Sea*, v. 3, Hill, M. N., ed., London: Interscience Publishers, p. 312–364.

Heezen, B. C., and Tharp, M., 1971, Arctic Ocean Floor, map accompanying *National Geographic Magazine*, October.

Heezen, B. C., Tharp, M., and Ewing, M., 1959, The floors of the oceans. I. The North Atlantic, *Geol. Soc. Am. Spec. Pap.* 65, 122 p.

Herman, Y., 1974, Arctic Ocean sediments, microfauna, and the climatic record in Late Cenozoic time, in: *Marine Geology and Oceanography of the Arctic Seas*, Herman, Y., ed., New York: Springer-Verlag, p. 283–348.

Herron, E. M., Dewey, J. F., and Pitman, W. C., III, 1974, Plate tectonics model for the evolution of the Arctic, *Geology*, v. 2, p. 377–380.

Hoffman, T. F., 1972, The origin and stratigraphic significance of pink layers in Late Cenozoic sediments of the Arctic Ocean, *Univ. Wisconsin Arctic Sed. Stud. Tech. Rep.* 11, 74 p.

Hope, E. R., 1959, Geotectonics of the Arctic Ocean and the great Arctic magnetic anomaly, *J. Geophy. Res.*, v. 64, p. 407–427.

Hunkins, K., 1958, The floor of the Arctic Ocean, American Association for the Advancement of Science, 125th Annual Meeting Program, Washington, D.C.

Hunkins, K., 1959, The floor of the Arctic Ocean, *Trans. Am. Geophys. Union*, v. 40, p. 159–162.

Hunkins, K., 1961, Seismic studies of the Arctic Ocean floor, in: *Geology of the Arctic*, Raasch, G. O., ed., Toronto: Toronto University Press, p. 645–665.

Hunkins, K., 1966, The Arctic continental shelf north of Alaska, in: *Symposium on Continental Margins and Island Arcs*, Poole, W. M., ed., *Geol. Surv. Can. Pap.* 66-15, p. 197–205.

Hunkins, K., 1968, Geomorphic provinces of the Arctic Ocean, in: *Arctic Drifting Stations*, Washington, D.C.: The Arctic Institute of North America, 365–376.

Hunkins, K., and Kutschale, H., 1967, Quaternary sedimentation in the Arctic Ocean, *Progr. Oceanogr.*, v. 4, p. 89–94.

Hunkins, K., Herron, T., Kutschale, H., and Peter, G., 1962, Geophysical studies of the Chukchi Cap, Arctic Ocean, *J. Geophys. Res.*, v. 67, p. 234–247.

Hunkins, K., Be, A. W. H., Opdyke, N. D., and Mathieu, G. 1971a, The Late Cenozoic history of the Arctic Ocean, in: *The Late Cenozoic Ice Ages*, Turekian, K., ed., New Haven: Yale University Press, p. 215–237.

Hunkins, K., Be, A. W. H., Opdyke, N. D., and Saito, T., 1971b, Arctic paleo-oceanography in Late Cenozoic time, *Science*, v. 174, p. 962.

Joy, J. A., and Clark, D. L., 1977, The distribution, ecology, and systematics of the benthonic Ostracoda of the central Arctic Ocean, *Micropaleontology*, v. 23, p. 129–154.

Kitchell, J. A., 1979, Deep-sea traces from the Central Arctic: An analysis of diversity, *Deep Sea Res.*, v. 26A, p. 1185–1198.

Kitchell, J. A., and Clark, D. L., 1979, A multivariate approach to biofacies analysis of deep-sea traces from the Central Arctic, *J. Paleontol.*, v. 53, p. 1045–1067.

Kutschale, H., 1966, Arctic Ocean geophysical studies: The southern half of the Siberia Basin, *Geophysics*, v. 31, p. 683–710.

Lachenbruch, A. H., and Marshall, B. V., 1966, Heat flow through the Arctic Ocean floor—The Canada Basin–Alpha Rise boundary, *J. Geophys. Res.*, v. 71, p. 1223–1248.

Lagoe, M. B., 1975, Recent benthic Foraminifera from the central Arctic Ocean, *Univ. Wisconsin Arctic Sed. Stud. Tech. Rep.* 20, 225 p.

Lagoe, M. B., 1976, Species diversity of deep-sea benthic Foraminifera from the central Arctic Ocean, *Geol. Soc. Am. Bull.*, v. 87, p. 1678–1683.

Ling, H. Y., McPherson, L. M., and Clark, D. L., 1973, Late Cretaceous (Maestrichtian) silicoflagellates from the Alpha Cordillera of the Arctic Ocean, *Science*, v. 180, p. 1360–1361.

Meyerhoff, A. A., 1973, Origin of Arctic and North Atlantic Oceans, in: *Arctic Geology*, Pitcher, M. G., ed., *Am. Assoc. Petrol. Geol. Mem.* 19, p. 562–582.

Mullen, R. E., Darby, D. A., and Clark, D. L., 1972, Sedimentary processes in the Arctic Ocean, *Geol. Soc. Am. Bull.*, v. 83, p. 205–211.

Nansen, F., 1904, *The Norwegian North Polar Expeditions, 1893–1896*, v. 4, London: Longmans, Green, 232 p.

Oliver, J., Ewing, M., and Press, F., 1955, Crustal structure of the Arctic regions from Lg phase, *Geol. Soc. Am. Bull.*, v. 66, p. 1063–74.

Ostenso, N. A., 1962, Geophysical investigations in the Arctic Ocean Basin, *Univ. Wisconsin Polar Res. Center Rep.* 62-4, 124 p.

Ostenso, N. A., 1963, Aeromagnetic survey of the Arctic Ocean Basin, Proc. Thirteenth Alaska Sci. Conf., Alaska Div. AAAS, Juneau, p. 115–148.

Ostenso, N. A., 1973, Sea-floor spreading and the origin of the Arctic Ocean Basin, in: *Implications of Continental Drift to the Earth Sciences*, London: Academic Press, p. 1965–1973.

Ostenso, N. A., and Wold., R. J., 1971, Aeromagnetic survey of the Arctic Ocean: Techniques and interpretations, *Mar. Geophys. Res.*, v. 1, p. 178–219.

Ostenso, N. A., and Wold, R. J., 1973, Aeromagnetic evidence for origin of the Arctic Ocean Basin, in: *Arctic Geology*, Pitcher, M. G., ed., *Am. Assoc. Petrol. Geol. Mem.* 19, p. 506–516.

Papanin, I. 1947, *Life on an Ice Floe*, London: Hutchinson, 240 p.

Pitman, W. C., III, and Talwani, M., 1972, Sea-floor spreading in the North Atlantic, *Geol. Soc. Am. Bull.*, v. 83, p. 619–646.

Ritchie, W., and Walker, H. J., 1974, River in the frozen north, *Geographical Magazine* (London), August 1974, p. 634–640.

Rutford, R. H., Craddock, C., White, C. M., and Armstrong, R. L., 1972, Tertiary glaciation in the Jones Mountains, *Antarctic Geology and Geophysics*, Universitetsforlaget, p. 239–243.

Savin, S. M., Douglas, R. G., and Stehli, F. G., 1975, Tertiary marine paleotemperatures, *Geol. Soc. Am. Bull.*, v. 86, p. 1494–1510.

Sclater, J. G., Anderson, R. N., Bell, H., and Lee, M., 1971, Elevation of ridges and evolution of the central eastern Pacific, *J. Geophys. Res.*, v. 76, p. 7889–7915.

Shaver, R., and Hunkins, K., 1964, Arctic Ocean geophysical studies: Chukchi Cap and Chukchi Abyssal Plain, *Deep Sea Res.*, v. 11, p. 905–916.

Somov, M. M., 1955, Observational data of the scientific research drifting station of 1950–1951, v. I–III, Morskoi Transport 1954–1955, *Trans. Am. Met. Soc. ASTIA*, p. 117–133.

Steuerwald, B. A. and Clark, D. L., 1972, *Globigerina pachyderma* in Pleistocene and recent Arctic Ocean sediment, *J. Paleontol*, v. 46, p. 573–580.

Tailleur, I. L., 1973, Probable rift origin of Canada Basin, Arctic Ocean, in: *Arctic Geology*, Pitcher, M. G., ed., *Am. Assoc. Petrol. Geol. Mem.* 19, p. 526–535.

Tibbs, J. F., 1967, On some planktonic Protozoa taken from the track of Drift Station Arlis I, 1960–61, *Arctic*, v. 20, p. 247–254.

Vogt, P. R., and Ostenso, N. A., 1970, Magnetic and gravity profiles across the Alpha Cordillera and their relation to Arctic sea-floor spreading, *J. Geophys. Res.*, v. p. 4925–4937.

Wilkins, G. H., 1928, Polar exploration by airplane, in: *Problems of Polar Research*, *Am. Geogr. Soc. Spec. Publ.*, v. 7, p. 396–417.

Wold, R. J., Woodzik, T. L., and Ostenso, N. A., 1970, Structure of the Beaufort Sea continental margin, *Geophysics*, v. 35, p. 849–861.

INDEX

that the Cretaceous material is a fossil "red tide," the concentration of organisms is similar to that found to latitudes of 55° or less in modern seas (Brongersma-Sanders, 1957). Certainly, the Cretaceous phytoplankton lived in warmer water than exists here today and paleomagnetic data indicate that the location of the sediment was 20° or more away from the Cretaceous north pole. Silicoflagellate abundance, present latitude distribution of taxa, and paleomagnetism all support the idea that the water of the Cretaceous Amerasian Basin was not ice covered but similar to that of the present North Atlantic in the Iceland and southern Greenland area (Fig. 20).

The early Cenozoic record also is based on a single core from the Alpha Ridge (Clark, 1974). This core consists of 160 cm of light orange–yellow tuffaceous sediment and 183 cm of dark brown to gray brown tuffaceous lutite. Montmorillonite, illite, kaolinite, and chlorite are the dominant clay minerals. Much of the tuff is amorphous. This core contains Eocene silicoflagellates and, like the Cretaceous core, is a slump mass resting in Plio-Pleistocene lutite.

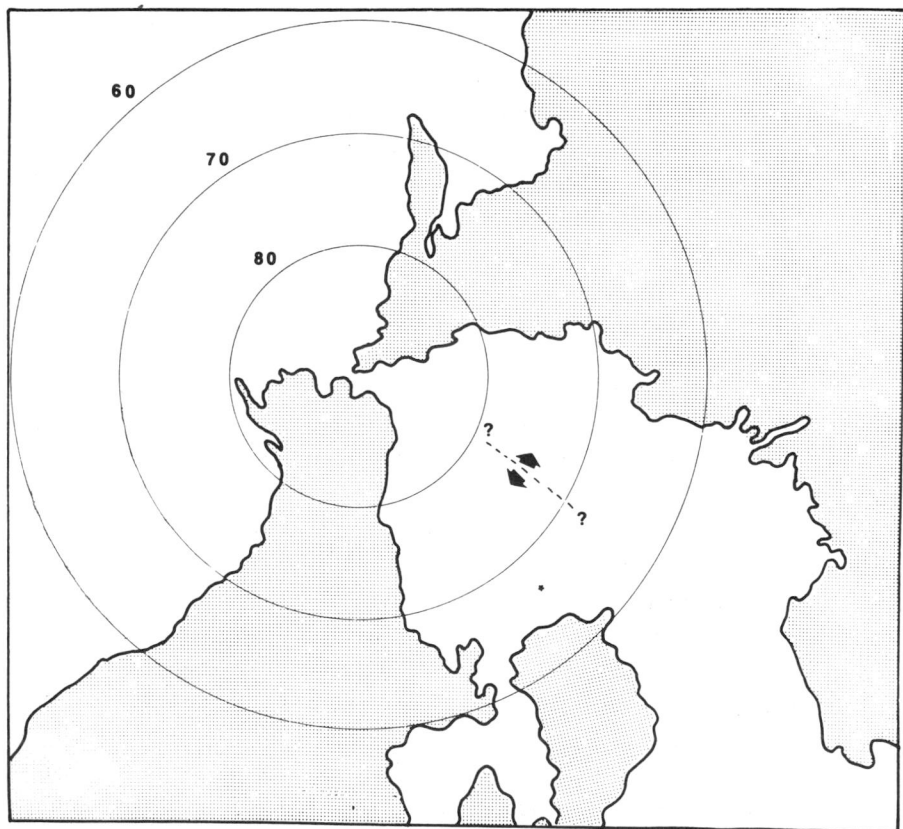

Fig. 20.   Possible late Cretaceous paleolatitudes and the Alpha Ridge as a spreading site during this time.